Stem Cell-Dependent Therapies

Stem Cell-Dependent Therapies

Editor: Mccoy Connor

FOSTER ACADEMICS
www.fosteracademics.com

www.fosteracademics.com

FOSTER ACADEMICS

Cataloging-in-Publication Data

Stem cell-dependent therapies / edited by Mccoy Connor.
p. cm.
Includes bibliographical references and index.
ISBN 978-1-64646-630-6
1. Stem cells--Therapeutic use. 2. Stem cells. 3. Pathology, Cellular. 4. Cytology. I. Connor, Mccoy.
QH588.S83 S743 2023
616.027 74--dc23

Foster Academics,
118-35 Queens Blvd., Suite 400,
Forest Hills, NY 11375, USA

ISBN 978-1-64646-630-6 (Hardback)

Contents

Preface

Stem cells refer to the type of cells that have the ability to develop into a wide types of cells in the body. They function as a repair system for the body. Adult stem cells and embryonic stem cells are the two categories of stem cells. The utilization of these cells to prevent or treat a condition or illness is known as stem cell therapy. It is a technique for the regeneration of damaged organs, tissues or functions through the transplantation of stem cells. It has the potential to treat cancer and degenerative diseases, and restoring damaged tissues. Bone marrow transplantation is a type of stem cell therapy that has been utilized for many years due to its effectiveness in clinical trials. Experts and students actively engaged in the research of stem cell-dependent therapies will find this book full of crucial and unexplored concepts. Those in search of information to further their knowledge will be greatly assisted by it.

After months of intensive research and writing, this book is the end result of all who devoted their time and efforts in the initiation and progress of this book. It will surely be a source of reference in enhancing the required knowledge of the new developments in the area. During the course of developing this book, certain measures such as accuracy, authenticity and research focused analytical studies were given preference in order to produce a comprehensive book in the area of study.

This book would not have been possible without the efforts of the authors and the publisher. I extend my sincere thanks to them. Secondly, I express my gratitude to my family and well-wishers. And most importantly, I thank my students for constantly expressing their willingness and curiosity in enhancing their knowledge in the field, which encourages me to take up further research projects for the advancement of the area.

Editor

1

Mesenchymal Stromal Cells for Transplant Tolerance

Manuel Alfredo Podestà [1,2], Giuseppe Remuzzi [1] and Federica Casiraghi [1]*

[1] Department of Molecular Medicine, Istituto di Ricerche Farmacologiche Mario Negri IRCCS, Bergamo, Italy, [2] Department of Health Sciences, Università degli Studi di Milano, Milan, Italy

***Correspondence:**
Federica Casiraghi
federica.casiraghi@marionegri.it

In solid organ transplantation lifelong immunosuppression exposes transplant recipients to life-threatening complications, such as infections and malignancies, and to severe side effects. Cellular therapy with mesenchymal stromal cells (MSC) has recently emerged as a promising strategy to regulate anti-donor immune responses, allowing immunosuppressive drug minimization and tolerance induction. In this review we summarize preclinical data on MSC in solid organ transplant models, focusing on potential mechanisms of action of MSC, including down-regulation of effector T-cell response and activation of regulatory pathways. We will also provide an overview of available data on safety and feasibility of MSC therapy in solid organ transplant patients, highlighting the issues that still need to be addressed before establishing MSC as a safe and effective tolerogenic cell therapy in transplantation.

Keywords: mesenchymal (stromal) stem cells, transplantation, T cells, macrophage, tolerance, B cells

Transplantation represents the treatment of choice for end-stage solid organ failure. The dissection of the mechanisms regulating the interplay between the host immune system and the transplanted graft have led to the introduction into clinical practice of effective T-cell immunosuppressive agents, which have abated the risk of acute rejection in the peri-transplant period, increasing the 1-year graft survival above 90%. However, current immunosuppressive agents failed to significantly affect the long-term outcome of solid organ transplantation, because these drugs are less effective in preventing chronic allograft rejection (1). Moreover, the lifelong systemic immunosuppression exposes transplant recipients to a significant risk of side effects, infections and malignancy (2, 3).

Since acute T-cell mediated rejection can be successfully managed in most cases, transplant research should focus on the identification of innovative strategies to achieve allograft tolerance, i.e., long-term stable graft function in the absence of immunosuppression. Therefore, the ideal strategy should target not only T cells, which are the main players of alloimmunity, but regulate in a concerted action also B cells, dendritic cells and macrophages, which all contribute to both the acute and chronic alloimmune response.

In this scenario, mesenchymal stromal cells (MSC) seem a very promising cellular therapy in the pursuit of transplantation tolerance induction, allowing minimization or even discontinuation of life-long immunosuppression. Indeed, MSC have a unique capability to inhibit the immune alloresponse at different levels and to dampen the activation of cells of both the adaptive and innate immune systems, reprogramming them into regulatory cells. MSC are a heterogeneous subset of non-hematopoietic cells, currently defined by standard criteria that include the ability to differentiate into tissues of mesodermal lineages *in vitro*, plastic adherence under standard culture conditions, expression of CD73, CD90 and CD105, and lack of CD45, CD34, CD14, or CD11b, CD79α or CD19, and HLA-DR surface antigens (4). Although MSC can be obtained from several adult and fetal tissues (including umbilical cord, Wharton's jelly, amniotic fluid, adipose tissue and dental pulp) (5–9), the bone marrow has been traditionally considered as their main source, and thus bone marrow-derived MSC are the ones best characterized both in humans and animal models.

In this review we will provide a summary of the alloimmune response events that accompany solid organ transplantation, discussing the multiple immune modulatory effects that MSC have demonstrated on both the innate and adaptive arms of the immune system and highlighting their tolerance-inducing properties in preclinical transplant models. We will also provide an overview of available data from clinical trials on MSC infusion in solid organ transplant patients, discussing future perspectives and issues that still need to be addressed.

THE ADAPTIVE IMMUNE SYSTEM: TERMINAL EFFECTOR OF THE ALLORESPONSE

The crucial step of the adaptive response is represented by the recognition of donor alloantigens (mostly non-self MHC molecules) by recipient lymphocytes. These cells are activated in the presence of costimulatory molecules, leading to a cascade of events that ultimately precipitates into graft rejection (10).

Transplant rejection results in the generation of cells with either regulatory or effector functions, with the latter prevailing during the alloresponse. Effector functions are mainly mediated by $CD8^+$ and $CD4^+$ T cells: cytotoxic $CD8^+$ T cells, activated by class I MHC-presented antigens, undergo clonal expansion, mature into effector cells and migrate into the graft, where they induce apoptosis and secrete cytotoxic molecules.

Class II MHC-restricted $CD4^+$ T lymphocytes differentiate into distinct subsets of helper T cells, depending on the cytokine milieu: IL-12 prompts the differentiation in T_H1 cells and promotes the secretion of IL-2 and IFNγ, which, respectively, sustain $CD8^+$ T-cell proliferation and induce a delayed-type hypersensitivity reaction through macrophage activation (11). T_H2 polarization is guided by IL-4, which results in the recruitment of graft-damaging eosinophils *via* IL-5 secretion (12). T_H17 induction is mediated by IL-21, IL-23, IL-6, and TGFβ stimulation. T_H17 cells play a role in both acute and chronic rejection through the recruitment of neutrophils into the transplanted graft (13). Finally, follicular helper T-cell (T_{FH}) differentiation is primarily mediated by IL-12, IL-6, and TGF-β signaling; T_{FH} cells promote B-cell differentiation into antibody-producing plasma cells (14, 15) and are the cells responsible for the development of donor-specific antibodies and antibody-mediated rejection (16, 17).

$CD4^+$ T cells may also differentiate into several regulatory subsets, which actively inhibit the alloresponse and therefore constitute an essential part of peripheral tolerance. Regulatory T cells (T_{REG}) express the transcription factor FoxP3 and are known to suppress the alloreaction through modulation of antigen presentation, production of anti-inflammatory cytokines, as well as competition and cytolysis of effector T cells (18, 19). T-regulatory type 1 cells (T_{R1}) can be induced from naïve T cells upon TCR stimulation in the presence of IL-10; these cells downregulate the alloresponse by producing anti-inflammatory soluble mediators, including IL-10 and TGFβ (20).

B-cell mediated humoral immune responses also play an important role in alloimmunity, which is reflected by the elevated incidence of chronic antibody-mediated rejection in long-term graft recipients. B-cell differentiation into antibody-producing plasma cells and memory B cells after alloantigen uptake depend on cognate interactions with T cells. Conversely, following uptake through their B-cell receptor, B cells can mount alloantigens on class-II MHC and present them to T cells. Specific B-cell subsets have displayed immunoregulatory properties, and have been termed regulatory B cells (B_{REG}). B_{REG} are a heterogenous population that promotes the development of T_{REG}, suppresses effector T cell differentiation and converts T_H1 effectors into T_R1 cells, mainly through secretion of the anti-inflammatory cytokines IL-10 and TGFβ (21, 22).

THE INNATE IMMUNE SYSTEM: OLD AND EMERGING EVIDENCE OF A CENTRAL ROLE IN TRANSPLANT REJECTION

Despite the high specificity of the adaptive system for non-self-antigens, the set of allogeneic responses that ultimately lead to transplant rejection derives from a complex interplay with the innate immune system, an interaction that is much more intertwined than originally hypothesized.

Innate immunity relies on the activation of pattern recognition receptors (PRRs), which is induced by evolutionarily preserved molecular motifs from pathogens. However, also self-molecules released in response to cell stress can activate PRRs, triggering the innate immunity. These molecules are termed DAMPs (damage-associated molecular patterns) and are massively released in the context of organ transplantation in response to brain/cardiac death and ischemia-reperfusion injury (23, 24).

The best characterized cellular PRRs are toll-like receptors (TLRs), localized on the outer membrane and on the surface of intracellular vesicles, NOD-like receptors, which are part of the inflammasome complex, and C-type lectin receptors. DAMP-induced signaling through these receptors activates the transcription of specific genes, leading to the secretion of inflammatory cytokines and upregulation of adhesion and costimulatory molecules (25, 26).

DAMP-induced TLR signaling promotes the maturation of dendritic cells (DCs), which represents the pivotal link between innate and adaptive immunity. This process determines a critical switch in the biological functions of DCs in solid

---

Abbreviations: APC, antigen presenting cell; B_{REG}, regulatory B cell; DAMP, damage-associated molecular pattern; DC, dendritic cell; HO-1, heme-oxygenase 1; IDO, indoleamine 2,3-dioxygenase; IFNγ, interferon gamma; MLR, mixed-lymphocyte reaction; MSC, mesenchymal stromal cell; NK, natural killer; PGE_2, prostaglandin-E_2; PRR, pattern recognition receptor; T_{FH}, follicular helper T-cell; T_{FR}, T-follicular regulatory cell; TGFβ, transforming growth factor β; T_H, helper T cell; TLR, toll-like receptor; TNFα, tumor necrosis factor α; T_{R1}, T-regulatory type 1 cell; T_{REG}, regulatory T cell.

organ transplantation: donor DCs migrate from the graft to the lymphoid organs of the recipient and present intact donor MHC-antigen complexes to host T cells ("direct presentation"), inducing the intense activation and rapid proliferation of effector T cells that underlies early rejection episodes. In addition, mature recipient DCs infiltrating the graft can also present processed donor antigens to recipient T cells ("indirect presentation"), providing a weaker but long-lasting stimulation that can eventually result in both acute and chronic rejection (27, 28). In the setting of acute rejection, most of the DCs found in the graft derive from infiltrating monocytes that locally differentiate into antigen-presenting cells (29, 30).

Monocytes can also contribute to transplant rejection by differentiating into macrophages, which frequently represent the majority of graft-infiltrating cells during rejection (31). Similarly to other cell types, macrophages can damage the allograft or have immunoregulatory functions depending on their state of activation. Classically-activated M_1 macrophages are pro-inflammatory cells that develop upon TLR engagement in the presence of IFNγ, damaging the graft *via* direct cytotoxicity or by inducing a delayed-type hypersensitivity reaction. Notably, these effector functions were shown to be triggered by alloantigens, and cytotoxicity proved to be allospecific after T-helper priming (32, 33). On the other end of the spectrum, alternatively-activated M_2 macrophages, whose differentiation is guided by IL-4 and IL-13, produce anti-inflammatory cytokines (IL-10, TGFβ), induce regulatory T-cell differentiation and promote tissue repair, mitigating graft damage (34, 35).

MECHANISTIC INSIGHTS ON MSC AND TOLERANCE INDUCTION

MSC infusion has not only shown encouraging results in controlling autoimmunity *in vivo* (36), but has also consistently proven effective in prolonging allograft survival in multiple animal models of solid organ transplantation (37–39) (**Table 1**). Compelling evidence demonstrated that during the alloresponse these cells can tip the balance between effector and regulatory functions in favor of the latter. MSC reduce the host-vs.-graft response in part through contact-dependent regulation (49, 50) and, most importantly, by secreting soluble factors with paracrine immunomodulatory effects (51, 52). A constantly growing number of soluble mediators have been implicated in MSC-induced functions on the immune system, including indoleamine 2,3-dioxygenase (IDO) (53, 54), nitric oxide (55, 56), TGF-β (49, 57–59), prostaglandin-E_2 (PGE_2) (50, 60), heme-oxygenase 1 (HO-1) (61), galectins (62) and HLA-G5 (63). Indeed, MSC-conditioned medium and MSC-derived extracellular vesicles, which both contain these soluble bioproducts, have been increasingly studied as cell-free alternatives to MSC administration (64). In addition, a different approach has been explored by some researchers, who engineered MSC to overexpress and secrete several different soluble mediators, including IDO (65), TGFβ (66), IL-10 (67), and HO-1 (68–70). Preliminary evidence from animal models of solid organ transplantation showed a modest but consistent increase in the pro-tolerogenic activity of genetically modified MSC compared to their standard counterparts (**Table 2**). Soluble mediators act on multiple cell targets, resulting in dramatic changes in their phenotype and functions.

T Cells

Autologous and allogeneic MSC regulate the T-cell response to the graft via modulation of alloantigen presentation and through direct, antigen-independent effects on T cell. In their seminal work, Bartholomew et al. demonstrated for the first time that MSC can prevent not only T-cell alloreactivity, but also polyclonal activation and proliferation of baboon T cells induced by mitogens *in vitro* (40). This dose-dependent effect was also confirmed in mice with a series of elegant experiments assessing single-antigen reactivity (71, 72), which showed that MSC inhibition was not limited to antigen-specific T-cell clones, but determined a generalized division arrest anergy in both naïve and memory T cells, eventually resulting in reduced $CD8^+$ T-cell proliferation and cytotoxicity. Similar results were obtained with human autologous and allogeneic MSC, which were able to suppress T-cell proliferation due to both antigen-priming and polyclonal activators (73–75). Findings that MSC are able to inhibit memory T cells, including $CD8^+$ memory T cells, is of particular relevance, since a high frequency of alloreactive memory T cells before transplantation represents a barrier to tolerance induction (76–78) and threatens allograft survival, especially in the context of T-cell depleting induction therapy (79–82).

MSC can also modify the activity of helper T cells by rewiring their polarization. Addition of MSC to T-cell cultures stimulated by mitogens, primed by allogenic T cells, or under T_H1-differentiating conditions, suppressed T_H1-cell proliferation and activation, and promoted a strong T_H2 polarization along with increased IL-10 production (74, 83, 84). The same effect was observed in a rat model of corneal transplantation, in which MSC injection shifted the balance between T_H1- and T_H2-specific cytokines in favor of the latter (46). Moreover, MSC displayed the capacity to inhibit T_H17 generation by suppressing the transcription factor RORγt and to induce suppressive T_{REG} from differentiated T_H17 cells, both *in vitro* and *in vivo* (44, 50, 85). Although data regarding MSC effect on T_{FH} cells in solid organ transplantation are currently lacking, MSC were shown to suppress the differentiation and proliferation of cultured T_{FH} cells obtained from healthy human donors (86).

Alongside effector T-cell inhibition, MSC expand several subsets of regulatory cells, making them a unique tool to modulate the alloresponse. In murine models, MSC polarized T cells toward a $FoxP3^+$ T_{REG} phenotype both *in vitro* and *in vivo* (41, 45, 83), an effect that was evident also in human T-cell cultures (49, 63) and in transplant recipients undergoing MSC infusion (87). Recently, MSC were also reported to promote IL-10 secretion and to expand the T_R1 regulatory subset in allogeneic mixed-lymphocyte reactions (MLRs), which mediated immunosuppression *in vitro* through the PGE_2 and IDO pathways (61, 88).

TABLE 1 | MSC in preclinical transplant models.

Transplant model	MSC			Outcome	Immunological mechanism	References
	Origin	Timing[a]	Dose/Route			
Skin tx in baboons	Donor BM	Day 0	1–2 × 10^7 /kg	Significant prolongation of graft survival		(40)
Liver tx in rats (LEW in BN rats)	BM from syngeneic, donor or TP (Wistar) rats	Days 0, +1, +2, +3, +8, +12, +16 (7 doses)	2 × 10^6/dose, IV	Significant prolongation of graft survival irrespective whether MSC were of syngeneic, donor or TP origin	$Foxp3^+$ Treg generation	(41)
Kidney tx in mice (C57 in BALB/c)	Donor BM	Day 1	1 × 10^6, IV	Indefinite graft survival	IDO-dependent $Foxp3^+$ Treg generation	(38)
Heart tx in mice (C57 in BALB/c mice)	Donor BM	Day +1	1 × 10^6, IV	Indefinite graft survival by MSC in combination with low-dose rapamycin	Tolerogenic DC and $Foxp3^+$ Treg generation	(39)
Kidney tx in rats (F344 in LEW rats)	TP (SD) BM	Week 11	0.5 × 10^6, IV	Prevention from chronic renal graft dysfunction and injury (IF/TA)	Anti-inflammatory effects	(42)
Kidney tx in mice (BALB/c in sensitized[b] C57 mice)	Syngeneic BM	Day−1 or day−7 or double pre-tx infusion (days−7 and−1) or at day +2	0.5 × 10^6, IV	Significant prolongation of graft survival when MSC were given pre-transplant, acute graft rejection when MSC were given post-transplantation	$Foxp3^+$ Treg generation	(43)
Heart tx in rats (Wistar in F344 rats)	Donor BM	Day−7, 0, +1, +2, +3 (5 doses)	2 × 10^6/dose, IV	Significant prolongation of graft survival	Reduced pro-inflammatory and increased anti-inflammatory cytokine expression	(37)
Heart tx in mice (C57 into BALB/c mice)	Donor adipose tissue	Day−4	1 × 10^6, IV	Significant prolongation of graft survival by MSC in combination with MMF	Conversion into $Foxp3^+$ Tregs (by MMF) of Th17 cells induced by MSC-educated MDSC	(44)
Heart tx in mice (B6C3 in C57 mice)	Syngeneic or donor BM	Days−7 and−1	0.5 × 10^6, IV (portal vein day−7, tail vein day−1)	Significant prolongation of graft survival with either syngeneic or donor-derived MSC	$Foxp3^+$ Treg generation	(45)
Corneal tx in rats (Wistar in LEW rats)	Donor BM	Days−3,−2 and−1 or days 0, 1 and 2	5 × 10^6, IV	Significant prolongation of graft survival when MSC are given post-transplant either alone or combined with CNI	$Foxp3^+$ Treg generation	(46)
Corneal tx in mice (C57 in BALB/c mice)	Human BM	Days−7 and−3	1 × 10^6, IV	Significant prolongation of graft survival	Conversion of lung monocyte/macrophage toward an immune regulatory phenotype in a TSG-6-depedendent manner	(47)
Corneal tx in rats (DA in sensitized[b] LEW rats)	TP (Wistar Furth) BM	Days−7 and−1	1 × 10^6, IV	30-day rejection free in 64% MSC-treated animals compared to 0% in the control group	Induction of PGE_2/TGFβ-producing and immunosuppressive $CD45^+CD11b^+B220^+$ lung monocytes and Foxp3+Treg generation	(48)

[a]*From day of transplant (Day 0);*

[b]*Donor-sensitization by donor splenocyte injection prior to transplantation.*

BM, bone marrow; BN, Brown Norway; CNI, calcineurin inhibitor; DA, Dark Agouti; DC, dendritic cells; IDO, indoleamine 2,3-dioxygenase; IF/TA, interstitial fibrosis/tubular atrophy; IV, intravenous; LEW, Lewis; MMF, mycophenolate mofetil; PGE2, prostaglandin E2; SD, Sprague-Dawley; TGFβ, transforming growth factor β; TP, third-party; TSG-6, tumor necrosis factor-inducible gene 6; Tx, transplant.

TABLE 2 | Preclinical data on genetically-engineered MSC in solid organ transplantation.

References	Type of graft	Donor/recipient	MSC source	Mediator/ vector	Outcome	Immunological effects
He et al. (65)	Kidney	NZ/Japanese white rabbits	NZ rabbit (BM)	IDO/ lentivirus	IDO-MSC prolonged graft survival compared to standard MSC, prevented rejection and induced donor-specific tolerance to skin grafts.	Inhibition of T-cell proliferation, Treg induction and increased levels of tolerance-related cytokines. All effects were greater with IDO-MSC compared to standard MSC.
Tang et al. (66)	Liver	DA/Lewis rats	Lewis rats (BM)	TGFβ/ lentivirus	TGFβ-MSC reduced graft rejection and increased survival compared to standard MSC.	Reduced effector T-cell proliferation *in vivo*, Treg induction (but reduction in natural Treg) and increased Treg/T_H17 ratio in graft-infiltrating cells.
Niu et al. (67)	Liver	DA/Lewis rats	DA rats (BM)	IL-10/ lentivirus	IL-10-MSC increased graft survival and reduced the histological rejection activity index compared to standard MSC.	Increased FoxP3 expression in intragraft $CD4^+CD25^+$ T cells, reduced pro-inflammatory and increased anti-inflammatory cytokines.
Wu et al. (68)	Liver	Lewis/BN rats	BN rats (BM)	HO-1/ adenovirus	HO-1-MSC increased survival rates and attenuated acute rejection compared to standard MSC.	Increased Treg fraction in splenocytes, induction of an anti-inflammatory cytokine profile.
Wang et al. (69)	Liver (reduced-size)	Lewis/BN rats	BN rats (BM)	HO-1/ adenovirus	HO-1-MSC attenuated the rejection activity index compared to standard MSC.	Reduction in mTOR/ERK levels, along with increased autophagy-related proteins (LC3 and Beclin-1).
Yang et al. (70)	Small bowel	BN/Lewis rats	Lewis rats (BM)	HO-1/ adenovirus.	HO-1-MSC improved survival rates, clinical manifestations and acute rejection grading compared to standard MSC	Reduction of the pro-inflammatory cytokine milieu, increased Treg fraction in splenocytes.

BN, Brown Norway; BM, bone marrow; DA, Dark Agouti; HO-1, heme oxygenase-1; IDO, indoleamine 2,3-dioxygenase; IL-10, interleukin-10; MSC, mesenchymal stromal cells; NZ, New-Zealand; TGFβ, transforming growth factor β.

B Cells

Different reports have shown that MSC may block B-cell proliferation through cell-cycle arrest into the G0/G1 phase (89, 90). This alteration results in a strong inhibition of proliferation and maturation of B cells into plasmablasts, causing a steep reduction in antibody secretion (91, 92). Of particular interest for the transplantation field, the addition of MSC to standard allogeneic MLRs was shown to inhibit the formation of donor-specific antibodies (92). Interestingly, *in vitro* B-cell proliferation was not suppressed when MSC and B cells were cultured without T cells, suggesting that this effect is at least in part mediated by T-cell help (91). On the other hand, absence of T cells did not interfere with MSC-mediated inhibition of B-cell maturation and with B_{REG} induction (93).

Inflammatory licensing by IFNγ significantly modifies MSC effect on B-cell subsets: MSC cultured in standard conditions were shown to increase the percentage of B_{REG} along with IL-10 production, while MSC exposed to high IFNγ concentrations did not promote B_{REG} expansion, but instead induced a greater inhibition of B-cell proliferation through IDO metabolic effects (94, 95).

Although, to date, there is a lack of studies exploring MSC effect on the B-cell compartment in animal models of solid organ transplantation, MSC infusion prevented the formation of circulating donor-specific antibodies in rats undergoing allogeneic kidney transplantation, suggesting that these cells are able to modulate humoral responses *in vivo* (42). Moreover, long term immunophenotyping of kidney allograft recipients that received MSC infusions revealed increased naïve and $CD24^{HI}CD38^{HI}$ ("transitional") B-cell subsets, a phenotypic signature that was associated with spontaneous operational tolerance (87, 96). Additional characterization will be needed to clarify whether MSC may have increased the frequency of IL-10 producing B cells (i.e., those associated with B_{REG} phenotype and function), especially in the "transitional" B-cell fraction.

Dendritic Cells

MSC modulate DC phenotype and function at multiple levels, inducing pro-tolerogenic functions. Exposure to mouse MSC interferes with DC maturation, downregulating class II MHC and the costimulatory molecules CD40 and CD86 (97, 98). Moreover, MSC also impair DC homing to secondary lymphoid organs by reducing the expression of CCR7 and CD49dβ1 (99).

These effects have direct consequences on alloantigen presentation by both donor and recipient DCs: since MLR reactivity mostly depends on direct antigen presentation, the inhibition of proliferation observed in these cultures after MSC addition suggests an inhibitory effect on "donor" DCs (97). This was confirmed by the hypo-responsiveness displayed by human allogeneic T-cell responders to MSC-primed DCs compared to mature DCs (100–102). *In vitro* experiments with ovalbumin-specific TCR-transgenic mouse T cells demonstrated that MSC can also exert regulatory effects on ovalbumin-pulsed "recipient" DCs, leading to defective indirect alloantigen presentation to

$CD4^+$ T cells and reduced cross-presentation of allopeptides to $CD8^+$ T cells (98, 99).

DC maturation block and inefficient antigen presentation, coupled with lack of co-stimulation, induces a regulatory phenotype from conventional T cells, thereby increasing the relative abundancy of T_{REG} compared to effector T cells and inhibiting the alloresponse. In an allogeneic kidney transplant model, MSC infusion was associated with high frequency of immature, tolerogenic DCs, along with impaired donor-specific T-cell proliferation and enrichment of suppressive T_{REG} in secondary lymphoid organs and into the graft (38). Similar results had been previously observed in mice receiving a cardiac allograft, in which tolerance was achieved by combining MSC infusion with low-dose rapamycin (39).

Macrophages

Several models have shown that MSC can increase macrophage proliferation and migration, while inducing a pro-tolerogenic polarization shift at the same time: MSC-reprogrammed M_2 macrophages show reduced TNFα, IFNγ, and IL-12 secretion and increased IL-10 production, which inhibit effector T-cell responses and promote T_{REG} proliferation (59, 103–105).

Several soluble mediators secreted by MSC have been implicated in macrophage reprogramming, including IDO and PGE_2 (105, 106). Recent reports, however, have shown that also phagocytosis of apoptotic MSC or MSC microparticles by macrophages is sufficient to induce an M_2 polarization shift, partly through IDO upregulation in these phagocytes (107–109). Moreover, macrophage uptake of the cell debris deriving from MSC-induced T-cell apoptosis also induced TGFβ secretion, which facilitated T_{REG} expansion (110).

In vivo, MSC infusion before allogeneic corneal transplantation redirected macrophages toward an M_2 phenotype, which conferred protection against allograft rejection (47). In this model, adoptive transfer and selective depletion experiments suggested that the monocyte lineage was responsible for tolerance induction (47). In pre-sensitized rats, allogeneic MSC administration before corneal transplant significantly increased rejection-free survival, which was associated with an early rise in alternatively-activated macrophages, followed by an increase in T_{REG} at later time points (48).

DUAL FUNCTION OF MSC AND CROSSTALK WITH THE MICROENVIRONMENT

Although MSC can act as potent immunomodulators of inflammation during the alloresponse, in resting conditions they mainly display homeostatic properties, supporting cells residing in their niche (111). Moreover, MSC can also acquire a completely opposite function, releasing pro-inflammatory cytokines and even acting as antigen-presenting cells following class II MHC upregulation (112–114). The acquisition of either of these specific functions strictly depends on the microenvironment MSC encounter: to become immunosuppressive, MSC need to "licensed" by a series of inflammatory stimuli that commonly occur in the setting of solid organ transplantation, deriving from both the innate and adaptive immune response.

Cytokines are the best studied mediators involved in MSC licensing: binding of IFNγ (chiefly secreted by T-cells after T_{H1} polarization) to its specific receptors on MSC represents a critical step of this process (115, 116). Other cytokines, including the macrophage-derived TNFα, IL-1α and IL-1β, have been shown to potentiate the licensing effect of IFNγ (55). Overall, this stimulation results in the production of chemokines (such as CXCR3 and CCR5) and in the induction of IDO, which are respectively necessary for T-cell recruitment and inhibition. However, below-threshold IFNγ and TNFα concentrations have been proven insufficient to induce IDO, therefore precluding the suppression of recruited T-cells and paradoxically increasing the inflammatory response (117).

MSC functional destiny is also influenced by the dynamics of DAMP-associated signaling that accompanies tissue injury. The presence of a wide array of TLRs has been confirmed on the surface of both human and mouse MSC (118–120), with variable expression influenced by environmental cues, including hypoxia and inflammatory cytokines (114, 121). TLR stimulation promotes MSC migration to the site of inflammation and enhances their survival (122, 123), but conflicting data have been reported regarding a DAMP-mediated licensing effect through TLR3 and TLR4 stimulation (124, 125). Recent evidence suggests that MSC can be differentially activated based on the type of TLR triggered: transient TLR4 engagement by low lipopolysaccharide concentrations polarizes MSC toward a proinflammatory phenotype, stimulating the secretion of IL-6 and IL-8, recruiting neutrophils and inducing T-cell activation in co-culture experiments (126, 127). On the other hand, sub-maximal triggering of TLR3 guides the differentiation into immunomodulating MSC, which upregulate IDO, secrete IL-10, CXCL10 and CCL5, and induce T-cell suppression *in vitro* (126).

In addition to its multiple pro-inflammatory effects, activation of the complement cascade also has direct effects on MSC activity. MSC are resistant to complement-mediated lysis due to expression and secretion of negative regulators (128, 129) but, at the same time, ligation of complement anaphylatoxins to their surface receptors enhances MSC resistance to oxidative stress and apoptosis (130). Although a direct licensing effect of complement has not been described, C3a and C5a act as potent chemotactic agents for MSC and recruit them at the site of inflammation, where functional polarization takes place (131, 132).

TIMING AS A KEY FACTOR FOR MSC-INDUCED TOLERANCE

MSC survival is limited both in culture systems and following *in vivo* administration (45, 133). This peculiar feature seems antithetic to the development of long-term tolerance after a single MSC infusion, unless one takes into account that these cells may determine a multi-level protolerogenic polarization within a limited time frame. Therefore, the reduced lifespan of MSC raises by itself the issue of optimal infusion timing in allograft recipients. This concern was further corroborated by the first evidence that the timing of MSC infusion dictates their pro-tolerogenic effect: in an allogeneic rat model of heart transplantation, donor MSC infused 4 days before the transplant induced acceptance of the graft, as opposed to infusions on the same day or after 3 days from the procedure (134).

Building upon data indicating that systemically infused MSC tend to migrate to damaged tissues (such as those exposed to ischemia/reperfusion injury) (135–137), we hypothesized that timing of infusion may impact on MSC localization, and confirmed that administration of MSC after transplantation results in migration of these cells inside the graft, rather than toward secondary lymphoid organs (43). In addition, the site at which MSC are recruited tightly correlates with their immunomodulatory properties: indeed, migration of MSC into kidney allografts following post-transplant infusion was associated with a proinflammatory phenotype and resulted in neutrophilic infiltration, complement deposition and acute kidney injury, both in animal models and humans (43, 87). The events leading to this polarization are still ill-defined, but the inflammatory microenvironment and the massive release of DAMPs are likely key factors. On the other hand, administration of MSC before renal transplantation led to localization of these cells into secondary lymphoid organs, where they promoted the formation of a pro-tolerogenic environment and prolonged graft survival (43).

Localization by itself, however, does not fully explain the opposing characteristics that MSC may acquire in the setting of solid organ transplantation. We recently showed that experimental inhibition of C3a and C5a receptors on MSC infused after kidney transplantation prevents intra-graft migration and allows homing of these cells to secondary lymphoid organs. However, despite correct localization and reduction of donor-specific T-cell alloreactivity, MSC failed to induce T_{REG} generation in this model and resulted only in limited prolongation of graft survival (132).

These experiments confirmed previous evidence suggesting that the effect of MSC on alloreactive T cells also depends on the degree of maturity these T cells display. Indeed, when MSC were added to MLRs after the primary stimulation step had already taken place, $CD8^+$ T-cell mediated cytotoxicity was unhindered, suggesting that a complete polarization favors the escape from MSC-mediated immunoregulation (138). Similarly, the *in vitro* regulatory effect of MSC on T_H1/T_H17 proliferation, activation and differentiation was progressively lost when MSC were added with increasing delay from the start of the polarization process (83).

Consistent with effects observed in mice, MSC added at later time points *in vitro* were also unable to convert terminally differentiated conventional T cells into T_{REG} (83, 139). Among mediators involved in T_{REG} induction, the trophic cytokine IL-2 seems to be critical for the whole process to occur (140, 141): since neither MSC nor T_{REG} secrete IL-2, the concentration of this cytokine relies on conventional T-cell production. Studies on the kinetics of IL-2 secretion during effector T-cell differentiation elucidated how terminal effectors progressively lose their ability to secrete this cytokine (142), resulting in a steep decline in IL-2 concentrations after 2 days in a conventional MLR (143). Thus, it is tempting to speculate that MSC may fail to induce T_{REG} from terminally differentiated conventional T cells due to low IL-2 concentrations, but other concomitant factors are likely to play a role. For instance, timing seems also involved in MSC-mediated induction of an immature, tolerogenic DC phenotype: addition of lipopolysaccharide to MSC-DC co-cultures, mimicking the DAMP-induced TLR triggering that accompanies graft transplantation, reduced the inhibitory effect of MSC on DC maturation (144, 145). As previously described, the acquisition of a mature DC phenotype prevents T_{REG} expansion, thus providing an additional plausible explanation of the reduced immunomodulating ability of MSC infused post-transplantation.

Failure to expand T_{REG} has been frequently associated with graft rejection in animal models (38, 43), suggesting a central role for suppression in MSC-induced tolerance. Thus, infusion protocols aimed at expanding the T_{REG} pool should have the highest potential of promoting a protolerogenic environment. Pre-transplant MSC administration has the best chances to achieve this aim, even though T_{REG} expansion in the absence of antigenic pressure from the graft produces a broad repertoire, which includes both donor-specific and non-alloreactive regulatory T cells.

Indeed, antigen-pressure is not necessary to induce suppressive T_{REG} from conventional T cells in MSC co-culture (146). However, FoxP3 expression in these cells is not stable, suggesting that polyclonal T_{REG} induction without alloantigen presentation does not provide the required survival and proliferation stimuli for a stable long-term expansion (146). *In vivo* administration of MSC to healthy, non-transplanted mice expands the T_{REG} pool (43), and adoptive transfer of splenocytes from these mice to syngeneic transplant recipients induces tolerance to the allograft (45), consistent with the notion that polyclonally-expanded T_{REG} can modulate the alloresponse. However, after graft tolerance has been established, T_{REG}-mediated suppression becomes donor-specific, without affecting the response to third-party alloantigen presentation *in vivo* (45). These findings suggest that MSC infused before the transplant and localizing to secondary lymphoid organs induce a polyclonal, antigen-independent expansion of T_{REG}. Antigenic pressure from the graft then provides the required survival and expansion stimuli to donor-specific T_{REG}, while non-alloreactive T_{REG} clones are progressively lost over time (**Figure 1**).

MSC AND SOLID ORGAN TRANSPLANTATION: OPEN ISSUES FOR CLINICAL TRANSLATION

Cellular therapy with MSC has been tested so far in phase 1 clinical studies in kidney, liver and lung transplantation (**Table 3**) using either autologous bone marrow- (BM-)derived (87, 147–151), allogeneic BM-derived (152, 155–158) or umbilical cord-derived MSC (154). These studies have indicated the safety, feasibility and tolerability of the procedure in all settings, including the intravenous injection of MSC in lung transplant

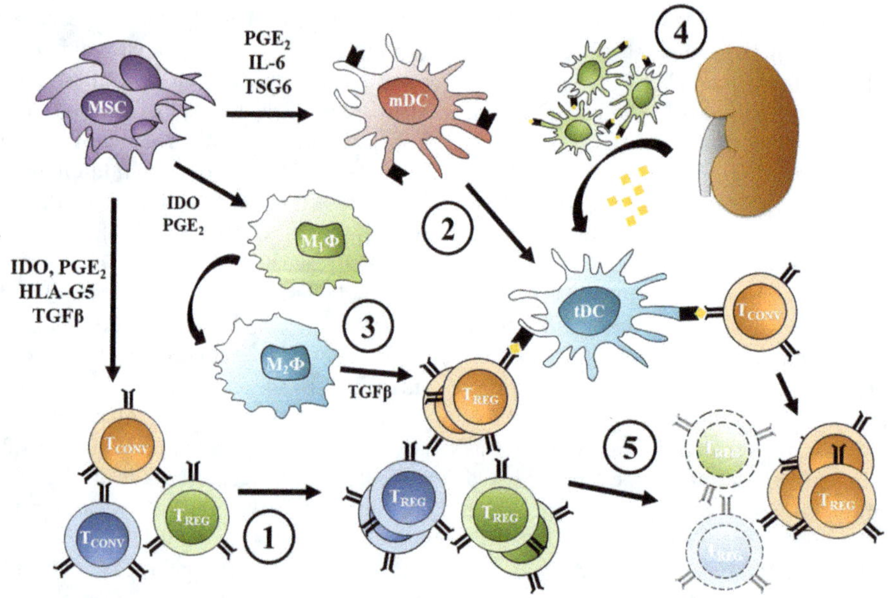

FIGURE 1 | Expansion of donor-specific T_{REG} following pre-transplant MSC infusion. (1) MSC infused before transplant induce T_{REG} from conventional T cells and promote a polyclonal, antigen-independent expansion; at the same time, (2) MSC inhibit dendritic cell maturation increasing the frequency of protolerogenic DCs, and (3) reprogram macrophages toward an M_2 phenotype. (4) Allogeneic transplantation causes migration of donor DCs from the graft to lymphoid organs and provides alloantigen for recipient DC uptake (other antigen presenting cells can act the same way). (5) Direct and indirect antigen presentation provide the survival stimuli necessary for donor-specific T_{REG} proliferation, while the non-alloreactive T_{REG} pool is progressively lost over time. Protolerogenic DCs can also induce T_{REG} from conventional alloreactive T cells, while alternatively-activated macrophage TGFβ secretion promotes T_{REG} expansion. MSC, mesenchymal stromal cell; T_{CONV}, conventional T cell; T_{REG}, regulatory T cell; mDC, mature dendritic cell; tDC, tolerogenic dendritic cell; $M_1 \phi$, classically-activated macrophage; $M_2 \phi$, alternatively-activated macrophage; PGE2, prostaglandin-E_2; IDO, indoleamine 2,3-dioxygenase; TGFβ, transforming growth factor β; HLA-G5, human leukocyte antigen G5; IL-6, interleukin 6; TSG6, tumor necrosis factor-inducible gene 6 protein.

recipients with moderate obstructive chronic lung allograft rejection (158). All MSC-treated patients showed good graft function over the 1–2 years-follow-up.

Even though an initial study in kidney transplantation reported a high incidence of opportunistic infections in MSC-treated kidney transplant recipients (150), subsequent studies allayed this concern. In kidney and liver transplant recipients MSC infusion was associated with a reduced (147) or unchanged (87, 155) incidence of opportunistic infections in the first 1–2 years post-transplant compared to controls. In our kidney transplant patients given autologous BM-derived MSC we did not observe increased susceptibility to infection or neoplasm even in the long term (5–7-years follow-up) (87).

The hypothetical increased risk of malignancy following MSC administration constitutes an additional matter of debate (159). This concern originally stemmed from the observation that murine MSC may undergo spontaneous malignant transformation during *ex vivo* cultures (160). However, this feature of murine MSC is not shared by their human counterparts, which are not prone to maldifferentiation even after long-term *in vitro* expansion (159, 161). Despite the lack of malignant potential, human MSC may still theoretically promote the growth of pre-existing undiagnosed tumors in transplant recipients. Preclinical evidence in this context is not univocal, with MSC effects varying from neoplastic facilitation to anti-tumoral activity (159, 162–164). Even though studies carried out so far did not detect any association between MSC infusion and malignancy (165), long-term surveillance systems need to be established in order to gauge a precise estimate of adverse events related to MSC administration.

A major concern regarding the use of third-party MSC therapy refers to possible recipient sensitization due to development of antibodies against allogeneic MSC, a phenomenon demonstrated in animal models (166, 167). This fundamental issue, however, has been considered only by one recent study in kidney transplantation (155): ten kidney transplant recipients from deceased donor were given third-party BM-derived MSC on day 3 ± 2 post-transplant. During the 12 months-follow up, four MSC-treated patients developed *de novo* antibodies against MSC or shared kidney donor-MSC HLA, albeit at very low levels and with only one patient showing *de novo* MSC-donor-specific antibodies with MFI > 1,500. The clinical significance of these alloantibodies is uncertain and further studies with longer follow-up are needed to definitely address the issue of possible sensitization after allogeneic MSC injection.

The available studies in kidney transplantation also provide some evidence of potential efficacy of the MSC therapy in enabling minimization of induction (147) and maintenance (152, 153) immunosuppressive drugs, in inducing a pro-tolerogenic environment (87, 148, 149) and in repairing chronic allograft damage (150).

TABLE 3 | Summary of clinical trials assessing MSC infusion in the setting of renal, liver and pulmonary transplantation.

References	Number of patients	Source of MSC	Timing	MSC dose	Clinical outcome	Immunological outcomes
KIDNEY TRANSPLANTATION						
Tan et al. (147)	105 MSC 51 controls	Autologous BM	Day 0, day +14	1–2·× 10^6/Kg	Safety and feasibility MSC can replace induction therapy with basiliximab and allow the reduction of CNI dose	N/A
Perico et al. (87, 148, 149)	5	Autologous BM	Day +7 (*n* = 2) Day−1 (*n* = 3)	1–2·× 10^6/Kg	Tolerability and feasibility Safetyproviding that MSC are infused at day−1	High ratio Treg/memory $CD8^+$ T cells, donor-specific $CD8^+$ T cell unresponsiveness, safe IS withdrawal in 1 patient
Reinders et al. (150)	6	Autologous BM	Month 6–10 post-transplant (2 infusions)	1·× 10^6/Kg	Safety, tolerability and feasibility Resolution of IF/TA in 2 patients Development of opportunistic infections in 3 patients	Decreased proliferation and cytokine production in response to donor cells
Mudrabettu et al. (151)	4	Autologous BM	Day−1 and day +30	0.2–0.3 (low dose) or 2.1–2.8 (high dose) ×·10^6/Kg	Safety, tolerability and feasibility	Trend of increased peripheral Treg percentages, reduced polyclonal $CD4^+$ T cell proliferation
Peng et al. (152), Pan et al. (153)	16	Allogeneic BM	Day 0, day +30	5·× 10^6 intragraft, then 2·× 10^6/Kg	Safety, tolerability and feasibility MSC allow 50% reduction of CNI dose	No significant changes in Treg percentages compared to controls or to basal values
Sun et al. (154)	21 MSC 21 controls	Umbilical cord	Day−1 and day 0	2·× 10^6/Kg pre-Tx, 5·× 10^6 intragraft	Safety, tolerability and feasibility	N/A
Erpicum et al. (155)	10 MSC 10 controls	Allogeneic BM	Day 3 ± 2	1.5–3.0·× 10^6/Kg	Safety, tolerability and feasibility Potential long-term immunization against MSC	Increased frequency of Treg
LIVER TRANSPLANTATION						
Soeder et al. (156)	1	Allogeneic BM	Day 0, day +2	150·× 10^6	Clinically diagnosed acute rejection at day 6 and biopsy proven acute rejection at day 219	Increased frequency of Tregs and reduced expression of HLA-DR on $CD14^+$ monocytes
Detry et al. (157)	10	Allogeneic BM	Day 3 ± 2	1.5–3.0·× 10^6/Kg	Safety, tolerability and feasibility Failure to withdraw IS	No significant changes in Treg counts or phenotype compared to controls
LUNG TRANSPLANTATION						
Keller et al. (158)	9	Allogeneic BM	–	1, 2 or 4·× 10^6/Kg (three groups)	Safety, tolerability and feasibility	N/A

BM, bone marrow; CNI, calcineurin inhibitor; IF/TA, interstitial fibrosis/tubular atrophy; IS, immunosuppression; MSC, mesenchymal stromal cells; N/A, not available.

In a large clinical trial conducted in China, autologous BM-MSC infusions (at reperfusion and 2 weeks later, 1–2 × 10^6/Kg each) could safely replace induction therapy with the anti-IL-2R antibody basiliximab and enable 20% CNI dose reduction in living donor kidney transplantation (147). In a subsequent study from China, allogeneic BM-derived MSC allowed a 50% reduction of tacrolimus dose used as maintenance therapy. Patients given MSC showed graft survival, incidence of acute rejection and graft function similar to the control patients receiving a full tacrolimus dose (153).

We are currently conducting clinical studies of MSC therapy in living (NCT00752479 and NCT02012153) (87, 148, 149) and deceased donor kidney (NCT 02565459) and liver (NCT02260375) transplantation. In our experience with autologous BM-derived MSC infusion in living donor kidney transplant recipients we showed that timing of cell infusion was crucial in establishing the eventual effect of MSC. MSC infusion at day 7 after transplantation on top of induction therapy with low dose thymoglobulin and basiliximab, resulted in transient renal insufficiency in the first two patients (148), likely caused by MSC migration toward the inflamed graft with the consequent activation of their pro-inflammatory phenotype. We therefore modified the protocol and the subsequent two patients were given infusion of autologous MSC the day before kidney transplantation (149). In this setting, we removed basiliximab from the induction therapy to avoid any deleterious effect of the anti-IL-2R antibody on T_{REG} possibly expanded by MSC. MSC infusion was no longer associated with acute renal insufficiency (149) but the basiliximab-free regimen exposed patients to an increased risk of acute rejection. The protocol was therefore further modified by re-introducing basiliximab induction therapy in the setting of pre-transplant (day−1) MSC infusion. Two additional patients have been treated so far with this protocol and results of 1 year follow up of the first patient have been recently published (87). The first four patients are now at 5–7-years follow-up with stable graft function and no major side effect (87). In these patients we performed extensive longitudinal immunological studies and were able to document a long lasting increase in the ratio between T_{REG} and memory $CD8^+$ T cells, along with a persistent reduction of donor-specific $CD8^+$ T-cell cytotoxicity *ex-vivo*. These parameters were particularly evident in two patients and were associated with high levels of circulating naïve and transitional B cells, parameters proposed as the B-cell signature of spontaneous and induced tolerance to kidney transplantation (96). In one of these two patients, who was also showing normal histology at 1-year protocol biopsy and no evidence of *de novo* donor-specific antibodies, we attempted immunosuppressive drug withdrawal. Cyclosporin was successfully withdrawn, mycophenolate mofetil was gradually tapered, and the patient is currently free from immunosuppression (Perico et al. manuscript submitted).

Analogously to our approach, the group of Detry et al. attempted immunosuppression withdrawal in liver transplant patients given allogeneic MSC infusion, despite the lack of laboratory evidence of a pro-tolerogenic environment (unchanged percentages of total T_{REG} or their naïve, resting, activated and proliferating subsets) in MSC-treated patients compared to controls. Immunosuppression withdrawal, attempted in nine MSC recipients, was unsuccessful and immunosuppressive drug discontinuation was not achieved (157). Ineffective MSC-mediated immunomodulation in these patients could be due, at least in part, to the absence of a peri-transplant T-cell depleting induction therapy, which would have likely enabled MSC to further expand T_{REG} while restraining memory T cells at the same time.

Overall, the available studies indicate the safety and feasibility of MSC infusion of both autologous and allogeneic source in solid organ transplant recipients. Some suggestions of possible therapeutic efficacy derived from the evidence that MSC allowed minimization of immunosuppression and could promote a T_{REG}-mediated pro-tolerogenic environment.

Many open issues such as source, number of infusions, extent of *in vitro* expansion and concomitant immunosuppression regimens still need to be addressed before MSC can become a standard treatment protocol for transplant recipients. In particular, large variations in the dose of MSC infusions are a common occurrence in clinical trials, since the choice is empirical. No direct comparison of MSC doses has been performed in solid organ transplantation so far, but a clinical trial on graft-vs.-host disease showed that amounts ranging from 0.4 to 9.0 $\times \cdot 10^6$/Kg were not only safe and well tolerated, but also had a similar immunomodulatory effect (168). On the other hand, lack of clinical trial uniformity regarding timing of MSC infusion likely reflects organizational issues rather than insufficient scientific evidence. In light of preclinical and clinical data available, MSC infusion should be always programmed before the transplantation procedure in kidney allograft recipients, in order to avoid potential graft damage resulting from infiltrating proinflammatory MSC. Similarly, the lack of tolerance induction in liver transplant recipients suggests that post-transplant MSC infusion does not efficiently modulate the alloresponse, an effect that might be obtained instead by a pre-transplant administration schedule.

Another fundamental question remains to be addressed, i.e., the final aim of using MSC cellular therapy in solid organ transplantation. We think that MSC should not be adopted to prevent acute allograft rejection, since this phenomenon is well controlled by conventional, less expensive and effective immunosuppressive drugs. Instead, MSC should be harnessed for complementing the tolerogenic potential of induction therapies, which should still be used for the prevention of acute graft rejection, allowing MSC to create a pro-tolerogenic environment at later stages. To this purpose, future efforts should focus on a deeper understanding of mechanism of action of MSC and their interplay with immune cell subsets, with the aim to identify biomarkers of response to MSC therapy that may allow to select patients amenable to safe immunosuppressive drug withdrawal.

AUTHOR CONTRIBUTIONS

MP and FC searched the literature and wrote the manuscript. GR critically revised the manuscript.

ACKNOWLEDGMENTS

This work was supported by the European Union's Horizon 2020 research and innovation programme under grant agreement No 634086. The materials presented and views expressed here are the responsibility of the authors only. The EU commission takes no responsibility for any use made of the information set out.

REFERENCES

1. Lodhi SA, Lamb KE, Meier-Kriesche HU. Solid organ allograft survival improvement in the United States: the long-term does not mirror the dramatic short-term success. *Am J Transplant.* (2011) 11:1226–35. doi: 10.1111/j.1600-6143.2011.03539.x
2. Kim SI. Bacterial infection after liver transplantation. *World J Gastroenterol.* (2014) 20:6211–20. doi: 10.3748/wjg.v20.i20.6211
3. Rama I, Grinyó JM. Malignancy after renal transplantation: the role of immunosuppression. *Nat Rev Nephrol.* (2010) 6:511–9. doi: 10.1038/nrneph.2010.102
4. Dominici M, Le Blanc K, Mueller I, Slaper-Cortenbach I, Marini F, Krause D, et al. Minimal criteria for defining multipotent mesenchymal stromal cells. The International Society for Cellular Therapy position statement. *Cytotherapy.* (2006) 8:315–7. doi: 10.1080/14653240600855905
5. Sarugaser R, Lickorish D, Baksh D, Hosseini MM, Davies JE. Human umbilical cord perivascular (HUCPV) cells: a source of mesenchymal progenitors. *Stem Cells.* (2005) 23:220–9. doi: 10.1634/stemcells.2004-0166
6. Batsali AK, Kastrinaki M-C, Papadaki HA, Pontikoglou C. Mesenchymal stem cells derived from Wharton's Jelly of the umbilical cord: biological properties and emerging clinical applications. *Curr Stem Cell Res Ther.* (2013) 8:144–55. doi: 10.2174/1574888X11308020005
7. In't Anker PS, Scherjon SA, Kleijburg-van der Keur C, Noort WA, Claas FHJ, Willemze R, et al. Amniotic fluid as a novel source of mesenchymal stem cells for therapeutic transplantation. *Blood.* (2003) 102:1548–9. doi: 10.1182/blood-2003-04-1291
8. Orbay H, Tobita M, Mizuno H. Mesenchymal stem cells isolated from adipose and other tissues: basic biological properties and clinical applications. *Stem Cells Int.* (2012) 2012:461718. doi: 10.1155/2012/461718
9. La Noce M, Paino F, Spina A, Naddeo P, Montella R, Desiderio V, et al. Dental pulp stem cells: state of the art and suggestions for a true translation of research into therapy. *J Dent.* (2014) 42:761–8. doi: 10.1016/j.jdent.2014.02.018
10. Wood KJ, Goto R. Mechanisms of rejection: current perspectives. *Transplantation.* (2012) 93:1–10. doi: 10.1097/TP.0b013e31823cab44
11. Jiang S, Herrera O, Lechler RI. New spectrum of allorecognition pathways: implications for graft rejection and transplantation tolerance. *Curr Opin Immunol.* (2004) 16:550–7. doi: 10.1016/j.coi.2004.07.011
12. Goldman M, Le Moine A, Braun M, Flamand V, Abramowicz D. A role for eosinophils in transplant rejection. *Trends Immunol.* (2001) 22:247–51. doi: 10.1016/S1471-4906(01)01893-2
13. Liu Z, Fan H, Jiang S. CD4+ T-cell subsets in transplantation. *Immunol Rev.* (2013) 252:183–91. doi: 10.1111/imr.12038
14. Yan L, de Leur K, Hendriks RW, van der Laan LJW, Shi Y, Wang L, et al. T Follicular Helper Cells as a new target for immunosuppressive therapies. *Front Immunol.* (2017) 8:1510. doi: 10.3389/fimmu.2017.01510
15. Chen M, Lin X, Cheukfai L, Olsen N, He X, Zheng SG. Advances in T follicular helper and T follicular regulatory cells in transplantation immunity. *Transplant Rev (Orlando).* (2018) 32:187–93. doi: 10.1016/j.trre.2018.07.002
16. Walters GD, Vinuesa CG. T Follicular Helper Cells in transplantation. *Transplantation.* (2016) 100:1650–5. doi: 10.1097/TP.0000000000001217
17. Chenouard A, Chesneau M, Bui Nguyen L, Le Bot S, Cadoux M, Dugast E, et al. Renal operational tolerance is associated with a defect of blood Tfh cells that exhibit impaired B cell help. *Am J Transplant.* (2017) 17:1490–501. doi: 10.1111/ajt.14142
18. Caridade M, Graca L, Ribeiro RM. Mechanisms underlying $CD4^+$ Treg immune regulation in the adult: from experiments to models. *Front Immunol.* (2013) 4:378. doi: 10.3389/fimmu.2013.00378
19. Waldmann H, Hilbrands R, Howie D, Cobbold S. Harnessing $FOXP3^+$ regulatory T cells for transplantation tolerance. *J Clin Invest.* (2014) 124:1439–45. doi: 10.1172/JCI67226
20. Battaglia M, Gregori S, Bacchetta R, Roncarolo M-G. Tr1 cells: from discovery to their clinical application. *Semin Immunol.* (2006) 18:120–7. doi: 10.1016/j.smim.2006.01.007
21. Wortel CM, Heidt S. Regulatory B cells: phenotype, function and role in transplantation. *Transpl Immunol.* (2017) 41:1–9. doi: 10.1016/j.trim.2017.02.004
22. Najar M, Raicevic G, Fayyad-Kazan H, Bron D, Toungouz M, Lagneaux L. Mesenchymal stromal cells and immunomodulation: a gathering of regulatory immune cells. *Cytotherapy.* (2016) 18:160–71. doi: 10.1016/j.jcyt.2015.10.011
23. Fung A, Zhao H, Yang B, Lian Q, Ma D. Ischaemic and inflammatory injury in renal graft from brain death donation: an update review. *J Anesth.* (2016) 30:307–16. doi: 10.1007/s00540-015-2120-y
24. van Werkhoven MB, Damman J, van Dijk MCRF, Daha MR, de Jong IJ, Leliveld A, et al. Complement mediated renal inflammation induced by donor brain death: role of renal C5a-C5aR interaction. *Am J Transplant.* (2013) 13:875–82. doi: 10.1111/ajt.12130
25. Todd JL, Palmer SM. Danger signals in regulating the immune response to solid organ transplantation. *J Clin Invest.* (2017) 127:2464–72. doi: 10.1172/JCI90594
26. Fukata M, Vamadevan AS, Abreu MT. Toll-like receptors (TLRs) and Nod-like receptors (NLRs) in inflammatory disorders. *Semin Immunol.* (2009) 21:242–53. doi: 10.1016/j.smim.2009.06.005
27. Hubo M, Trinschek B, Kryczanowsky F, Tüttenberg A, Steinbrink K, Jonuleit H. Costimulatory molecules on immunogenic versus tolerogenic human dendritic cells. *Front Immunol.* (2013) 4:82. doi: 10.3389/fimmu.2013.00082
28. Sagoo P, Lombardi G, Lechler RI. Relevance of regulatory T cell promotion of donor-specific tolerance in solid organ transplantation. *Front Immunol.* (2012) 3:184. doi: 10.3389/fimmu.2012.00184
29. Zuidwijk K, de Fijter JW, Mallat MJK, Eikmans M, van Groningen MC, Goemaere NN, et al. Increased influx of myeloid dendritic cells during acute rejection is associated with interstitial fibrosis and tubular atrophy and predicts poor outcome. *Kidney Int.* (2012) 81:64–75. doi: 10.1038/ki.2011.289
30. Zhuang Q, Liu Q, Divito SJ, Zeng Q, Yatim KM, Hughes AD, et al. Graft-infiltrating host dendritic cells play a key role in organ transplant rejection. *Nat Commun.* (2016) 7:12623. doi: 10.1038/ncomms12623
31. Hancock WW, Thomson NM, Atkins RC. Composition of interstitial cellular infiltrate identified by monoclonal antibodies in renal biopsies of rejecting human renal allografts. *Transplantation.* (1983) 35:458–63. doi: 10.1097/00007890-198305000-00013
32. Liu W, Xiao X, Demirci G, Madsen J, Li XC. Innate NK cells and macrophages recognize and reject allogeneic nonself *in vivo* via different mechanisms. *J Immunol.* (2012) 188:2703–11. doi: 10.4049/jimmunol.1102997
33. Zecher D, van Rooijen N, Rothstein DM, Shlomchik WD, Lakkis FG. An innate response to allogeneic nonself mediated by monocytes. *J Immunol.* (2009) 183:7810–6. doi: 10.4049/jimmunol.0902194
34. Salehi S, Reed E. The divergent roles of macrophages in solid organ transplantation. *Curr Opin Organ Transpl.* (2015) 20:446–53. doi: 10.1097/MOT.0000000000000209
35. Schmidt A, Zhang X-M, Joshi RN, Iqbal S, Wahlund C, Gabrielsson S, et al. Human macrophages induce CD4(+)Foxp3(+) regulatory T cells via binding and re-release of TGF-β. *Immunol Cell Biol.* (2016) 94:747–62. doi: 10.1038/icb.2016.34
36. Pistoia V, Raffaghello L. Mesenchymal stromal cells and autoimmunity. *Int Immunol.* (2017) 29:49–58. doi: 10.1093/intimm/dxx008
37. Zhou HP, Yi DH, Yu SQ, Sun GC, Cui Q, Zhu HL, et al. Administration of donor-derived mesenchymal stem cells can prolong the survival

of rat cardiac allograft. *Transplant Proc.* (2006) 38:3046–51. doi: 10.1016/j.transproceed.2006.10.002

38. Ge W, Jiang J, Arp J, Liu W, Garcia B, Wang H. Regulatory T-cell generation and kidney allograft tolerance induced by mesenchymal stem cells associated with indoleamine 2,3-dioxygenase expression. *Transplantation.* (2010) 90:1312–20. doi: 10.1097/TP.0b013e3181fed001
39. Ge W, Jiang J, Baroja ML, Arp J, Zassoko R, Liu W, et al. Infusion of mesenchymal stem cells and rapamycin synergize to attenuate alloimmune responses and promote cardiac allograft tolerance. *Am J Transplant.* (2009) 9:1760–72. doi: 10.1111/j.1600-6143.2009.02721.x
40. Bartholomew A, Sturgeon C, Siatskas M, Ferrer K, McIntosh K, Patil S, et al. Mesenchymal stem cells suppress lymphocyte proliferation *in vitro* and prolong skin graft survival *in vivo*. *Exp Hematol.* (2002) 30:42–8. doi: 10.1016/S0301-472X(01)00769-X
41. Wang Y, Zhang A, Ye Z, Xie H, Zheng S. Bone marrow-derived mesenchymal stem cells inhibit acute rejection of rat liver allografts in association with regulatory T-cell expansion. *Transplant Proc.* (2009) 41:4352–6. doi: 10.1016/j.transproceed.2009.08.072
42. Franquesa M, Herrero E, Torras J, Ripoll E, Flaquer M, Gomà M, et al. Mesenchymal stem cell therapy prevents interstitial fibrosis and tubular atrophy in a rat kidney allograft model. *Stem Cells Dev.* (2012) 21:3125–35. doi: 10.1089/scd.2012.0096
43. Casiraghi F, Azzollini N, Todeschini M, Cavinato RA, Cassis P, Solini S, et al. Localization of mesenchymal stromal cells dictates their immune or proinflammatory effects in kidney transplantation. *Am J Transplant.* (2012) 12:2373–83. doi: 10.1111/j.1600-6143.2012.04115.x
44. Obermajer N, Popp FC, Soeder Y, Haarer J, Geissler EK, Schlitt HJ, et al. Conversion of Th17 into IL-17Aneg regulatory T cells: a novel mechanism in prolonged allograft survival promoted by mesenchymal stem cell-supported minimized immunosuppressive therapy. *J Immunol.* (2014) 193:4988–99. doi: 10.4049/jimmunol.1401776
45. Casiraghi F, Azzollini N, Cassis P, Imberti B, Morigi M, Cugini D, et al. Pretransplant infusion of mesenchymal stem cells prolongs the survival of a semiallogeneic heart transplant through the generation of regulatory T cells. *J Immunol.* (2008) 181:3933–46. doi: 10.4049/jimmunol.181.6.3933
46. Jia Z, Jiao C, Zhao S, Li X, Ren X, Zhang L, et al. Immunomodulatory effects of mesenchymal stem cells in a rat corneal allograft rejection model. *Exp Eye Res.* (2012) 102:44–9. doi: 10.1016/j.exer.2012.06.008
47. Ko JH, Lee HJ, Jeong HJ, Kim MK, Wee WR, Yoon S-O, et al. Mesenchymal stem/stromal cells precondition lung monocytes/macrophages to produce tolerance against allo- and autoimmunity in the eye. *Proc Natl Acad Sci USA.* (2016) 113:158–63. doi: 10.1073/pnas.15229 05113
48. Lohan P, Murphy N, Treacy O, Lynch K, Morcos M, Chen B, et al. Third-party allogeneic mesenchymal stromal cells prevent rejection in a pre-sensitized high-risk model of corneal transplantation. *Front Immunol.* (2018) 9:2666. doi: 10.3389/fimmu.2018.02666
49. English K, Ryan JM, Tobin L, Murphy MJ, Barry FP, Mahon BP. Cell contact, prostaglandin E(2) and transforming growth factor beta 1 play non-redundant roles in human mesenchymal stem cell induction of CD4+CD25(High) forkhead box P3^{+} regulatory T cells. *Clin Exp Immunol.* (2009) 156:149–60. doi: 10.1111/j.1365-2249.2009.03874.x
50. Duffy MM, Pindjakova J, Hanley SA, McCarthy C, Weidhofer GA, Sweeney EM, et al. Mesenchymal stem cell inhibition of T-helper 17 cell- differentiation is triggered by cell-cell contact and mediated by prostaglandin E2 via the EP4 receptor. *Eur J Immunol.* (2011) 41:2840–51. doi: 10.1002/eji.201141499
51. Gao F, Chiu SM, Motan DAL, Zhang Z, Chen L, Ji H-L, et al. Mesenchymal stem cells and immunomodulation: current status and future prospects. *Cell Death Dis.* (2016) 7:e2062. doi: 10.1038/cddis.20 15.327
52. Burr SP, Dazzi F, Garden OA. Mesenchymal stromal cells and regulatory T cells: the Yin and Yang of peripheral tolerance? *Immunol Cell Biol.* (2013) 91:12–8. doi: 10.1038/icb.2012.60
53. Meisel R, Zibert A, Laryea M, Göbel U, Däubener W, Dilloo D. Human bone marrow stromal cells inhibit allogeneic T-cell responses by indoleamine 2,3-dioxygenase-mediated tryptophan degradation. *Blood.* (2004) 103:4619–21. doi: 10.1182/blood-2003-11-3909
54. Reading JL, Yang JHM, Sabbah S, Skowera A, Knight RR, Pinxteren J, et al. Clinical-grade multipotent adult progenitor cells durably control pathogenic T cell responses in human models of transplantation and autoimmunity. *J Immunol.* (2013) 190:4542–52. doi: 10.4049/jimmunol.1202710
55. Ren G, Zhang L, Zhao X, Xu G, Zhang Y, Roberts AI, et al. Mesenchymal stem cell-mediated immunosuppression occurs via concerted action of chemokines and nitric oxide. *Cell Stem Cell.* (2008) 2:141–50. doi: 10.1016/j.stem.2007.11.014
56. Sato K, Ozaki K, Oh I, Meguro A, Hatanaka K, Nagai T, et al. Nitric oxide plays a critical role in suppression of T-cell proliferation by mesenchymal stem cells. *Blood.* (2007) 109:228–34. doi: 10.1182/blood-2006-02-002246
57. Nemeth K, Keane-Myers A, Brown JM, Metcalfe DD, Gorham JD, Gorham JD, et al. Bone marrow stromal cells use TGF-beta to suppress allergic responses in a mouse model of ragweed-induced asthma. *Proc Natl Acad Sci USA.* (2010) 107:5652–7. doi: 10.1073/pnas.0910720107
58. Zhao W, Wang Y, Wang D, Sun B, Wang G, Wang J, et al. TGF-beta expression by allogeneic bone marrow stromal cells ameliorates diabetes in NOD mice through modulating the distribution of CD4+ T cell subsets. *Cell Immunol.* (2008) 253:23–30. doi: 10.1016/j.cellimm.2008.06.009
59. Melief SM, Schrama E, Brugman MH, Tiemessen MM, Hoogduijn MJ, Fibbe WE, et al. Multipotent stromal cells induce human regulatory T cells through a novel pathway involving skewing of monocytes toward anti-inflammatory macrophages. *Stem Cells.* (2013) 31:1980–91. doi: 10.1002/stem.1432
60. Aggarwal S, Pittenger MF. Human mesenchymal stem cells modulate allogeneic immune cell responses. *Blood.* (2005) 105:1815–22. doi: 10.1182/blood-2004-04-1559
61. Mougiakakos D, Jitschin R, Johansson CC, Okita R, Kiessling R, Blanc KL. The impact of inflammatory licensing on heme oxygenase-1-mediated induction of regulatory T cells by human mesenchymal stem cells. *Blood.* (2011) 117:4826–35. doi: 10.1182/blood-2010-12-3 24038
62. Gieseke F, Böhringer J, Bussolari R, Dominici M, Handgretinger R, Müller I. Human multipotent mesenchymal stromal cells use galectin-1 to inhibit immune effector cells. *Blood.* (2010) 116:3770–9. doi: 10.1182/blood-2010-02-270777
63. Selmani Z, Naji A, Zidi I, Favier B, Gaiffe E, Obert L, et al. Human leukocyte Antigen-G5 secretion by human mesenchymal stem cells is required to suppress T lymphocyte and natural killer function and to induce CD4+CD25highFOXP3+ regulatory T cells. *Stem Cells.* (2008) 26:212–22. doi: 10.1634/stemcells.2007-0554
64. Ferreira JR, Teixeira GQ, Santos SG, Barbosa MA, Almeida-Porada G, Gonçalves RM. Mesenchymal stromal cell secretome: influencing therapeutic potential by cellular pre-conditioning. *Front Immunol.* (2018) 9:2837. doi: 10.3389/fimmu.2018.02837
65. He Y, Zhou S, Liu H, Shen B, Zhao H, Peng K, et al. Indoleamine 2, 3-Dioxgenase transfected mesenchymal stem cells induce kidney allograft tolerance by increasing the production and function of regulatory T cells. *Transplantation.* (2015) 99:1829–38. doi: 10.1097/TP.0000000000000856
66. Tang J, Yang R, Lv L, Yao A, Pu L, Yin A, et al. Transforming growth factor-β-expressing mesenchymal stem cells induce local tolerance in a rat liver transplantation model of acute rejection. *Stem Cells.* (2016) 34:2681–92. doi: 10.1002/stem.2437
67. Niu J, Yue W, Song Y, Zhang Y, Qi X, Wang Z, et al. Prevention of acute liver allograft rejection by IL-10-engineered mesenchymal stem cells. *Clin Exp Immunol.* (2014) 176:473–84. doi: 10.1111/cei.12283
68. Wu B, Song H-L, Yang Y, Yin M-L, Zhang B-Y, Cao Y, et al. Improvement of liver transplantation outcome by Heme Oxygenase-1-transduced bone marrow mesenchymal stem cells in rats. *Stem Cells Int.* (2016) 2016:9235073. doi: 10.1155/2016/9235073
69. Wang R, Shen Z, Yang L, Yin M, Zheng W, Wu B, et al. Protective effects of heme oxygenase-1-transduced bone marrow-derived mesenchymal stem cells on reduced-size liver transplantation: role of autophagy regulated by the ERK/mTOR signaling pathway. *Int J Mol Med.* (2017) 40:1537–48. doi: 10.3892/ijmm.2017.3121
70. Yang Y, Song HL, Zhang W, Wu BJ, Fu NN, Dong C, et al. Heme oxygenase-1-transduced bone marrow mesenchymal stem cells in reducing acute rejection and improving small bowel transplantation outcomes in rats. *Stem Cell Res Ther.* (2016) 7:164. doi: 10.1186/s13287-016-0427-8

71. Krampera M, Glennie S, Dyson J, Scott D, Laylor R, Simpson E, et al. Bone marrow mesenchymal stem cells inhibit the response of naive and memory antigen-specific T cells to their cognate peptide. *Blood.* (2003) 101:3722–9. doi: 10.1182/blood-2002-07-2104
72. Glennie S, Soeiro I, Dyson PJ, Lam EW-F, Dazzi F. Bone marrow Mesenchymal stem cells induce division arrest anergy of activated T cells. *Blood.* (2005) 105:2821–7. doi: 10.1182/blood-2004-09-3696
73. Di Nicola M, Carlo-Stella C, Magni M, Milanesi M, Longoni PD, Matteucci P, et al. Human bone marrow stromal cells suppress T-lymphocyte proliferation induced by cellular or nonspecific mitogenic stimuli. *Blood.* (2002) 99:3838–43. doi: 10.1182/blood.V99.10.3838
74. Mareschi K, Castiglia S, Sanavio F, Rustichelli D, Muraro M, Defedele D, et al. Immunoregulatory effects on T lymphocytes by human mesenchymal stromal cells isolated from bone marrow, amniotic fluid, and placenta. *Exp Hematol.* (2016) 44:138–50 e1. doi: 10.1016/j.exphem.2015.10.009
75. Tse WT, Pendleton JD, Beyer WM, Egalka MC, Guinan EC. Suppression of allogeneic T-cell proliferation by human marrow stromal cells: implications in transplantation. *Transplantation.* (2003) 75:389–97. doi: 10.1097/01.TP.0000045055.63901.A9
76. Nadazdin O, Boskovic S, Murakami T, Tocco G, Smith R-N, Colvin RB, et al. Host alloreactive memory T cells influence tolerance to kidney allografts in nonhuman primates. *Sci Transl Med.* (2011) 3:86ra51. doi: 10.1126/scitranslmed.3002093
77. Bestard O, Cruzado JM, Lucia M, Crespo E, Casis L, Sawitzki B, et al. Prospective assessment of antidonor cellular alloreactivity is a tool for guidance of immunosuppression in kidney transplantation. *Kidney Int.* (2013) 84:1226–36. doi: 10.1038/ki.2013.236
78. Heeger PS, Greenspan NS, Kuhlenschmidt S, Dejelo C, Hricik DE, Schulak JA, et al. Pretransplant frequency of donor-specific, IFN-gamma-producing lymphocytes is a manifestation of immunologic memory and correlates with the risk of posttransplant rejection episodes. *J Immunol.* (1999) 163:2267–75.
79. Rosenblum JM, Kirk AD. Recollective homeostasis and the immune consequences of peritransplant depletional induction therapy. *Immunol Rev.* (2014) 258:167–82. doi: 10.1111/imr.12155
80. Moxham VF, Karegli J, Phillips RE, Brown KL, Tapmeier TT, Hangartner R, et al. Homeostatic proliferation of lymphocytes results in augmented memory-like function and accelerated allograft rejection. *J Immunol.* (2008) 180:3910–8. doi: 10.4049/jimmunol.180.6.3910
81. Ayasoufi K, Yu H, Fan R, Wang X, Williams J, Valujskikh A. Pretransplant antithymocyte globulin has increased efficacy in controlling donor-reactive memory T cells in mice. *Am J Transplant.* (2013) 13:589–99. doi: 10.1111/ajt.12068
82. Neujahr DC, Chen C, Huang X, Markmann JF, Cobbold S, Waldmann H, et al. Accelerated memory cell homeostasis during T cell depletion and approaches to overcome it. *J Immunol.* (2006) 176:4632–9. doi: 10.4049/jimmunol.176.8.4632
83. Luz-Crawford P, Kurte M, Bravo-Alegría J, Contreras R, Nova-Lamperti E, Tejedor G, et al. Mesenchymal stem cells generate a CD4+CD25+Foxp3+ regulatory T cell population during the differentiation process of Th1 and Th17 cells. *Stem Cell Res Ther.* (2013) 4:65. doi: 10.1186/scrt216
84. Aksu AE, Horibe E, Sacks J, Ikeguchi R, Breitinger J, Scozio M, et al. Co-infusion of donor bone marrow with host mesenchymal stem cells treats GVHD and promotes vascularized skin allograft survival in rats. *Clin Immunol.* (2008) 127:348–58. doi: 10.1016/j.clim.2008.02.003
85. Ghannam S, Pène J, Moquet-Torcy G, Torcy-Moquet G, Jorgensen C, Yssel H. Mesenchymal stem cells inhibit human Th17 cell differentiation and function and induce a T regulatory cell phenotype. *J Immunol.* (2010) 185:302–12. doi: 10.4049/jimmunol.0902007
86. Liu R, Li X, Zhang Z, Zhou M, Sun Y, Su D, et al. Allogeneic mesenchymal stem cells inhibited T follicular helper cell generation in rheumatoid arthritis. *Sci Rep.* (2015) 5:12777. doi: 10.1038/srep12777
87. Perico N, Casiraghi F, Todeschini M, Cortinovis M, Gotti E, Portalupi V, et al. Long-term clinical and immunological profile of kidney transplant patients given mesenchymal stromal cell immunotherapy. *Front Immunol.* (2018) 9:1359. doi: 10.3389/fimmu.2018.01359
88. Hsu W-T, Lin C-H, Chiang B-L, Jui H-Y, Wu KK-Y, Lee C-M. Prostaglandin E2 potentiates mesenchymal stem cell-induced IL-10+IFN-γ+CD4+ regulatory T cells to control transplant arteriosclerosis. *J Immunol.* (2013) 190:2372–80. doi: 10.4049/jimmunol.1202996
89. Corcione A, Benvenuto F, Ferretti E, Giunti D, Cappiello V, Cazzanti F, et al. Human mesenchymal stem cells modulate B-cell functions. *Blood.* (2006) 107:367–72. doi: 10.1182/blood-2005-07-2657
90. Tabera S, Pérez-Simón JA, Díez-Campelo M, Sánchez-Abarca LI, Blanco B, López A, et al. The effect of mesenchymal stem cells on the viability, proliferation and differentiation of B-lymphocytes. *Haematologica.* (2008) 93:1301–9. doi: 10.3324/haematol.12857
91. Rosado MM, Bernardo ME, Scarsella M, Conforti A, Giorda E, Biagini S, et al. Inhibition of B-cell proliferation and antibody production by mesenchymal stromal cells is mediated by T cells. *Stem Cells Dev.* (2015) 24:93–103. doi: 10.1089/scd.2014.0155
92. Comoli P, Ginevri F, Maccario R, Avanzini MA, Marconi M, Groff A, et al. Human mesenchymal stem cells inhibit antibody production induced *in vitro* by allostimulation. *Nephrol Dial Transplant.* (2008) 23:1196–202. doi: 10.1093/ndt/gfm740
93. Franquesa M, Mensah FK, Huizinga R, Strini T, Boon L, Lombardo E, et al. Human adipose tissue-derived mesenchymal stem cells abrogate plasmablast formation and induce regulatory B cells independently of T helper cells. *Stem Cells.* (2015) 33:880–91. doi: 10.1002/stem.1881
94. Luk F, Carreras-Planella L, Korevaar SS, de Witte SFH, Borràs FE, Betjes MGH, et al. Inflammatory conditions dictate the effect of mesenchymal stem or stromal cells on B cell function. *Front Immunol.* (2017) 8:1042. doi: 10.3389/fimmu.2017.01042
95. Hermankova B, Zajicova A, Javorkova E, Chudickova M, Trosan P, Hajkova M, et al. Suppression of IL-10 production by activated B cells via a cell contact-dependent cyclooxygenase-2 pathway upregulated in IFN-γ-treated mesenchymal stem cells. *Immunobiology.* (2016) 221:129–36. doi: 10.1016/j.imbio.2015.09.017
96. Newell KA, Asare A, Kirk AD, Gisler TD, Bourcier K, Suthanthiran M, et al. Identification of a B cell signature associated with renal transplant tolerance in humans. *J Clin Invest.* (2010) 120:1836–47. doi: 10.1172/JCI39933
97. Djouad F, Charbonnier L-M, Bouffi C, Louis-Plence P, Bony C, Apparailly F, et al. Mesenchymal stem cells inhibit the differentiation of dendritic cells through an interleukin-6-dependent mechanism. *Stem Cells.* (2007) 25:2025–32. doi: 10.1634/stemcells.2006-0548
98. English K, Barry FP, Mahon BP. Murine mesenchymal stem cells suppress dendritic cell migration, maturation and antigen presentation. *Immunol Lett.* (2008) 115:50–8. doi: 10.1016/j.imlet.2007.10.002
99. Chiesa S, Morbelli S, Morando S, Massollo M, Marini C, Bertoni A, et al. Mesenchymal stem cells impair *in vivo* T-cell priming by dendritic cells. *Proc Natl Acad Sci USA.* (2011) 108:17384–9. doi: 10.1073/pnas.1103650108
100. Zhang W, Ge W, Li C, You S, Liao L, Han Q, et al. Effects of mesenchymal stem cells on differentiation, maturation, and function of human monocyte-derived dendritic cells. *Stem Cells Dev.* (2004) 13:263–71. doi: 10.1089/154732804323099190
101. Nauta AJ, Kruisselbrink AB, Lurvink E, Willemze R, Fibbe WE. Mesenchymal stem cells inhibit generation and function of both CD34+-derived and monocyte-derived dendritic cells. *J Immunol.* (2006) 177:2080–7. doi: 10.4049/jimmunol.177.4.2080
102. Zhao Z-G, Xu W, Sun L, You Y, Li F, Li Q-B, et al. Immunomodulatory function of regulatory dendritic cells induced by mesenchymal stem cells. *Immunol Invest.* (2012) 41:183–98. doi: 10.3109/08820139.2011.607877
103. Gao S, Mao F, Zhang B, Zhang L, Zhang X, Wang M, et al. Mouse bone marrow-derived mesenchymal stem cells induce macrophage M2 polarization through the nuclear factor-κB and signal transducer and activator of transcription 3 pathways. *Exp Biol Med (Maywood).* (2014) 239:366–75. doi: 10.1177/1535370213518169
104. Kim J, Hematti P. Mesenchymal stem cell-educated macrophages: a novel type of alternatively activated macrophages. *Exp Hematol.* (2009) 37:1445–53. doi: 10.1016/j.exphem.2009.09.004
105. Maggini J, Mirkin G, Bognanni I, Holmberg J, Piazzón IM, Nepomnaschy I, et al. Mouse bone marrow-derived mesenchymal stromal cells turn activated macrophages into a regulatory-like profile. *PLoS ONE.* (2010) 5:e9252. doi: 10.1371/journal.pone.0009252
106. François M, Romieu-Mourez R, Li M, Galipeau J. Human MSC suppression correlates with cytokine induction of indoleamine 2,3-dioxygenase and

bystander M2 macrophage differentiation. *Mol Ther.* (2012) 20:187–95. doi: 10.1038/mt.2011.189

107. Galleu A, Riffo-Vasquez Y, Trento C, Lomas C, Dolcetti L, Cheung TS, et al. Apoptosis in mesenchymal stromal cells induces in vivo recipient-mediated immunomodulation. *Sci Transl Med.* (2017) 9:7828. doi: 10.1126/scitranslmed.aam7828
108. de Witte SFH, Luk F, Sierra Parraga JM, Gargesha M, Merino A, Korevaar SS, et al. Immunomodulation by therapeutic mesenchymal stromal cells (MSC) is triggered through phagocytosis of MSC by monocytic cells. *Stem Cells.* (2018) 36:602–15. doi: 10.1002/stem.2779
109. Gonçalves F da C, Luk F, Korevaar SS, Bouzid R, Paz AH, et al. Membrane particles generated from mesenchymal stromal cells modulate immune responses by selective targeting of pro-inflammatory monocytes. *Sci Rep.* (2017) 7:12100. doi: 10.1038/s41598-017-12121-z
110. Akiyama K, Chen C, Wang D, Xu X, Qu C, Yamaza T, et al. Mesenchymal-stem-cell-induced immunoregulation involves FAS-ligand-/FAS-mediated T cell apoptosis. *Cell Stem Cell.* (2012) 10:544–55. doi: 10.1016/j.stem.2012.03.007
111. Krampera M. Mesenchymal stromal cell "licensing": a multistep process. *Leukemia.* (2011) 25:1408–14. doi: 10.1038/leu.2011.108
112. Chan JL, Tang KC, Patel AP, Bonilla LM, Pierobon N, Ponzio NM, et al. Antigen-presenting property of mesenchymal stem cells occurs during a narrow window at low levels of interferon-gamma. *Blood.* (2006) 107:4817–24. doi: 10.1182/blood-2006-01-0057
113. François M, Romieu-Mourez R, Stock-Martineau S, Boivin M-N, Bramson JL, Galipeau J. Mesenchymal stromal cells cross-present soluble exogenous antigens as part of their antigen-presenting cell properties. *Blood.* (2009) 114:2632–8. doi: 10.1182/blood-2009-02-207795
114. Raicevic G, Rouas R, Najar M, Stordeur P, Boufker HI, Bron D, et al. Inflammation modifies the pattern and the function of Toll-like receptors expressed by human mesenchymal stromal cells. *Hum Immunol.* (2010) 71:235–44. doi: 10.1016/j.humimm.2009.12.005
115. Krampera M, Cosmi L, Angeli R, Pasini A, Liotta F, Andreini A, et al. Role for interferon-gamma in the immunomodulatory activity of human bone marrow mesenchymal stem cells. *Stem Cells.* (2006) 24:386–98. doi: 10.1634/stemcells.2005-0008
116. English K, Barry FP, Field-Corbett CP, Mahon BP. IFN-gamma and TNF-alpha differentially regulate immunomodulation by murine mesenchymal stem cells. *Immunol Lett.* (2007) 110:91–100. doi: 10.1016/j.imlet.2007.04.001
117. Li W, Ren G, Huang Y, Su J, Han Y, Li J, et al. Mesenchymal stem cells: a double-edged sword in regulating immune responses. *Cell Death Differ.* (2012) 19:1505–13. doi: 10.1038/cdd.2012.26
118. Opitz CA, Litzenburger UM, Lutz C, Lanz TV, Tritschler I, Köppel A, et al. Toll-like receptor engagement enhances the immunosuppressive properties of human bone marrow-derived mesenchymal stem cells by inducing indoleamine-2,3-dioxygenase-1 via interferon-beta and protein kinase R. *Stem Cells.* (2009) 27:909–19. doi: 10.1002/stem.7
119. Pevsner-Fischer M, Morad V, Cohen-Sfady M, Rousso-Noori L, Zanin-Zhorov A, Cohen S, et al. Toll-like receptors and their ligands control mesenchymal stem cell functions. *Blood.* (2007) 109:1422–32. doi: 10.1182/blood-2006-06-028704
120. Le Blanc K, Mougiakakos D. Multipotent mesenchymal stromal cells and the innate immune system. *Nat Rev Immunol.* (2012) 12:383–96. doi: 10.1038/nri3209
121. Hwa Cho H, Bae YC, Jung JS. Role of toll-like receptors on human adipose-derived stromal cells. *Stem Cells.* (2006) 24:2744–52. doi: 10.1634/stemcells.2006-0189
122. Tomchuck SL, Zwezdaryk KJ, Coffelt SB, Waterman RS, Danka ES, Scandurro AB. Toll-like receptors on human mesenchymal stem cells drive their migration and immunomodulating responses. *Stem Cells.* (2008) 26:99–107. doi: 10.1634/stemcells.2007-0563
123. Wang Z, Zhang F, Wang L, Yao Y, Zhao Q, Gao X. Lipopolysaccharides can protect mesenchymal stem cells (MSCs) from oxidative stress-induced apoptosis and enhance proliferation of MSCs via Toll-like receptor(TLR)-4 and PI3K/Akt. *Cell Biol Int.* (2009) 33:665–74. doi: 10.1016/j.cellbi.2009.03.006
124. Liotta F, Angeli R, Cosmi L, Filì L, Manuelli C, Frosali F, et al. Toll-like receptors 3 and 4 are expressed by human bone marrow-derived mesenchymal stem cells and can inhibit their T-cell modulatory activity by impairing Notch signaling. *Stem Cells.* (2008) 26:279–89. doi: 10.1634/stemcells.2007-0454
125. Rashedi I, Gómez-Aristizábal A, Wang X-H, Viswanathan S, Keating A. TLR3 or TLR4 activation enhances mesenchymal stromal cell-mediated Treg induction via notch signaling. *Stem Cells.* (2017) 35:265–75. doi: 10.1002/stem.2485
126. Waterman RS, Tomchuck SL, Henkle SL, Betancourt AM. A new mesenchymal stem cell (MSC) paradigm: polarization into a pro-inflammatory MSC1 or an immunosuppressive MSC2 phenotype. *PLOS ONE.* (2010) 5:e10088. doi: 10.1371/journal.pone.0010088
127. Brandau S, Jakob M, Hemeda H, Bruderek K, Janeschik S, Bootz F, et al. Tissue-resident mesenchymal stem cells attract peripheral blood neutrophils and enhance their inflammatory activity in response to microbial challenge. *J Leukoc Biol.* (2010) 88:1005–15. doi: 10.1189/jlb.0410207
128. Tu Z, Li Q, Bu H, Lin F. Mesenchymal stem cells inhibit complement activation by secreting factor H. *Stem Cells Dev.* (2010) 19:1803–9. doi: 10.1089/scd.2009.0418
129. Moll G, Jitschin R, von Bahr L, Rasmusson-Duprez I, Sundberg B, Lönnies L, et al. Mesenchymal stromal cells engage complement and complement receptor bearing innate effector cells to modulate immune responses. *PLoS ONE.* (2011) 6:e21703. doi: 10.1371/journal.pone.0021703
130. Le Blanc K, Davies LC. Mesenchymal stromal cells and the innate immune response. *Immunol Lett.* (2015) 168:140–6. doi: 10.1016/j.imlet.2015.05.004
131. Schraufstatter IU, DiScipio RG, Zhao M, Khaldoyanidi SK. C3a and C5a are chemotactic factors for human mesenchymal stem cells, which cause prolonged ERK1/2 phosphorylation. *J Immunol.* (2009) 182:3827–36. doi: 10.4049/jimmunol.0803055
132. Casiraghi F, Todeschini M, Azzollini N, Cravedi P, Cassis P, Solini S, et al. Effect of timing and complement receptor antagonism on intragraft recruitment and pro-tolerogenic effects of mesenchymal stromal cells in murine kidney transplantation. *Transplantation.* (2019) doi: 10.1097/TP.0000000000002611. [Epub ahead of print].
133. Eggenhofer E, Benseler V, Kroemer A, Popp FC, Geissler EK, Schlitt HJ, et al. Mesenchymal stem cells are short-lived and do not migrate beyond the lungs after intravenous infusion. *Front Immunol.* (2012) 3:297. doi: 10.3389/fimmu.2012.00297
134. Popp FC, Eggenhofer E, Renner P, Slowik P, Lang SA, Kaspar H, et al. Mesenchymal stem cells can induce long-term acceptance of solid organ allografts in synergy with low-dose mycophenolate. *Transpl Immunol.* (2008) 20:55–60. doi: 10.1016/j.trim.2008.08.004
135. Li Q, Zhou X, Shi Y, Li J, Zheng L, Cui L, et al. *In vivo* tracking and comparison of the therapeutic effects of MSCs and HSCs for liver injury. *PLoS ONE.* (2013) 8:e62363. doi: 10.1371/journal.pone.0062363
136. Herrera MB, Bussolati B, Bruno S, Morando L, Mauriello-Romanazzi G, Sanavio F, et al. Exogenous mesenchymal stem cells localize to the kidney by means of CD44 following acute tubular injury. *Kidney Int.* (2007) 72:430–41. doi: 10.1038/sj.ki.5002334
137. Dong F, Harvey J, Finan A, Weber K, Agarwal U, Penn MS. Myocardial CXCR4 expression is required for mesenchymal stem cell mediated repair following acute myocardial infarction. *Circulation.* (2012) 126:314–24. doi: 10.1161/CIRCULATIONAHA.111.082453
138. Rasmusson I, Ringdén O, Sundberg B, Le Blanc K. Mesenchymal stem cells inhibit the formation of cytotoxic T lymphocytes, but not activated cytotoxic T lymphocytes or natural killer cells. *Transplantation.* (2003) 76:1208–13. doi: 10.1097/01.TP.0000082540.43730.80
139. Carrión F, Nova E, Luz P, Apablaza F, Figueroa F. Opposing effect of mesenchymal stem cells on Th1 and Th17 cell polarization according to the state of CD4+ T cell activation. *Immunol Lett.* (2011) 135:10–6. doi: 10.1016/j.imlet.2010.09.006
140. Bayer AL, Yu A, Adeegbe D, Malek TR. Essential role for interleukin-2 for CD4+CD25+ T regulatory cell development during the neonatal period. *J Exp Med.* (2005) 201:769–77. doi: 10.1084/jem.20041179
141. Amado IF, Berges J, Luther RJ, Mailhé M-P, Garcia S, Bandeira A, et al. IL-2 coordinates IL-2-producing and regulatory T cell interplay. *J Exp Med.* (2013) 210:2707–20. doi: 10.1084/jem.20122759

142. Seder RA, Darrah PA, Roederer M. T-cell quality in memory and protection: implications for vaccine design. *Nat Rev Immunol.* (2008) 8:247–58. doi: 10.1038/nri2274
143. Cayeux S, Meuer S, Pezzutto A, Körbling M, Haas R, Schulz R, et al. Allogeneic mixed lymphocyte reactions during a second round of ontogeny: normal accessory cells did not restore defective interleukin-2 (IL-2) synthesis in T cells but induced responsiveness to exogeneous IL-2. *Blood.* (1989) 74:2278–84.
144. Beyth S, Borovsky Z, Mevorach D, Liebergall M, Gazit Z, Aslan H, et al. Human mesenchymal stem cells alter antigen-presenting cell maturation and induce T-cell unresponsiveness. *Blood.* (2005) 105:2214–9. doi: 10.1182/blood-2004-07-2921
145. Spaggiari GM, Abdelrazik H, Becchetti F, Moretta L. MSCs inhibit monocyte-derived DC maturation and function by selectively interfering with the generation of immature DCs: central role of MSC-derived prostaglandin E2. *Blood.* (2009) 113:6576–83. doi: 10.1182/blood-2009-02-203943
146. Khosravi M, Bidmeshkipour A, Cohen JL, Moravej A, Hojjat-Assari S, Naserian S, et al. Induction of CD4+CD25+FOXP3+ regulatory T cells by mesenchymal stem cells is associated with modulation of ubiquitination factors and TSDR demethylation. *Stem Cell Res Ther.* (2018) 9:273. doi: 10.1186/s13287-018-0991-1
147. Tan J, Wu W, Xu X, Liao L, Zheng F, Messinger S, et al. Induction therapy with autologous mesenchymal stem cells in living-related kidney transplants: a randomized controlled trial. *JAMA.* (2012) 307:1169–77. doi: 10.1001/jama.2012.316
148. Perico N, Casiraghi F, Introna M, Gotti E, Todeschini M, Cavinato RA, et al. Autologous mesenchymal stromal cells and kidney transplantation: a pilot study of safety and clinical feasibility. *Clin J Am Soc Nephrol.* (2011) 6:412–22. doi: 10.2215/CJN.04950610
149. Perico N, Casiraghi F, Gotti E, Introna M, Todeschini M, Cavinato RA, et al. Mesenchymal stromal cells and kidney transplantation: pretransplant infusion protects from graft dysfunction while fostering immunoregulation. *Transpl Int.* (2013) 26:867–78. doi: 10.1111/tri.12132
150. Reinders MEJ, de Fijter JW, Roelofs H, Bajema IM, de Vries DK, Schaapherder AF, et al. Autologous bone marrow-derived mesenchymal stromal cells for the treatment of allograft rejection after renal transplantation: results of a phase I study. *Stem Cells Transl Med.* (2013) 2:107–11. doi: 10.5966/sctm.2012-0114
151. Mudrabettu C, Kumar V, Rakha A, Yadav AK, Ramachandran R, Kanwar DB, et al. Safety and efficacy of autologous mesenchymal stromal cells transplantation in patients undergoing living donor kidney transplantation: a pilot study. *Nephrology (Carlton).* (2015) 20:25–33. doi: 10.1111/nep.12338
152. Peng Y, Ke M, Xu L, Liu L, Chen X, Xia W, et al. Donor-derived mesenchymal stem cells combined with low-dose tacrolimus prevent acute rejection after renal transplantation: a clinical pilot study. *Transplantation.* (2013) 95:161–8. doi: 10.1097/TP.0b013e3182754c53
153. Pan G-H, Chen Z, Xu L, Zhu J-H, Xiang P, Ma J-J, et al. Low-dose tacrolimus combined with donor-derived mesenchymal stem cells after renal transplantation: a prospective, non-randomized study. *Oncotarget.* (2016) 7:12089–101. doi: 10.18632/oncotarget.7725
154. Sun Q, Huang Z, Han F, Zhao M, Cao R, Zhao D, et al. Allogeneic mesenchymal stem cells as induction therapy are safe and feasible in renal allografts: pilot results of a multicenter randomized controlled trial. *J Transl Med.* (2018) 16:52. doi: 10.1186/s12967-018-1422-x
155. Erpicum P, Weekers L, Detry O, Bonvoisin C, Delbouille M-H, Grégoire C, et al. Infusion of third-party mesenchymal stromal cells after kidney transplantation: a phase I-II, open-label, clinical study. *Kidney Int.* (2019) 95:693–707. doi: 10.1016/j.kint.2018.08.046
156. Soeder Y, Loss M, Johnson CL, Hutchinson JA, Haarer J, Ahrens N, et al. First-in-human case study: multipotent adult progenitor cells for immunomodulation after liver transplantation. *Stem Cells Transl Med.* (2015) 4:899–904. doi: 10.5966/sctm.2015-0002
157. Detry O, Vandermeulen M, Delbouille M-H, Somja J, Bletard N, Briquet A, et al. Infusion of mesenchymal stromal cells after deceased liver transplantation: a phase I-II, open-label, clinical study. *J Hepatol.* (2017) 67:47–55. doi: 10.1016/j.jhep.2017.03.001
158. Keller CA, Gonwa TA, Hodge DO, Hei DJ, Centanni JM, Zubair AC. Feasibility, safety, and tolerance of mesenchymal stem cell therapy for obstructive chronic lung allograft dysfunction. *Stem Cells Transl Med.* (2018) 7:161–7. doi: 10.1002/sctm.17-0198
159. Casiraghi F, Remuzzi G, Abbate M, Perico N. Multipotent mesenchymal stromal cell therapy and risk of malignancies. *Stem Cell Rev.* (2013) 9:65–79. doi: 10.1007/s12015-011-9345-4
160. Miura M, Miura Y, Padilla-Nash HM, Molinolo AA, Fu B, Patel V, et al. Accumulated chromosomal instability in murine bone marrow mesenchymal stem cells leads to malignant transformation. *Stem Cells.* (2006) 24:1095–103. doi: 10.1634/stemcells.2005-0403
161. Bernardo ME, Zaffaroni N, Novara F, Cometa AM, Avanzini MA, Moretta A, et al. Human bone marrow derived mesenchymal stem cells do not undergo transformation after long-term *in vitro* culture and do not exhibit telomere maintenance mechanisms. *Cancer Res.* (2007) 67:9142–9. doi: 10.1158/0008-5472.CAN-06-4690
162. Souza LEB, Almeida DC, Yaochite JNU, Covas DT, Fontes AM. Intravenous administration of bone marrow-derived multipotent mesenchymal stromal cells enhances the recruitment of CD11b(±) myeloid cells to the lungs and facilitates B16-F10 melanoma colonization. *Exp Cell Res.* (2016) 345:141–9. doi: 10.1016/j.yexcr.2015.05.021
163. Cousin B, Ravet E, Poglio S, De Toni F, Bertuzzi M, Lulka H, et al. Adult stromal cells derived from human adipose tissue provoke pancreatic cancer cell death both *in vitro* and *in vivo*. *PLoS ONE.* (2009) 4:e6278. doi: 10.1371/journal.pone.0006278
164. Khakoo AY, Pati S, Anderson SA, Reid W, Elshal MF, Rovira II, et al. Human mesenchymal stem cells exert potent antitumorigenic effects in a model of Kaposi's sarcoma. *J Exp Med.* (2006) 203:1235–47. doi: 10.1084/jem.20051921
165. Lalu MM, McIntyre L, Pugliese C, Fergusson D, Winston BW, Marshall JC, et al. Safety of cell therapy with mesenchymal stromal cells (SafeCell): a systematic review and meta-analysis of clinical trials. *PLoS ONE.* (2012) 7:e47559. doi: 10.1371/journal.pone.0047559
166. Zangi L, Margalit R, Reich-Zeliger S, Bachar-Lustig E, Beilhack A, Negrin R, et al. Direct imaging of immune rejection and memory induction by allogeneic mesenchymal stromal cells. *Stem Cells.* (2009) 27:2865–74. doi: 10.1002/stem.217
167. Eliopoulos N, Stagg J, Lejeune L, Pommey S, Galipeau J. Allogeneic marrow stromal cells are immune rejected by MHC class I- and class II-mismatched recipient mice. *Blood.* (2005) 106:4057–65. doi: 10.1182/blood-2005-03-1004
168. Le Blanc K, Frassoni F, Ball L, Locatelli F, Roelofs H, Lewis I, et al. Mesenchymal stem cells for treatment of steroid-resistant, severe, acute graft-versus-host disease: a phase II study. *Lancet.* (2008) 371:1579–86. doi: 10.1016/S0140-6736(08)60690-X

2

Preclinical Toxicity Evaluation of Clinical Grade Placenta-Derived Decidua Stromal Cells

Behnam Sadeghi[1*], *Gianluca Moretti*[1], *Fabian Arnberg*[2,3], *Erik Samén*[2,3,4], *Bita Kohein*[1], *Rusan Catar*[5,6], *Julian Kamhieh-Milz*[7], *Sven Geissler*[8,9], *Guido Moll*[5,8,10], *Staffan Holmin*[2,3] *and Olle Ringdén*[1]

[1] *Translational Cell Therapy Research (TCR), Department of Clinical Science Intervention and Technology (CLINTEC), Karolinska Institutet and Karolinska University Hospital, Stockholm, Sweden,* [2] *Department of Clinical Neuroscience, Karolinska Institutet and Karolinska University Hospital, Stockholm, Sweden,* [3] *Department of Neuroradiology, Karolinska Institutet and Karolinska University Hospital, Stockholm, Sweden,* [4] *Department of Radiopharmacy, Karolinska Institutet and Karolinska University Hospital, Stockholm, Sweden,* [5] *Department of Nephrology and Internal Intensive Care Medicine, Charité-Universitätsmedizin Berlin, Corporate Member of Freie-Universität Berlin, Humboldt-Universität zu Berlin, and Berlin Institute of Health (BIH), Berlin, Germany,* [6] *Berlin Institute of Health (BIH), Berlin, Germany,* [7] *Department of Transfusion Medicine, Charité-Universitätsmedizin Berlin, Corporate Member of Freie-Universität Berlin, Humboldt-Universität zu Berlin, and Berlin Institute of Health (BIH), Berlin, Germany,* [8] *BIH Center for Regenerative Therapies (BCRT), Charité-Universitätsmedizin Berlin, Corporate Member of Freie-Universität Berlin, Humboldt-Universität zu Berlin, and Berlin Institute of Health (BIH), Berlin, Germany,* [9] *Julius Wolff Institute (JWI), Charité-Universitätsmedizin Berlin, Corporate Member of Freie-Universität Berlin, Humboldt-Universität zu Berlin, and Berlin Institute of Health (BIH), Berlin, Germany,* [10] *Berlin-Brandenburg School for Regenerative Therapies (BSRT), Charité-Universitätsmedizin Berlin, Corporate Member of Freie-Universität Berlin, Humboldt-Universität zu Berlin, and Berlin Institute of Health (BIH), Berlin, Germany*

Correspondence:
Behnam Sadeghi
behnam.sadeghi@ki.se

Placenta-derived decidua stromal cells (DSCs) are being investigated as an alternative to other sources of mesenchymal stromal cells (MSCs) for cellular therapy. DSCs are more effective in treating acute inflammatory diseases in human and this is our preclinical safety study of human DSCs in Sprague-Dawley rats and Balb/c mice. Human DSCs were cultured and expanded from fetal membranes obtained from placentas following cesarean section. In rats, 0.5×10^6 cells/kg were injected intravenously ($n = 4$) or intra-aortal ($n = 4$). In mice, DSCs were given intravenously at doses ranging from $4–40 \times 10^6$ cells/kg (total of $n = 120$ mice). *In vivo* tracking of human cells in mice was performed by using transduced DSC with luciferin gene, and in rats by using ^{18}F-FDG PET. Clotting parameters were determined *in vitro* and *in vivo*. All intra-arterially DSC-treated rats had normal motility and behavior and histological examination was normal for liver, spleen kidneys and thigh muscles. Mice treated with DSCs showed no immediate or long-term side effects. None of the mice died or showed acute toxicity or adverse reactions 3 and 30 days after DSC infusion. Murine blood biochemistry profiles related to liver, kidney, heart, and inflammatory indices was not influenced by DSC infusion and complete blood counts were normal. *In vivo* tracking of infused DSCs detected a signal in the lungs for up to 4 days post infusion. Compared to bone marrow derived MSCs, the DSCs had better viability, smaller size, but stronger clotting in human blood and plasma. Both MSC- and DSC-induced coagulation and complement activation markers, thrombin-anti-thrombin complex (TAT) and C3a, and *in vitro* clotting parameters were decreased by heparin supplementation. In conclusion, DSCs are safe with almost no side effects even with doses 40 times higher than are used clinically, particularly when supplemented with low-dose heparin.

Keywords: placenta-derived decidua stromal cells, mesenchymal stromal cells, toxicity, side effects, cellular therapy

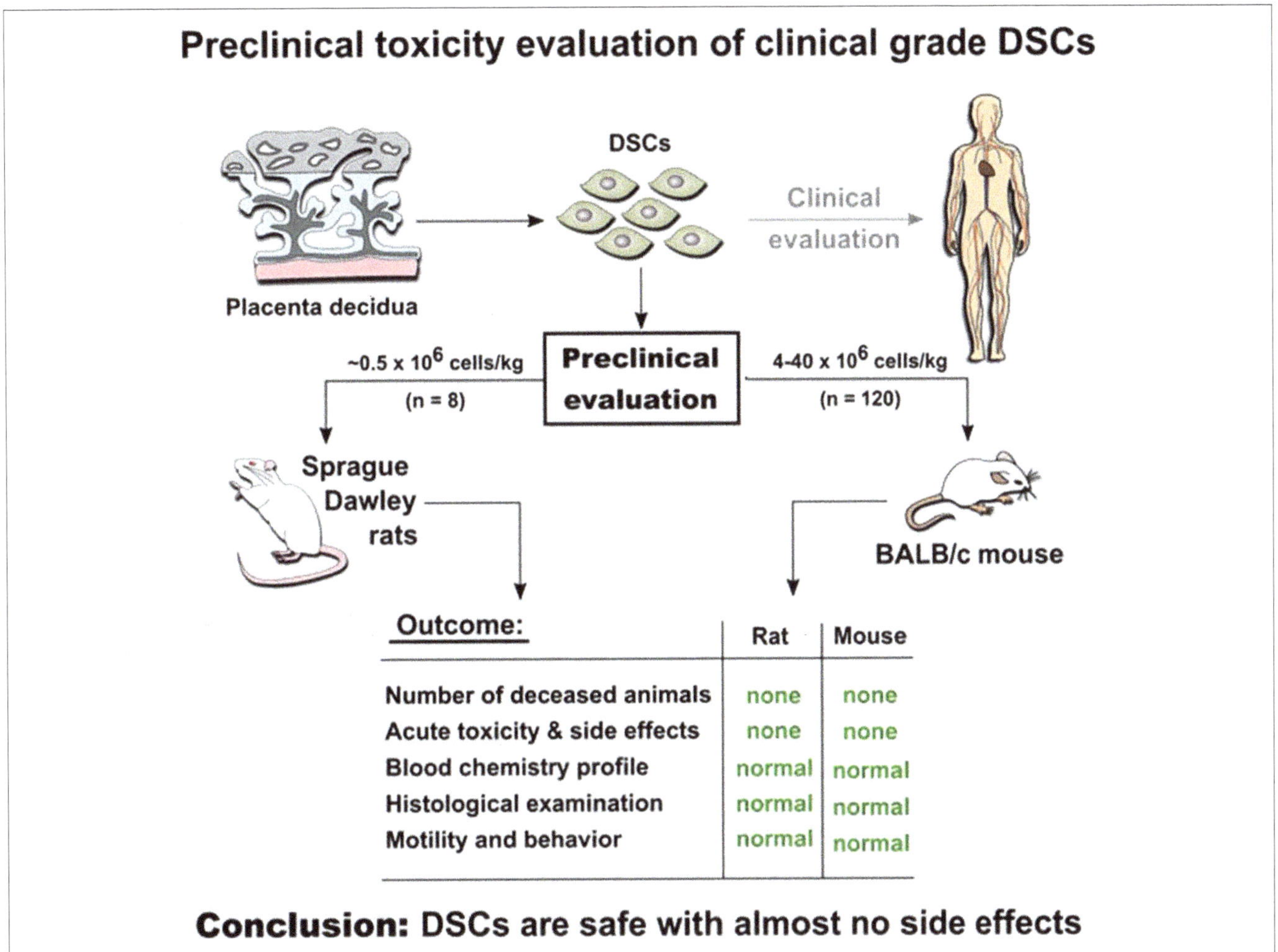

GRAPHICAL ABSTRACT| Experimental *in vivo* studies indicate safety of DSCs infusion in two animal models.

INTRODUCTION

Mesenchymal stromal cells (MSCs), first described by Friedenstein et al. (1), have the potential to differentiate into several mesenchymal lineages and are found in many vascularized human tissues (2, 3). MSCs have multiple beneficial properties; e.g., they support hematopoiesis and have potent immunomodulatory property, and have therefore been in experimental clinical use for treatment of a series of inflammatory diseases, including graft-vs.-host disease (GvHD) and hemorrhagic cystitis following hematopoietic stem cell transplantation (HSCT), autoimmune diseases and in regenerative medicine (4–10). Galleu et al. demonstrated that infused MSCs are actively induced to undergo perforin-dependent apoptosis by recipient cytotoxic cells (11) and this process appears to be required for MSC-induced immune suppression (8, 12–14). Galipeau and Sensébé reasoned that the clearance of apoptotic MSC-like cells and in particular lung-embolized placental stromal material leads to reprograming of lung macrophages by efferocytosis, thus promoting fetomaternal tolerance (8).

Infusions of placenta-derived decidual stromal cells (DSCs) may thus mimic a highly conserved biological process in mammals that induces systemic immunomodulation and feto-maternal tolerance during pregnancy (8, 15–17). Placental DSCs differ from bone marrow (BM)-MSCs in several aspects. Compared to MSCs, the DSCs are only half the size, show less differentiation into chondrocytes and osteocytes, have a stronger inhibitory effect on allo-reactive T-cells, and promote stronger coagulation (18–20).

Systemic or local administration of clinical grade MSCs derived from various adult and perinatal tissue sources have been used in both the autologous and allogeneic transplantation setting for many decades (21). Numerous preclinical and clinical studies have evaluated the safety and side effects of therapeutic MSCs (15, 22–24). Nonetheless, some reports on potential adverse events highlight a general need for better MSC characterization and handling (15, 24, 25). Multiple research and clinical groups recently reported that heparin improves both the safety and efficacy of MSC therapy (18, 26, 27).

Our initial two clinical reports showed that intravenous infusion of human BM-MSCs and DSCs triggers an innate immune attack, termed the instant blood-mediated inflammatory reaction (IBMIR) (15, 18, 28). Liao et al. recently confirmed this finding demonstrating that BM-MSCs are not fully compatible with blood due to their intrinsic Tissue Factor (TF/CD142) expression, particularly after extensive expansion, which was furthermore found to be conserved among different species of mammals (27).

Liao et al. found that large doses of MSCs induced symptoms of respiratory and/or heart failure attributed to the triggering of intravascular thrombosis promoting cell embolization in the lungs (27). In contrast, clinically more relevant MSC doses induced only mild and reversible coagulation, but anticoagulation with heparin (400 U/kg) effectively prevented MSC-induced coagulation and concomitant adverse events of large cell doses.

The most common cell dose infused in patients is $1–2 \times 10^6$ cells/kg, but does up to $10–20 \times 10^6$ cells/kg have also been tested (15). Thus, a major bottleneck is the need for robust *ex vivo* expansion of GMP grade cell product to generate clinically relevant cell doses (25). A practical solution to overcome these restrictions may be the use of MSCs generated from other tissue sources with a more favorable amount of starting material and better growth characteristics during *ex vivo* expansion, such as placenta-derived DSCs.

We previously reported on the good safety and efficacy of DSCs in treatment of GvHD and HC following HSCT (29, 30) as well as in experimental setting (31, 32). When employed at the typical low clinical cell doses, DSCs demonstrated a safe toxicity profile and no side effects in the clinical setting (33) and in an animal model of BMT and GvHD (31). This is in analogy to very recent reports by Perlee et al. that the infusion of clinical-grade TF-bearing adipose tissue (AT)-MSCs is safe at the typically employed clinical doses (34, 35).

We here report the comprehensive preclinical toxicity study of systemic DSC infusion in two animal models (rat and mouse) and in series of *in vitro* assays employing human blood.

MATERIALS AND METHODS

Isolation, Quality Testing, and Reconstitution of Stromal Cells

The BM-MSCs and placenta DSCs were isolated and characterized as described previously (18, 29). The MSCs were obtained from BM aspirates of adult healthy volunteer donors and the placentas were collected following elective cesarean section delivery after obtaining informed consent. Donors were negative for HIV, hepatitis B and C infection and syphilis. The cell culture was done in clean rooms and culture media and cell suspensions were negative for bacteria and fungi and polymerase chain reaction (PCR)-negative for Mycoplasma. The cells were expanded for up to four passages in a medium containing 10% fetal calf serum (FCS; Hyclone, Logan, UT).

Flow cytometric analysis was conducted on cells labeled with the antibodies outlined in supporting information, **Table S1**. The cells were labeled with antibodies, fixed with 1% paraformaldehyde, and analyzed on a FACS Aria (Becton Dickinson, Franklin Lakes, NJ); 5,000 gated events were quantified and analyzed with Summit v.4.1 (Dako, Glostrup, Denmark) as previously described (36). The adipogenic and the osteogenic differentiation capacities were evaluated as described previously (37). DSCs did not differentiate well to bone and fat in contrast to BM-MSCs as previously reported in detail (19). DSCs were positive for typical MSC markers CD105, CD166, CD73, CD90, and CD29. They were negative for hematopoietic markers CD34, CD45, and CD14. A detailed FACS analysis of DSCs was previously published (37).

The cells were stored in liquid nitrogen and thawed for intravenous infusion or the *in vitro* assays and washed twice in buffer containing 5% human serum albumin (HSA; CSL Behring, Marburg, Germany). Cell viability, cell size, and other morphological parameter of the reconstituted products were assessed by trypan blue exclusion dye, flow cytometry, and automated electrical impedance-based Cell Counter (CASY-TT; Roche, Germany), as previously described (18, 36).

Intra-Arterial/Venous Infusion Toxicity Evaluation in Rats

All animal experiments were conducted with ethical approval given by the local authority (Stockholms Norra djurförsöksetiska nämnd, ethical approval #N138/10). Adult male Sprague-Dawley rats (mean body weight = 340 ± 4.9 grams) were permitted to have access food and water *ad libitum* until surgery. Anesthesia was induced using 4% isoflurane mixed with 100% O_2 and subsequently maintained at 2% isoflurane. The animals were kept normothermic by means of a rectal thermistor coupled with a heating pad. For intravenous cell administration the tail vein was cannulated ($n = 4$).

For intra-aortal administration a midline incision (5 mm) was made proximally on the ventral side of the tail ($n = 4$). The fascia covering the ventral artery was cut, and the exposed artery was ligated distally. Next, a 7-0 silk ligature was tied loosely around the proximal part of the exposed artery, and a microvascular clip was placed over the ventral artery. The artery was cut and a "0.0157" "Sonic" hydrophilic micro-catheter

(Balt Extrusion, France) carrying a micro wire was advanced to a tip position in the thoracic aorta With the catheter in this position, cell suspensions containing 0.5×10^6 DSC dispersed in 0.5 ml saline with 10% fetal calf serum were infused during 1 min. After infusion, the micro-catheter was retracted and the proximal ligature on the tail artery was tightened and the incision was closed.

The animals were returned to their cages with food and water *ad libitum*. The animals were weighed before and 24 h after cell injection. Activity, gait, grooming and motility were evaluated at 2, 4, and 24 h for gross evaluation of the animals' conditions.

Histopathology Evaluation of Rat Organs

Twenty four hours after the transplantation, all animals were sacrificed by decapitation under identical anesthesia as during the initial surgery. The left kidney, liver, spleen and left thigh muscles were snap-frozen and stored at −80°C. Next, 14 μm sagittal cryo-sections were cut serially throughout the organs and stored at −20°C. The sections were rehydrated, fixed in formaldehyde and stained with Hematoxylin-Eosin according to Mayer's protocol. Microscopical analysis was performed on a sample of 10 sections at even intervals throughout the organs to assess presence of tissue infarction.

Short- and Long-Term Toxicity Assay Following DSC Administration in Mice

Female Balb/c mice, 10 to 12 weeks old, were purchased from Scanbur (Sollentuna). The mice were maintained under pathogen-free conditions with controlled humidity (55% ± 5%), 12 h of alternating light and dark, controlled temperature (21 ± 2°C) and high efficiency particulate air (HEPA)-filtered air. Groups of 5–10 mice were kept in individually ventilated cages and were fed autoclaved mouse chow and tap water *ad libitum*. The South Stockholm Ethics Committee for Animal Research approved all the experiments described here (No 8/16).

The DSCs were thawed and washed as outlined above, passed through a 70 μm cell strainer and suspended in sodium chloride 0.9% containing 5% HSA (infusion buffer). Four different doses of DSC ranging from 0.1–2–10^6 cells/mouse were reconstituted in 200 μl of infusion buffer. In some experiments to evaluate the effect of anticoagulation, 10 Units/animal of heparin were added to the infusion buffer. In the toxicity study each mouse received a single infusion of DSC via a lateral tail vein. The mice were humanly killed 3 or 30 days after cell infusion.

Blood Analysis and Serum Biochemistry Analysis in Mice

At designated time (2, 8, 24 h, +3 and +30 days after cell infusion) and after giving anesthesia to the animals, peripheral blood was collected in different type of tubes (EDTA, citrate or heparin tubes). Blood samples were centrifuged immediately and serum or plasma was transferred into new tubes and placed in −80°C until analysis. Frozen serum or plasma was transported with dry ice to the research section of the central blood- and biochemistry lab at the Karolinska University Hospital. Different biochemical indices were analyzed using various instruments according to the manufacturer instruction. Final data was sent back to us in Excel files. Formation of the blood activation markers thrombin-anti-thrombin complex (TAT) and complement component C3 activation fragment a (C3a) in murine plasma was measured with ELISA (Cusabio Biotech LTD, Wuhan, China), at 2, 8, and 24 h after the DSC infusion.

DSC Labeling and *in vivo* Cell Tracking in Mice and Rats

Transfection and transduction of DSCs with luciferin gene containing green fluorescent protein (GFP), was performed as explained elsewhere (38). Lentiviral vector co-expressed GFP and luciferin was kindly provided by Joseph C. Wu (University of California San Francisco, Institute for Regenerative Medicine). Briefly, the virus supernatant was harvested 24 and 48 h after transfection and concentrated by centrifugation at 6,000 g for 16 h at 4°C. The DSCs were infected with the virus supernatant overnight in the presence of polybrene (8 mg/mL). Transduced cells were selected by adding puromycin (1.5 mg/mL) for 48 h Transduced. GFP+ cells were confirmed by use of fluorescence microscopy. Transfected DSCs were sorted using flow cytometry and GFP+ cells were isolated. The purity of GFP+ cell was more than 99%. The surface markers expression profile of DSC-Lu+ was tested with flow cytometry. The DSC-Lu+ immunomodulatory function was evaluated using the mixed lymphocyte reaction (MLR) assays (31).

For the *in vivo* bio-distribution study, a group of Balb/c mice received 1×10^6 DSC-Lu+ cell via a lateral tail vein. At different time points, +1, +4, +6 h, and 1, 2, 3, 4, and 7 days after cell infusions the mice were imaged using a Xenogen IVIS100 imaging system, using the manufacturer's directions. For *in vivo* imaging of the cells, mice were anesthetized with isoflurane and injected intraperitoneally with 50 mg/kg D-luciferin (Caliper, Hopkintown, MA) at 5–10 min was given before imaging. To control the background photon emission, the obtained data were subjected to average background subtraction, using data from control animals that were only injected with identical doses of luciferin.

For *in vivo* bio-distribution studies in rats, Sprague-Dawley rats were anesthetized and catheterized as described above. Next, the animals received intra-aortal injections ($n = 2$) and tail-vein injections ($n = 2$) with 1×10^6 DSCs labeled with [2-^{18}F]-2-fluoro-2-deoxy-D-glucose (^{18}F-FDG) for dynamic *in-vivo* positron emission tomography of cell trafficking immediately after injection. The ^{18}F-FDG used was an aliquot obtained from daily productions for clinical PET at the Karolinska University Hospital and had passed all quality requirements for administrations in humans. ^{18}F-FDG (10–20 MBq, 500 μL) followed by a saline flush of 300 μL.

Histopathology Evaluation of Organs in Mice

Sample preparation was conducted as explained previously (39). Briefly, at the appropriate time point (day +3 and +30 after cell infusion), mice ($N = 6$ in each group/time point) were given general anesthesia and killed by cervical dislocation. Lung, liver, kidney, spleen, and GI tract were harvested and placed in a sealed container with 4% paraformaldehyde. After harvesting, the lungs were perfused with 4% paraformaldehyde through the trachea, and then transferred to the sealed container. After 24 h the tissues were transferred to 70% ethanol and were thereafter dehydrated using increasing concentrations of ethanol, cleared with xylene and embedded in paraffin. The tissues were cut 5 μm thicknesses and deparaffinized with tissue clear reagent, rehydrated in series of ethanol in decreased concentrations. The sections were stained with hematoxylin and eosin, dehydrated in ethanol. Tissue Tek Prisma (Sakura Finetek Inc., Torrance, CA, USA). An automated slide-stainer was used for the staining procedure according to the manufacturer's recommendations. Samples from different treatment as well as the control group were observed and scored by a blinded animal histopathologist.

Clotting Analysis of DSCs in Human Blood and Plasma *in vitro*

The clotting time (in seconds) was recorded on a semi-automatic 10-channel ball coagulometer (MC10plus; Merlin Medical ABW Medizin und Technik GmbH, Lemgo, Germany), as reported earlier. Frozen aliquots of DSCs were thawed, washed twice, and re-suspended in buffer containing 10% HSA, with or without supplementation of low-dose heparin, as indicated in the figure legends. Sodium citrate anti-coagulated human blood was obtained from healthy volunteers who had not received any medication for at least 10 days, and citrated plasma was collected after centrifugation at 1,000 × *g* for 10 min. The cuvette was filled with either 100 μl of citrated blood diluted 1:1 in PBS, or 100 μl of citrated normal plasma. Blood or plasma was then supplemented with 50 μl of buffer with or without 3,000 stromal cells to a final concentration of 15,000 cells/ml corresponding to a dose of 1–2 × 10^6 cells/kg as typically used in clinical trials. To initiate clotting, 50 μl of 40 mM Ca^{2+} solution was added, to a final concentration of 10 mM.

Statistical Analysis

All data are expressed as mean ± standard error (SE) unless otherwise stated. Differences between groups were analyzed by ANOVA and the Student's *t*-test was used. If the data did not fit a normal distribution, the Mann-Whitney test or the Wilcoxon matched-pairs test was used (two-tailed, 95% confidence intervals). Any $P < 0.05$ were considered statistically significant. Prism software 5.0 was used for analysis and making the graphs (GraphPad Software, La Jolla, CA).

RESULTS

The Effect of IA/IV Infusion of DSC in Rat and Biodistribution and Safety Evaluation Following Intra-Arterial DSC Infusion

In a first exploratory experiment in rats, a dose of 0.5 × 10^6 cells/animal was slowly infused either by cannulation of the tail-vein or by assistance of a micro-catheter into the abdominal aorta to assess principal DSC toxicity. All DSC-treated rats were healthy and no short time adverse effects were observed.

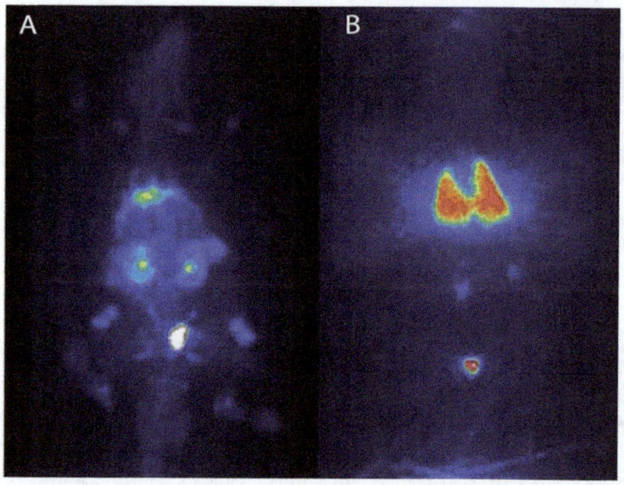

FIGURE 1 | Migration pattern of i.a and i.v. infused human DSCs in the rat. Naïve Sprague-Dawley rats received intra-aortal (i.a) injections ($n = 2$) and tail-vein (i.v) injections ($n = 2$) with 1 × 10^6 DSCs labeled with [2-^{18}F]-2-fluoro-2-deoxy-D-glucose (^{18}F-FDG). Dynamic *in-vivo* positron emission tomography of cell trafficking was done immediately after injection. **(A,B)**, are cell distribution patterns following i.a and i.v infusion of DSC, respectively.

No negative effect on body weight was seen in any of the animals at 24 h. All animals showed normal motility and exploratory behavior at 2, 4, and 24 h post cell infusion. Macroscopic and histological analysis of internal organs did not reveal any evidence for infarction, hemorrhage or pathologic lesion (data not shown).

Injections of ^{18}F-FDG-labeled DSCs and micro-PET imaging 90 min following DSC i.a. injection in rats showed radioactivity distribution primarily in the abdominal organs whereas i.v. infusion resulted in radioactivity distribution almost exclusively in the lungs (**Figure 1**).

To examine possible effects on organs if the DSCs were administered intra-arterially, we performed intra-aortic infusions of cells proximal to the renal arteries by endovascular technique and analyzed clinical behavior and possible ischemic events in the liver, kidney, spleen and muscles in the lower extremities. During the 24 h follow up, the animals did not lose weight, showed normal activity, gait, grooming and motility. H&E staining of liver, kidney, spleen, and muscle sample were normal and without evidence of ischemic- or other tissue injury (data not shown).

Short- and Long-Term Toxicity in Mice

All animals tolerated the intravenous cell infusion even by the dose of 1 × 10^6 cells/mouse ($N = 5$–9 in each group/time point), which is equal to 40 × 10^6 cell/kg. None of them showed any immediate side effects including restless, breathing problem (dyspnea) or any changes in grooming and activity.

However, the dose of 2 × 10^6 cell/mouse (equal to 80 × 10^6 cells/kg) should be infused slowly and carefully otherwise

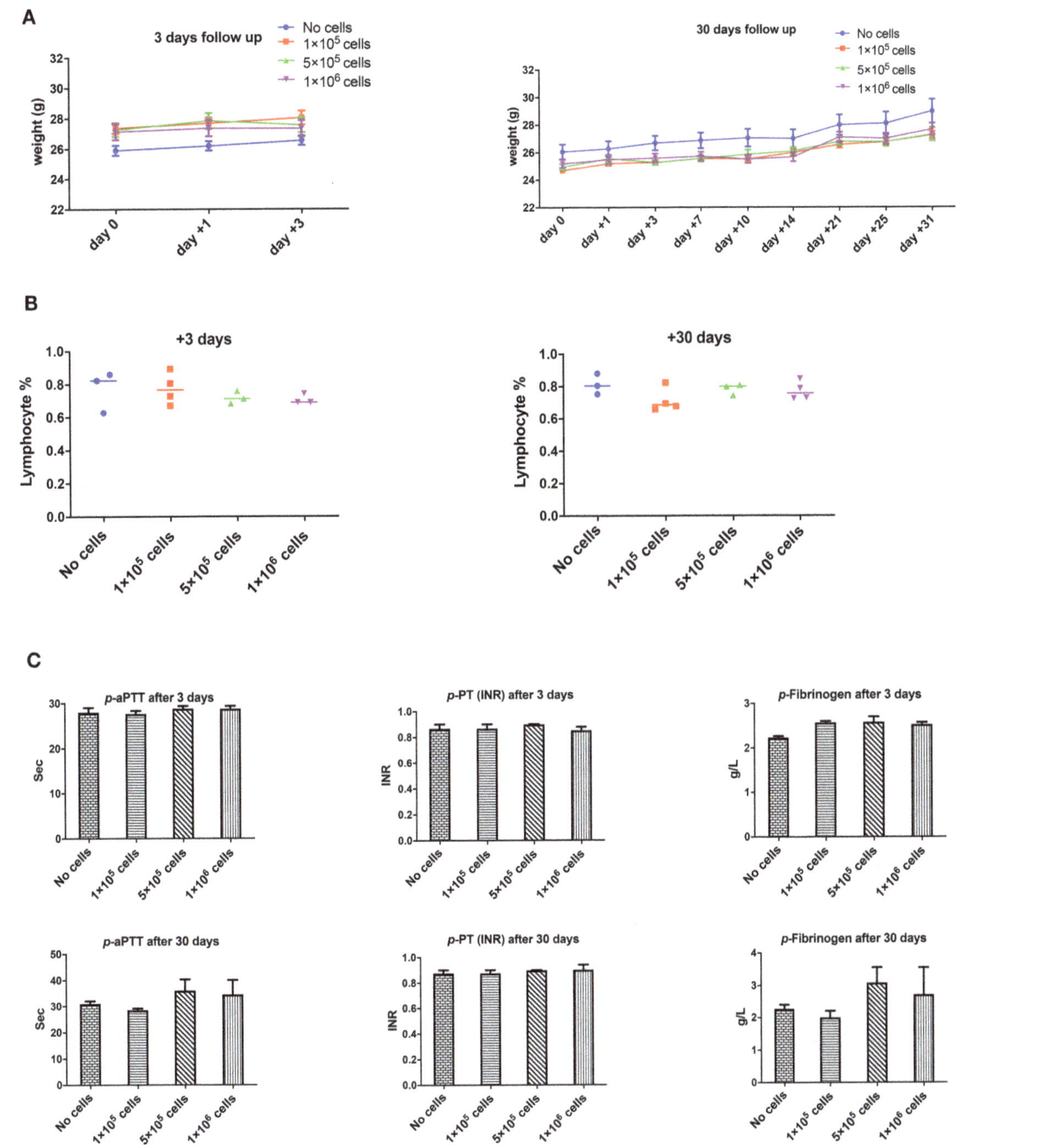

FIGURE 2 | The effect of human DSCs infusion on the mice weight, blood and coagulation system. To evaluate the clinical and biological changes in animals, human DSCs were infused i.v. at different doses (0–1 × 10^6 cell/mouse). The outcome was measured in short (+3 days after) or long term (+30 days after) follow up (N = 3–5 mouse per dose/timepoint). **(A)** Weight and general health of treated animals was not affected by any dose in short and long term. **(B)** The frequency of peripheral blood lymphocyte did not change after human cell infusion. **(C)** Plasma level of coagulation factors was not influenced by the cell infusion as well.

rapid infusion might induce restless, dyspnea and finally death (2 out of 5 animals). It seems that the problem could be due to transient lung emboli (observed in few numbers of healthy mice), as observed by Perlee et al. in their mouse model (35), but which resolved within 24 h. No changes in body weight were seen in any of the mice 3 or 30 days after DSC infusion (**Figure 2A**).

The Effect of DSC Infusion on Blood Count and Coagulation

None of the complete blood count (CBC) items changed in short- or long-term follow-up after different doses of intravenous DSC infusion in mice. Peripheral blood lymphocyte frequency did not alter by cell dose at different time points (**Figure 2B**).

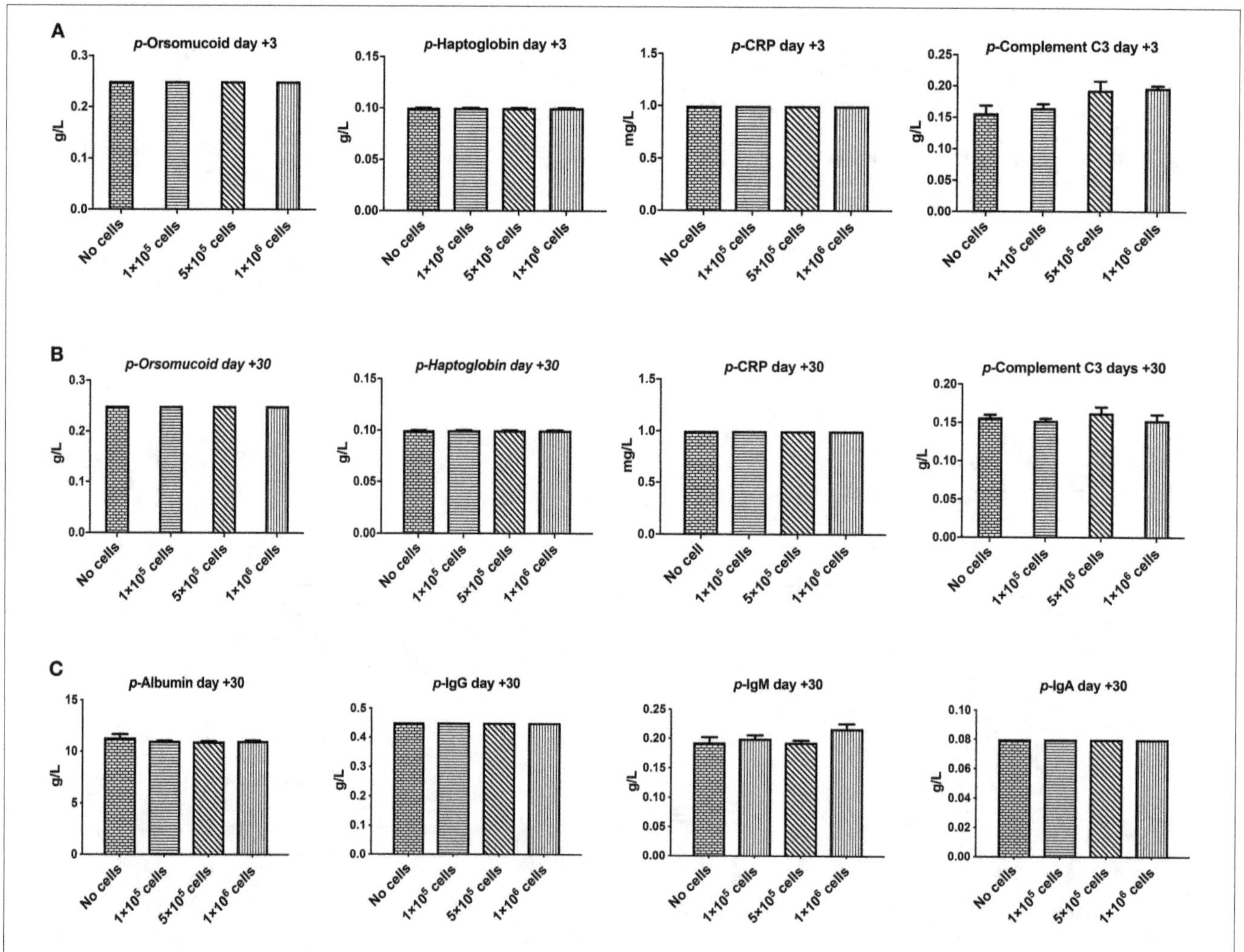

FIGURE 3 | The effect of human DSCs infusion on the inflammatory markers, serum proteins level. Early and delayed inflammatory reaction could be an important and relevant side effect following systemic cell therapy. Healthy mice received human DSCs i.v. at different doses (0–1 × 10^6 cell/mouse) ($N = 3$–5 mouse per dose/timepoint). **(A–C)** The acute inflammatory reactions' protein as well as serum level of immunoglobulin and albumin were measured and showed no significant changes in the short (+3 days after) or the long term (+30 days after) follow up.

One of the most important issues with intravascular cell therapy procedures is the evaluation of coagulopathy and the blood clotting system. We evaluated the effect of DSC infusion on the mice coagulation system. As shown in **Figure 2C**, coagulation cascade proteins or the protease was not affected by the DSC infusion and different doses of DSCs did not change activated partial thromboplastin time (aPTT), prothrombin time/international normalized ratio (PT-INR) and Fibrinogen level neither in short nor long term.

Plasma Inflammatory and Hemolysis Markers and DSC Infusion

Orosomucoid, Haptoglobin, and C-reactive protein that are indicators for acute inflammation response and hemolysis were evaluated at 3 and 30 days after different doses of DSC infusion in mice. As shown in **Figures 3A,B** none of these indices were changed following cell infusion. The major plasma complement C3 levels did not show any significant deviation after different cell dose infusions (**Figure 3B**).

We also evaluated other plasma proteins especially immunoglobulin and albumin levels. DSC infusion did not induce any changes in plasma protein levels, and it seemed that up to 1 × 10^6 cells/mouse are safe for plasma protein balance both in the short and in the long term follow up (**Figure 3C**).

Liver and Kidney Function Was Not Affected by DSC Infusion

Liver and kidney functions tests are crucial toxicity evaluation for any new medication. As shown in **Figure 4**, plasma levels of liver parameters ALAT, ASAT, ALk P as well as creatinine and Urea were not affected by any cell dose at short or long-term assays (**Figures 4A,B**). On the contrary, a high dose of DSCs infusion significantly decreased the ALAT level in manipulated animals. DSC infusion did not affect the plasma lipid profile, although some non-significant decrement in the

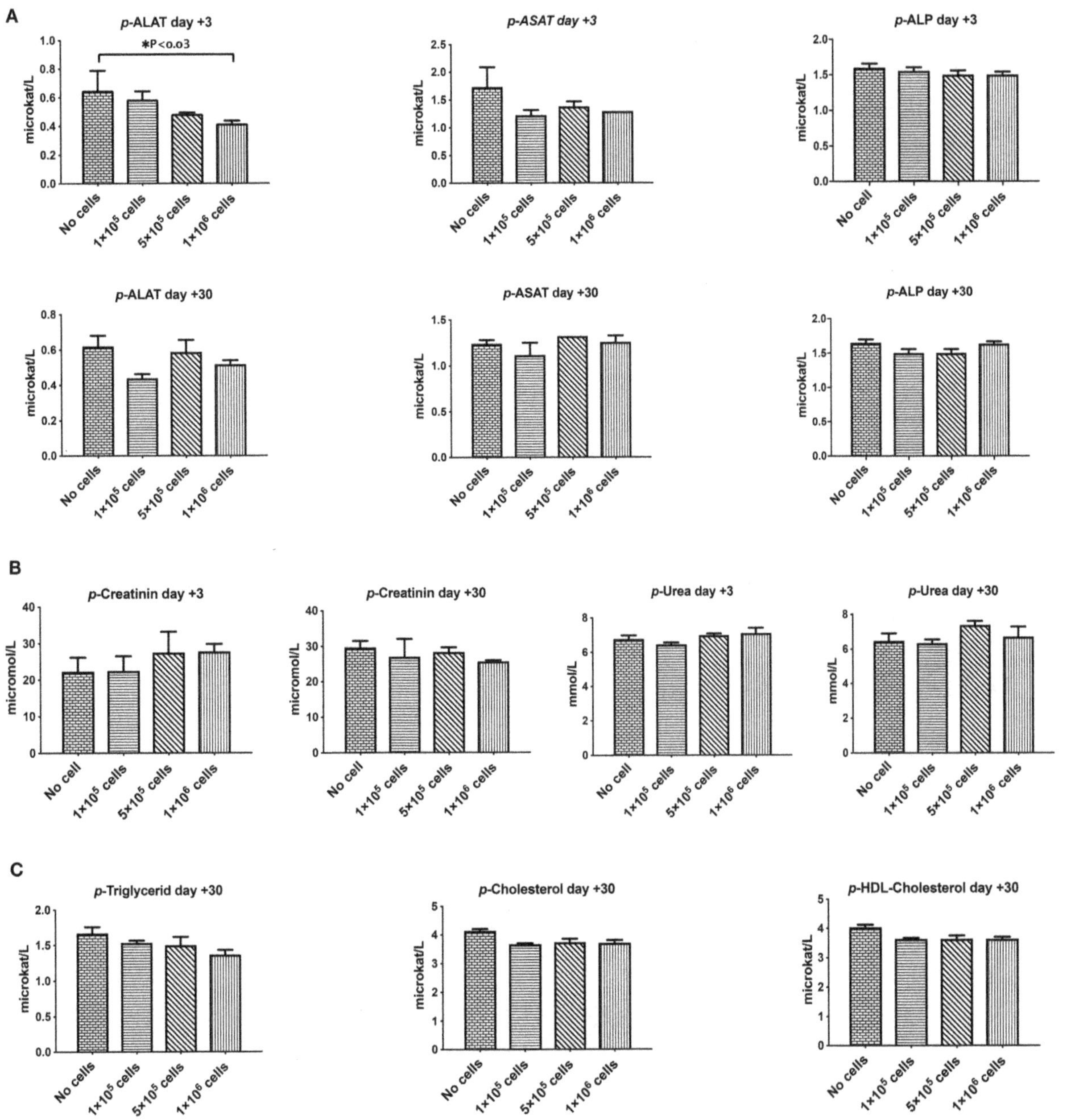

FIGURE 4 | The effect of human DSCs infusion on the liver and kidney function. Organ toxicity is one of the most important concerns in all cell based therapy protocols. Healthy mice received human DSCs i.v. at different doses (0–1 $\times$ 10^6 cell/mouse) (N = 3–5 mouse per dose/timepoint). **(A)** Liver enzymes **(B)** kidney function biomarkers and **(C)** serum lipid profile were measured and compared among different group in short and long term following cell infusion. Interestingly, human cell infusion ranged from 0–40 $\times$ 10^6 cell/kg and did not show any toxicity relevant to these organs.

plasma levels of cholesterol and triglyceride were observed at day +30 (**Figure 4C**).

In vivo Tracking of the Infused Therapeutic Cell Products in Mice

We next evaluated the migration pattern of injected cells in healthy animals. The infused DSC first moved to the lung (**Figure 5**), and the signal continuously faded with time. Signals in the lungs were detected up to 4 days after cell infusion, but not thereafter. We could not detect any significant signals in liver or spleen at any evaluated time period (**Figure 5**).

Histopathology Evaluation of Liver, Kidney, and Lung Following DSC Infusion

Histopathology evaluation of the liver at short (+3) or long term (+30) follow up did not show any significant abnormality related to cell infusion. Mild diffuse glycogenesis, occasional individual cell with micro vesicular lipidosis, moderate cell

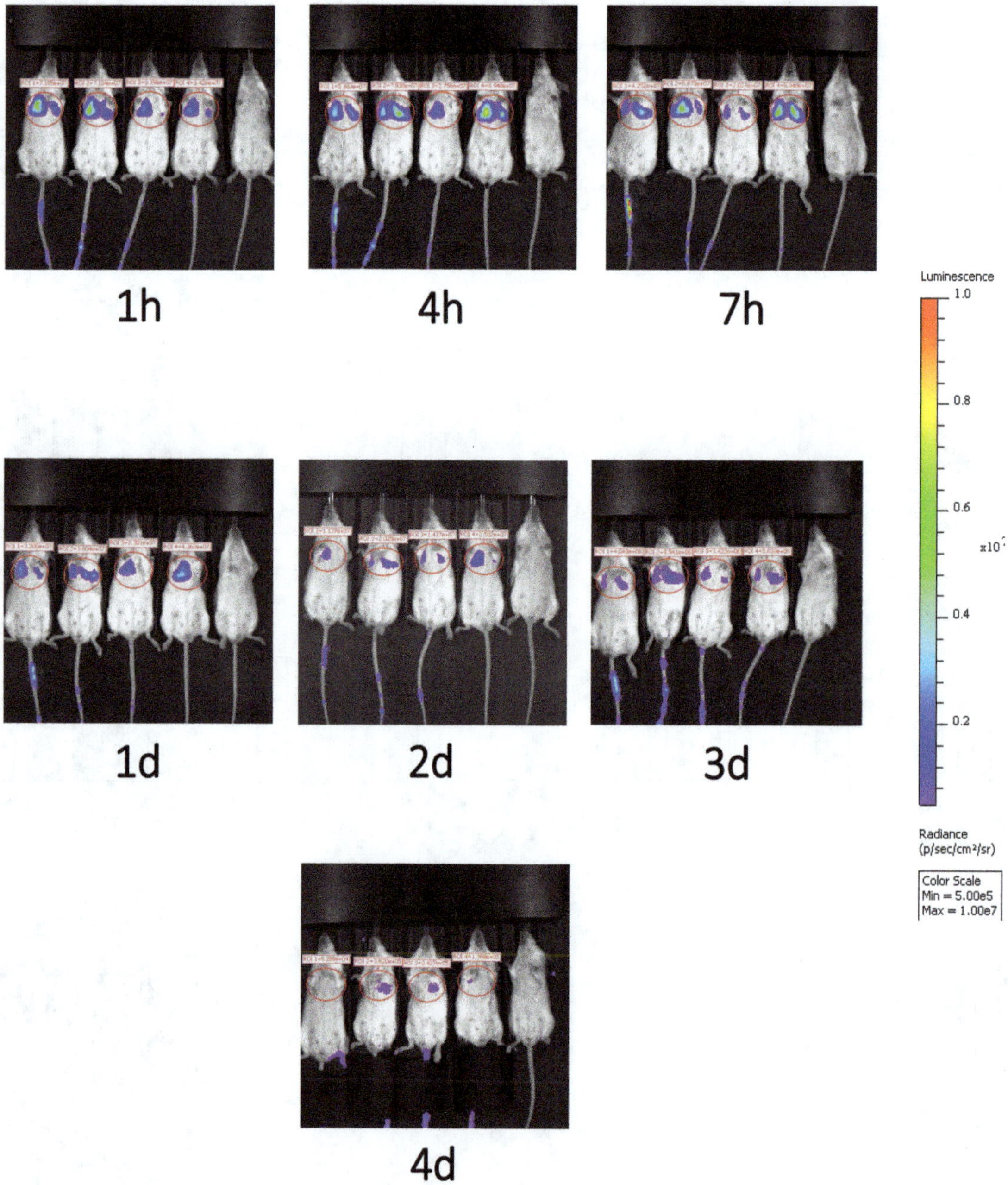

FIGURE 5 | Migration pattern of i.v. infused human DSCs in healthy mice. Migration pattern of systemic infused stromal cells, in live animals, was done by using labeled cells. Human DSCs were transduced with Luciferin gene containing GFP using lentivirus vector (methods). 1×10^6 cell/mouse were infused i.v. and after light anesthesia, live animals ($N = 2$) were kept in IVIS CCD camera and at different time points whole body scan were recorded. Infused cells first pass to the lung and stayed there up to 4 days. In the Luciferin tracking model, just live cells will be recorded. It means that the human DCs mainly reside it the lung.

vacuolization; moderate karyomegaly/anisokaryosis; scattered gray-green pigment in kupffer cells among others were observed in liver samples. However, none of this observations were related to cell infusion or even cell dose. Heparin infusion did have any effect on histopathological manifestation (**Figure 6A**). Kidney tissue was not essentially affected by any dose of cell infusion with or without heparin at any time point (data not shown).

The lungs are the first organs in which the infused cells will arrest. Thus, histopathology signs for thromboembolic features, hemorrhage and granulomatous formation were carefully evaluated among different groups. DSC infusion did not induce any of the mentioned pathologic features in the short term (+3) period. Heparin also did not change the histopathology pattern in short term.

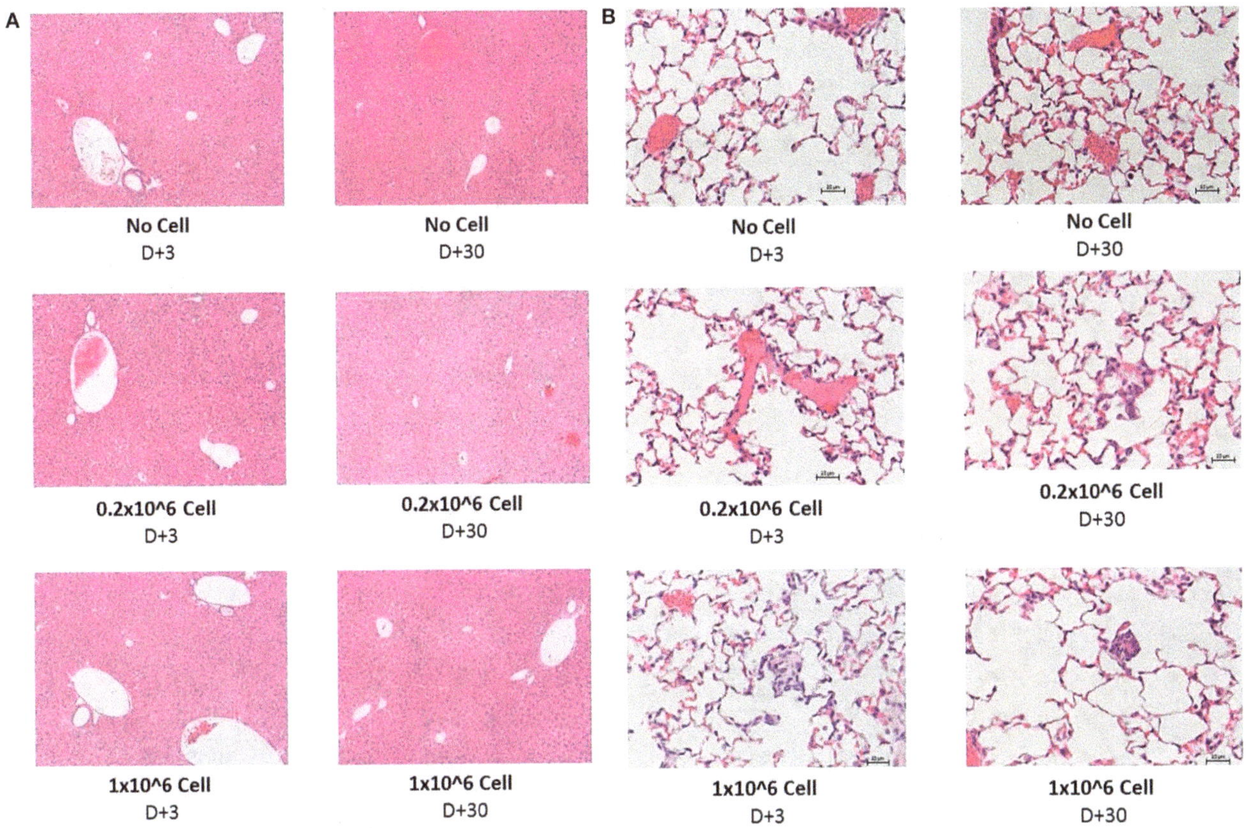

FIGURE 6 | Histopathology evaluation of lung and liver after i.v. infusion of human DSCs in healthy mice. To assess the effect of cell infusion on vital organs liver, lung and kidney were prepared for histopathology evaluation 3 and 30 days after different doses of cell infusion ($N = 3$ mouse per dose/timepoint). **(A)** Liver tissue with no cell, 0, 2, and 1×10^6 DSC IV injection (+heparin) 3 days (left column) and 30 days after cell infusion. **(B)** Lung tissue with no cell, 0, 2, and 1×10^6 DSC IV injection (+heparin) 3 days (left column) and 30 days after cell infusion.

Moderate diffuse hyperemia; moderately frequent interstitial foci of unidentified/inflammatory cells; scattered foamy macrophages and occasional megakaryocytes were seen more often in the 30 days follow up. Few vessels contain apparently coagulated material. However, it did not have any correlation to cell dose or heparin treatment. This finding was also observed in the control group (**Figure 6B**).

Cell Morphology, Immunophenotype, and *in vitro* Clotting Analysis of Stromal Cells With and Without Addition of Heparin

Compared to BM-MSCs, the DSCs had 16% better viability post thawing (Mean 69% vs. 85%, $P = 0.05$, **Figure 7A**). The DSCs had a smaller size than MSCs (Mean peak diameter 16 vs. 20 μm, $P = 0.05$, **Figure 7B**) and a smaller cell volume (Mean peak volume 2,000 vs. 6,500 fl, $P < 0.05$, **Figure 7C**). CD142 expression was higher for DSCs than BM-MSCs (Mean 43% vs. 5% positive, $P < 0.001$, **Figure 7D**). All parameters were unchanged by the addition of heparin (**Figures 7A–D**).

There was a stronger clotting of DSCs compared to BM-MSCs in human blood and plasma *in vitro* ($P < 0.001$, **Figures 7E,F**). Clotting was stronger for DSCs (75% reduction) than MSCs (50% reduction) when compared to the buffer controls ($P < 0.0001$ and $P < 0.001$, respectively). Clotting in human blood and plasma was abrogated for both cell types by the addition of low-dose heparin (**Figures 7E,F**).

In vivo Monitoring of Systemic Blood Parameters After Stromal Cell Infusion With and Without Heparin in Mice

The peak for the generation of coagulation activation marker TAT was found at 8 h post BM-MSC and DSC infusion compared to the buffer control ($P < 0.01$ and $P < 0.001$, respectively, **Figure 8A**). TAT was still detected following DSCs, but not following BM-MSCs infusion after 24 h ($P < 0.05$). TAT generation was decreased by the supplementation of heparin (10 U/animal), with minor TAT formation at 8 h post DSCs infusion ($P < 0.05$) (**Figure 8A**).

Similarly, complement activation marker C3a demonstrated a peak at 8 h post BM-MSC and DSC infusion compared to the buffer control ($P = 0.06$ and $P < 0.05$, respectively, **Figure 8B**). This was still evident for DSCs but not BM-MSCs at 24 h post cell infusion ($P < 0.01$), C3a generation was antagonized by

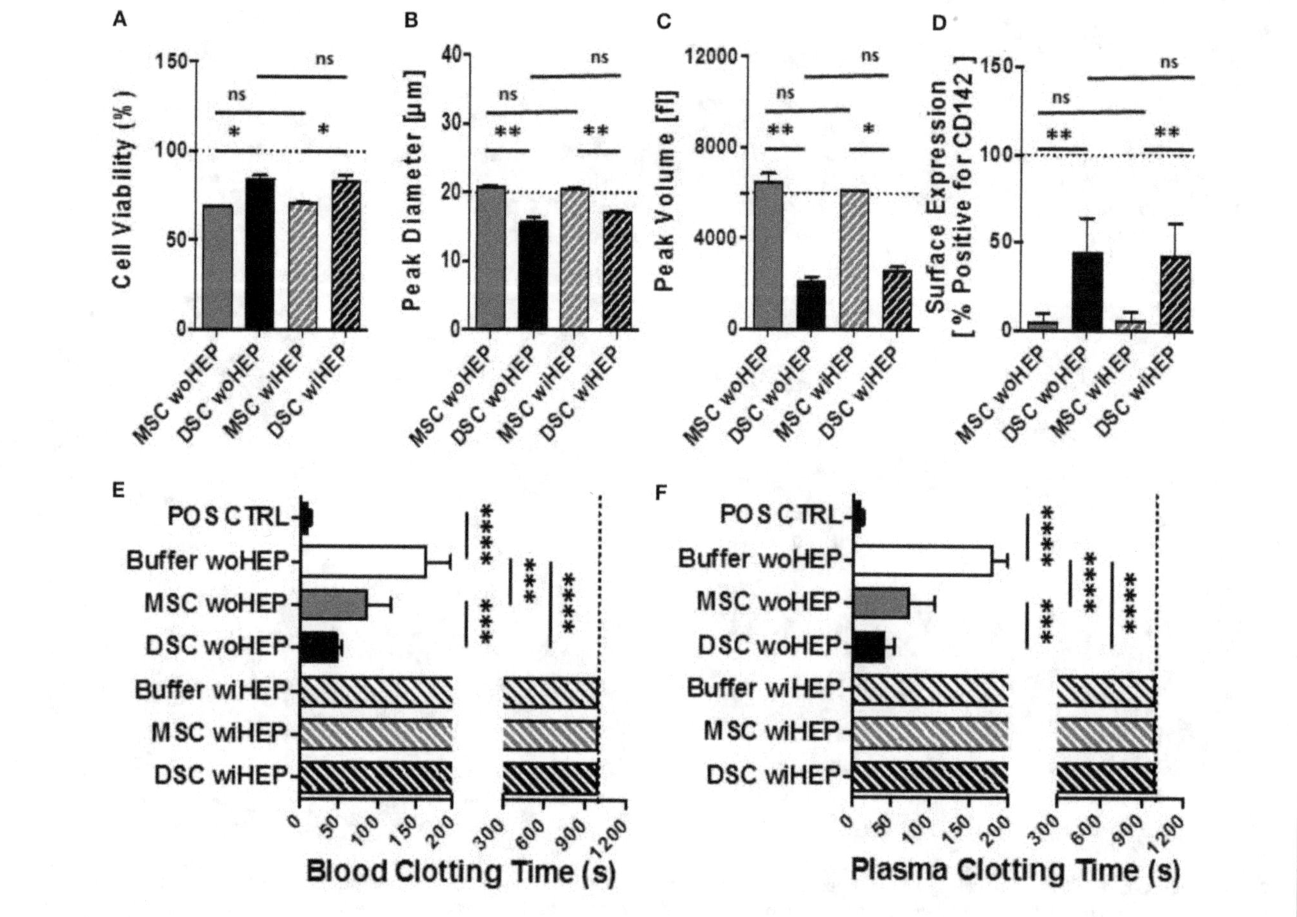

FIGURE 7 | Cell morphology, immunophenotype, and *in vitro* clotting analysis of stromal cells prepared with and without addition of heparin. In order to simulate the preparation and formulation of our clinical products, aliquots of cryobanked therapeutic bone marrow mesenchymal stromal cells (MSC) and placenta-derived decidua stromal cells (DSC) were thawed, reconstituted and supplemented in buffer supplemented with or without low-dose heparin (HEP) as commonly done during clinical procedures, and then evaluated for viability, phenotype, and clotting time in human blood and plasma. **(A–C)** Cell suspensions of stromal cell batches from different donors ($n = 4$ each cell type) were analyzed with automated electrical impedance-based CASY counter for quantification of parameters of importance for systemic cell infusion: **(A)** Cell viability (%), **(B)** Peak diameter (um: micrometers), and **(C)** Peak volume (fL: femtoliters), **(D)** Flow cytometric analysis for cell surface expression of Tissue Factor (CD142, % of cells positive, $n = 8$) compared to isotype control, and **(E,F)** Analysis of clotting time (in seconds, $n = 14$ tests) after exposure of stromal cells (15,000 cells/mL) resuspended with and without heparin (10 U/mL) to fresh recalcified human blood **(E)** or plasma **(F)**, using a semiautomatic 10-channel ball coagulometer. Mean ± SD. $^{*}P < 0.05$, $^{**}P < 0.01$, $^{***}P < 0.001$, and $^{****}P < 0.0001$, and ns, not significant.

heparin, with only weak and non-significant C3a formation at the observed time points (**Figure 8B**).

DISCUSSION

There is a bulk of data regarding the safety of BM-MSCs (22, 40). All showed that BM-MSCs are safe to infuse with few if any side effects (15). DSCs differ in many ways from stromal cells from other tissues such as bone marrow (18, 19). Therefore, comprehensive safety analysis is needed for DSCs.

Before treating patients, we injected DSCs to rats, as reported in this article and in rabbits (32) and found it to be safe with no side effects. Subsequently, we treated patients with acute and chronic GvHD, hemorrhagic cystitis and acute respiratory distress syndrome (30, 33, 41, 42). Side effects were minimal and severe adverse events were those commonly seen in patients undergoing allogeneic HSCT, such as infections, hemorrhages, graft failure and multi-organ failure (29, 33).

DSCs like other cells first go to the lung after i.v. infusion (41, 43, 44). These findings are confirmed in mice and rats in the present study, where homing to the lungs is seen up to several days after i.v. infusion. Special attention should be made to evaluate possible DSCs induced side-effects in the lungs such as thrombosis or pneumonias. We found no pulmonary embolism in the present study where DSC were infused to mice in 40 times higher dose than used in humans, which is in agreement with studies on AT-MSCs by Perlee et al. (34, 35).

We previously reported that mice, treated with total body irradiation prior to high dose DSCs infusion (20×10^6/kg), died from pulmonary embolism (31). Using BM-MSCs, an increased risk for pneumonia-related death after HSCT was found at our center (45). DSCs have a stronger effect on coagulation compared to BM-MSCs (18). However, DSCs are only half the size of bone

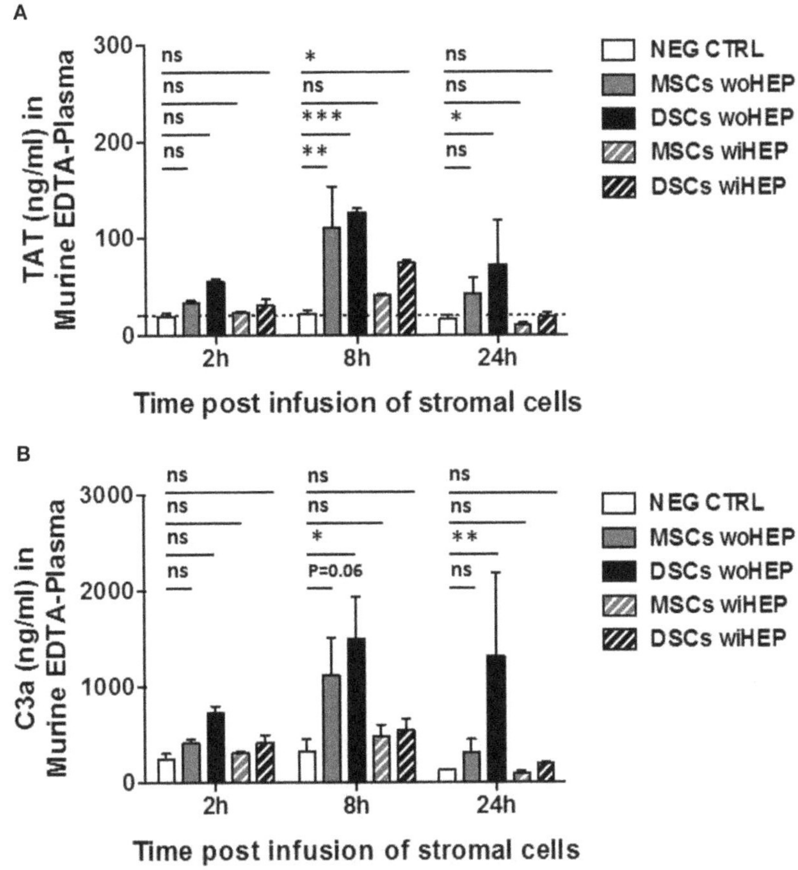

FIGURE 8 | Analysis of systemic coagulation and complement activation markers in mice after systemic infusion of stromal cells prepared with and without heparin. Aliquots of cryobanked therapeutic bone marrow mesenchymal stromal cells (MSCs) and placenta-derived decidua stromal cells (DSCs) were thawed, washed and reconstituted in buffer supplemented with or without low-dose heparin (HEP, final dose 10 units/animal), and infused at 500,000 cells/animal ($n = 2$ animals each bar), and kinetics of coagulation and complement activation markers monitored after 2, 8, and 24 h. Quantification of: **(A)** Coagulation activation marker TAT (ng/mL) and **(B)** Complement activation marker C3a (ng/mL) in murine EDTA-plasma. Mean ± SD. $^{*}P < 0.05$, $^{**}P < 0.01$, and $^{***}P < 0.001$, and ns, not significant. Dashed lines indicate background levels of C3a and TAT. TAT, thrombin-anti-thrombin complex, and C3a, complement component 3 activation fragment a.

marrow MSCs as confirmed in the present study. Furthermore, there was no primary toxicity seen with DSCs doses up to 40×10^6/kg.

One way of overcoming pulmonary trapping of cells and facilitate homing is to administer cells intra-arterially upstream of target tissue. In this study we found that intra-aortal injections of DSCs in rats does not cause animal morbidity or tissue injury and result in DSC distribution in the abdominal organs instead of the lungs.

In patients infused with I-111-marked DSCs, signals were first found in the lung and after 48 h in liver and spleen. This is in contrast to the present study in mice where Luciferase-marked DSCs were found in the lungs of the animals up to 4 days, but in no other organ (**Figure 4**). This difference may be due to that Luciferase is a marker for live cells, whereas I-111 will also appear on dying cells. This seems reasonable due to the xeno-reactive mice immune system, which may distract the human DSCs already in the lung.

In this study, treating animals with doses up to 40×10^6/kg we did not see any abnormal changes in laboratory values, including lymphocytes, hemoglobin, antithrombin, coagulation parameters, haptoglobin, complement C3, albumin, creatinine, liver enzymes, triglycerides, cholesterol, and HDL-cholesterol. This confirms that even extremely high doses of DSCs seem safe in accordance with clinical safety studies using doses around 1×10^6 DSCs/kg. Histology of all examined organs did not show any adverse effects by infusion of DSCs.

The results from the clotting experiments confirmed previous data (18). DSCs had stronger clotting than BM-MSCs. By the addition of low dose heparin, clotting was entirely abrogated (**Figure 7**). Kinetics for *in vivo* monitoring of systemic coagulation and complement activation markers using DSCs and MSCs infusion in mice, revealed a weak triggering of the coagulation and complement cascades. We found similarly to our results from *in vitro* blood exposure, that DSCs without heparin were slightly more pro-coagulant *in vivo* than BM-MSCs. These data support the use of heparin when infusing stromal cells to patients for clinical use. Heparin also prevented BM-MSCs induced coagulation and the acute adverse events in experimental colitis (27).

In the comparison between BM-MSCs and DSCs, the former donors were slightly older. BM-MSCs were taken from bone marrow transplant donors with an age ranging from 30 to 50 years of age. The DSCs were from fertile women ranging in age between 20 and 35 years of age. The difference in donor age, most probably had no effect on the differences seen with coagulation and other parameters. We previously found that donor age had no impact on GvHD response to MSC therapy (46). We also reported that DSCs were more effective to treat acute GvHD than BM-MSCs (29). The differences in efficacy are more likely due to differences in source of MSC, than to MSC donor age.

As outlined in the introduction, it was demonstrated in mice that BM-MSCs following contact with activated T cells or NK cells, activate caspase and undergo apoptosis, which appears to be critical for MSC-induced systemic immunosuppression *in vivo* (11). Apoptosis is induced by the bystander release of cytotoxic granules by the activated immune cells. The killing is contact-dependent but not antigen-specific and does not require the engagement of the immunological synapses. MSCs, which undergo apoptosis, promote the chemotaxis of monocytes and macrophages that phagocytose the MSCs (8, 12–14). By these mechanisms, the macrophages produce anti-inflammatory activities *in vivo*, which are dependent, among others, on host indoleamine deoxygenase. If similar mechanisms are also induced by DSCs remains to be elucidated.

The present study demonstrates the safety of infusing DSCs i.v. and i.a. with normal appearing in rats and mice, no side effects and no toxicity to any organs. Minor effects on coagulation were normalized by heparin infusion.

AUTHOR CONTRIBUTIONS

BS, GM, SH, and OR: concept and design. SG, GM, SH, and OR: financial support. BS and OR: administrative support. BS, GM, FA, ES, BK, RC, JK-M, SG, GM, SH, and OR: collection and assembly of data and final approval of manuscript. BS, GM, FA, GM, SH, and OR: data analysis and interpretation.

FUNDING

We thank Gunilla Tillinger for help in the preparation of this manuscript. This study was supported by grants to OR from the Swedish Cancer Foundation (CAN 2013/671), the Swedish Research Council (K2014-64X-05971-34-4), the Children's Cancer Foundation (PR2013-0045), the Cancer Society in Stockholm (111293) and to SH from the Söderberg Foundations, Karolinska Institutet, and the Erling Persson Family foundation. OR was also supported by a distinguished professor award from Karolinska Institutet. GM's contributions were supported by German Research Foundation (DFG) and German Federal Ministry of Education and Research (BMBF) funding through the BSRT (GSC203) and BCRT. SG gratefully acknowledge the German Research Foundation (FOR2165, GE2512/2-2), and the German Federal Ministry of Education, Research (BMBF, DIMEOS: 01EC1402B) for funding his contribution. This study has received support under the European Union's Horizon 2020 research and innovation program (Horizon 2020 Framework Program) under the grant agreements no. 733006 (PACE) and no. 779293 (HIPGEN).

REFERENCES

1. Friedenstein AJ, Petrakova KV, Kurolesova AI, Frolova GP. Heterotopic of bone marrow. Analysis of precursor cells for osteogenic and hematopoietic tissues. *Transplantation.* (1968) 6:230–47. doi: 10.1097/00007890-196803000-00009
2. Le Blanc K, Ringdén O. Immunomodulation by mesenchymal stem cells and clinical experience. *J Intern Med.* (2007) 262:509–25. doi: 10.1111/j.1365-2796.2007.01844.x
3. Pittenger MF, Mackay AM, Beck SC, Jaiswal RK, Douglas R, Mosca JD, et al. Multilineage potential of adult human mesenchymal stem cells. *Science.* (1999) 284:143–7. doi: 10.1126/science.284.5411.143
4. Ringdén O, Uzunel M, Rasmusson I, Remberger M, Sundberg B, Lönnies H, et al. Mesenchymal stem cells for treatment of therapy-resistant graft-versus-host disease. *Transplantation.* (2006) 81:1390–7. doi: 10.1097/01.tp.0000214462.63943.14
5. Melmed GY, Pandak WM, Casey K, Abraham B, Valentine J, Schwartz D, et al. Human Placenta-derived Cells (PDA-001) for the treatment of moderate-to-severe Crohn's disease: a phase 1b/2a study. *Inflamm Bowel Dis.* (2015) 21:1809–16. doi: 10.1097/MIB.0000000000000441
6. Darlington PJ, Boivin MN, Bar-Or A. Harnessing the therapeutic potential of mesenchymal stem cells in multiple sclerosis. *Expert Rev Neurother.* (2011) 11:1295–303. doi: 10.1586/ern.11.113
7. Morando S, Vigo T, Esposito M, Casazza S, Novi G, Principato MC, et al. The therapeutic effect of mesenchymal stem cell transplantation in experimental autoimmune encephalomyelitis is mediated by peripheral and central mechanisms. *Stem Cell Res Ther.* (2012) 3:3. doi: 10.1186/scrt94
8. Galipeau J, Sensébé L. Mesenchymal stromal cells: clinical challenges and therapeutic opportunities. *Cell Stem Cell.* (2018) 22:824–33. doi: 10.1016/j.stem.2018.05.004
9. Mathiasen AB, Qayyum AA, Jørgensen E, Helqvist S, Fischer-Nielsen A, Kofoed KF, et al. Bone marrow-derived mesenchymal stromal cell treatment in patients with severe ischaemic heart failure: a randomized placebo-controlled trial (MSC-HF trial). *Eur Heart J.* (2015) 36:1744–53. doi: 10.1093/eurheartj/ehv136
10. Singh A, Singh A, Sen D. Mesenchymal stem cells in cardiac regeneration: a detailed progress report of the last 6 years (2010–2015). *Stem Cell Res Ther.* (2016) 7:82. doi: 10.1186/s13287-016-0341-0

11. Galleu A, Riffo-Vasquez Y, Trento C, Lomas C, Dolcetti L, Cheung TS, et al. Apoptosis in mesenchymal stromal cells induces *in vivo* recipient-mediated immunomodulation. *Sci Transl Med.* (2017) 9:eaam7828. doi: 10.1126/scitranslmed.aam7828
12. Moll G, Geißler S, Catar R, Ignatowicz L, Hoogduijn MJ, Strunk D, et al. Cryopreserved or fresh mesenchymal stromal cells: only a matter of taste or key to unleash the full clinical potential of MSC therapy? *Adv Exp Med Biol.* (2016) 951:77–98. doi: 10.1007/978-3-319-454 57-3_7
13. Weiss ARR, Dahlke MH. Immunomodulation by Mesenchymal Stem Cells (MSCs): mechanisms of action of living, apoptotic, dead MSCs. *Front Immunol.* (2019) 10:1191. doi: 10.3389/fimmu.2019.01191
14. Weiss DJ, English K, Krasnodembskaya A, Isaza-Correa JM, Hawthorne IJ, Mahon BP. The necrobiology of mesenchymal stromal cells affects therapeutic efficacy. *Front Immunol.* (2019) 10:1228. doi: 10.3389/fimmu.2019.01228
15. Moll G, Ankrum JA, Kamhieh-Milz J, Bieback K, Ringdén O, Volk HD, et al. Intravascular mesenchymal stromal/stem cell therapy product diversification: time for new clinical guidelines. *Trends Mol Med.* (2019) 25:149–63. doi: 10.1016/j.molmed.2018.12.006
16. Munoz-Suano A, Hamilton AB, Betz AG. Gimme shelter: the immune system during pregnancy. *Immunol Rev.* (2011) 241:20–38. doi: 10.1111/j.1600-065X.2011.01002.x
17. Vento-Tormo R, Efremova M, Botting RA, Turco MY, Vento-Tormo M, Meyer KB, et al. Single-cell reconstruction of the early maternal-fetal interface in humans. *Nature.* (2018) 563:347–53. doi: 10.1038/s41586-018-0698-6
18. Moll G, Ignatowicz L, Catar R, Luecht C, Sadeghi B, Hamad O, et al. Different procoagulant activity of therapeutic mesenchymal stromal cells derived from bone marrow and placental decidua. *Stem Cells Dev.* (2015) 24:2269–79. doi: 10.1089/scd.2015.0120
19. Erkers T, Nava S, Yosef J, Ringdén O, Kaipe H. Decidual stromal cells promote regulatory T cells and suppress alloreactivity in a cell contact-dependent manner. *Stem Cells Dev.* (2013) 22:2596–605. doi: 10.1089/scd.2013.0079
20. Karlsson H, Erkers T, Nava S, Ruhm S, Westgren M, Ringdén O. Stromal cells from term fetal membrane are highly suppressive in allogeneic settings *in vitro*. *Clin Exp Immunol.* (2012) 167:543–55. doi: 10.1111/j.1365-2249.2011.04540.x
21. Sharma RR, Pollock K, Hubel A, McKenna D. Mesenchymal stem or stromal cells: a review of clinical applications and manufacturing practices. *Transfusion.* (2014) 54:1418–37. doi: 10.1111/trf.12421
22. Lalu MM, McIntyre L, Pugliese C, Fergusson D, Winston BW, Marshall JC, et al. Safety of cell therapy with mesenchymal stromal cells (SafeCell): a systematic review and meta-analysis of clinical trials. *PLoS ONE.* (2012) 7:e47559. doi: 10.1371/journal.pone.0047559
23. Lalu MM, Sullivan KJ, Mei SH, Moher D, Straus A, Fergusson DA, et al. Evaluating mesenchymal stem cell therapy for sepsis with preclinical meta-analyses prior to initiating a first-in-human trial. *Elife.* (2016) 5:e17850. doi: 10.7554/eLife.17850
24. Caplan H, Olson SD, Kumar A, George M, Prabhakara KS, Wenzel P, et al. Mesenchymal stromal cell therapeutic delivery: translational challenges to clinical application. *Front Immunol.* (2019) 10:1645. doi: 10.3389/fimmu.2019.01645
25. Marks PW, Witten CM, Califf RM. Clarifying stem-cell therapy's benefits and risks. *N Engl J Med.* (2017) 376:1007–9. doi: 10.1056/NEJMp1 613723
26. Gleeson BM, Martin K, Ali MT, Kumar AH, Pillai MG, Kumar SP, et al. Bone marrow-derived mesenchymal stem cells have innate procoagulant activity and cause microvascular obstruction following intracoronary delivery: amelioration by antithrombin therapy. *Stem Cells.* (2015) 33:2726–37. doi: 10.1002/stem.2050
27. Liao L, Shi B, Chang H, Su X, Zhang L, Bi C, et al. Heparin improves BMSC cell therapy: anticoagulant treatment by heparin improves the safety and therapeutic effect of bone marrow-derived mesenchymal stem cell cytotherapy. *Theranostics.* (2017) 7:106–16. doi: 10.7150/thno.16911
28. Moll G, Rasmusson-Duprez I, von Bahr L, Connolly-Andersen AM, Elgue G, Funke L, et al. Are therapeutic human mesenchymal stromal cells compatible with human blood? *Stem Cells.* (2012) 30:1565–74. doi: 10.1002/stem.1111
29. Ringden O, Baygan A, Remberger M, Gustafsson B, Winiarski J, Khoein B, et al. Placenta-derived decidua stromal cells for treatment of severe acute graft-versus-host disease. *Stem Cells Transl Med.* (2018) 7:325–31. doi: 10.1002/sctm.17-0167
30. Aronsson-Kurttila W, Baygan A, Moretti G, Remberger M, Khoein B, Moll G, et al. Placenta-derived decidua stromal cells for hemorrhagic cystitis after stem cell transplantation. *Acta Haematol.* (2018) 139:106–14. doi: 10.1159/000485735
31. Sadeghi B, Heshmati Y, Khoein B, Kaipe H, Uzunel M, Walfridsson J, et al. Xeno-immunosuppressive properties of human decidual stromal cells in mouse models of alloreactivity *in vitro* and *in vivo*. *Cytotherapy.* (2015) 17:1732–45. doi: 10.1016/j.jcyt.2015.09.001
32. Arnberg F, Lundberg J, Olsson A, Samén E, Jaff N, Jussing E, et al. Intra-arterial administration of placenta-derived decidual stromal cells to the superior mesenteric artery in the rabbit: distribution of cells, feasibility, and safety. *Cell Transplant.* (2016) 25:401–10. doi: 10.3727/096368915X 688191
33. Baygan A, Aronsson-Kurttila W, Moretti G, Tibert B, Dahllöf G, Klingspor L, et al. Safety and side effects of using placenta-derived decidual stromal cells for graft-versus-host disease and hemorrhagic cystitis. *Front Immunol.* (2017) 8:795. doi: 10.3389/fimmu.2017.00795
34. Perlee D, van Vught LA, Scicluna BP, Maag A, Lutter R, Kemper EM, et al. Intravenous infusion of human adipose mesenchymal stem cells modifies the host response to lipopolysaccharide in humans: a randomized, single-blind, parallel group, placebo controlled trial. *Stem Cells.* (2018) 36:1778–88. doi: 10.1002/stem.2891
35. Perlee D, de Vos AF, Scicluna BP, Maag A, Mancheño P, de la Rosa O, et al. Role of tissue factor in the procoagulant and antibacterial effects of human adipose-derived mesenchymal stem cells during pneumosepsis in mice. *Stem Cell Res Ther.* (2019) 10:286. doi: 10.1186/s13287-019-1391-x
36. Moll G, Alm JJ, Davies LC, von Bahr L, Heldring N, Stenbeck-Funke L, et al. Do cryopreserved mesenchymal stromal cells display impaired immunomodulatory and therapeutic properties? *Stem Cells.* (2014) 32:2430–42. doi: 10.1002/stem.1729
37. Ringdén O, Erkers T, Nava S, Uzunel M, Iwarsson E, Conrad R, et al. Fetal membrane cells for treatment of steroid-refractory acute graft-versus-host disease. *Stem Cells.* (2013) 31:592–601. doi: 10.1002/stem.1314
38. Cao F, Xie X, Gollan T, Zhao L, Narsinh K, Lee RJ, et al. Comparison of gene-transfer efficiency in human embryonic stem cells. *Mol Imaging Biol.* (2010) 12:15–24. doi: 10.1007/s11307-009-0236-x
39. Sadeghi B, Aghdami N, Hassan Z, Forouzanfar M, Rozell B, Abedi-Valugerdi M, et al. GVHD after chemotherapy conditioning in allogeneic transplanted mice. *Bone Marrow Transplant.* (2008) 42:807–18. doi: 10.1038/bmt.2008.261
40. Yin F, Battiwalla M, Ito S, Feng X, Chinian F, Melenhorst JJ, et al. Bone marrow mesenchymal stromal cells to treat tissue damage in allogeneic stem cell transplant recipients: correlation of biological markers with clinical responses. *Stem Cells.* (2014) 32:1278–88. doi: 10.1002/stem.1638
41. Erkers T, Kaipe H, Nava S, Molldén P, Gustafsson B, Axelsson R, et al. Treatment of severe chronic graft-versus-host disease with decidual stromal cells and tracing with (111)indium radiolabeling. *Stem Cells Dev.* (2015) 24:253–63. doi: 10.1089/scd.2014.0265
42. Ringden O, Solders M, Erkers T, Nava S, Molldén P, Hultcrantz M, et al. Successful reversal of acute lung injury using placenta-derived decidual stromal cells. *J Stem Cell Res Ther.* (2014) 4:1–5. doi: 10.4172/2157-7633.1000244
43. Devine SM, Cobbs C, Jennings M, Bartholomew A, Hoffman R. Mesenchymal stem cells distribute to a wide range of tissues following systemic infusion into nonhuman primates. *Blood.* (2003) 101:2999–3001. doi: 10.1182/blood-2002-06-1830

44. Gao J, Dennis JE, Muzic RF, Lundberg M, Caplan AI. The dynamic *in vivo* distribution of bone marrow-derived mesenchymal stem cells after infusion. *Cells Tissues Organs.* (2001) 169:12–20. doi: 10.1159/000047856
45. Forslöw U, Blennow O, LeBlanc K, Ringdén O, Gustafsson B, Mattsson J, et al. Treatment with mesenchymal stromal cells is a risk factor for pneumonia-related death after allogeneic hematopoietic stem cell transplantation. *Eur J Haematol.* (2012) 89:220–7. doi: 10.1111/j.1600-0609.2012.01824.x
46. Le Blanc K, Frassoni F, Ball L, Locatelli F, Roelofs H, Lewis I, et al. Mesenchymal stem cells for treatment of steroid-resistant, severe, acute graft-versus-host disease: a phase II study. *Lancet.* (2008) 371:1579–86. doi: 10.1016/S0140-6736(08)60690-X

3

Mesenchymal Stromal Cell Therapeutic Delivery: Translational Challenges to Clinical Application

Henry Caplan [1†], Scott D. Olson [1†], Akshita Kumar [1], Mitchell George [1], Karthik S. Prabhakara [1], Pamela Wenzel [1], Supinder Bedi [1], Naama E. Toledano-Furman [1], Fabio Triolo [1], Julian Kamhieh-Milz [2], Guido Moll [3] and Charles S. Cox Jr. [1*]*

[1] Department of Pediatric Surgery, McGovern Medical School, University of Texas Health Science Center at Houston, Houston, TX, United States, [2] Department of Transfusion Medicine, Charité Universitätsmedizin Berlin, Corporate Member of Freie Universität Berlin, Humboldt-Universität zu Berlin, and Berlin Institute of Health (BIH), Berlin, Germany, [3] BIH Center for Regenerative Therapies (BCRT), Charité Universitätsmedizin Berlin, Corporate Member of Freie Universität Berlin, Humboldt-Universität zu Berlin, and Berlin Institute of Health (BIH), Berlin, Germany

***Correspondence:**
Scott D. Olson
scott.D.olson@uth.tmc.edu
Charles S. Cox Jr.
charles.s.cox@uth.tmc.edu

† These authors have contributed equally to this work

For several decades, multipotent mesenchymal stromal cells (MSCs) have been extensively studied for their therapeutic potential across a wide range of diseases. In the preclinical setting, MSCs demonstrate consistent ability to promote tissue healing, down-regulate excessive inflammation and improve outcomes in animal models. Several proposed mechanisms of action have been posited and demonstrated across an array of *in vitro* models. However, translation into clinical practice has proven considerably more difficult. A number of prominent well-funded late-phase clinical trials have failed, thus calling out for new efforts to optimize product delivery in the clinical setting. In this review, we discuss novel topics critical to the successful translation of MSCs from pre-clinical to clinical applications. In particular, we focus on the major routes of cell delivery, aspects related to hemocompatibility, and potential safety concerns associated with MSC therapy in the different settings.

Keywords: cellular therapy, mesenchymal stromal cell, clinical translation, safety, cell delivery, hemocompatibility, complement, coagulation

INTRODUCTION

The study of multipotent mesenchymal stromal cells (MSCs) and their usefulness for treating human injury and disease is almost 40 years old and has evolved through a number of phases roughly defined by the most commonly proposed mechanisms of action (MoA) at work. Interest in MSCs began as a study of bone marrow stromal cells in the 1970s by Friedenstein and contemporaries (1). This grew into an interest in their osteogenic differentiation potential in the late 1980s, and later broadened to trilineage differentiation (bone, fat, cartilage) (2). Cell fusion was briefly considered as a possible mechanism of repair (3–6). Aside from MSCs, a multitude of other stem cell populations with distinct properties have been isolated from adult rodent and human tissue, including multipotent adult progenitor cells (MAPCs) (7). Today, many studies focus on paracrine growth and immunomodulatory factors as key mediators of MSC's therapeutic effect, identified to protect injured tissue and to encourage endogenous repair mechanisms (8). In 2006, the International Society for Cellular Therapy (ISCT) published their first position statement on defining minimal criteria for MSCs (9), followed by several updates mainly focusing on the refinement of standards for therapeutic efficacy (10–12). Efforts to further refine cell pharmacology and drug delivery are ongoing (13). Thus, it becomes apparent that MSC research has undergone numerous advancements over time, in order to understand and benefit from the interesting properties of these cells. In this review, we will discuss key translational hurdles to clinical applications of MSCs. We will first outline popular cell delivery methods specific to their clinical application and then address newly identified efficacy and safety concerns regarding specific delivery methods (14).

ROUTES OF THERAPEUTIC CELL DELIVERY

A number of notable efforts have been made to compare the efficacy of different routes of MSC administration, which has become increasingly difficult with the large number of pre-clinical and clinical studies that are being published daily (14–18). Despite a number of direct comparisons in animal models and efforts to compare specific routes in a limited number of clinical trials, there is no consensus on the optimal method for MSC delivery (**Figure 1**), with specific limitations or advantages being associated with either method in clinical situations. Numerous methods for delivery of MSCs exist today and it stands to reason that different clinical indications and pathologies will require different delivery routes for optimal therapeutic efficacy (19, 20). Notably, a number of efforts are being made to develop MSC-derived exosomes or extracellular vesicles as a new "cell-free" way to recapitulate MSC activity with unique challenges and considerations (21).

Many investigators and industry driven studies rely on practical or logistical considerations (22). Nonetheless, new data-driven approaches for product optimization are currently being tested that integrate desired therapeutic MSC properties with representative simulated microenvironment interactions *in vitro* in an effort to determine optimal delivery and improve clinical outcomes (23–25). Selecting a suitable delivery route for future studies should also include a consideration of the desired MoA, and whether MSC culture techniques sufficiently highlight that mechanism, and if MSCs can be better primed by an alternative method (26).

Here, we will briefly discuss the rationale behind the most common delivery methods -topical application, intra-muscular (IM) or direct injection (DI) into tissues/organs, intra-venous (IV) infusion, and intra-arterial (IA) infusion, followed by notable considerations and translational challenges from preclinical to clinical application.

Topical Application and Local Injection

Classically, localized topical application or injection of a cell therapy into a specific site or target tissue, e.g., intramuscular or penumbral area of an injury, has been shown to be very useful for precision delivery of MSCs and to increase the engraftment of therapeutic cells at a specific site of interest (14, 27). These strategies are often associated with a tissue replacement strategy or direct paracrine support as a MoA and can be particularly useful when combined with specifically tailored exogenous support systems and biomaterials to guide MSC-host interaction and encourage endogenous therapeutic actions (28, 29).

While the direct differentiation and replacement of host tissue by MSCs has been challenged as a result of a different activity, such as cell fusion or transfer of genetic material (6, 30–34), there are also some notable recent reports of MSCs directly contributing to tissue regeneration, such as in recent work in trachea and esophageal replacement (35, 36).

Topical application of MSCs is the least invasive method of delivery and has demonstrated great potential in the fields of burn medicine and wound care. Topically applied MSCs have improved outcomes, wound healing, and skin graft survival in burn wounds, diabetic-related wounds, and other chronic wounds (37, 38). Using a fibrin polymer spray system, Falanga et al. demonstrated that topically applied MSCs improved wound closure rates in a preclinical model, as well as in patients with chronic non-healing lower extremity wounds (39).

Intra-muscular (IM) delivery of MSCs, like topical application, presents a safe and simple method for cell delivery and, furthermore, leads to improved dwell time compared to other routes such as IV, intra-peritoneal (IP) and subcutaneous cell delivery (40). In the study by Braid et al. IM delivery of MSCs in a mouse model led to survival of human MSCs for up to 5 months after injection. In addition to extended dwell time, IM skeletal muscle fibers provide a highly vascular conduit for local and systemic release of trophic factors and support for MSC paracrine actions (27).

In critical limb ischemia (CLI), for example, MSCs may exert their restorative effects via promotion of angiogenesis and revascularization of ischemic tissue (41). A recent Cochrane analysis of autologous cells treatments, including bone marrow (BM)-MSCs, for CLI found no differences between IA and IM deliveries (42). Furthermore, Soria et al. found that IM delivery may be superior to IA delivery regarding the mitigation of adipose tissue (AT)-derived MSCs prothrombotic properties (43). Interestingly, work by Lataillade et al. has also shown promising effects of local IM-injections of MSCs in dosimetry-guided surgery treatment of radiations burns (44), while both, local IM and systemic IV delivery of MSCs and MSC-like cells has led to rescue from lethal radiation in animal models (45, 46).

In addition to topical and IM delivery, early investigative efforts often focused on the potential of MSCs to repair tissues by local engraftment and/or differentiation via direct injection (DI) into the target tissue or organ. Pre-clinical studies in neurological disease, such as stroke, attributed the beneficial effects of MSCs to their ability to engraft and differentiate into neurons and/or glia (47). However, the notion that MSCs can differentiate into functioning neuronal cells was subsequently challenged and appears unlikely (48–50).

Others contended that MSC engraftment facilitated endogenous neurorestorative mechanisms such as promotion of host neural and glial cell remodeling (51, 52). Regardless of the MoA, DI has potential advantage of bypassing the blood-brain barrier to increase delivery of cells into the central nervous system (CNS). For example, a recent Phase 1/2a clinical trial investigated intra-cerebral implantation of the SB623 MSC

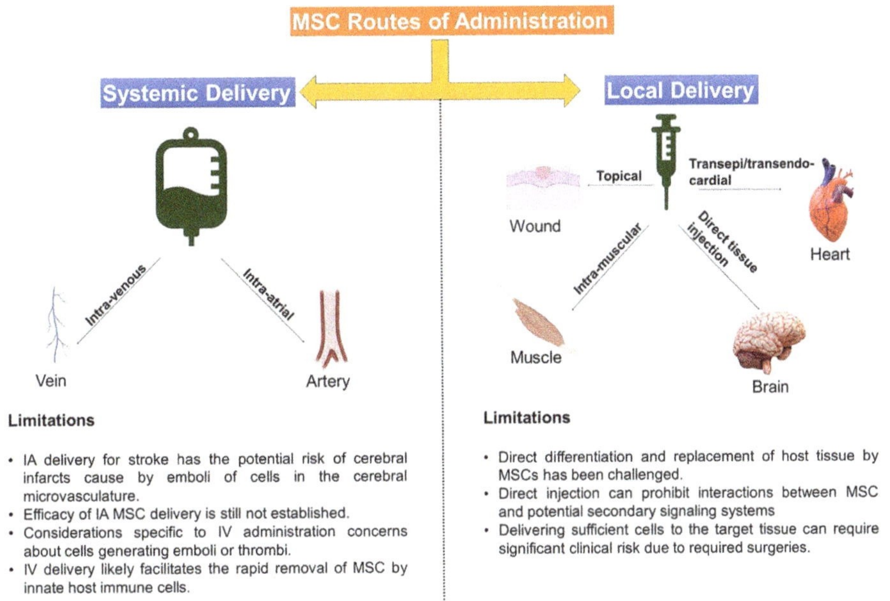

FIGURE 1 | Delivery Routes Common for MSC therapies. Depicted above are the main methods that MSC are administered to target tissues, accompanied by some limitations of each approach.

cell line in adults with chronic, non-hemorrhagic stroke via magnetic resonance imaging (MRI) stereotactic guidance into the peri-infarct area (53). The authors concluded that MSC implantation was safe, feasible and also improved neurologic outcomes at 12 months. However, important to note, is that the study was limited by patient selection (only 4.7% of screened patients were enrolled) and a lack of a control/placebo group.

Direct injection (DI) of MSCs has also been widely utilized for the treatment of cardiac disease—both acute myocardial infarction (AMI) and ischemic heart failure (IHF)—via open trans-epicardial and catheter-based trans-endocardial injection. There have been numerous preclinical and clinical trials in recent years, but here we will highlight only a few selected studies. Important to note, the notion that MSCs can differentiate into cardiomyocytes or promote cardiac stem cell proliferation and differentiation have largely been abandoned in the past decade. Furthermore, even the existence of resident adult cardiac stem cells has recently been challenged and appears unlikely (54, 55).

The PROMETHEUS trial investigated the use of intra-myocardial MSC injections into non-revascularized ischemic myocardium in patients undergoing coronary artery bypass grafting (CABG) for IHF and found improvements in regional myocardial and global left ventricular (LV) function (56). The authors rationalized the use of intra-myocardial injections based on the theory the MSCs exert their effects predominately at the injection site via release of anti-fibrotic matrix metaloproteases and stimulation of neovascularization.

Indeed, the authors found that the effect of MSC injection dropped off as a function of distance from the injection site. Of note, this study involved only 6 patients and no control group. However, a recently published, randomized trial of intra-myocardial injection of mesenchymal precursor cells (MPCs) in 159 patients with advanced heart failure undergoing left ventricular assist device (LVAD) placement found that MPC therapy did not demonstrate improvement in the primary outcome, weaning from LVAD support within 6 months (57). The authors also noted that one potential factor for the lack of efficacy may have been the use of trans-epicardial injections, which can lead to significant cell loss.

A systematic review and meta-analysis of MSC delivery methods in preclinical and clinical AMI found that trans-endocardial injections produced more favorable results in swine models in comparison to direct implantation (intra-myocardial) (58). Furthermore, the trans-endocardial approach allows for a minimally invasive, catheter-based direct implantation of cells into the myocardium and avoids the invasive thoracotomy, and thus additional risks for patient harm, required for trans-epicardial delivery.

There are several significant risks and considerations unique to DI delivery of MSC. Among them are reports of MSC differentiating into problematic tissue/ectopic tissue formation (59, 60), particularly heterotopic ossification into ectopic bone (61, 62). Additionally, localized DI may prohibit interactions between therapeutic MSCs and potential host secondary signaling systems in the lung, spleen, and peripheral blood, thus limiting their repertoire of therapeutic MoA. Consideration must also be given to the logistics and feasibility of various DI approaches, as delivering sufficiently high cell numbers to the selected target tissue can create significant clinical risk due to required surgeries, such as laminectomies to treat spinal cord injuries (63).

Intra-arterial Infusion

There is substantial evidence that MSCs exert their effects largely via direct cell contact and local paracrine effects, as opposed to engraftment and differentiation within the target organ (64, 65). Additionally, upon systemic infusion, the interaction with endogenous inflammatory and tissue repair signals, such as immune cells and the innate immune cascade systems in the bloodstream likely influences MSCs responses, bio-distribution and homing to injured or diseased tissues (14, 66).

Intra-arterial (IA) delivery may prove the most efficacious method in one treatment indication, but may be potentially harmful in another. IA delivery of MSCs allows for infusion of cells within the local vascular system of the target organ without the physical risks of direct implantation and pitfalls of IV administration, especially the trapping of cells within the lung microvasculature, and may thus allow more cells to reach the intended target tissue (17, 65, 67).

Importantly, based on a survey of published results (68), the IA delivery of MSCs for stroke entails the potential risk of cerebral infarcts, caused by emboli of cells in the cerebral microvasculature. Factors such as vascular access, cell size, cell dosage and delivery speed must be considered, especially when delivering cells into coronary or cerebral arteries (69, 70).

In AMI several clinical trials have demonstrated safety and improvements in functional outcomes with the use of intracoronary infusion of MSCs and other BM cell populations (71–73). Here, IA delivery is likely a valid option, as it avoids invasive procedures (which are not part of the usual care in AMI) and ensures delivery of cells to the area of focal tissue injury and hypoxia.

As mentioned above, the SafeCell Heart study demonstrated significant improvements in LVEF with intracoronary MSC delivery (74). IA cell delivery has also been utilized in other pathologies including, but not limited to intra-carotid delivery in stroke, intra-renal delivery for renovascular disease, and intra-hepatic delivery for cirrhosis (75–77). However, the efficacy of IA MSC delivery in the above examples is still under investigation.

Intra-venous Infusion

The most commonly used method to apply MSCs is via IV infusion, due to the relative ease and limited risk (14). This method results in a large number of MSCs accumulating in the lungs, but also distributing throughout the body and other organs, such as the spleen, throughout 24–48 h (67, 78–80). Similar to intra-arterial delivery of cell therapies, IV administration is most often associated with mechanisms involving secondary signaling effector cell systems and interactions with the host immune system (64, 80, 81).

Most of the attempts at developing a commercial cell therapy have used IV administration in order to facilitate use at multiple centers. Perhaps most widely known, the systemically IV-infused Remestemcel-L (Prochymal) has been developed by Osiris Therapeutics (Now Mesoblast), to primarily treat graft-vs.-host disease (GvHD) (82). Other notable efforts also include the use of Multistem, a multipotent adult progenitor cell that is somewhat similar to MSCs, to treat ischemic stroke (83) and our own use of cell therapies to treat traumatic brain injury (84). Common among these approaches is the use of cell therapies to modulate inflammation and activation of the immune system in order to decrease inflammation-related secondary injuries and restore homeostasis.

Among the considerations specific to IV administration are the same concerns about cells generating emboli or thrombi, however, with the advantage of the lung capturing potential vascular obstructions before they can disrupt other organ function. IV infusion certainly may result in limited numbers of cells reaching target tissues, a more transient persistence of cells, and a dilution of paracrine factors reaching target tissues. Finally, while it may be part of MSCs MoA in modulating immune responses (85), IV delivery likely facilitates the rapid removal of clinical MSCs by innate host immune cells (14, 86).

Additional Routes of Administration

Several additional routes of MSC administration are notable for consideration for some specific applications. Intra-nasal (IN) administration of MSCs is promising route to treat neurologic pathologies that avoids the risks associated with direct injection of cells into the CNS. Preclinical data has demonstrated efficacy of IN MSCs across a spectrum of CNS disorders, including perinatal ischemic brain injury, subarachnoid hemorrhage, and neurodegenerative disorders such as Parkinson's disease (20, 87–90).

Intrathecal (IT) administration of MSCs, often through lumbar puncture, has shown benefit in preclinical and clinical studies across a wide array of neurologic disorders, including chronic neuropathic pain secondary to spinal cord injury (91–94), amyotrophic lateral sclerosis (ALS) (95–97), and epilepsy (98, 99).

Intravitreal administration of MSCs has demonstrated improvement in outcomes in several animal models of retinal injury and dysfunction (100–104); however, the safety of intravitreal injection of MSCs in humans is still an area of clinical concern (105).

Furthermore, the administration of MSCs during *ex vivo* perfusion of solid organs, e.g., in kidney transplantation or injury, avoids the trapping of cells in the lung and spleen and permits direct interaction of MSCs with the target tissue (106). Gregorini et al. demonstrated that delivering MSCs during hypothermic machine perfusion improved outcomes in a rat model of ischemic kidney injury (107). Recently, Sierra Parraga et al. demonstrated that machine perfusion of MSCs supports their function and survival, although more work will be required to determine the optimal conditions for perfusion (108). Notably, these strategies may reduce the need for immunosuppression to prevent organ rejection (109).

TRANSLATIONAL HURDLES WITH SYSTEMIC AND LOCAL DELIVERY

Any therapeutic modality, whether a new pharmacologic agent or surgical procedure, must not only be efficacious, but also safe for the patient. While the vast majority of preclinical and clinical studies have found MSCs to be safe and well-tolerated (110), the rise of unregulated and unproven stem cell interventions have resulted in several reported clinical adverse events (14, 60). Here we will highlight mechanisms by which MSCs may cause unwanted or adverse reactions, including interactions with the host's innate and adaptive immune system, as well as their tumorigenic potential (**Figure 2**).

Triggering of Innate Immune Responses by MSCs

Systemically infused MSCs activate the host innate immune cascade systems, such as complement and coagulation, termed the instant blood-mediated inflammatory reaction (IBMIR) (14, 66, 111). IBMIR was first used to describe the procoagulant activity of pancreatic islet cells and hepatocytes in relation to their expression of procoagulant tissue factor (TF/CD142) (112, 113).

MSCs demonstrate similar effects when they contact human blood with variable amounts of procoagulant activity when isolated from different issues (14). They normally reside in the perivascular space around blood vessels, are predominantly excluded from direct contact with blood, and express variable amounts of procoagulant TF, first described on placenta-derived decidual stromal cells (DSCs) (114).

The effects of *ex vivo* expanded MSCs administered IV into patients include a significant increase in complement C3 activation fragment a (C3a) and the coagulation activation marker thrombin-antithrombin complex (TAT), which was accompanied by a decrease in platelet counts and a significant increase in fibrinolysis marker D-dimer (111, 115). Importantly, MSC's TF expression was shown to increase as cell passage increases, and the procoagulant effect of MSCs increases as their TF expression increases (14).

Furthermore, in a study investigating human BM- and AT-derived MSCs, TF expression and procoagulant activity were measured using flow cytometry and calibrated automated thrombogram and thromboelastography, respectively, demonstrating large donor and tissue variability (116). A causal relationship between MSC-associated TF with clot formation has also been demonstrated in human blood using flow cytometry to measure the density of TF on different types of MSCs (24), finding that MSC induce a dose-dependent change in clotting time and thrombin production based on TF expression.

Several authors demonstrated a reduction in procoagulant activity in MSCs when the cells were diluted or treated with TF pathway blocking reagents (24, 111, 117). Multiple authors also found that heparin can nullify accelerated clotting time due to MSC associated TF *in vitro* (115, 117–119). These data highlight the importance of monitoring MSC's procoagulant activity and provide a possible clinical solution. Thus, MSC-associated TF must be considered as a safety release criterion prior to administration in patients (14).

While a substantial number of *in vitro* investigations have demonstrated that MSCs exert a procoagulant effect after blood contact, there are limited reports of MSC-associated thrombotic events in humans in the literature, specifically in established clinical trials (60). One of the first reports involved the use of human placenta-derived MSC-like cells in a Phase 1b/2a study in Crohn's disease. The authors reported that two patients suffered from venous thrombosis after infusion (120), and posit TF expression on clinical-grade therapeutic MSCs as a possible cause (111).

Furthermore, subsequent *in vitro* studies comparing placenta-derived DSCs to BM-MSCs found a 15-fold higher expression of TF in the DSCs (115), which may help to explain the results from the above study. The need for increased caution with perinatal tissue (PT)-derived MSC products is further substantiated by a recent report of thromboembolism in two patients treated with umbilical cord MSC products (121).

Another key report of thrombotic events was in a trial that evaluated the use of autologous AT-derived MSCs to treat patients with critical limb ischemia (CLI) (122). The investigators found that two of the patients, both of whom also had diabetes, developed distal microthrombi after infusion. Interestingly, the investigators found no cases of thrombotic events when using autologous AT-MSCs in non-diabetic patients or autologous BM mononuclear cells (BM-MNCs) from diabetic patients.

It was subsequently demonstrated that AT-MSCs from diabetic patients release higher levels of plasminogen activator inhibitor type 1, reduced levels of tissue plasminogen activator, and lower d-dimer formation in comparison to non-diabetic AT-MSCs, all of which might lead to blunted fibrinolytic activity. Furthermore, these diabetic-derived AT-MSCs upregulated TF expression and displayed altered platelet-derived growth factor (PDGF) signaling, which was abrogated using PDGF-BB treatment (123).

There are other case reports of MSC-associated thrombotic events. A man who presented with chest pain was found to have small bilateral pulmonary emboli 1 month after receiving the last of multiple systemic infusions of AT–derived MSC for herniated cervical discs (124). His parents were reported to also have had MSC infusion for osteoarthritis, and both were found to have small pulmonary emboli, although neither was symptomatic. No evidence of hypercoagulable disease was found in any family member. There was no clear mention of the exact source of the clinical MSCs infused, or if these patients received MSCs in conjunction with a clinical trial. There is an additional media report of a 73 year old man who died from a pulmonary embolism after receiving an infusion of AT-derived stromal cells in Japan (125). It appears that this patient was not enrolled in any published clinical trial, and the company involved has come under intense scrutiny. In both of the latter two cases, there are few details regarding the source, manufacturing process and testing of the cells, as well as details surrounding the treating parties.

While the evidence for MSC-associated thrombotic events is nominal, there is a need to ensure the safety of MSC therapy, including their procoagulant effects (14). Furthermore, many patients who may benefit from MSC therapy—those with inflammatory mediated disorders, diabetes, cancer, cardiac dysfunction, or trauma-related injuries—are likely to have an acquired hypercoagulable state or are at high risk of a thrombotic event secondary to their primary disease process. Therefore, we must continue to monitor the pro-thrombotic effects of MSCs as part of release criteria and in clinical trials.

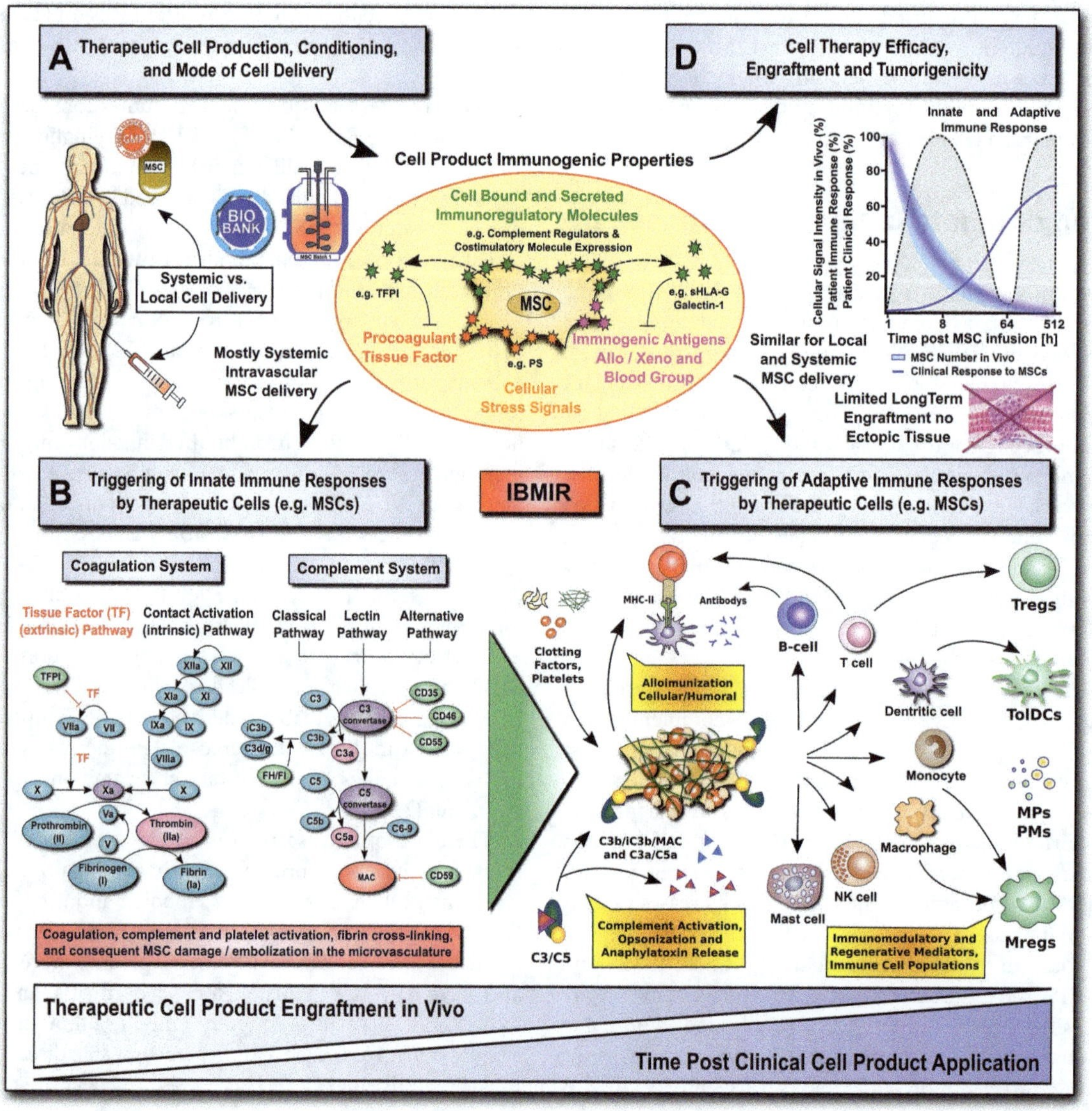

FIGURE 2 | Translational challenges with systemic and local cell delivery. **(A-D)** Therapeutic cell production / conditioning (e.g., 2D vs. 3D culture and cytokine priming) and the mode of cell delivery (e.g., systemic intravascular infusion vs. local injection) have a major impact on the cell product's immunogenic properties (shown in **A**), and consequent rapid triggering of innate and adaptive immune responses (shown in **B,C**), thus affecting its therapeutic efficacy, engraftment and tumorigenicity (shown in **D**). The MSC product's immunogenic properties are affected by numerous cell-bound and secreted immunoregulatory mediators (e.g., complement regulators, coagulation regulator TFPI, or regulators of the adaptive immune response, such as co-stimulatory molecule expression, sHLA-G and galectin-1). The cells can also exhibit a number of immunogenic features, such as procoagulant TF-expression, cellular stress signals (e.g., PS), and immunogenic antigens (e.g., allo, xeno, and blood groups). **(B)** The innate coagulation and complement cascade systems are two of the major effector arms of the instant blood-mediated inflammatory reaction (IBMIR) that can recognize blood-incompatible therapeutic cell features and thus trigger the detrimental thromboinflammation compromising cellular therapeutics. The innate immune cascade systems employ multiple sophisticated molecular sensors (e.g., FVII and FXII, or C1q and MBL, respectively), to recognize aberrant cell surface molecular features on infused therapeutic cells (e.g., TF and PS, respectively), which can trigger innate immune cascade activation and amplification by effector cells (e.g., platelets, PMNs, monocytes/macrophages, and T/B cells), potentially leading to adverse reactions (e.g., cell lysis, inflammation, sequestration and rejection). **(C)** Innate and adaptive effector cell modulation: triggering of IBMIR and therapeutic cell injury/disintegration promotes the release of various bioactive molecules, in itself and from dying MSCs, upon crosstalk with the responsive host immune system, such as activated clotting factors (e.g., thrombin), anaphylatoxins (C3a and C5a), opsonins (iC3b, and C3d/g), and MSC-derived constituents (e.g., microparticles, cytokines and growth factors) in a highly conditional manner, thus greatly amplifying the initial signal, leading to modulation of multiple effector cell types. This can result in alloimmunization and consecutive cellular and humoral responses (e.g., T-cells and B-cells alloantibodies), but also in the induction and release of multiple immunoregulatory and regenerative cell types and mediators (e.g., Tregs, Mregs, TolDCs, MPs, and PMs). **(D)** A large fraction of the infused therapeutic cells is lost within the first hours to days of infusion due to the triggering of instant innate immune responses, which can be furthermore aggravated by triggering of adaptive immune responses in case of allogeneic cell products. Studies on MSC persistence *in vivo* have shown prolonged survival, dwell-time, and engraftment by alternative routes of delivery (e.g., local injection in conjunction with biomaterials), although long-term engraftment is very limited and ectopic tissue formation rarely reported. Currently, patient clinical

(Continued)

FIGURE 2 | responses are still sub-optimal for many MSC therapeutics leaving room for improvement in long-term survival. AT, antithrombin; FI-FXII / FIa-FXIIa; native and activated coagulation factors I-XII; TF / TFPI, tissue factor and tissue factor pathway inhibitor; C3/5-9, complement component 3 and 5 to 9; C3a/C5a, activation fragment a of complement component 3 and 5; C3b/iC3b/C3d(g), complement component 3 sequential degradation fragments b, inactivated fragment b and d(g); complement regulatory molecules: CD35, complement receptor 1, CD46/MCP, membrane cofactor protein; CD55/DAF, decay accelerating factor; CD59, protectin; FI and FH, complement factor I and H; PS, phosphatidyl-serine; MAC, membrane attack complex; and MHC-II, major histocompatibility complex class-II; sHLA-G, soluble human leukocyte antigen G; MPs and PMs, MSC and immune cell-derived micro-particles and paracrine mediators; Tregs, Mregs and TolDCs, immunogerulatory T-cells, myeloid cells, and tolerogenic dendritic cells.

The complement system, another major part of the innate immune cascade systems, is one of the first lines of defense against foreign pathogens. It is not surprising that systemically infused MSCs activate and interact with the complement system (126). In fact, the complement system appears to play a critical role in initiating the immunomodulatory reaction between infused therapeutic MSCs and host cells (127). What role the complement system plays in MSC's MoA and how it affects their efficacy is currently under investigation.

Early studies by Tu et al. revealed that MSCs constitutively express factor H, a complement inhibitor (128); however, even when augmented by the presence of pro-inflammatory cytokines TNF-α and IFN-γ, the overall production of factor H by MSCs proved relatively insignificant in comparison to overall systemic levels. Subsequently, the same group and others demonstrated that although MSCs express cell-surface complement regulators, MSCs activate the complement system upon contact with human sera *in vitro*, leading to cytotoxicity in a dose-dependent manner (126, 129, 130).

Blockade of these complement inhibitors led to increased cytotoxicity, while both upregulation of CD55 (a cell-surface complement inhibitor) on MSCs via transfection with recombinant adenovirus, and addition of anti-C5 immunoglobulin significantly reduced MSCs cytotoxicity after contact with serum. Li et al. also determined allogeneic MSCs caused increased complement activation, with associated increased cytotoxicity, in comparison to autologous MSCs (126).

In order to prevent complement-mediated cytotoxicity against therapeutic cell products, investigators have engineered complement-resistant MSCs. Heparin is known to inhibit complement activity, and systemic administration of heparin reduced MSC cell damage after infusion (131). In order to prevent unwanted anti-coagulation effects of systemic heparin administration, the authors then demonstrated that incubation of MSCs with activated heparin led to binding of heparin to the surface of MSCs and furthermore, these "heparin-painted" MSCs showed less surface deposition of complement C3 activation fragment b (C3b/iC3b) and suffered less cell damage after contact with serum.

The proposed mechanism of heparin-mediated protection involves recruitment of factor H binding to the MSC cell-surface. Subsequent "painting" of factor H onto MSCs via pre-incubation led to decreased complement deposition onto MSCs surface, reduced cell damage, and increased cell survival *in vitro* and *in vivo* (132). In addition, MSCs with factor H incubation attenuated C5a release, which significantly reduced complement-mediated activation of neutrophils and led to improvement in MSC function and reduced cell damage.

Investigators have also engineered human MSCs via transduction of a retrovirus encoding genes from human cytomegalovirus (HCMV), which downregulated HLA1 expression on MSCs without increased vulnerability to NK killing (133). Furthermore, HCMV is known to incorporate the host-encoded complement inhibitor proteins and upregulate host-encoded CD55 and CD46 in order to evade the innate immune system. Soland et al demonstrated that induction of MSCs with the specific HCMV US2 protein led to upregulated expression of complement inhibitors CD46, CD55, and CD59, and a reduction in complement mediated MSC lysis (129).

While further investigation into engineered MSCs as described above is certainly necessary, the ability to evade or attenuate complement-mediated cell damage and lysis may prove crucial to the efficacy of future MSC therapeutics (86). Further investigation has demonstrated a much more complex relation between MSCs and the complement system, with discrepancies between pre-clinical and clinical results. In a clinical trial of patients receiving MSC infusions for treatment-resistant GvHD, the investigators examined the relationship between complement activation, immunosuppressive potential of the MSCs *in vitro* and the clinical effectiveness of MSC therapy *in vivo* (127). MSCs activation of the complement system was found to mediate effector cell activation and modulate their immunomodulatory activity in a multifactorial manner.

This finding was mechanistically substantiated by using *in vitro* inhibition of complement function, which resulted in decreased CD11b upregulation on effector cells. Furthermore, MSC's ability to activate the complement system was found to correlate with its immunosuppressive potential *in vitro*: MSCs with increased complement activating properties demonstrated increased suppression of peripheral blood monocyte (PBMC) proliferation in mixed lymphocyte reactions (MLRs) and increased ability to trigger CD11b+ effector cells in whole blood. Surprisingly, the authors found an inverse correlation between the immunosuppressive potential of the MSCs *in vitro* and their clinical effectiveness in human patients, with average suppressing cells yielding the most beneficial therapeutic effects *in vivo*.

Importantly, it appears that substantial differences exist with respect to complement activation and potential efficacy of fresh vs. freeze-thawed MSCs upon systemic infusion (86). Freeze thawed MSCs were found to demonstrate increased activation of the IBMIR and susceptibility to complement-mediate cell lysis (130). Similar changes in the immunomodulatory capacity of MSC due to cryopreservation have also been reported elsewhere: such as the alteration of MSC-mediated attenuation of T cell activation, inflammatory cytokine concentrations, and an increased susceptibility of MSC to lysis by mixed immune cells (134, 135).

In the clinical sample evaluated (130), the majority of GvHD patients were treated with freeze-thawed MSCs with a small number of fresh culture-derived MSCs being available for comparison; while the study lacked sufficient power and thus should be considered with caution, as also emphasized in the discussion, the authors noted an interesting trend of improved clinical outcomes with fresh MSCs, especially those delivered at early passage, in comparison to freeze-thawed MSCs (130, 136).

Overall, it is now evident that MSCs activate the complement system upon contact with blood, which appears to be positively correlated with their immunosuppressive ability *in vitro*. Furthermore, freeze-thawed MSCs, in comparison to fresh MSCs, demonstrated an increased triggering of the IBMIR and associated complement-mediated lysis *in vitro*, leading to a significant reduction in viable cells (130, 137). What remains unanswered is: (1) how and to what extent complement activation influences the clinical efficacy of systemic MSC therapy, and (2) whether fresh MSCs are subject to decreased complement-mediated lysis and, as a result, are more effective clinically than freeze-thawed MSCs *in vivo*.

Triggering of Adaptive Immune Responses: Autologous vs. Allogeneic MSCs

A comprehensive review of the immunobiology of MSCs and the many ways that MSCs interact with the local and systemic immune system in both normal and activated systems is beyond the scope of this review, and has previously been covered in several notable efforts to consolidate the literature (138–141). It would not be an understatement to say that MSCs have extensive possible interactions with every major component of the immune system through a combination of paracrine activity, extracellular matrix remodeling, direct contact-based signaling, and more recently, through the use of extracellular vesicles.

These wide-ranging putative molecular mechanisms have made it incredibly difficult for the field to come to a meaningful consensus regarding the effects of self vs. donor antigens, further complicated by the additional xenogenic antigens introduced during common cell culture techniques (such as expansion in FBS) (137, 142, 143). Until recently, MSCs were widely reported to be immune privileged, enabling their use as an allogeneic therapy without concurrent immunosuppression.

As the field increasingly focused on the immunobiology of MSCs, there was a correlating rise in the number of studies that found that MSCs were not exempt from immune recognition. As summarized in a number of more comprehensive literature reviews (140, 144–146), allogeneic MSCs with poorly matched HLA can and do generate both innate and humoral responses from the immune system, albeit responses that appear to be dependent upon the conditional expression and balance of both immune-activating antigens (such as MHCs) and immune-modifying cytokines, molecules and metabolites, like tumor necrosis factor-inducible gene 6 (TSG-6), galectin-1, prostaglandin E2 (PGE_2), and indoleamine 2,3-dioxygenase (IDO).

The eventual immune recognition of allogeneic and HLA-mismatched MSCs has become increasingly implicated as a barrier to clinical efficacy (140). On the one hand, the huge number of pre-clinical studies in both xenogeneic and allogeneic systems and clinical studies using allogeneic cells without regard to conventional graft-vs.-host compatibility considerations indicate that MSC efficacy is often independent of the eventual immune "rejection" of donor MSCs, either by virtue of MSC activity occurring prior to immune recognition or perhaps even due to a MoA that includes the host immune system.

In a recent report it was found that MSCs can modulate the immune system by being engulfed by antigen presenting cells, and that the subsequent display of MSC antigens by antigen-presenting cells (APCs) results in a chain of anti-inflammatory activity and downstream therapeutic outcomes (85). On the other hand, the recognition and removal of MSCs by the host immune system may also limit the duration and possible efficacy of a number of MSC MoAs.

Among a number of strategies to evade immune recognition, there is a large amount of interest in the use of biomaterials and engineering techniques to shield MSCs from immune activity to prolong paracrine activity (29, 147, 148), efforts to further decrease the immunogenicity of cells (149–151), sophisticated banking strategies to allow for autologous or HLA-matched cells to treat acute injuries in a timely fashion (152, 153), and conventional pharmaceutical immunosuppression given temporarily, all to provide a larger window for MSC activity *in vivo* (154, 155).

It is our opinion and that of others (86, 136, 140, 156), that reducing the activity of the host immune system is likely to also reduce the therapeutic efficacy of MSCs, as we feel that many of the pleiotropic effects of MSCs require the participation of the host immune system.

Cell Engraftment and Tumorigenicity

The potential for malignant transformation of MSCs is of obvious concern. Because MSC therapy involves *ex-vivo* production and expansion of cell lines, and even allogeneic MSCs have the capacity to escape elements of immune recognition, it is crucial to ensure that transplanted or infused cells do not contain transformed, potentially tumorigenic cells (157).

Concerns for the tumorigenic potential initially came from studies of mouse-derived MSCs transplanted into a mouse model (158). The murine-derived BM-MSCs used in the study were reported to spontaneously transform into malignant fibrosarcomas in multiple organs *in vivo* after systemic infusion into immunocompromised mice. Of note, this same study also evaluated human MSCs and found no evidence of malignant potential *in vitro*.

Another study demonstrated that the injection of mouse MSCs with spontaneous p53 mutations led to development of fibrosarcomas at the site of injection in immunocompetent mice (159). Yet, there was no evidence that the transfer of MSCs without such mutations led to tumor formation. Transformed MSCs have also been identified for other non-human species.

The authenticity of such findings is difficult to confirm in non-human cells, as many of such studies lack true verification by modern stringent methods, such as short tandem repeat (STR) profiling (160).

Additional reports of spontaneous malignant transformation of human MSCs further intensified concerns for the tumorigenic potential of MSCs. Two separate studies identified spontaneous malignant transformations of human MSCs in culture, and injection of these transformed cells led the development of tumors in mice (161, 162). However, the findings from both of these studies have since been retracted, as the MSC cultures in both instances were found to have been contaminated with established malignant human cell lines (163).

While two more recent studies have demonstrated, and confirmed using STR analysis, the development of malignant transformation of MSCs in both cynomolgus macaques and human cell lines (164–166), there are, in contrast, far more studies that have demonstrated that *ex vivo* expanded human MSCs are rather resistant to malignant transformation, even after long-term culture, development of chromosomal aberrations, and application of physical and chemical stress (167–169), and that they undergo senescence rather than becoming tumorigenic.

Further reports concerning the genetic instability of MSCs appear to be grossly overstated and even misleading (170, 171). In addition, studies of Good Manufacturing Practices (GMP) grade MSCs have also demonstrated a lack of malignant potential *in vitro* and *in vivo* (172, 173). While the possibility of rare tumorigenic transformations in MSCs cannot be ignored, careful monitoring of cell cultures, minimization of *in vitro* expansion length and evaluation for cytogenic aberrations when concerns rise should be considered (174).

Human MSCs have been used clinically for more than two decades, the majority from BM, but also increasingly also from other sources. To date, there are no clinical studies that have attributed MSC therapy to the development of tumors or malignancy. A meta-analysis of 36 studies, including 8 randomized control trials (RCTs), involving 1,012 patients found no evidence of association between MSCs and tumor formation (110). A 2013 report from the ISCT noted that although the risk of tumorigenicity of MSCs had yet to be confirmed or denied, no tumors have been diagnosed in patients that would originate from administered MSCs (175). While it appears that MSC therapy is safe and well-tolerated in human subjects, the risk of tumorigenicity must continue to be studied both in clinical trials with long-term follow up and during the culture/expansion process prior to any therapeutic infusion.

COMPARATIVE STUDIES FOR OPTIMAL DELIVERY

Two important considerations exist in determining optimal cell delivery. The first consideration is patient safety. The second consideration is the efficacy of a therapy. If DI, IA, or IV demonstrate similar clinical effectiveness, then the least invasive method of cell delivery is preferred. But do MSCs work differently when delivered DI v. IA v. IV? Here we have selected a few examples of preclinical and clinical studies that have compared cell delivery methods head to head in order to demonstrate potential advantages/disadvantages of one method over another.

In heart disease, DI into myocardium may not provide improved MSC engraftment rates or outcomes over IA infusion. In a model of porcine model of ischemic cardiomyopathy, animals received either surgical implantation, trans-endocardial injection, or intracoronary infusion of autologous MSCs and were euthanized 4 h after infusion (176). DI of MSCs via surgical implantation or trans-endocardial injection led to only 16 and 11% retention of MSCs within the myocardium, respectively. The majority of the cells, around 45%, accumulated in non-target organs for all three delivery methods. Intracoronary infusion led to similar rates of intra-myocardial MSC retention as DI (both 11%). IA delivery necessitates the need for patent vasculature and, therefore, may not provide benefit in pathologies such as AMI or ischemic stroke secondary to occluded internal carotid or intra-cerebral arteries.

There have also been investigations of IA vs. IV cell delivery, especially in stroke, as endovascular treatments have become increasingly utilized (70). Byun et al. demonstrated that IA delivery leads to improved cerebral engraftment and outcomes over IV delivery in a rat model of cerebral infarction (177). A meta-analysis of preclinical studies of MSCs in ischemic stroke models found that although DI provided the greatest benefit, all 3 methods of delivery—DI, IA, and IV— consistently demonstrated significant improvement in outcomes (178).

Direct comparison of delivery methods is often lacking in clinical trials due to logistical concerns. Furthermore, pooled meta-analysis and systematic reviews often combine different cell types, related pathologies and delivery methods. Therefore, the following small selection of studies is limited and should not be considered a representative sample.

In stroke, a pooled analysis of clinical trials using multiple cells types, largely BM-MNCs and MSCs, Jeong et al. determined that IA provided increased benefit over DI or IV (179). Furthermore, the SafeCell Heart study, a systematic review and meta-analysis of cell therapy in heart disease, found that IA and intra-myocardial (catheter-directed trans-endocardial) infusions provided significant improvements in LVEF, which were not seen with trans-epicardial or IV cell delivery (74).

In another meta-analysis of preclinical and clinical studies of MSCs in cardiac disease, Kanelidis et al. found that cell delivery method did have an effect on outcomes in AMI, and that trans-endocardial and IV delivery improved outcomes in both swine models and clinical trials, while IA infusion with subsequent intracoronary delivery did not demonstrate significant benefit (58).

Clearly, the currently available data, based on few preclinical studies and limited clinical trials, which are often contradictory, are not sufficient to make any major conclusions as to whether one delivery method is superior to another. However, clinical trials directly comparing cell delivery methods will likely not happen until MSC therapy, via any delivery method, is proven efficacious for a particular pathology.

CONCLUSION

During the past decades, MSC therapeutics have undergone a continuous transition from proof-of-concept to clinically approved therapies. In order to improve our ability to utilize MSCs therapies, great efforts are ongoing to refine potency assessment, cell pharmacology and drug delivery. Compared to advancements in cell sorting, manufacturing and biobanking, the importance of cell delivery methods and the *in vivo* effects of MSCs on the human immune and hematologic systems are still largely underappreciated today. Thus, we here discussed key aspects related to the effective and safe delivery of MSCs, in the context of recent clinical studies with focus on different methods of MSC administration. As the growth of MSC-based therapeutics accelerates in private, public, and fringe applications, it is vitally important to remember historical safety concerns, recognize modern clinical risks, and use methodology and delivery consistent with the intended MoA, in order to yield the most effective and safest economically viable therapeutic approaches. We encourage our colleagues to careful consider their assumptions and commonly used practices to ensure that their long-held views about MSC biology are supported by modern studies.

AUTHOR CONTRIBUTIONS

HC, SO, AK, MG, and PW wrote the manuscript. KP, JK-M, and GM worked on illustrations. HC, SO, AK, SB, NT-F, FT, GM, and CC edited the manuscript.

ACKNOWLEDGMENTS

This study was supported by the NIH 2T32 grant GM0879201-11A1, the Clare A. Glassell family Pedriatic Surgery Research Fund, the Grace Reynolds Wall Research Fund, and the Bentsen Stroke Center. GM's contributions were made possible by German Research Foundation (DFG) and German Federal Ministry of Education and Research (BMBF) funding through the BSRT (GSC203) and BCRT.

REFERENCES

1. Friedenstein AJ. Precursor cells of mechanocytes. *Int Rev Cytol.* (1976) 47:327–59. doi: 10.1016/S0074-7696(08)60092-3
2. Pittenger MF, Mackay AM, Beck SC, Jaiswal RK, Douglas R, Mosca JD, et al. Multilineage potential of adult human mesenchymal stem cells. *Science.* (1999) 284:143–7. doi: 10.1126/science.284.5411.143
3. Terada N, Hamazaki T, Oka M, Hoki M, Mastalerz DM, Nakano Y, et al. Bone marrow cells adopt the phenotype of other cells by spontaneous cell fusion. *Nature.* (2002) 416:542–5. doi: 10.1038/nature730
4. Alison MR, Poulsom R, Otto WR, Vig P, Brittan M, Direkze NC, et al. Recipes for adult stem cell plasticity: fusion cuisine or readymade? *J Clin Pathol.* (2004) 57:113–20. doi: 10.1136/jcp.2003.010074
5. Gunn WG, Conley A, Deininger L, Olson SD, Prockop DJ, Gregory CA. A crosstalk between myeloma cells and marrow stromal cells stimulates production of DKK1 and interleukin-6: a potential role in the development of lytic bone disease and tumor progression in multiple myeloma. *Stem Cells.* (2006) 24:986–91. doi: 10.1634/stemcells.2005-0220
6. Spees JL, Olson SD, Whitney MJ, Prockop DJ. Mitochondrial transfer between cells can rescue aerobic respiration. *Proc Natl Acad Sci USA.* (2006) 103:1283–8. doi: 10.1073/pnas.0510511103
7. Jacobs SA, Roobrouck VD, Verfaillie CM, Van Gool SW. Immunological characteristics of human mesenchymal stem cells and multipotent adult progenitor cells. *Immunol Cell Biol.* (2013) 91:32–9. doi: 10.1038/icb.2012.64
8. Griffin MD, Elliman SJ, Cahill E, English K, Ceredig R, Ritter T. Concise review: Adult mesenchymal stromal cell therapy for inflammatory diseases: How well are we joining the dots? *Stem Cells.* (2013) 31:2033–41. doi: 10.1002/stem.1452
9. Dominici M, Le Blanc K, Mueller I, Slaper-Cortenbach I, Marini F, Krause D, et al. Minimal criteria for defining multipotent mesenchymal stromal cells. *Int Soc Cell Ther Position Statement Cytother.* (2006) 8:315–7. doi: 10.1080/14653240600855905
10. Krampera M, Galipeau J, Shi Y, Tarte K, Sensebe L, Therapy, MSCCOTISFC. Immunological characterization of multipotent mesenchymal stromal cells–The International Society for Cellular Therapy (ISCT) working proposal. *Cytotherapy.* (2013) 15:1054–61. doi: 10.1016/j.jcyt.2013.02.010
11. Galipeau J, Krampera M. The challenge of defining mesenchymal stromal cell potency assays and their potential use as release criteria. *Cytotherapy.* (2015) 17:125–7. doi: 10.1016/j.jcyt.2014.12.008
12. Galipeau J, Krampera M, Barrett J, Dazzi F, Deans RJ, Debruijn J, et al. International Society for Cellular Therapy perspective on immune functional assays for mesenchymal stromal cells as potency release criterion for advanced phase clinical trials. *Cytotherapy.* (2016) 18:151–9. doi: 10.1016/j.jcyt.2015.11.008
13. Aijaz A, Vaninov N, Allen A, Barcia RN, Parekkadan B. Convergence of cell pharmacology and drug delivery. *Stem Cells Transl Med.* (2019) doi: 10.1002/sctm.19-0019. [Epub ahead of print].
14. Moll G, Ankrum JA, Kamhieh-Milz J, Bieback K, Ringden O, Volk HD, et al. Intravascular mesenchymal stromal/stem cell therapy product diversification: time for new clinical guidelines. *Trends Mol Med.* (2019) 25:149–63. doi: 10.1016/j.molmed.2018.12.006
15. Prockop DJ, Olson SD. Clinical trials with adult stem/progenitor cells for tissue repair: let's not overlook some essential precautions. *Blood.* (2007) 109:3147–51. doi: 10.1182/blood-2006-03-013433
16. Liu S, Zhou J, Zhang X, Liu Y, Chen J, Hu B, et al. Strategies to optimize adult stem cell therapy for tissue regeneration. *Int J Mol Sci.* (2016) 17:982. doi: 10.3390/ijms17060982
17. Watanabe M, Yavagal DR. Intra-arterial delivery of mesenchymal stem cells. *Brain Circ.* (2016) 2:114–7. doi: 10.4103/2394-8108.192522
18. Detante O, Rome C, Papassin J. How to use stem cells for repair in stroke patients. *Rev Neurol.* (2017) 173:572–6. doi: 10.1016/j.neurol.2017.09.003
19. Galipeau J, Sensebe L. Mesenchymal stromal cells: clinical challenges and therapeutic opportunities. *Cell Stem Cell.* (2018) 22:824–33. doi: 10.1016/j.stem.2018.05.004
20. Nijboer CH, Kooijman E, Van Velthoven CT, Van Tilborg E, Tiebosch IA, Eijkelkamp N, et al. Intranasal stem cell treatment as a novel therapy for subarachnoid hemorrhage. *Stem Cells Dev.* (2018) 27:313–25. doi: 10.1089/scd.2017.0148
21. Phinney DG, Pittenger MF. Concise Review: MSC-Derived Exosomes for Cell-Free Therapy. *Stem Cells.* (2017) 35:851–8. doi: 10.1002/stem.2575
22. Abou-El-Enein M, Elsanhoury A, Reinke P. Overcoming challenges facing advanced therapies in the EU market. *Cell Stem Cell.* (2016) 19:293–7. doi: 10.1016/j.stem.2016.08.012
23. Chinnadurai R, Rajan D, Qayed M, Arafat D, Garcia M, Liu Y, et al. Potency analysis of mesenchymal stromal cells using a combinatorial assay matrix approach. *Cell Rep.* (2018) 22:2504–17. doi: 10.1016/j.celrep.2018.02.013
24. George MJ, Prabhakara K, Toledano-Furman NE, Wang YW, Gill BS, Wade CE, et al. Clinical cellular therapeutics accelerate clot formation. *Stem Cells Transl Med.* (2018) 7:731–9. doi: 10.1002/sctm.18-0015
25. Li M, Khong D, Chin LY, Singleton A, Parekkadan B. Therapeutic delivery specifications identified through compartmental analysis of a

mesenchymal stromal cell-immune reaction. *Sci Rep.* (2018) 8:6816. doi: 10.1038/s41598-018-24971-2
26. Yin JQ, Zhu J, Ankrum JA. Manufacturing of primed mesenchymal stromal cells for therapy. *Nat Biomed Eng.* (2019) 3:90–104. doi: 10.1038/s41551-018-0325-8
27. Hamidian Jahromi S, Davies JE. Concise review: skeletal muscle as a delivery route for mesenchymal stromal cells. *Stem Cells Transl Med.* (2019) 8:456–65. doi: 10.1002/sctm.18-0208
28. Qazi TH, Mooney DJ, Duda GN, Geissler S. *Biomaterials* that promote cell-cell interactions enhance the paracrine function of MSCs. *Biomaterials.* (2017) 140:103–14. doi: 10.1016/j.biomaterials.2017.06.019
29. Qazi TH, Duda GN, Ort MJ, Perka C, Geissler S, Winkler T. Cell therapy to improve regeneration of skeletal muscle injuries. *J Cachexia Sarcopenia Muscle.* (2019) 10:501–16. doi: 10.1002/jcsm.12416
30. Alvarez-Dolado M, Pardal R, Garcia-Verdugo JM, Fike JR, Lee HO, Pfeffer K, et al. Fusion of bone-marrow-derived cells with Purkinje neurons, cardiomyocytes and hepatocytes. *Nature.* (2003) 425:968–73. doi: 10.1038/nature02069
31. Spees JL, Olson SD, Ylostalo J, Lynch PJ, Smith J, Perry A, et al. Differentiation, cell fusion, and nuclear fusion during *ex vivo* repair of epithelium by human adult stem cells from bone marrow stroma. *Proc Natl Acad Sci USA.* (2003) 100:2397–402. doi: 10.1073/pnas.0437997100
32. Kemp K, Gordon D, Wraith DC, Mallam E, Hartfield E, Uney J, et al. Fusion between human mesenchymal stem cells and rodent cerebellar Purkinje cells. *Neuropathol Appl Neurobiol.* (2011) 37:166–78. doi: 10.1111/j.1365-2990.2010.01122.x
33. Freeman BT, Kouris NA, Ogle BM. Tracking fusion of human mesenchymal stem cells after transplantation to the heart. *Stem Cells Transl Med.* (2015) 4:685–94. doi: 10.5966/sctm.2014-0198
34. Melzer C, Von Der Ohe J, Hass R. *In vivo* cell fusion between mesenchymal stroma/stem-like cells and breast cancer cells. *Cancers.* (2019) 11:185. doi: 10.3390/cancers11020185
35. Petrella F, Spaggiari L, Acocella F, Barberis M, Bellomi M, Brizzola S, et al. Airway fistula closure after stem-cell infusion. *N Engl J Med.* (2015) 372:96–7. doi: 10.1056/NEJMc1411374
36. La Francesca S, Aho JM, Barron MR, Blanco EW, Soliman S, Kalenjian L, et al. Long-term regeneration and remodeling of the pig esophagus after circumferential resection using a retrievable synthetic scaffold carrying autologous cells. *Sci Rep.* (2018) 8:4123. doi: 10.1038/s41598-018-22401-x
37. Hanson SE, Bentz ML, Hematti P. Mesenchymal stem cell therapy for nonhealing cutaneous wounds. *Plast Reconstr Surg.* (2010) 125:510–6. doi: 10.1097/PRS.0b013e3181c722bb
38. Magne B, Lataillade JJ, Trouillas M. Mesenchymal stromal cell preconditioning: the next step toward a customized treatment for severe burn. *Stem Cells Dev.* (2018) 27:1385–405. doi: 10.1089/scd.2018.0094
39. Falanga V, Iwamoto S, Chartier M, Yufit T, Butmarc J, Kouttab N, et al. Autologous bone marrow-derived cultured mesenchymal stem cells delivered in a fibrin spray accelerate healing in murine and human cutaneous wounds. *Tissue Eng.* (2007) 13:1299–312. doi: 10.1089/ten.2006.0278
40. Braid LR, Wood CA, Wiese DM, Ford BN. Intramuscular administration potentiates extended dwell time of mesenchymal stromal cells compared to other routes. *Cytotherapy.* (2018) 20:232–44. doi: 10.1016/j.jcyt.2017.09.013
41. Han KH, Kim AK, Kim DI. Therapeutic potential of human mesenchymal stem cells for treating ischemic limb diseases. *Int J Stem Cells.* (2016) 9:163–8. doi: 10.15283/ijsc16053
42. Abdul Wahid SF, Ismail NA, Wan Jamaludin WF, Muhamad NA, Abdul Hamid MKA, Harunarashid H, et al. Autologous cells derived from different sources and administered using different regimens for 'no-option' critical lower limb ischaemia patients. *Cochr Database Syst Rev.* (2018) 8:CD010747. doi: 10.1002/14651858.CD010747.pub2
43. Soria-Juan B, Escacena N, Capilla-González V, Aguilera Y, Llanos L, Tejedo JR, et al. Cost-effective, safe, and personalized cell therapy for critical limb ischemia in type 2 diabetes mellitus. *Front Immunol.* 10:01151. doi: 10.3389/fimmu.2019.01151
44. Lataillade JJ, Doucet C, Bey E, Carsin H, Huet C, Clairand I, et al. New approach to radiation burn treatment by dosimetry-guided surgery combined with autologous mesenchymal stem cell therapy. *Regen Med.* (2007) 2:785–94. doi: 10.2217/17460751.2.5.785
45. Lange C, Brunswig-Spickenheier B, Cappallo-Obermann H, Eggert K, Gehling UM, Rudolph C, et al. Radiation rescue: mesenchymal stromal cells protect from lethal irradiation. *PLoS One.* (2011) 6:e14486. doi: 10.1371/journal.pone.0014486
46. Pinzur L, Akyuez L, Levdansky L, Blumenfeld M, Volinsky E, Aberman Z, et al. Rescue from lethal acute radiation syndrome (ARS) with severe weight loss by secretome of intramuscularly injected human placental stromal cells. *J Cachexia Sarcopenia Muscle.* (2018) 9:1079–92. doi: 10.1002/jcsm.12342
47. Zhao LR, Duan WM, Reyes M, Keene CD, Verfaillie CM, Low WC. Human bone marrow stem cells exhibit neural phenotypes and ameliorate neurological deficits after grafting into the ischemic brain of rats. *Exp Neurol.* (2002) 174:11–20. doi: 10.1006/exnr.2001.7853
48. Castro RF. Failure of bone marrow cells to transdifferentiate into neural cells *in vivo*. *Science.* (2002) 297:1299. doi: 10.1126/science.297.5585.1299
49. Wagers AJ, Sherwood RI, Christensen JL, Weissman IL. Little evidence for developmental plasticity of adult hematopoietic stem cells. *Science.* (2002) 297:2256–9. doi: 10.1126/science.1074807
50. Coyne TM, Marcus AJ, Woodbury D, Black IB. Marrow stromal cells transplanted to the adult brain are rejected by an inflammatory response and transfer donor labels to host neurons and glia. *Stem Cells.* (2006) 24:2483–92. doi: 10.1634/stemcells.2006-0174
51. Li Y, Chopp M. Marrow stromal cell transplantation in stroke and traumatic brain injury. *Neurosci Lett.* (2009) 456:120–3. doi: 10.1016/j.neulet.2008.03.096
52. Tajiri N, Duncan K, Antoine A, Pabon M, Acosta SA, De La Pena I, et al. Stem cell-paved biobridge facilitates neural repair in traumatic brain injury. *Front Syst Neurosci.* (2014) 8:116. doi: 10.3389/fnsys.2014.00116
53. Steinberg GK, Kondziolka D, Wechsler LR, Lunsford LD, Coburn ML, Billigen JB, et al. Clinical outcomes of transplanted modified bone marrow-derived mesenchymal stem cells in Stroke: A Phase 1/2a Study. *Stroke.* (2016) 47:1817–24. doi: 10.1161/STROKEAHA.116.012995
54. Lee RT. Adult cardiac stem cell concept and the process of science. *Circulation.* (2018) 138:2940–2. doi: 10.1161/CIRCULATIONAHA.118.036407
55. Li Y, He L, Huang X, Bhaloo SI, Zhao H, Zhang S, et al. Genetic lineage tracing of nonmyocyte population by dual recombinases. *Circulation.* (2018) 138:793–805. doi: 10.1161/CIRCULATIONAHA.118.034250
56. Karantalis V, Difede DL, Gerstenblith G, Pham S, Symes J, Zambrano JP, et al. Autologous mesenchymal stem cells produce concordant improvements in regional function, tissue perfusion, and fibrotic burden when administered to patients undergoing coronary artery bypass grafting: The Prospective Randomized Study of Mesenchymal Stem Cell Therapy in Patients Undergoing Cardiac Surgery (PROMETHEUS) trial. *Circ Res.* (2014) 114:1302–10. doi: 10.1161/CIRCRESAHA.114.303180
57. Yau TM, Pagani FD, Mancini DM, Chang HL, Lala A, Woo YJ, et al. Intramyocardial injection of mesenchymal precursor cells and successful temporary weaning from left ventricular assist device support in patients with advanced heart failure: a randomized clinical trial. *JAMA.* (2019) 321:1176–86. doi: 10.1001/jama.2019.2341
58. Kanelidis AJ, Premer C, Lopez J, Balkan W, Hare JM. Route of delivery modulates the efficacy of mesenchymal stem cell therapy for myocardial infarction: a meta-analysis of preclinical studies and clinical trials. *Circ Res.* (2017) 120:1139–50. doi: 10.1161/CIRCRESAHA.116.309819
59. Kan C, Chen L, Hu Y, Lu H, Li Y, Kessler JA, et al. Microenvironmental factors that regulate mesenchymal stem cells: lessons learned from the study of heterotopic ossification. *Histol Histopathol.* (2017) 32:977–85. doi: 10.14670/HH-11-890
60. Bauer G, Elsallab M, Abou-El-Enein M. Concise review: a comprehensive analysis of reported adverse events in patients receiving unproven stem cell-based interventions. *Stem Cells Transl Med.* (2018) 7:676–85. doi: 10.1002/sctm.17-0282
61. Agarwal S, Loder S, Cholok D, Li J, Breuler C, Drake J, et al. Surgical excision of heterotopic ossification leads to re-emergence of mesenchymal stem cell populations responsible for recurrence. *Stem Cells Transl Med.* (2017) 6:799–806. doi: 10.5966/sctm.2015-0365
62. Kan C, Chen L, Hu Y, Ding N, Li Y, Mcguire TL, et al. Gli1-labeled adult mesenchymal stem/progenitor cells and hedgehog signaling

contribute to endochondral heterotopic ossification. *Bone*. (2018) 109:71–9. doi: 10.1016/j.bone.2017.06.014

63. Mendonca MV, Larocca TF, De Freitas Souza BS, Villarreal CF, Silva LF, Matos AC, et al. Safety and neurological assessments after autologous transplantation of bone marrow mesenchymal stem cells in subjects with chronic spinal cord injury. *Stem Cell Res Ther*. (2014) 5:126. doi: 10.1186/scrt516
64. Walker PA, Letourneau PA, Bedi S, Shah SK, Jimenez F, Cox CS Jr. Progenitor cells as remote "bioreactors": neuroprotection via modulation of the systemic inflammatory response. *World J Stem Cells*. (2011) 3:9–18. doi: 10.4252/wjsc.v3.i2.9
65. Kean TJ, Lin P, Caplan AI, Dennis JE. MSCs: delivery routes and engraftment, cell-targeting strategies, and immune modulation. *Stem Cells Int*. (2013) 2013:732742. doi: 10.1155/2013/732742
66. Leibacher J, Henschler R. Biodistribution, migration and homing of systemically applied mesenchymal stem/stromal cells. *Stem Cell Res Ther*. (2016) 7:7. doi: 10.1186/s13287-015-0271-2
67. Fischer UM, Harting MT, Jimenez F, Monzon-Posadas WO, Xue H, Savitz SI, et al. Pulmonary passage is a major obstacle for intravenous stem cell delivery: the pulmonary first-pass effect. *Stem Cells Dev*. (2009) 18:683–92. doi: 10.1089/scd.2008.0253
68. Argibay B, Trekker J, Himmelreich U, Beiras A, Topete A, Taboada P, et al. Intraarterial route increases the risk of cerebral lesions after mesenchymal cell administration in animal model of ischemia. *Sci Rep*. (2017) 7:40758. doi: 10.1038/srep40758
69. Janowski M, Lyczek A, Engels C, Xu J, Lukomska B, Bulte JW, et al. Cell size and velocity of injection are major determinants of the safety of intracarotid stem cell transplantation. *J Cereb Blood Flow Metab*. (2013) 33:921–7. doi: 10.1038/jcbfm.2013.32
70. Guzman R, Janowski M, Walczak P. Intra-arterial delivery of cell therapies for stroke. *Stroke*. (2018) 49:1075–82. doi: 10.1161/STROKEAHA.117.018288
71. Chen SL, Fang WW, Ye F, Liu YH, Qian J, Shan SJ, et al. Effect on left ventricular function of intracoronary transplantation of autologous bone marrow mesenchymal stem cell in patients with acute myocardial infarction. *Am J Cardiol*. (2004) 94:92–5. doi: 10.1016/j.amjcard.2004.03.034
72. Katritsis DG, Sotiropoulou PA, Karvouni E, Karabinos I, Korovesis S, Perez SA, et al. Transcoronary transplantation of autologous mesenchymal stem cells and endothelial progenitors into infarcted human myocardium. *Catheter Cardiovasc Interv*. (2005) 65:321–9. doi: 10.1002/ccd.20406
73. Leistner DM, Fischer-Rasokat U, Honold J, Seeger FH, Schachinger V, Lehmann R, et al. Transplantation of progenitor cells and regeneration enhancement in acute myocardial infarction (TOPCARE-AMI): final 5-year results suggest long-term safety and efficacy. *Clin Res Cardiol*. (2011) 100:925–34. doi: 10.1007/s00392-011-0327-y
74. Lalu MM, Mazzarello S, Zlepnig J, Dong YYR, Montroy J, Mcintyre L, et al. Safety and efficacy of adult stem cell therapy for acute myocardial infarction and ischemic heart failure (safecell heart): a systematic review and meta-analysis. *Stem Cells Transl Med*. (2018) 7:857–66. doi: 10.1002/sctm.18-0120
75. Moniche F, Gonzalez A, Gonzalez-Marcos JR, Carmona M, Pinero P, Espigado I, et al. Intra-arterial bone marrow mononuclear cells in ischemic stroke: a pilot clinical trial. *Stroke*. (2012) 43:2242–4. doi: 10.1161/STROKEAHA.112.659409
76. Suk KT, Yoon JH, Kim MY, Kim CW, Kim JK, Park H, et al. Transplantation with autologous bone marrow-derived mesenchymal stem cells for alcoholic cirrhosis: phase 2 trial. *Hepatology*. (2016) 64:2185–97. doi: 10.1002/hep.28693
77. Saad A, Dietz AB, Herrmann SMS, Hickson LJ, Glockner JF, Mckusick MA, et al. Autologous mesenchymal stem cells increase cortical perfusion in renovascular disease. *J Am Soc Nephrol*. (2017) 28:2777–85. doi: 10.1681/ASN.2017020151
78. Gao J, Dennis JE, Muzic RF, Lundberg M, Caplan AI. The dynamic *in vivo* distribution of bone marrow-derived mesenchymal stem cells after infusion. *Cells Tissues Organs*. (2001) 169:12–20. doi: 10.1159/000047856
79. Schrepfer S, Deuse T, Reichenspurner H, Fischbein MP, Robbins RC, Pelletier MP. Stem cell transplantation: the lung barrier. *Transplant Proc*. (2007) 39:573–6. doi: 10.1016/j.transproceed.2006.12.019
80. Lee RH, Pulin AA, Seo MJ, Kota DJ, Ylostalo J, Larson BL, et al. Intravenous hMSCs improve myocardial infarction in mice because cells embolized in lung are activated to secrete the anti-inflammatory protein TSG-6. *Cell Stem Cell*. (2009) 5:54–63. doi: 10.1016/j.stem.2009.05.003
81. Walker PA, Shah SK, Jimenez F, Gerber MH, Xue H, Cutrone R, et al. Intravenous multipotent adult progenitor cell therapy for traumatic brain injury: preserving the blood brain barrier via an interaction with splenocytes. *Exp Neurol*. (2010) 225:341–52. doi: 10.1016/j.expneurol.2010.07.005
82. Locatelli F, Algeri M, Trevisan V, Bertaina A. Remestemcel-L for the treatment of graft versus host disease. *Expert Rev Clin Immunol*. (2017) 13:43–56. doi: 10.1080/1744666X.2016.1208086
83. Mays RW, Savitz SI. Intravenous cellular therapies for acute ischemic stroke. *Stroke*. (2018) 49:1058–65. doi: 10.1161/STROKEAHA.118.018287
84. Cox CS, Jr, Juranek J, Bedi S. Clinical trials in traumatic brain injury: cellular therapy and outcome measures. *Transfusion*. (2019) 59:858–68. doi: 10.1111/trf.14834
85. De Witte SFH, Luk F, Sierra Parraga JM, Gargesha M, Merino A, Korevaar SS, et al. Immunomodulation by therapeutic mesenchymal stromal cells (MSC) is triggered through phagocytosis of MSC by monocytic cells. *Stem Cells*. (2018) 36:602–15. doi: 10.1002/stem.2779
86. Moll G, Geissler S, Catar R, Ignatowicz L, Hoogduijn MJ, Strunk D, et al. Cryopreserved or fresh mesenchymal stromal cells: only a matter of taste or key to unleash the full clinical potential of MSC therapy? *Adv Exp Med Biol*. (2016) 951:77–98. doi: 10.1007/978-3-319-45457-3_7
87. Danielyan L, Schäfer R, Von Ameln-Mayerhofer A, Bernhard F, Verleysdonk S, Buadze M, et al. Therapeutic efficacy of intranasally delivered mesenchymal stem cells in a rat model of Parkinson disease. *Rejuvenation Res*. (2011) 14:3–16. doi: 10.1089/rej.2010.1130
88. Danielyan L, Beer-Hammer S, Stolzing A, Schafer R, Siegel G, Fabian C, et al. Intranasal delivery of bone marrow-derived mesenchymal stem cells, macrophages, and microglia to the brain in mouse models of Alzheimer's and Parkinson's disease. *Cell Transplant*. (2014) 23(Suppl. 1):S123–39. doi: 10.3727/096368914X684970
89. Oppliger B, Joerger-Messerli M, Mueller M, Reinhart U, Schneider P, Surbek DV, et al. Intranasal delivery of umbilical cord-derived mesenchymal stem cells preserves myelination in perinatal brain damage. *Stem Cells Dev*. (2016) 25:1234–42. doi: 10.1089/scd.2016.0027
90. Salama M, Sobh M, Emam M, Abdalla A, Sabry D, El-Gamal M, et al. Effect of intranasal stem cell administration on the nigrostriatal system in a mouse model of Parkinson's disease. *Exp Ther Med*. (2017) 13:976–82. doi: 10.3892/etm.2017.4073
91. Schafer S, Berger JV, Deumens R, Goursaud S, Hanisch UK, Hermans E. Influence of intrathecal delivery of bone marrow-derived mesenchymal stem cells on spinal inflammation and pain hypersensitivity in a rat model of peripheral nerve injury. *J Neuroinflammation*. (2014) 11:157. doi: 10.1186/s12974-014-0157-8
92. Liu L, Hua Z, Shen J, Yin Y, Yang J, Cheng K, et al. Comparative efficacy of multiple variables of mesenchymal stem cell transplantation for the treatment of neuropathic pain in rats. *Mil Med*. (2017) 182:175–84. doi: 10.7205/MILMED-D-16-00096
93. Vaquero J, Zurita M, Rico MA, Bonilla C, Aguayo C, Fernandez C, et al. Repeated subarachnoid administrations of autologous mesenchymal stromal cells supported in autologous plasma improve quality of life in patients suffering incomplete spinal cord injury. *Cytotherapy*. (2017) 19:349–59. doi: 10.1016/j.jcyt.2016.12.002
94. Vaquero J, Zurita M, Rico MA, Aguayo C, Bonilla C, Marin E, et al. Intrathecal administration of autologous mesenchymal stromal cells for spinal cord injury: safety and efficacy of the 100/3 guideline. *Cytotherapy*. (2018) 20:806–19. doi: 10.1016/j.jcyt.2018.03.032
95. Petrou P, Gothelf Y, Argov Z, Gotkine M, Levy YS, Kassis I, et al. Safety and clinical effects of mesenchymal stem cells secreting neurotrophic factor transplantation in patients with amyotrophic lateral sclerosis: results of phase 1/2 and 2a clinical trials. *JAMA Neurol*. (2016) 73:337–44. doi: 10.1001/jamaneurol.2015.4321
96. Sykova E, Rychmach P, Drahoradova I, Konradova S, Ruzickova K, Vorisek I, et al. Transplantation of mesenchymal stromal cells in patients with amyotrophic lateral sclerosis: results of phase I/IIa clinical trial. *Cell Transplant*. (2017) 26:647–58. doi: 10.3727/096368916X693716

97. Oh KW, Noh MY, Kwon MS, Kim HY, Oh SI, Park J, et al. Repeated intrathecal mesenchymal stem cells for amyotrophic lateral sclerosis. *Ann Neurol.* (2018) 84:361–73. doi: 10.1002/ana.25302
98. Hlebokazov F, Dakukina T, Ihnatsenko S, Kosmacheva S, Potapnev M, Shakhbazau A, et al. Treatment of refractory epilepsy patients with autologous mesenchymal stem cells reduces seizure frequency: An open label study. *Adv Med Sci.* (2017) 62:273–9. doi: 10.1016/j.advms.2016.12.004
99. Milczarek O, Jarocha D, Starowicz-Filip A, Kwiatkowski S, Badyra B, Majka M. Multiple autologous bone marrow-derived CD271(+) mesenchymal stem cell transplantation overcomes drug-resistant epilepsy in children. *Stem Cells Transl Med.* (2018) 7:20–33. doi: 10.1002/sctm.17-0041
100. Junyi L, Na L, Yan J. Mesenchymal stem cells secrete brain-derived neurotrophic factor and promote retinal ganglion cell survival after traumatic optic neuropathy. *J Craniofac Surg.* (2015) 26:548–52. doi: 10.1097/SCS.0000000000001348
101. Ezquer M, Urzua CA, Montecino S, Leal K, Conget P, Ezquer F. Intravitreal administration of multipotent mesenchymal stromal cells triggers a cytoprotective microenvironment in the retina of diabetic mice. *Stem Cell Res Ther.* (2016) 7:42. doi: 10.1186/s13287-016-0299-y
102. Mathew B, Poston JN, Dreixler JC, Torres L, Lopez J, Zelkha R, et al. Bone-marrow mesenchymal stem-cell administration significantly improves outcome after retinal ischemia in rats. *Graefes Arch Clin Exp Ophthalmol.* (2017) 255:1581–92. doi: 10.1007/s00417-017-3690-1
103. Park SS, Moisseiev E, Bauer G, Anderson JD, Grant MB, Zam A, et al. Advances in bone marrow stem cell therapy for retinal dysfunction. *Prog Retin Eye Res.* (2017) 56:148–65. doi: 10.1016/j.preteyeres.2016.10.002
104. Wang JD, An Y, Zhang JS, Wan XH, Jonas JB, Xu L, et al. Human bone marrow mesenchymal stem cells for retinal vascular injury. *Acta Ophthalmol.* (2017) 95:e453–e461. doi: 10.1111/aos.13154
105. Satarian L, Nourinia R, Safi S, Kanavi MR, Jarughi N, Daftarian N, et al. Intravitreal injection of bone marrow mesenchymal stem cells in patients with advanced retinitis pigmentosa; a safety study. *J Ophthalmic Vis Res.* (2017) 12:58–64. doi: 10.4103/2008-322X.200164
106. Sierra-Parraga JM, Eijken M, Hunter J, Moers C, Leuvenink H, Moller B, et al. Mesenchymal stromal cells as anti-inflammatory and regenerative mediators for donor kidneys during normothermic machine perfusion. *Stem Cells Dev.* (2017) 26:1162–70. doi: 10.1089/scd.2017.0030
107. Gregorini M, Corradetti V, Pattonieri EF, Rocca C, Milanesi S, Peloso A, et al. Perfusion of isolated rat kidney with Mesenchymal Stromal Cells/Extracellular Vesicles prevents ischaemic injury. *J Cell Mol Med.* (2017) 21:3381–93. doi: 10.1111/jcmm.13249
108. Sierra Parraga JM, Rozenberg K, Eijken M, Leuvenink HG, Hunter J, Merino A, et al. Effects of normothermic machine perfusion conditions on mesenchymal stromal cells. *Front Immunol.* (2019) 10:765. doi: 10.3389/fimmu.2019.00765
109. Reinders MEJ, Van Kooten C, Rabelink TJ, De Fijter JW. Mesenchymal stromal cell therapy for solid organ transplantation. *Transplantation.* (2018) 102:35–43. doi: 10.1097/TP.0000000000001879
110. Lalu MM, Mcintyre L, Pugliese C, Fergusson D, Winston BW, Marshall JC, et al. Safety of cell therapy with mesenchymal stromal cells (SafeCell): a systematic review and meta-analysis of clinical trials. *PLoS ONE.* (2012) 7:e47559. doi: 10.1371/journal.pone.0047559
111. Moll G, Rasmusson-Duprez I, Von Bahr L, Connolly-Andersen AM, Elgue G, Funke L, et al. Are therapeutic human mesenchymal stromal cells compatible with human blood? *Stem Cells.* (2012) 30:1565–74. doi: 10.1002/stem.1111
112. Moberg L, Johansson H, Lukinius A, Berne C, Foss A, Kallen R, et al. Production of tissue factor by pancreatic islet cells as a trigger of detrimental thrombotic reactions in clinical islet transplantation. *Lancet.* (2002) 360:2039–45. doi: 10.1016/S0140-6736(02)12020-4
113. Gustafson EK, Elgue G, Hughes RD, Mitry RR, Sanchez J, Haglund U, et al. The instant blood-mediated inflammatory reaction characterized in hepatocyte transplantation. *Transplantation.* (2011) 91:632–8. doi: 10.1097/TP.0b013e31820ae459
114. Drake TA, Morrissey JH, Edgington TS. Selective cellular expression of tissue factor in human tissues. Implications for disorders of hemostasis and thrombosis. *Am J Pathol.* (1989) 134:1087–97.
115. Moll G, Ignatowicz L, Catar R, Luecht C, Sadeghi B, Hamad O, et al. Different procoagulant activity of therapeutic mesenchymal stromal cells derived from bone marrow and placental decidua. *Stem Cells Dev.* (2015) 24:2269–79. doi: 10.1089/scd.2015.0120
116. Christy BA, Herzig MC, Montgomery RK, Delavan C, Bynum JA, Reddoch KM, et al. Procoagulant activity of human mesenchymal stem cells. *J Trauma Acute Care Surg.* (2017) 83:S164–9. doi: 10.1097/TA.0000000000001485
117. Liao L, Shi B, Chang H, Su X, Zhang L, Bi C, et al. Heparin improves BMSC cell therapy: anticoagulant treatment by heparin improves the safety and therapeutic effect of bone marrow-derived mesenchymal stem cell cytotherapy. *Theranostics.* (2017) 7:106–16. doi: 10.7150/thno.16911
118. Gleeson BM, Martin K, Ali MT, Kumar AH, Pillai MG, Kumar SP, et al. Bone marrow-derived mesenchymal stem cells have innate procoagulant activity and cause microvascular obstruction following intracoronary delivery: amelioration by antithrombin therapy. *Stem Cells.* (2015) 33:2726–37. doi: 10.1002/stem.2050
119. George MJ, Gill B, Prabhakara K, Toledano Furman NE, Olson SD, Cotton BA, et al. Procoagulant *in vitro* effects of human mesenchymal stem cells and mononuclear cells in a severely injured trauma population. *J Am College Surg.* (2018) 227:S273. doi: 10.1016/j.jamcollsurg.2018.07.563
120. Melmed GY, Pandak WM, Casey K, Abraham B, Valentine J, Schwartz D, et al. Human Placenta-derived Cells (PDA-001) for the treatment of moderate-to-severe crohn's disease: a phase 1b/2a study. *Inflamm Bowel Dis.* (2015) 21:1809–16. doi: 10.1097/MIB.0000000000000441
121. Wu Z, Zhang S, Zhou L, Cai J, Tan J, Gao X, et al. Thromboembolism induced by umbilical cord mesenchymal stem cell infusion: a report of two cases and literature review. *Transplant Proc.* (2017) 49:1656–8. doi: 10.1016/j.transproceed.2017.03.078
122. Acosta L, Hmadcha A, Escacena N, Perez-Camacho I, De La Cuesta A, Ruiz-Salmeron R, et al. Adipose mesenchymal stromal cells isolated from type 2 diabetic patients display reduced fibrinolytic activity. *Diabetes.* (2013) 62:4266–9. doi: 10.2337/db13-0896
123. Capilla-Gonzalez V, Lopez-Beas J, Escacena N, Aguilera Y, De La Cuesta A, Ruiz-Salmeron R, et al. PDGF restores the defective phenotype of adipose-derived mesenchymal stromal cells from diabetic patients. *Mol Ther.* (2018) 26:2696–709. doi: 10.1016/j.ymthe.2018.08.011
124. Jung JW, Kwon M, Choi JC, Shin JW, Park IW, Choi BW, et al. Familial occurrence of pulmonary embolism after intravenous, adipose tissue-derived stem cell therapy. *Yonsei Med J.* (2013) 54:1293–6. doi: 10.3349/ymj.2013.54.5.1293
125. Cyranoski D. Korean deaths spark inquiry. *Nature.* (2010) 468:485. doi: 10.1038/468485a
126. Li Y, Lin F. Mesenchymal stem cells are injured by complement after their contact with serum. *Blood.* (2012) 120:3436–43. doi: 10.1182/blood-2012-03-420612
127. Moll G, Jitschin R, Von Bahr L, Rasmusson-Duprez I, Sundberg B, Lonnies L, et al. Mesenchymal stromal cells engage complement and complement receptor bearing innate effector cells to modulate immune responses. *PLoS ONE.* (2011) 6:e21703. doi: 10.1371/journal.pone.0021703
128. Tu Z, Li Q, Bu H, Lin F. Mesenchymal stem cells inhibit complement activation by secreting factor H. *Stem Cells Dev.* (2010) 19:1803–9. doi: 10.1089/scd.2009.0418
129. Soland MA, Bego M, Colletti E, Zanjani ED, St. Jeor S, Porada CD, et al. Mesenchymal stem cells engineered to inhibit complement-mediated damage. *PLoS ONE.* (2013) 8:e60461. doi: 10.1371/journal.pone.0060461
130. Moll G, Alm JJ, Davies LC, Von Bahr L, Heldring N, Stenbeck-Funke L, et al. Do cryopreserved mesenchymal stromal cells display impaired immunomodulatory and therapeutic properties? *Stem Cells.* (2014) 32:2430–42. doi: 10.1002/stem.1729
131. Li Y, Fung J, Lin F. Local inhibition of complement improves mesenchymal stem cell viability and function after administration. *Mol Ther.* (2016) 24:1665–74. doi: 10.1038/mt.2016.142
132. Li Y, Qiu W, Zhang L, Fung J, Lin F. Painting factor H onto mesenchymal stem cells protects the cells from complement- and neutrophil-mediated damage. *Biomaterials.* (2016) 102:209–19. doi: 10.1016/j.biomaterials.2016.05.055
133. Soland MA, Bego MG, Colletti E, Porada CD, Zanjani ED, St. Jeor S, et al. Modulation of human mesenchymal stem cell immunogenicity through

forced expression of human cytomegalovirus us proteins. *PLoS ONE*. (2012) 7:e36163. doi: 10.1371/journal.pone.0036163

134. François M, Copland IB, Yuan S, Romieu-Mourez R, Waller EK, Galipeau J. Cryopreserved mesenchymal stromal cells display impaired immunosuppressive properties as a result of heat-shock response and impaired interferon-γ licensing. *Cytotherapy*. (2012) 14:147–52. doi: 10.3109/14653249.2011.623691
135. Chinnadurai R, Rajan D, Ng S, Mccullough K, Arafat D, Waller EK, et al. Immune dysfunctionality of replicative senescent mesenchymal stromal cells is corrected by IFNgamma priming. *Blood Adv*. (2017) 1:628–43. doi: 10.1182/bloodadvances.2017006205
136. Moll G, Le Blanc K. Engineering more efficient multipotent mesenchymal stromal (stem) cells for systemic delivery as cellular therapy. *ISBT Science Series*. (2015) 10:357–65. doi: 10.1111/voxs.12133
137. Moll G, Hult A, Von Bahr L, Alm JJ, Heldring N, Hamad OA, et al. Do ABO blood group antigens hamper the therapeutic efficacy of mesenchymal stromal cells? *PLoS ONE*. (2014) 9:e85040. doi: 10.1371/journal.pone.0085040
138. Kode JA, Mukherjee S, Joglekar MV, Hardikar AA. Mesenchymal stem cells: immunobiology and role in immunomodulation and tissue regeneration. *Cytotherapy*. (2009) 11:377–91. doi: 10.1080/14653240903080367
139. Singer NG, Caplan AI. Mesenchymal stem cells: mechanisms of inflammation. *Annu Rev Pathol*. (2011) 6:457–78. doi: 10.1146/annurev-pathol-011110-130230
140. Ankrum JA, Ong JF, Karp JM. Mesenchymal stem cells: immune evasive, not immune privileged. *Nat Biotechnol*. (2014) 32:252–60. doi: 10.1038/nbt.2816
141. Stroncek DF, Butterfield LH, Cannarile MA, Dhodapkar MV, Greten TF, Grivel JC, et al. Systematic evaluation of immune regulation and modulation. *J Immunother Cancer*. (2017) 5:21. doi: 10.1186/s40425-017-0223-8
142. Spees JL, Gregory CA, Singh H, Tucker HA, Peister A, Lynch PJ, et al. Internalized antigens must be removed to prepare hypoimmunogenic mesenchymal stem cells for cell and gene therapy. *Mol Ther*. (2004) 9:747–56. doi: 10.1016/j.ymthe.2004.02.012
143. Sundin M, Ringden O, Sundberg B, Nava S, Gotherstrom C, Le Blanc K. No alloantibodies against mesenchymal stromal cells, but presence of anti-fetal calf serum antibodies, after transplantation in allogeneic hematopoietic stem cell recipients. *Haematologica*. (2007) 92:1208–15. doi: 10.3324/haematol.11446
144. Consentius C, Reinke P, Volk HD. Immunogenicity of allogeneic mesenchymal stromal cells: what has been seen *in vitro* and *in vivo*? *Regen Med*. (2015) 10:305–15. doi: 10.2217/rme.15.14
145. Berglund AK, Fortier LA, Antczak DF, Schnabel LV. Immunoprivileged no more: measuring the immunogenicity of allogeneic adult mesenchymal stem cells. *Stem Cell Res Ther*. (2017) 8:288. doi: 10.1186/s13287-017-0742-8
146. Lohan P, Treacy O, Griffin MD, Ritter T, Ryan AE. Anti-donor immune responses elicited by allogeneic mesenchymal stem cells and their extracellular vesicles: are we still learning? *Front Immunol*. (2017) 8:1626. doi: 10.3389/fimmu.2017.01626
147. Elliott Donaghue I, Tam R, Sefton MV, Shoichet MS. Cell and biomolecule delivery for tissue repair and regeneration in the central nervous system. *J Control Release*. (2014) 190:219–27. doi: 10.1016/j.jconrel.2014.05.040
148. Qazi TH, Mooney DJ, Pumberger M, Geissler S, Duda GN. *Biomaterials* based strategies for skeletal muscle tissue engineering: existing technologies and future trends. *Biomaterials*. (2015) 53:502–21. doi: 10.1016/j.biomaterials.2015.02.110
149. Ding DC, Chou HL, Chang YH, Hung WT, Liu HW, Chu TY. Characterization of HLA-G and related immunosuppressive effects in human umbilical cord stroma-derived stem cells. *Cell Transplant*. (2016) 25:217–28. doi: 10.3727/096368915X688182
150. Kobolak J, Dinnyes A, Memic A, Khademhosseini A, Mobasheri A. Mesenchymal stem cells: Identification, phenotypic characterization, biological properties and potential for regenerative medicine through biomaterial micro-engineering of their niche. *Methods*. (2016) 99:62–8. doi: 10.1016/j.ymeth.2015.09.016
151. Miki T, Triolo F. *Perinatal Tissue-Derived Stem Cells: Alternative Sources of Fetal Stem Cells*. New York, NY: Springer Science+Business Media (2016).
152. Sensebe L. Clinical grade production of mesenchymal stem cells. *Biomed Mater Eng*. (2008) 18:S3–10.
153. De Soure AM, Fernandes-Platzgummer A, Da Silva CL, Cabral JM. Scalable microcarrier-based manufacturing of mesenchymal stem/stromal cells. *J Biotechnol*. (2016) 236:88–109. doi: 10.1016/j.jbiotec.2016.08.007
154. Pischiutta F, D'amico G, Dander E, Biondi A, Biagi E, Citerio G, et al. Immunosuppression does not affect human bone marrow mesenchymal stromal cell efficacy after transplantation in traumatized mice brain. *Neuropharmacology*. (2014) 79:119–26. doi: 10.1016/j.neuropharm.2013.11.001
155. Hajkova M, Hermankova B, Javorkova E, Bohacova P, Zajicova A, Holan V, et al. Mesenchymal stem cells attenuate the adverse effects of immunosuppressive drugs on distinct T cell subopulations. *Stem Cell Rev*. (2017) 13:104–15. doi: 10.1007/s12015-016-9703-3
156. Galleu A, Riffo-Vasquez Y, Trento C, Lomas C, Dolcetti L, Cheung TS, et al. Apoptosis in mesenchymal stromal cells induces *in vivo* recipient-mediated immunomodulation. *Sci Transl Med*. (2017) 9:eaam7828. doi: 10.1126/scitranslmed.aam7828
157. von Bahr L, Batsis I, Moll G, Hagg M, Szakos A, Sundberg B, et al. Analysis of tissues following mesenchymal stromal cell therapy in humans indicates limited long-term engraftment and no ectopic tissue formation. *Stem Cells*. (2012) 30:1575–8. doi: 10.1002/stem.1118
158. Miura M, Miura Y, Padilla-Nash HM, Molinolo AA, Fu B, Patel V, et al. Accumulated chromosomal instability in murine bone marrow mesenchymal stem cells leads to malignant transformation. *Stem Cells*. (2006) 24:1095–103. doi: 10.1634/stemcells.2005-0403
159. Li H, Fan X, Kovi RC, Jo Y, Moquin B, Konz R, et al. Spontaneous expression of embryonic factors and p53 point mutations in aged mesenchymal stem cells: a model of age-related tumorigenesis in mice. *Cancer Res*. (2007) 67:10889–98. doi: 10.1158/0008-5472.CAN-07-2665
160. Torsvik A, Rosland GV, Bjerkvig R. Comment to: "Spontaneous transformation of adult mesenchymal stem cells from cynomolgus macaques *in vitro*" by Z. Ren et al. Exp. Cell Res. (2011) 317:2950–7: spontaneous transformation of mesenchymal stem cells in culture: facts or fiction? *Exp Cell Res*. (2012) 318:441–3. doi: 10.1016/j.yexcr.2011.12.009
161. Rubio D, Garcia-Castro J, Martin MC, De La Fuente R, Cigudosa JC, Lloyd AC, et al. Spontaneous human adult stem cell transformation. *Cancer Res*. (2005) 65:3035–9. doi: 10.1158/0008-5472.CAN-04-4194
162. Rosland GV, Svendsen A, Torsvik A, Sobala E, Mccormack E, Immervoll H, et al. Long-term cultures of bone marrow-derived human mesenchymal stem cells frequently undergo spontaneous malignant transformation. *Cancer Res*. (2009) 69:5331–9. doi: 10.1158/0008-5472.CAN-08-4630
163. Torsvik A, Rosland GV, Svendsen A, Molven A, Immervoll H, Mccormack E, et al. Spontaneous malignant transformation of human mesenchymal stem cells reflects cross-contamination: putting the research field on track - letter. *Cancer Res*. (2010) 70:6393–6. doi: 10.1158/0008-5472.CAN-10-1305
164. Ren Z, Wang J, Zhu W, Guan Y, Zou C, Chen Z, et al. Spontaneous transformation of adult mesenchymal stem cells from cynomolgus macaques *in vitro*. *Exp Cell Res*. (2011) 317:2950–7. doi: 10.1016/j.yexcr.2011.09.008
165. Ren Z, Zhang YA, Chen Z. Spontaneous transformation of cynomolgus mesenchymal stem cells *in vitro*: further confirmation by short tandem repeat analysis. *Exp Cell Res*. (2012) 318:435–40. doi: 10.1016/j.yexcr.2011.12.012
166. Pan Q, Fouraschen SM, De Ruiter PE, Dinjens WN, Kwekkeboom J, Tilanus HW, et al. Detection of spontaneous tumorigenic transformation during culture expansion of human mesenchymal stromal cells. *Exp Biol Med*. (2014) 239:105–15. doi: 10.1177/1535370213506802
167. Bernardo ME, Zaffaroni N, Novara F, Cometa AM, Avanzini MA, Moretta A, et al. Human bone marrow derived mesenchymal stem cells do not undergo transformation after long-term *in vitro* culture and do not exhibit telomere maintenance mechanisms. *Cancer Res*. (2007) 67:9142–9. doi: 10.1158/0008-5472.CAN-06-4690
168. Conforti A, Starc N, Biagini S, Tomao L, Pitisci A, Algeri M, et al. Resistance to neoplastic transformation of *ex-vivo* expanded human mesenchymal stromal cells after exposure to supramaximal physical and chemical stress. *Oncotarget*. (2016) 7:77416–29. doi: 10.18632/oncotarget.12678

169. Stultz BG, Mcginnis K, Thompson EE, Lo Surdo JL, Bauer SR, Hursh DA. Chromosomal stability of mesenchymal stromal cells during *in vitro* culture. *Cytotherapy.* (2016) 18:336–43. doi: 10.1016/j.jcyt.2015.11.017
170. Ben-David U, Mayshar Y, Benvenisty N. Large-scale analysis reveals acquisition of lineage-specific chromosomal aberrations in human adult stem cells. *Cell Stem Cell.* (2011) 9:97–102. doi: 10.1016/j.stem.2011.06.013
171. Sensebe L, Tarte K, Galipeau J, Krampera M, Martin I, Phinney DG, et al. Limited acquisition of chromosomal aberrations in human adult mesenchymal stromal cells. *Cell Stem Cell.* (2012) 10:9–10; author reply 10–11. doi: 10.1016/j.stem.2011.12.005
172. Luetzkendorf J, Nerger K, Hering J, Moegel A, Hoffmann K, Hoefers C, et al. Cryopreservation does not alter main characteristics of Good Manufacturing Process-grade human multipotent mesenchymal stromal cells including immunomodulating potential and lack of malignant transformation. *Cytotherapy.* (2015) 17:186–98. doi: 10.1016/j.jcyt.2014.10.018
173. Guess AJ, Daneault B, Wang R, Bradbury H, La Perle KMD, Fitch J, et al. Safety profile of good manufacturing practice manufactured interferon gamma-primed mesenchymal stem/stromal cells for clinical trials. *Stem Cells Transl Med.* (2017) 6:1868–79. doi: 10.1002/sctm.16-0485
174. Wuchter P, Bieback K, Schrezenmeier H, Bornhauser M, Muller LP, Bonig H, et al. Standardization of Good Manufacturing Practice-compliant production of bone marrow-derived human mesenchymal stromal cells for immunotherapeutic applications. *Cytotherapy.* (2015) 17:128–39. doi: 10.1016/j.jcyt.2014.04.002
175. Barkholt L, Flory E, Jekerle V, Lucas-Samuel S, Ahnert P, Bisset L, et al. Risk of tumorigenicity in mesenchymal stromal cell-based therapies–bridging scientific observations and regulatory viewpoints. *Cytotherapy.* (2013) 15:753–9. doi: 10.1016/j.jcyt.2013.03.005
176. Van Der Spoel TI, Vrijsen KR, Koudstaal S, Sluijter JP, Nijsen JF, De Jong HW, et al. Transendocardial cell injection is not superior to intracoronary infusion in a porcine model ofischaemic cardiomyopathy: a study on delivery efficiency. *J Cell Mol Med.* (2012) 16:2768–76. doi: 10.1111/j.1582-4934.2012.01594.x
177. Byun JS, Kwak BK, Kim JK, Jung J, Ha BC, Park S. Engraftment of human mesenchymal stem cells in a rat photothrombotic cerebral infarction model: comparison of intra-arterial and intravenous infusion using MRI and histological analysis. *J Korean Neurosurg Soc.* (2013) 54:467–76. doi: 10.3340/jkns.2013.54.6.467
178. Vu Q, Xie K, Eckert M, Zhao W, Cramer SC. Meta-analysis of preclinical studies of mesenchymal stromal cells for ischemic stroke. *Neurology.* (2014) 82:1277–86. doi: 10.1212/WNL.0000000000000278
179. Jeong H, Yim HW, Cho YS, Kim YI, Jeong SN, Kim HB, et al. Efficacy and safety of stem cell therapies for patients with stroke: a systematic review and single arm meta-analysis. *Int J Stem Cells.* (2014) 7:63–9. doi: 10.15283/ijsc.2014.7.2.63

4

Dissecting Allo-Sensitization After Local Administration of Human Allogeneic Adipose Mesenchymal Stem Cells in Perianal Fistulas of Crohn's Disease Patients

Alvaro Avivar-Valderas[1†], Cristina Martín-Martín[2†], Cristina Ramírez[1], Borja Del Río[1], Ramón Menta[1], Pablo Mancheño-Corvo[1], Maitane Ortiz-Virumbrales[1], Ángel Herrero-Méndez[1], Julián Panés[3], Damián García-Olmo[4], José Luís Castañer[2], Itziar Palacios[1], Eleuterio Lombardo[1], Wilfried Dalemans[5] and Olga DelaRosa[1*]*

[1] Takeda Madrid, Cell Therapy Technology Center—Cell Therapies, Madrid, Spain, [2] Department of Immunology, University Hospital Ramon y Cajal, Madrid, Spain, [3] Department of Gastroenterology, Hospital Clínic, IDIBAPS, CIBERehd, Barcelona, Spain, [4] Department of Surgery, Hospital U. Fundación Jiménez Díaz, Madrid, Spain, [5] TiGenix NV, Leuven, Belgium

***Correspondence:**
Alvaro Avivar-Valderas
alvaro.avivar@takeda.com
Olga DelaRosa
olga.delarosa@takeda.com

†These authors have contributed equally to this work

Adipose mesenchymal stem cells (ASC) are considered minimally immunogenic. This is due to the low expression of human leukocyte antigens I (HLA-I), lack of HLA-II expression and low expression of co-stimulatory molecules such as CD40 and CD80. The low rate of observed immunological rejection as well as the immunomodulatory qualities, position ASC as a promising cell-based therapy for the treatment of a variety of inflammatory indications. Yet, few studies have addressed relevant aspects of immunogenicity such as ASC donor-to-patient HLA histocompatibility or assessment of immune response triggered by ASC administration, particularly in the cases of presensitization. The present study aims to assess allo-immune responses in a cohort of Crohn's disease patients administered with allogeneic ASC (darvadstrocel formerly Cx601) for the treatment of complex perianal fistulas. We identified donor-specific antibodies (DSA) generation in a proportion of patients and observed that patients showing preexisting immunity were prone to generating DSA after allogeneic therapy. Noteworthy, naïve patients generating DSA at week 12 (W12) showed a significant reduction in DSA titer at week 52 (W52), whereas DSA titer was reduced in pre-sensitized patients only with no specificities against the donor administered. Remarkably, we did not observe any correlation of DSA generation with ASC therapeutic efficacy. *In vitro* complement-dependent cytotoxicity (CDC) studies have revealed limited cytotoxic levels based upon HLA-I expression and binding capacity even in pro-inflammatory conditions. We sought to identify CDC coping mechanisms contributing to the limited cytotoxic killing observed in ASC *in vitro*. We found that ASC express membrane-bound complement regulatory proteins (mCRPs) CD55, CD46, and CD59 at basal levels, with CD46 more actively expressed in pro-inflammatory conditions. We demonstrated that CD46 is a main driver of CDC signaling; its depletion significantly enhances sensitivity of ASC to CDC. In summary, despite relatively high clearance, DSA generation may represent a major challenge for allogeneic cell therapy management. Sensitization may be a significant concern when evaluating re-treatment or multi-donor trials. It is still unknown whether DSA generation could potentially be the consequence of donor-to-patient interaction and, therefore, subsequently link to efficacy or biological activity. Lastly, we propose that CDC modulators such as CD46 could be used to ultimately link CDC specificity with allogeneic cell therapy efficacy.

Keywords: ASC, cell therapy and immunotherapy, allogeneic, allo-sensitization, CD46, crispr gene editing, HLA class 1, complement dependent cytotoxicity (CDC)

INTRODUCTION

The use of autologous and allogeneic mesenchymal stem cells (MSC) for treating a variety of conditions has been evaluated across several clinical trials (1), including those focused on the treatment inflammatory bowel disease (2), sepsis (3), and graft vs. host disease (GvHD) (4), as well as several autoimmune diseases (5). MSC therapeutic effects are largely attributed to their capacity to modulate immune response. MSC interact with both innate and adaptive arms of the immune system, inhibiting proliferation of T, B and natural killer (NK) cells and fueling expansion of regulatory T cells (Treg) (6, 7). MSC are able to hinder a series of immune cell functions, including: reducing cytokine secretion of T and NK cells; impairing maturation and activation of antigen-presenting cells (APC) and maturation and antibody (Ab) secretion of B cells (8, 9). Lastly, MSC have been shown to promote monocyte to macrophage transition and facilitate M1 (pro-inflammatory) to M2 (anti-inflammatory) macrophage skewing (10). Specifically, the immune modulatory capacities and therapeutic potential of allogeneic Adipose mesenchymal stem cells (ASC) have been extensively characterized by us and others (11–14).

The prevailing paradigm is that MSC in general, and ASC in particular, are not considered to induce a strong immunogenic response. This is mainly due to low human leukocyte antigen (HLA)-I expression, an absence of HLA-II expression, and a low expression of co-stimulatory molecules, such as CD40 and CD80 (7, 11). In fact, in the majority of clinical trials, patients have been administered with allogeneic MSC without the need of major histocompatibility complex (MHC) matching prior to treatment and without utilizing immunosuppressive therapies (15, 16), in contrast to other allogeneic cell-based therapies. Recent data has indicated that although well-tolerated, allogeneic MSC therapy has shown some signals of immune response, such as the generation of donor-specific antibodies (DSA) (17–21). Moreover, during allogeneic administration HLA-mediated cytotoxicity can potentially contribute to the elimination of ASC, eventually compromising efficacy. There are few studies investigating the immunogenicity of administered allogeneic ASC. The current study aims to address this issue.

Crohn's disease (CD) is a chronic inflammation of the digestive tract characterized by transmural inflammation and fistula formation (22). Perianal fistulas are a complication of CD that affects ~28% of patients in the first two decades following diagnosis (23). They are extremely difficult to treat as there are not any recognized effective treatment options, and the condition is associated with a high rate of relapse following the withdrawal of antibiotic or anti-tumor necrosis factor alpha (TNFα) therapy (22). ADMIRE CD (global randomized, double-blind, parallel-group, placebo-controlled trial) demonstrated that local allogeneic darvadstrocel (Cx601) administration can be an effective and safe treatment for complex perianal fistulas in adult patients with non- or mildly active luminal CD (21, 24). A cohort of ADMIRE CD patients was screened for generation of DSA, and it was determined that there were no safety signals associated with the development of DSA (24). A cohort of patients was screened for generation of DSA through Luminex solid-phase assays (SPA) using recombinant HLA molecules bound to beads.

The complement system is a key component of the innate immune response and serves as a nexus with the adaptive immune response, but it is primarily involved in inflammatory processes (25). During allogeneic recognition, complement-dependent cytotoxicity (CDC) is initiated when C1q, the initiating mediator of the classical complement pathway, is fixed to the Fc portion of HLA-I antigen-bound antibodies resulting in its activation and subsequent complement cascade signaling (26). Complement activation can be also triggered in damaged tissues where anaphylatoxins act as chemotactic factors to amplify the immune response. MSC express functional anaphylatoxin receptors (27) which facilitates their recruitment to the injured site and subsequent classic C1q CDC pathway activation. During the ADMIRE CD1 clinical trial, ASC were directly delivered into all fistula tracts and internal openings, which are essentially injured tissues (24). The prominent role of CDC in safeguarding homeostasis within damaged tissues (i.e., fistula) prompted us to examine the impact of the complement activation pathways in the fate of ASC.

Complement signaling is buffered by regulatory molecules (i.e., CD46, CD55, CD59) that are found in the membrane of different cell types, including ASC (28). These molecules function as decay accelerators and are known as membrane complement regulatory proteins (mCRP) (29). Despite mCRP expression being reported in MSC of varying origins, very little is known about their contribution to complement mediated cytotoxicity (30–32). Our objective is to evaluate the differential cytotoxic capacity among plasma samples collected from ADMIRE CD1 patients, using conventional functional assays against donor ASC *in vitro*.

MATERIALS AND METHODS

Monitoring DSA Generation in ADMIRE CD Patients

We focused in a subgroup of 123 patients enrolled in the randomized, double-blind, parallel-group, placebo-controlled study ADMIRE CD (24). All of them were adult patients (≥18 years) with CD and treatment-refractory, draining complex perianal fistulas and selected to a single local injection of 120 million ASC or 24 mL saline solution (control). A total of 60 and 63 patients received control or infusion of allogeneic ASC, respectively, from which 105 (58 ASC, 47 control) were successfully followed up 52 weeks after administration. Note, that clinical remission data was missing for 10 naïve patients which explains why number of clinical remission data-points (33) differs to total number of naïve patients in the study (34).

ASC Donors and Bone Marrow (BM)-MSC

Human adipose tissue aspirates from healthy donors were processed as described elsewhere (35). For the present study we used ASC from seven different donors, DonA was the donor used for ADMIRE CD clinical trial. Additionally, we tested mCRP expression and CDC sensitivity to the following donors: DonB, DonC, DonD, DonE, DonF, and DonG, manufactured by Tigenix (Takeda Pharmaceuticals). All ASC donors comply with International Federation for Adipose Therapeutics (IFATS) and the International Society for Cellular Therapy (ISCT) identity and purity criteria (36). ASC culture has been described elsewhere (11). BM-MSC were purchased to Lonza and cultured following manufacturer instructions.

HLA Antibodies (Abs) Detection

A plasma sample was obtained by centrifugation of a peripheral blood tube with ethylenediaminetetraacetic acid (Vacutainer® spry-coated K2EDTA tubes, BD™), collected from all patients, at baseline, 12 and 52 weeks after control or ASC administration. HLA Abs were detected in a Luminex platform using a LabscreenMixed™ kit (One Lambda Inc.® Canoga Park, CA, US) according to manufacturer instructions. All samples with a signal >800 units of median fluorescence intensity (MFI) were considered positive, and donor specificities for HLA Abs were determined using Labscreen Single Antigen™ kit (One Lambda Inc.® Canoga Park, CA, US). All signals were normalized according to Quantiplex™ beads fluorescence and specificities > 20,000 units of standard fluorescent intensity were considered relevant. Qualitatively, we defined the HLA antibody titer as the resulting MFI sum of all the determinant beads of the HLA class I molecules included in the Labscreen Mixed kit. We will refer to pre-existing HLA Abs detected in patients before ASC administration as HLA Abs whereas donor ASC-induced HLA Abs will be referred as DSA.

HLA Typing

The assignment of HLA allele expressed in patients or ASC donors was determined over DNA samples obtained from peripheral blood sample or ASC pellets using chemagic DNA Blood250 KIT (PerkinElmer). After checking purity via examination of the A260/280 absorbance ratio, all samples with a DNA concentration of 20 ng/μl or more were tested by LABType®SSO assay (One Lambda, Canoga Park, CA) specifically for loci A, B, and C of HLA according manufacturer instructions. The characterization of the incompatibilities between patient and donor ASC were defined as an unshared, unique chain of polymorphic residues, using the algorithm HLA matchmaker hereafter referred to as *eplets* (37).

Standardization of Flow Cytometry Crossmatch (FCXM) Binding With Recombinant HLA Abs (rHLA)

Standard Curves

We established the level of class I (DonA and DonB) and class II HLA (DonA) expression in the indicated ASC donors used, under basal conditions and pre-activated with interferon gamma (IFNγ) (3 ng/mL for 48 h). We stained 50,000 ASC with the PE (R-phycoerythrin) anti-human class I HLA Ab (clone W6/32) and Peridinin Chlorophyll Protein Complex (PerCP) anti-human class II HLA Ab (clone L243) (Becton Dickinson, Franklin Lakes, New Jersey, US) in increasing concentrations (from 0 to 15 ng/ml for clone W6/32, and 0 to 3 ng/ml for clone L243) and incubated 30 min (min) in the dark at room temperature (RT).

Plasma Samples FCXM Binding Strength and CDC

We tested pre-treatment, week 12 (W12) and week 52 (W52) plasma samples of all patients who had received the ASC administration, previously de-complemented at 56°C for 30 min and washed once with autoMACS Running Buffer (Miltenyi). We incubated 50 μl of de-complemented plasma with 50,000 ASC in a final volume of 100 μl during 30 min at RT. Without washing we added 250 μl of rabbit serum as source of complement anti-human class I HLA (CABC-1D, One Lambda Inc.® Canoga Park, CA, US) for 1 h. Then cells were washed twice and incubated with 20 μl FITC anti-human IgG during 20 min and once washed, adding 5 μl of viability dye 7-Aminoactinomycin D (7-AAD) by acquisition in a LSR Fortessa flow cytometer (BD™). The presence of HLA antibodies and the recognition of ASC was determined by the increase of MFI with respect to the control serum and the cytotoxic capacity by the percentage of 7-AAD+ cells acquiring 10,000 events in P1 gate (total population of ASC) per sample. For analysis we used FlowJo software version 9.7.5.

mCRP Quantification *via* FACS

ASC were grown in normal or 3 ng/mL IFNγ conditions for 48 h. ASC were then trypsinized and counted for a final concentration of 50,000 ASC per 100 μL autoMACS Running Buffer (Miltenyi). For antibody staining we used CD46 (564253, BD), CD55 (MCA1614PE, Serotec), and CD59 (BRA-10G, Novus Biologicals) Abs and their respective isotypes as controls (IgG2a-APC, IgG1-PE, and IgG2b-PE from BD). After 20 min ice incubation ASC were washed with autoMACS Running Buffer (Miltenyi) and centrifuged 500 × g for 4 min. Finally, ASC were resuspended in 100 μL autoMACS Running Buffer (Miltenyi) transferred to cytometry tubes acquired in a LSR Fortessa flow cytometer (BD) and analyzed with BD FacsDiva™ (BD).

Generation of CD46^{KO} ASC

Guide RNA was designed to target CD46 exon 3 using the following public genomic tools: https://genome.ucsc.edu/, https://www.ncbi.nlm.nih.gov/gene. For CRISPR RNA (crRNA) delivery we used the Alt-R® CRISPR-Cas9 System (IDT Integrated DNA Technologies) as per manufacturer instructions. Briefly, ASC were thawed and left overnight. Following this, we prepared and delivered ribonucleoprotein (RNP) complexes using Lipofectamine™ RNAiMAX (Thermo-Fisher). We mixed crRNA and trans-activating crRNA (tracrRNA) in equimolar concentration in a sterile micro-centrifuge tube at a final oligo duplex working concentration of 1 μM. Following 20 min at RT mix incubation, we added the transfection complexes to the culture plate before adding the ASC suspension. After 24 h we replaced the ASC medium and verified lipofection efficacy of labeled tracrRNA-ATTO^{550} ASC with fluorescence microscope.

RESULTS

Long-Term DSA Presence in ADMIRE CD Treated Patients

Blood samples were collected from 123 patients (63 ASC and 60 control) at baseline and 12 weeks after treatment administration. At 52 weeks after treatment administration, 105 patients (58 ASC and 47 placebo) provided blood samples (**Figure 1A**). Analysis by solid phase assay using Luminex technology revealed that 23 patients generated DSA 12 weeks after treatment. As expected, no patients receiving placebo generated DSA (**Figure 1A,** right chart). Additionally, results indicated that 16% (10/63) of treatment-ASC group and 15% (9/60) of placebo group had pre-existing HLA abs (pre-sensitized patients) at baseline. Out of the 53 naïve patients, 17 generated DSA at W12, and 6 of the 10 pre-sensitized patients generated DSA at W12 (**Figure 1A,** left chart). In all cases specificities of DSA were only detected against class I of HLA and not against HLA class II. Long term follow up revealed that no additional generation of DSA was detected 52 weeks after treatment and 7 out of the 23 sensitized patients (30%) cleared DSA at W52. Interestingly, the group of naïve patients generating DSA at W12 showed 35% clearance (6/17), whereas pre-sensitized patients generating DSA at W12 showed 17% (1/6) clearance at W52. Further, pre-sensitized patients were prone to sustained humoral response at W52 post-treatment, indicative for a sort of secondary response in pre-sensitized patients. In contrast, naïve patients generating DSA showed a trend to return to the initial baseline status as naïve patients, suggesting a primary immune response kinetic. The level of DSA positivity for a given sample was selected using the most restrictive threshold of the single antigen results, interpreted by categorical values (yes or no over a given cut-off). Alternatively, the amount of antibody bound relative to the total antigen present on the purified HLA-coated beads can be also quantified as the sum of MFI HLA class I LSM microspheres. We calculated time-course curves measuring plasma DSA titer throughout time (before treatment and 12–52 weeks post-treatment) illustrating the response kinetics and determining the likelihood of reducing their DSA levels (**Figure 1B**). Patients from the treatment group were clustered in the following groups: Naïve patients that did not generate DSA (where baseline levels will be used for comparisons), naïve patients that generated DSA after allo-ASC administration, pre-sensitized patients sharing specificities of the donor ASC administered and pre-sensitized patients not sharing specificities against the donor used. As expected, naïve patients not generating DSA did not modify their antibody titer throughout the course of the study (**Figure 1B,** upper left panel) whereas naïve patients that generated DSA exhibited an increase of antibody titer at W12 that reduces at W52, mimicking a primary immune response kinetic. In addition, baseline MFI values to W12 differed among patients, with certain patients exhibiting a particularly intense response, patient 92 (Pat92) being the most allo-reactive (**Figure 1B,** blue circle). Interestingly, the kinetics of the pre-sensitized group of patients that shared specificities with donor administered did not show a consistent reduction of antibody titer at W52 (**Figure 1B,** lower left panel), whereas those pre-sensitized patients that did not share specificities with the donor administered showed increase of DSA titer at W12 that is reduced at W52 (**Figure 1B,** lower right panel) like the primary immune response kinetics observed in the upper left. We examined a possible connection between donor–patient HLA matching grade and the probability to generate DSA (38, 39). We aimed to identify polymorphic residues present in the ASC donor HLA type (*eplets*) but absent in patients as main precursors of the allogeneic recognition. Each patient's HLA allele was aligned with ASC donor HLA allele for mismatch quantification (37). In our study, allo-sensitization arose mainly against HLA class I (*data not shown*), therefore we focused in characterizing loci A and B of HLA class I. Total number of *eplets* correlated with patients' susceptibility to generate DSA (**Figure 1C**).

Finally, we focused on investigating a potential connection of DSA generation and ASC therapeutic effect. We correlated DSA generation with clinical remission (closure of all treated external openings draining at baseline) at W12 and W52 (**Supplementary Figure 1A**). DSA presence did not affect clinical remission ratios in naïve patients (66.7 against 67.6%) at W12 clinical end-points. Clinical remission levels were slightly higher in naïve patients with DSA vs. patients that did not generated DSA (55.6 against 50%) at W52 end-point. However, no statistical difference was observed at this clinical end-point. Due to the low number of pre-sensitized patients, no statistical analysis could be performed in this sub-group.

To summarize, the above data suggest that although ASC could trigger allo-sensitization in a segment of patients, in most cases we observed a reduction in the antibody titer over time. Although we identified a link between DSA generation and donor-to-patient histocompatibility (based on HLA class I A and B loci mismatch) no correlation with clinical efficacy was observed.

In vitro Binding of HLA Abs to ASC

Although MSC found in different tissues share common hallmarks including immuno-modulatory properties or identity markers, differential immunogenic responses have been reported in *in vivo* models (40–44). To characterize immunogenic response of ASC we first sought to quantify the expression of HLA-I and -II molecules on DonA ASC and their ability

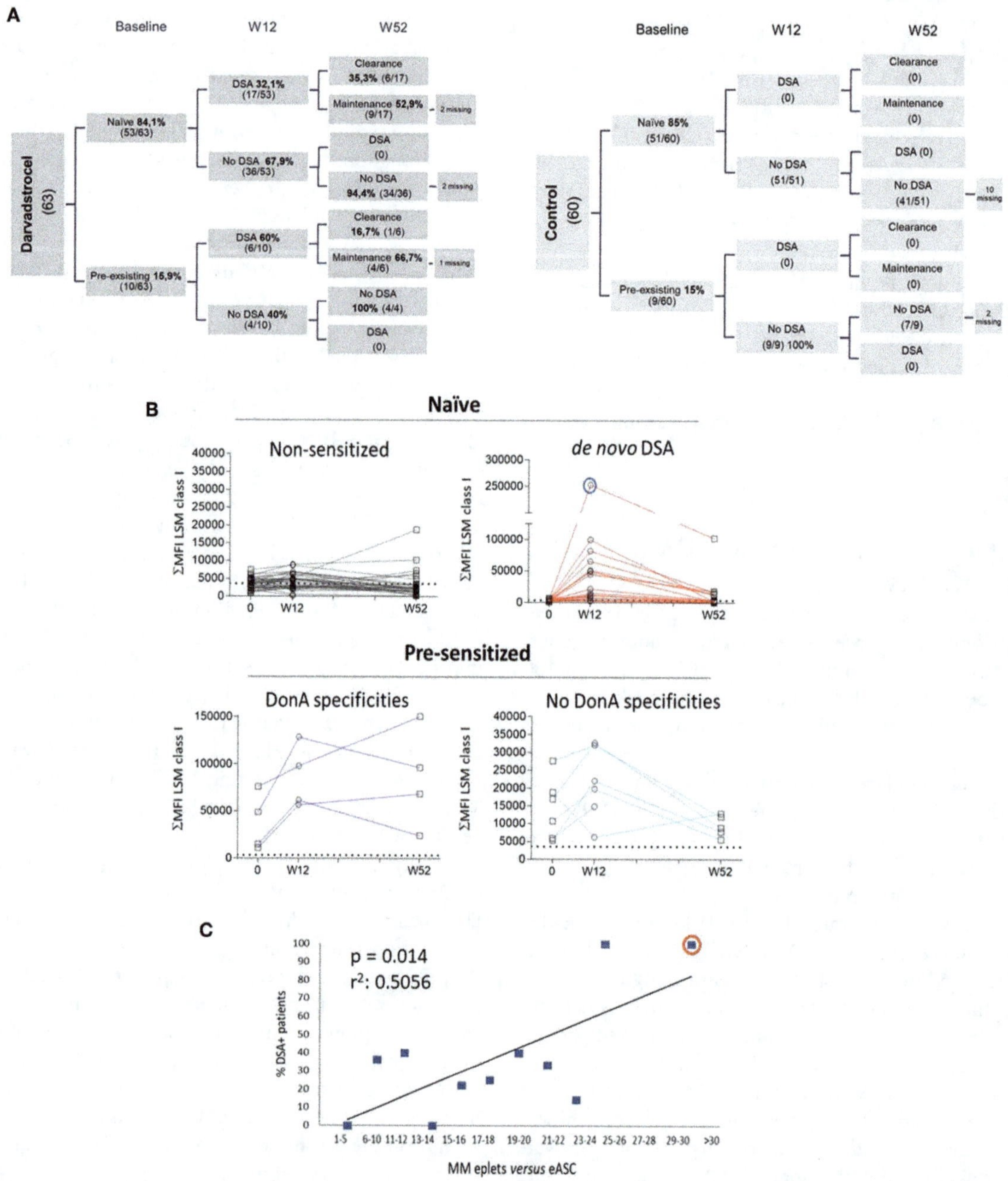

FIGURE 1 | Characterization of DSA generation in ADMIRE CD. **(A)** Distribution of Labscreen Mixed/Labscreen Single Antigen (LSM/LSA) results for indicated visits: pre-treatment (baseline), week 12 (W12) and week 52 (W52) in both control (right chart) and darvadstrocel (left chart) arms of the study. A total of 5 (darvadstrocel arm) and 12 (control arm) patients withdrew from the study and, therefore, no LSM/LSA data were available. **(B)** Kinetic curves illustrating HLA Abs titer represented as the sum of MFI (ΣMFI) of each micro-sphere measured with the LSM assay at the indicated time points. The dotted line in each graph indicates the MFI > 3000 threshold applied for positivity. Pat92 is evidenced in a blue line circle. **(C)** Graph representing HLA incompatibility between each patient and ASC. For the correlation of incompatibility, the percentage of individuals that generated DSA was plotted vs. the number number (range) of mismatched *eplets*. Highlighted with a red circle is Pat92. For linear regression we applied Pearson test (r^2). *P*-values were determined by the Student's *t*-test.

to bind patient's HLA Abs *via* FACS analysis (**Figure 2A**). We incubated ASC, pre-stimulated or not with IFNγ, with increasing concentrations of fluorescence-labeled HLA-I (W6/32) or HLA-II (L243) recombinant Abs and quantified the MFI of the staining. As expected, ASC expressed HLA-I at basal level exhibiting a strong increase after IFNγ stimulation (**Figure 2A**). Conversely, HLA-II levels were negative at baseline and were increased following IFNγ stimulation.

Among all plasma samples analyzed, one interesting high HLA titer sample brought our attention to a detailed analysis. This plasma sample, Pat92 (**Figure 1B,** blue circle), was carrying the largest DSA titer post-treatment. Following IFNγ stimulation, we observed a significant increase of ASC binding strength exclusively in positive control (pool of hyper-immunized samples, HI *pool*) and Pat92 sample (**Figure 2B,** lower left panel). The increase in the binding was accompanied by high percentage of cytotoxic killing, specifically 34.5% in Pat92 (**Figure 2B,** light blue). This percentage was significantly higher than the percentage of killing quantified in the rest of 22 patients tested (ranging from 3.3 to 9.3%, *data not shown*), confirming that Pat92 was the most allo-reactive. Interestingly, Pat92 was the one showing highest level of mismatching with the ASC administered (37 mismatched *eplets*) and did clear antibody 52 weeks after treatment. Further studies will allow us to better understand whether there is a correlation between pre-sensitization and lack of efficacy due to the immediate ASC elimination.

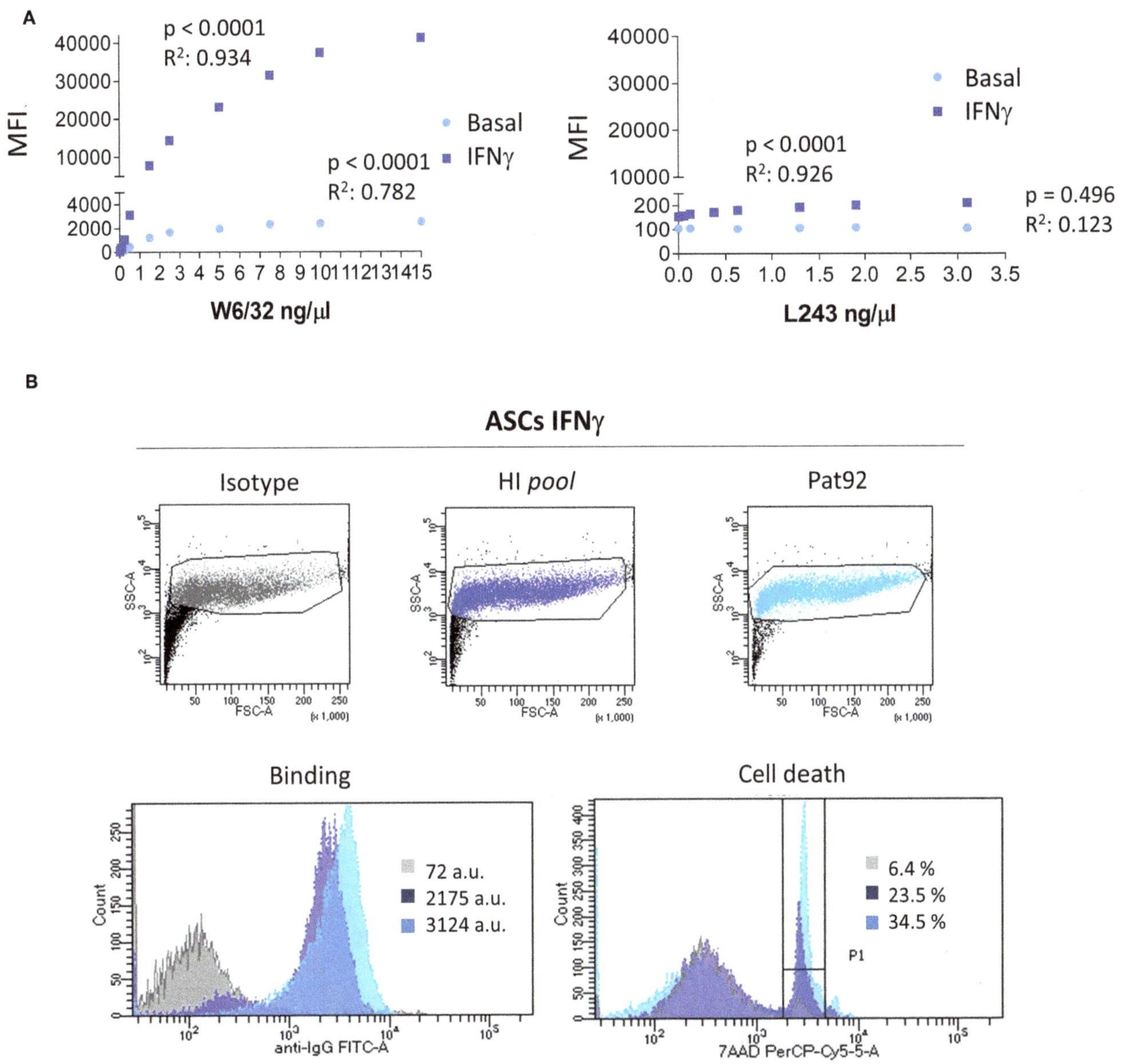

FIGURE 2 | HLA expression in ASC and *in vitro* binding HLA Abs. **(A)** Graphs showing the correlation between MFI increase and each concentration of class I HLA (W6/32) Ab and class II HLA (L243) directed against untreated (light blue) or pre-activated with IFNγ (dark blue) ASC. **(B)** Plots of FcTox (complement-dependent cytotoxicity by flow cytometry assay) representing negative control (isotype), positive control (hyper-immunized samples, HI pool) and patient 92 (Pat92). Lower panels show histogram analysis of binding strength (left) and percentage of cell death, 7-AAD % (right panel). *P*-values were determined by the Student's *t*-test and r^2 by Pearson test.

Plasma DSA Binds ASC Inducing Moderate Killing *in vitro*

In the results above we have shown that from a cohort of 63 ADMIRE CD patients, 10 had pre-existing HLA-I Ab and 17 generated DSA *de novo*. We have also demonstrated that ASC express HLA-I antigen and bind to rHLA-I Ab; however, it is unknown whether patients' DSA have the ability to bind and, subsequently, induce cytotoxic killing of ASC. To test that, we strove to quantify differential affinities of pre-sensitized and *de novo* DSA+ groups to bind HLA class I antigens in ASC *in vitro*. In addition to the original donor, DonA (administered in the ADMIRE CD trial), we included an additional donor, DonB (**Supplementary Figure 1B**), to function as a control for DSA specificity since it was not administered to the patients (**Figure 3A**). We analyzed baseline and W12 samples *via* FCXM and measured HLA-I binding strength to ASC cultured in basal conditions or pre-stimulated with IFNγ (**Figure 3A,** upper panels). As expected, we did not observe high binding capacity in pre-sensitized nor in *de novo* DSA+ patients samples for both ASC donors in basal conditions. When ASC were pre-stimulated with IFNγ, we observed that pre-sensitized patients showed high and comparable binding affinities in samples from W0 and W12 visits only with DonA (**Figure 3A,** upper panels). As expected, basal conditions and thus low HLA-I antigen expression in the membrane of ASC donors, correlated with low binding affinity both in DonA and DonB (**Figure 3A,** lower panels). Similarly, we detected a significant increase the binding affinity when we compared W12 vs. baseline in *de novo* DSA+ patient's samples only with DonA (**Figure 3A,** lower panels). These results agree with solid transplant flow cytometry cross-matching observations (33), where only patients with high DSA levels that also share immune-specificities with donors resulted in significant HLA-I binding.

Next, we aimed to understand whether pre-existing HLA-I Abs and DSA generated after allogeneic administration of ASC were able to fix complement, and therefore trigger *in vitro* CDC. In the cohort of pre-sensitized patients, W12 samples induced modest cytotoxic killing in DonB (0–5% 7-AAD+ cells) both in basal and IFNγ conditions. As expected, incubation of same samples induced higher cytotoxic killing in basal (2 patients) and IFNγ (3 patients) exclusively in DonA ASC (**Figure 3B**). Similarly, patients with high cytotoxic levels in basal conditions increased their threshold when stimulated with IFNγ at W12 (**Supplementary Figure 2**) exclusively in DonA ASC. *De novo* DSA+ W12 samples were able to fix complement and induce cytotoxic killing in basal conditions. As anticipated, a higher percentage of cell death was reached exclusively in DonA when ASC were stimulated with IFNγ (**Figure 3B**). At W12, IFNγ stimulation could increase cytotoxic levels even to patients with already high basal levels (**Supplementary Figure 2**) specially in DonA. This data suggests that pre-existing HLA-I Abs and DSA can bind ASC and induce modest CDC specifically in DonA ASC. Interestingly, although DonB did not induce significant cytotoxic percentages neither in pre-sensitized nor in *DSA*+ cohorts, we observed a trend of increased death levels in W12 vs. baseline. One possible explanation could be the existence of shared HLA polymorphic alleles between the two donors used in this study. To evaluate that we performed HLA typing and observed indeed that DonB shares the same HLA-A allele with DonA (**Supplementary Figure 1B**). We believe this shared HLA allele might be responsible for the trend toward increased cytotoxic levels in W12 samples in DonB.

High Expression of mCRP in ASC

To understand the moderate killing of ASC impinged by pre-existing HLA Abs and DSA we sought to identify complement inhibition strategies that might enable ASC to cope and/or evade cytotoxic killing. One classical mechanism for complement signaling inhibition is the induction of mCRP proteins CD46, CD55, and CD59 (29, 45, 46). While some authors have shown that MSC express low levels of CD46 and CD55, and high CD59 (30), others suggest that MSC express moderate levels of all mCRP (31, 32). To address this controversy, we analyzed CD46, CD55, and CD59 expression levels in seven ASC donors stimulated or not with IFNγ, and compared expression levels with commercial BM-MSC (**Figure 4A**). We observed that basal levels of CD46, CD55, and CD59 were higher in ASC compared to BM-MSC. To recreate a physiologically relevant scenario, we tested mCRP levels in the presence of IFNγ (pro-inflammatory environment), which is also a critical mediator of ASC immune-modulatory response (11). We did not observe significant modulation of mCRPs in BM-MSC, whereas ASC appeared to potently induce mCRP following IFNγ stimulation. mCRP induction was particularly prominent in CD46 with a ~2.14-fold increase following IFNγ stimulation (**Figure 4A**). While the expression pattern of mCRP in different ASC donors was comparable, some donors exhibited differential expression of specific mCRPs (**Figure 4B**). Specifically, DonC exhibited higher CD46 and CD55 levels than rest of donors; DonC, DonE and DonF preferentially over-expressed CD55 (**Figure 4B**). This prompted us to investigate whether the differential expression of mCRP could impinge differential sensitivities to CDC. To answer this, we investigated the kinetics of HLA-I antigen expression and binding affinities among this panel of ASC donors. ASC donors were incubated with increased concentrations of recombinant HLA-I antibody (rHLA-I) W6/32 and measured its binding affinity with FACS (**Figure 4C**). We observed different binding affinities among donors in basal conditions (i.e., a 12-fold difference when comparing DonC with DonG with 10 ng/mL W6/32), suggesting differential antigen HLA-I expression. As expected, following IFNγ stimulation HLA-I antigen was induced in ASC donors as indicated by increased W6/32 binding (**Figure 4C**). Again, we observed varying binding affinities among donors (a 6.25-fold difference at 10 ng/mL comparing DonB with DonG). In parallel, we studied the sensitivity of the different ASC donors to the CDC assay. Under basal conditions DonE exhibited ~25% cell death at the highest W6/32 concentrations, in the rest of the donors this was ~15%, except DonD and DonG which exhibited a lower percentage (**Figure 4C**). As predicted, following IFNγ stimulation CDC sensitivity levels increased dramatically in all ASC donors, but

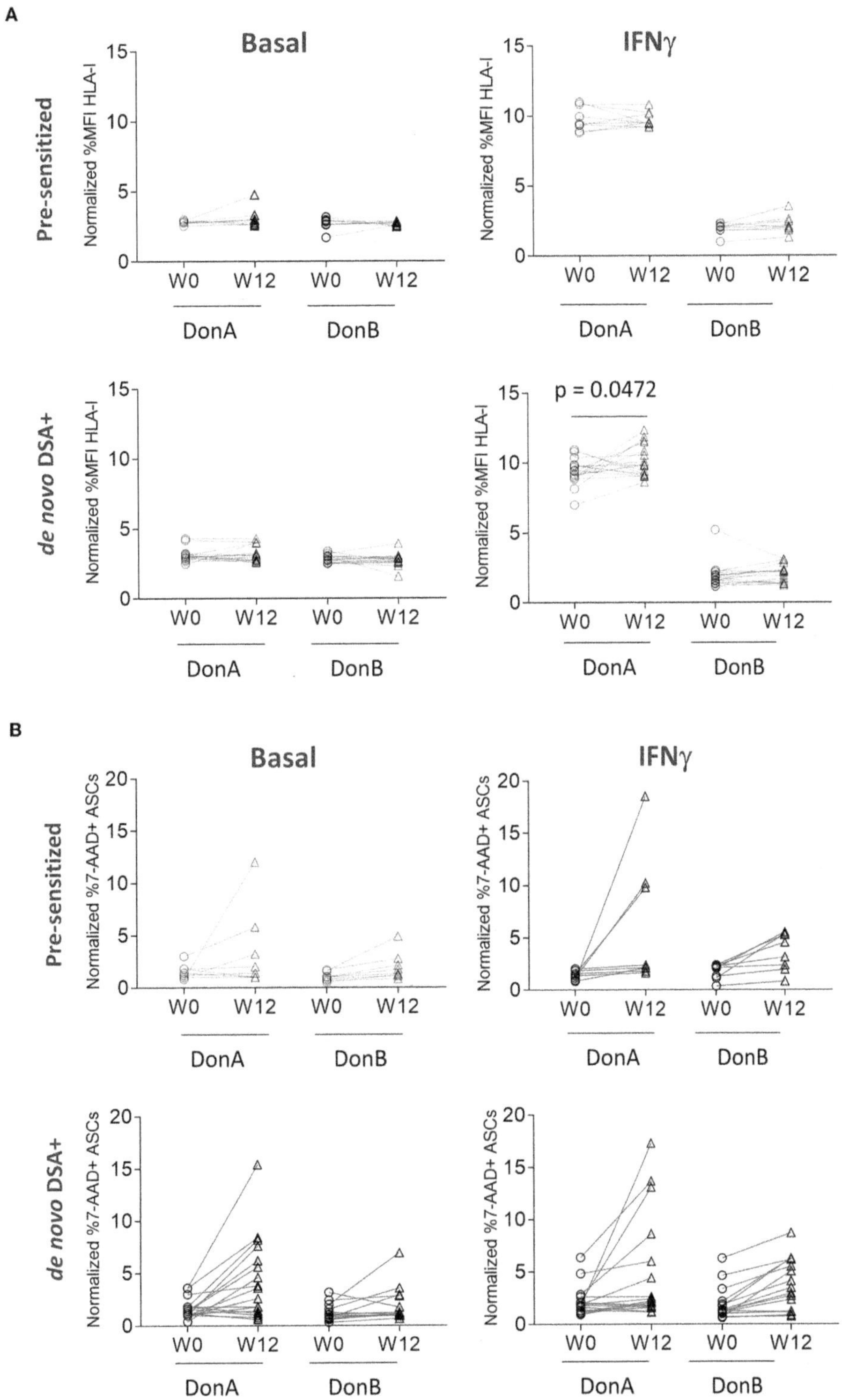

FIGURE 3 | ADMIRE CD plasma samples induce low cytotoxic killing in ASC *in vitro*. **(A)** Graphs showing normalized percent values of HLA-I binding in 10 pre-sensitized (upper panels) and 17 *de novo* DSA+ patients (lower panels) at the indicated time-points (W0 pre-treatment and W12 post-treatment). Prior to binding assay DonA (donor administered in the ADMIRE CD trial) and DonB ASC were grown in normal (basal) conditions or in presence of 3 ng/mL IFNγ (IFNγ) for 48 h. **(B)** Graphs showing normalized percent values of 7-AAD positive ASC in 10 pre-sensitized (upper panels) and 17 *de novo* DSA+ patients (lower panels) at the indicated time-points. *P*-values were determined by the Student's.

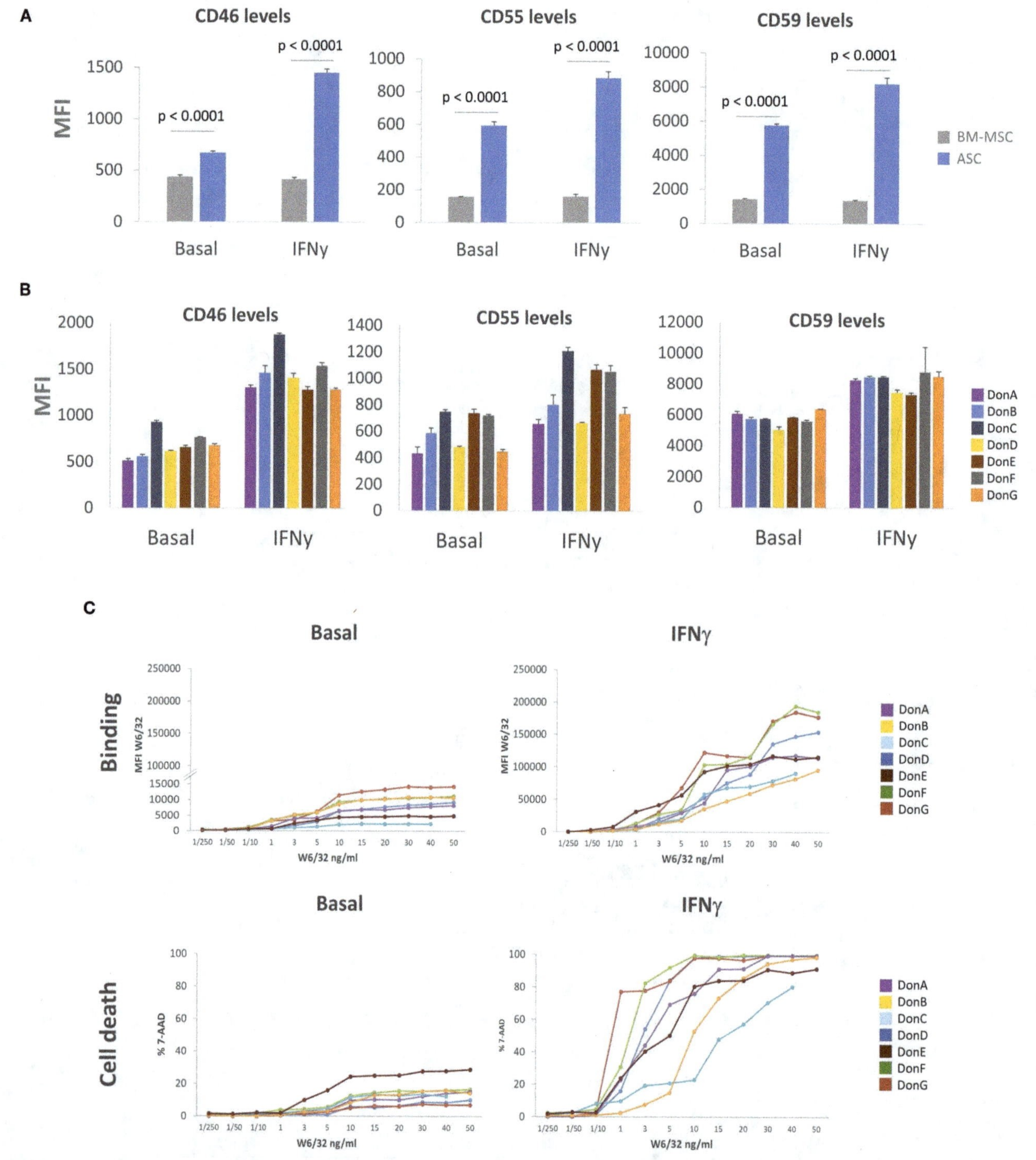

FIGURE 4 | ASC express high levels of mCRP. (A) Graphs showing MFI values of CD46, CD55, and CD59 in seven ASC donors (blue bars) and one BM-MSC donor (gray bars) *via* FACS analysis. Cells were grown in the presence of 3 ng/mL (IFNγ) for 48 h or left untreated (basal). **(B)** Graphs showing differential MFI values of CD46, CD55, and CD59 in the seven ASC donors *via* FACS analysis. Cells were grown in the presence of 3 ng/mL IFNγ for 48 h (IFNγ) of left untreated (basal). **(C)** *P*-values were determined by the Student's *t*-test.

to a lesser extent for DonB and DonC, that remained relatively resistant to CDC-mediated cell death (**Figure 4C**).

Next, we sought to determine whether high W6/32 binding affinity would correlate with enhanced CDC sensitivity. For performing this correlation analysis, we set W6/32 concentration to 10ng/mL as this was the Ab amount driving the transitional phase of the curves (after exponential and before plateau). We did not observe any significant correlation in basal conditions nor in the IFNγ stimulated scenario (**Supplementary Figure 3A**). Remarkably, we identified a group of ASC donors that despite expressing relatively low W6/32 Ab MFI levels (low binding) exhibited high sensitivity to CDC. This suggested that there is not an absolute correlation between W6/32 Ab binding and sensitivity to CDC. We hypothesized that ASC donor-specific expression of mCRP might be the main driver of CDC sensitivity among donors. To determine which of the three mCRPs is the major contributor to CDC inhibition, we correlated cell death levels reached with 10 ng/mL of W6/32 with MFI expression levels of the three mCRPs both in basal and IFNγ conditions (**Supplementary Figure 3B**). We did not observe a positive correlation in basal conditions with any of the tested mCRP molecules. However, we noticed that following IFNγ stimulation, lower CD46 and CD55 levels significantly correlated with higher cell death levels. Finally, CD46 slope significance was slightly higher compared with CD55.

CD46 Depletion Increases CDC Sensitivity of ASC *in vitro*

The robust induction of CD46 following IFNγ stimulation compared to CD55 or CD59 and the higher significance of cytotoxicity correlation together with the reduced inter-donor variability of CD46 vs. CD55, prompted us to perform an in-depth analysis of CD46 and its potential impact in CDC sensitivity in ASC. Using public genome browsers, we identified top gRNA sequences to knock-down *CD46* (ncbi.nlm.nih.gov/gene and crispr.mit.edu). We selected optimal gRNA sequences based on two parameters, high specificity and low off-target score. Two optimal crRNAs (crRNA1 blue and crRNA2 yellow) targeting exon 3 were selected for efficacy screening (**Supplementary Figure 3C**). Delivery of crRNA:tracrRNA-ATTO^{550}:Cas9 complexes was examined under fluorescence microscope. We observed that after 24 h of lipotransfection the vast majority of ASC had incorporated the RNP complexes which would correlate with high Cas9-mediated double-strand breaks events (**Supplementary Figure 3C**). To check CRISPR-mediated knock-down efficacy, we cultured ASC in the presence or absence of IFNγ and analyzed CD46 expression *via* FACS. The efficacy of crRNA1 was comparable to crRNA2 both in normal and IFNγ conditions, thus we selected crRNA1 for the generation of the ASC-CD46^{KO} clones (**Supplementary Figure 3D**).

We then selected DonB having low CDC sensitivity to test whether selective depletion of CD46 could sensitize it to cytotoxic killing. The W6/32-mediated cytotoxic assays was performed in basal and IFNγ conditions (**Figure 5A**). As expected, CD46 knock-down induced modest increase in the cytotoxic killing in basal conditions (**Figure 5A,** left graph), which was likely due to the low HLA-I binding and subsequent complement fixation. Following IFNγ stimulation we observed a significant boost in the percentage of cytotoxic killing in parental ASC and this was further enhanced in CD46^{KO} ASC (**Figure 5A,** right graph). At a physiological dose of 10 ng/mL W6/32 we obtained ∼50% 7-AAD positive cells in parental ASC and CD46 knock-down increased the percentage of killing up to ∼95%, suggesting that CD46 plays a critical function in preventing cytotoxic killing.

To test whether CD46 cytotoxic inhibitory functions are effective in other ASC donors, we performed cytotoxicity analysis in the panel of seven donors and plotted mean curves in basal conditions (**Figure 5B,** left graph) and after ASC IFNγ stimulation (**Figure 5B,** right graph). Under both testing conditions CD46^{KO} ASC donors exhibited enhanced sensitivity to CDC than parental controls. A shift to the left of the half maximal effective concentration (EC_{50}) curves implies a decrease in the concentration of the W6/32 Ab required to induce CDC, which correlates with enhanced sensitivity to CDC. Next, to quantify the shift in the curves we calculated of W6/32 Ab (**Figure 5C**). In basal conditions, CD46^{KO} donors exhibited decreased EC_{50} compared to parental donors, suggesting higher sensitivity to W6/32 (**Figure 5C,** fourth column). We observed a similar effect in IFNγ conditions (**Figure 5C,** seventh column), confirming our original hypothesis that *CD46* expression confers CDC resistance to ASC *in vitro*. This data confirms that *CD46* is a key mediator of CDC and that its depletion in ASC correlates with enhanced CDC sensitivity.

DISCUSSION

There are high expectations for utilizing allogeneic MSC as therapeutic tools for treating numerous diseases; however, further investigations are required to assess safety and potential toxicity. Few studies have examined their immunogenicity by characterizing the immune responses induced by their therapeutic administration. Noteworthy, a significant number of studies have found that MSC administered in several animal models with mismatching MHC are in fact rejected (40–42). These observations allow us to strongly argue against the canonical notion that MSC are immuno-privileged. Nevertheless, several clinical studies have demonstrated that allogeneic MSC treatment induces DSA without having negative consequences on safety or efficacy (19, 20, 47). In general, potential immunological risk has been downplayed based on general features observed *in vitro:* i.e., lack of expression of classic co-stimulatory molecules and ability to downregulate NK and T-cells proliferation (48, 49), which balances the risk–benefit profile toward being a safe therapy. The generation of DSA are the consequence of indirect allo-recognition of HLAs from allogeneic MSCs by patient APC. As a result, the induction of allo-specific $CD4^{+}$ T cells will activate the HLA-specific-IgG producing B cells (18).

Investigation of DSA induced by ASC was an important focus in the ADMIRE CD study. In ADMIRE CD we devoted efforts to understand the immunogenic impact of ASC in pre-sensitized and naïve patients. The present study sheds light on the persistence and function of DSA, and provides *in vitro* mechanistic insights in HLA Abs and DSA interactions with ASC. Here we confirm that patients can be primed by allo-MHC molecules and mount a DSA response against allogeneic ASC after a single local injection. Pre-sensitization to allo-antigens can result from priming due to the transfusion of blood components

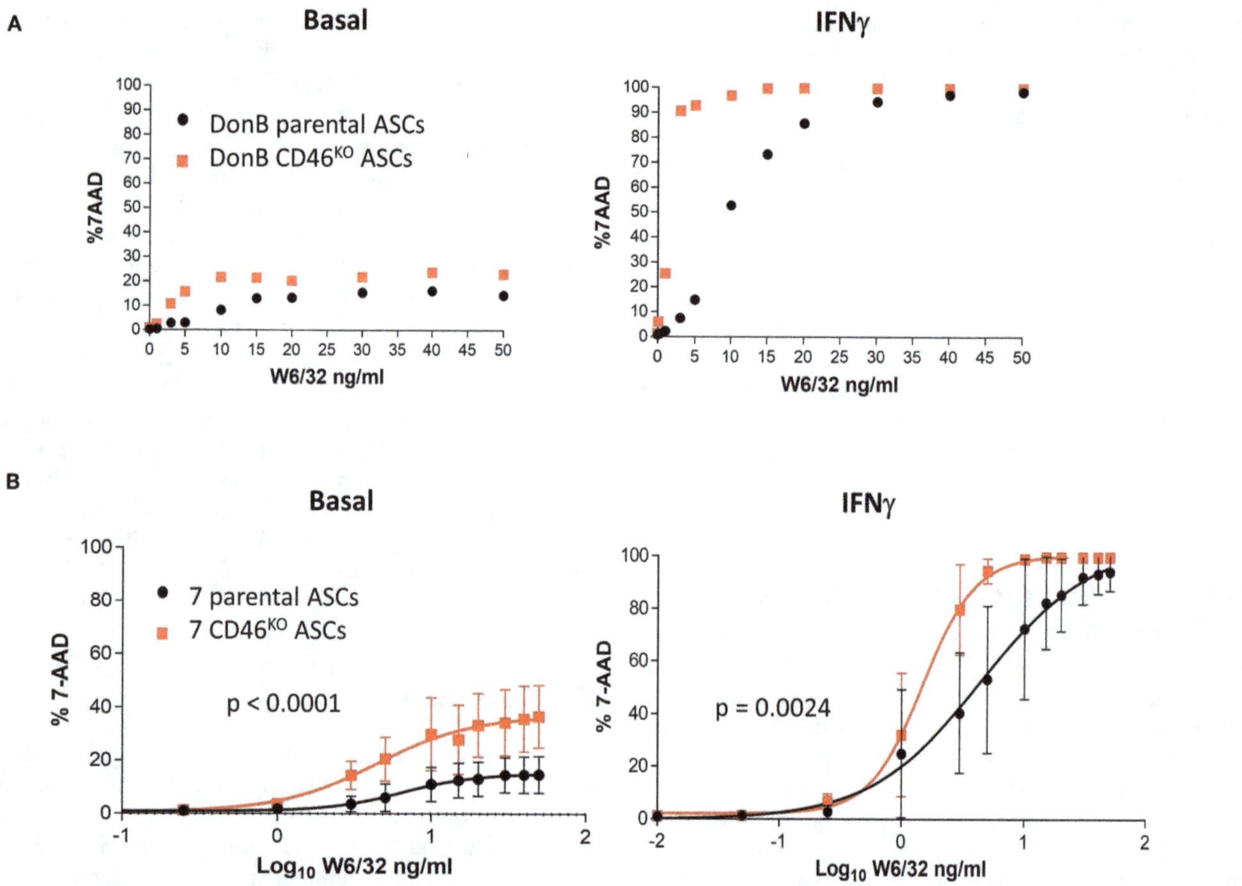

C

W6/32 EC_{50} (ng/mL)	Basal Parental	Basal $CD46^{KO}$	Basal Fold change	IFNγ Parental	IFNγ $CD46^{KO}$	IFNγ Fold change
DonA	8.2	4.3	0.5	3.5	1.6	0.4
DonB	10.3	3.3	0.3	9.7	1.4	0.1
DonC	6.0	3.9	0.6	24.7	1.5	0.1
DonD	17.0	3.4	0.2	2.8	3.1	1.1
DonE	4.5	3.6	0.8	3.7	0.6	0.2
DonF	6.9	5.1	0.7	1.5	1.4	0.9
DonG	7.7	5.4	0.7	0.6	2.1	3.8

Effect of CD46 knockdown:

Enhanced sensitivity | No change | Reduced sensitivity

FIGURE 5 | CD46 is a key mediator of complement cytotoxicity in ASC. **(A)** Graph showing percentage of 7-AAD positive parental and $CD46^{KO}$ DonB ASC against increased concentration levels of W6/32 Ab. Prior analysis parental and $CD46^{KO}$ DonB ASC were grown in the presence of 3 ng/mL IFNγ for 48 h (IFNγ) or left untreated (basal). **(B)** Sigmoidal curves displaying percentage of 7-AAD positive parental and $CD46^{KO}$ ASC against EC_{50} of W6/32 Ab (transformed from lineal to log_{10}). *P*-values were determined by the two-way ANOVA test. **(C)** Table showing half-maximal effective concentration of W6/32 Ab (ng/mL) of seven parental and the corresponding $CD46^{KO}$ ASC in basal and IFNγ conditions. In columns 4 and 7, we calculated fold-change differences ($CD46^{KO}$ EC_{50}/parental EC_{50}) and applied the color code showed in the right.

(i.e., thrombocytes, leukocytes), organ or cell transplantation, pregnancy, or simply by unspecific cross-reactivity (50). In fact, HLA-specific memory T-cells could be detected in 6–12% of healthy donors, yet >50% of patients on transplant waiting lists (51, 52). Lessons from allogeneic solid-organ and hematopoietic stem cell transplantation indicate that HLA Abs are a major limitation of effective tissue and organ transplantation (34, 53–55). In line with this, we envision that HLA Abs are key contributors to the immediate elimination of allogeneic MSC before they exert modulating effect on inflammation.

The present data demonstrates that a higher proportion of pre-sensitized patients generate and maintain DSA over time, compared with the naïve population. As expected, pre-sensitized patients sharing specificities with the administered product were prone to generate, maintain or increase the DSA titer, indicating a boost effect analogous to a secondary response. However, there is clear evidence showing that pre-existing anti-HLA abs can also lead to a primary allogeneic response when HLA Abs do not share specificities with administered ASC.

Even if it appears not to be a correlation between DSA generation and safety and efficacy of the treatment (24 and **Supplementary Figure 1A**), the potential impact on a future administration of an allogeneic cell therapy product or transplantation is still unknown. To evaluate the risk-benefit of allogeneic treatment in pre-sensitized patients, the consequences of the presence of HLA Abs *in vivo* should be also evaluated.

Allo-antibodies can bind to HLA and subsequently initiate antibody-dependent cytotoxicity because of their interaction with innate immune cells *via* Fc receptors (56, 57). Complement can also bind Fc region of the HLA Abs antibody resulting in CDC (52, 56, 58). Cancer cells have developed different strategies for CDC evasion, for instance over-expression of mCRPs (29, 59, 60). Our data indicates that to some extent ASC are also well-equipped for complement system evasion. We observed an approximate $\sim$3.49-fold increased expression of CD46 in ASC compared to BM-MSC in pro-inflammatory conditions, suggesting ASC have the ability to induce CD46 expression. The canonical role of CD46 is to bind and inhibit the opsonin factors C3b and C4b and therefore its function has mainly been linked to innate immunity. However, recent investigations have highlighted the role of CD46 in adaptive immunity (61, 62). CD46 has been directly implicated in the modulation of Th1 IFNγ to interleukin-10, switching to and regulating an adaptive T-cell response (61). Such evidence implicates CD46 as playing a pivotal role in immunogenic response, further investigations are required to characterize the extent at which it may be relevant for influencing the immunomodulatory properties of ASC. Finally, we cannot exclude that some other immune-evasive strategies, such as the over-expression of heat-shock proteins or complement inhibitors, contribute to ASC immune-evasive capabilities (29, 63). Several groups are evaluating the modulation of complement evasive molecules to cope with allogeneic MSC cytotoxicity and subsequently prolong MSC persistence *in vivo*. Herein, we propose the examination of CDC/HLA^{MFI} curves and HLA binding kinetics for assessing donors' differential susceptibility to complement dependent toxicity.

The capacity of HLA-I to transduce signals is dependent on the degree of molecular aggregation of the HLA-I molecules which relies on the level of HLA-I expression and HLA antibody titer. In solid organ transplants, high antibody titer induces death of endothelial cells; a low titer promotes their survival and CDC resistance (64, 65). In addition, the migration of monocytes/macrophages in response to Fc-dependent effector factors has also been proposed to have a relevant role (66, 67). In this case, the presence of DSA could potentially also promote monocyte macrophages trafficking toward ASC. We could show though that pre-existing HLA Ab from patients' serum bind and trigger CDC onto ASC *in vitro*; only in the presence of high levels of DSA and when ASC are preactivated could the killing effect of ASC be observed. Although the present data supports darvadstrocel efficacy and long-term DSA clearance in naïve patients, there are outstanding questions that remain unanswered. First, is whether the indirect allorecognition of the foreign HLA molecule could be defined as the consequence of donor-to-patient crosstalk (i.e., NK cells, monocytes, macrophages) and whether the survival and modulatory mechanisms of ASC are causing a delay or reduction of this allo-response. Secondly, it remains to be determined whether MSC elimination is directly associated with a lack of efficacy or, as proposed in other studies, that apoptosis of MSC and the engulfment of phagocytes with apoptotic MSC is in fact essential for MSC modulation (68). These outstanding questions need to be addressed not only in experimental models but also *via* extensive monitoring in clinical studies in which therapeutic efficacy, TEAEs and immunogenicity are interconnected and correlated.

As a summary, we have found that despite allo-ASC generate DSA, those levels are not sufficient to bind ASC and subsequently execute antibody-dependent cytotoxicity *in vitro*. Only particular cases where the number of mismatched *eplets* is extremely high, the DSA titer could potentially be detrimental for ASC survival, in particular if ASC are activated. To reduce the DSA titer, we are inclined to propose donor screening for optimal eligibility rather than donor-to-patient HLA-matching oriented strategies as proposed by others (69). An interesting scheme, similar to what has been proposed for allogeneic induced pluripotent stem cells banks, is the generation of ASC banks from HLA-homozygous donors (70), minimizing the number of allo-antigens exposed to the patient. An interesting approach to minimize immediate rejection in pre-sensitized patients might be the fine-tune modulation of mCRPs such as CD46 in ASC donors. We propose CD46^{HIGH} as a biomarker of CDC resistance that could be used for screening ASC donors suitable for the treatment of pre-sensitized patients and/or facilitating re-treatment options.

ETHICS STATEMENT

Current study involve use of frozen plasma samples belonging to ADMIRE CD clinical trial. The study was done in accordance with the 2008 Declaration of Helsinki and all relevant international, national, and local rules and regulations (https://www.clinicaltrialsregister.eu/ctr-search/trial/2011-006064-43/results). The protocol was approved by the local ethics committee of participating centers. All patients gave written informed consent before enrolment.

AUTHOR CONTRIBUTIONS

OD and AA-V: conceptualization, visualization, and supervision. AA-V, CM-M, CR, BD, RM, PM-C, and MO-V: methodology. AA-V and CM-M: investigation. JC and OD: resources. AA-V, OD, and CM-M: writing. AA-V, OD, WD, EL, JP, CM-M, MO-V, ÁH-M, PM-C, DG-O, and IP: revision. OD: funding resources.

ACKNOWLEDGMENTS

ADMIRE CD study was sponsored by TiGenix. We thank the study participants and investigators involved in this clinical trial. We also thank Inmaculada Gilaberte and Marie Paul Richard for their contribution to the analysis of clinical aspects. Editing for language was provided by Syneos Health Medical Communications and funded by Takeda Pharmaceuticals U.S.A. Inc.

SUPPLEMENTARY MATERIAL

Supplementary Figure 1 | Effect of DSA generation at W12 and W52 in the proportions of patients with clinical remission (closure of all treated external openings that were draining at baseline despite gentle finger compression). (N) is the number of patients in analysis population and (n) number of patients with observation. There was no clinical remission data for 10 naïve patients which explains why number of clinical remission data-points (43 patients) differs to total number of naïve patients in the study (53 patients) **Figure 1A**. Percentages were calculated based on HLA Abs status at baseline and patients exhibiting DSA both at W12 and W52. **(B)** DNA from DonA and DonB was purified and tested by LABTypeSSO assay for HLA allele characterization. In bold, we highlight HLA-A allele shared by DonA and DonB.

Supplementary Figure 2 | ADMIRE CD plasma samples induce low cytotoxic killing in ASC *in vitro*. Graphs showing normalized percent values of 7-AAD positive ASC in 10 pre-sensitized (upper panels) and 17 *de novo* DSA+ patients (lower panels) before and after INFγ stimulation at the indicated time-points (week 0 and week 12).

Supplementary Figure 3 | We correlated MFI values of W6/32 **(A)** and CD46, CD55, and CD59 **(B)** of ASC donors grown in the presence of 3 ng/mL IFNγ for 48h (red dots) or basal conditions (black dots). *P*-values shows slope significance of linear regression. The R-squared value for the significant of the slopes in IFNγ conditions in CD46 and CD55 **(B)** was 0.74 and 0.71, respectively. **(C)** Left image, annotation of human full-length messenger RNA (mRNA) of CD46. Highlighted in blue (crRNA1) and yellow (crRNA2) are the specific guide RNA sequences tested and green are the protospacer adjacent motif (PAM) sites. Lastly, red lines correspond to CAS9 cleavage sites. The right image shows the merged picture of phase contrast and fluorescence PE channel of ASC transfected with ribo-nucleo-protein complex including tracrRNA-ATTO^{550} (red). Scale bar = 20 μm. **(D)** Graph showing CD46 MFI values in parental (black) and CD46^{KO} (dark and light red) ASC. Prior analysis parental and CD46^{KO} ASC were grown in the presence of 3 ng/mL IFNγ for 48 h (IFNγ) or left untreated (basal). *P*-values were determined by the Student's *t*-test.

REFERENCES

1. Galipeau J, Sensebe L. Mesenchymal stromal cells: clinical challenges and therapeutic opportunities. *Cell Stem Cell.* (2018) 22:824–33. doi: 10.1016/j.stem.2018.05.004
2. Semont A, Francois S, Mouiseddine M, Francois A, Sache A, Frick J, et al. Mesenchymal stem cells increase self-renewal of small intestinal epithelium and accelerate structural recovery after radiation injury. *Adv Exp Med Biol.* (2006) 585:19–30. doi: 10.1007/978-0-387-34133-0_2
3. Lombardo E, Van Der Poll T, Delarosa O, Dalemans W. Mesenchymal stem cells as a therapeutic tool to treat sepsis. *World J Stem Cells.* (2015) 7:368–79. doi: 10.4252/wjsc.v7.i2.368
4. Munneke JM, Spruit MJ, Cornelissen AS, Van Hoeven V, Voermans C, Hazenberg MD. The potential of mesenchymal stromal cells as treatment for severe steroid-refractory acute graft-versus-host disease: a critical review of the literature. *Transplantation.* (2016) 100:2309–14. doi: 10.1097/TP.0000000000001029
5. Wang S, Qu X, Zhao RC. Clinical applications of mesenchymal stem cells. *J Hematol Oncol.* (2012) 5:19. doi: 10.1186/1756-8722-5-19
6. Singer NG, Caplan AI. Mesenchymal stem cells: mechanisms of inflammation. *Annu Rev Pathol.* (2011) 6:457–78. doi: 10.1146/annurev-pathol-011110-130230
7. Le Blanc K, Davies LC. Mesenchymal stromal cells and the innate immune response. *Immunol Lett.* (2015) 168:140–6. doi: 10.1016/j.imlet.2015.05.004
8. Bernardo ME, Fibbe WE. Mesenchymal stromal cells: sensors and switchers of inflammation. *Cell Stem Cell.* (2013) 13:392–402. doi: 10.1016/j.stem.2013.09.006
9. Chiossone L, Conte R, Spaggiari GM, Serra M, Romei C, Bellora F, et al. Mesenchymal stromal cells induce peculiar alternatively activated macrophages capable of dampening both innate and adaptive immune responses. *Stem Cells.* (2016) 34:1909–21. doi: 10.1002/stem.2369
10. Vasandan AB, Jahnavi S, Shashank C, Prasad P, Kumar A, Prasanna SJ. Human Mesenchymal stem cells program macrophage plasticity by altering their metabolic status via a PGE2-dependent mechanism. *Sci Rep.* (2016) 6:38308. doi: 10.1038/srep38308
11. Delarosa O, Lombardo E, Beraza A, Mancheno-Corvo P, Ramirez C, Menta R, et al. Requirement of IFN-gamma-mediated indoleamine 2,3-dioxygenase expression in the modulation of lymphocyte proliferation by human adipose-derived stem cells. *Tissue Eng Part A.* (2009) 15:2795–806. doi: 10.1089/ten.tea.2008.0630
12. Delarosa O, Dalemans W, Lombardo E. Toll-like receptors as modulators of mesenchymal stem cells. *Front Immunol.* (2012) 3:182. doi: 10.3389/fimmu.2012.00182
13. Franquesa M, Mensah FK, Huizinga R, Strini T, Boon L, Lombardo E, et al. Human adipose tissue-derived mesenchymal stem cells abrogate plasmablast formation and induce regulatory B cells independently of T helper cells. *Stem Cells.* (2015) 33:880–91. doi: 10.1002/stem.1881
14. Gao F, Chiu SM, Motan DA, Zhang Z, Chen L, Ji HL, et al. Mesenchymal stem cells and immunomodulation: current status and future prospects. *Cell Death Dis.* (2016) 7:e2062. doi: 10.1038/cddis.2015.327
15. Heathman TR, Nienow AW, Mccall MJ, Coopman K, Kara B, Hewitt CJ. The translation of cell-based therapies: clinical landscape and manufacturing challenges. *Regen Med.* (2015) 10:49–64. doi: 10.2217/rme.14.73
16. Abbasalizadeh S, Pakzad M, Cabral JMS, Baharvand H. Allogeneic cell therapy manufacturing: process development technologies and facility design options. *Expert Opin Biol Ther.* (2017) 17:1201–19. doi: 10.1080/14712598.2017.1354982
17. Hare JM, Fishman JE, Gerstenblith G, Difede Velazquez DL, Zambrano JP, Suncion VY, et al. Comparison of allogeneic vs autologous bone marrow-derived mesenchymal stem cells delivered by transendocardial injection in patients with ischemic cardiomyopathy: the POSEIDON randomized trial. *JAMA.* (2012) 308:2369–79. doi: 10.1001/jama.2012.25321
18. Griffin MD, Ryan AE, Alagesan S, Lohan P, Treacy O, Ritter T. Anti-donor immune responses elicited by allogeneic mesenchymal stem cells: what have we learned so far? *Immunol Cell Biol.* (2013) 91:40–51. doi: 10.1038/icb.2012.67

19. Alvaro-Gracia JM, Jover JA, Garcia-Vicuna R, Carreno L, Alonso A, Marsal S, et al. Intravenous administration of expanded allogeneic adipose-derived mesenchymal stem cells in refractory rheumatoid arthritis (Cx611): results of a multicentre, dose escalation, randomised, single-blind, placebo-controlled phase Ib/IIa clinical trial. *Ann Rheum Dis.* (2017) 76:196–202. doi: 10.1136/annrheumdis-2015-208918
20. Garcia-Sancho J, Sanchez A, Vega A, Noriega DC, Nocito M. Influence of HLA matching on the efficacy of allogeneic mesenchymal stromal cell therapies for osteoarthritis and degenerative disc disease. *Transplant Direct.* (2017) 3:e205. doi: 10.1097/TXD.0000000000000724
21. Panes J, Garcia-Olmo D, Van Assche G, Colombel JF, Reinisch W, Baumgart DC, et al. Long-term efficacy and safety of stem cell therapy (Cx601) for complex perianal fistulas in patients with Crohn's Disease. *Gastroenterology.* (2018) 154:1334–42 e1334. doi: 10.1053/j.gastro.2017.12.020
22. Baumgart DC, Sandborn WJ. Crohn's disease. *Lancet.* (2012) 380:1590–605. doi: 10.1016/S0140-6736(12)60026-9
23. Eglinton TW, Barclay ML, Gearry RB, Frizelle FA. The spectrum of perianal Crohn's disease in a population-based cohort. *Dis Colon Rectum.* (2012) 55:773–7. doi: 10.1097/DCR.0b013e31825228b0
24. Panes J, Garcia-Olmo D, Van Assche G, Colombel JF, Reinisch W, Baumgart DC, et al. Expanded allogeneic adipose-derived mesenchymal stem cells (Cx601) for complex perianal fistulas in Crohn's disease: a phase 3 randomised, double-blind controlled trial. *Lancet.* (2016) 388:1281–90. doi: 10.1016/S0140-6736(16)31203-X
25. Medzhitov R, Janeway CAJr. Decoding the patterns of self and nonself by the innate immune system. *Science.* (2002) 296:298–300. doi: 10.1126/science.1068883
26. Meyer S, Leusen JH, Boross P. Regulation of complement and modulation of its activity in monoclonal antibody therapy of cancer. *MAbs.* (2014) 6:1133–44. doi: 10.4161/mabs.29670
27. Schraufstatter IU, Discipio RG, Zhao M, Khaldoyanidi SK. C3a and C5a are chemotactic factors for human mesenchymal stem cells, which cause prolonged ERK1/2 phosphorylation. *J Immunol.* (2009) 182:3827–36. doi: 10.4049/jimmunol.0803055
28. Donnenberg AD, Meyer EM, Rubin JP, Donnenberg VS. The cell-surface proteome of cultured adipose stromal cells. *Cytometry A.* (2015) 87:665–74. doi: 10.1002/cyto.a.22682
29. Gancz D, Fishelson Z. Cancer resistance to complement-dependent cytotoxicity (CDC): problem-oriented research and development. *Mol Immunol.* (2009) 46:2794–800. doi: 10.1016/j.molimm.2009.05.009
30. Moll G, Jitschin R, Von Bahr L, Rasmusson-Duprez I, Sundberg B, Lonnies L, et al. Mesenchymal stromal cells engage complement and complement receptor bearing innate effector cells to modulate immune responses. *PLoS ONE.* (2011) 6:e21703. doi: 10.1371/journal.pone.0021703
31. Soland MA, Bego M, Colletti E, Zanjani ED, St. Jeor S, Porada CD, et al. Mesenchymal stem cells engineered to inhibit complement-mediated damage. *PLoS ONE.* (2013) 8:e60461. doi: 10.1371/journal.pone.0060461
32. Atkinson K. *The Biology and Therapeutic Application of Mesenchymal Cells.* Hoboken, NJ: John Wiley and Sons (2017). doi: 10.1002/9781118907474
33. Lentine KL, Graff RJ, Xiao H, Modanlou KA, Salvalaggio PR, Brennan DC, et al. Flow cytometry crossmatch before kidney transplantation in contemporary practice: target cell utilization, results patterns, and associated long-term graft survival. *Clin Transpl.* (2008) 253–66.
34. Brugiere O, Suberbielle C, Thabut G, Lhuillier E, Dauriat G, Metivier AC, et al. Lung transplantation in patients with pretransplantation donor-specific antibodies detected by Luminex assay. *Transplantation.* (2013) 95:761–5. doi: 10.1097/TP.0b013e31827afb0f
35. Lopez-Santalla M, Mancheno-Corvo P, Menta R, Lopez-Belmonte J, Delarosa O, Bueren JA, et al. Human adipose-derived mesenchymal *stem cells* modulate experimental autoimmune arthritis by modifying early adaptive T cell responses. *Stem Cells.* (2015) 33:3493–503. doi: 10.1002/stem.2113
36. Bourin P, Bunnell BA, Casteilla L, Dominici M, Katz AJ, March KL, et al. Stromal cells from the adipose tissue-derived stromal vascular fraction and culture expanded adipose tissue-derived stromal/stem cells: a joint statement of the International Federation for Adipose Therapeutics and Science (IFATS) and the International Society for Cellular Therapy (ISCT). *Cytotherapy.* (2013) 15:641–8. doi: 10.1016/j.jcyt.2013.02.006
37. Duquesnoy RJ. HLAMatchmaker: a molecularly based algorithm for histocompatibility determination. I Description of the algorithm. *Hum Immunol.* (2002) 63:339–52. doi: 10.1016/S0198-8859(02)00382-8
38. Dankers MK, Witvliet MD, Roelen DL, De Lange P, Korfage N, Persijn GG, et al. The number of amino acid triplet differences between patient and donor is predictive for the antibody reactivity against mismatched human leukocyte antigens. *Transplantation.* (2004) 77:1236–9. doi: 10.1097/01.TP.0000120385.03278.28
39. Williams RC, Opelz G, Weil EJ, Mcgarvey CJ, Chakkera HA. The risk of transplant failure with hla mismatch in first adult kidney Allografts 2: living donors, summary, guide. *Transpl Direct.* (2017) 3:e152. doi: 10.1097/TXD.0000000000000664
40. Eliopoulos N, Stagg J, Lejeune L, Pommey S, Galipeau J. Allogeneic marrow stromal cells are immune rejected by MHC class I- and class II-mismatched recipient mice. *Blood.* (2005) 106:4057–65. doi: 10.1182/blood-2005-03-1004
41. Nauta AJ, Westerhuis G, Kruisselbrink AB, Lurvink EG, Willemze R, Fibbe WE. Donor-derived mesenchymal stem cells are immunogenic in an allogeneic host and stimulate donor graft rejection in a nonmyeloablative setting. *Blood.* (2006) 108:2114–20. doi: 10.1182/blood-2005-11-011650
42. Zangi L, Margalit R, Reich-Zeliger S, Bachar-Lustig E, Beilhack A, Negrin R, et al. Direct imaging of immune rejection and memory induction by allogeneic mesenchymal stromal cells. *Stem Cells.* (2009) 27:2865–74. doi: 10.1002/stem.217
43. Isakova IA, Lanclos C, Bruhn J, Kuroda MJ, Baker KC, Krishnappa V, et al. Allo-reactivity of mesenchymal stem cells in rhesus macaques is dose and haplotype dependent and limits durable cell engraftment *in vivo. PLoS ONE.* (2014) 9:e87238. doi: 10.1371/journal.pone.0087238
44. Pezzanite LM, Fortier LA, Antczak DF, Cassano JM, Brosnahan MM, Miller D, et al. Equine allogeneic bone marrow-derived mesenchymal stromal cells elicit antibody responses *in vivo. Stem Cell Res Ther.* (2015) 6:54. doi: 10.1186/s13287-015-0053-x
45. Ricklin D, Hajishengallis G, Yang K, Lambris JD. Complement: a key system for immune surveillance and homeostasis. *Nat Immunol.* (2010) 11:785–97. doi: 10.1038/ni.1923
46. Tegla CA, Cudrici C, Patel S, Trippe R III, Rus V, Niculescu F, Rus H. Membrane attack by complement: the assembly and biology of terminal complement complexes. *Immunol Res.* (2011) 51:45–60. doi: 10.1007/s12026-011-8239-5
47. Hare JM, Difede DL, Rieger AC, Florea V, Landin AM, El-Khorazaty J, et al. Randomized comparison of allogeneic versus autologous mesenchymal stem cells for nonischemic dilated cardiomyopathy: POSEIDON-DCM trial. *J Am Coll Cardiol.* (2017) 69:526–37. doi: 10.1016/j.jacc.2016.11.009
48. Lombardo E, Delarosa O, Mancheno-Corvo P, Menta R, Ramirez C, Buscher D. Toll-like receptor-mediated signaling in human adipose-derived stem cells: implications for immunogenicity and immunosuppressive potential. *Tissue Eng Part A.* (2009) 15:1579–89. doi: 10.1089/ten.tea.2008.0340
49. Laranjeira P, Pedrosa M, Pedreiro S, Gomes J, Martinho A, Antunes B, et al. Effect of human bone marrow mesenchymal stromal cells on cytokine production by peripheral blood naive, memory, and effector T cells. *Stem Cell Res Ther.* (2015) 6:3. doi: 10.1186/scrt537
50. Hricik DE, Poggio ED, Woodside KJ, Sarabu N, Sanchez EQ, Schulak JA, et al. Effects of cellular sensitization and donor age on acute rejection and graft function after deceased-donor kidney transplantation. *Transplantation.* (2013) 95:1254–8. doi: 10.1097/TP.0b013e31828ad866
51. Andree H, Nickel P, Nasiadko C, Hammer MH, Schonemann C, Pruss A, et al. Identification of dialysis patients with panel-reactive memory T cells before kidney transplantation using an allogeneic cell bank. *J Am Soc Nephrol.* (2006) 17:573–80. doi: 10.1681/ASN.2005030299
52. Brand A, Doxiadis IN, Roelen DL. On the role of HLA antibodies in hematopoietic stem cell transplantation. *Tissue Antigens.* (2013) 81:1–11. doi: 10.1111/tan.12040
53. Bradley JA, Baldwin WM, Bingaman A, Ellenrieder C, Gebel HM, Glotz D, et al. Antibody-mediated rejection–an ounce of prevention is worth a pound of cure. *Am J Transplant.* (2011) 11:1131–9. doi: 10.1111/j.1600-6143.2011.03581.x

54. Ruggeri A, Rocha V, Masson E, Labopin M, Cunha R, Absi L, et al. Impact of donor-specific anti-HLA antibodies on graft failure and survival after reduced intensity conditioning-unrelated cord blood transplantation: a Eurocord, Societe Francophone d'Histocompatibilite et d'Immunogenetique (SFHI) and Societe Francaise de Greffe de Moelle et de Therapie Cellulaire (SFGM-TC) analysis. *Haematologica.* (2013) 98:1154–60. doi: 10.3324/haematol.2012.077685
55. Thomas KA, Valenzuela NM, Reed EF. The perfect storm: HLA antibodies, complement, FcgammaRs, and endothelium in transplant rejection. *Trends Mol Med.* (2015) 21:319–29. doi: 10.1016/j.molmed.2015.02.004
56. Wehner J, Morrell CN, Reynolds T, Rodriguez ER, Baldwin WM III. Antibody and complement in transplant vasculopathy. *Circ Res.* (2007) 100:191–203. doi: 10.1161/01.RES.0000255032.33661.88
57. Resch T, Fabritius C, Ebner S, Ritschl P, Kotsch K. The Role of Natural Killer Cells in Humoral Rejection. *Transplantation.* (2015) 99:1335–40. doi: 10.1097/TP.0000000000000757
58. Loupy A, Lefaucheur C, Vernerey D, Prugger C, Duong Van Huyen JP, Mooney N, et al. Complement-binding anti-HLA antibodies and kidney-allograft survival. *N Engl J Med.* (2013) 369:1215–26. doi: 10.1056/NEJMoa1302506
59. Maciejczyk A, Szelachowska J, Szynglarewicz B, Szulc R, Szulc A, Wysocka T, et al. CD46 Expression is an unfavorable prognostic factor in breast cancer cases. *Appl Immunohistochem Mol Morphol.* (2011) 19:540–6. doi: 10.1097/PAI.0b013e31821a0be9
60. Kesselring R, Thiel A, Pries R, Fichtner-Feigl S, Brunner S, Seidel P, et al. The complement receptors CD46, CD55 and CD59 are regulated by the tumour microenvironment of head and neck cancer to facilitate escape of complement attack. *Eur J Cancer.* (2014) 50:2152–61. doi: 10.1016/j.ejca.2014.05.005
61. Cardone J, Le Friec G, Vantourout P, Roberts A, Fuchs A, Jackson I, et al. Complement regulator CD46 temporally regulates cytokine production by conventional and unconventional T cells. *Nat Immunol.* (2010) 11:862–71. doi: 10.1038/ni.1917
62. Cardone J, Le Friec G, Kemper C. CD46 in innate and adaptive immunity: an update. *Clin Exp Immunol.* (2011) 164:301–11. doi: 10.1111/j.1365-2249.2011.04400.x
63. Li Y, Qiu W, Zhang L, Fung J, Lin F. Painting factor H onto mesenchymal stem cells protects the cells from complement- and neutrophil-mediated damage. *Biomaterials.* (2016) 102:209–19. doi: 10.1016/j.biomaterials.2016.05.055
64. Narayanan K, Jaramillo A, Phelan DL, Mohanakumar T. Pre-exposure to sub-saturating concentrations of HLA class I antibodies confers resistance to endothelial cells against antibody complement-mediated lysis by regulating Bad through the phosphatidylinositol 3-kinase/Akt pathway. *Eur J Immunol.* (2004) 34:2303–12. doi: 10.1002/eji.200324843
65. Jindra PT, Zhang X, Mulder A, Claas F, Veale J, Jin YP, et al. Anti-HLA antibodies can induce endothelial cell survival or proliferation depending on their concentration. *Transplantation.* (2006) 82:S33–35. doi: 10.1097/01.tp.0000231447.34240.3c
66. Tinckam KJ, Djurdjev O, Magil AB. Glomerular monocytes predict worse outcomes after acute renal allograft rejection independent of C4d status. *Kidney Int.* (2005) 68:1866–74. doi: 10.1111/j.1523-1755.2005.00606.x
67. Valenzuela NM, Mulder A, Reed EF. HLA class I antibodies trigger increased adherence of monocytes to endothelial cells by eliciting an increase in endothelial P-selectin and, depending on subclass, by engaging FcgammaRs. *J Immunol.* (2013) 190:6635–50. doi: 10.4049/jimmunol.1201434
68. Galleu A, Riffo-Vasquez Y, Trento C, Lomas C, Dolcetti L, Cheung TS, et al. Apoptosis in mesenchymal stromal cells induces *in vivo* recipient-mediated immunomodulation. *Sci Transl Med.* (2017) 9:7828. doi: 10.1126/scitranslmed.aam7828
69. Hocine HR, Costa HE, Dam N, Giustiniani J, Palacios I, Loiseau P, et al. Minimizing the risk of allo-sensitization to optimize the benefit of allogeneic cardiac-derived stem/progenitor cells. *Sci Rep.* (2017) 7:41125. doi: 10.1038/srep44720
70. Saito KM, Matsunaga A, Takasu N, and Yamanaka S. Donor recruitment and eligibility criteria for HLA-homozygous iPS cell bank in Japan. In: Ilic D, editor. *Stem Cell Banking. Stem Cell Biology and Regenerative Medicine.* New York, NY: Springer (2014) 67–76.

5

Immunomodulation by Mesenchymal Stem Cells (MSCs): Mechanisms of Action of Living, Apoptotic and Dead MSCs

Andreas Robert Rudolf Weiss [1*] *and Marc Hendrik Dahlke* [2]

[1] *Department of Surgical Sciences, Dunedin School of Medicine, University of Otago, Dunedin, New Zealand,* [2] *Department of Surgery, Robert Bosch Hospital, Stuttgart, Germany*

Correspondence:
Andreas Robert Rudolf Weiss
andi-weiss@outlook.com

Expectations on mesenchymal stem cell (MSC) treatment are high, especially in the fields of sepsis, transplant medicine, and autoimmune diseases. Various pre-clinical studies have been conducted with encouraging results, although the mechanisms of action behind the observed immunomodulatory capacity of mesenchymal stem cells have not been fully understood. Previous studies have demonstrated that the immunomodulatory effect of MSCs is communicated via MSC-secreted cytokines and has been proven to rely on the local microenvironment as some of the observed effects depend on a pre-treatment of MSCs with inflammatory cytokines. Nonetheless, recent findings indicate that the cytokine-mediated effects are only one part of the equation as apoptotic, metabolically inactivated, or even fragmented MSCs have been shown to possess an immunomodulatory potential as well. Both cytokine-dependent and cytokine-independent mechanisms suggest a key role for regulatory T cells and monocytes in the overall pattern, but the principle as to why viable and non-viable MSCs have similar immunomodulatory capacities remains elusive. Here we review the current knowledge on cellular and molecular mechanisms involved in MSC-mediated immunomodulation and focus on the viability of MSCs, as there is still uncertainty concerning the tumorigenic potential of living MSCs.

Keywords: mesenchymal stem cell (MSC), immunomodulation, immunogenicity, tumor induction, apoptosis, HI-MSC, monocytes, regulatory T cells

INTRODUCTION

Mesenchymal stem cell (MSC) therapy offers a promising treatment option for autoimmune diseases, sepsis, and in transplant surgery (1–7). However, the underlying cellular and molecular mechanisms of MSC-mediated immunomodulation have not been fully clarified. Studies have demonstrated various immunomodulatory changes following the administration of MSCs, although a clear picture is still missing and study results are often inconsistent. This might partially be explained by the fact that MSCs from different sources and under different culture conditions express different surface markers, show varying cytokine secretion profiles and differ in telomere-length and methylation patterns (8–15).

However, comparing the available data is complicated by a lack of standardization for the isolation, culture, and characterization of MSCs (16). MSCs can be harvested from various adult tissues such as bone marrow, adipose tissue, inner organs, and peripheral blood as well as from neonatal tissues (e.g., umbilical cord, placenta, amniotic fluid, amniotic membrane). In clinical studies, adipose tissue, and umbilical cord-derived MSCs have regularly been used due to their accessibility. The broad range of potential sources makes a comparison of study results challenging, as MSCs display varying features *in vitro* and *in vivo* depending on the tissue they originate from (17–19). In most study protocols MSCs were administered intravenously, yet in others they were delivered via an intraarterial, intraportal, intraperitoneal, or topical route or were administered directly into the damaged tissue (20–24).

Furthermore, freshly thawed MSCs seem to have an impaired immunomodulatory capacity compared to continuously cultured MSCs (25). The fact that MSCs act differently depending on the local microenvironment contributes even more to the complexity of understanding MSC-mediated immunomodulation (26–28). MSCs have a short half-life and cannot pass through the lung capillary network after IV administration, which appears to contradict the observed long-term immunomodulatory effects, particularly in transplant settings (29, 30).

Nevertheless, there are certain patterns and pathways that seem to be consistent and have been repeatedly demonstrated. MSC-mediated immunomodulation operates through a synergy of cell contact-dependent mechanisms and soluble factors (8, 31). MSCs reveal their immunomodulatory potential via functional changes of monocytes/macrophages, dendritic cells, T cells, B cells, and natural killer cells (6, 27, 32–36). In particular, anti-inflammatory monocytes/macrophages and regulatory T cells (Tregs) play a prominent role as they unfold their full immunomodulatory potential in a complex interaction catalyzed by MSCs (32, 37, 38). The interaction between MSCs, monocytes, and Tregs have often been attributed to MSC-secreted cytokines, although there is increasing evidence for mechanisms that rely on a direct cell-cell interaction, which—in the case of MSCs—does not necessarily require an intact cell metabolism (27, 31, 39, 40). Recent studies could demonstrate that apoptotic, metabolically inactivated, or even fragmented MSCs possess immunomodulatory capacities (21, 39, 41). As there are still ongoing concerns as to what extent living MSCs might contribute to tumorigenesis, the option to use dead cells or even cell fragments could be a promising alternative. This review summarizes the current knowledge on cellular and molecular interactions in MSC-derived immunomodulation by highlighting the different immune responses to living, apoptotic, and dead MSCs and provides an overview of the potential risks of MSC treatment in terms of tumor induction.

IMMUNOMODULATION BY LIVING MSCs

Effect on Monocytes/Macrophages and Dendritic Cells

MSC were shown to promote the polarization of monocytes/macrophages toward an anti-inflammatory/immune-regulatory (type 2) phenotype and to directly inhibit the differentiation into the type 1 phenotype and dendritic cells (DCs) (10, 42–45). MSC-secreted Interleukin 1 Receptor Antagonist (IL1-RA) can promote the polarization of macrophages toward the type 2 phenotype (36). Anti-inflammatory monocytes secret high levels of IL-10 and have decreased levels of IL-12p70, TNF-a, and IL-17 expression—a process that is mediated by MSC-produced IL-6 and hepatocyte growth factor (HGF) (10, 40). A key role for the MSC-mediated, increased production of IL-10 has been demonstrated in a sepsis model in mice where IL-10 neutralization reversed the beneficial effect of bone marrow-derived MSCs on overall survival after induction of sepsis via cecal ligation and puncture (CLP) (6). Monocyte-derived IL-10 prevents monocyte differentiation into DCs and shifts monocytes toward an anti-inflammatory, IL-10-secreting subtype in terms of a positive-feedback loop (10). Apart from IL-10, MCS-primed monocytes express high levels of MHC class II, CD45R, and CD11b and seem to be able to suppress T-cell activity regardless of FoxP3$^+$ Tregs (46). The supernatants of type 2 macrophages induce the formation of FoxP3$^+$ Tregs from naïve CD4$^+$ T cells, which emphasizes the role of soluble factors in MSC-mediated immunomodulation (47). The monocyte-induced Treg-formation is mediated by monocyte-produced CCL-18 and monocyte-released transforming growth factor beta 1 (TGF- β1) (45, 47). Macrophages bind and re-release TGF-β1 during their differentiation into type 2 macrophages and might thereby contribute to the MSC-induced formation of Tregs as MSCs have been shown to secrete TGF-β1 (45, 47). The neutralization of CCL-18 leads to a significant reduction in MSC-induced Treg formation (45). CCL-18 can turn memory CD4$^+$ T cells into to CD4$^+$CD25$^+$Foxp3$^+$ Tregs with an increased IL-10 and TGF- β1 production. CCL-18-pretreated Tregs inhibit CD4$^+$CD25$^-$ effector T cell proliferation via the activation of G-protein-coupled receptors (48). Macrophage type 2-derived CCL-18 can differentiate DCs into tolerogenic DCs, which are in turn able to prime Tregs (45, 48, 49) (**Figure 1A**). Interestingly, high concentrations of CCL-18 producing antigen-presenting cells can be found in the lungs, where MSCs become caught in the capillary system after IV application (50–52) (**Figure 1B**).

MSCs also suppress the migration and maturation of DCs (32). In the presence of MSCs, DCs are less capable to support antigen specific CD4$^+$ T cell proliferation and to display an MHC class II-peptide complex (54). Mature type 1 DCs secrete significantly less TNF-α, if co-cultured with MSCs and anti-inflammatory mature type 2 DCs show an increased IL-10 secretion (55). Furthermore, Sca-1$^+$CD117$^-$Lin$^-$ bone marrow-derived MSCs have been shown to generate regulatory DCs with immune regulatory function from hematopoietic stem cells in mice (56).

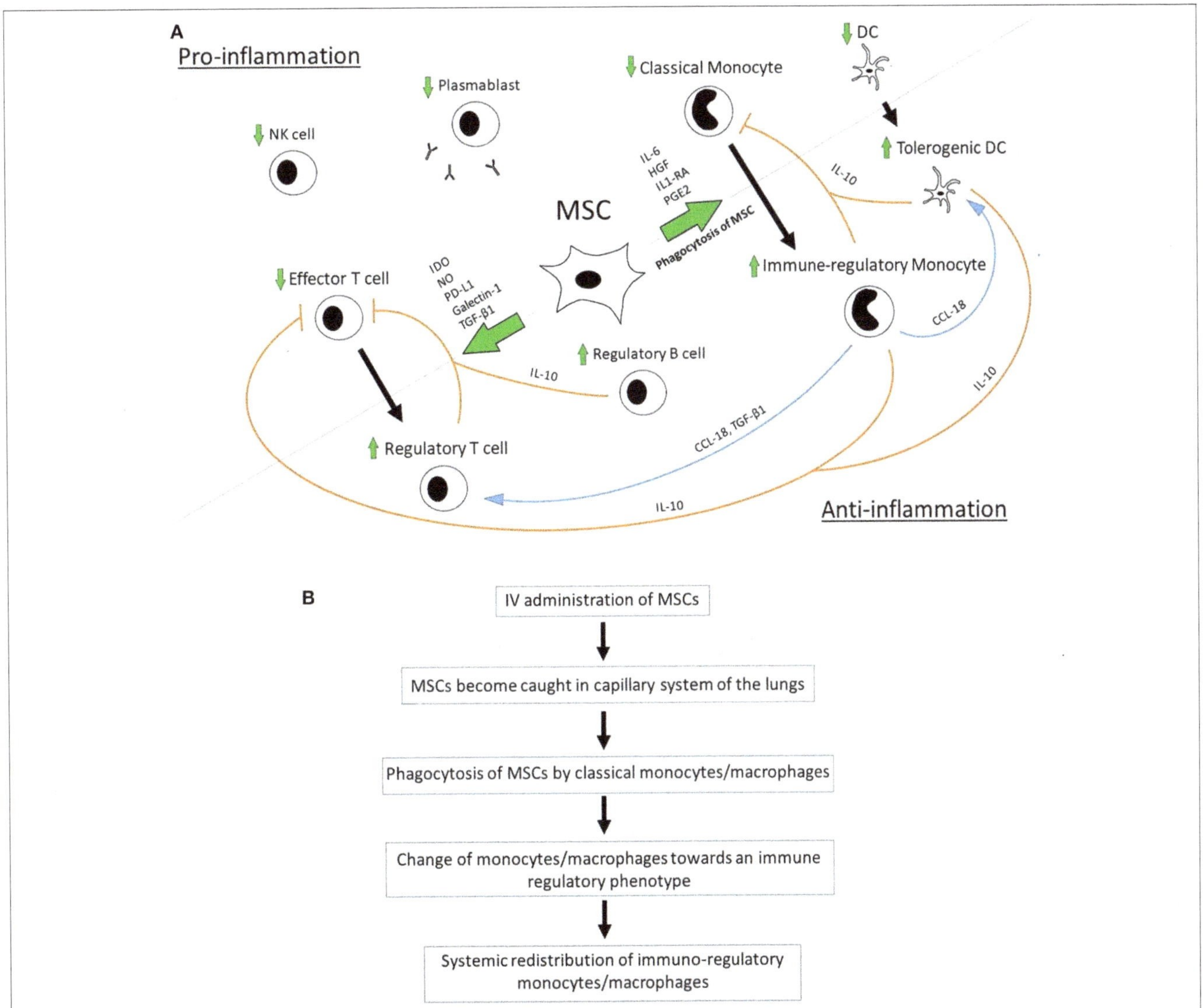

FIGURE 1 | (A) Proposed interaction of MSCs with host immune cells. **(B)** Proposed pathway of MSC-mediated anti-inflammation via phagocytosis of MSCs [in accordance with De Witte et al. (31) and Braza et al. (53)].

Recently, a cytokine-independent pathway for the MSC-induced polarization of monocytes/macrophages has been revealed. In a mouse model of asthma, MSCs were phagocytosed by lung macrophages. The phagocytosis of MSCs caused monocytes to turn into a type 2 immunosuppressive phenotype (53). A previous study from 2012 observed a similar pattern in a p. aeruginosa peritonitis model in mice (57). The intravenous application of bone marrow-derived MSCs was followed by an increased phagocytic activity in blood monocytes compared with the PBS control group. Furthermore, an increase in alternately activated $CD163^{+}CD206^{+}$ monocytes/macrophages in the spleen of the MSC-treated mice could be observed (57). De Witte et al. (31) could show that phagocytosed MSCs are mainly found in non-classical $Ly6C^{low}$ monocytes. The phagocytosis of MSCs caused $CD14^{++}CD16^{-}$ classical monocytes to polarize toward a $CD14^{++}CD16^{+}CD206^{+}$ immune regulatory intermediate subtype with anti-inflammatory properties and increased expression of IL-10 and programmed death ligand-1 (PD-L1). After phagocytosis of MSCs, these primed monocytes were able to induce $CD4^{+}CD25^{hi}$ Treg formation *in vitro* to a significantly higher extent than un-primed monocytes (31). Likewise, an increase in anti-inflammatory $Ly6C^{low}$ monocytes in the blood, heart, and spleen was observed after IV administration of MSCs in a Coxsackievirus B3-induced myocarditis model in mice (58).

The depletion of phagocytic cells demonstrates their indispensable role in MSC-mediated immunomodulation as the absence of monocytes/macrophages and dendritic cells abrogates the ability of MSCs to suppress T cell proliferation *in vitro* and their immunomodulatory effect in *in vivo* transplant models (59).

Effect on T Cells

MSCs suppress T cell proliferation ($CD4^+$ and $CD8^+$ T cell subsets) in mixed lymphocyte reactions in a dose-dependent manner (39, 60, 61). In interaction with DCs, MSCs cause a shift from pro-inflammatory Th1 to anti-inflammatory Th2 cells including a change in the cytokine profile toward anti-inflammation (32, 62–64). Moreover, MSCs facilitate the formation of Tregs *in vitro* and *in vivo* (3, 32, 55). Tregs are essential for the immune homeostasis by preventing autoimmunity (65, 66). The induction of $CD4^+CD25^+FoxP3^+$ Tregs is a mainstay in MSC-mediated immunomodulation and was shown to be essential for tolerance induction in a kidney allograft transplantation model (32). The MSC-induced upregulation of Tregs does not result from an expansion of pre-existing natural Tregs but via induction of Tregs from conventional T cells (67, 68). TGF-β1-neutralization studies have shown that the generation of Tregs is TGF-β1-mediated and that MSCs constitutively secrete TGF-β1. However, TGF-β1 seems not to be sufficient on its own, as the presence of monocytes was proven to be essential for the formation of Tregs (45). In co-treatment with MMF, MSCs have been shown to promote a direct conversion of IL-$17A^+$ cells into IL-$17A^{neg}Foxp3^+$ Tregs (35). MSCs also constitutively secrete indoleamine 2,3-dioxygenase (IDO) and the secretion is increased upon stimulation by INF-γ. The consecutive tryptophan depletion leads to an inhibition of allogeneic T cell responses, stimulates the secretion of IL-4 in Th2 cells and decreases the IFN-γ production by Th1 cells (27, 32, 55, 69, 70). Gieseke et al. have shown that MSCs can directly inhibit the proliferation of alloreactive $CD4^+$ and $CD8^+$ T cells without the presence of other immune cells and that this process is partially mediated by MSC-derived galectin-1 (60). Via the secretion of PD-L1, MSCs can suppress T cell activation and induce an irreversible T cell hypo-responsiveness and apoptosis (71, 72) (**Figure 1A**).

Effect on B Cells

MSCs directly interact with B cells and are able to reduce plasmablast formation as well as to promote the induction of regulatory B cells (Bregs) (73). Bregs have immunosuppressive properties through which they provide immunological tolerance (74). IL-10-producing Bregs were shown to convert effector $CD4^+$ T cells into $Foxp3^+$ Tregs (75). The stimulatory effect of MSCs on Breg formation and IL-10 production is not mediated via soluble factors but seems to be dependent on direct cell-cell contact or at least on a close proximity of the corresponding cells (27). However, it was shown that the stimulatory effect of MSCs on Breg formation and their suppressive effect on T cell proliferation requires an active cell metabolism (27, 41). Luz-Crawford et al. revealed a cytokine-triggered mechanism, by which MSC-secreted IL1-RA inhibits B cell differentiation (36). In the presence of T cells, MSCs also inhibit the proliferation of B cells, which could be due to T cell-secreted IFN-γ as IFN-γ pre-treated MSCs are able to inhibit B cell proliferation (27).

TABLE 1 | Overview of important immunomodulatory effects of living, apoptotic, and dead MSCs.

	Living MSCs	Apoptotic MSCs	Dead MSCs*
Inhibition of T cell proliferation in MLR	+	n.a.	–
Induction of regulatory T cells *in vivo*	+	n.a.	n.a.
Modulation of monocyte function *in vivo*	+	(+)**	+
Attenuation of sepsis *in vivo*	+	+	+

**Heat-inactivated MSCs without active cell metabolism according to Luk et al. (41).*
***Indirect evidence in study from Galleu et al. (80).*
n.a. information not available.

Effect on Natural Killer Cells

MSC are also strong inhibitors of natural killer cell (NK cell) proliferation. NK cells show an impaired cytotoxic activity and cytokine production after co-culture with MSCs. There is evidence that the inhibitory effect of MSCs on NK cells involves MSC-secreted prostaglandin E2 (PGE2), IDO, TGF-β1, IL-6 and nitric oxide (NO) (10, 33, 76) (**Figure 1A**).

IMMUNOMODULATION BY APOPTOTIC AND DEAD MSCs

Effect of Apoptotic MSCs

The viability of MSCs does not appear to be a prerequisite for some of their immunomodulatory effects. Apoptotic adipose tissue-derived MSCs (A-ADMSCs) have been shown to reduce mortality in rats after sepsis induction via CLP (21) (**Table 1**). Mortality, circulating TNF-α level as well as circulating and splenic levels of T helper cells and cytotoxic T cells following CLP were significantly lower in the group treated with A-ADMSCs compared to CLP alone (21). A study group around Chen et al. provided similar results in a CLP-induced model of acute kidney injury in mice with a reduced splenic level of T helper cells and cytotoxic T cells as well as a lower circulating TNF-α level in the group treated with intravenous A-ADMSCs compared to the CLP control group (77). Interestingly, the study by Chang et al. could not prove any benefit and even a trend toward a reduced survival after administration of living MSCs (21). In keeping with the results of Chang et al. it was demonstrated that an IV A-ADMSC treatment is superior to a treatment with living MSCs in a CLP-induced sepsis model in rats (78). The parameters for sepsis-induced acute lung injury (ALI) and acute kidney injury were significantly lower in the group treated with apoptotic MSCs. Furthermore, A-ADMSC treatment was more effective in reducing inflammation, oxidative stress, and apoptosis as well as sepsis-induced histopathological alterations in the lungs and kidneys compared to living MSCs (78). Likewise, A-ADMSCs were shown to be superior to living MSCs for the treatment of an acute lung ischemia-reperfusion injury in rats if administered along with melatonin (79).

However, there are several studies that demonstrated a significant positive effect of living MSCs in the attenuation of sepsis in different animal models (22, 81–83). Interestingly, a recent study demonstrated that recipient cytotoxic cells cause perforin-induced apoptosis in infused MSCs (80). The apoptosis of MSCs was the prerequisite for MSCs to unfold their immune-regulatory effect in a murine graft vs. host model. Hence, the cytotoxic activity against MSCs was demonstrated to be a crucial part in MSC-mediated immunomodulation. The recognition by cytotoxic cells in this model was shown to be MHC-independent and non-antigen-specific. Moreover, the use of apoptotic MSCs skipped the need for recipient cytotoxic cells. Apoptotic MSCs were immunosuppressive in a TH2-type inflammation model and induced the IDO production in recipient phagocytes (80) (**Table 1**). Another study provided evidence that the supernatants of macrophages that phagocytized apoptotic mesenchymal stem cells improve the survival of hypoxic cardiomyocytes (84). These findings are in keeping with the "dying stem cell hypothesis" published by Thum in 2005, which stated that the apoptosis of MSCs causes a modulation of the local immune response with a down-regulation of the innate and adaptive immunity (85).

Contrary findings were provided in an endotoxin-induced ALI model in mice. The intrapulmonary administration of apoptotic MSCs did not improve survival or decrease the severity of an endotoxin-induced ALI. Moreover, no decrease in TNF-α levels and no increase in IL-10 levels could be observed, neither in plasma nor in fluid from bronchoalveolar lavage (22). Compared with the above mentioned studies, the intrapulmonary application via trachea/bronchi was a unique characteristic of this study compared to the commonly used IV route and might explain the differing results.

Effect of Metabolically Inactivated MSCs (HI-MSCs) and MSC Cell Fragments

In 2016, a heat-inactivation protocol for MSCs was introduced, in which human MSCs were heated for 30 min to 50°C (41). Heat-inactivation causes an irreversible cessation of the metabolic and proliferative activity of MSCs. HI-MSCs do not secrete cytokines but their cell integrity remains largely intact. Over the course of time, HI-MSCs are subject to physical disintegration rather than to apoptosis, as they do not overexpress heat shock proteins Hsp27 and Hsp70 and pro-apoptotic Bax (41). Therefore, in this review heat-inactivated MSCs (HI-MSCs) are referred to as being "dead." In contrast to living MSCs, HI-MSCs do not inhibit T cell proliferation and do not induce Breg formation in mixed lymphocyte reactions. However, HI-MSCs are still able to attenuate the inflammatory response in mice following LPS-administration. After administration of HI-MSCs, serum levels of IFN-γ were significantly reduced whereas IL-10 serum levels were increased (41). MSCs and HI-MSCs show similar effects on monocyte function with significantly reduced production of TNF-α in response to lipopolysaccharide (41). Even MSC-derived membrane particles seem to possess immunomodulatory properties. Goncalves et al. used MSC-derived membrane particles with a size ranging between 63 and 700 nm (>95% smaller than 200 nm). These MSC-derived membrane particles were shown to be enzymatically active but did not suppress T cell proliferation in mixed lymphocyte reactions. The MSC membrane particles were taken up by monocytes and became bound to their plasma membranes thus inducing selective apoptosis of pro-inflammatory $CD14^{+}CD16^{+}$ monocytes (39) (**Table 1**).

SCENE OF THE EVENT

There is an ongoing discussion, as to whether MSCs are able to migrate to the site of inflammation/tissue damage or to a transplanted organ. In that context, it is worthwhile to differentiate between endogenous MSCs and exogenously administered MSCs. The data concerning the migratory capacity of endogenous MSCs is controversial and there is no convincing evidence that endogenous MSCs find their way to the site of inflammation/tissue damage via the bloodstream (86). Nevertheless, endogenous MSCs seem to be able to migrate within the tissues and might thereby reach the damaged or inflamed areas (50). A recently published study revealed that exogenous MSCs, if transplanted directly into the tissue, could survive up to 4 months at the effector site with few of these transplanted MSCs beginning to develop a resident cell tissue phenotype (23).

However, in most clinical and pre-clinical models MSCs were administered intravenously. Interestingly, the majority of IV-administered MSCs does not pass the capillary system of the lungs, which contradicts the suggestion that MSCs exert their immunomodulatory effect by migrating to the sites of harm. Within 24 h after being trapped in the pulmonary system a major decrease in the number of viable MSCs can be observed (29, 30). The fate of these MSCs was further unveiled by demonstrating that MSCs become phagocytosed by blood-derived monocytes and neutrophils after their initial entrapment in the lungs and thereby become redistributed systemically with a major accumulation in the liver and the spleen (31, 87) (**Figure 1B**).

The change in the monocytes' profile toward an immuno-regulatory phenotype after phagocytosis of MSCs followed by their systemic redistribution might be the key mechanism to explain the contradictory findings with regards to the short half-life of MSCs and the long lasting immunomodulatory effects that have been observed (31, 35).

It remains unclear, whether living MSCs are subject to phagocytosis by host-innate immune cells or if MSCs have to undergo apoptosis in order to become phagocytosed. Galleu et al. have shown that infused living MSCs are subject to perforin-induced apoptosis through recipient cytotoxic cells (80). These findings emphasize the importance of apoptosis in MSC-mediated immunomodulation and might explain previous study results, in which apoptotic MSCs were shown to be superior compared to living MSCs (21, 78, 79). Nevertheless, apoptosis cannot explain by which means dead HI-MSCs—which are subject to physical disintegration rather than to apoptosis—unfold their immunomodulatory potential (27, 41).

AUTOLOGOUS VS. ALLOGENIC—EVIDENCE FOR IMMUNOGENICITY OF MSCs

Allogeneic MSCs are frequently used in clinical studies even though their immunogenic potential has not always been taken into account (88). In the past, MSCs were thought to be immune privileged. Several pre-clinical studies could show that both, autologous and allogenic MSCs suppress T cell proliferation in mixed lymphocyte reactions in a dose-dependent manner (61, 89). In animal models, donor-derived MSCs prolonged the survival of semi-allogeneic heart transplants in mice via the generation of Tregs and a shift in the Th1/Th2 balance toward Th2 (90, 91). Interestingly, the IV administration of MSCs is followed by a systemic inflammatory response with an increase in macrophages in lung tissue 2 h after infusion and increased serum levels of pro-inflammatory IL-6, CXCL1, and monocyte chemoattractant protein-1. The phase of acute inflammatory response seems to be followed by a phase with reduced immune reactivity, which might partially explain the increased allograft survival in animal models, in which MSCs were administered prior to transplantation (34, 59, 92).

However, donor-derived MSCs given prior to transplantation only prolonged the survival of allogenic heart grafts if a short course of mycophenolate mofetil (MMF) was administered from the day of transplantation onwards. The administration of donor-derived MSCs alone caused a prompt T cell infiltration of the grafts with consecutive graft rejection, which supports the assumption that allogeneic MSCs are immunogenic and sensitize the recipient (34, 59). It has been demonstrated that allogenic MSCs can activate T cells (89, 93). The additional treatment with MMF seems to eliminate these activated T cells with a consecutive reduction of T cell infiltration in the respective allografts (59).

Further evidence that MSCs are not immune privileged has been provided in a study, in which syngeneic, erythropoietin-releasing MSCs persisted for more than 200 days, whereas allogeneic MSCs were rapidly rejected (94). Zangi et al. provided similar findings with luciferase-labeled MSCs. The survival of allogeneic MSCs was significantly shorter compared to syngeneic MSCs. Moreover, allogeneic MSCs seem to induce an immune memory represented by an increase in T cells with a memory phenotype (95).

It can be assumed that MSCs are not immune privileged, but rather that allogenic MSCs have a lower immunogenic potential as other allogeneic cell types (88). As the cytotoxic cell-induced apoptosis of MSCs was described to be essential for the MSC-mediated immunomodulation and because this effect was shown to be MHC-independent and non–antigen-specific, the allogenic component might be of secondary importance (80).

POTENTIAL BENEFIT OF DEAD MSCs IN TERMS OF REDUCED CANCER RISK

Previous studies have suggested that MSCs could favor tumor growth *in vivo* (96, 97). It has been reported that implanted MSCs cause an earlier onset of syngeneic tumors and allow B16 melanoma cells to form tumors in allogenic mice (98). Furthermore, TNFα-activated MSCs can facilitate tumor growth and promote cancer metastasis via $CXCR2^{+}$ neutrophil recruitment (99).

The mechanism behind MSC-induced tumor growth involves the formation of carcinoma associated fibroblasts (CAFs). Human bone marrow-derived MSCs (BM-MSCs) were shown to adopt a CAF-like phenotype with similar functional properties after prolonged exposure to tumor-conditioned medium (100, 101). CAFs and CAF-like MSCs produce growth factors, cytokines, and chemokines, and thereby provide the microvascularization and the stromal network for tumor progression (102). In a mouse model of inflammation-induced gastric cancer, at least 20% of CAFs derived from the bone marrow and developed from MSCs (103).

Furthermore, MSCs seem to have a distinct tropism for tumors as BM-MSCs were shown to accumulate in brain tumors after intracarotid injection, whereas fibroblasts and U87 glioma cells did not (20). This tropism was further elucidated in a study from 2013 that unveiled an active recruitment of MSCs to prostate cancer via prostate cancer-secreted CXCL-16. CXCL-16 binds to CXCR6 expressed by MSCs. The CXCL16/CXCR6 signaling induces the conversion of MSCs into CAFs (104). Ren et al. could show that under inflammatory conditions CAF-like MSCs stimulate tumor growth via the recruitment of monocytes and macrophages. The essential role monocytes/macrophages in MSC-mediated immunomodulation was demonstrated once more as their depletion abrogated the promotion of tumor growth by lymphoma isolated-MSCs (105).

Whilst MSCs can exhibit immunosuppressive or immune-enhancing properties depending on the presence or absence of certain inflammatory or anti-inflammatory stimuli, the tumorigenic potential of living MSCs poses a risk that cannot be neglected (26, 28, 106). As dead MSCs have no active cell metabolism, it can be assumed that they do not differentiate into CAF-like cells with the corresponding secretion of growth factors, cytokines, and chemokines. However, a clear discrimination between CAFs, CAF-like cells, and MSCs is still missing in the current literature. Furthermore, dead MSCs may still provoke changes in the tumor microenvironment due to their immunomodulatory properties. Therefore, the question whether dead or fragmented MSCs are beneficial in terms of a reduced cancer risk cannot be answered, yet.

CONCLUSION

It remains a challenge to connect the dots between the various MSC-mediated immunomodulatory effects, especially as MSCs are very heterogenic and subject to significant changes upon inflammatory or anti-inflammatory stimuli. Viable

MSCs might provoke more complex immunomodulatory mechanisms due to their intact secretome. However, since the discussion about a universal donor for MSC therapy has not been finally answered, the possibility to use dead MSCs should also be considered. HI-MSCs or fragmented MSCs are most likely not subject to changes in their immunomodulatory characteristics upon different environmental stimuli. Hence, their immunomodulatory effects might be more predictable, which would allow a better comparison of future study results.

AUTHOR CONTRIBUTIONS

AW drafted and wrote the manuscript. MD drafted and critically revised the manuscript. Both authors have approved the manuscript for publication.

ACKNOWLEDGMENTS

The authors want to thank Prof. Schlitt and his team in the surgical department at the University Hospital Regensburg for their support.

REFERENCES

1. Le Blanc K, Frassoni F, Ball L, Locatelli F, Roelofs H, Lewis I, et al. Mesenchymal stem cells for treatment of steroid-resistant, severe, acute graft-versus-host disease: a phase II study. *Lancet.* (2008) 371:1579–86. doi: 10.1016/S0140-6736(08)60690-X
2. Forbes GM, Sturm MJ, Leong RW, Sparrow MP, Segarajasingam D, Cummins AG, et al. A phase 2 study of allogeneic mesenchymal stromal cells for luminal Crohn's disease refractory to biologic therapy. *Clin Gastroenterol Hepatol.* (2014) 12:64–71. doi: 10.1016/j.cgh.2013.06.021
3. Gonzalez MA, Gonzalez-Rey E, Rico L, Buscher D, Delgado M. Adipose-derived mesenchymal stem cells alleviate experimental colitis by inhibiting inflammatory and autoimmune responses. *Gastroenterology.* (2009) 136:978–89. doi: 10.1053/j.gastro.2008.11.041
4. Reinders ME, de Fijter JW, Roelofs H, Bajema IM, de Vries DK, Schaapherder AF, et al. Autologous bone marrow-derived mesenchymal stromal cells for the treatment of allograft rejection after renal transplantation: results of a phase I study. *Stem Cells Transl Med.* (2013) 2:107–11. doi: 10.5966/sctm.2012-0114
5. Bartholomew A, Sturgeon C, Siatskas M, Ferrer K, McIntosh K, Patil S, et al. Mesenchymal stem cells suppress lymphocyte proliferation *in vitro* and prolong skin graft survival *in vivo*. *Exp Hematol.* (2002) 30:42–8. doi: 10.1016/S0301-472X(01)00769-X
6. Nemeth K, Leelahavanichkul A, Yuen PS, Mayer B, Parmelee A, Doi K, et al. Bone marrow stromal cells attenuate sepsis via prostaglandin E(2)-dependent reprogramming of host macrophages to increase their interleukin-10 production. *Nat Med.* (2009) 15:42–9. doi: 10.1038/nm.1905
7. Hu J, Yu X, Wang Z, Wang F, Wang L, Gao H, et al. Long term effects of the implantation of Wharton's jelly-derived mesenchymal stem cells from the umbilical cord for newly-onset type 1 diabetes mellitus. *Endocr J.* (2013) 60:347–57. doi: 10.1507/endocrj.EJ12-0343
8. Wu Y, Hoogduijn MJ, Baan CC, Korevaar SS, de Kuiper R, Yan L, et al. Adipose Tissue-derived mesenchymal stem cells have a heterogenic cytokine secretion profile. *Stem Cells Int.* (2017) 2017:4960831. doi: 10.1155/2017/4960831
9. Nasef A, Mathieu N, Chapel A, Frick J, Francois S, Mazurier C, et al. Immunosuppressive effects of mesenchymal stem cells: involvement of HLA-G. *Transplantation.* (2007) 84:231–7. doi: 10.1097/01.tp.0000267918.07906.08
10. Deng Y, Zhang Y, Ye L, Zhang T, Cheng J, Chen G, et al. Umbilical cord-derived mesenchymal stem cells instruct monocytes towards an IL10-producing phenotype by secreting IL6 and HGF. *Sci Rep.* (2016) 6:37566. doi: 10.1038/srep37566
11. De Witte SFH, Peters FS, Merino A, Korevaar SS, Van Meurs JBJ, O'Flynn L, et al. Epigenetic changes in umbilical cord mesenchymal stromal cells upon stimulation and culture expansion. *Cytotherapy.* (2018) 20:919–29. doi: 10.1016/j.jcyt.2018.05.005
12. Bonab MM, Alimoghaddam K, Talebian F, Ghaffari SH, Ghavamzadeh A, Nikbin B. Aging of mesenchymal stem cell *in vitro*. *BMC Cell Biol.* (2006) 7:14. doi: 10.1186/1471-2121-7-14
13. Schellenberg A, Lin Q, Schuler H, Koch CM, Joussen S, Denecke B, et al. Replicative senescence of mesenchymal stem cells causes DNA-methylation changes which correlate with repressive histone marks. *Aging.* (2011) 3:873–88. doi: 10.18632/aging.100391
14. Baxter MA, Wynn RF, Jowitt SN, Wraith JE, Fairbairn LJ, Bellantuono I. Study of telomere length reveals rapid aging of human marrow stromal cells following *in vitro* expansion. *Stem Cells.* (2004) 22:675–82. doi: 10.1634/stemcells.22-5-675
15. de Witte SFH, Lambert EE, Merino A, Strini T, Douben HJCW, O'Flynn L, et al. Aging of bone marrow- and umbilical cord-derived mesenchymal stromal cells during expansion. *Cytotherapy.* (2017) 19:798–807. doi: 10.1016/j.jcyt.2017.03.071
16. Yin JQ, Zhu J, Ankrum JA. Manufacturing of primed mesenchymal stromal cells for therapy. *Nat Biomed Eng.* (2019) 28:90–104. doi: 10.1038/s41551-018-0325-8
17. Elahi KC, Klein G, Avci-Adali M, Sievert KD, MacNeil S, Aicher WK. Human Mesenchymal stromal cells from different sources diverge in their expression of cell surface proteins and display distinct differentiation patterns. *Stem Cells Int.* (2016) 2016:5646384. doi: 10.1155/2016/5646384
18. Hass R, Kasper C, Bohm S, Jacobs R. Different populations and sources of human mesenchymal stem cells (MSC): a comparison of adult and neonatal tissue-derived MSC. *Cell Commun Signal.* (2011) 9:12. doi: 10.1186/1478-811X-9-12
19. da Silva Meirelles L, Chagastelles PC, Nardi NB. Mesenchymal stem cells reside in virtually all post-natal organs and tissues. *J Cell Sci.* (2006) 119(Pt 11):2204–13. doi: 10.1242/jcs.02932
20. Nakamizo A, Marini F, Amano T, Khan A, Studeny M, Gumin J, et al. Human bone marrow-derived mesenchymal stem cells in the treatment of gliomas. *Cancer Res.* (2005) 65:3307–18. doi: 10.1158/0008-5472.CAN-04-1874
21. Chang CL, Leu S, Sung HC, Zhen YY, Cho CL, Chen A, et al. Impact of apoptotic adipose-derived mesenchymal stem cells on attenuating organ damage and reducing mortality in rat sepsis syndrome induced by cecal puncture and ligation. *J Transl Med.* (2012) 10:244. doi: 10.1186/1479-5876-10-244
22. Gupta N, Su X, Popov B, Lee JW, Serikov V, Matthay MA. Intrapulmonary delivery of bone marrow-derived mesenchymal stem cells improves survival and attenuates endotoxin-induced acute lung injury in mice. *J Immunol.* (2007) 179:1855–63. doi: 10.4049/jimmunol.179.3.1855
23. Munoz MF, Arguelles S, Guzman-Chozas M, Guillen-Sanz R, Franco JM, Pintor-Toro JA, et al. Cell tracking, survival, and differentiation capacity of adipose-derived stem cells after engraftment in rat tissue. *J Cell Physiol.* (2018) 233:6317–28. doi: 10.1002/jcp.26439
24. Avritscher R, Abdelsalam ME, Javadi S, Ensor J, Wallace MJ, Alt E, et al. Percutaneous intraportal application of adipose tissue-derived mesenchymal stem cells using a balloon occlusion catheter in a porcine model of liver fibrosis. *J Vasc Interv Radiol.* (2013) 24:1871–8. doi: 10.1016/j.jvir.2013.08.022
25. Moll G, Geissler S, Catar R, Ignatowicz L, Hoogduijn MJ, Strunk D, et al. Cryopreserved or fresh mesenchymal stromal cells: only a matter of taste or key to unleash the full clinical potential of MSC therapy? *Adv Exp Med Biol.* (2016) 951:77–98. doi: 10.1007/978-3-319-45457-3_7
26. Wang Y, Chen X, Cao W, Shi Y. Plasticity of mesenchymal stem cells in immunomodulation: pathological and therapeutic implications. *Nat Immunol.* (2014) 15:1009–16. doi: 10.1038/ni.3002

27. Luk F, Carreras-Planella L, Korevaar SS, de Witte SFH, Borras FE, Betjes MGH, et al. Inflammatory conditions dictate the effect of mesenchymal stem or stromal cells on B cell function. *Front Immunol.* (2017) 8:1042. doi: 10.3389/fimmu.2017.01042
28. de Witte SFH, Merino AM, Franquesa M, Strini T, van Zoggel JAA, Korevaar SS, et al. Cytokine treatment optimises the immunotherapeutic effects of umbilical cord-derived MSC for treatment of inflammatory liver disease. *Stem Cell Res Ther.* (2017) 8:140. doi: 10.1186/s13287-017-0590-6
29. Eggenhofer E, Benseler V, Kroemer A, Popp FC, Geissler EK, Schlitt HJ, et al. Mesenchymal stem cells are short-lived and do not migrate beyond the lungs after intravenous infusion. *Front Immunol.* (2012) 3:297. doi: 10.3389/fimmu.2012.00297
30. Fischer UM, Harting MT, Jimenez F, Monzon-Posadas WO, Xue H, Savitz SI, et al. Pulmonary passage is a major obstacle for intravenous stem cell delivery: the pulmonary first-pass effect. *Stem Cells Dev.* (2009) 18:683–91. doi: 10.1089/scd.2008.0253
31. de Witte SFH, Luk F, Sierra Parraga JM, Gargesha M, Merino A, Korevaar SS, et al. Immunomodulation by therapeutic mesenchymal stromal cells (MSC) is triggered through phagocytosis of msc by monocytic cells. *Stem Cells.* (2018) 36:602–15. doi: 10.1002/stem.2779
32. Ge W, Jiang J, Arp J, Liu W, Garcia B, Wang H. Regulatory T-cell generation and kidney allograft tolerance induced by mesenchymal stem cells associated with indoleamine 2,3-dioxygenase expression. *Transplantation.* (2010) 90:1312–20. doi: 10.1097/TP.0b013e3181fed001
33. Spaggiari GM, Capobianco A, Abdelrazik H, Becchetti F, Mingari MC, Moretta L. Mesenchymal stem cells inhibit natural killer-cell proliferation, cytotoxicity, and cytokine production: role of indoleamine 2,3-dioxygenase and prostaglandin E2. *Blood.* (2008) 111:1327–33. doi: 10.1182/blood-2007-02-074997
34. Popp FC, Eggenhofer E, Renner P, Slowik P, Lang SA, Kaspar H, et al. Mesenchymal stem cells can induce long-term acceptance of solid organ allografts in synergy with low-dose mycophenolate. *Transpl Immunol.* (2008) 20:55–60. doi: 10.1016/j.trim.2008.08.004
35. Obermajer N, Popp FC, Soeder Y, Haarer J, Geissler EK, Schlitt HJ, et al. Conversion of Th17 into IL-17A(neg) regulatory T cells: a novel mechanism in prolonged allograft survival promoted by mesenchymal stem cell-supported minimized immunosuppressive therapy. *J Immunol.* (2014) 193:4988–99. doi: 10.4049/jimmunol.1401776
36. Luz-Crawford P, Djouad F, Toupet K, Bony C, Franquesa M, Hoogduijn MJ, et al. Mesenchymal stem cell-derived interleukin 1 receptor antagonist promotes macrophage polarization and inhibits B cell differentiation. *Stem Cells.* (2016) 34:483–92. doi: 10.1002/stem.2254
37. Eggenhofer E, Popp FC, Mendicino M, Silber P, Van't Hof W, Renner P, et al. Heart grafts tolerized through third-party multipotent adult progenitor cells can be retransplanted to secondary hosts with no immunosuppression. *Stem Cells Transl Med.* (2013) 2:595–606. doi: 10.5966/sctm.2012-0166
38. Riquelme P, Haarer J, Kammler A, Walter L, Tomiuk S, Ahrens N, et al. TIGIT(+) iTregs elicited by human regulatory macrophages control T cell immunity. *Nat Commun.* (2018) 9:2858. doi: 10.1038/s41467-018-05167-8
39. Goncalves FDC, Luk F, Korevaar SS, Bouzid R, Paz AH, Lopez-Iglesias C, et al. Membrane particles generated from mesenchymal stromal cells modulate immune responses by selective targeting of pro-inflammatory monocytes. *Sci Rep.* (2017) 7:12100. doi: 10.1038/s41598-017-12121-z
40. Melief SM, Geutskens SB, Fibbe WE, Roelofs H. Multipotent stromal cells skew monocytes towards an anti-inflammatory interleukin-10-producing phenotype by production of interleukin-6. *Haematologica.* (2013) 98:888–95. doi: 10.3324/haematol.2012.078055
41. Luk F, de Witte SF, Korevaar SS, Roemeling-van Rhijn M, Franquesa M, Strini T, et al. Inactivated mesenchymal stem cells maintain immunomodulatory capacity. *Stem Cells Dev.* (2016) 25:1342–54. doi: 10.1089/scd.2016.0068
42. Zheng G, Huang R, Qiu G, Ge M, Wang J, Shu Q, et al. Mesenchymal stromal cell-derived extracellular vesicles: regenerative and immunomodulatory effects and potential applications in sepsis. *Cell Tissue Res.* (2018) 374:1–15. doi: 10.1007/s00441-018-2871-5
43. Ramasamy R, Fazekasova H, Lam EW, Soeiro I, Lombardi G, Dazzi F. Mesenchymal stem cells inhibit dendritic cell differentiation and function by preventing entry into the cell cycle. *Transplantation.* (2007) 83:71–6. doi: 10.1097/01.tp.0000244572.24780.54
44. Jiang XX, Zhang Y, Liu B, Zhang SX, Wu Y, Yu XD, et al. Human mesenchymal stem cells inhibit differentiation and function of monocyte-derived dendritic cells. *Blood.* (2005) 105:4120–6. doi: 10.1182/blood-2004-02-0586
45. Melief SM, Schrama E, Brugman MH, Tiemessen MM, Hoogduijn MJ, Fibbe WE, et al. Multipotent stromal cells induce human regulatory T cells through a novel pathway involving skewing of monocytes toward anti-inflammatory macrophages. *Stem Cells.* (2013) 31:1980–91. doi: 10.1002/stem.1432
46. Ko JH, Lee HJ, Jeong HJ, Kim MK, Wee WR, Yoon SO, et al. Mesenchymal stem/stromal cells precondition lung monocytes/macrophages to produce tolerance against allo- and autoimmunity in the eye. *Proc Natl Acad Sci USA.* (2016) 113:158–63. doi: 10.1073/pnas.1522905113
47. Schmidt A, Zhang XM, Joshi RN, Iqbal S, Wahlund C, Gabrielsson S, et al. Human macrophages induce CD4(+)Foxp3(+) regulatory T cells via binding and re-release of TGF-β. *Immunol Cell Biol.* (2016) 94:747–62. doi: 10.1038/icb.2016.34
48. Chang Y, de Nadai P, Azzaoui I, Morales O, Delhem N, Vorng H, et al. The chemokine CCL18 generates adaptive regulatory T cells from memory CD4+ T cells of healthy but not allergic subjects. *FASEB J.* (2010) 24:5063–72. doi: 10.1096/fj.10-162560
49. Azzaoui I, Yahia SA, Chang Y, Vorng H, Morales O, Fan Y, et al. CCL18 differentiates dendritic cells in tolerogenic cells able to prime regulatory T cells in healthy subjects. *Blood.* (2011) 118:3549–58. doi: 10.1182/blood-2011-02-338780
50. Eggenhofer E, Luk F, Dahlke MH, Hoogduijn MJ. The life and fate of mesenchymal stem cells. *Front Immunol.* (2014) 5:148. doi: 10.3389/fimmu.2014.00148
51. Kodelja V, Muller C, Politz O, Hakij N, Orfanos CE, Goerdt S. Alternative macrophage activation-associated CC-chemokine-1, a novel structural homologue of macrophage inflammatory protein-1 alpha with a Th2-associated expression pattern. *J Immunol.* (1998) 160:1411–8.
52. Vulcano M, Struyf S, Scapini P, Cassatella M, Bernasconi S, Bonecchi R, et al. Unique regulation of CCL18 production by maturing dendritic cells. *J Immunol.* (2003) 170:3843–9. doi: 10.4049/jimmunol.170.7.3843
53. Braza F, Dirou S, Forest V, Sauzeau V, Hassoun D, Chesne J, et al. Mesenchymal stem cells induce suppressive macrophages through phagocytosis in a mouse model of asthma. *Stem Cells.* (2016) 34:1836–45. doi: 10.1002/stem.2344
54. English K, Barry FP, Mahon BP. Murine mesenchymal stem cells suppress dendritic cell migration, maturation and antigen presentation. *Immunol Lett.* (2008) 115:50–8. doi: 10.1016/j.imlet.2007.10.002
55. Aggarwal S, Pittenger MF. Human mesenchymal stem cells modulate allogeneic immune cell responses. *Blood.* (2005) 105:1815–22. doi: 10.1182/blood-2004-04-1559
56. Liu X, Ren S, Ge C, Cheng K, Zenke M, Keating A, et al. Sca-1+Lin-CD117- mesenchymal stem/stromal cells induce the generation of novel IRF8-controlled regulatory dendritic cells through Notch-RBP-J signaling. *J Immunol.* (2015) 194:4298–308. doi: 10.4049/jimmunol.1402641
57. Krasnodembskaya A, Samarani G, Song Y, Zhuo H, Su X, Lee JW, et al. Human mesenchymal stem cells reduce mortality and bacteremia in gram-negative sepsis in mice in part by enhancing the phagocytic activity of blood monocytes. *Am J Physiol Lung Cell Mol Physiol.* (2012) 302:L1003–13. doi: 10.1152/ajplung.00180.2011
58. Miteva K, Pappritz K, El-Shafeey M, Dong F, Ringe J, Tschope C, et al. Mesenchymal stromal cells modulate monocytes trafficking in coxsackievirus b3-induced myocarditis. *Stem Cells Transl Med.* (2017) 6:1249–61. doi: 10.1002/sctm.16-0353
59. Eggenhofer E, Steinmann JF, Renner P, Slowik P, Piso P, Geissler EK, et al. Mesenchymal stem cells together with mycophenolate mofetil inhibit antigen presenting cell and T cell infiltration into allogeneic heart grafts. *Transpl Immunol.* (2011) 24:157–63. doi: 10.1016/j.trim.2010.12.002
60. Gieseke F, Bohringer J, Bussolari R, Dominici M, Handgretinger R, Muller I. Human multipotent mesenchymal stromal cells use galectin-1 to inhibit immune effector cells. *Blood.* (2010) 116:3770–9. doi: 10.1182/blood-2010-02-270777
61. Duffy MM, Ritter T, Ceredig R, Griffin MD. Mesenchymal stem cell effects on T-cell effector pathways. *Stem Cell Res Ther.* (2011) 2:34. doi: 10.1186/scrt75

62. Wang Q, Sun B, Wang D, Ji Y, Kong Q, Wang G, et al. Murine bone marrow mesenchymal stem cells cause mature dendritic cells to promote T-cell tolerance. *Scand J Immunol.* (2008) 68:607–15. doi: 10.1111/j.1365-3083.2008.02180.x
63. Bai L, Lennon DP, Eaton V, Maier K, Caplan AI, Miller SD, et al. Human bone marrow-derived mesenchymal stem cells induce Th2-polarized immune response and promote endogenous repair in animal models of multiple sclerosis. *Glia.* (2009) 57:1192–203. doi: 10.1002/glia.20841
64. Fiorina P, Jurewicz M, Augello A, Vergani A, Dada S, La Rosa S, et al. Immunomodulatory function of bone marrow-derived mesenchymal stem cells in experimental autoimmune type 1 diabetes. *J Immunol.* (2009) 183:993–1004. doi: 10.4049/jimmunol.0900803
65. Marson A, Kretschmer K, Frampton GM, Jacobsen ES, Polansky JK, MacIsaac KD, et al. Foxp3 occupancy and regulation of key target genes during T-cell stimulation. *Nature.* (2007) 445:931–5. doi: 10.1038/nature05478
66. Sakaguchi S, Sakaguchi N, Asano M, Itoh M, Toda M. Immunologic self-tolerance maintained by activated T cells expressing IL-2 receptor alpha-chains (CD25). Breakdown of a single mechanism of self-tolerance causes various autoimmune diseases. *J Immunol.* (1995) 155:1151–64.
67. Khosravi M, Karimi MH, Hossein Aghdaie M, Kalani M, Naserian S, Bidmeshkipour A. Mesenchymal stem cells can induce regulatory T cells via modulating miR-126a but not miR-10a. *Gene.* (2017) 627:327–36. doi: 10.1016/j.gene.2017.06.012
68. Engela AU, Hoogduijn MJ, Boer K, Litjens NHR, Betjes MGH, Weimar W, et al. Human adipose-tissue derived mesenchymal stem cells induce functional *de-novo* regulatory T cells with methylated FOXP3 gene DNA. *Clin Exp Immunol.* (2013) 173:343–54. doi: 10.1111/cei.12120
69. Meisel R, Zibert A, Laryea M, Gobel U, Daubener W, Dilloo D. Human bone marrow stromal cells inhibit allogeneic T-cell responses by indoleamine 2,3-dioxygenase-mediated tryptophan degradation. *Blood.* (2004) 103:4619–21. doi: 10.1182/blood-2003-11-3909
70. Cutler AJ, Limbani V, Girdlestone J, Navarrete CV. Umbilical cord-derived mesenchymal stromal cells modulate monocyte function to suppress T cell proliferation. *J Immunol.* (2010) 185:6617–23. doi: 10.4049/jimmunol.1002239
71. Davies LC, Heldring N, Kadri N, Le Blanc K. Mesenchymal stromal cell secretion of programmed death-1 ligands regulates T cell mediated immunosuppression. *Stem Cells.* (2017) 35:766–76. doi: 10.1002/stem.2509
72. Beyth S, Borovsky Z, Mevorach D, Liebergall M, Gazit Z, Aslan H, et al. Human mesenchymal stem cells alter antigen-presenting cell maturation and induce T-cell unresponsiveness. *Blood.* (2005) 105:2214–9. doi: 10.1182/blood-2004-07-2921
73. Franquesa M, Mensah FK, Huizinga R, Strini T, Boon L, Lombardo E, et al. Human adipose tissue-derived mesenchymal stem cells abrogate plasmablast formation and induce regulatory B cells independently of T helper cells. *Stem Cells.* (2015) 33:880–91. doi: 10.1002/stem.1881
74. Rosser EC, Mauri C. Regulatory B cells: origin, phenotype, and function. *Immunity.* (2015) 42:607–12. doi: 10.1016/j.immuni.2015.04.005
75. Carter NA, Vasconcellos R, Rosser EC, Tulone C, Munoz-Suano A, Kamanaka M, et al. Mice lacking endogenous IL-10-producing regulatory B cells develop exacerbated disease and present with an increased frequency of Th1/Th17 but a decrease in regulatory T cells. *J Immunol.* (2011) 186:5569–79. doi: 10.4049/jimmunol.1100284
76. Sato K, Ozaki K, Oh I, Meguro A, Hatanaka K, Nagai T, et al. Nitric oxide plays a critical role in suppression of T-cell proliferation by mesenchymal stem cells. *Blood.* (2007) 109:228–34. doi: 10.1182/blood-2006-02-002246
77. Chen HH, Lin KC, Wallace CG, Chen YT, Yang CC, Leu S, et al. Additional benefit of combined therapy with melatonin and apoptotic adipose-derived mesenchymal stem cell against sepsis-induced kidney injury. *J Pineal Res.* (2014) 57:16–32. doi: 10.1111/jpi.12140
78. Sung PH, Chang CL, Tsai TH, Chang LT, Leu S, Chen YL, et al. Apoptotic adipose-derived mesenchymal stem cell therapy protects against lung and kidney injury in sepsis syndrome caused by cecal ligation puncture in rats. *Stem Cell Res Ther.* (2013) 4:155. doi: 10.1186/scrt385
79. Yip HK, Chang YC, Wallace CG, Chang LT, Tsai TH, Chen YL, et al. Melatonin treatment improves adipose-derived mesenchymal stem cell therapy for acute lung ischemia-reperfusion injury. *J Pineal Res.* (2013) 54:207–21. doi: 10.1111/jpi.12020
80. Galleu A, Riffo-Vasquez Y, Trento C, Lomas C, Dolcetti L, Cheung TS, et al. Apoptosis in mesenchymal stromal cells induces *in vivo* recipient-mediated immunomodulation. *Sci Transl Med.* (2017) 9:eaam7828. doi: 10.1126/scitranslmed.aam7828
81. Sun CK, Yen CH, Lin YC, Tsai TH, Chang LT, Kao YH, et al. Autologous transplantation of adipose-derived mesenchymal stem cells markedly reduced acute ischemia-reperfusion lung injury in a rodent model. *J Transl Med.* (2011) 9:118. doi: 10.1186/1479-5876-9-118
82. Gonzalez-Rey E, Anderson P, Gonzalez MA, Rico L, Buscher D, Delgado M. Human adult stem cells derived from adipose tissue protect against experimental colitis and sepsis. *Gut.* (2009) 58:929–39. doi: 10.1136/gut.2008.168534
83. Johnson CL, Soeder Y, Dahlke MH. Concise review: mesenchymal stromal cell-based approaches for the treatment of acute respiratory distress and sepsis syndromes. *Stem Cells Transl Med.* (2017) 6:1141–51. doi: 10.1002/sctm.16-0415
84. Lu W, Fu C, Song L, Yao Y, Zhang X, Chen Z, et al. Exposure to supernatants of macrophages that phagocytized dead mesenchymal stem cells improves hypoxic cardiomyocytes survival. *Int J Cardiol.* (2013) 165:333–40. doi: 10.1016/j.ijcard.2012.03.088
85. Thum T, Bauersachs J, Poole-Wilson PA, Volk HD, Anker SD. The dying stem cell hypothesis: immune modulation as a novel mechanism for progenitor cell therapy in cardiac muscle. *J Am Coll Cardiol.* (2005) 46:1799–802. doi: 10.1016/j.jacc.2005.07.053
86. Hoogduijn MJ, Verstegen MM, Engela AU, Korevaar SS, Roemeling-van Rhijn M, Merino A, et al. No evidence for circulating mesenchymal stem cells in patients with organ injury. *Stem Cells Dev.* (2014) 23:2328–35. doi: 10.1089/scd.2014.0269
87. Kidd S, Spaeth E, Dembinski JL, Dietrich M, Watson K, Klopp A, et al. Direct evidence of mesenchymal stem cell tropism for tumor and wounding microenvironments using *in vivo* bioluminescent imaging. *Stem Cells.* (2009) 27:2614–23. doi: 10.1002/stem.187
88. Ankrum JA, Ong JF, Karp JM. Mesenchymal stem cells: immune evasive, not immune privileged. *Nat Biotechnol.* (2014) 32:252–60. doi: 10.1038/nbt.2816
89. Le Blanc K, Tammik L, Sundberg B, Haynesworth SE, Ringden O. Mesenchymal stem cells inhibit and stimulate mixed lymphocyte cultures and mitogenic responses independently of the major histocompatibility complex. *Scand J Immunol.* (2003) 57:11–20. doi: 10.1046/j.1365-3083.2003.01176.x
90. Casiraghi F, Azzollini N, Cassis P, Imberti B, Morigi M, Cugini D, et al. Pretransplant infusion of mesenchymal stem cells prolongs the survival of a semiallogeneic heart transplant through the generation of regulatory T cells. *J Immunol.* (2008) 181:3933–46. doi: 10.4049/jimmunol.181.6.3933
91. Zhou HP, Yi DH, Yu SQ, Sun GC, Cui Q, Zhu HL, et al. Administration of donor-derived mesenchymal stem cells can prolong the survival of rat cardiac allograft. *Transplant Proc.* (2006) 38:3046–51. doi: 10.1016/j.transproceed.2006.10.002
92. Hoogduijn MJ, Roemeling-van Rhijn M, Engela AU, Korevaar SS, Mensah FK, Franquesa M, et al. Mesenchymal stem cells induce an inflammatory response after intravenous infusion. *Stem Cells Dev.* (2013) 22:2825–35. doi: 10.1089/scd.2013.0193
93. Klyushnenkova E, Mosca JD, Zernetkina V, Majumdar MK, Beggs KJ, Simonetti DW, et al. T cell responses to allogeneic human mesenchymal stem cells: immunogenicity, tolerance, and suppression. *J Biomed Sci.* (2005) 12:47–57. doi: 10.1007/s11373-004-8183-7
94. Eliopoulos N, Stagg J, Lejeune L, Pommey S, Galipeau J. Allogeneic marrow stromal cells are immune rejected by MHC class I- and class II-mismatched recipient mice. *Blood.* (2005) 106:4057–65. doi: 10.1182/blood-2005-03-1004
95. Zangi L, Margalit R, Reich-Zeliger S, Bachar-Lustig E, Beilhack A, Negrin R, et al. Direct imaging of immune rejection and memory induction by allogeneic mesenchymal stromal cells. *Stem Cells.* (2009) 27:2865–74. doi: 10.1002/stem.217
96. Djouad F, Plence P, Bony C, Tropel P, Apparailly F, Sany J, et al. Immunosuppressive effect of mesenchymal stem cells favors

tumor growth in allogeneic animals. *Blood.* (2003) 102:3837–44. doi: 10.1182/blood-2003-04-1193

97. Zhu W, Xu W, Jiang R, Qian H, Chen M, Hu J, et al. Mesenchymal stem cells derived from bone marrow favor tumor cell growth *in vivo. Exp Mol Pathol.* (2006) 80:267–74. doi: 10.1016/j.yexmp.2005.07.004
98. Djouad F, Bony C, Apparailly F, Louis-Plence P, Jorgensen C, Noel D. Earlier onset of syngeneic tumors in the presence of mesenchymal stem cells. *Transplantation.* (2006) 82:1060–6. doi: 10.1097/01.tp.0000236098.13804.0b
99. Yu PF, Huang Y, Han YY, Lin LY, Sun WH, Rabson AB, et al. TNFα-activated mesenchymal stromal cells promote breast cancer metastasis by recruiting CXCR2(+) neutrophils. *Oncogene.* (2017) 36:482–90. doi: 10.1038/onc.2016.217
100. Mishra PJ, Mishra PJ, Humeniuk R, Medina DJ, Alexe G, Mesirov JP, et al. Carcinoma-associated fibroblast-like differentiation of human mesenchymal stem cells. *Cancer Res.* (2008) 68:4331–9. doi: 10.1158/0008-5472.CAN-08-0943
101. Spaeth EL, Dembinski JL, Sasser AK, Watson K, Klopp A, Hall B, et al. Mesenchymal stem cell transition to tumor-associated fibroblasts contributes to fibrovascular network expansion and tumor progression. *PLoS ONE.* (2009) 4:e4992. doi: 10.1371/journal.pone.0004992
102. Shi Y, Du L, Lin L, Wang Y. Tumour-associated mesenchymal stem/stromal cells: emerging therapeutic targets. *Nat Rev Drug Discov.* (2017) 16:35–52. doi: 10.1038/nrd.2016.193
103. Quante M, Tu SP, Tomita H, Gonda T, Wang SS, Takashi S, et al. Bone marrow-derived myofibroblasts contribute to the mesenchymal stem cell niche and promote tumor growth. *Cancer Cell.* (2011) 19:257–72. doi: 10.1016/j.ccr.2011.01.020
104. Jung Y, Kim JK, Shiozawa Y, Wang J, Mishra A, Joseph J, et al. Recruitment of mesenchymal stem cells into prostate tumours promotes metastasis. *Nat Commun.* (2013) 4:1795. doi: 10.1038/ncomms2766
105. Ren G, Zhao X, Wang Y, Zhang X, Chen X, Xu C, et al. CCR2-dependent recruitment of macrophages by tumor-educated mesenchymal stromal cells promotes tumor development and is mimicked by TNFα. *Cell Stem Cell.* (2012) 11:812–24. doi: 10.1016/j.stem.2012.08.013
106. Zamarron BF, Chen W. Dual roles of immune cells and their factors in cancer development and progression. *Int J Biol Sci.* (2011) 7:651–8. doi: 10.7150/ijbs.7.651

6

Metabolism in Human Mesenchymal Stromal Cells: A Missing Link Between hMSC Biomanufacturing and Therapy?

Xuegang Yuan[1], Timothy M. Logan[2,3] and Teng Ma[1,2]*

[1] Department of Chemical and Biomedical Engineering, Florida State University, Tallahassee, FL, United States, [2] Institute of Molecular Biophysics, Florida State University, Tallahassee, FL, United States, [3] Department of Chemistry and Biochemistry, Florida State University, Tallahassee, FL, United States

***Correspondence:**
Teng Ma
teng@eng.fsu.edu

Human mesenchymal stem cells (hMSCs) are the most commonly-tested adult stem cells in cell therapy. While the initial focus for hMSC clinical applications was to exploit their multi-potentiality for cell replacement therapies, it is now apparent that hMSCs empower tissue repair primarily by secretion of immuno-regulatory and pro-regenerative factors. A growing trend in hMSC clinical trials is the use of allogenic and culture-expanded cells because they are well-characterized and can be produced in large scale from specific donors to compensate for the donor pathological condition(s). However, donor morbidity and large-scale expansion are known to alter hMSC secretory profile and reduce therapeutic potency, which are significant barriers in hMSC clinical translation. Therefore, understanding the regulatory mechanisms underpinning hMSC phenotypic and functional property is crucial for developing novel engineering protocols that maximize yield while preserving therapeutic potency. hMSC are heterogenous at the level of primary metabolism and that energy metabolism plays important roles in regulating hMSC functional properties. This review focuses on energy metabolism in regulating hMSC immunomodulatory properties and its implication in hMSC sourcing and biomanufacturing. The specific characteristics of hMSC metabolism will be discussed with a focus on hMSC metabolic plasticity and donor- and culture-induced changes in immunomodulatory properties. Potential strategies of modulating hMSC metabolism to enhance their immunomodulation and therapeutic efficacy in preclinical models will be reviewed.

Keywords: MSCs (mesenchymal stromal cells), immunomodulation, metabolic plasticity, biomanufacturing, therapeutic potentials

BACKGROUND

Human mesenchymal stem or stromal cells (hMSCs) are the most commonly-tested adult stem cells in experimental cell therapy and have been used in more than 50% of clinical trials using stem cells since 2000. Clinically, the most beneficial aspects of hMSCs are their multilineage differentiation for damage tissue replacement and their ability to empower tissue repair by secretion of immuno-regulatory and pro-regenerative factors. Clinical applications of hMSC-based therapy initially exploited their multi-potentiality but increasingly focused on their secretion of immunomodulatory and trophic factors. In this immunoregulatory scenario, hMSCs promote tissue regeneration by coordinating an anti-inflammatory response through communication with the host's inflammatory microenvironment, making hMSC logical candidates for the treatment of immune disorders and inflammatory diseases. In contrast to the promising results from preclinical studies and small-scale clinical trials, the clinical outcomes using manufactured hMSC have been inconsistent and suboptimal (1–3). The close scrutiny of the discrepant outcome from these studies suggests that culture expansion, cryopreservation, and inappropriate delivery routes and dosage, are major factors that adversely influence hMSC's therapeutic efficacy (1, 4). For example, in graft-vs.-host disease, 1-year survival for patients receiving hMSC at passages 1–2 was 75% in contrast to 21% using hMSC at passages 3–4 (5). In clinical application, hMSC therapeutics are often cryobanked as "off-the-shelf" products prior to transfusion. However, cryopreservation is known to reduce hMSC immunomodulatory properties as a result of cellular stress such as cellular acidosis and metabolic uncoupling induced during freezing and thaw cycles (6, 7). It is worth noting that many of these functional changes are not readily reflected in the assessment of the minimal criterial and potency assays, suggesting the need for the identification of additional surrogate markers and regulatory pathways that can be readily assessed, and modulated to restore hMSC therapeutic potency.

This immunoregulatory function is achieved through rapid hMSC phenotype polarization and sustained production of immunoregulatory factors in response to inflammatory stimuli, which requires cellular plasticity and metabolic fitness to enable this response (8). The metabolic fitness of hMSCs is dependent on donor age and morbidity, and can be significantly altered by culture conditions imposed on the cells during *in vitro* expansion. Each of these factors, and probably other, currently unknown factors, can reduce hMSC immunomodulatory capacity and, therefore, reduce their therapeutic potency. A typical clinical dose of hMSCs is on the order of tens- to hundreds-of millions of cells per patient and, with hMSCs being approved for use in an ever-expanding number of clinical indications, it is estimated 300 trillion (300×10^{12}) hMSCs will be needed annually by 2030 (9, 10). Current engineering protocols isolate hMSCs from adult donors and expand them under nutrient-rich conditions that significantly promote proliferation but ultimately and inadvertently reduce stemness and therapeutic potency. To compensate for the culture-induced decline of hMSC therapeutics potency, non-genetic preconditioning such as hypoxia pretreatment or 3D culture has demonstrated significant potential in restoring hMSC properties (1). To better preserve hMSC property during cryopreservation, hydrogen peroxide preconditioning has also been shown to enhance adipose derived stem cells (ASCs) resistance and survival under oxidative stress (7). While many of these preconditioning strategies have demonstrated effectiveness in restoring hMSC functional properties, the regulatory mechanism remains to be fully understood for widespread implementation in hMSC manufacturing and translation (11).

Among the core pathways to improving hMSC function, metabolism has emerged as an important hub. In their native environment, hMSCs are present in a quiescent state characterized by low proliferation and high multi-potentiality, which is maintained throughout adult life. In this state, stem cells appear to be primarily glycolytic, with "young" mitochondria maintained by active autophagy and mitophagy (12, 13). However, numerous studies show that transferring hMSC into the nutrient-rich artificial culture environment that promotes rapid proliferation reconfigures their central energy metabolism to become significantly more dependent on oxidative phosphorylation (OXPHOS) to support the rapid proliferation (14, 15). The high OXPHOS-fueled metabolic profile also results in accumulation of cytotoxic metabolic byproducts, including reactive oxygen species (ROS) (16), that reduce the basal autophagy and mitophagy rate while simultaneously increasing the fraction of senescent cells with reduced clinical potency (17). Similar metabolic alterations have been reported for hMSC undergoing large-scale expansion in bioreactor systems (18). The influence of these metabolic changes on hMSC functional properties has just begun to be revealed. Beyond providing cells with building blocks and energy source to power cellular processes, the energy metabolism and intermediate metabolites play important roles in shaping cellular functional properties (19). Therefore, a clearer understanding of how hMSC metabolism is affected by long-term culture and how specific metabolic states impact immunomodulatory function may be the missing link between engineering practices for expansion and consistent, predictable outcomes in hMSC-based therapies. This review focuses on the role of energy metabolism in regulating hMSC immunomodulatory properties and discusses its implication for hMSC large-scale manufacturing and therapy. The specific characteristics of hMSC metabolism, culture-induced metabolic changes, and the metabolism underpinning hMSC's immunomodulatory properties will be discussed. We will also review and discuss recent studies on hMSC metabolic modulation of their immunomodulatory properties and therapeutic efficacy.

HMSCS METABOLIC PLASTICITY AND CULTURE-INDUCED METABOLIC CHANGES

Once viewed as a mere consequence of the cellular state, energy metabolism not only provides energy and substrates for cell growth but is also intimately linked to cell signaling and control of cell fate (16, 20, 21). As depicted in **Figure 1**, freshly isolated hMSCs mainly compressed a clonogenic subset with high glycolytic activity. Expansion under currently adopted engineering protocols reduces the fraction of clonogenic/glycolytic cells and increases the fraction of mature/OXPHOS-based cells. In fact, this phenotypic and metabolic heterogeneity exists even at the clonal level of hMSCs (22). It has been suggested that changes in the microenvironment

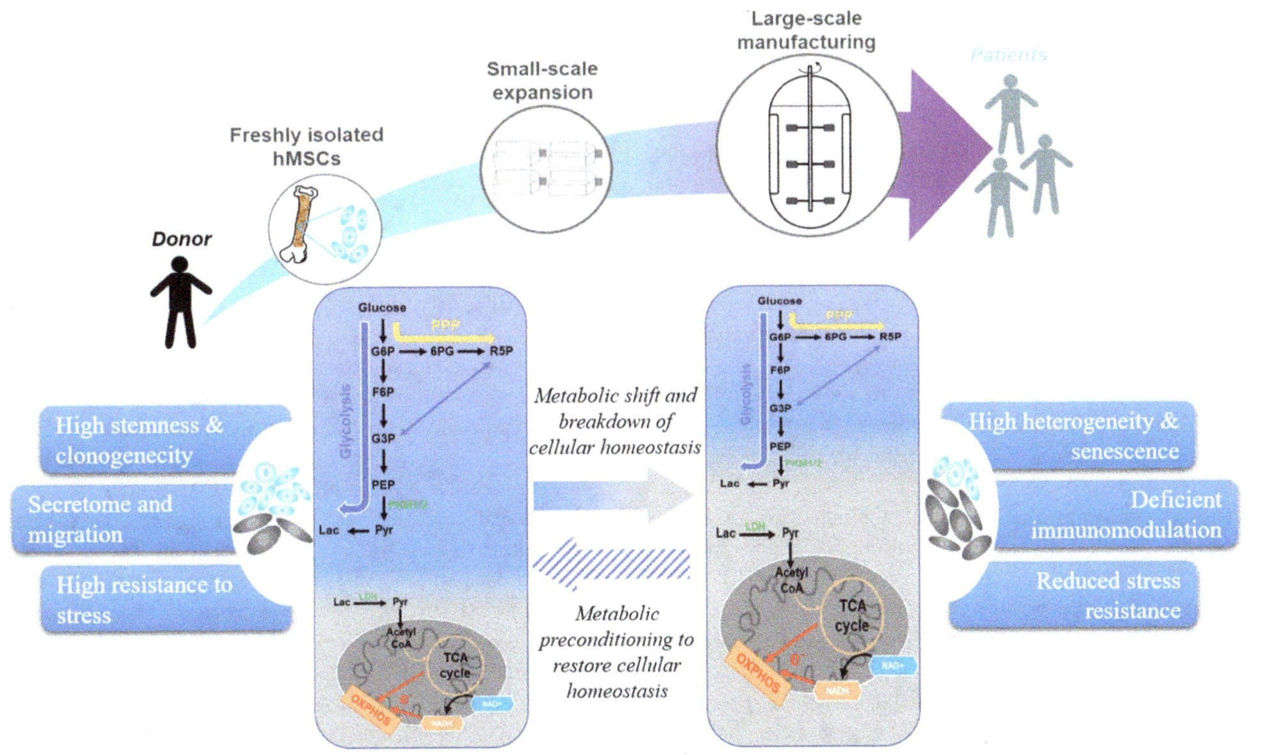

FIGURE 1 | Current hMSC manufacturing practices lead to metabolic shift that reduces therapeutic properties. hMSC manufacturing utilizes freshly isolated hMSCs and expand under artificial environment to obtain sufficient cell number for clinical application. However, external stresses during replicative expansion, and cryopreservation shift hMSC metabolism from glycolysis toward OXPHOS, which increases senescent subset and contributes to a breakdown of cellular homeostasis. Metabolic preconditioning targeting specific pathways can restore hMSC cellular homeostasis and enhance their therapeutic potency.

and accumulated replicative stress experienced during extended *in vitro* expansion accelerate hMSC cellular and metabolic heterogeneity by reconfiguring energy metabolism (21, 22). As mentioned above, hMSCs exhibit a quiescent, glycolytic phenotype in a hypoxic *in vivo* niche such as bone marrow (16, 21). This particular metabolic state may serve to preserve hMSC cellular homeostasis by minimizing ROS production (16, 21, 23), because the high glycolytic flux is also cytoprotective due to increased generation of antioxidant precursors from the pentose phosphate pathway (PPP) (24). Upon removal from this hypoxic, *in vivo* niche and transfer to the oxygenated, nutrient-rich environment, this quiescent, glycolytic phenotype is no longer of benefit to hMSC (11, 22). During early passages, hMSCs still maintain to an aerobic glycolysis profile, despite its low efficiency for ATP production (25). However, protracted expansion of hMSCs under this nutrient-rich environment induces a metabolic shift from glycolysis toward OXPHOS (22, 26). This metabolic shift is associated with increased coupling between glycolysis and TCA cycle and significantly increased production of ROS and dysfunctional mitochondria (21). A consequence of this culture-induced metabolic shift is a breakdown of cellular homeostasis, characterized by reduced autophagy/mitophagy activity and increased senescence (22, 27–29). A recent proteomic analysis identified several proteins involved in energy metabolism, mitochondrial dysfunction, OXPHOS, and nuclear factor erythroid 2-related factor 2 -mediated oxidative stress response as being among the top canonical pathways that are altered in large-scale production of hMSC in bioreactor system (18). In addition, intrinsic biological factors such as donor disease or chronological age also alter hMSC metabolic profile. For example, hMSCs from obese donors have a higher number of defective mitochondria with reduced dependence on glycolysis and an altered metabolic profile compared to cells obtained from healthy controls (30, 31). hMSCs from aged donors also exhibit greater population heterogeneity with lower mitochondrial-to-cytoplasm area ratio, reduced level of manganese superoxide dismutase (MnSOD) expression, and accumulated ROS compared with cells obtained from young donors (32). Furthermore, hMSCs from donors with age-related atherosclerosis exhibited impaired mitochondrial function that also contributed to metabolic alterations compared to cells obtained from healthy, young donors (33). Culture-based interventions aimed at restoring the mitochondrial function of atherosclerotic-hMSCs by treatment with ROS scavengers effectively restored their immunosuppressive ability to that of healthy donors (33). This study directly linked metabolic alteration due to donor morbidity to reduced hMSC therapeutic potency and demonstrated that metabolic treatment can restore hMSCs to a state capable of delivering the desired therapeutic outcomes. In contrast, the extent of influence of culture-induced metabolic changes on hMSC therapeutic potency is largely unknown because few studies have characterized the metabolic profiles of hMSC used in therapeutic applications. Beyond supplying energy and anabolic production of macromolecules, metabolic circuits engage genetic programs to regulate cellular events and phenotypic and functional properties, reflecting the metabolic substrate, specific pathways, and environmental conditions (23). Although large scale hMSC biomanufacturing often entails significant changes of culture conditions such as media composition and substrates (e.g., sugar *vs.* fatty acids), supplements (e.g., fetal bovine serum *vs.* platelet lysate), expansion protocols (e.g., multi-well plates vs. bioreactor), and cryopreservation (e.g., cryogen and freeze-thaw cycles), few studies have elucidated the influence of these changes on hMSC functional properties and clinical outcome (1, 6, 34–38).

Preclinical studies on the mechanistic connections between metabolism and hMSC phenotype provide specific molecular targets and pathways that can be modulated to maintain, or remodel metabolic profiles of culture expanded hMSCs. For example, hypoxia culture has been widely used in hMSC expansion as it better preserves the clonogenic subset during hMSC expansion by maintaining glycolysis and suppressing TCA cycle and OXPHOS activity (15, 26, 39, 40), most likely through activation of hypoxia-inducible factor (HIF) related genes (41). As a consequence of the glycolytic metabolic profile, hypoxia culture also activates hMSC autophagy while reducing senescence and preserving hMSC functional properties by maintaining cellular homeostasis *in vitro* (40, 42). However, hMSCs are not only sensitive to the absolute oxygen level but also to the fluctuation in oxygen tension, which significantly complicates implementing hypoxia as an engineering protocol for maintaining stem cell properties in large scale manufacturing (43, 44). More recently, the benefits of hMSC hypoxia culture have been recapitulated at ambient oxygen tensions by treatment of small molecule modulators that target specific metabolic regulatory pathways identified in hypoxia studies. Since HIF is a downstream effector of mTOR, inhibition of Akt/mTOR signaling pathway with rapamycin and LY294002 reduced mitochondrial activity, and glycolysis–TCA coupling, prevented culture-induced senescence (20, 45). These studies highlight the importance for a balance between AKT/mTOR activity and intracellular signaling for maintaining glycolytic metabolism to preserve stem cell functions. Non-hypoxic stabilization of HIF-1α using hypoxia mimetics such as desferoxamine (DFO), ciclopirox olamine, the HIF-prolyl hydroxylase (PHD) inhibitor FG-4497, or cobalt chloride (CoCl2) have also been shown to overcome the challenge of controlling ambient oxygen tension to effectively maintain MSC properties (45–47).

hMSCs have intriguing properties of self-assembly into three-dimensional (3D) aggregates mediated by cell-cell and cell-ECM interaction, which better preserve hMSC phenotypic properties compared to their 2D counterparts (48). The benefits of 3D aggregation culture in preserving hMSC stemness and enhancing secretion of immunomodulatory cytokines can also be attributed to aggregation-induced metabolic reconfiguration, which inhibits mitochondrial activity and increases glycolysis with increased anaplerotic flux (28, 48–50). While the metabolic reprograming in 3D aggregates has been commonly attributed to oxygen diffusion limitation, our recent studies reveal that actin-mediated cellular compaction activates PI3K/Akt pathway and induces metabolic shift toward glycolysis (48, 51), highlighting energy metabolism as a signaling hub in regulating hMSC functions during *in vitro* culture.

THE ROLE OF METABOLISM IN THE HMSC IMMUNE RESPONSE AND IMMUNOMODULATION

As mentioned earlier in this review, an attractive feature of hMSC as a cell therapy product is their immunomodulatory properties in response to environmental stimuli from surrounding tissues resulting in the secretion of beneficial cytokines and cellular components such as microRNA and extracellular vesicles; as might be expected, these properties are significantly influenced by metabolism. As shown in **Figure 2**, in the presence of inflammatory cytokines such as interferon-γ (IFN-γ) alone or in combination with tumor necrosis factor-α (TNF-α) or interlukin-1 (IL-1), MSCs secrete chemokines such as CXC-chemokine receptor 3 (CXCR3), CC-chemokine receptor 5 (CCR5) ligands, CXC-chemokine ligand 9 (CXCL9), CXCL10, and CXCL11 (52, 53), which attract immune cells via chemotaxis (52, 54–56). Recruited T cells are inhibited by activated hMSCs through the secretion of indoleamine 2,3-dioxygenase (IDO), a catabolic enzyme that regulates tryptophan metabolism (54). Besides IDO, hMSC is also a potent source of other soluble immunosuppressive factors, such as nitric oxide (NO, in rodent MSCs), prostaglandin E2 (PGE2), transforming growth factor-β1 (TGF-β1), hepatocyte growth factor (HGF), interleukins and cyclooxygenase 2 (COX-2). MSCs promote the polarization of macrophages from a pro-inflammatory M1 phenotype to an anti-inflammatory M2 phenotype and suppress IL-6 and TNF-α production in macrophages through secretion of PGE2 and IDO (57–61). Similarly, co-culture with MSCs inhibits the maturation and activation of antigen-presenting dendritic cells (DCs) and reduces B cell proliferation by increased production of IL-10, chemokine receptors, and immunoglobulins (62–68). Compared to B cells, MSCs inhibit T cell proliferation regardless of their lineage difference (naïve, CD4+ or CD8+ lineage) (20, 69). Moreover, MSCs from multiple sources exhibit similar effects on inducing apoptosis of T cells (70). IDO, HGF, TGF-β, PGE2, and PD-1/PD-L1 ligation from MSCs all contribute to the immunosuppressive effect (20, 71–75). As the hMSC secretome is central to their immunomodulatory properties, preserving the secretory properties of hMSC during long-term expansion has become an important engineering challenge in hMSC translation (4).

The profile of MSC secretome is tightly regulated by intrinsic (e.g., donor morbidity and aging) or external (e.g., culture conditions) factors through metabolic regulation. Adipose-derived stem cells showed a higher secretion of immunomodulatory cytokines, such as IL-6 and TGF compared to bone marrow (BM)-MSCs due to higher metabolic activity (76). Conversely, compared to cells obtained from lean

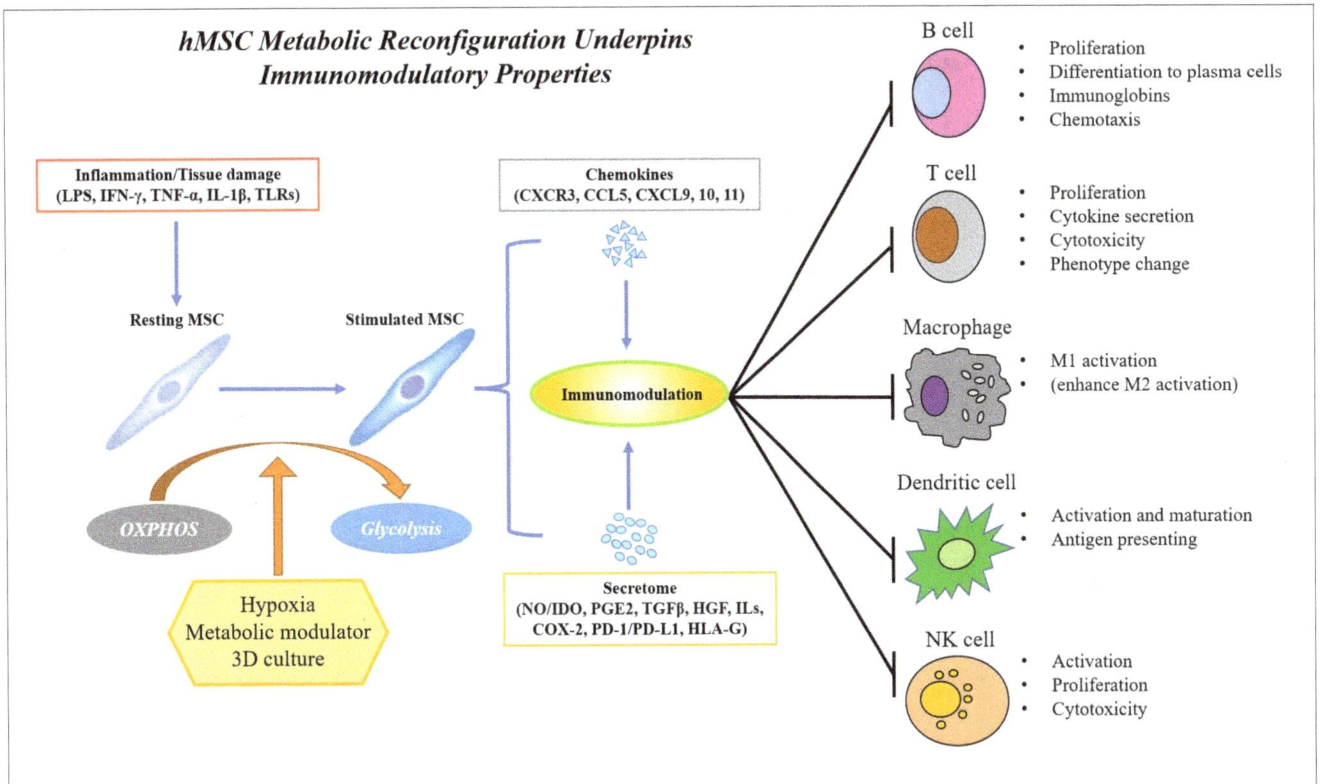

FIGURE 2 | hMSC immunomodulation requires polarization by inflammatory environment and is achieved by the secretion of immunomodulatory factors such as chemokines and cytokines, extracellular vesicles and exosome, and direct cell-cell contact. hMSC's immunomodulatory property requires a metabolic reconfiguration toward aerobic glycolysis to sustain the production of secretome. hMSC's immunomodulatory capacity can be enhanced by modulation of hMSC metabolism via hypoxia, small molecule metabolic mediators, or 3D aggregation.

patients, ASCs from obese patients exhibiting reduced glycolytic activity and upregulated expression of inflammatory genes and increased secretion of inflammatory cytokines such as IL-6 and IL-8 (30, 77, 78), making them less effective in suppressing lymphocyte proliferation and activating the M2 macrophage phenotype (79). The expression of inflammation-response genes has been reported to decline in hMSC from aged donors as they have altered metabolic profile, although conflicting results were also reported (80–82). Interestingly, recent clinical study has shown that hMSCs isolated from patient with atherosclerosis have impaired mitochondrial functional properties, contributing to the reduced suppression of T cell proliferation (33). To identify the specific role of metabolism in sustaining hMSC secretion of immunomodulatory cytokines, we observed a pronounced shift in hMSC energy metabolism toward glycolysis upon activation by IFN-γ treatment (20) and showed that inhibition of these metabolic changes prevented the production of the key immunosuppressive cytokines such as IDO and PGE2. We also demonstrated that mitochondrial ROS and Akt/mTOR signaling play a critical role in initiating this metabolic remodeling in response to inflammatory stimuli (20, 55, 83). It is not surprising that culture conditions favoring glycolysis also potentiate MSC immunomodulatory properties. As mentioned above, hypoxia culture upregulates MSC secretion of IDO, PGE2, PD-L1, ILs, etc., and enhances inhibition of CD4+/CD8+ T lymphocyte proliferation while increasing the regulatory T cells (T-reg) populations (84–86). In addition to the MSC secretome, direct cell-cell contact could directly interact with immune cells (87–92), but the influence of specific MSC metabolic profiles on these interactions remain to be investigated.

An important subset of the MSC secretome is extracellular vehicles (EVs) and exosomes, which are small membrane vesicles (ranging from 50 to 1,000 nm) derived from multi-vesicular bodies or from the plasma membrane and are enriched with proteins, lipids, and nucleic acids for intercellular communication including immune regulation (93, 94). MSC-derived EVs and exosomes have been tested in several disease models such as acute kidney injury, experimental autoimmune encephalomyelitis, type-1 diabetes mellitus and myocardial ischemic injury through microRNA (miRNA) regulation (95–99). Mechanistically, MSC-derived EVs and exosomes are enriched with miRNA such as miR-15a, miR-15b, miR-16 that inhibit CXC ligands and suppress chemotaxis of macrophage (95, 100). Recent studies have shown that metabolism regulates EV biogenesis and cargo composition through regulation of endosomal secretion pathways. In tumor cells, pyruvate kinase type M2 (PKM2), the rate limiting enzyme in glycolysis, acts as a protein kinase to promote exosome release via phosphorylating snaptosome-associated protein 23 (SNAP-23), which mediates the fusion of intracellular vesicles with membrane compartments

TABLE 1 | Metabolic enhancement of mesenchymal stem cell-mediated immunomodulation in preclinical studies.

MSC source	Pre-treatment	Metabolic targets	Functional enhancement	*In vivo* study	References
Rats adipose	Hypoxia	HIF	Enhance secretion of angiogenesis and neuroprotection cytokines	Improve functional recovery of DED rat; enhanced eNOS expression; increased expression of endothelial and smooth muscle markers;	(107)
Mice bone marrow	Tetrahydrocannabinol or with AM630	Mitochondrial respiration	Increased MSC IL-10 production; activated MSC ERK signaling pathway; enhance immunomodulation of microglia;	Reduced thermal hyperalgesia and mechanical allodynia response; reduced inflammation in chronic constriction injury model;	(108)
Human bone marrow	Hypoxia	HIF	Upregulated mRNA levels of IL-1β, IL-6, IL-8, and TGF β-1; mitogenic, chemoattractive and angiogenic paracrine effects	Enhance Balb/c nude mouse skin wounds healing process; Increased macrophage recruitment at wound site;	(109)
Human umbilical cord blood	Hypoxia and calcium ion	HIF	Reduced secretion of IL-6 and IL-8 and increased secretion of PGE2. Improve the inhibition of T cell proliferation.	Improve survival of humanized GVHD mouse model; decreased immune cell infiltration and characteristic tissue injuries	(110)
Human umbilical cords	IL-1β	Glycolysis	Upregulated COX-2, IL-6, IL-8 gene expression; Enhanced COX-2 protein expression;	Modulate the balance of macrophage polarization; reduced local inflammation and improve migration to DSS-induced murine colitis	(111)
Human bone marrow or umbilical cord blood	IFN-γ	Glycolysis and mitochondria	Increased gene expression of CXCL9, CXCL10, CCL8, and IDO. Enhanced secretion of IDO and inhibition of hPBMCs proliferation.	Reduced the symptoms of graft-versus-host disease (GVHD) in NOD-SCID mice and improve survival rate;	(112)
Human bone marrow	3D aggregation	Mitochondria	Increased gene expression of CXCR4, TSG-6, STC-1, IL-24, TRAIL, and LIF; elevated expression of TSG-6, LIF, and STC-1; Decreased macrophage activation	Decreased the protein content of the lavage fluid and neutrophil activity; reduced levels of the proinflammatory molecules in mouse model for peritonitis	(113)

(101). HIF-1α overexpression in hMSCs significantly enhanced exosome secretion and also upregulated Notch ligand Jagged-1, which induce dendritic cell maturation and regulatory T cell proliferation (102, 103). Hypoxia is known to mediate the expression of Rab22 and Rab20 and ceramide production, which are associated with EV formation and secretion (104). MSCs also manage intracellular oxidative stress by targeting depolarized mitochondria to the plasma membrane and unload partially depolarized mitochondria as EVs to enhance cell survival (105). Conversely, secreted exosomes regulate metabolism of recipient cells. For example, MSC exosome carry a cargo rich in active glycolytic enzymes and promote ischemic myocardium repair by enhancing glycolytic flux to compensate for the reduced OXPHOS in defective mitochondria (106). Future studies are needed to establish the mechanistic connection between hMSC metabolism and biogenesis and cargo composition of hMSC-derived EV.

TARGETING HMSC METABOLISM TO ENHANCE IMMUNOMODULATORY PROPERTIES IN PRECLINICAL STUDIES

Investigating the mechanistic connections between metabolism and immunomodulation has identified specific molecular targets that can be modulated to overcome metabolic deficiency due to donor age and morbidity, and to enhance the therapeutic potency of culture-expanded hMSCs. **Table 1** summarized preclinical studies of enhanced MSC immunomodulation via metabolic regulation. As discussed above, hypoxia treatment is commonly used to enhance hMSC immunoregulatory properties by increasing hMSC anti-inflammatory properties while attenuating the secretion of pro-inflammatory cytokines, both *in vitro* and *in vivo* (84–86). In a preclinical study, hypoxic pre-treatment of hMSC enhanced the secretion of IL-10 and Fas ligand, thereby reducing recruitment of inflammatory cells, resulting in a more organized granulation tissue at the wound site in an excisional skin-healing mouse model (85). Transplantation of hMSC expanded under hypoxia in a humanized mouse graft vs. host disease (GVHD) model improved animal survival and weight loss, and reduced histopathologic injuries in GVHD target organs, presumably due to enhanced PGE-2 secretion and reduced IL-6/IL-8 secretion (110). To overcome the complexity and inconsistency associated with *in vitro* hypoxia culture, overexpression of HIF1-α and hypoxia mimetics targeting HIF pathway are being actively pursued to enhance hMSC immunomodulatory properties (114, 115). Oxidative preconditioning of MSCs by ROS leads to redox-dependent HIF-1α stabilization and reduced apoptosis in inflammatory environment (116, 117). Pretreatment with N-acetylcysteine (NAC), which stabilizes HIF-1, improved MSC anti-inflammation and cell retention in

Box 1 | Current knowledge and future directions of metabolic perturbation in hMSC biomanufacturing and immunotherapy.

Energy metabolism in hMSC biomanufacturing and immunotherapy

Current knowledge

- hMSC's are metabolically "plastic" and reconfigure metabolism to match divergent demands of cellular events;
- hMSC's metabolic "fitness" or the ability to adapt metabolically influences their functional properties;
- Donor morbidity and *in vitro* bioprocessing such as extended expansion and cryopreservation reduce hMSC metabolic fitness by preconditioning is an effective strategy to enhance hMSC therapeutic potency.

Open question and future direction

- What is the mechanistic link between hMSC metabolic profile and therapeutic outcome in a given disease?
- Can hMSC metabolism profile be standardized as a potency indicator as part of Critical Quality Attributes (CQA)?
- How the media composition and culture supplement such as human platelet lysate and lipid contents used in hMSC expansion influence metabolic fitness and functional properties?
- How to maintain a desired metabolic profile during large-scale expansion, harvesting, and cryopreservation (distribution and administration) in hMSC biomanufacturing?
- How does a given metabolic profile influence specific aspect of hMSC functional properties such as angiogenesis, or immunomodulation?

bleomycin-induced lung injury model by improving antioxidant capacity (118).

Pre-activation of MSCs with immunomodulatory cytokines has been widely reported to enhance MSC immunomodulatory effect in various preclinical models. Interestingly, many such cytokines are also metabolic regulators and the extent of their effects is influenced by oxygen levels and MSC's metabolic plasticity (119, 120). Pretreatment of hMSC with IFN-γ activates hMSC's anti-inflammatory properties by enhancing the secretion of anti-inflammatory cytokines and inhibits the proliferation of NK cells and CD4+/CD8+ T cells (20, 111, 121, 122). Infusion of IFN-γ pretreated MSCs in an immunodeficient mouse model significantly reduced the symptoms of GVHD and improved survival (112). As mentioned above, IFN-γ treatment reconfigures hMSC metabolism toward a glycolytic phenotype, generating a metabolic profile that enhances cell survival and sustains the secretion of immunomodulatory factors (20). Fan et al. reported that MSCs preconditioned by IL-1β exposure significantly attenuated the development of dextran sulfate sodium (DSS)-induced colitis in mice by enhancing MSC migration to the inflammatory site via upregulating CXCR4 expression (111). The IL-1 cytokine family members are important regulators of metabolism and upregulate glycolysis in various cell types (119, 123). TGF- β, a pleiotropic cytokine involved in immune regulation, is also a potent regulator of hMSC immunomodulatory properties and can display either anti-inflammatory or proinflammatory effects depending on the cell niche. TGF-β activates MSC to promote immune response and altered hASC secretory profile (124–126). Interestingly, TGF-β signaling in tumor growth is compartment-specific and induces a "Warburg-like" metabolism in cancer-associated fibroblasts that fuels tumor growth; a similar metabolic shift toward glycolysis is also observed in human chondrocytes and plays important role in maintaining cartilage homeostasis (127, 128). The effects of TGF-β on hMSC metabolic properties during immune activation remains to be elucidated.

As mentioned above, 3D aggregation reconfigures hMSC metabolic state toward glycolysis and this approach has emerged as a novel engineering strategy to potentiate MSC secretory and immunomodulatory functions. Bartosh et al. reported *in vivo* aggregation of MSCs into 3D spheroid enhanced macrophage activation in mice with regulated expression of inflammation-modulating factors TSG-6, STC-1, and COX-2 (129). Zimmerman et al. demonstrated the enhanced suppression of T cells by MSC spheroids together with IFN-γ via the secretion of IDO from 3D MSC aggregates (130). Increased secretion of PGE2 and COX2 from 3D MSC aggregates converted LPS-activated macrophages into M2 phenotype with reduced production of TNF-α, IL-6, IL-23, and CXCL2 (113, 131). Intraluminal injection of MSC aggregates in mice with DSS-induced colitis reduced local inflammatory cytokines including TNF-α, IFN-γ, IL-6, and the system inflammation marker serum amyloid A (132), whereas the local expression of PGE-2 and COX-2 in mice distal colons were increased, resulting in less body weight loss and lower disease activity score (132). Inhibition of this metabolic reconfiguration in hMSC 3D aggregates reduces secretion of IDO, PGE2, COX2, IL-6, TGF-β and other anti-inflammatory cytokines (113, 129, 131, 133). These studies demonstrate the potential of 3D aggregation to potentiate hMSC immunomodulatory properties.

CONCLUSIONS

This review shows that energy metabolism has emerged as a central hub connecting hMSC sourcing and biomanufacturing practices to core signaling pathways that regulate hMSC phenotypic properties and clinical outcome. Modulation of hMSC metabolism by specific engineering practices or metabolic modulators is an effective strategy for enhancing hMSC functional properties and improving therapeutic potency. hMSC metabolic characteristics or fitness can also be used as defining criteria to determine cellular quality in hMSC sourcing and large-scale manufacturing. As summarized in **Box 1**, many open questions remain in the implementation of metabolic strategies to enhance hMSC therapeutic potency in large scale manufacturing. Future studies that elucidate the signaling roles of other intermediate metabolites are needed to identify novel targets to improve hMSC clinical outcomes.

AUTHOR CONTRIBUTIONS

XY and TM organized the structure, collected references, and wrote the manuscript. TL organized and revised the manuscript. All authors contributed to manuscript revision, read, and approved the submitted version.

FUNDING

This work is supported by the National Science Foundation (CBET 1743426) and National Institutes of Health (NIH) (R01 NS102395). The content is solely the responsibility of the authors and does not necessarily represent the official views of the NSF and NIH.

REFERENCES

1. Yin JQ, Zhu J, Ankrum JA. Manufacturing of primed mesenchymal stromal cells for therapy. *Nat Biomed Eng.* (2019) 3:90–104. doi: 10.1038/s41551-018-0325-8
2. Castro-Manrreza ME, Montesinos JJ. Immunoregulation by mesenchymal stem cells: biological aspects and clinical applications. *J Immunol Res.* (2015) 2015:394917. doi: 10.1155/2015/394917
3. Galipeau J, Krampera M, Barrett J, Dazzi F, Deans RJ, DeBruijn J, et al. International Society for Cellular Therapy perspective on immune functional assays for mesenchymal stromal cells as potency release criterion for advanced phase clinical trials. *Cytotherapy.* (2016) 18:151–9. doi: 10.1016/j.jcyt.2015.11.008
4. Galipeau J, Sensebe L. Mesenchymal stromal cells: clinical challenges and therapeutic opportunities. *Cell Stem Cell.* (2018) 22:824–33. doi: 10.1016/j.stem.2018.05.004
5. von Bahr L, Sundberg B, Lonnies L, Sander B, Karbach H, Hagglund H, et al. Long-term complications, immunologic effects, and role of passage for outcome in mesenchymal stromal cell therapy. *Biol Blood Marrow Transplant.* (2012) 18:557–64. doi: 10.1016/j.bbmt.2011.07.023
6. Moll G, Alm JJ, Davies LC, von Bahr L, Heldring N, Stenbeck-Funke L, et al. Do cryopreserved mesenchymal stromal cells display impaired immunomodulatory and therapeutic properties? *Stem Cells.* (2014) 32:2430–42. doi: 10.1002/stem.1729
7. Castro B, Martinez-Redondo D, Gartzia I, Alonso-Varona A, Garrido P, Palomares T. Cryopreserved H2 O2 -preconditioned human adipose-derived stem cells exhibit fast post-thaw recovery and enhanced bioactivity against oxidative stress. *J Tissue Eng Regen Med.* (2019) 13:328–41. doi: 10.1002/term.2797
8. Waterman RS, Tomchuck SL, Henkle SL, Betancourt AM. A new mesenchymal stem cell (MSC) paradigm: polarization into a pro-inflammatory MSC1 or an Immunosuppressive MSC2 phenotype. *PLoS ONE.* (2010) 5:e10088. doi: 10.1371/journal.pone.0010088
9. Olsen TR, Ng KS, Lock LT, Ahsan T, Rowley JA. Peak MSC-are we there yet? *Front Med.* (2018) 5:178. doi: 10.3389/fmed.2018.00178
10. Robb KP, Fitzgerald JC, Barry F, Viswanathan S. Mesenchymal stromal cell therapy: progress in manufacturing and assessments of potency. *Cytotherapy.* (2018) 21:289–306. doi: 10.1016/j.jcyt.2018.10.014
11. Bara JJ, Richards RG, Alini M, Stoddart MJ. Concise review: bone marrow-derived mesenchymal stem cells change phenotype following *in vitro* culture: implications for basic research and the clinic. *Stem Cells.* (2014) 32:1713–23. doi: 10.1002/stem.1649
12. Sbrana FV, Cortini M, Avnet S, Perut F, Columbaro M, De Milito A, et al. The role of autophagy in the maintenance of stemness and differentiation of mesenchymal stem cells. *Stem Cell Rev.* (2016) 12:621–33. doi: 10.1007/s12015-016-9690-4
13. Katajisto P, Dohla J, Chaffer CL, Pentinmikko N, Marjanovic N, Iqbal S, et al. Stem cells. Asymmetric apportioning of aged mitochondria between daughter cells is required for stemness. *Science.* (2015) 348:340–3. doi: 10.1126/science.1260384
14. Sherr CJ, DePinho RA. Cellular senescence: mitotic clock or culture shock? *Cell.* (2000) 102:407–10. doi: 10.1016/S0092-8674(00)00046-5
15. Pattappa G, Heywood HK, de Bruijn JD, Lee DA. The metabolism of human mesenchymal stem cells during proliferation and differentiation. *J Cell Physiol.* (2011) 226:2562–70. doi: 10.1002/jcp.22605
16. Shyh-Chang N, Daley GQ, Cantley LC. Stem cell metabolism in tissue development and aging. *Development.* (2013) 140:2535–47. doi: 10.1242/dev.091777
17. Ho TT, Warr MR, Adelman ER, Lansinger OM, Flach J, Verovskaya EV, et al. Autophagy maintains the metabolism and function of young and old stem cells. *Nature.* (2017) 543:205–10. doi: 10.1038/nature21388
18. Cunha B, Aguiar T, Carvalho SB, Silva MM, Gomes RA, M.Carrondo JT, et al. Bioprocess integration for human mesenchymal stem cells: from up to downstream processing scale-up to cell proteome characterization. *J Biotechnol.* (2017) 248:87–98. doi: 10.1016/j.jbiotec.2017.01.014
19. Green DR, Galluzzi L, Kroemer G. Cell biology. Metabolic control of cell death. *Science.* (2014) 345:1250256. doi: 10.1126/science.1250256
20. Liu Y, Yuan X, Munoz N, Logan TM, Ma T. Commitment to aerobic glycolysis sustains immunosuppression of human mesenchymal stem cells. *Stem Cells Transl Med.* (2019) 8:93–106. doi: 10.1002/sctm.18-0070
21. Liu Y, Ma T. Metabolic regulation of mesenchymal stem cell in expansion and therapeutic application. *Biotechnol Prog.* (2015) 31:468–81. doi: 10.1002/btpr.2034
22. Liu Y, Munoz N, Bunnell BA, Logan TM, Ma T. Density-dependent metabolic heterogeneity in human mesenchymal stem cells. *Stem Cells.* (2015) 33:3368–81. doi: 10.1002/stem.2097
23. Folmes CD, Dzeja PP, Nelson TJ, Terzic A. Metabolic plasticity in stem cell homeostasis and differentiation. *Cell Stem Cell.* (2012) 11:596–606. doi: 10.1016/j.stem.2012.10.002
24. Galluzzi L, Kepp O, Vander Heiden MG, Kroemer G. Metabolic targets for cancer therapy. *Nat Rev Drug Discov.* (2013) 12:829–46. doi: 10.1038/nrd4145
25. Vander Heiden MG, Cantley LC, Thompson CB. Understanding the Warburg effect: the metabolic requirements of cell proliferation. *Science.* (2009) 324:1029–33. doi: 10.1126/science.1160809
26. Estrada JC, Albo C, Benguría A, Dopazo A, López-Romero P, Carrera-Quintanar L, et al. Culture of human mesenchymal stem cells at low oxygen tension improves growth and genetic stability by activating glycolysis. *Cell Death Different.* (2011) 19:743. doi: 10.1038/cdd.2011.172
27. Gao L, Cen S, Wang P, Xie Z, Liu Z, Deng W, et al. Autophagy improves the immunosuppression of $CD4^+$ T cells by mesenchymal stem cells through transforming growth factor-beta1. *Stem Cells Transl Med.* (2016) 5:1496–505. doi: 10.5966/sctm.2015-0420
28. Liu Y, Munoz N, Tsai AC, Logan TM, Ma T. Metabolic reconfiguration supports reacquisition of primitive phenotype in human mesenchymal stem cell aggregates. *Stem Cells.* (2017) 35:398–410. doi: 10.1002/stem.2510
29. Oh J, Lee YD, Wagers AJ. Stem cell aging: mechanisms, regulators and therapeutic opportunities. *Nat Med.* (2014) 20:870–80. doi: 10.1038/nm.3651
30. Perez LM, Bernal A, de Lucas B, San Martin N, Mastrangelo A, Garcia A, et al. Altered metabolic and stemness capacity of adipose tissue-derived stem cells from obese mouse and human. *PLoS ONE.* (2015) 10:e0123397. doi: 10.1371/journal.pone.0123397
31. Meng Y, Eirin A, Zhu XY, Tang H, Chanana P, Lerman A, et al. Obesity-induced mitochondrial dysfunction in porcine adipose tissue-derived mesenchymal stem cells. *J Cell Physiol.* (2018) 233:5926–36. doi: 10.1002/jcp.26402

32. Pietila M, Palomaki S, Lehtonen S, Ritamo I, Valmu L, Nystedt J, et al. Mitochondrial function and energy metabolism in umbilical cord blood- and bone marrow-derived mesenchymal stem cells. *Stem Cells Dev.* (2012) 21:575–88. doi: 10.1089/scd.2011.0023
33. Kizilay Mancini O, Lora M, Cuillerier A, Shum-Tim D, Hamdy R, Burelle Y, et al. Mitochondrial oxidative stress reduces the immunopotency of mesenchymal stromal cells in adults with coronary artery disease. *Circ Res.* (2018) 122:255–66. doi: 10.1161/CIRCRESAHA.117.311400
34. Fillmore N, Huqi A, Jaswal JS, Mori J, Paulin R, Haromy A, et al. Effect of fatty acids on human bone marrow mesenchymal stem cell energy metabolism and survival. *PLoS ONE.* (2015) 10:e0120257. doi: 10.1371/journal.pone.0120257
35. Heathman TR, Glyn VA, Picken A, Rafiq QA, Coopman K, Nienow AW, et al. Expansion, harvest and cryopreservation of human mesenchymal stem cells in a serum-free microcarrier process. *Biotechnol Bioeng.* (2015) 112:1696–707. doi: 10.1002/bit.25582
36. Becherucci V, Piccini L, Casamassima S, Bisin S, Gori V, Gentile F, et al. Human platelet lysate in mesenchymal stromal cell expansion according to a GMP grade protocol: a cell factory experience. *Stem Cell Res Ther.* (2018) 9:124. doi: 10.1186/s13287-018-0863-8
37. Moya A, Paquet J, Deschepper M, Larochette N, Oudina K, Denoeud C, et al. Human mesenchymal stem cell failure to adapt to glucose shortage and rapidly use intracellular energy reserves through glycolysis explains poor cell survival after implantation. *Stem Cells.* (2018) 36:363–76. doi: 10.1002/stem.2763
38. Savelli S, Trombi L, D'Alessandro D, Moscato S, Pacini S, Giannotti S, et al. Pooled human serum: a new culture supplement for bioreactor-based cell therapies. Preliminary results. *Cytotherapy.* (2018) 20:556–63. doi: 10.1016/j.jcyt.2017.12.013
39. Dos Santos F, Andrade PZ, Boura JS, Abecasis MM, da Silva CL, Cabral JM. *Ex vivo* expansion of human mesenchymal stem cells: a more effective cell proliferation kinetics and metabolism under hypoxia. *J Cell Physiol.* (2010) 223:27–35. doi: 10.1002/jcp.21987
40. Munoz N, Kim J, Liu Y, Logan TM, Ma T. Gas chromatography-mass spectrometry analysis of human mesenchymal stem cell metabolism during proliferation and osteogenic differentiation under different oxygen tensions. *J Biotechnol.* (2014) 169:95–102. doi: 10.1016/j.jbiotec.2013.11.010
41. Ma T, Grayson WL, Frohlich M, Vunjak-Novakovic G. Hypoxia and stem cell-based engineering of mesenchymal tissues. *Biotechnol Prog.* (2009) 25:32–42. doi: 10.1002/btpr.128
42. Liu J, Hao H, Huang H, Tong C, Ti D, Dong L, et al. Hypoxia regulates the therapeutic potential of mesenchymal stem cells through enhanced autophagy. *Int J Low Extrem Wounds.* (2015) 14:63–72. doi: 10.1177/1534734615573660
43. Wenger RH, Kurtcuoglu V, Scholz CC, Marti HH, Hoogewijs D. Frequently asked questions in hypoxia research. *Hypoxia.* (2015) 3:35–43. doi: 10.2147/HP.S92198
44. Grayson WL, Zhao F, Bunnell B, Ma T. Hypoxia enhances proliferation and tissue formation of human mesenchymal stem cells. *Biochem Biophys Res Commun.* (2007) 358:948–53. doi: 10.1016/j.bbrc.2007.05.054
45. Gharibi B, Farzadi S, Ghuman M, Hughes FJ. Inhibition of Akt/mTOR attenuates age-related changes in mesenchymal stem cells. *Stem Cells.* (2014) 32:2256–66. doi: 10.1002/stem.1709
46. Liu XB, Wang JA, Ogle ME, Wei L. Prolyl hydroxylase inhibitor dimethyloxalylglycine enhances mesenchymal stem cell survival. *J Cell Biochem.* (2009) 106:903–11. doi: 10.1002/jcb.22064
47. Najafi R, Sharifi AM. Deferoxamine preconditioning potentiates mesenchymal stem cell homing *in vitro* and in streptozotocin-diabetic rats. *Expert Opin Biol Ther.* (2013) 13:959–72. doi: 10.1517/14712598.2013.782390
48. Tsai AC, Liu Y, Yuan X, Ma T. Compaction, fusion, and functional activation of three-dimensional human mesenchymal stem cell aggregate. *Tissue Eng Part A.* (2015) 21:1705–19. doi: 10.1089/ten.tea.2014.0314
49. Sart S, Tsai AC, Li Y, Ma T. Three-dimensional aggregates of mesenchymal stem cells: cellular mechanisms, biological properties, and applications. *Tissue Eng Part B Rev.* (2014) 20:365–80. doi: 10.1089/ten.teb.2013.0537
50. Yuan X, Tsai AC, Farrance I, Rowley J, Ma T. Aggregation of culture expanded human mesenchymal stem cells in microcarrier-based bioreactor. *Biochem Eng J.* (2018) 131:39–46. doi: 10.1016/j.bej.2017.12.011
51. Bijonowski BM, Daraiseh SI, Yuan X, Ma T. Size-dependent cortical compaction induces metabolic adaptation in mesenchymal stem cell aggregates. *Tissue Eng Part A.* (2018) 25:575–87. doi: 10.1089/ten.TEA.2018.0155
52. Ren G, Zhang L, Zhao X, Xu G, Zhang Y, Roberts AI, et al. Mesenchymal stem cell-mediated immunosuppression occurs via concerted action of chemokines and nitric oxide. *Cell Stem Cell.* (2008) 2:141–50. doi: 10.1016/j.stem.2007.11.014
53. Gao F, Chiu SM, Motan DA, Zhang Z, Chen L, Ji HL, et al. Mesenchymal stem cells and immunomodulation: current status and future prospects. *Cell Death Dis.* (2016) 7:e2062. doi: 10.1038/cddis.2015.327
54. Li W, Ren G, Huang Y, Su J, Han Y, Li J, et al. Mesenchymal stem cells: a double-edged sword in regulating immune responses. *Cell Death Differ.* (2012) 19:1505–13. doi: 10.1038/cdd.2012.26
55. Turner MD, Nedjai B, Hurst T, Pennington DJ. Cytokines and chemokines: at the crossroads of cell signalling and inflammatory disease. *Biochim Biophys Acta.* (2014) 1843:2563–82. doi: 10.1016/j.bbamcr.2014.05.014
56. Hemeda H, Jakob M, Ludwig AK, Giebel B, Lang S, Brandau S. Interferon-gamma and tumor necrosis factor-alpha differentially affect cytokine expression and migration properties of mesenchymal stem cells. *Stem Cells Dev.* (2010) 19:693–706. doi: 10.1089/scd.2009.0365
57. Cho DI, Kim MR, Jeong HY, Jeong HC, Jeong MH, Yoon SH, et al. Mesenchymal stem cells reciprocally regulate the M1/M2 balance in mouse bone marrow-derived macrophages. *Exp Mol Med.* (2014) 46:e70. doi: 10.1038/emm.2013.135
58. Melief SM, Geutskens SB, Fibbe WE, Roelofs H. Multipotent stromal cells skew monocytes towards an anti-inflammatory function: the link with key immunoregulatory molecules. *Haematologica.* (2013) 98:e121–2. doi: 10.3324/haematol.2013.093864
59. Yen BL, Chang CJ, Liu KJ, Chen YC, Hu HI, Bai CH, et al. Brief report–human embryonic stem cell-derived mesenchymal progenitors possess strong immunosuppressive effects toward natural killer cells as well as T lymphocytes. *Stem Cells.* (2009) 27:451–6. doi: 10.1634/stemcells.2008-0390
60. Sotiropoulou PA, Perez SA, Gritzapis AD, Baxevanis CN, Papamichail M. Interactions between human mesenchymal stem cells and natural killer cells. *Stem Cells.* (2006) 24:74–85. doi: 10.1634/stemcells.2004-0359
61. Thomas H, Jager M, Mauel K, Brandau S, Lask S, Flohe SB. Interaction with mesenchymal stem cells provokes natural killer cells for enhanced IL-12/IL-18-induced interferon-gamma secretion. *Mediators Inflamm.* (2014) 2014:143463. doi: 10.1155/2014/143463
62. Spaggiari GM, Moretta L. Interactions between mesenchymal stem cells and dendritic cells. *Adv Biochem Eng Biotechnol.* (2013) 130:199–208. doi: 10.1007/10_2012_154
63. Spaggiari GM, Capobianco A, Becchetti S, Mingari MC, Moretta L. Mesenchymal stem cell-natural killer cell interactions: evidence that activated NK cells are capable of killing MSCs, whereas MSCs can inhibit IL-2-induced NK-cell proliferation. *Blood.* (2006) 107:1484–90. doi: 10.1182/blood-2005-07-2775
64. Nauta AJ, Kruisselbrink AB, Lurvink E, Willemze R, Fibbe WE. Mesenchymal stem cells inhibit generation and function of both CD34+-derived and monocyte-derived dendritic cells. *J Immunol.* (2006) 177:2080–7. doi: 10.4049/jimmunol.177.4.2080
65. Wang Y, Chen X, Cao W, Shi Y. Plasticity of mesenchymal stem cells in immunomodulation: pathological and therapeutic implications. *Nat Immunol.* (2014) 15:1009–16. doi: 10.1038/ni.3002
66. Franquesa M, Hoogduijn MJ, Bestard O, Grinyo JM. Immunomodulatory effect of mesenchymal stem cells on B cells. *Front Immunol.* (2012) 3:212. doi: 10.3389/fimmu.2012.00212
67. Tabera S, Perez-Simon JA, Diez-Campelo M, Sanchez-Abarca LI, Blanco B, Lopez A, et al. The effect of mesenchymal stem cells on the viability, proliferation and differentiation of B-lymphocytes. *Haematologica.* (2008) 93:1301–9. doi: 10.3324/haematol.12857
68. Corcione A, Benvenuto F, Ferretti E, Giunti D, Cappiello V, Cazzanti F, et al. Human mesenchymal stem cells modulate B-cell functions. *Blood.* (2006) 107:367–72. doi: 10.1182/blood-2005-07-2657

69. Wang LT, Ting CH, Yen ML, Liu KJ, Sytwu HK, Wu KK, et al. Human mesenchymal stem cells (MSCs) for treatment towards immune- and inflammation-mediated diseases: review of current clinical trials. *J Biomed Sci.* (2016) 23:76. doi: 10.1186/s12929-016-0289-5
70. Carrade Holt DD, Wood JA, Granick JL, Walker NJ, Clark KC, Borjesson DL. Equine mesenchymal stem cells inhibit T cell proliferation through different mechanisms depending on tissue source. *Stem Cells Dev.* (2014) 23:1258–65. doi: 10.1089/scd.2013.0537
71. Hsu WT, Lin CH, Chiang BL, Jui HY, Wu KK, Lee CM. Prostaglandin E2 potentiates mesenchymal stem cell-induced IL-10^{+}IFN-gamma^{+}CD4^{+} regulatory T cells to control transplant arteriosclerosis. *J Immunol.* (2013) 190:2372–80. doi: 10.4049/jimmunol.1202996
72. Luz-Crawford P, Noel D, Fernandez X, Khoury M, Figueroa F, Carrion F, et al. Mesenchymal stem cells repress Th17 molecular program through the PD-1 pathway. *PLoS ONE.* (2012) 7:e45272. doi: 10.1371/journal.pone.0045272
73. Wang WB, Yen ML, Liu KJ, Hsu PJ, Lin MH, Chen PM, et al. Interleukin-25 mediates transcriptional control of PD-L1 via STAT3 in multipotent Human Mesenchymal Stromal Cells (hMSCs) to suppress Th17 responses. *Stem Cell Rep.* (2015) 5:392–404. doi: 10.1016/j.stemcr.2015.07.013
74. Augello, Tasso R, Negrini SM, Amateis A, Indiveri F, Cancedda R, et al. Bone marrow mesenchymal progenitor cells inhibit lymphocyte proliferation by activation of the programmed death 1 pathway. *Eur J Immunol.* (2005) 35:1482–90. doi: 10.1002/eji.200425405
75. English K, Barry FP, Field-Corbett CP, Mahon BP. IFN-gamma and TNF-alpha differentially regulate immunomodulation by murine mesenchymal stem cells. *Immunol Lett.* (2007) 110:91–100. doi: 10.1016/j.imlet.2007.04.001
76. Melief SM, Zwaginga JJ, Fibbe WE, Roelofs H. Adipose tissue-derived multipotent stromal cells have a higher immunomodulatory capacity than their bone marrow-derived counterparts. *Stem Cells Transl Med.* (2013) 2:455–63. doi: 10.5966/sctm.2012-0184
77. Silva KR, Liechocki S, Carneiro JR, Claudio-da-Silva C, Maya-Monteiro CM, Borojevic R, et al. Stromal-vascular fraction content and adipose stem cell behavior are altered in morbid obese and post bariatric surgery ex-obese women. *Stem Cell Res Ther.* (2015) 6:72. doi: 10.1186/s13287-015-0029-x
78. Perez LM, de Lucas B, Lunyak VV, Galvez BG. Adipose stem cells from obese patients show specific differences in the metabolic regulators vitamin D and Gas5. *Mol Genet Metab Rep.* (2017) 12:51–6. doi: 10.1016/j.ymgmr.2017.05.008
79. Serena C, Keiran N, Ceperuelo-Mallafre V, Ejarque M, Fradera R, Roche K, et al. Obesity and type 2 diabetes alters the immune properties of human adipose derived stem cells. *Stem Cells.* (2016) 34:2559–73. doi: 10.1002/stem.2429
80. Baker N, Boyette LB, Tuan RS. Characterization of bone marrow-derived mesenchymal stem cells in aging. *Bone.* (2015) 70:37–47. doi: 10.1016/j.bone.2014.10.014
81. Killer MC, Nold P, Henkenius K, Fritz L, Riedlinger T, Barckhausen C, et al. Immunosuppressive capacity of mesenchymal stem cells correlates with metabolic activity and can be enhanced by valproic acid. *Stem Cell Res Ther.* (2017) 8:100. doi: 10.1186/s13287-017-0553-y
82. Siegel G, Kluba T, Hermanutz-Klein U, Bieback K, Northoff H, Schafer R. Phenotype, donor age and gender affect function of human bone marrow-derived mesenchymal stromal cells. *BMC Med.* (2013) 11:146. doi: 10.1186/1741-7015-11-146
83. Cham CM, Gajewski TF. Glucose availability regulates IFN-gamma production and p70S6 kinase activation in CD8+ effector T cells. *J Immunol.* (2005) 174:4670–7. doi: 10.4049/jimmunol.174.8.4670
84. Kadle RL, Abdou SA, Villarreal-Ponce AP, Soares MA, Sultan DL, David JA, et al. Microenvironmental cues enhance mesenchymal stem cell-mediated immunomodulation and regulatory T-cell expansion. *PLoS ONE.* (2018) 13:e0193178. doi: 10.1371/journal.pone.0193178
85. Jiang CM, Liu J, Zhao JY, Xiao L, An S, Gou YC, et al. Effects of hypoxia on the immunomodulatory properties of human gingiva-derived mesenchymal stem cells. *J Dent Res.* (2015) 94:69–77. doi: 10.1177/0022034514557671
86. Bobyleva PI, Andreeva ER, Gornostaeva AN, Buravkova LB. Tissue-related hypoxia attenuates proinflammatory effects of allogeneic PBMCs on adipose-derived stromal cells *in vitro*. *Stem Cells Int.* (2016) 2016:4726267. doi: 10.1155/2016/4726267
87. Di Nicola M, Carlo-Stella C, Magni M, Milanesi M, Longoni PD, Matteucci P, et al. Human bone marrow stromal cells suppress T-lymphocyte proliferation induced by cellular or nonspecific mitogenic stimuli. *Blood.* (2002) 99:3838–43. doi: 10.1182/blood.V99.10.3838
88. Sheng H, Wang Y, Jin Y, Zhang Q, Zhang Y, Wang L, et al. A critical role of IFNgamma in priming MSC-mediated suppression of T cell proliferation through up-regulation of B7-H1. *Cell Res.* (2008) 18:846–57. doi: 10.1038/cr.2008.80
89. Majumdar MK, Keane-Moore M, Buyaner D, Hardy WB, Moorman MA, McIntosh KR, et al. Characterization and functionality of cell surface molecules on human mesenchymal stem cells. *J Biomed Sci.* (2003) 10:228–41. doi: 10.1007/BF02256058
90. Han KH, Ro H, Hong JH, Lee EM, Cho B, Yeom HJ, et al. Immunosuppressive mechanisms of embryonic stem cells and mesenchymal stem cells in alloimmune response. *Transpl Immunol.* (2011) 25:7–15. doi: 10.1016/j.trim.2011.05.004
91. Raicevic G, Najar M, Najimi M, El Taghdouini A, van Grunsven LA, Sokal E, et al. Influence of inflammation on the immunological profile of adult-derived human liver mesenchymal stromal cells and stellate cells. *Cytotherapy.* (2015) 17:174–85. doi: 10.1016/j.jcyt.2014.10.001
92. Najar M, Raicevic G, Fayyad-Kazan H, De Bruyn C, Bron D, Toungouz M, et al. Immune-related antigens, surface molecules and regulatory factors in human-derived mesenchymal stromal cells: the expression and impact of inflammatory priming. *Stem Cell Rev.* (2012) 8:1188–98. doi: 10.1007/s12015-012-9408-1
93. Mathieu M, Martin-Jaular L, Lavieu G, Thery C. Specificities of secretion and uptake of exosomes and other extracellular vesicles for cell-to-cell communication. *Nat Cell Biol.* (2019) 21:9–17. doi: 10.1038/s41556-018-0250-9
94. Robbins PD, Morelli AE. Regulation of immune responses by extracellular vesicles. *Nat Rev Immunol.* (2014) 14:195–208. doi: 10.1038/nri3622
95. Zou X, Zhang G, Cheng Z, Yin D, Du T, Ju G, et al. Microvesicles derived from human Wharton's Jelly mesenchymal stromal cells ameliorate renal ischemia-reperfusion injury in rats by suppressing CX3CL1. *Stem Cell Res Ther.* (2014) 5:40. doi: 10.1186/scrt428
96. Nojehdehi S, Soudi S, Hesampour A, Rasouli S, Soleimani M, Hashemi SM. Immunomodulatory effects of mesenchymal stem cell-derived exosomes on experimental type-1 autoimmune diabetes. *J Cell Biochem.* (2018) 119:9433–43. doi: 10.1002/jcb.27260
97. Bai L, Shao H, Wang H, Zhang Z, Su C, Dong L, et al. Effects of mesenchymal stem cell-derived exosomes on experimental autoimmune uveitis. *Sci Rep.* (2017) 7:4323. doi: 10.1038/s41598-017-04559-y
98. Yu B, Kim HW, Gong M, Wang J, Millard RW, Wang Y, et al. Exosomes secreted from GATA-4 overexpressing mesenchymal stem cells serve as a reservoir of anti-apoptotic microRNAs for cardioprotection. *Int J Cardiol.* (2015) 182:349–60. doi: 10.1016/j.ijcard.2014.12.043
99. Aghajani N, Lerman LO, Eirin A. Mesenchymal stem cell-derived extracellular vesicles for kidney repair: current status and looming challenges. *Stem Cell Res Ther.* (2017) 8:273. doi: 10.1186/s13287-017-0727-7
100. Toh WS, Zhang B, Lai RC, Lim SK. Immune regulatory targets of mesenchymal stromal cell exosomes/small extracellular vesicles in tissue regeneration. *Cytotherapy.* (2018) 20:1419–26. doi: 10.1016/j.jcyt.2018.09.008
101. Wei Y, Wang D, Jin F, Bian Z, Li L, Liang H, et al. Pyruvate kinase type M2 promotes tumour cell exosome release via phosphorylating synaptosome-associated protein 23. *Nat Commun.* (2017) 8:14041. doi: 10.1038/ncomms14041
102. Gonzalez-King H, Garcia NA, Ontoria-Oviedo I, Ciria M, Montero JA, Sepulveda P. Hypoxia inducible factor-1alpha potentiates jagged 1-mediated angiogenesis by mesenchymal stem cell-derived exosomes. *Stem Cells.* (2017) 35:1747–59. doi: 10.1002/stem.2618
103. Cahill EF, Tobin LM, Carty F, Mahon BP, English K. Jagged-1 is required for the expansion of CD4^{+} CD25^{+} FoxP3^{+} regulatory T cells and tolerogenic dendritic cells by murine mesenchymal stromal cells. *Stem Cell Res Ther.* (2015) 6:19. doi: 10.1186/s13287-015-0021-5
104. Zhang W, Zhou X, Yao Q, Liu Y, Zhang H, Dong Z. HIF-1-mediated production of exosomes during hypoxia is protective

in renal tubular cells. *Am J Physiol Renal Physiol.* (2017) 313:F906–13. doi: 10.1152/ajprenal.00178.2017

105. Phinney DG, Di Giuseppe M, Njah J, Sala E, Shiva S, St Croix CM, et al. Mesenchymal stem cells use extracellular vesicles to outsource mitophagy and shuttle microRNAs. *Nat Commun.* (2015) 6:8472. doi: 10.1038/ncomms9472
106. Arslan F, Lai RC, Smeets MB, Akeroyd L, Choo A, Aguor EN, et al. Mesenchymal stem cell-derived exosomes increase ATP levels, decrease oxidative stress and activate PI3K/Akt pathway to enhance myocardial viability and prevent adverse remodeling after myocardial ischemia/reperfusion injury. *Stem Cell Res.* (2013) 10:301–12. doi: 10.1016/j.scr.2013.01.002
107. Wang X, Liu C, Li S, Xu Y, Chen P, Liu Y, et al. Hypoxia precondition promotes adipose-derived mesenchymal stem cells based repair of diabetic erectile dysfunction via augmenting angiogenesis and neuroprotection. *PLoS ONE.* (2015) 10:e0118951. doi: 10.1371/journal.pone.0118951
108. Xie J, Xiao D, Xu Y, Zhao J, Jiang L, Hu X, et al. Up-regulation of immunomodulatory effects of mouse bone-marrow derived mesenchymal stem cells by tetrahydrocannabinol pre-treatment involving cannabinoid receptor CB2. *Oncotarget.* (2016) 7:6436–47. doi: 10.18632/oncotarget.7042
109. Chen L, Xu Y, Zhao J, Zhang Z, Yang R, Xie J, et al. Conditioned medium from hypoxic bone marrow-derived mesenchymal stem cells enhances wound healing in mice. *PLoS ONE.* (2014) 9:e96161. doi: 10.1371/journal.pone.0096161
110. Kim Y, Jin HJ, Heo J, Ju H, Lee HY, Kim S, et al. Small hypoxia-primed mesenchymal stem cells attenuate graft-versus-host disease. *Leukemia.* (2018) 32:2672–84. doi: 10.1038/s41375-018-0151-8
111. Fan H, Zhao G, Liu L, Liu F, Gong W, Liu X, et al. Pre-treatment with IL-1beta enhances the efficacy of MSC transplantation in DSS-induced colitis. *Cell Mol Immunol.* (2012) 9:473–81. doi: 10.1038/cmi.2012.40
112. Kim DS, Jang IK, Lee MW, Ko YJ, Lee DH, Lee JW, et al. Enhanced immunosuppressive properties of human mesenchymal stem cells primed by interferon-gamma. *EBio Med.* (2018) 28:261–73. doi: 10.1016/j.ebiom.2018.01.002
113. Bartosh TJ, Ylostalo JH, Mohammadipoor A, Bazhanov N, Coble K, Claypool K, et al. Aggregation of human mesenchymal stromal cells (MSCs) into 3D spheroids enhances their antiinflammatory properties. *Proc Natl Acad Sci USA.* (2010) 107:13724–9. doi: 10.1073/pnas.1008117107
114. Martinez VG, Ontoria-Oviedo I, Ricardo CP, Harding SE, Sacedon R, Varas A, et al. Overexpression of hypoxia-inducible factor 1 alpha improves immunomodulation by dental mesenchymal stem cells. *Stem Cell Res Ther.* (2017) 8:208. doi: 10.1186/s13287-017-0659-2
115. Fujisawa K, Takami T, Okada S, Hara K, Matsumoto T, Yamamoto N, et al. Analysis of metabolomic changes in mesenchymal stem cells on treatment with desferrioxamine as a hypoxia mimetic compared with hypoxic conditions. *Stem Cells.* (2018) 36:1226–36. doi: 10.1002/stem.2826
116. Calvani M, Comito G, Giannoni E, Chiarugi P. Time-dependent stabilization of hypoxia inducible factor-1alpha by different intracellular sources of reactive oxygen species. *PLoS ONE.* (2012) 7:e38388. doi: 10.1371/journal.pone.0038388
117. Li S, Deng Y, Feng J, Ye W. Oxidative preconditioning promotes bone marrow mesenchymal stem cells migration and prevents apoptosis. *Cell Biol Int.* (2009) 33:411–8. doi: 10.1016/j.cellbi.2009.01.012
118. Wang Q, Zhu H, Zhou WG, Guo XC, Wu MJ, Xu ZY, et al. N-acetylcysteine-pretreated human embryonic mesenchymal stem cell administration protects against bleomycin-induced lung injury. *Am J Med Sci.* (2013) 346:113–22. doi: 10.1097/MAJ.0b013e318266e8d8
119. Tan Q, Huang Q, Ma YL, Mao K, Yang G, Luo P, et al. Potential roles of IL-1 subfamily members in glycolysis in disease. *Cytokine Growth Factor Rev.* (2018) 44:18–27. doi: 10.1016/j.cytogfr.2018.11.001
120. Andreeva ER, Udartseva OO, Zhidkova OV, Buravkov SV, Ezdakova MI, Buravkova LB. IFN-gamma priming of adipose-derived stromal cells at "physiological" hypoxia. *J Cell Physiol.* (2018) 233:1535–47. doi: 10.1002/jcp.26046
121. Krampera M, Cosmi L, Angeli R, Pasini A, Liotta F, Andreini A, et al. Role for interferon-gamma in the immunomodulatory activity of human bone marrow mesenchymal stem cells. *Stem Cells.* (2006) 24:386–98. doi: 10.1634/stemcells.2005-0008
122. Li H, Wang W, Wang G, Hou Y, Xu F, Liu R, et al. Interferon-gamma and tumor necrosis factor-alpha promote the ability of human placenta-derived mesenchymal stromal cells to express programmed death ligand-2 and induce the differentiation of CD4(+)interleukin-10(+) and CD8(+)interleukin-10(+)Treg subsets. *Cytotherapy.* (2015) 17:1560–71. doi: 10.1016/j.jcyt.2015.07.018
123. Ben-Shlomo, Kol S, Roeder LM, Resnick CE, Hurwitz A, Payne DW, et al. Interleukin (IL)-1beta increases glucose uptake and induces glycolysis in aerobically cultured rat ovarian cells: evidence that IL-1beta may mediate the gonadotropin-induced midcycle metabolic shift. *Endocrinology.* (1997) 138:2680–8. doi: 10.1210/endo.138.7.5229
124. Zhao G, Miao H, Li X, Chen S, Hu Y, Wang Z, et al. TGF-beta3-induced miR-494 inhibits macrophage polarization via suppressing PGE2 secretion in mesenchymal stem cells. *FEBS Lett.* (2016) 590:1602–13. doi: 10.1002/1873-3468.12200
125. Xu C, Yu P, Han X, Du L, Gan J, Wang Y, et al. TGF-beta promotes immune responses in the presence of mesenchymal stem cells. *J Immunol.* (2014) 192:103–9. doi: 10.4049/jimmunol.1302164
126. Rodriguez TM, Saldias A, Irigo M, Zamora JV, Perone MJ, Dewey RA. Effect of TGF-beta1 stimulation on the secretome of human adipose-derived mesenchymal stromal cells. *Stem Cells Transl Med.* (2015) 4:894–8. doi: 10.5966/sctm.2015-0012
127. Guido C, Whitaker-Menezes D, Capparelli C, Balliet R, Lin Z, Pestell RG, et al. Metabolic reprogramming of cancer-associated fibroblasts by TGF-beta drives tumor growth: connecting TGF-beta signaling with "Warburg-like" cancer metabolism and L-lactate production. *Cell Cycle.* (2012) 11:3019–35. doi: 10.4161/cc.21384
128. Wang C, Silverman RM, Shen J, O'Keefe RJ. Distinct metabolic programs induced by TGF-beta1 and BMP2 in human articular chondrocytes with osteoarthritis. *J Orthop Translat.* (2018) 12:66–73. doi: 10.1016/j.jot.2017.12.004
129. Bartosh TJ, Ylostalo JH, Bazhanov N, Kuhlman J, Prockop DJ. Dynamic compaction of human mesenchymal stem/precursor cells into spheres self-activates caspase-dependent IL1 signaling to enhance secretion of modulators of inflammation and immunity (PGE2, TSG6, and STC1). *Stem Cells.* (2013) 31:2443–56. doi: 10.1002/stem.1499
130. Zimmermann JA, Hettiaratchi MH, McDevitt TC. Enhanced immunosuppression of T cells by sustained presentation of bioactive interferon-gamma within three-dimensional mesenchymal stem cell constructs. *Stem Cells Transl Med.* (2017) 6:223–37. doi: 10.5966/sctm.2016-0044
131. Ylostalo JH, Bartosh TJ, Coble K, Prockop DJ. Human mesenchymal stem/stromal cells cultured as spheroids are self-activated to produce prostaglandin E2 that directs stimulated macrophages into an anti-inflammatory phenotype. *Stem Cells.* (2012) 30:2283–96. doi: 10.1002/stem.1191
132. Molendijk I, Barnhoorn MC, de Jonge-Muller ES, Mieremet-Ooms MA, van der Reijden JJ, van der Helm D, et al. Intraluminal injection of mesenchymal stromal cells in spheroids attenuates experimental colitis. *J Crohns Colitis.* (2016) 10:953–64. doi: 10.1093/ecco-jcc/jjw047
133. Zimmermann JA, McDevitt TC. Pre-conditioning mesenchymal stromal cell spheroids for immunomodulatory paracrine factor secretion. *Cytotherapy.* (2014) 16:331–45. doi: 10.1016/j.jcyt.2013.09.004

7

Multi-Parameter Analysis of Biobanked Human Bone Marrow Stromal Cells Shows Little Influence for Donor Age and Mild Comorbidities on Phenotypic and Functional Properties

Anastazja Andrzejewska[1,2,3], *Rusan Catar*[1,4], *Janosch Schoon*[1,2,3], *Taimoor Hasan Qazi*[1,2,3], *Frauke Andrea Sass*[1,2,3], *Dorit Jacobi*[1,3], *Antje Blankenstein*[1,3], *Simon Reinke*[1,3], *David Krüger*[3], *Mathias Streitz*[1,5], *Stephan Schlickeiser*[1,5], *Sarina Richter*[1,5], *Naima Souidi*[1,2,5], *Christien Beez*[1,2,5], *Julian Kamhieh-Milz*[6], *Ulrike Krüger*[7], *Tomasz Zemojtel*[7], *Karsten Jürchott*[1], *Dirk Strunk*[8], *Petra Reinke*[1,5,9], *Georg Duda*[1,2,3], *Guido Moll*[1,2,3,4*†] *and Sven Geissler*[1,2,3*†]

[1] BIH Center for Regenerative Therapies (BCRT), Charité Universitätsmedizin Berlin, Corporate Member of Freie Universität Berlin, Humboldt-Universität zu Berlin, Berlin Institute of Health (BIH), Berlin, Germany, [2] Berlin-Brandenburg School for Regenerative Therapies, Charité Universitätsmedizin Berlin, Corporate Member of Freie Universität Berlin, Humboldt-Universität zu Berlin, BIH, Berlin, Germany, [3] Julius Wolff Institute, Charité Universitätsmedizin Berlin, Corporate Member of Freie Universität Berlin, Humboldt-Universität zu Berlin, BIH, Berlin, Germany, [4] Department of Nephrology and Internal Intensive Care Medicine, Charité Universitätsmedizin Berlin, Corporate Member of Freie Universität Berlin, Humboldt-Universität zu Berlin, BIH, Berlin, Germany, [5] Institute of Medical Immunology, Charité Universitätsmedizin Berlin, Corporate Member of Freie Universität Berlin, Humboldt-Universität zu Berlin, BIH, Berlin, Germany, [6] Department of Transfusion Medicine, Charité Universitätsmedizin Berlin, Corporate Member of Freie Universität Berlin, Humboldt-Universität zu Berlin, BIH, Berlin, Germany, [7] BIH Core Unit Genomics Charité Universitätsmedizin Berlin, Corporate Member of Freie Universität Berlin, Humboldt-Universität zu Berlin, Berlin Institute of Health, Berlin, Germany, [8] Berlin Center for Advanced Therapies, Charité Universitätsmedizin Berlin, Corporate Member of Freie Universität Berlin, Humboldt-Universität zu Berlin, BIH, Berlin, Germany, [9] Spinal Cord Injury and Tissue Regeneration Center, Experimental and Clinical Cell Therapy Institute, Paracelsus Medical University, Salzburg, Austria

***Correspondence:**
Guido Moll
guido.moll@charite.de
Sven Geissler
sven.geissler@charite.de

† These authors have contributed equally to this work and share senior authorship

Heterogeneous populations of human bone marrow-derived stromal cells (BMSC) are among the most frequently tested cellular therapeutics for treating degenerative and immune disorders, which occur predominantly in the aging population. Currently, it is unclear whether advanced donor age and commonly associated comorbidities affect the properties of *ex vivo*-expanded BMSCs. Thus, we stratified cells from adult and elderly donors from our biobank ($n = 10$ and $n = 13$, mean age 38 and 72 years, respectively) and compared their phenotypic and functional performance, using multiple assays typically employed as minimal criteria for defining multipotent mesenchymal stromal cells (MSCs). We found that BMSCs from both cohorts meet the standard criteria for MSC, exhibiting similar morphology, growth kinetics, gene expression profiles, and pro-angiogenic and immunosuppressive potential and the capacity to differentiate toward adipogenic, chondrogenic, and osteogenic lineages. We found no substantial differences between cells from the adult and elderly cohorts. As positive controls, we studied the impact of *in vitro* aging and inflammatory cytokine stimulation. Both conditions clearly affected the cellular properties, independent of donor age. We conclude that *in vitro* aging rather than *in vivo* donor aging influences BMSC characteristics.

Keywords: cellular therapy, bone marrow stromal cell, mesenchymal stromal cell, *in vivo* and *in vitro* aging, comorbidity, *in vitro* potency assay

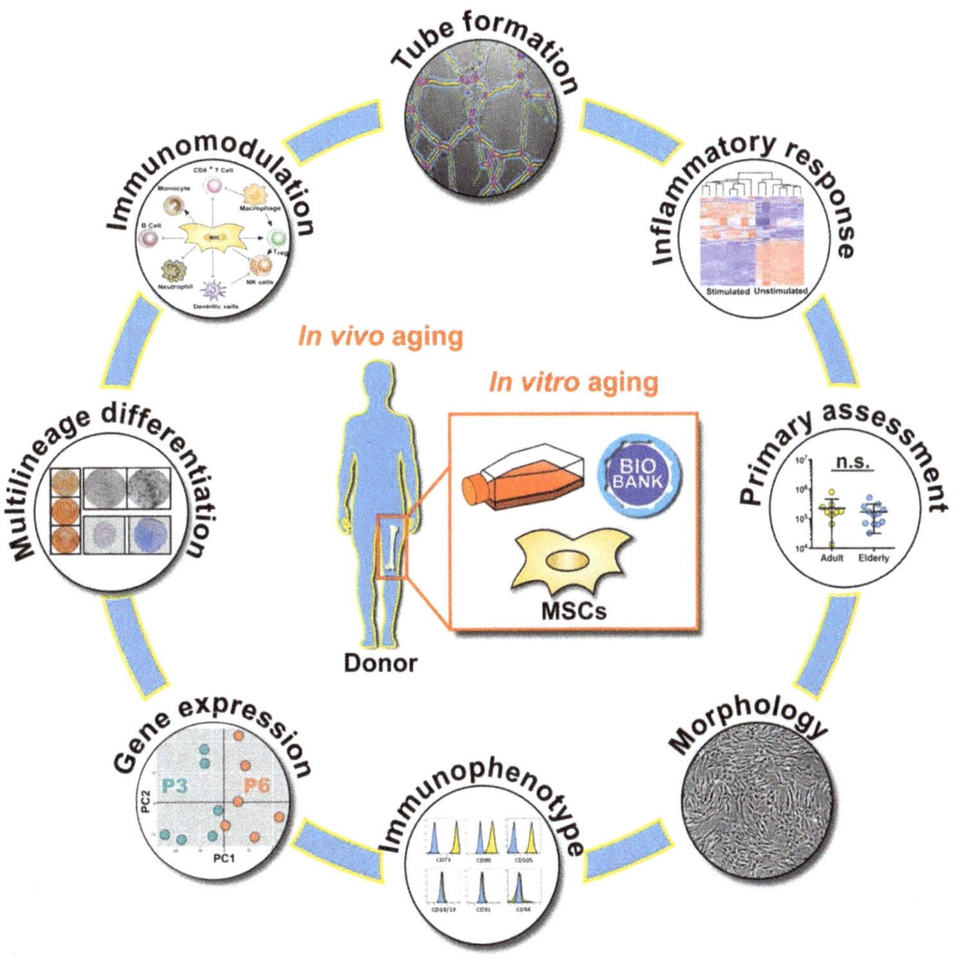

GRAPHICAL ABSTRACT | Overview on the molecular and functional assays used for the characterization of biobanked bone marrow stromal cells (BMSC) with respect to *in vivo* and *in vitro* aging, with primary assessment of starting material composition, cell morphology, immunophenotype, gene expression profile, multilineage differentiation capacity, immunomodulation, endothelial tube formation and inflammatory response.

INTRODUCTION

Qualifying adult regenerative cell sources in biobanking approaches is an essential task in order to overcome major pitfalls in regenerative medicine (1). Donor-intrinsic variation between different cell batches may influence the safety and efficacy of bone-marrow stromal cells (BMSCs) (2–4). Our previous work suggests that multiple parameters, such as tissue origin (5–7), culture time (8, 9), media supplementation (7, 10), and mode of cell delivery (4, 9, 11–13) can substantially affect cellular therapeutic properties. In addition, advanced donor age and the commonly associated comorbidities are thought to be another substantial confounder of potentially compromising BMSC phenotype and function (14–22).

Previous studies investigating the impact of donor age on BMSCs reported variable or partly inconclusive outcomes considering their *in vivo* frequency, their gene expression profile, and many of their functional parameters, such as antioxidant defense, cytoskeleton dynamics, migration behavior, differentiation capacity, and immunomodulatory and paracrine activity (**Table 1**) (14–22). These discrepancies may result from differences in experimental parameters such as donor species, cell isolation, and culture protocols and from small sample size or limited functional characterization. Potential age-dependent impairments by chronological *in vivo* aging may be further aggravated by the process of *in vitro* aging during serial expansion in tissue culture (14, 16). Thus, the true impact of advanced donor age on the therapeutic value of BMSCs is still rather unclear.

We hypothesized that donor age, in combination with age-related comorbidities, contributes to the perceived large phenotypic and functional heterogeneity between individual donor-derived cellular specimens. Surprisingly, we found no substantial association between donor age or comorbidities and BMSC characteristics. In contrast, our analysis revealed that *in vitro* aging and inflammatory cytokine stimulation clearly alter cellular properties.

MATERIALS AND METHODS

Isolation and Culture of Bone Marrow Stromal Cells (BMSCs)

BMSCs were received from the Core-Facility "Cell Harvesting" of the BIH Center for Regenerative Therapies (BCRT). The cells were isolated from metaphyseal bone marrow (BM) biopsies from patients undergoing hip replacement at Charité University Hospital, as previously stated (1, 66–68). Written informed consent was given, and ethics approval was obtained from the local ethics committee/institutional review board (IRB) of the Charité University Hospital.

Briefly, the BM mononuclear cell fraction (BM-MNC) in primary BM and the BMSC fraction post Ficoll-density gradient centrifugation (Histopaque 1077; Sigma-Aldrich) were quantified with an automated electrical impedance-based CASY® Cell Counter (Schaerfe System GmbH). The BMSC-containing interphase was plated in a 300 cm^2 tissue culture flask (ThermoFischer) and cultured under standard conditions (37°C, 5% CO_2) in an expansion medium (Dulbecco's Modified Eagle Medium-Low Glucose [DMEM-LG; Sigma-Aldrich] containing 10% fetal calf serum [FCS; Biochrom AG], 100 U/mL penicillin, and 100 µg/mL streptomycin [Biochrom AG], and 2 mM L-alanyl-L-glutamine [GlutaMAX; Gibco]). The non-adherent fraction was removed by washing with PBS (Gibco), the medium was changed every 72 h, and the cells were allowed to reach about 80% confluence before passaging.

The BMSCs were then expanded for several passages and were characterized with multiple functional and molecular assays, in line with the minimal criteria of the International Society for Cellular Therapy (ISCT) (69), at passage three (P3, early passage) and six (P6, late passage), respectively, as also shown in overview in the Graphical Abstract.

Cell Morphology, Viability, Growth Kinetics, and Immunophenotyping

Cell morphology was determined at regular intervals by using bright field light microscopy. Cell number, viability, size, and volume were determined at each culture passage by using the CASY® Cell Counter as outlined previously (6, 9). BMSC growth kinetics were quantified by calculating population doublings at each passage based on the following equation: $PD = \log(N/N_0)/\log(2)$. In this formula, N stands for the total number of viable cells at harvest, and N_0 is the initial number of cells seeded.

Flow cytometric immunophenotyping was conducted as described earlier (6, 9) using a BMSC Duraclone-panel (DURAClone SC Mesenchymal Tube; Beckman Coulter) containing the following antibodies: CD14, CD19, CD31, CD34, and CD45 (as negative markers) and CD73, CD90, CD105, and CD146 (as positive markers), or unlabeled control cells. Upon antibody labeling, the cells were washed with PBS, fixed with 1% paraformaldehyde, and analyzed on a Cytoflex flow cytometer (Beckman Coulter), and 5,000–10,000 gated events were quantified and analyzed with FlowJo v10.3.1 (FlowJo LLC).

Gene Expression Analysis by RNA Sequencing

The mRNA transcripts of resting or cytokine-stimulated BMSCs were studied by seeding the cells at 2,000–4,000 cells/cm^2 in 75 cm^2 culture flasks and expanding them for 1 week. Before harvest, the sub-confluent cells were washed twice with PBS and lysed with 1 mL of RLT-buffer (Qiagen). Total RNA was extracted by using the Qiagen RNeasy Plus Mini Kit (Qiagen), according to the manufacturer's instructions. A total of $n = 37$ BMSC samples were analyzed ($n = 24$ samples for P3 and $n = 6$ for P6, which were matched to $n = 6$ of the P3 donors). A subset of $n = 7$ donors was treated for 24 h with or without cytokines (TNF-alpha and IFN-gamma, both 10 ng/mL), which were matched to the corresponding unstimulated cells and processed in parallel.

Total RNA was extracted by using the RNeasy Plus Mini Kit, and the quality was assessed by Bioanalyzer RNA 6000 Nano assay (Agilent). Only high-quality RNA with RIN scores > 7 was used for library preparations. The RNA (1 µg of total RNA) from each sample was converted to complementary DNA (cDNA) using an iScript™ cDNA Synthesis Kit (Bio-Rad). Sequencing library preparation was performed using the NEBNext® Ultra™ RNA Library Prep Kit for Illumina® and PolyA mRNA selected from 500 ng of total RNA with a NEBNextPoly(A) mRNA Magnetic Isolation Module (both New England Biolabs) followed by library preparation. Libraries were quantified with a Qubit® dsDNA HS Assay Kit (Thermo Fischer) and sequenced on a HiSeq 4000 System (Illumina) in single-read mode with a 50-cycle read length.

FASTQ-files were quality-controlled with "fastQC" and trimmed for residual adapter sequences and low-quality reads with "AdapterRemoval" (70). Reads were aligned to the GRCh38 human genome using "tophat" and "bowtie2" (71, 72). Counts per gene were calculated using the "featureCounts" algorithm implemented in the "Rsubread" package in R (73). Genes were annotated with the "biomaRt" package and Ensembl-Version 94. Protein-coding genes were selected, expression values normalized, and variance stabilizing transformed using the "DESeq2" package in R (74). Principle component analysis (PCA) was performed for the 1,000 genes with the highest variance

TABLE 1 | Literature study on *in vivo* and *in vitro* aging and/or comorbidities of mesenchymal stromal cells.

Cell type (Tissue)	Type of aging	Model, donor source, donor number	Age range (Years)	Passage number	Parameters changed by aging and/or comorbidity	Ref.
BMSCs	*In vivo*	Human; Healthy individuals; N=8	16–32, 69–77	N.A.	(–) Proliferation (–) Immunophenotype (–) Metabolic activity (–) Trophic factor secretion *Therapeutic efficacy in C57/BL6 mouse model* ↓ Wound healing ↓ Neovascularization ↓ Trophic factor secretion ↓ Expression of genes involved in regeneration	(23)
BMSCs	*In vivo* *In vitro*	Human; Healthy individuals; N=12	21–25, 44–55, 80–92	P2–P11	*In vivo* Altered gene expression (–) Cell size (–) Immunophenotype (–) Osteogenic, adipogenic, chondrogenic potential *In vitro* ↓ Adipogenic potential ↑ Osteogenic potential ↑ Senescence	(24)
BMSCs	*In vivo*	Human; Healthy individuals, Patients with hip arthroplasty; *N* = 16	≤23, ≥65	P1–P2	↓ CFU-F ↑ Senescence ↑ Cell size ↑ SASP- cytokine production (–) Cell viability (–) Immunophenotype	(25)
BMSCs	*In vivo*	Human; Patients with hip OA; *N* = 19	19–70	P2	↓ Proliferation ↓ Osteogenic potential ↑ Apoptosis	(26)
BMSCs	*In vivo* *In vitro*	Human; Healthy individuals; *N* = 25	2–13, 20–50	≤ P27	*In vivo* ↓ Proliferation (–) Immunophenotype (–) Telomere length *In vitro* ↓ Telomere length	(27)
BMSCs	*In vivo*	Human; Healthy individuals; *N* = 30	0–60	N.A.	↓ Proliferation ↓ Adipogenic potential ↑ Osteogenic potential Altered gene expression	(28)
BMSCs	*In vivo*	Human; Healthy individuals; *N* = 33	5–55	P1–P5	↓ CFU-F ↓ Proliferation ↓ Osteogenic potential ↓ Chondrogenic potential ↑ Cell size ↑ Apoptosis Altered immunophenotype (CD44, CD90, CD105, Stro-1) (–) Adipogenic potential	(29)
BMSCs	*In vivo*	Human; Healthy individuals; *N* = 36	41–86		↓ Proliferation	(30)
BMSCs	*In vivo*	Healthy individuals; *N* = 41	3–70	N.A.	↓ Number of osteoprogenitors	(31)
BMSCs	*In vivo*	Human; Healthy individuals; *N* = 46	≥18	P1–P3	(–) Cell number/sample weight (–) Immunophenotype (–) Proliferation (–) Osteogenic, adipogenic, chondrogenic potential	(32)
BMSCs	*In vivo*	Human Healthy individuals; *N* = 53	13–80	P1	Altered immunophenotype, (–) Proliferation (–) Adipogenic, osteogenic, chondrogenic potential (–) Immunomodulatory activity (–) Trophic factor secretion	(33)

(Continued)

TABLE 1 | Continued

Cell type (Tissue)	Type of aging	Model, donor source, donor number	Age range (Years)	Passage number	Parameters changed by aging and/or comorbidity	Ref.
BMSCs	*In vivo*	Human; Patients with cardiac complications; N.A.	1–5, 50–70	N.A.	↓ CFU-F ↓ Proliferation (–) Immunophenotype	(34)
BMSCs	*In vitro*	Human; Healthy individuals; $N = 3$	N.A.	P4, P8, P12	↓ Proliferation ↓ Immunomodulatory activity (–) Immunophenotype, telomere length (–) Metabolic activity	(35)
BMSCs	*In vitro*	Human; Healthy individuals; $N = 6$	20–40	P1–P9	↓ Proliferation ↑ Cell size ↑ Senescence ↑ Telomere length (–) Immunophenotype	(36)
BMSCs	*In vitro*	Human; Healthy individuals; $N = 11$	23–63	≤P10	↓ Proliferation ↓ Adipogenic potential ↓ Osteogenic potential (–) Immunophenotype	(37)
BMSCs	*In vitro*	Human; Healthy individuals; $N = 3$	9,27,36	Early, late passage (≥ 38 PD)	Altered gene expression (–) Immunophenotype (–) Adipogenic and osteogenic potential	(38)
BMSCs	*In vivo*	Mouse; (C57Bl/6); $N = 3$	6 to 8-week-old ≥24-week-old	P2–P3	↓ Proliferation ↓ Osteogenic potential ↓ Immunomodulatory activity	(39)
BMSCs	*In vivo*	Mouse; (C57BL/6J); $N = 6$	3-month-old 16-month-old	N.A.	↓ Osteogenic potential ↑ Senescence ↑ Adipogenic potential	(40)
BMSCs	*In vivo*	Mouse; (C57BL/6); N.A.	6 to 8-month-old 20 to 26-month-old	N.A.	↓ Osteogenic potential ↑ Adipogenic potential	(41)
BMSCs	*In vivo*	Mouse; (SAMP6, SAMR1); N.A.	3 to 5-month-old	N.A.	↓ Osteogenic potential ↑ Adipogenic potential	(42)
BMSCs	*In vivo*	Mouse; (C57BL/6); N.A.	4 to 5-month-old 22 to 25-month-old	N.A.	↓ CFU-F	(43)
BMSCs	*In vivo* *In vitro*	Mouse; (C57Bl/6); $N = 3$	6 day-old 6 week-old 1-year-old	P1–P6	*In vivo* ↓ Proliferation ↓ Adipogenic potential ↓ Osteogenic potential ↓ Chondrogenic potential *In vitro* ↓ Adipogenic potential ↓ Osteogenic potential ↓ Chondrogenic potential	(44)
BMSCs	*In vivo* *In vitro*	Mouse; (BALB/c); $N = 20$	<4-week-old, 5 to 12-week-old 13 to 34-week-old	P3 - P24	*In vivo* ↓ CFU-F (–) Cell size (–) Proliferation (–) Immunophenotype (except CD73) (–) Adipogenic and osteogenic potential (–) Immunomodulatory activity *In vitro* ↓ Cell size ↑ CFU-F ↑ Proliferation ↑ Osteogenic potential (–) Immunophenotype (except Sca-1) (–) Adipogenic potential (–) Immunomodulatory activity,	(45)
BMSCs AT-MSCs	*In vivo*	Rat; (Lewis, Brown Norway); $N = 12$	4-week-old 15-month-old	N.A.	Altered immunophenotype (CD29, CD90, CD11, CD45)	(46)
BMSCs	*In vivo*	Rat; (Wistar); N.A.	12-month-old 24-month-old	N.A.	↓ Osteogenic potential	(47)

(Continued)

TABLE 1 | Continued

Cell type (Tissue)	Type of aging	Model, donor source, donor number	Age range (Years)	Passage number	Parameters changed by aging and/or comorbidity	Ref.
BMSCs	*In vivo* *In vitro*	Rat; (Sprague-Dawley); N.A.	3-week-old 12-month-old	≤P100	*In vivo* ↓ Migration potential (–) Proliferation, osteogenic and adipogenic potential, cell size, *In vitro* ↓ Cell size ↓ Adipogenic potential ↓ Osteogenic potential ↓ Metabolic activity ↓ Gene expression involved in differentiation and mitochondrial functions (–) Proliferation	(48)
AT-MSCs	*In vitro*	Human; Healthy individuals; $N = 3$	N.A.	P5, P10, P15	↓ Proliferation ↑ Cell size ↑ Morphological heterogeneity (–) Osteogenic, adipogenic, chondrogenic potential (–) Immunophenotype (except CD105)	(49)
AT-MSCs	*In vivo*	Human; Healthy individuals; $N = 8$	0–1 70–80	P3–P8	↓ Proliferation ↓ Osteogenic potential ↓ Adipogenic potential ↑ Senescence	(50)
AT-MSCs	*In vivo*	Human; Healthy individuals; $N = 24$	6–12 22–27 60–73	P1–P5	↓ CFU-F ↓ Proliferation ↓ Osteogenic potential ↓ Adipogenic potential ↓ Migration potential ↑ Senescence (–) Cell viability (–) Immunophenotype	(51)
AT-MSCs	*In vivo*	Human; Patients with CAD & healthy individuals; $N = 95$	2–82	P2	↓ Angiogenic potential ↓ Telomerase activity (–) Immunophenotype;	(52)
AT-MSCs	*In vivo*	Human; Healthy individuals; $N = 260$	5–97	P0, P5	↓ Adipogenic potential (–) CFU-F (–) Proliferation (–) Osteogenic and chondrogenic potential	(53)
UC-MSCs	*In vitro*	Human; Healthy individuals; N.A.	>37 pregnancy week	P0–P16	Altered gene expression	(54)

Cell type (Tissue)	Type of comorbidity	Model, donor source, donor number	Age range (Years)	Passage number	Parameters changed by comorbidity	Ref.
BMSCs	T1D with renal failure	Human; T1D Patients & non-diabetic individuals; $N = 31$	18–70	P1–P5	Altered expression of genes involved in wound healing and stress response (–) CFU-F (–) Immunophenotype (–) Proliferation (–) Migration potential (–) Immunomodulatory activity	(55)
BMSCs	T1D	Human; T1D Patients and non-diabetic individuals; N.A.	23, 31	N.A.	(–) Cell size (–) Immunophenotype (–) Adipogenic differentiation (–) Immunomodulatory activity (–) Gene expression	(56)
BMSCs	DM,CLI, CAD	Human; Ischemic Patients (+DM) and healthy individuals $N = 12$	N.A.	P3–P6	↓ Proliferation in later passages (–) Immunophenotype (–) Angiogenic potential	(57)

(Continued)

TABLE 1 | Continued

Cell type (Tissue)	Type of comorbidity	Model, donor source, donor number	Age range (Years)	Passage number	Parameters changed by comorbidity	Ref.
AT-MSCs	T2D with CLI	Human; T2D Patients; N.A.	N.A.	N.A.	↓ Proliferation ↓ Migration potential ↓ CFU ↓ PDFG signaling ↓ Osteogenic potential ↑ Adipogenic potential ↑ Prothrombotic phenotype (–) Immunophenotype	(58–60)
AT-MSCs	T2D	Human; T2D and non-diabetic patients; $N = 40$	N.A.	N.A.	Altered immunophenotype (CD90, CD105) ↑ Expression of stemness markers (NANOG, OCT4) ↑ Oxidative stress ↑ Production of pro-inflammatory cytokines	(61)
AT-MSCs	T2D	Human; T2D and non-diabetic patients; $N = 40$	60–76	N.A.	↓ CFU ↑ Apoptosis ↑ Senescence (–) Proliferation	(62)
AT-MSCs	ATH T2D	Human; Patients with/without ATH; $N = 50$	<65 ≥65	P2–P3	↓ Immunomodulatory activity	(62)
AT-MSCs	Obesity T2D	Human; Healthy individuals, Patients with Obesity & T2D $N = 12$	30–55	P3–P7	↓ Immunomodulatory activity ↑ Metabolic activity ↑ Migration potential ↑ Expression of inflammatory markers	(63)
BM-ECs	T1D	Mouse; T1D and non-diabetic (CD1); N.A.	N.A.	N.A.	↓ Angiocrine activity, migration ↓ Angiogenic potential ↑ Transendothelial migration ↑ Permeability	(64)
BM-ECs	T1D	Mouse; T1D and non-diabetic (CD1); N.A.	N.A.	N.A.	↓ Hematopoietic fraction in bone ↓ Migration ↓ Angiogenic potential ↑ Osteopenia in bone ↑ Fat cells in bone ↑ Senescence ↑ Oxidative stress	(65)

N.A., not available; ***Cell types****: BMSCs, bone marrow stromal cells; MSC, mesenchymal stromal cells;* ***Tissue sources****: AT, adipose tissue; BM, bone marrow; UC, umbilical cord; P, passage number;* ***Parameters****: PD, population doubling; CFU-F, colony-forming unit fibroblast; SASP, senescence-associated secretory phenotype; PDGF, platelet-derived growth factor;* ***Comorbidities****: CLI, critical limb ischemia; T1D/T2D, type 1 and 2 diabetes mellitus; ATH, atherosclerosis; CAD, coronary artery disease; OA, osteoarthritis.*

across all samples. Differentially expressed genes between groups were determined using negative binomial distribution models as implemented in the "DESeq2" package. Raw p-values were adjusted for multiple testing with Bonferoni correction, and an adjusted p-value below 0.05 was used for the selection of significant genes. Functional annotation and enrichment analysis were carried out using "DAVID" with the "clusterProfiler" package in R (75). False discovery rates were used to adjust raw p-values for multiple testing, and a threshold of $p < 0.05$ was used for the selection of significant results. Dotplots of top-ranking results were created with the function implemented in the "clusterProfiler" package. GOcirc plots were created using the algorithm in the "GOplot" package in R (76). The raw data on expression are available at the Gene Expression Omnibus under the GEO-Accession-ID (GSE139073).

Multilineage Differentiation Analysis

Adipogenic, osteogenic, and chondrogenic differentiation of BMSCs at P3 and P6 were induced by using specific differentiation media and evaluated as previously described (48, 66–68, 77–79). Briefly, BMSCs were plated in 24-well plates at specific densities for adipogenic (1.44×10^4 cells/well) or osteogenic (1.28×10^4 cells/well) differentiation or in V-bottom 96-well plates at higher density (3.0×10^5 cells/well) for chondrogenic differentiation. Control cells were exposed to normal culture media, and all cultures were sustained for up to 22 days. To compare the differentiation potential of BMSCs among age groups and comorbidities and between passages, the differentiation responses from each individual were normalized to their respective controls.

Adipogenic Induction

BMSCs were cultured with complete DMEM-HG (High Glucose) supplemented with 10 μM dexamethasone, 50 μM indomethacin, 10 mM 3-isobutyl-1-methylxanthine, and 0.1 μM insulin (all from Sigma-Aldrich, St. Louis, MI, USA). Adipogenic differentiation was demonstrated by performing Nile Red staining (Sigma-Aldrich) to visualize lipid droplet formation. Quantification was achieved by measurement of Nile Red fluorescence (Ex/Em 485/540), which was normalized to the cell number quantified by staining with Hoechst 33258 dye (Life Technologies) and consecutive readout on a multimode microplate reader (TECAN M200 PRO) (67).

Osteogenic Induction

BMSCs were cultured with complete DMEM-LG supplemented with 0.1 μM dexamethasone, 50 μM ascorbic acid, and 10 mM beta-glycerol-phosphate disodium salt hydrate (all Sigma-Aldrich). Osteogenesis was assessed by Alizarin Red S (Merck) staining to determine mineralized matrix deposition, which was quantified at days 14, 18, and 22 by measuring the absorbance of Alizarin Red S and then normalized to the cell number determined by Hoechst staining, with consecutive readout of absorbance on the TECAN reader. The ALP activity level was quantified by measuring the consumption of p-nitrophenyl phosphate (pNPP; Sigma-Aldrich), which was normalized to the amount of viable cell metabolic activity as measured by PrestoBlue® assay (Life Technologies) (67, 78).

Chondrogenesis

BMSCs (only at passage 6) were placed in V-bottom 96-well plates, centrifuged and subsequently cultured for up to 21 days in a chondrogenic medium [FBS-free DMEM-HG supplemented with 6.25 μg/mL insulin-transferrin-selenium, 0.1 μM dexamethasone, 50 μg/mL L-Ascorbic acid 2-phosphate sesquimagnesium salt hydrate, 1 mM sodium pyruvate, 0.35 mM L-proline, 1.25 mg/mL bovine serum albumin, 5.35 mg/mL linoleic acid (all from Sigma-Aldrich), and 10 ng/mL transforming growth factor-beta-3 (TGF-beta; Peprotech)] (78), and to quantify chondrogenesis, proteoglycan production was normalized to total protein amount (67).

Immunomodulation, Endothelial Tube Formation, and Cytokine Measurements

The immunomodulatory effects of BMSCs were assayed as described previously (80). Human peripheral blood mononuclear cells (PBMCs) were stained with 5 μM carboxy-fluorescein-succinimidyl ester (CFSE; Life Technologies), stimulated with anti-CD3 and anti-CD28 0.25 μg/mL (Biolegend) or with phytohemagglutinin (PHA; 0.5 μg/mL; Sigma Aldrich), and co-cultured with or without BMSCs at a ratio of 10:1. After 5 days, the CFSE-labeled PBMCs were harvested, stained with antibodies specific for CD4 and CD8 (anti-CD4-APC; anti-CD8-PE; both Miltenyi), and subjected to analysis with flow cytometry.

For tube formation assay (81), human umbilical vein endothelial cells (HUVECs) were plated in 96-well plates coated with Matrigel (Corning) and co-cultured for 16 h with conditioned culture medium derived from BMSCs or with unconditioned blank control. Bright-field microscopic images of each well were taken for computer-assisted quantification of multiple parameters associated with endothelial network formation, e.g., total master segment length (TMSL/field) (ImageJ 1.51; Bethesda, USA). For the generation of BMSC-conditioned media, the cells were seeded at a density of 1×10^4 cells/cm^2 in 24-well plates (DMEM + 10%FCS), allowed to adhere overnight, washed once to remove residual protein, and cultured for 24 h with media containing 0.5% FCS to collect the cells secretome. The conditioned a medium was collected and centrifuged to remove cell debris and supernatants, filtered, and stored at −80°C until assayed. Levels of interleukin 6 (IL-6) and vascular endothelial growth factor (VEGF) were assayed by using enzyme-linked immunosorbent assays (ELISA; R&D Systems).

Exploratory and Descriptive Statistics

Statistical analyses were performed using ANOVA and the Student's *t*-test. All data sets from individual experiments were tested for normal distribution with the Shapiro-Wilks test prior to testing for statistical significance. When performing multiple pair-wise comparisons, one way or two-way ANOVA was used, and Bonferroni *post-hoc* corrections were performed to adjust the *p*-values. For single-group testing, statistical significance was tested by Student's *t*-test (paired or unpaired, two-tailed). If the data did not fit a normal distribution, the Mann-Whitney test or the Wilcoxon matched-pairs test was used (two-tailed confidence intervals, 95%; $P < 0.05$ was considered statistically significant; Prism 5.0; Graphpad Software, USA).

RESULTS

Donor Stratification, Phenotypic, and Growth Characteristics of BMSCs

This study included BMSC preparations from 10 adult (38.2 ± 11.1 years) and 13 elderly (72.2 ± 7.5 years) donors, selected upon stratification of clinical background, to dissect the influence of donor age and common comorbidities (**Table 2**). We found that 60% of the adult cohort and 100% of the elderly cohort presented with comorbidities, with diabetes mellitus in 10% (1/10) and 54% (7/13) of cases (mainly early-stage, 6 non-insulin, and 2 insulin-dependent), respectively, followed by hypertension and other cardiovascular complications.

For both the adult and elderly cohort, the BM used as the starting material for cell isolation had a similar sample weight (**Figure 1A**, left panel), content of BM-MNCs (**Figure 1B**, left panel), and BMSC frequencies (**Figure 1C**, left panel). Stratification according to non-diabetic and diabetic donors also resulted in a comparable sample weight (**Figure 1A**, right panel), number of BM-MNCs (**Figure 1B**, right panel), and BMSC content (**Figure 1C**, right panel). Independent of donor age or diabetic status, the isolated BMSCs exhibited similar growth kinetics, as quantified by cumulative population doublings at passages 3 to 6 (**Figure 1D**). We observed a trend of reduced cell growth with increasing passage number in some of the BMSC preparations generated from the elderly diabetic donors with multiple comorbidities (see below).

TABLE 2 | Characteristics of bone marrow donors used for isolation of MSCs.

Donor ID	Sex (M/F)	Age (Years)	Comorbidities (Number)	Diabetes mellitus	Other types of potential comorbidities diagnosed
P127	F	16	None	None	None
P264	F	25	None	None	None
P276	F	48	Yes (2)	Yes (NID)	Bone cyst
P285	F	35	None	None	None
P289	F	45	None	None	None
P293	M	48	Yes (1)	None	Hypertension
P308	F	47	Yes (1)	None	Hypertension
P357	M	37	Yes (1)	None	Hyperuraemia
P784	M	33	Yes (1)	None	Hypertension
P819*	F	48	Yes (1)	None	Hypertension
Adults	**3 / 7**	**38.2 ± 11.1**	**6 / 10 (60%)**	**1 / 10 (10%)**	**6 / 10 (60%)**
P237	M	61	Yes (4)	Yes (NID)	Hypertension, HPLA, and RA
P265	M	63	Yes (1)	None	Hypertension
P278	F	82	Yes (2)	None	Hypertension and Hyperuraemia
P316	F	71	Yes (2)	Yes (NID)	Hypertension
P336	M	80	Yes (3)	Yes (NID)	Haematuria and Bradycardia
P354	M	85	Yes (2)	Yes (NID)	Hypertension
P374	M	68	Yes (1)	None	Hypertension
P378	M	78	Yes (1)	None	Hypertension
P386	F	68	Yes (1)	None	Hypertension
P651	F	69	Yes (1)	None	Hypertension
P660	F	65	Yes (2)	Yes (NID)	Hypertension
P777	F	73	Yes (4)	Yes (ID)	Hypertension, CKD3, and HVI
P821*	F	75	Yes (6)	Yes (ID)	Hypertension, CKD3, and HVI/DVT/PE
Elderly	**6 / 7**	**72.2 ± 7.5**	**12 / 12 (100%)**	**7 / 13 (54%)**	**13 / 13 (100%)**

*The age range is defined as follows: Adults younger than 50 years, and Elderly older than 60 years, and age values are presented as mean ± SD. The main type of comorbidity studied is diabetes mellitus. (*)Patients 819 and 821 were only included in **Figures 5, 6**. CKD3, chronic kidney disease grade 3; HLPA, hyperlipoproteinemia; HVI, heart valve insufficiency; DVT, deep-vein thrombosis; PE, pulmonary embolism; RA, rheumatoid polyarthritis; and NID/ID, non-insulin-/insulin-dependent diabetes mellitus.*

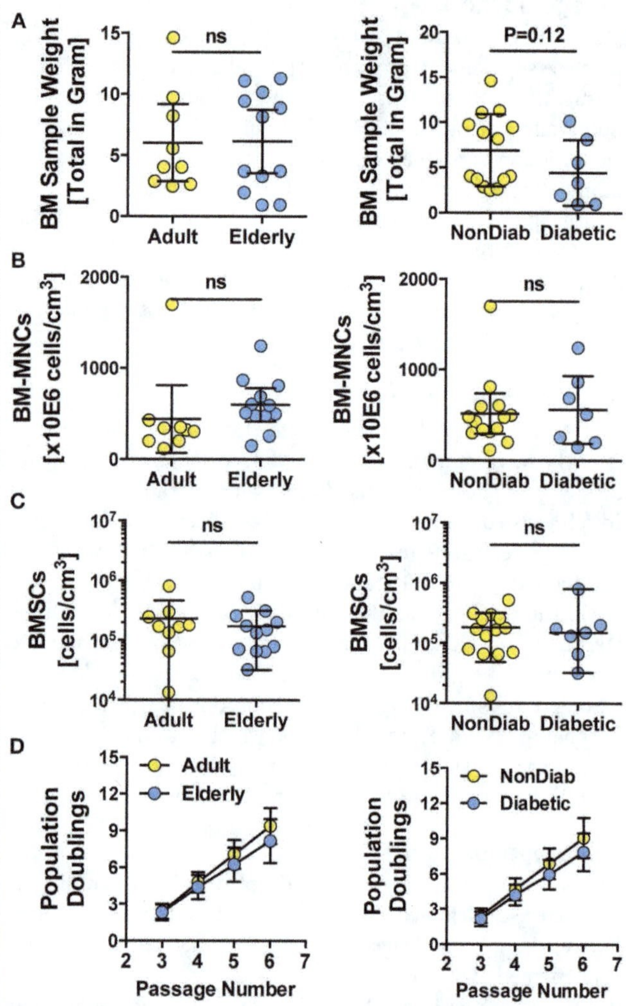

FIGURE 1 | Primary isolation and growth kinetics of BMSCs. **(A–C)** Quality control of BM samples: **(A)** bone marrow sample weight (in grams) and **(B)** number of BM-MNCs per sample (cells/cm^3), **(C)** number of BMSCs per sample (cells/cm^3), and **(D)** growth kinetics of BMSCs, with population doublings determined at different passages (P3-6), were quantified for BMSC preparations from adult vs. elderly ($n = 9$ vs. 12) and for non-diabetic vs. predominantly non-insulin-dependent early-stage diabetic donors ($n = 14$ vs. $n = 7$). Data are shown as mean ± SD, and the statistics were evaluated with a Student's t-test.

Phenotypic profiling revealed that all of the isolated BMSC cultures had a typical fibroblast-like morphology that was preserved during *in vitro* expansion up to passage 6 (**Figure 2A** and **Figure S1**). Regardless of donor age, trypsin-detached spheroid BMSCs had similar cell diameter and volume values at passage 3 (**Figure 2B** and **Figure S1**). Cells from adult and elderly donors, however, exhibited an increase in cell diameter and volume with increasing culture time, though this difference only reaches static significance between BMSCs derived from elderly donors at passages 3 and 6 ($P < 0.001$). BMSCs from diabetic donors showed a similar trend of increased cell size and volume, especially at higher passages, but this was not significant (**Figure 2B**).

Analysis of the cell surface marker pattern revealed that all of the BMSC preparations exhibited a similar surface marker phenotype at passage 3, as defined by the ISCT criteria (69), independent of donor age and disease status (**Figure 2C**). The isolated cells express typical MSC markers (CD73, CD90, CD105, and CD146) while being negative for contaminating cell populations (CD14, CD19, CD31, CD34, and CD45) such as cells of myeloid, B-cell, endothelial, and hematopoietic origin, respectively.

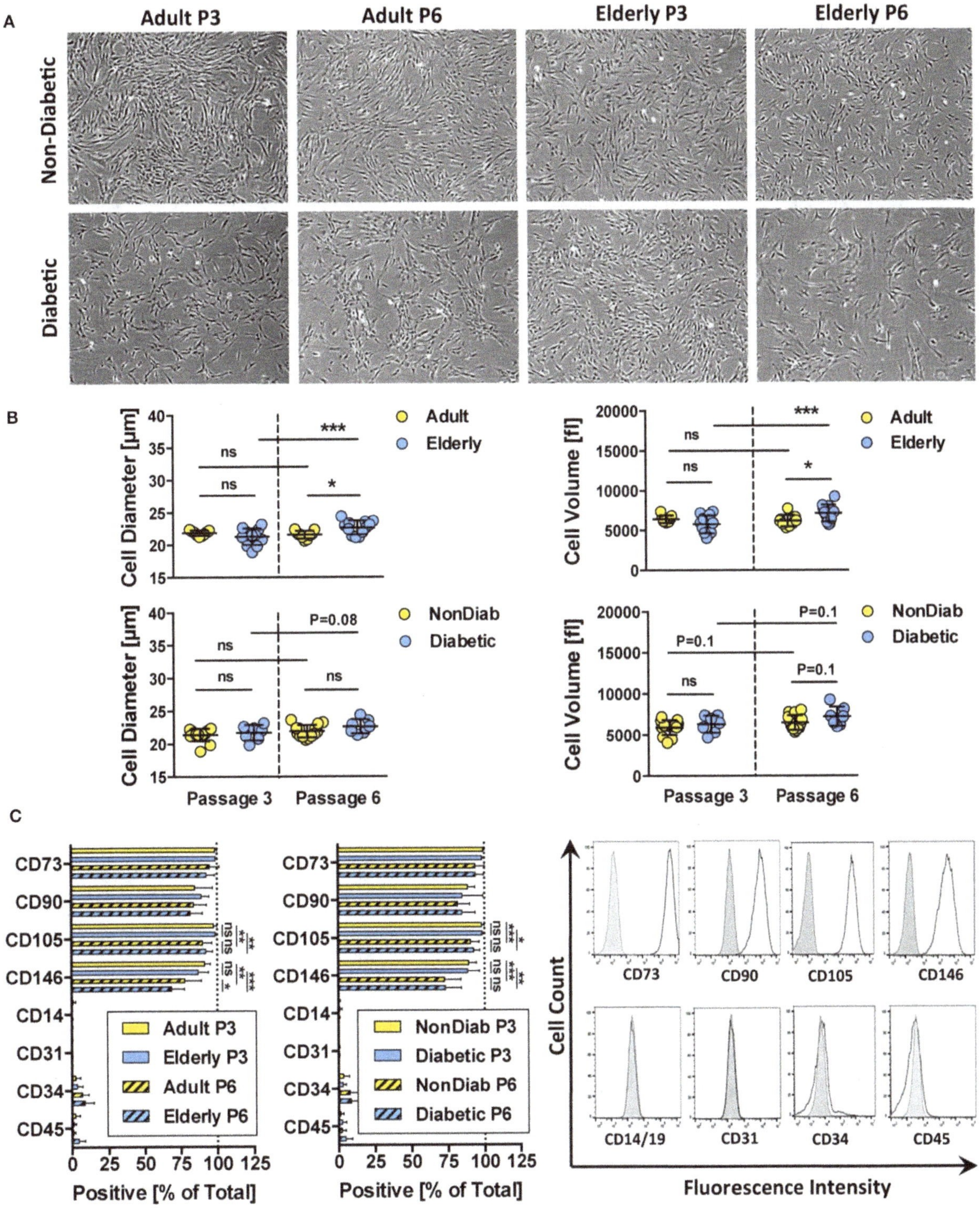

FIGURE 2 | Morphology and immunophenotype of BMSCs. **(A)** Representative bright-field microscopy images of BMSC cultures at passage 3 and 6, comparing adult vs. elderly and non-diabetic vs. diabetic donors, showing typical fibroblast-like morphology with a trend toward more irregular morphology in diabetic donors; **(B)** analysis of cell size and cell volume of trypsin-detached BMSCs from adult vs. elderly ($n = 9$ vs. 12) and non-diabetic vs. diabetic donors ($n = 14$ vs. 7). Adult and elderly have similar cell size and volume at passage 3, and the cell size of adult donor-derived BMSCs does not increase with passages, while elderly donor-derived BMSCs display cell enlargement with increasing passage; and **(C)** flow cytometry analysis of BMSCs (% positive cells; $n = 6$ random adult or elderly donors at early passage 1–3) with representative histograms shown to the right (unlabeled controls are shown in solid gray). The cells highly express typical BMSC-associated markers CD44, CD73, CD90, and CD105 while exhibiting no/weak expression of non-MSC-associated markers CD14, CD31, CD34, and CD45. Data are shown as mean ± SD, and statistical evaluation was performed by Student's t-test ($^{*}P < 0.05$ and $^{***}P < 0.001$).

Interestingly, we noted a weak decline in the expression of CD105 ($P < 0.05$ to $P < 0.001$) and CD146 ($P < 0.01$ to $P < 0.001$) upon extended culture up to P6 in all BMSC preparations, suggesting that long-term culture had a negative impact on the MSC phenotype of the cells. Indeed, a weak decline in CD105 and a stronger decline in CD106- and CD146-expression after extended culture, particularly in DMEM-media, or upon repeated passaging, have been reported previously (82). Functionally, the altered expression of CD105 was found to be associated with decreased osteogenic potential and altered Notch signaling (83).

In vitro Aging, but Not Donor Aging Alters the Transcriptome of Biobanked BMSCs

Next, we studied whether subgroups of BMSCs under resting conditions differed in their gene expression profiles by performing global transcriptome analysis. Multivariate statistical analysis using principal component analysis (PCA) was performed to study the variability between the groups with subsequent visualization of significant differences by hierarchical clustering heat maps and gene-ontology (GO) term enrichment analysis.

Our PCA-analysis found no clear separation between unstimulated BMSCs at P3 either for the comparison of adult and elderly ($n = 9$ vs. $n = 12$ donors) or non-diabetic and diabetic ($n = 14$ vs. $n = 7$ donors) donors (**Figure 3A,** left and central panel). Our diabetic group consisted of one adult and six elderly donors, and thus the comparison between non-diabetics (8 adults and 6 elderly) and diabetics is mainly a comparison with elderly diabetics (86% of the group), which could potentially weaken this analysis. To clarify this point, we carried out a sub-stratified comparison only between elderly diabetic and elderly non-diabetic donors ($n = 6$ each) (**Figure S2**). However, this analysis came to the same conclusion as the prior comparison and revealed no apparent differences in transcriptome between the two groups.

This result could suggest either, that the RNA sequencing-obtained gene-expression profiles of our isolated and *in vitro* expanded BMSCs are not affected consistently enough by the parameters age or diabetic status to allow for a robust multivariate statistical separation of these groups under resting conditions or, alternatively, that the studied *in vivo* imprint of the donor (e.g., aging and mild comorbidities) is simply not strong enough or is not maintained after isolation and *in vitro* expansion for several weeks.

Thus, as a positive control, we compared the gene expression pattern of selected donors from the same cohorts at P3 and P6 ($n = 6$ per group). We detected a clear difference between the two groups (**Figure 3A,** right panel) that confirms earlier reports (16, 38, 54). Our PCA showed a 32% variance in PC1 and 24% variance in PC2, indicating that PC1 (influence of culture time) accounts for most of the observed differences in expression pattern between the two groups. Accordingly, a subsequent hierarchical clustering analysis separated the transcriptome of BMSC preparations by their number of passages into two distinct P3 and P6 groups (**Figure 3B**).

The "Top 10 Results" from GO-term enrichment analysis of the biological processes predominantly involved identified highly significant changes in gene signatures associated with cell cycle, nuclear cell division, and chromosome segregation (**Figure 3C**). Concomitant in-depth delineation of the biological process and molecular functions involved using a combined DAVID and GO-database "Top 8 Results" analysis (**Figure 3D**) revealed that the altered biological processes were associated with changes in cytoskeletal proteins and microtubule binding, tubulin, fibronectin, enzyme, and protein kinase binding and kinase activity.

In summary, we found that the process of BMSC *in vitro* expansion had a strong impact on molecular phenotype, which may mask any rather weak *in vivo* imprint from donor aging and associated mild comorbidities after expansion over several passages in culture, as is typically done for cell production and biobanking. This does not generally exclude potential differences becoming apparent for the assessment of larger age differences (e.g., when comparing very young vs. elderly donors) or when studying the impact of stronger comorbidities.

In order to understand the functional heterogeneity of BMSC preparations, we next analyzed their functional performance in multiple *in vitro* assays typically employed for BMSC characterization (84).

In vitro Aging, but Not Donor Aging Affects BMSC Differentiation Capacity

In accordance with the prior transcriptome analysis setup, we studied the differentiation capacity of BMSCs depending on donor age and diabetic status with adjunct comparison of passages 3 and 6 (**Figure 4** and **Figures S3, S4**). In line with the prior results, we did not find any major differences in the differentiation capacity of BMSC preparations with respect to donor age and the presence of early-stage mild diabetes in the donor cohort under standard culture conditions (**Figures 4A–C** and **Figure S4**).

Along with the transcriptome changes observed during replicative *in vitro* aging, we compared the differentiation capacity of BMSC toward osteogenic and adipogenic lineages between P3 and P6 in order to evaluate whether the substantial transcriptional changes may also reflect alterations in functional behavior (**Figures 4A,B**). Indeed, BMSCs from adult and elderly donors exhibited significantly diminished osteogenic differentiation at passage 6 compared with passage 3 ($P < 0.01$ and $P < 0.001$, **Figure 4A**), which was also evident for comparison of non-diabetic and diabetic donors at both passages ($P < 0.001$ and $P < 0.05$). In contrast, adipogenic differentiation potential only showed minor changes, mainly reduced lipid formation, when comparing early and late passages or at later readout ($P < 0.05$, **Figure 4B**). The BMSCs from both adult and elderly donors displayed a similar increase in proteoglycan production upon chondrogenic differentiation with an induction medium containing transforming growth factor-beta (TGF-β) ($P < 0.01$ and P

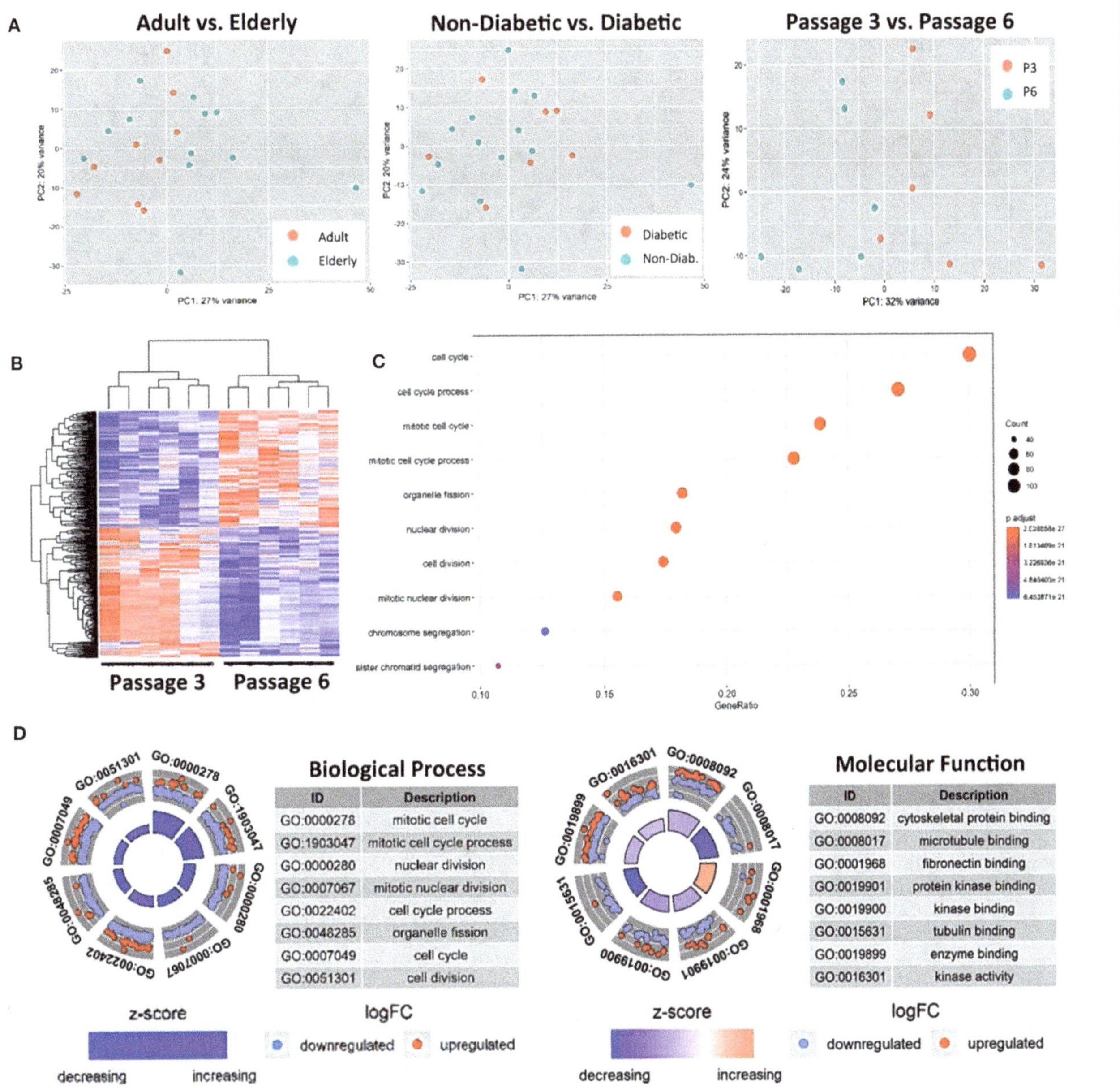

FIGURE 3 | Gene-expression-profiling of BMSCs with RNA-sequencing. **(A)** Principle component analysis (PCA) was employed for the visualization of group separation. Groups were stratified either according to donor age (left panel; $n = 9$ adult vs. $n = 12$ elderly donors), health status (central panel: $n = 14$ non-diabetic vs. $n = 7$ diabetic donors), and passage number (right panel; $n = 6$ random donors at passage 3 vs. 6). The PCA showed clear separation when comparing passage 3 vs. 6 (right panel), but no separation for the comparison of adult vs. elderly or diabetic vs. non-diabetic donors, with a random dot distribution throughout the graph. **(B–D)** passage 3 vs. 6 comparisons: **(B)** hierarchical clustering heat map with expression values sorted according to donor (rows) and gene (columns), where low expression is denoted by blue and high expression by red; **(C)** gene ontology (GO) enrichment analysis, with the "Top 10 Results" for changes in biological process (e.g., cell cycle, nuclear and cell division, and chromosome segregation) shown on the left y-axis and the size of the dots representing the counts of genes involved, while the color of the circle (scaled blue/lowest to red/highest) indicates significance expressed as adjusted p-value. **(D)** Combined David and GO "Top 8 Results" database analysis for changes in biological process (left panel; e.g., mitotic cell cycle and nuclear division) and molecular function (right panel; e.g., cytoskeletal protein binding and microtubule binding), with the z-score indicating the overall decrease or increase in expression for certain GO terms and the log-FC analysis indicating the actual number of down- or up-regulated genes within a certain GO term.

< 0.001, **Figure 4C**). BMSCs from non-diabetic and early-stage diabetic donors displayed similar chondrogenic differentiation capacity upon induction with the specific differentiation medium, thus excluding any major detectable influence of advanced BMSC donor age or diabetic status on chondrogenic differentiation capacity.

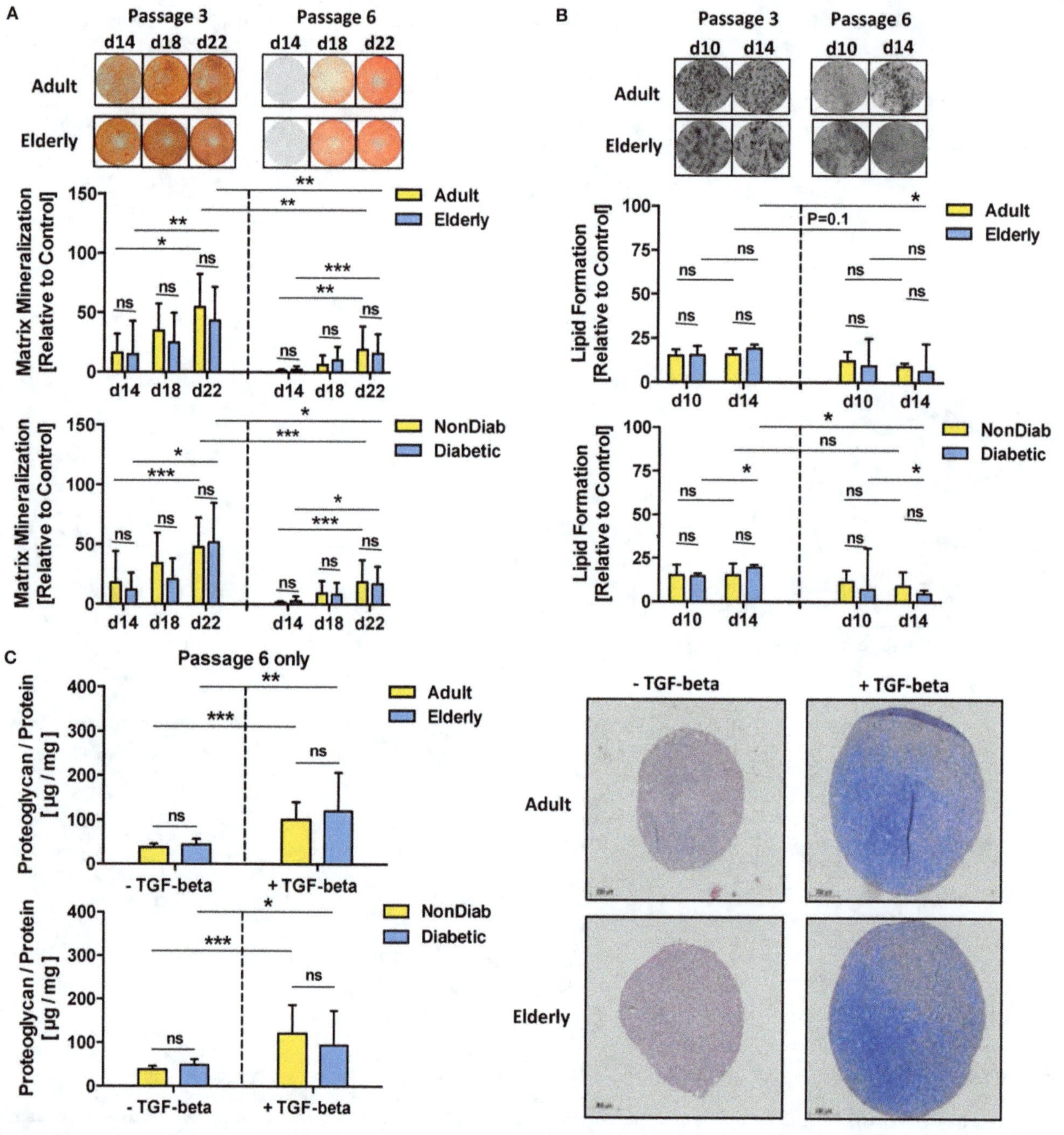

FIGURE 4 | Multilineage differentiation potential of BMSCs. The differentiation capacity of MSCs from adult and elderly ($n = 9$ vs. 12) and non-diabetic and predominantly non-insulin-dependent early-stage diabetic donors ($n = 14$ vs. 7) was assessed with different *in vitro* assays. **(A)** Osteogenic differentiation was assessed by quantification of BMSC matrix mineralization upon osteogenic induction for 14, 18, and 22 days, with representative images for Alizarin red staining shown at the top. Quantification revealed a time-dependent increase in mineralization for all groups (day 14 vs. day 22) both at passages 3 and 6, while there was no difference between adult vs. elderly or non-diabetic vs. diabetic donors at either time point or passage but a strong reduction in mineralization for higher vs. lower passage cells (P6 vs. P3). **(B)** Adipogenic differentiation of BMSCs upon *in vitro* adipogenic induction for 10 and 14 days was quantified by staining of lipid-rich vacuoles with Nile Red, with representative images for vacuole formation shown at the top. Quantification revealed similar vacuole formation for both time points (day 10 vs. day 14), with a trend of modest reduction when comparing higher vs. lower passage cells (P6 vs. P3). There was again no difference between adult vs. elderly or non-diabetic vs. diabetic donors at either time point but a stronger passage-dependent reduction for cells from elderly or diabetic donor donors (P3 vs. P6), though this was less notable for cells from adult or non-diabetic donors. **(C)** Chondrogenic differentiation of BMSCs upon induction with TGF-beta was quantified as the ratio of proteoglycan synthesis relative to protein content, with representative histology images for alcian blue proteoglycan staining of pellet sections shown to the right. There was no difference between BMSCs from adult vs. elderly or non-diabetic vs. diabetic donors. Data are shown as mean ± SD, and statistical evaluation was performed by using a Student's t-test or ANOVA followed by post-tests ($^{*}P < 0.05$, $^{**}P < 0.01$, and $^{***}P < 0.001$).

We found that assay readout-time had a considerable confounding influence on the obtained results, e.g., when comparing mineralization at day 14 to day 22 for cells at passage 3 and 6 ($P < 0.05$ to $P < 0.001$, **Figure 4A**) or when studying the optimal time point for the assessment of ALP-activity, which was found to be highest on day 5 for both adult and elderly and both non-diabetic and diabetic cohorts at passage 3 and 6 (**Figure S3**). This assay-readout time-dependence was less evident for the assessment of adipogenic differentiation (**Figure 4B**), e.g., when comparing day 10 and 14 at passages 3 and 6. Interestingly, we could detect weakly compromised lipid formation for BMSCs from elderly donors ($P < 0.05$, **Figure 4B**).

Cytokine Challenge *in vitro* Reveals Altered Gene Regulation in BMSCs From Select Elderly Donors With Multiple Comorbidities

Many clinical applications of BMSCs involve the therapeutic delivery of the cells into challenging *in vivo* environments characterized by inflammation or anoxia (13). Thus, it is indeed advisable to conduct cell-profiling approaches under resting and stimulating conditions (**Figure 5A**) (85). Stimulation with a cytokine cocktail resulted in a clear separation into two distinct groups in the hierarchical clustering heat map (**Figure 5A**), with multiple changes in biological processes such as immune system process or immune and inflammatory responses, as typically observed upon cytokine challenge of MSCs, e.g., during potency analysis of MSC products (85, 86).

Substratified expression analysis focusing only on stimulated samples found (**Figure 5B**), that BMSCs from the two elderly donors with insulin-dependent diabetes and multiple other comorbidities (P777 and P821, $n = 4$ and $n = 6$ comorbidities, respectively) clustered differently from the elderly and adult donors with fewer comorbidities (P265 and P374; and P264, P784, and P819, $n = 0$–1 comorbidities each). However, this has to be interpreted with caution since the number of donors with multiple comorbidities in this analysis was very limited due to their rare occurrence in our biobank. Indeed, changes in BMSC transcriptome upon disease progression to a more advanced stage (e.g., in advanced insulin-dependent diabetes and renal failure) have been reported earlier (55, 58, 60).

In our study, this was associated with a decline in cell proliferation and progression from a regular to a more irregular cell morphology in culture, particularly for BMSCs from elderly donors with insulin-dependent diabetes and multiple comorbidities (**Figure 5B,** lower panel). Analysis of the biological processes that differ in BMSC specimens from the two elderly donors with multiple comorbidities (**Figure 5B,** right panel) identified the downregulation of processes associated with cell proliferation (e.g., mitotic cell cycle, cell division, chromosome organization, and organelle fission). In contrast, upregulated processes entailed categories associated with cell differentiation (e.g., multicellular organism process, anatomic structure development and morphogenesis, and ECM organization), thus potentially indicating a progressive loss of the MSC "stem cell" phenotype over time in the presence of multiple strong comorbidities, although any conclusions from this analysis are limited due to the small sample size.

Inflammatory Challenge Affects BMSC Paracrine Activity

The prior assays involving advanced donor age and the predominant early-stage diabetic status of the included patients showed only minor effects on the *in vitro* expansion and differentiation capacity, as well as the transcriptome of unstimulated BMSCs. The subsequent experiment of *in vitro* cytokine challenge demonstrated a degree of altered responsiveness on the transcriptome level in BMSCs with advanced donor comorbidities.

In line with these transcriptome changes, we observed more pronounced effects on the secretome and paracrine activity of cytokine-activated BMSCs upon environmental challenge by an inflammatory environment (**Figure 6**), e.g., upon stimulation with pro-inflammatory cytokines and consecutive readout of angiogenesis in response to their secretome (conditioned media). We found that their immunomodulatory activity to suppress T-cell proliferation was not altered, thus confirming that this is a rather well-preserved process, even if cells are obtained from donors with advanced diabetes and renal failure (55). Nonetheless, the overall impact of these donor parameters on BMSC function was rather modest.

BMSC preparations from both adult and elderly donors and both non-diabetic and diabetic donors effectively suppressed CD8 and CD4 T-cell proliferation in either anti-CD3/CD28 or PHA-stimulated PBMC cultures (**Figure 6A**). The average remaining proliferation of CD8-responses for adult and elderly BMSCs was 15.9 ± 7.2% and 10.3 ± 6.0% for anti-CD3/CD28-stimulated cultures ($P < 0.05$ and $P < 0.001$) and 17.7 ± 9.0% and 12.1 ± 8.7% for PHA-stimulated cultures compared to control ($P < 0.05$ and $P < 0.001$), with no significant differences between adult and elderly groups for the two different stimuli ($P = 0.06$ and $P = 0.16$, respectively).

The average suppression of CD4-responses was generally weaker, with a remaining proliferation of 46.7 ± 28.7% and 27.9 ± 20.8% for anti-CD3/CD28-stimulated cultures ($P < 0.05$ and $P < 0.001$) and 38.0 ± 23.3% and 27.6 ± 19.8% for PHA-stimulated cultures compared to control ($P < 0.05$ and $P < 0.001$), with no significant differences between adult and elderly groups for the two stimuli ($P = 0.09$ and $P = 0.3$, respectively). A similar suppression pattern of CD8 and CD4 T-cell proliferation in anti-CD3/CD28- or PHA-stimulated co-cultures was found for BMSCs from non-diabetic and diabetic donors (All $P < 0.01$ and $P < 0.05$), while again no significant differences were found between non-diabetic vs. diabetic donors.

Next, we analyzed the secretion of IL-6 and VEGF, two key paracrine mediators associated with BMSC function (**Figure 6B**). Both, adult and elderly donor-derived BMSCs demonstrated strong secretion of IL-6 compared to the negative control medium ($P < 0.01$ and $P < 0.001$, mean 806 vs. 1,300 pg/mL), with higher IL-6 secretion in BMSC-conditioned media from elderly compared to adult BMSC donors ($P < 0.05$). Similarly, we detected a strong secretion of VEGF by both types of BMSCs

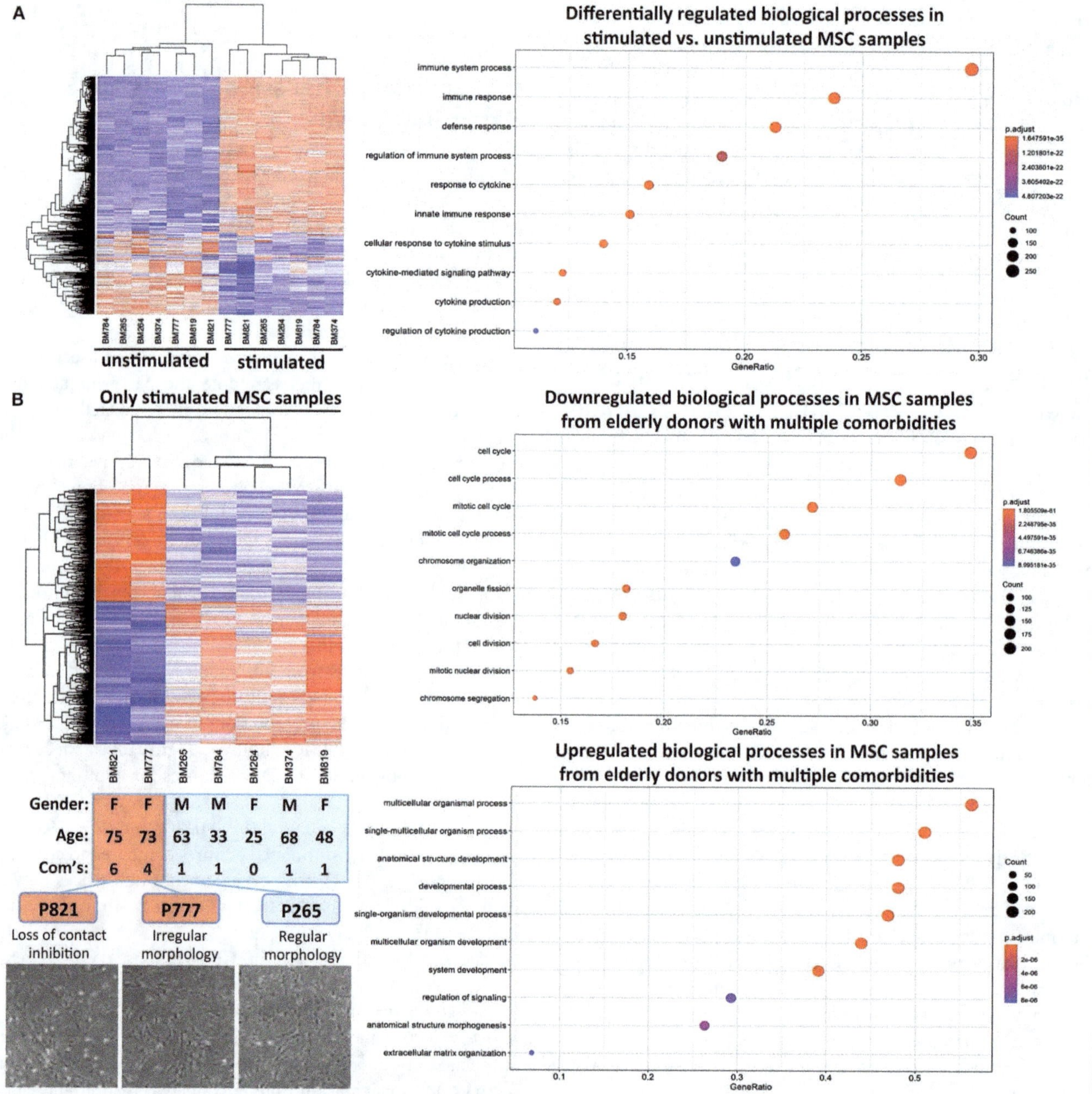

FIGURE 5 | Cytokine challenge reveals altered gene regulation in BMSCs from elderly donors with multiple comorbidities. **(A)** Comparison of unstimulated vs. cytokine-stimulated BMSCs ($n = 7$ donors each): hierarchical clustering heat map with expression values sorted according to donors (rows) and genes (columns), where low expression is denoted by blue and high expression by red, and corresponding gene ontology (GO) enrichment analysis with the "Top 10 Results" for changes in biological process (e.g., immune system process and immune and defense response) shown on the left y-axis, and the size of the dots representing the counts of genes involved, while the color of the circle (scaled blue/lowest to red/highest) indicating significance expressed as adjusted p-value. **(B)** Substratified gene expression analysis focusing only on stimulated samples ($n = 7$ donors): BMSCs from elderly donors with multiple comorbidities (P777 and P821, with $n = 4$ and $n = 6$ comorbidities, respectively) cluster separately from both elderly and adult donors with few comorbidities (P265 and P374; and P264, P784, and P819, respectively, $n = 0$–1 comorbidities), indicating that the accumulation of multiple comorbidities during advanced age results in a detectable *in vivo* imprint in the transcriptome of BMSCs. This was accompanied by a decline in cell proliferation and a progression from regular to irregular morphology in culture (representative images at the bottom), resulting in a progressive loss of contact inhibition and cell aggregation. Analysis of the biological processes, showing that P777 and P821 differ under stimulating conditions from the other donors, identified significant downregulation of processes associated with cell proliferation (e.g., mitotic cell cycle, cell division, and chromosome organization), while upregulated processes entailed categories associated with cell differentiation (e.g., multicellular organism process, anatomic structure development, morphogenesis, and ECM organization).

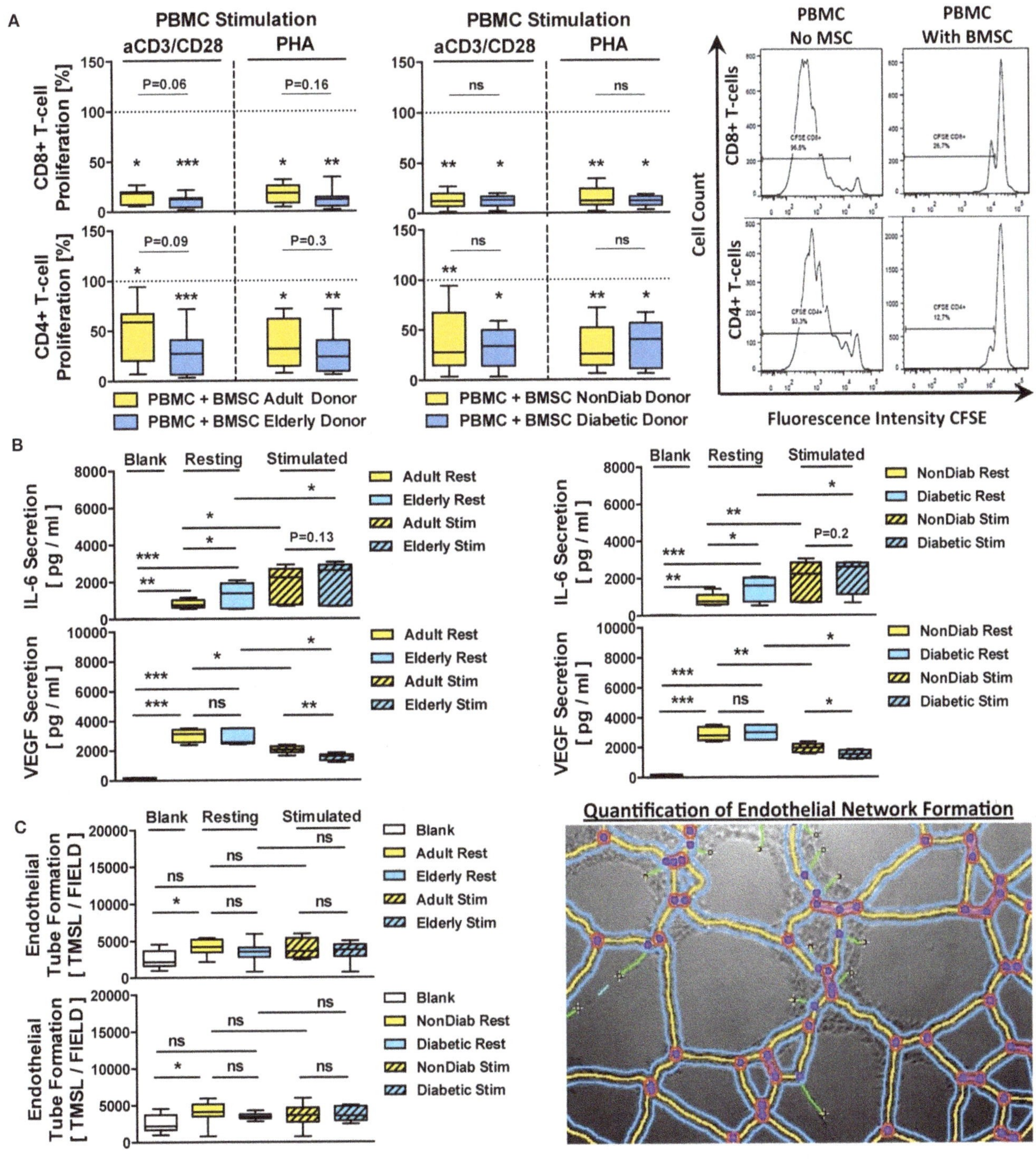

FIGURE 6 | Immunomodulatory and paracrine activity of BMSCs. **(A)** Immunomodulatory activity of BMSCs ($n = 9$ adult vs. $n = 12$ elderly and $n = 14$ non-diabetic vs. $n = 7$ diabetic donors, passage 3) to suppress anti-CD3/CD28- or phytohemagglutinin (PHA)-stimulated peripheral blood mononuclear cell (PBMC)-proliferation. The PBMCs were labeled with the cell proliferation-tracker dye CFSE, activated with either of the two different stimuli, and cocultured for 5 days with BMSCs from adult or elderly donors, in order to assess their capacity to inhibit the proliferation of CD4+ and CD8+ T-cells with flow cytometry, with representative histograms shown to the right. The quantitative assessment of T-cell proliferation is expressed as the percentage proliferation of CD4+ and CD8+ T cells relative to the positive control without BMSCs. Both adult vs. elderly and non-diabetic vs. diabetic donor-derived MSCs are equally potent in inhibiting CD4 and CD8 T-cell proliferation. **(B,C)** Paracrine and proangiogenic activity of BMSCs ($n = 9$ adult vs. $n = 12$ elderly and $n = 14$ non-diabetic vs. $n = 7$ diabetic donors) with and without cytokine stimulation (10 ng/mL of TNF-alpha and IFN-gamma for 24 h): **(B)** secretion of IL-6 and VEGF (pg/mL) in BMSC-conditioned culture media was assessed with ELISA, detecting elevated levels of IL-6 secretion by BMSCs obtained from elderly or diabetic donors, and **(C)** proangiogenic activity of BMSC-conditioned media compared to blank group (medium only) in an endothelial tube formation assay, with a representative image for the quantification of endothelial network formation by quantification of the total master segment length (TMSL/field, with 3–5 images assessed per test condition); TMSL/field was slightly increased for adult and non-diabetic donors. Results are given as box plot ± min-max whiskers. Statistical analysis was performed using either a Student's t-test or ANOVA followed by post-tests ($^{*}P < 0.05$, $^{**}P < 0.01$, and $^{***}P < 0.001$).

compared to negative controls (Both $P < 0.001$, mean 3,033 vs. 2,840 pg/mL), although there was no significant difference between the age groups.

Interestingly, IL-6 and VEGF were inversely regulated upon pro-inflammatory challenge with tumor necrosis factor-alpha (TNF-a) and interferon-gamma (IFN-g), with significant upregulation of IL-6 ($P < 0.05$) but downregulation of VEGF ($P < 0.05$) in stimulated cells compared to resting controls, thus demonstrating an inverse relationship between the two factors under stimulating conditions. Importantly, both BMSCs from adult vs. elderly donors and from non-diabetic vs. diabetic donors showed differential secretion of VEGF upon cytokine stimulation ($P < 0.01$ and $P < 0.05$), which was not the case for IL-6.

When exposing endothelial cells (ECs) to BMSC-conditioned culture medium from adult vs. elderly or non-diabetic vs. diabetic donors (**Figure 6C**), we found increased proangiogenic activity with media from adult donors compared to negative control ($P < 0.05$, mean 2,551 vs. 4,131 TMSL/field), while conditioned media from elderly donors and the comparison of adult and elderly donors did not reveal any differences. BMSC conditioned medium from non-diabetic donors showed the most profound proangiogenic activity compared to control ($P < 0.05$).

These results taken together indicate a weak but notable influence of advanced donor age and early-stage diabetes on BMSC regulation of its paracrine activity in response to cytokine challenge, while this was less evident for its immunomodulatory activity to suppress T-cell proliferation *in vitro*.

DISCUSSION

The goal of this study was to understand heterogeneity among BMSC specimens, which are frequently considered for autologous therapy approaches. We asked whether the donor-specific variability in morphological and functional parameters could be explained by intrinsic cell-donor attributes such as variations in donor age and common comorbidities. We thus conducted donor stratification and multi-parameter analysis in a defined patient cohort to allow for a robust readout of individual assay parameters.

Previous studies investigating the impact of donor age and comorbidities on BMSC properties reported partially inconclusive or contradictory outcomes (**Table 1**). We hypothesized that BMSCs from elderly donors (>60 years), who commonly suffer from mild comorbidities, display reduced regenerative function compared to adult donors (<50 years), who were found to have a much lower burden of common comorbidities. To our surprise, we found that for our prospective stratification, both adult and elderly donors demonstrated on average rather similar performance in most assays and that *in vitro* aging rather than *in vivo* aging and the typically associated mild comorbidities predominantly affected BMSC properties.

Clinical and Biological Relevance of Donor Aging for BMSC Biobanking Approaches

Many of the treatment indications targeted by allogeneic and autologous BMSC therapies are associated with advanced age, thus making elderly patients with multiple comorbidities one of the high-demand groups for cell-based therapies. They are thus frequently found among the cell donors of our biobank (1). The donor age of BMSC donors is one of the most clearly defined and readily accessible parameters and has been widely investigated, while the impact of certain comorbidities associated with advanced age is more difficult to assess.

Both, autologous and allogeneic BMSC therapies are widely studied, and it is not clear yet which approach will be favored for specific treatment indications (87). Allogeneic approaches provide the great advantage of being able to choose a well-defined starting material, e.g., from developmentally young tissue such as placenta or umbilical cord blood. However, they may come at the cost of immunological incompatibility, which may compromise efficacy and lead to allo-sensitization of the patient. Thus, a major advantage of autologous approaches is their neutral immune profile, but they may be limited by compromised bioactivity of cells sourced from elderly donors with multiple comorbidities, complex diseases, and pharmacological regimens (18, 19, 88).

Many studies focusing on BMSCs in the context of aging have compared various parameters either in cells from younger vs. elderly donors or the impact on cell expansion. These were mainly: (1) Cellular phenotype and proliferation capacity, (2) Gene expression profile, and (3) Various functional parameters, such as mesodermal differentiation capacity and immunosuppressive and paracrine properties. A problem with BMSC characterization is the great number of potentially confounding experimental variables that may impede the readout (85, 89).

BM-Sample Cellularity, BMSC Growth, Morphology, and Immunophenotype

Multiple studies have reported a decline of BM cellularity (e.g., BM-MNC and BMSC), CFU-F capacity, and growth kinetics with advanced donor age (17, 27, 29–31, 34). BMSCs have been shown to demonstrate a logarithmic decline with increasing age (31, 90), most evidently in the first years of life (e.g., newborns 1/10.000 and teens 1/100.000 BMSCs per BM-MNC), but this was less evident in later life, e.g., when comparing adults vs. elderly donors (e.g., 35-year old 1/250.000, 50-year old 1/400.000, and 60-year old 1/2-million BMSCs per BM-MNC). In our study, we could not detect an age-related difference in cellularity, showing that BM aspirates from adult donors had similar cell content as those from elderly donors (mean 38 vs. 72 years). We could thereby minimize a major confounding sampling bias in the starting material that may have caused initial disparities between the stratified groups.

In prior studies, differences in cell growth were most evident when comparing pediatric vs. elderly donors (27, 34), e.g., defined as the age ranges of 1–5 years vs. 50–70 years, respectively (34). In our study, we found no significant differences in the proliferation of BMSCs from adult and elderly donors, which may be explained by the different time-windows of analysis (Mean 38 vs. 72 years), since the donor population from our biobank does not include pediatric patients and contains few young adults. Our findings are in line with a report by Siegel et al. (33), who also found no correlation between the growth rate of BMSCs and donor age. Nonetheless, we could observe a trend of declining proliferation with successive passages, in particular in BMSCs from elderly donors with multiple comorbidities, although, surprisingly, this was not significant for the entire cohort.

Several reviews have summarized the impact of patient-specific aging and comorbidities on the morphological parameters of BMSCs (18, 19, 91), highlighting cell enlargement, decreased proliferation and replicative quiescence- and senescence-associated ß-galactosidase, as also discussed earlier (26). Siegel et al. reported an impact of donor aging on cell size (BMSCs at P1 from younger donors were smaller) and surface marker expression profile (e.g., increased levels of integrins, PDGFR-beta, and CD90 in younger donors).

In contrast to Siegel et al., who assessed the cell size at passage one, we assessed both cell size and volume of BMSCs at passage three and found that chronological aging had a rather minor influence, which may be due to the longer culture period (two passages longer). Nonetheless, we observed that BMSCs from elderly individuals, but not the younger donors, were more prone to an increase in cell size and volume at later passage (P6). This may support the notion of the so-called "Hayflick limit," the earlier reaching of proliferative senescence in cells from aged donors that is commonly observed in primary cells (16, 49).

Furthermore, a recent review by Baker et al. suggested that extended *in vitro* cultivation might alter the immunophenotype of BMSCs (17). We analyzed the cellular surface markers proposed by ISCT (69), and found that BMSCs from adult and elderly donors at both passages express typical MSC-associated markers, CD73, CD90, CD105, and CD146 and were negative for CD14, CD31, CD34, and CD45, with a limited influence of passage (decrease of CD105 and CD146). However, we observed that the expression of CD105 and CD146 declines with increasing passage number, suggesting a potential relationship between altered cell surface marker pattern and reduced functional capacity.

Indeed, reduced expression of CD105, CD106, and CD146 after extended culture and repeated passaging, particularly when the cells are cultured in DMEM-media, has been reported previously (82). Surface expression of CD105 can also be affected by the mode of culture (e.g., flasks vs. bioreactor or enzymatic detachment) (92). Downregulation of the TGF-beta receptor endoglin (CD105) and its associated signaling pathways (e.g., the Notch pathway) may also partly explain the evident decline in osteogenic differentiation capacity in BMSCs at increasing passage number (83).

BMSC Gene Expression Studies

So far, only a small number of studies have comprehensively studied the influence of advanced donor age and comorbidities on the transcriptome and methylome of BMSC products, mainly focusing on the effects of *in vitro* expansion before entering the senescent state (38, 93).

Our PCA of RNA-sequencing-derived gene expression profiles of unstimulated resting BMSC samples at early passage (P3) showed no clear separation between adult vs. elderly or non-diabetic vs. early-stage diabetic donors (both analysis of the whole cohort $n = 14$ vs. 7 or for the sub-stratified analysis of elderly non-diabetics vs. elderly diabetics $n = 6$ each). We compared non-diabetic vs. diabetic donors, since this was the most common and well-defined comorbidity in our cohort and is also frequently studied by others (60).

It should be noted that the majority of our diabetic donors suffered from non-insulin-dependent early-stage type 2 diabetes and that only two of the donors had more advanced insulin-dependent type 2 diabetes and multiple other notable comorbidities, owing to their rare presence in our biobank. However, the two donors with a more advanced disease stage in particular demonstrated differential gene expression upon cytokine challenge, corresponding to gene pathways associated with cellular decline. Davies et al. reported previously that BMSCs from late-stage type 1 diabetic donors show significant alterations in transcriptome compared to healthy controls under resting conditions (55). Indeed, patients who are affected by advanced late-stage type 1 or type 2 diabetes often present with renal failure and other more serious comorbidities and were thus on average sicker than the donors included in our analysis, which may explain the difference (60).

Furthermore, a recent methylome analysis of young and aged adults ($n = 5$ each, mean 22 vs. 75 years, range 20–24 vs. 62–82 years, respectively) at early and late passage (P4 vs. P8) was prospectively designed to distinguish between donor-age- and culture-induced changes (93). The authors found that a larger number of CpGs were differentially methylated in aged donors during culture and biological aging while there were fewer changes in young donors across genic elements, indicating that younger donors are more refractory to culture-induced changes. Furthermore, it was found that the majority of methylation changes appeared to be specific to either young or aged donors, with a subset being specific to long-term culture irrespective of adult donor age.

We also found in our second set of experiments under cytokine-challenge that particular elderly donors with multiple comorbidities (e.g., insulin-dependent diabetes and renal failure) demonstrated differential gene-expression profiles to healthy adults and healthy elderly donors, although these results have to be interpreted with great caution due to the small number of available samples that could be included in this analysis. Analysis of the underlying gene expression pathways indicated that this was associated with a loss of stemness and increased differentiation, going hand in hand with the observed methylome changes in the study above. This may indicate that, while methylome changes between adult and elderly donors are already evident in resting cells, transcriptome changes and their functional impact may become more evident under challenging environmental conditions.

As a positive control, we also analyzed how replicative aging upon *in vitro* expansion influences the expression profiles of BMSCs by comparing the gene expression pattern of six donors at P3 and P6. We found a clear separation between the two groups, thus confirming earlier findings on the matter and demonstrating the validity of our approach (16, 38, 54). Hierarchical clustering heat maps showed distinct gene expression patterns, with a passage-dependent decline in the expression of genes associated with cell cycle and cell proliferation. Indeed, earlier studies have demonstrated continuous and progressive gene-expression changes in BMSCs upon long-term culture expansion (16, 38). When comparing BMSCs at culture intervals from P2 up to P11, gene expression changes accumulated with each additional passage, with P2-3 vs. P4-5 vs. P6-11 being distinguishable, demonstrating the great analytic power of the method.

Similarly, a recent study documented accumulating transcriptome drift in UC-MSCs cultured until replicative senescence, with transcriptome changes becoming evident at P5, with a greater increase when reaching senescence at P9-12 (54), thus making it possible to distinguish between early passage (P2-4), medium passage (P6-8), pre-senescent (P10-12), and senescent (P14) cells. Both of these examples clearly illustrated that *in vitro* expansion influences BMSC gene expression signatures, which can dilute or mask any consistent *in vivo* signatures associated with donor aging and comorbidities.

Functional Assessment: Differentiation, Paracrine, and Immunomodulatory Activity

Since the key report by Pittenger et al. (84), the majority of studies on BMSCs have assessed multilineage differentiation potential as part of the minimal criteria for their characterization (69). Importantly, those that have studied the impact of aging and comorbidities often reported a negative impact of advanced donor age, with differential effects on the individual lineages (e.g., osteogenic, chondrogenic, and adipogenic potential). A review by Baker et al. pointed out that this is disputable and that approximately half of the studies do not find differences (17).

Multiple reports have shown no effect or that osteogenic differentiation decreases with increasing donor age (26, 29, 31, 33, 94). D'Ippolito et al. found a reduced ALP-positive CFU-F number and osteogenic potential in younger vs. elderly donors (3–36 vs. 41–70 years, respectively) (31). Müller et al. also reported a strong donor-age related decline in the osteogenic potential of BMSCs isolated from total hip arthroplasty patients ($\leq$50 years 7/11 or 63% positive, $\geq$60 years 5/19 or 26% positive) (94).

Furthermore, Stolzing et al. reported that osteogenic and chondrogenic potential were diminished with advanced age, while adipogenic differentiation was not (29). However, others found no age-dependent differences in differentiation capacity for either lineage (24, 33, 45). This inconsistency between study results may be explained by differences in methodology, e.g., using cells at different passages (26), as also indicated by our transcriptome analysis.

We thus analyzed the differentiation potential of adult vs. elderly and non-diabetic vs. early-stage diabetic donor-derived BMSCs at both early and late passage (P3 vs. P6). We found that the differentiation of BMSCs toward osteogenic, adipogenic, and chondrogenic lineages was mainly independent of the age and mild comorbidities of the donor and that *in vitro* aging, rather than *in vivo* aging, had a notable impact.

A large share of the therapeutic activity of BMSCs is attributed to their secretion of trophic and immunomodulatory factors (95, 96), which can be modulated by the environment the cells persist in or are brought into. Several reviews have summarized how these properties are potentially altered in the context of donor aging and its associated comorbidities (18, 21, 97).

Siegel et al. reported increased expression of IL-6 by BMSCs in association with aging (33), while Efimenko et al. reported reduced expression of VEGF and the loss of angiogenic potential in elderly donors with cardiovascular complications who more frequently presented with diabetes. We also found that BMSCs from elderly donors produced higher levels of IL-6, while both elderly and diabetic donors showed a stronger decline in VEGF-production under stimulating conditions, which was also reflected in lower *in vitro* angiogenic activity.

Considering their immunomodulatory activity, Siegel et al. did not find a correlation between increased donor age and a loss of immunomodulatory activity in BMSCs in a large cohort of more than 50 donors aged 13–80 years (33). Furthermore, two studies did not find a negative impact of diabetes when comparing BMSCs from healthy donors either to early- and late-stage type 1 diabetic donors (55) or to newly diagnosed type 1 diabetics (56). In contrast, Manchini et al. (n = 27 adult vs. n = 23 elderly donors, cut-off 65 years) and Serena et al. (n = 4 donors each) reported a negative impact of age, atherosclerosis, obesity and type 2 diabetes on the immunomodulatory properties of adipose-derived MSCs.

Similarly to Siegel, Davies, and Yaochite and colleagues, we did not observe any differences in the capacity of BMSC to suppress CD4 or CD8 T-cell proliferation in anti-CD3/CD28- or PHA-stimulated cultures with respect to age and comorbidities. This may be explained by the rather weak comorbidities in our patient cohort and the different age cut-offs for stratification of the groups or by differences in methodology.

STUDY LIMITATIONS

Qualifying adult stem cell sources in biobanking approaches is of major importance for understanding their behavior in preclinical and clinical studies (1). Importantly, the results obtained in most of the studies by other groups and also our own studies are shaped by the starting material, cell isolation and expansion protocols, and consecutive analysis methods. In the following sections, we discuss a few prominent limitations that are important for the interpretation of the results of this and other studies.

Starting Material and Baseline Characteristics of Diabetic Patients

Different outcomes considering an *in vivo* imprint of the cell donor and tissue source may be obtained when using different starting materials (e.g., adipose or placental tissue instead of BM) (98). We have predominantly banked BM-MSCs at our facility so far, and we can thus not extend our analysis to

MSCs derived from other tissue sources in the same depth. Our study is limited by the absence of control cells from young and healthy volunteer donors from a commercially available source, e.g., American Type Culture Collection (ATCC). However, multiple MSC batches from young, healthy donors isolated at our facility were included, and we here focused on the clinically relevant patient population, who are most in need of autologous BMSC-based therapies. Another important aspect is the baseline characteristics of the patients and their comorbidities, in particular the elderly diabetic patients. As discussed above, the majority of the diabetic donors were non-insulin-dependent early-stage diabetic (where few differences were found), while two elderly donors with more progressed insulin-dependent type 2 diabetes and multiple comorbidities demonstrated more substantial phenotypic and functional alterations in line with results from other groups (60), which should be anticipated when interpreting the results for clinical use of BMSCs.

Cell Isolation, Expansion, and Enzymatic Detachment Protocols

Our protocol is based on one of the most commonly used methods, separation of the BM-MNCs with density-gradient centrifugation and plastic adherence, with culture and expansion employing DMEM-LG containing 10% FCS. It is noteworthy that alternative culture media (e.g., chemically defined serum-/xeno-free MesenCult-XF medium or StemPro XF and SF media) and growth supplementation (e.g., MSCGM SingleQuots instead of 10% FCS) have become popular and may yield different results when considering *in vitro* aging and donor-specific *in vivo* imprint. Furthermore, our cell detachment and passaging protocol is based on the most commonly used protease trypsin, but other protocols (e.g., employing more gentle cell dissociation with dispase instead of trypsin), may lead to different observations in terms of *in vitro* aging, while it is rather unlikely that any *in vivo* donor imprint is better preserved by using alternative enzymatic detachment.

Transcriptome Analysis Method

Gene regulation at the post-transcriptional level is of great importance in (adult) stem cells, and differences in transcriptome or the lack thereof have to be interpreted with caution and should not be equaled with proteome or functional conclusions. We thus paralleled our transcriptome analysis with multiple functional *in vitro* assays, partly under challenging inflammatory conditions, to mimic the cells' responsiveness in a disease context (85, 86). We also used the known *in vitro* aging effect during successive passaging as a positive control to align transcriptional with functional outcomes. Novel methods such as epigenetic methylome analysis and Ribo-profiling are of great interest (93, 99). The latter attempts to better reflect the "active" proteome by employing analysis of mRNA associated with polysomes, which may more closely reflect the true levels of the actively translated transcripts in the cells, although the method is technically more challenging than the already established RNAseq (99).

SUMMARY AND CONCLUSION

Our study demonstrates that donor age and its typically associated mild comorbidities may exert less influence on the phenotype and functionality of BMSCs than previously assumed and that these two parameters do not explain the inherent donor variation in our biobank. In fact, *in vitro* aging rather than *in vivo* aging exerted a strong influence on the cellular properties in our setting, with prolonged culture expansion impairing the regenerative functions of BMSCs at later passages, which should therefore be strongly controlled for in preclinical and clinical studies. Therapeutic approaches would best require a large number of minimally expanded cells with optimum potency. Since sufficient BMSC numbers can only be obtained by extensive expansion, this might be a limiting factor for using BMSCs in cellular therapy, unless a sufficient amount of starting material allowing for limited expansion can be obtained. Alternative tissue sources with better expansion capacity (e.g., perinatal tissue sources such as placenta or umbilical cord), may thus offer certain advantages, but, similarly to other MSC sources, bear an additional risk of thromboembolic complications when applied systemically (4). Therefore, the ideal source of therapeutic MSCs still needs to be defined, and therapeutic approaches utilizing BMSCs should critically review *in vitro* expansion protocols.

ETHICS STATEMENT

BMSCs were received from the Core-Facility "Cell Harvesting" of the BIH Center for Regenerative Therapies (BCRT). The cells were isolated from metaphyseal bone marrow (BM) biopsies from patients undergoing hip replacement at Charité University Hospital, as previously stated (1, 66–68). Written informed consent was given, and ethics approval was obtained from the local ethics committee/institutional review board (IRB) of the Charité University Hospital.

AUTHOR CONTRIBUTIONS

GM, SG, and AA: conception and design, administrative support, data analysis and interpretation. GM, SG, GD, PR, DS, and TZ: financial support. GM, SG, AA, RC, JS, TQ, FS, DJ, AB, SRe, DK, MS, SS, SRi, NS, CB, JK-M, UK, TZ, and KJ: collection and assembly of data. GM, SG, DS, and AA: manuscript writing. AA, RC, JS, TQ, FS, DJ, AB, SRe, DK, MS, SS, SRi, NS, CB, JK-M, UK, TZ, KJ, DS, PR, GD, GM, and SG: final approval of manuscript.

FUNDING

The contributions of AA, JS, TQ, NS, CB, and GM were supported by the German Research Foundation (DFG) and German Federal Ministry of Education and Research (BMBF) funding through the BSRT (GSC203) and BCRT. SG was supported by grants from the BMBF (DIMEOS, 01EC1402B) and DFG (FOR2165, GE2512/2-2). This study has received support under the European Union's Horizon 2020 research and innovation program (Horizon 2020 Framework Program) under grant agreement nos. 733006 (PACE) and 779293 (HIPGEN). We acknowledge support from the DFG and the Open Access Publication Fund of Charité - Universitätsmedizin Berlin.

ACKNOWLEDGMENTS

The authors would like to thank the Cell Harvesting Unit (BCRTCH), the Genomics Core Unit, and the Flow Cytometry and Mass Cytometry Core Unit of the Berlin Institute of Health and the BCRT, Charité - Universitätsmedizin Berlin, for their excellent technical assistance and support.

Figure S1 | Morphological heterogeneity in BMSC preparations from individual adult and elderly donors. Representative bright-field microscopy images (40x magnification) of $n = 21$ BMSC preparations at passages 2–3 vs. 5–6, generated either from adult ($n = 9$) or elderly donors ($n = 12$) with typical fibroblast-like morphology and some minor phenotypic variation being evident in both cohorts. The BMSCs derived from elderly donors with diabetes mellitus [DIAB; Elderly P6; e.g., P237, P316, P336, P354, P660, and P777)] in particular appear to show a trend of deterioration in morphology and slower growth at higher passages, which also appears to be the case for the one diabetic donor in the adult group (Adult P6; P276).

Figure S2 | Focused gene-expression profiling of biobanked resting BMSCs from elderly donors with or without predominantly non-insulin-dependent early-stage diabetes. **(A,B)** Principle component analysis (PCA) was employed to study group separation comparing elderly donors with or without diabetes ($n = 6$ each), with a random dot distribution throughout the graph.

Figure S3 | Heterogeneity in osteogenic and adipogenic differentiation potential for BMSCs from individual donors. **(A)** Osteogenic differentiation: Alizarin red staining of BMSC matrix mineralization for individual donors ($n = 9$ adult vs. $n = 12$ elderly donors) at passages 3 and 6 upon *in vitro* osteogenic induction for 14, 18, and 21 days. **(B)** Adipogenic differentiation: Nile Red staining of lipid-rich vacuoles for BMSCs from individual donors ($n = 9$ adult vs. $n = 12$ elderly donors) at passages 3 and 6 upon *in vitro* adipogenic induction for 10 and 14 days. In general, both adult and elderly BMSCs display a very large time-dependent heterogeneity in osteogenic and adipogenic differentiation, which makes any predictions of functional performance difficult.

Figure S4 | ALP-activity during osteogenic differentiation of adult vs. elderly and non-diabetic vs. diabetic donors. Alkaline phosphatase (ALP) enzyme activity was assessed upon *in vitro* osteogenic induction of BMSCs for 0, 5, and 10 days at P3 and P6 either for: **(A)** Adult vs. elderly donors ($n = 9$ and $n = 12$) or **(B)** non-diabetic vs. diabetic donors ($n = 14$ vs. $n = 7$, respectively). For all comparisons, ALP activity peaks at day 5, with similar values at P3 and P6. Data are shown as mean ± SD and statistical evaluation was performed by using a Student's *t*-test or ANOVA followed by post-tests ($^{*}P < 0.05$, $^{**}P < 0.01$, and $^{***}P < 0.001$).

REFERENCES

1. Reinke S, Dienelt A, Blankenstein A, Duda GN, Geissler S. Qualifying stem cell sources: how to overcome potential pitfalls in regenerative medicine? *J Tissue Eng Regen Med.* (2016) 10:3–10. doi: 10.1002/term.1923
2. Moll G, Le Blanc K. Engineering more efficient multipotent mesenchymal stromal (stem) cells for systemic delivery as cellular therapy. *ISBT Sci Ser.* (2015) 10:357–65. doi: 10.1111/voxs.12133
3. Moll G, Geissler S, Catar R, Ignatowicz L, Hoogduijn MJ, Strunk D, et al. Cryopreserved or fresh mesenchymal stromal cells: only a matter of taste or key to unleash the full clinical potential of MSC therapy? *Adv Exp Med Biol.* (2016) 951:77–98. doi: 10.1007/978-3-319-45457-3_7
4. Moll G, Ankrum JA, Kamhieh-Milz J, Bieback K, Ringden O, Volk HD, et al. Intravascular mesenchymal stromal/stem cell therapy product diversification: time for new clinical guidelines. *Trends Mol Med.* (2019) 25:149–63. doi: 10.1016/j.molmed.2018.12.006
5. Ketterl N, Brachtl G, Schuh C, Bieback K, Schallmoser K, Reinisch A, et al. A robust potency assay highlights significant donor variation of human mesenchymal stem/progenitor cell immune modulatory capacity and extended radio-resistance. *Stem Cell Res Ther.* (2015) 6:236. doi: 10.1186/s13287-015-0233-8
6. Moll G, Ignatowicz L, Catar R, Luecht C, Sadeghi B, Hamad O, et al. Different procoagulant activity of therapeutic mesenchymal stromal cells derived from bone marrow and placental decidua. *Stem Cells Dev.* (2015) 24:2269–79. doi: 10.1089/scd.2015.0120
7. Oeller M, Laner-Plamberger S, Hochmann S, Ketterl N, Feichtner M, Brachtl G, et al. Selection of tissue factor-deficient cell transplants as a novel strategy for improving hemocompatibility of human bone marrow stromal cells. *Theranostics.* (2018) 8:1421–34. doi: 10.7150/thno.21906
8. Von Bahr L, Sundberg B, Lonnies L, Sander B, Karbach H, Hagglund H, et al. Long-term complications, immunologic effects, and role of passage for outcome in mesenchymal stromal cell therapy. *Biol Blood Marrow Transplant.* (2012) 18:557–64. doi: 10.1016/j.bbmt.2011.07.023
9. Moll G, Alm JJ, Davies LC, Von Bahr L, Heldring N, Stenbeck-Funke L, et al. Do cryopreserved mesenchymal stromal cells display impaired immunomodulatory and therapeutic properties? *Stem Cells.* (2014) 32:2430–42. doi: 10.1002/stem.1729
10. Moll G, Hult A, Von Bahr L, Alm JJ, Heldring N, Hamad OA, et al. Do ABO blood group antigens hamper the therapeutic efficacy of mesenchymal stromal cells? *PLoS ONE.* (2014) 9:e85040. doi: 10.1371/journal.pone.0085040
11. Pumberger M, Qazi TH, Ehrentraut MC, Textor M, Kueper J, Stoltenburg-Didinger G, et al. Synthetic niche to modulate regenerative potential of MSCs and enhance skeletal muscle regeneration. *Biomaterials.* (2016) 99:95–108. doi: 10.1016/j.biomaterials.2016.05.009
12. Qazi TH, Mooney DJ, Duda GN, Geissler S. Biomaterials that promote cell-cell interactions enhance the paracrine function of MSCs. *Biomaterials.* (2017) 140:103–14. doi: 10.1016/j.biomaterials.2017.06.019
13. Caplan H, Olson SD, Kumar A, George M, Prabhakara KS, Wenzel P, et al. Mesenchymal stromal cell therapeutic delivery: translational challenges to clinical application. *Front Immunol.* (2019) 10:1645. doi: 10.3389/fimmu.2019.01645
14. Rubin H. Cell aging *in vivo* and *in vitro*. *Mech Ageing Dev.* (1997) 98:1–35. doi: 10.1016/S0047-6374(97)00067-5
15. Dimmeler S, Leri A. Aging and disease as modifiers of efficacy of cell therapy. *Circ Res.* (2008) 102:1319–30. doi: 10.1161/CIRCRESAHA.108.175943

16. Wagner W, Bork S, Lepperdinger G, Joussen S, Ma N, Strunk D, et al. How to track cellular aging of mesenchymal stromal cells? *Aging.* (2010) 2:224–30. doi: 10.18632/aging.100136
17. Baker N, Boyette LB, Tuan RS. Characterization of bone marrow-derived mesenchymal stem cells in aging. *Bone.* (2015) 70:37–47. doi: 10.1016/j.bone.2014.10.014
18. Efimenko AY, Kochegura TN, Akopyan ZA, Parfyonova YV. Autologous stem cell therapy: how aging and chronic diseases affect stem and progenitor cells. *Biores Open Access.* (2015) 4:26–38. doi: 10.1089/biores. 2014.0042
19. Schimke MM, Marozin S, Lepperdinger G. Patient-Specific Age: The other side of the coin in advanced mesenchymal stem cell therapy. *Front Physiol.* (2015) 6:362. doi: 10.3389/fphys.2015.00362
20. Charif N, Li YY, Targa L, Zhang L, Ye JS, Li YP, et al. Aging of bone marrow mesenchymal stromal/stem cells: implications on autologous regenerative medicine. *Biomed Mater Eng.* (2017) 28:S57–63. doi: 10.3233/BME-171624
21. Van De Vyver M. Intrinsic mesenchymal stem cell dysfunction in diabetes mellitus: implications for autologous cell therapy. *Stem Cells Dev.* (2017) 26:1042–53. doi: 10.1089/scd.2017.0025
22. Yang YK. Aging of mesenchymal stem cells: implication in regenerative medicine. *Regen Ther.* (2018) 9:120–2. doi: 10.1016/j.reth.2018.09.002
23. Khong SML, Lee M, Kosaric N, Khong DM, Dong Y, Hopfner U, et al. Single-cell transcriptomics of human mesenchymal stem cells reveal age-related cellular subpopulation depletion and impaired regenerative function. *Stem Cells.* (2019) 37:240–6. doi: 10.1002/stem.2934
24. Wagner W, Bork S, Horn P, Krunic D, Walenda T, Diehlmann A, et al. Aging and replicative senescence have related effects on human stem and progenitor cells. *PLoS ONE.* (2009) 4:e5846. doi: 10.1371/journal.pone.0005846
25. Block TJ, Marinkovic M, Tran ON, Gonzalez AO, Marshall A, Dean DD, et al. Restoring the quantity and quality of elderly human mesenchymal stem cells for autologous cell-based therapies. *Stem Cell Res Ther.* (2017) 8:239. doi: 10.1186/s13287-017-0688-x
26. Zhou S, Greenberger JS, Epperly MW, Goff JP, Adler C, Leboff MS, et al. Age-related intrinsic changes in human bone-marrow-derived mesenchymal stem cells and their differentiation to osteoblasts. *Aging Cell.* (2008) 7:335–43. doi: 10.1111/j.1474-9726.2008.00377.x
27. Mareschi K, Ferrero I, Rustichelli D, Aschero S, Gammaitoni L, Aglietta M, et al. Expansion of mesenchymal stem cells isolated from pediatric and adult donor bone marrow. *J Cell Biochem.* (2006) 97:744–54. doi: 10.1002/jcb. 20681
28. Laschober GT, Brunauer R, Jamnig A, Singh S, Hafen U, Fehrer C, et al. Age-specific changes of mesenchymal stem cells are paralleled by upregulation of CD106 expression as a response to an inflammatory environment. *Rejuvenation Res.* (2011) 14:119–31. doi: 10.1089/rej.2010.1077
29. Stolzing A, Jones E, Mcgonagle D, Scutt A. Age-related changes in human bone marrow-derived mesenchymal stem cells: consequences for cell therapies. *Mech Ageing Dev.* (2008) 129:163–73. doi: 10.1016/j.mad.2007.12.002
30. Mendes SC, Tibbe JM, Veenhof M, Bakker K, Both S, Platenburg PP, et al. Bone tissue-engineered implants using human bone marrow stromal cells: effect of culture conditions and donor age. *Tissue Eng.* (2002) 8:911–20. doi: 10.1089/107632702320934010
31. D'ippolito G, Schiller PC, Ricordi C, Roos BA, Howard GA. Age-related osteogenic potential of mesenchymal stromal stem cells from human vertebral bone marrow. *J Bone Miner Res.* (1999) 14:1115–22. doi: 10.1359/jbmr.1999.14.7.1115
32. Prall WC, Saller MM, Scheumaier A, Tucholski T, Taha S, Bocker W, et al. Proliferative and osteogenic differentiation capacity of mesenchymal stromal cells: influence of harvesting site and donor age. *Injury.* (2018) 49:1504–12. doi: 10.1016/j.injury.2018.06.024
33. Siegel G, Kluba T, Hermanutz-Klein U, Bieback K, Northoff H, Schafer R. Phenotype, donor age and gender affect function of human bone marrow-derived mesenchymal stromal cells. *BMC Med.* (2013) 11:146. doi: 10.1186/1741-7015-11-146
34. Fan M, Chen W, Liu W, Du GQ, Jiang SL, Tian WC, et al. The effect of age on the efficacy of human mesenchymal stem cell transplantation after a myocardial infarction. *Rejuvenation Res.* (2010) 13:429–38. doi: 10.1089/rej.2009.0986
35. De Witte SFH, Lambert EE, Merino A, Strini T, Douben H, O'flynn L, et al. Aging of bone marrow- and umbilical cord-derived mesenchymal stromal cells during expansion. *Cytotherapy.* (2017) 19:798–807. doi: 10.1016/j.jcyt.2017.03.071
36. Oja S, Komulainen P, Penttila A, Nystedt J, Korhonen M. Automated image analysis detects aging in clinical-grade mesenchymal stromal cell cultures. *Stem Cell Res Ther.* (2018) 9:6. doi: 10.1186/s13287-017-0740-x
37. Bonab MM, Alimoghaddam K, Talebian F, Ghaffari SH, Ghavamzadeh A, Nikbin B. Aging of mesenchymal stem cell *in vitro*. *BMC Cell Biol.* (2006) 7:14. doi: 10.1186/1471-2121-7-14
38. Schallmoser K, Bartmann C, Rohde E, Bork S, Guelly C, Obenauf AC, et al. Replicative senescence-associated gene expression changes in mesenchymal stromal cells are similar under different culture conditions. *Haematologica.* (2010) 95:867–74. doi: 10.3324/haematol.2009.011692
39. Yin Y, Wu RX, He XT, Xu XY, Wang J, Chen FM. Influences of age-related changes in mesenchymal stem cells on macrophages during *in-vitro* culture. *Stem Cell Res Ther.* (2017) 8:153. doi: 10.1186/s13287-017-0608-0
40. Ma Y, Qi M, An Y, Zhang L, Yang R, Doro DH, et al. Autophagy controls mesenchymal stem cell properties and senescence during bone aging. *Aging Cell.* (2018) 17. doi: 10.1111/acel.12709. [Epub ahead of print].
41. Moerman EJ, Teng K, Lipschitz DA, Lecka-Czernik B. Aging activates adipogenic and suppresses osteogenic programs in mesenchymal marrow stroma/stem cells: the role of PPAR-gamma2 transcription factor and TGF-beta/BMP signaling pathways. *Aging Cell.* (2004) 3:379–89. doi: 10.1111/j.1474-9728.2004.00127.x
42. Kajkenova O, Lecka-Czernik B, Gubrij I, Hauser SP, Takahashi K, Parfitt AM, et al. Increased adipogenesis and myelopoiesis in the bone marrow of SAMP6, a murine model of defective osteoblastogenesis and low turnover osteopenia. *J Bone Miner Res.* (1997) 12:1772–9. doi: 10.1359/jbmr.1997.1 2.11.1772
43. Chen TL. Inhibition of growth and differentiation of osteoprogenitors in mouse bone marrow stromal cell cultures by increased donor age and glucocorticoid treatment. *Bone.* (2004) 35:83–95. doi: 10.1016/j.bone.2004.03.019
44. Kretlow JD, Jin YQ, Liu W, Zhang WJ, Hong TH, Zhou G, et al. Donor age and cell passage affects differentiation potential of murine bone marrow-derived stem cells. *BMC Cell Biol.* (2008) 9:60. doi: 10.1186/1471-2121-9-60
45. Katsara O, Mahaira LG, Iliopoulou EG, Moustaki A, Antsaklis A, Loutradis D, et al. Effects of donor age, gender, and *in vitro* cellular aging on the phenotypic, functional, and molecular characteristics of mouse bone marrow-derived mesenchymal stem cells. *Stem Cells Dev.* (2011) 20:1549–61. doi: 10.1089/scd.2010.0280
46. Wu LW, Wang YL, Christensen JM, Khalifian S, Schneeberger S, Raimondi G, et al. Donor age negatively affects the immunoregulatory properties of both adipose and bone marrow derived mesenchymal stem cells. *Transpl Immunol.* (2014) 30:122–7. doi: 10.1016/j.trim.2014.03.001
47. Santos LF, Singulani MP, Stringhetta-Garcia CT, Oliveira SHP, Chaves-Neto AH, Dornelles RCM. Oxytocin effects on osteoblastic differentiation of Bone Marrow Mesenchymal Stem Cells from adult and aging female Wistar rats. *Exp Gerontol.* (2018) 113:58–63. doi: 10.1016/j.exger.2018. 09.023
48. Geissler S, Textor M, Kuhnisch J, Konnig D, Klein O, Ode A, et al. Functional comparison of chronological and *in vitro* aging: differential role of the cytoskeleton and mitochondria in mesenchymal stromal cells. *PLoS ONE.* (2012) 7:e52700. doi: 10.1371/journal.pone.0052700
49. Truong NC, Bui KH, Van Pham P. Characterization of senescence of human adipose-derived stem cells after long-term expansion. *Adv Exp Med Biol.* (2018) 1084:109–28. doi: 10.1007/5584_2018_235
50. Khanh VC, Zulkifli AF, Tokunaga C, Yamashita T, Hiramatsu Y, Ohneda O. Aging impairs beige adipocyte differentiation of mesenchymal stem cells via the reduced expression of Sirtuin 1. *Biochem Biophys Res Commun.* (2018) 500:682–90. doi: 10.1016/j.bbrc.2018.04.136
51. Liu M, Lei H, Dong P, Fu X, Yang Z, Yang Y, et al. Adipose-derived mesenchymal stem cells from the elderly exhibit decreased migration and differentiation abilities with senescent properties. *Cell Transplant.* (2017) 26:1505–19. doi: 10.1177/0963689717721221
52. Efimenko A, Dzhoyashvili N, Kalinina N, Kochegura T, Akchurin R, Tkachuk V, et al. Adipose-derived mesenchymal stromal cells from aged patients with

coronary artery disease keep mesenchymal stromal cell properties but exhibit characteristics of aging and have impaired angiogenic potential. *Stem Cells Transl Med.* (2014) 3:32–41. doi: 10.5966/sctm.2013-0014
53. Kawagishi-Hotta M, Hasegawa S, Igarashi T, Yamada T, Takahashi M, Numata S, et al. Enhancement of individual differences in proliferation and differentiation potentials of aged human adipose-derived stem cells. *Regen Ther.* (2017) 6:29–40. doi: 10.1016/j.reth.2016.12.004
54. Wiese DM, Ruttan CC, Wood CA, Ford BN, Braid LR. Accumulating transcriptome drift precedes cell aging in human umbilical cord-derived mesenchymal stromal cells serially cultured to replicative senescence. *Stem Cells Transl Med.* (2019) 8:945–58. doi: 10.1002/sctm.18-0246
55. Davies LC, Alm JJ, Heldring N, Moll G, Gavin C, Batsis I, et al. Type 1 diabetes mellitus donor mesenchymal stromal cells exhibit comparable potency to healthy controls *in vitro*. *Stem Cells Transl Med.* (2016) 5:1485–95. doi: 10.5966/sctm.2015-0272
56. Yaochite JN, De Lima KW, Caliari-Oliveira C, Palma PV, Couri CE, Simoes BP, et al. Multipotent mesenchymal stromal cells from patients with newly diagnosed type 1 diabetes mellitus exhibit preserved *in vitro* and *in vivo* immunomodulatory properties. *Stem Cell Res Ther.* (2016) 7:14. doi: 10.1186/s13287-015-0261-4
57. Brewster L, Robinson S, Wang R, Griffiths S, Li H, Peister A, et al. (2017). Expansion and angiogenic potential of mesenchymal stem cells from patients with critical limb ischemia. *J Vasc Surg.* 65, 826–38 e821. doi: 10.1016/j.jvs.2015.02.061
58. Acosta L, Hmadcha A, Escacena N, Perez-Camacho I, De La Cuesta A, Ruiz-Salmeron R, et al. Adipose mesenchymal stromal cells isolated from type 2 diabetic patients display reduced fibrinolytic activity. *Diabetes.* (2013) 62:4266–9. doi: 10.2337/db13-0896
59. Capilla-Gonzalez V, Lopez-Beas J, Escacena N, Aguilera Y, De La Cuesta A, Ruiz-Salmeron R, et al. PDGF restores the defective phenotype of adipose-derived mesenchymal stromal cells from diabetic patients. *Mol Ther.* (2018) 26:2696–709. doi: 10.1016/j.ymthe.2018.08.011
60. Soria-Juan,B, Escacena N, Capilla-González V, Aguilera Y, Llanos L, Tejedo JR, et al. Cost-effective, safe, and personalized cell therapy for critical limb ischemia in type 2 diabetes mellitus. *Front Immunol.* (2019) 10:1151. doi: 10.3389/fimmu.2019.01151
61. Dentelli P, Barale C, Togliatto G, Trombetta A, Olgasi C, Gili M, et al. A diabetic milieu promotes OCT4 and NANOG production in human visceral-derived adipose stem cells. *Diabetologia.* (2013) 56:173–84. doi: 10.1007/s00125-012-2734-7
62. Nawrocka D, Kornicka K, Szydlarska J, Marycz K. Basic fibroblast growth factor inhibits apoptosis and promotes proliferation of adipose-derived mesenchymal stromal cells isolated from patients with type 2 diabetes by reducing cellular oxidative stress. *Oxid Med Cell Longev.* (2017) 2017:3027109. doi: 10.1155/2017/1083618
63. Kizilay Mancini O, Shum-Tim D, Stochaj U, Correa JA, Colmegna I. Age, atherosclerosis and type 2 diabetes reduce human mesenchymal stromal cell-mediated T-cell suppression. *Stem Cell Res Ther.* (2015) 6:140. doi: 10.1186/s13287-015-0127-9
64. Mangialardi G, Katare R, Oikawa A, Meloni M, Reni C, Emanueli C, et al. Diabetes causes bone marrow endothelial barrier dysfunction by activation of the RhoA-Rho-associated kinase signaling pathway. *Arterioscler Thromb Vasc Biol.* (2013) 33:555–64. doi: 10.1161/ATVBAHA.112.300424
65. Oikawa A, Siragusa M, Quaini F, Mangialardi G, Katare RG, Caporali A, et al. Diabetes mellitus induces bone marrow microangiopathy. *Arterioscler Thromb Vasc Biol.* (2010) 30:498–508. doi: 10.1161/ATVBAHA.109.200154
66. Reinke S, Geissler S, Taylor WR, Schmidt-Bleek K, Juelke K, Schwachmeyer V, et al. Terminally differentiated CD8(+) T cells negatively affect bone regeneration in humans. *Sci Transl Med.* (2013) 5:177ra136. doi: 10.1126/scitranslmed.3004754
67. Rakow A, Schoon J, Dienelt A, John T, Textor M, Duda G, et al. Influence of particulate and dissociated metal-on-metal hip endoprosthesis wear on mesenchymal stromal cells *in vivo* and *in vitro*. *Biomaterials.* (2016) 98:31–40. doi: 10.1016/j.biomaterials.2016.04.023
68. Schoon J, Geissler S, Traeger J, Luch A, Tentschert J, Perino G, et al. Multi-elemental nanoparticle exposure after tantalum component failure in hip arthroplasty: in-depth analysis of a single case. *Nanomedicine.* (2017) 13:2415–23. doi: 10.1016/j.nano.2017.08.004
69. Dominici M, Le Blanc K, Mueller I, Slaper-Cortenbach I, Marini F, Krause D, et al. Minimal criteria for defining multipotent mesenchymal stromal cells. *Int Soc Cell Ther Posit Statem Cytother.* (2006) 8:315–7. doi: 10.1080/14653240600855905
70. Schubert M, Lindgreen S, Orlando L. AdapterRemoval v2: rapid adapter trimming, identification, and read merging. *BMC Res Notes.* (2016) 9:88. doi: 10.1186/s13104-016-1900-2
71. Langmead B, Salzberg SL. Fast gapped-read alignment with Bowtie 2. *Nat Methods.* (2012) 9:357–9. doi: 10.1038/nmeth.1923
72. Kim D, Pertea G, Trapnell C, Pimentel H, Kelley R, Salzberg SL. TopHat2: accurate alignment of transcriptomes in the presence of insertions, deletions and gene fusions. *Genome Biol.* (2013) 14:R36. doi: 10.1186/gb-2013-14-4-r36
73. Liao Y, Smyth GK, Shi W. The Subread aligner: fast, accurate and scalable read mapping by seed-and-vote. *Nucleic Acids Res.* (2013) 41:e108. doi: 10.1093/nar/gkt214
74. Love MI, Huber W, Anders S. Moderated estimation of fold change and dispersion for RNA-seq data with DESeq2. *Genome Biol.* (2014) 15:550. doi: 10.1186/s13059-014-0550-8
75. Yu G, Wang LG, Han Y, He QY. clusterProfiler: an R package for comparing biological themes among gene clusters. *OMICS.* (2012) 16:284–7. doi: 10.1089/omi.2011.0118
76. Walter W, Sanchez-Cabo F, Ricote M. GOplot: an R package for visually combining expression data with functional analysis. *Bioinformatics.* (2015) 31:2912–4. doi: 10.1093/bioinformatics/btv300
77. Geissler S, Textor M, Schmidt-Bleek K, Klein O, Thiele M, Ellinghaus A, et al. In serum veritas-in serum sanitas? *Cell* non-autonomous aging compromises differentiation and survival of mesenchymal stromal cells via the oxidative stress pathway. *Cell Death Dis.* (2013) 4:e970. doi: 10.1038/cddis.2013.501
78. Ode A, Schoon J, Kurtz A, Gaetjen M, Ode JE, Geissler S, et al. CD73/5'-ecto-nucleotidase acts as a regulatory factor in osteo-/chondrogenic differentiation of mechanically stimulated mesenchymal stromal cells. *Eur Cell Mater.* (2013) 25:37–47. doi: 10.22203/eCM.v025a03
79. Zickler D, Luecht C, Willy K, Chen L, Witowski J, Girndt M, et al. Tumour necrosis factor-alpha in uraemic serum promotes osteoblastic transition and calcification of vascular smooth muscle cells via extracellular signal-regulated kinases and activator protein 1/c-FOS-mediated induction of interleukin 6 expression. *Nephrol Dial Transplant.* (2018) 33:574–85. doi: 10.1093/ndt/gfx316
80. Flemming A, Schallmoser K, Strunk D, Stolk M, Volk HD, Seifert M. Immunomodulative efficacy of bone marrow-derived mesenchymal stem cells cultured in human platelet lysate. *J Clin Immunol.* (2011) 31:1143–56. doi: 10.1007/s10875-011-9581-z
81. Catar R, Witowski J, Zhu N, Lucht C, Derrac Soria A, Uceda Fernandez J, et al. IL-6 trans-signaling links inflammation with angiogenesis in the peritoneal membrane. *J Am Soc Nephrol.* (2016) 28:1188–99. doi: 10.1681/ASN.2015101169
82. Yang YK, Ogando CR, Wang See C, Chang TY, Barabino GA. Changes in phenotype and differentiation potential of human mesenchymal stem cells aging *in vitro*. *Stem Cell Res Ther.* (2018) 9:131. doi: 10.1186/s13287-018-0876-3
83. Na T, Liu J, Zhang K, Ding M, Yuan BZ. The notch signaling regulates CD105 expression, osteogenic differentiation and immunomodulation of human umbilical cord mesenchymal stem cells. *PLoS ONE.* (2015) 10:e0118168. doi: 10.1371/journal.pone.0118168
84. Pittenger MF, Mackay AM, Beck SC, Jaiswal RK, Douglas R, Mosca JD, et al. Multilineage potential of adult human mesenchymal stem cells. *Science.* (1999) 284:143–7. doi: 10.1126/science.284.5411.143
85. Galipeau J, Krampera M, Barrett J, Dazzi F, Deans RJ, Debruijn J, et al. International Society for Cellular Therapy perspective on immune functional assays for mesenchymal stromal cells as potency release criterion for advanced phase clinical trials. *Cytotherapy.* (2016) 18:151–9. doi: 10.1016/j.jcyt.2015.11.008
86. Chinnadurai R, Rajan D, Qayed M, Arafat D, Garcia M, Liu Y, et al. Potency analysis of mesenchymal stromal cells using a combinatorial assay matrix approach. *Cell Rep.* (2018) 22:2504–17. doi: 10.1016/j.celrep.2018.02.013
87. Ankrum JA, Ong JF, Karp JM. Mesenchymal stem cells: immune evasive, not immune privileged. *Nat Biotechnol.* (2014) 32:252–60. doi: 10.1038/nbt.2816

88. Dimmeler S, Ding S, Rando TA, Trounson A. Translational strategies and challenges in regenerative medicine. *Nat Med.* (2014) 20:814–21. doi: 10.1038/nm.3627
89. Viswanathan S, Keating A, Deans R, Hematti P, Prockop D, Stroncek DF, et al. Soliciting strategies for developing cell-based reference materials to advance mesenchymal stromal cell research and clinical translation. *Stem Cells Dev.* (2014) 23:1157–67. doi: 10.1089/scd.2013.0591
90. Caplan AI. The mesengenic process. *Clin Plast Surg.* (1994) 21:429–35.
91. Fehrer C, Lepperdinger G. Mesenchymal stem cell aging. *Exp Gerontol.* (2005) 40:926–30. doi: 10.1016/j.exger.2005.07.006
92. Das R, Roosloot R, Van Pel M, Schepers K, Driessen M, Fibbe WE, et al. Preparing for cell culture scale-out: establishing parity of bioreactor- and flask-expanded mesenchymal stromal cell cultures. *J Transl Med.* (2019) 17:241. doi: 10.1186/s12967-019-1989-x
93. Pasumarthy KK, Doni Jayavelu N, Kilpinen L, Andrus C, Battle SL, Korhonen M, et al. Methylome analysis of human bone marrow MSCs reveals extensive age- and culture-induced changes at distal regulatory elements. *Stem Cell Rep.* (2017) 9:999–1015. doi: 10.1016/j.stemcr.2017.07.018
94. Mueller SM, Glowacki J. Age-related decline in the osteogenic potential of human bone marrow cells cultured in three-dimensional collagen sponges. *J Cell Biochem.* (2001) 82:583–90. doi: 10.1002/jcb.1174
95. Doorn J, Moll G, Le Blanc K, Van Blitterswijk C, De Boer J. Therapeutic applications of mesenchymal stromal cells: paracrine effects and potential improvements. *Tissue Eng Part B Rev.* (2012) 18:101–15. doi: 10.1089/ten.teb.2011.0488
96. Ferreira JR, Teixeira GQ, Santos SG, Barbosa MA, Almeida-Porada G, Goncalves RM. Mesenchymal stromal cell secretome: influencing therapeutic potential by cellular pre-conditioning. *Front Immunol.* (2018) 9:2837. doi: 10.3389/fimmu.2018.02837
97. Lepperdinger G. Inflammation and mesenchymal stem cell aging. *Curr Opin Immunol.* (2011) 23:518–24. doi: 10.1016/j.coi.2011.05.007
98. Menard C, Dulong J, Roulois D, Hebraud B, Verdiere L, Pangault C, et al. Integrated transcriptomic, phenotypic, and functional study reveals tissue-specific immune properties of mesenchymal stromal cells. *Stem Cells.* (2019). doi: 10.1002/stem.3077. [Epub ahead of print].
99. Erhard F, Halenius A, Zimmermann C, L'hernault A, Kowalewski DJ, Weekes MP, et al. Improved Ribo-seq enables identification of cryptic translation events. *Nat Methods.* (2018) 15:363–6. doi: 10.1038/nmeth.4631

8

Mesenchymal Stromal Cell Secretome: Influencing Therapeutic Potential by Cellular Pre-Conditioning

Joana R. Ferreira[1,2,3], *Graciosa Q. Teixeira*[1,2†], *Susana G. Santos*[1,2], *Mário A. Barbosa*[1,2,3], *Graça Almeida-Porada*[4] *and Raquel M. Gonçalves*[1,2,3*]

[1] *Instituto de Investigação e Inovação em Saúde (i3S), Universidade do Porto, Porto, Portugal,* [2] *Instituto de Engenharia Biomédica, Universidade do Porto, Porto, Portugal,* [3] *Instituto de Ciências Biomédicas Abel Salazar, Universidade do Porto, Porto, Portugal,* [4] *Wake Forest Institute for Regenerative Medicine, Winston-Salem, NC, United States*

***Correspondence:**
Raquel M. Gonçalves
raquelg@ineb.up.pt

Mesenchymal stromal cells (MSCs) are self-renewing, culture-expandable adult stem cells that have been isolated from a variety of tissues, and possess multipotent differentiation capacity, immunomodulatory properties, and are relatively non-immunogenic. Due to this unique set of characteristics, these cells have attracted great interest in the field of regenerative medicine and have been shown to possess pronounced therapeutic potential in many different pathologies. MSCs' mode of action involves a strong paracrine component resulting from the high levels of bioactive molecules they secrete in response to the local microenvironment. For this reason, MSCs' secretome is currently being explored in several clinical contexts, either using MSC-conditioned media (CM) or purified MSC-derived extracellular vesicles (EVs) to modulate tissue response to a wide array of injuries. Rather than being a constant mixture of molecular factors, MSCs' secretome is known to be dependent on the diverse stimuli present in the microenvironment that MSCs encounter. As such, the composition of the MSCs' secretome can be modulated by preconditioning the MSCs during *in vitro* culture. This manuscript reviews the existent literature on how preconditioning of MSCs affects the therapeutic potential of their secretome, focusing on MSCs' immunomodulatory and regenerative features, thereby providing new insights for the therapeutic use of MSCs' secretome.

Keywords: MSCs (Mesenchymal Stromal Cells), pre-conditioning, regeneration, immunomodulation, therapeutic potential, secretome

INTRODUCTION

Mesenchymal stromal cells (MSCs), defined by the International Society for Stem Cell Research (ISSCR) as fibroblast-like non-hematopoietic cells, have been explored in recent years due to the clinical promise they hold for tissue repair in regenerative medicine (1, 2). They present a capacity to differentiate into multiple lineages, which was on the basis of the high number of clinical trials using MSCs. By 2015, 493 MSC-based clinical trials were reported (2), a number that greatly increased in the next 2 years, reaching a total of 861 trials in 2018 according to the official database of the US National Institutes of Health. In an effort to address this fast-increasing knowledge base, several reviews have been published to provide a thorough analysis of the evolution of MSC-based clinical trials (3, 4). Perhaps one of the best documented properties of these cells is their ability to promote regeneration in a variety of tissues and to be a major contributor to the positive results achieved in many published papers (5, 6). Indeed, up to 2015 most of the studies with MSCs had focused on their use to treat disorders of the musculoskeletal system, namely in their application to repair bone or cartilage (2). Looking beyond their potential in tissue repair and regeneration, MSCs have also been used extensively for their immunomodulatory properties, for example to treat graft-vs.-host disease (GVHD) (7) and auto-immune diseases such as lupus (8, 9), or Crohn's disease (10). Furthermore, MSCs' clinical potential has been extended to treat myocardial infarction (11, 12), stroke (13), multiple sclerosis (14, 15), liver cirrhosis (16, 17), diabetes (18, 19), lung injuries (20), among others. MSCs are known as relatively immune-inert cells (21), but depending on the context can have immunosuppressive (22–24), or immune-stimulating capacity (25, 26) (see **Figure 1**).

Despite this great promise, however, their therapeutic benefits are not limited solely to their regenerative abilities. MSCs have also been referred to as trophic "factories" due to the large number of bioactive molecules they secrete in response to the local environment, which then exert paracrine effects upon neighboring cells and tissues (27). Indeed, an increasing number of authors have come to consider these paracrine or trophic properties to be the primary means by which MSCs conduct many of their therapeutic effects (28–30). This conclusion has been furthered by the observation that, in many cases, the number of differentiated cells is far too small to explain the observed response (27). Nevertheless, this paracrine action is known to be influenced by the microenvironment surrounding the cells (31). Therefore, there's a need to understand how *in vitro* culture conditions affect the regenerative and immunomodulatory potential of MSCs' secretome, with the ultimate goal of defining an optimal "cocktail" to precondition MSCs for a given therapeutic application. While the fast pace of research in this field is providing a large amount of data related to MSCs' therapeutic potential, an integrated investigation into how preconditioning can specifically influence the MSC secretome is lacking. To address this deficiency, we performed a comprehensive literature search on the following databases: clinicaltrials.gov, Google Scholar, Scopus, and PubMed, using either direct word-correspondence search or MESH integrated search, with several combinations of the following words: mesenchymal stem cells, hypoxia, inflammatory, pretreatment, preconditioning, stimulation, stimulus, priming, regeneration, immunomodulation, secretome, conditioned medium (CM), paracrine, therapeutic, brain, nervous system, bone, cartilage, kidney, liver, lung, pancreas, cancer, tumor, diabetes, skin, heart, cardiovascular, and intervertebral disc. The compilation of database outputs (~20,000 papers) was analyzed according to the focus of the study and relevance of the results obtained. From these results, articles found within reference lists were also screened and included when relevant to this article, considering the focus on MSCs preconditioning.

MSCS SECRETOME: PRECLINICAL AND CLINICAL EVIDENCES OF ITS THERAPEUTIC POTENTIAL

The MSCs-derived cell-free secretome appears to be able to recapitulate many of the properties/effects that have been described for the MSCs themselves. MSCs secretome is enriched in several soluble factors including cytokines, chemokines, immunomodulatory molecules, and growth factors (32). Additionally, paracrine factors produced by cells can be found encapsulated in cell-secreted vesicles. These Extracellular Vesicles (EV) are usually divided according to their size and origin in the cell into exosomes, microvesicles and apoptotic bodies. The smaller nanosized vesicle populations have deserved the most attention. Microvesicles (100–1,000 nm) originate on the plasma membrane, and exosomes (30–120 nm) that are formed in the multivesicular endosomes, have overlapping size ranges and when their separation cannot be completely ascertained are collectively designated EV (33, 34). EV content is thought to mimic that of the cells (35). The exact composition of MSCs' secretome has been investigated to identify the key molecules responsible for MSCs therapeutic potential, with the final goal being the substitution of a cell-free product to achieve the desired therapeutic effect (see **Table 1**) (32, 36–38, 40–43). Pro-regenerative effects of MSCs secretome have been observed in many different systems, acting by modulating the immune system (44), inhibiting cell death and fibrosis (45, 46), stimulating vascularization (44), promoting tissue remodeling, and recruiting other cells (47).

Preclinical Evidence

Preclinical evidence of the regenerative potential of MSCs secretome will be briefly described. ASC (adipose tissue-derived MSCs)-CM was able to regenerate/repair mandible lesions in rabbits. In the ASC-CM obtained from 24 h culture in serum-free medium under hypoxic conditions, the authors detected 43 angiogenic factors, 11 of which also appeared to be involved in bone regeneration: IGF-1, TGF-β1, VEGF, Angiogenin, IL-6, PDGF-BB, basic FGF (bFGF), EGF, RANTES, MCP-1, and MCP3 (38). This repertoire of secreted factors seemed to be in accordance with the BM-derived MSCs-CM composition reported by other authors (36, 41, 43, 48), with the remarkable exception that the BM-derived MSCs-CM also contained HGF (41) and BMP-1 (36). HGF in particular seems to be a key factor in MSCs-mediated reversal of hepatic fibrosis (49). Other studies exploring the effect of locally administered MSCs to degenerated tissues found evidence to support the notion that the soluble factors produced in response to the injury played a decisive role in the observed benefits of MSCs administration (50–52).

In the context of intervertebral disc (IVD) injury, MSCs also seem to act via a paracrine role through crosstalk with IVD cells (53–55). In an *ex-vivo* bovine model of pro-inflammatory/degenerated IVDs, MSCs in co-culture were able to immunomodulate the inflammatory reaction mediated by the nucleus pulposus (NP), even though few cells were found to

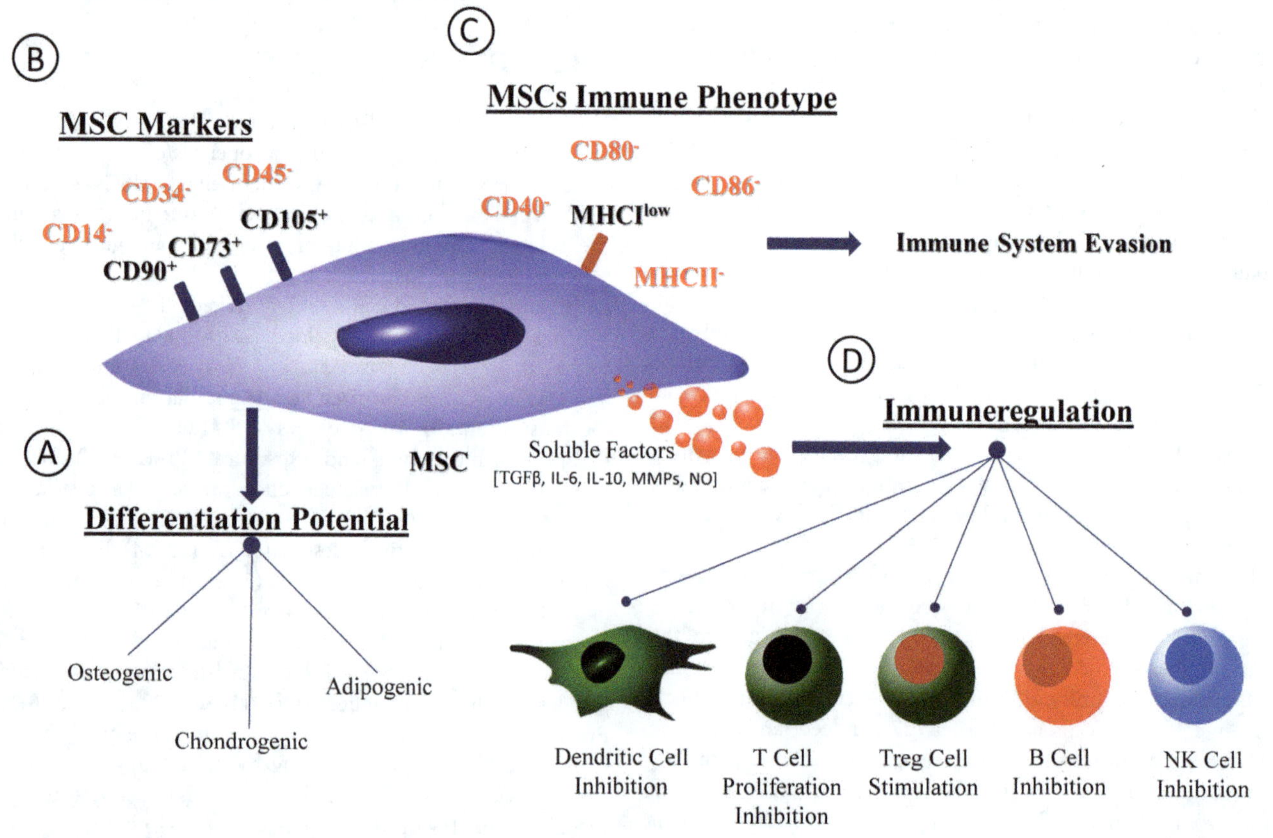

FIGURE 1 | MSCs phenotype, differentiation potential, and immunological properties. Schematic representation of MSCs phenotype and immunological profile. **(A)** MSCs capacity of differentiation into osteogenic, chondrogenic and adipogenic lineages. **(B)** MSCs phenotype accordingly with the International Society for Stem Cell Research (ISSCR). **(C)** MSCs immunological profile. **(D)** Soluble factors families produced by MSCs and profile of interaction with immune cells.

have actually migrated to the disc (56). Zheng et al. further analyzed MSCs-CM effect on the gene expression of NP-like cells, and found an upregulation of KRT19 and downregulation of MMP12 and MGP (57). As MMP12, KRT19, and MGP have been associated with IVD degeneration, the authors suggested that a healthy NP-like phenotype could be restored by MSCs-CM. In fact, it was further proposed that the MSCs' secretome was stimulating IVD progenitor cells activity (54) and the communication mechanism between MSCs and NP cells was at least partially via secretion of microvesicles (58).

Evidence for the pivotal role of MSCs paracrine activity in injured tissues continues to arise in many different systems and pathologic conditions. In 2007, Dai et al. observed that, in myocardial infarction, using MSCs-CM had a similar, albeit less intense, effect to what had been reported earlier for MSCs *per se*, indicating that at least part of the effect that had been observed following MSCs injection could be attributed to soluble factors (59). In the context of neuronal damage, a local injection of MSCs to the lesion site in a stroke model improved coordinated function, inhibited scar tissue formation and cell apoptosis, and stimulated angiogenesis (60). Despite these marked improvements, no neural differentiation of the transplanted MSCs was observed, reinforcing the key role of their paracrine mode of action. Moreover, it has been established that the presence of BDNF, Glial Cell Line-derived Neurotrophic Factor (GDNF), Nerve Growth Factor (NGF), and IGF in the MSCs secretome is necessary to observe the MSCs-induced neuronal survival and differentiation both *in vivo* and *in vitro* (61). Other models in which MSCs-CM has demonstrated therapeutic efficacy include chronic kidney disease, in which administration of MSCs-CM partially rescued kidney function, mainly by attracting endothelial cells, which led to neo-angiogenesis and stimulated wound closure (62). In this study, the authors concluded that the renal-protective paracrine factors present within the MSCs-CM were likely to be VEGF, HGF, and IGF.

MSCs-derived EV, particularly exosomes, have been increasingly shown to contribute to or even completely replicate the therapeutic effects observed with the use of the entire secretome (63). They were shown to improve cardiac function after a porcine myocardial infarction, reducing infarct size and maintaining the systolic and diastolic performance, as a result of inducing neo-revascularization and modulating the inflammatory response (64). Similarly, hBM-MSCs-derived exosomes injected locally 24 h after an induced focal

TABLE 1 | Main factors detected in the MSCs secretome.

Factors	References
BMP	(36)
CCL5/RANTES	(37, 38)
EGF	(38)
FGF	(38)
G-CSF	(39)
GM-CSF	(39)
HGF	(40, 41)
ICAM	(37)
IDO	(37)
IGF	(38, 40–43)
IL-10	(37)
IL-6	(38, 39, 42)
IL-8	(39, 42)
LIF	(42)
MCP-1	(38, 39, 42)
MMP-1	(36)
MMP-2	(36)
MMP-3	(36)
MMP-7	(36)
PDGF	(38)
PGE2	(37)
TGF-β	(37–39, 41)
TIMP-1	(36, 42)
TIMP-2	(36, 42)
VEGF	(38, 40–43)

cerebral ischemia were able to reduce the resulting functional impairments through an increase of angioneurogenesis and the modulation of the peripheral immune response (65). Additionally, the treatment seemed to also induce long-term neuroprotection. Other studies reported that MSCs-derived exosomes could mediate the transfer of the micro RNA (miRNA)-133b to neuronal cells, which induced neurite outgrowth and functional recovery after stroke (66), hinting to the importance of this mechanism in the neuronal protective capacity exhibited by MSCs. These effects were also observed by others in different models of ischemic injury (67, 68), even though their ability to modulate the local inflammatory reaction has not been observed by all (67). In another study, a single administration of MSCs-derived microvesicles inhibited apoptosis and stimulated tubular epithelial cell proliferation, thus protecting animals from acute kidney injury (69). Bruno et al. has demonstrated as well that the treatment of acute kidney injury with MSC-EVs leads to functional improvements and reduced mortality through an inhibition of the apoptotic cascade (70). Moreover, treatment with multiple administrations was shown to be significantly more effective than a single administration of the EVs. In a similar fashion, MSCs-derived exosomes were shown to protect hepatocytes and reduce both hepatic inflammation and collagen deposition (45). Indeed, MSCs-derived vesicles have consistently been reported to play a key role in the paracrine activity of these cells.

Clinical Trials

While the preclinical evidence showing the regenerative and immunomodulatory potential of the MSCs secretome continues to expand rapidly, the clinical studies revolving around this hypothesis are still scarce. Even so, the few clinical trials performed using the product of the MSCs paracrine activity seem to have already established the safety and feasibility of this method, as none of them reported related adverse effects (71–75). Furthermore, the use of the secretome seemed to be effective in improving the clinical outcomes of the involved patients. In the case of alveolar bone regeneration, conditioned media from commercially available BM-MSCs was administered to 8 patients suffering from severe alveolar bone atrophy and needing bone augmentation (75). These patients received either porous pure beta-tricalcium phosphate (β-TCP) or shell-shaped atelocollagen sponge (ACS) scaffold grafts soaked in the CM. After the surgery, minor inflammation of the local tissues was observed with less infiltration of inflammatory cells recorded. The scaffold was gradually replaced by newly formed bone, with no records of bone resorption in any of the cases and early mineralization observed in the augmented bone. IGF-1, VEGF, TGF-β, and HGF were present in the CM, even though molecules typically involved in bone homeostasis, like BMP-2, were not detected by the methodology used.

Clinical trials addressing alopecia (73) and Female Pattern Hair Loss (74) were able to increase hair density after injecting patients not with the MSCs secretome but with a commercially available product containing its protein components. Furthermore, the treatment of one treatment-refractory GVHD patient with MSCs-derived exossomes yielded a pronounced clinical improvement shortly after the administration with a decrease in more than 50% of the IL-1β/INF-γ/TNF-α-producing peripheral blood mononuclear cells (PBMCs) (72). After 4 months, the clinical condition of the patient was still stable, indicating a long-lasting therapeutic effect of the exossomes. Currently, allogeneic MSC-derived exosomes, enriched for miR-124, are reported in a registered clinical trial, directed to stroke patients (http://clinicaltrials.gov).

THE EFFECT OF PRE-CONDITIONING ON MSCs SECRETOME

Although MSCs have an innate potential to induce and/or contribute to regeneration, this potential is now known to be greatly influenced by diverse extrinsic factors such as the tissue source of the MSCs, the health status and age of the MSCs donor, the batch/lot of serum used for the *in vitro* culture of the MSCs, passage number, oxygen concentration, and the presence/absence of a pro-inflammatory environment when the MSCs are infused (76–82). Thus, *in vitro* preconditioning of MSCs with a variety of different factors has been explored to enhance the therapeutic capacity/potential of MSCs, which included: 3D culture (83–85), pharmacological compounds (86–88), inflammatory cytokines (89, 90), and hypoxia (91, 92) (**Table 2**). Considering MSCs main mechanism of action upon transplantation might be via paracrine signaling, it is somewhat surprising that only a few groups have studied how preconditioning of MSCs affects their secretory profile. This is particularly relevant when the MSCs' secretome may ultimately prove to be an extremely valuable therapeutic tool. The influence of these factors on MSCs' secretome will be reviewed in the following section.

Hypoxia

Normoxic oxygen tension, as used for standard cell culture, is the atmospheric pressure (21% O_2). The term hypoxia, when employed in the context of cell culture is routinely used to refer to oxygen tensions ranging from 0 to 10% (108). The physiological oxygen tension in tissues can vary from 1% in cartilage and bone marrow to 12% in peripheral blood (109). Thus, the 21% O_2 routinely used for MSCs culture is far higher than the oxygen found physiologically.

In general, hypoxic preconditioning enhances MSCs' regenerative and cytoprotective effects (82, 91–98). Moreover, culturing MSCs in hypoxic conditions has been shown to maintain MSCs' multipotency (110), enhance MSCs proliferation (111), and increase their levels of cytoprotective molecules (98) (**Table 3**), thereby improving the ability of MSCs to survive in the harsh environment found within injury sites upon transplantation. The beneficial effects of hypoxic culture preconditioning can likely be explained by the fact that MSCs exist *in vivo* in hypoxic environments (131) and hence have the ability to respond to a hypoxic microenvironment through the upregulation of the transcription factor HIF-1α (132). When stabilized due to the lack of oxygen, and dependent upon the increase of phosphorylated Akt and p38 mitogen-activated protein kinase (p38MAPK), this factor binds to the promoter regions of genes responsive to hypoxia, leading to an increase in available glucose (109). As MSCs are capable of switching from aerobic to anaerobic metabolic pathways, they are then able to endure very low oxygen tension values in their microenvironment (133). Therefore, using these culture conditions to precondition MSCs enhances their capacity to survival for longer periods, increases their proliferation rate, and maintains them in an undifferentiated state (109, 131). Small differences, however, in the oxygen tension used to culture MSCs, and in the culture protocol itself, can influence both their ability to differentiate into each of the different mesenchymal lineages (134) and their paracrine production (109). This extreme sensitivity to oxygen tension is an important factor to bear in mind when analyzing results from studies using different preconditioning protocols. The various studies to-date that have used hypoxia as a means of preconditioning MSCs used a concentration up to 2% O2, for a time period of 4–72 h (**Table 4**). Unfortunately, a high degree of variability exists between the protocols that have been employed, and this must be considered when assessing the MSCs' therapeutic function.

HIF-1α activation due to preconditioning MSCs with hypoxia leads to the induction of factors such as VEGF and Angiotensin, promoters of vascularization (136, 137). As neovascularization is a key factor in the regenerative process of damaged tissues, this may account, in itself, for the better therapeutic capacity that has been seen with MSCs pretreated with hypoxia. This hypothesis is supported by a growing number of publications identifying VEGF as a crucial molecule for the observed pro-regenerative effects of MSCs (47, 121, 138, 139). Liu et al. described a direct impact of the hypoxia-preconditioned MSCs treatment on endothelial cell proliferation with a simultaneous reduction in apoptosis (139). In addition, infusion of hypoxia-preconditioned BM-MSCs into the portal vein of rats subject to hepatectomy promoted hepatocyte proliferation and survival and improved serum albumin levels after surgery through a TGF-β dependent mechanism (95). Again, increased production of VEGF was observed. Hypoxia-preconditioning induced MSCs to express higher levels of HIF-1α, and the growth factors GDNF, BDNF, VEGF, Ang-1, and SDF-1, as well as its receptor CXCR4, all of which have been linked to neovascularization, as well as EPO and its receptor EPOR, a neuroprotective and pro-angiogenic molecule (120). Also, when using specifically hypoxia-preconditioned MSCs-derived EVs to treat acute myocardial infarction, authors reported the importance of the increased vascularization in the therapeutic effects. Bian et al. observed that EVs derived from BM-MSCs preconditioned with hypoxia for 72 h were able to significantly improve cardiac function after acute myocardial infarction, mainly through the promotion of angiogenesis (140). Indeed, a comprehensive proteomic analysis of exosomes derived from hypoxia-exposed MSCs showed that these exosomes induce angiogenesis in endothelial cells via the activation of the NFκB pathway (141). However, in another study exosomes derived from hypoxia-preconditioned MSCs contributed to the attenuation of the injury resulting from an ischemia/reperfusion episode via the Wnt signaling pathway (142). Beyond that, hypoxia seems to increase exosome secretion in general (141). Also, in a fat graft model, co-transplantation of exosomes from hypoxia pre-conditioned adipose-derived MSC improved vascularization and graft survival (143) (see **Table 5**).

Nevertheless, other growth factors are also upregulated in response to this stimulus (43, 46, 147) (**Table 3**), and these factors likely contribute to the specificity of tissue regeneration in a variety of scenarios. An analysis of the hypoxia-preconditioned MSCs' CM used to treat wounds in diabetic rats, revealed higher levels of VEGF, IGF-1, and bFGF (94), while another study reported increased production of VEGF-1α and Bcl-2, with upregulation of HIF-1α, HGF, bFGF, MMP9, and PDGF in MSCs pretreated with hypoxia (101). In agreement with these aforementioned studies, Zhang and colleagues observed that pre-treatment with hypoxia led to increased levels of VEGF, bFGF, and Akt that were implicated in the enhancement of MSCs' anti-oxidative, anti-apoptotic, and pro-angiogenic effects in a rat model of acute kidney injury (96). Additionally, other studies showed that both hypoxia and, to an even greater degree, forced overexpression of *Akt,* upregulated expression of VEGF, bFGF, HGF, IGF, and TB4, molecules associated with tissue repair and regeneration (113). The Akt signaling pathway was also reported to play a role in the enhanced wound healing observed in mice treated with the secretome from hypoxia-preconditioned MSCs (121). The effect of this secretome was related to increased levels of fibronectin, AKT, PI3K, and SMAD2 in the injured tissue; molecules that are all involved in cell proliferation and migration. The hypoxia-preconditioned MSCs secretome was

TABLE 2 | MSCs preconditioning parameters diversity.

Pre-conditioning treatment	Treatment conditions	MSCs Sources	References
Hypoxia	Anoxia to 2% O2, 4–72 h	Placenta, Gingiva, Bone marrow, Adipose tissue, Umbilical Cord Blood	(82, 91–98)
Cytokines, growth factors and hormones	SDF-1, TGF-α, Angiotensin II, INF-γ, TNF-α, Melatonin, Oxytocin 30 min–7 days	Bone marrow, Umbilical blood cord	(89, 90, 99–101)
3D	Aggregates/spheroids, 24 h–4 days	Bone marrow, Adipose tissue, Synovium	(76, 83–85, 102–104)
Pharmacological agents	Atorvastatin, Diazoxide, LPS, Paclitaxel, Curcumin, S1P, Valproate, Lithium, 30 min–48 h	Bone marrow, Adipose tissue, Cell line	(86–88, 101, 105–107)

also shown to contain higher levels of VEGF and TGF-β, which led to increased cell proliferation and migration of dermal fibroblasts, via the TGF-β/SMAD2 and PI3K/AKT signaling pathways (121). These results were further validated by Chen et al. who demonstrated that the hypoxia-preconditioned MSCs secretome significantly increased proliferation and migration of keratinocytes, fibroblasts, endothelial cells, and monocytes *in vitro*, and that skin wound contraction was accelerated in an *in vivo* mouse model (47). The secretome produced by hypoxia-preconditioned placenta-derived MSCs was also shown to reduce scar formation and inhibit proliferation and migration of skin fibroblasts *in vitro* (126). In this case, IL-10 was identified as the key player in the process. In agreement with all these results, Lan and colleagues reported increased expression of anti-apoptotic (HGF, Bcl-2), anti-oxidative (catalase, HO-1), and pro-angiogenic (VEGF) factors in hypoxia-treated BM-MSCs infused with the goal of improving the respiratory function of mice suffering from pulmonary fibrosis (98). Chen et al. observed an increase in the MSCs production levels of not only VEGF-A and bFGF, but also IL-6 and IL-8 (molecules involved in the inflammatory response) under hypoxic conditions (47).

The cytoprotective effect of the hypoxia-pretreatment of MSCs, along with changes in metabolism and maintenance of their differentiation potential, have now been repeatedly demonstrated by a variety of authors, despite differences in the hypoxic conditions used (138, 148). From these studies, it has been concluded that hypoxia-preconditioning increases MSCs' survival in harsh environments (148) and enhances their angiogenic capacity, which together boost MSCs' regenerative and immunomodulatory abilities, contributing to the regulation of excessive fibrosis and cell death due to uncontrolled inflammation (96, 98, 101).

Cytokines, Growth Factors, and Hormones

When considering a significant amount of experimental data regarding MSCs preconditioning with inflammatory cytokines, it is readily apparent that such a stimulus seems to predominantly promote an increase in the production of factors involved in the regulation of the immune response (see **Table 3**). This includes chemoattraction of most immune cells, modulation of inflammation, and even enhancing migration and homing of transplanted MSCs to sites with higher concentrations of such inflammatory molecules. Their immunoregulatory abilities encompass the inhibition of the complement system activation, the inhibition of NK cells, the guidance of monocyte differentiation toward anti-inflammatory macrophages (M2 phenotype), the suppression of cytotoxic T cell proliferation, and the increase in the numbers of regulatory T cells (149). Many of these outcomes are explained by the large number of chemokines produced by the MSCs that effectively attract numerous immune cells to resolve an inflammatory response (150). Specifically, IL-6, PGE2, and IDO all seem to be major effector molecules in the immunoregulatory effects MSCs mediate (123, 151, 152). The production of this potent triad of immunomodulatory molecules is stimulated by the presence of pro-inflammatory factors such as IL-1β, TNF-α, IFN-γ, and LPS (22, 42, 123, 153, 154), that induce MSCs to adopt an immunomodulatory phenotype and to trigger the production of a cocktail of growth factors. These studies thus collectively indicate the close relationship that exists between inflammation and regeneration. In agreement with this supposition, the therapeutic effects that were observed with TNF-α treated MSCs in a wound closure model, mainly mediated by increased angiogenesis and immune cells infiltration, were observed to be dependent on increased levels of IL-6 and IL-8 (116). Indeed, a recent publication featuring an extensive proteomic analysis of the secretome from BM-MSCs preconditioned with pro-inflammatory factors (IL-1β, IL-6, and TNF-α) clearly demonstrates how a pro-inflammatory stimulus mainly increases MSCs production of proteins involved in inflammation and angiogenesis (155). Moreover, the authors also explore the idea that MSCs role in regulating the proteolytic activity in tissues is key for the regulation of these processes.

Still, the mechanism by which these factors seem to influence MSCs is still largely undefined. There is evidence that MSCs immunomodulatory abilities are mediated by both cell-to-cell contact-derived mechanisms (156–158) and paracrine communication (159–161). Also, some authors believe that MSCs are not naturally immunosuppressive and thus, need licensing at the site of inflammation to become so (162–165). This theory is supported by results demonstrating that molecules such as IFN-γ, TNF-α, or IL-1β are necessary to activate the MSCs immunomodulatory activity (166, 167). One study exploring the effect of the preconditioning with TNF-α on MSCs-derived exosomes demonstrated that the effect the stimulatory

TABLE 3 | Dynamics of MSCs secretome composition with cells pre-conditioning.

	Molecule	Preconditioning factors			References
		Hypoxia	Inflammatory stimuli	3D culture	
Adhesion	Gal-9		+		(32)
	VCAM-1		+		(112)
	ICAM-1		+		(112)
	ICAM-4		+		(112)
Antioxidation	Catalase	+			(98)
	HO-1	+			(98)
Apoptosis	IL-24			+	(32)
	TRAIL			+	(32)
	CD82			+	(32)
Cell proliferation and differentiation	IGF	+	+	+	(32, 38, 40–43, 46, 85, 101, 113)
	EGF	+		+	(38)
	G-CSF			+	(114)
	TB4	+			(113)
Chemoattraction	CCL2 (MCP-1)	+	+	+	(37, 38, 48, 114)
	CCL5 (RANTES)	+	+		(37, 38, 48, 112)
	CCL7 (MCP-3)	+		+	(38, 48, 114)
	CCL20		+		(112)
	CXCL1		+		(112)
	CXCL2		+		(115)
	CXCL3		+		(112)
	CXCL5		+		(112, 115)
	CXCL6		+		(112, 115)
	CXCL8 (IL-8)	+	+		(42, 47, 112, 115, 116)
	CXCL9		+		(117)
	CXCL10		+		(112, 115, 117)
	CXCL11		+		(112, 115, 117)
	CXCL12 (SDF-1)	+		+	(41, 43, 114)
	CXCR4	+	+	+	(32, 93, 118–120)
	CXCR7	+			(118)
Immunoregulation	TGF-β	+	+	+	(32, 37, 38, 41, 42, 114, 121)
	IDO	+	+		(32, 37, 117, 122–124)
	Factor H		+		(32, 125)
	IL-10	−/+			(37, 126, 127)
	PD-L1		+		(117)
	HLA-G		+		(117)
	IL-1Ra			+	(114)
	PD-L2		+		(117)
	TSG-6			+	(32, 85)
Inflammation	IL-6	+	+	+	(37, 38, 42, 47, 48, 112, 114–116, 119, 128)
	PGE_2		+	+	(32, 37, 122–124)
	PTX3		+		(115)
	Complement factor B		+		(115)
	Complement factor D		+		(115)
	COX-2		+		(119)
	TNF-α		+		(112)
	IL-23		+		(112)
	IL-16			+	(114)

(Continued)

TABLE 3 | Continued

	Molecule	Preconditioning factors			References
		Hypoxia	Inflammatory stimuli	3D culture	
	IL-7			+	(114)
	IL-11			+	(129)
	IL-2Rα			+	(114, 129)
Metabolism	STC-1			+	(32)
	Cathepsin L1		+		(115)
	Procathepsin B			+	(129)
Migration	MMP-1		+		(36, 115)
	MMP12	+			(130)
Migration Inhibition	PAI-1		+		(115)
	PAI-2		+		(115)
Neuroprotection	BDNF	+			(46, 120)
	GDNF	+			(46)
Osteogenesis	BMP		+	+	(32, 36, 41, 43, 129)
Pluripotency	Oct4	+			(32)
	Rex1	+			(32)
	LIF			+	(42, 114)
Survival	HGF	+	+	+	(32, 40, 41, 85, 98, 113, 130)
	Bcl-2	+		+	(85, 98, 101)
	Akt	+	+		(89, 91, 98, 120)
	HIF-1α	+			(93, 101, 120)
Vascularization	Angiogenin	+		+	(38, 93, 120, 129)
	FGF	+		+	(32, 38, 41–43, 47, 85, 101, 113, 120, 129, 130)
	PDGF	+			(38, 41, 43)
	VEGF	+	+	+	(32, 37, 38, 40–43, 47, 85, 90, 93–95, 101, 113, 114, 120, 121, 129, 130)
	EPO	+			(93, 120)
	EPOR	+			(93)

effects these vesicles had on human osteoblasts was potentiated through increase of Wnt-3a content in ASC-exosomes (168). Conversely, IFN-γ priming of MSC before EV isolation was reported not to influence the immunomodulatory capacity of exosomes or microparticles, which displayed dose-dependent immunomodulatory effects in inflammatory animal models (169). Additionally, TLRs (Toll-Like Receptor) have also been implicated as important mediators of this activation. Optiz et al. reported that activation of TLR3 and TLR4 lead to the induction of IDO which, in turn, mediated the immunosuppressive actions of the MSCs (170). Activation of TLR-2 was shown to cause an increase in the production of galectin-3 by MSCs and, thus, potentiate their capacity to suppress T-cell activation (171). Nonetheless, contradictory reports have also been published. Liotta et al. demonstrated that TLR3 and TLR4 activation not only increased the production of pro-inflammatory molecules but also reduced their inhibitory effect on the proliferation of T-cells (172). Furthermore, they observed that the activation of these TLRs didn't have any effect on levels of IDO. More recently, along with the demonstration that priming with IFN-γ enhanced MSCs immunosuppressive abilities, mainly through the induction of IDO, it was also shown that TLR3 activation did not affect IDO levels and did not influence the cells immunosuppressive activity (165).

Preconditioning with a myriad of other soluble factors, such as growth factors or hormones, seems to also potentiate MSCs regenerative capacity, mainly by stimulating angiogenesis and inhibiting fibrosis. For example, intracardiac transplantation of SDF-1-preconditioned MSCs increased angiogenesis and reduced fibrosis in the ischemic area of a post-infarct heart (89). The effects observed were attributed to the activation of the Akt signaling pathway, similarly to what was described for hypoxia-preconditioned MSCs. TGF-α-preconditioned MSCs enhanced cardiac function mainly through increased VEGF production via a p38 MAPK-dependent mechanism (90). TNF-α or hypoxia were then combined with the TGF-α during prestimulation, and this led to a further improvement in cardiac function. Once more, VEGF seemed to play a key role in MSCs' mode

TABLE 4 | Effect of preconditioning on therapeutic potential of MSCs secretome.

Pre-conditioning treatment	Animal	Study model	MSCs source	Treatment conditions	Main identified mediators	Major conclusions	References
Hypoxia	Rat	*In vitro* ischemic heart	BM	0.5% O_2 for 12 h	CM	Cytoprotection of ARVCs to hypoxia	(135)
	Mouse	Acute kidney injury	AT	0.5% O_2 for 48 h	CM	Enhancement of tissue regeneration and renal function. Decrease in levels of IL-1β and IL-6	(130)
	Mouse	Scald skin wound	Placenta	1–5% O_2 for 72 h	CM (IL-10)	Reduction in scar formation. Inhibition of proliferation and migration of skin fibroblasts	(126)
	Mouse	Excisional skin wound	AT	1/5% O_2 for 72 h	CM (VEGF, TGF-β1, via TGF-β/SMAD and PI3K/Akt)	Increase in MSCs and skin fibroblasts proliferation. Acceleration of wound closure	(121)
	Mouse	Excisional skin wound	BM	2% O_2 for 48 h	CM (bFGF, VEGF, IL-6, IL-8)	Enhancemente of proliferation/ migration of fibroblasts, keratinocytes and enthelial cells. Neovascularization and recruitment of macrophages. Acceleration of wound contraction	(47)
Cytokines, growth factors and hormones	Rat	Cutaneous wound	AT	TNF-α (10 ng/mL) for 48 h	CM (IL-6, IL-8)	Acceleration of wound closure. Increase in angiogenesis and infiltration of immune cells into the wound	(116)

TABLE 5 | Effect of preconditioning on therapeutic potential of MSCs-derived exosomes.

Pre-conditioning treatment	Animal	Study model	MSCs source	Treatment conditions	Main identified mediators	Major conclusions	References
Hypoxia	Rat	Acute myocardial infarction	BM	1%O_2 for 72 h	EVs	Increased angiogenesis and improved cardiac function	(140)
	Rat	I/R cardiac injury	?	?	EVs (miRNA26a)	Attenuation of the injured area and arrythmias	(142)
	Mouse	Acute myocardial infarction	BM	Anoxia + reoxygenation	EVs (miRNA-22)	Reduction of post-infarction fibrosus	(144)
Cytokines, Growth Factors and Hormones	Rat	Kidney ischemia/reperfusion injury	UCB	IFN-γ (100 ng/mL) for 24–48 h	Evs	Loss of cytoprotective effect. Loss of complement factors and lipid binding proteins and gain of tetraspanins, a more complete proteasome complex and MHCI	(145)
Pharmacological agents	Rat	Local cerebral ischemia	Cell line	BYHWD (2,4 g'mL) for 48 h	Evs (VEGF)	Attenuation of ischemic injury by an increase in vascularization	(146)

of action. Preconditioning MSCs with a cocktail of growth-factors (FGF-2, IGF-1, and BMP-2) was also attempted, and this was found to yield protective effects on cardiomyocytes and to improve left ventricular systolic function in a rat myocardial infarction model (173). Another soluble molecule that has been used to precondition MSCs is melatonin, which activates the ERK 1/2 signaling pathway, and consequently enhances cell survival under oxidative stress (100). Thus, melatonin-preconditioned MSCs increased angiogenesis and neurogenesis, reduced infarct size, and improved neurobehavioral outcome in a rat cerebral ischemia model, and once more this seems to have been related to increased VEGF levels (100). Melatonin-preconditioned MSCs also exhibited significantly higher survival rates after intraparenchymal injection in a rat kidney ischemia model (99). This effect was attributed to an upregulation of the enzymes catalase and superoxide dismutase-1 that imbued MSCs with greater antioxidant capacity. Lastly, H_2O_2 has been used to precondition MSCs whose exosomes were used to treat and ischemia/reperfusion injury in a rat model (174). The treatment lead to increased vascularization, which led to higher survival rates, and a reduced inflammatory reaction.

3-Dimensional (3D) Culture

MSCs culture in a 3-dimensional (3D) environment is another type of preconditioning that aims to more closely mimic the physiological conditions which the cells would see *in vivo*. 3D culture of MSCs, namely as spheroids, induces an increase in the production of factors associated with cell survival and proliferation and vascularization (129, 175) (**Table 3**). This, in turn, has been shown to increase these cells' immunomodulatory, angiogenic, anti-fibrotic, and anti-apoptotic activities (83, 85, 104, 176). MSCs spheroids decreased neutrophil activity, the levels of the pro-inflammatory molecules TNF-α, IL-1β, CXCL2/MIP-2, PGE2, and plasmin activity in a mouse peritonitis model (83). Other studies demonstrated that 3D culture of MSCs in spheroids seems to stimulate the cells' pro-angiogenic ability, as demonstrated by the implantation of AT-MSCs aggregates leading to an improvement in renal function in a rat model of acute renal ischemia/reperfusion (85). As previously mentioned, the 3D culture of MSCs has also been demonstrated to increase their anti-fibrotic potential. MSCs spheroids were shown to decrease tissue fibrosis in a mouse model of hepatic fibrosis (104). However, in another model system, MSCs' anti-fibrotic activity was shown to be dependent on the dose of MSCs aggregates, with higher density aggregates being unable to regenerate the cartilage in rabbits suffering from full-thickness osteochondral defects (84). Administration of 3D cultures of MSCs aggregates (with more than 125,000 cells) during 24 h increased wound healing rate in mice with full-thickness diabetic wounds. Lower cell doses yielded results similar to those observed in the vehicle-treated mice (76), supporting the notion that the dose used is key.

The spheroid 3D culture creates a microenvironment where inner layers are exposed to much lower levels of oxygen and nutrients, originating an hypoxic environment (177). Although, as a consequence, MSCs express higher

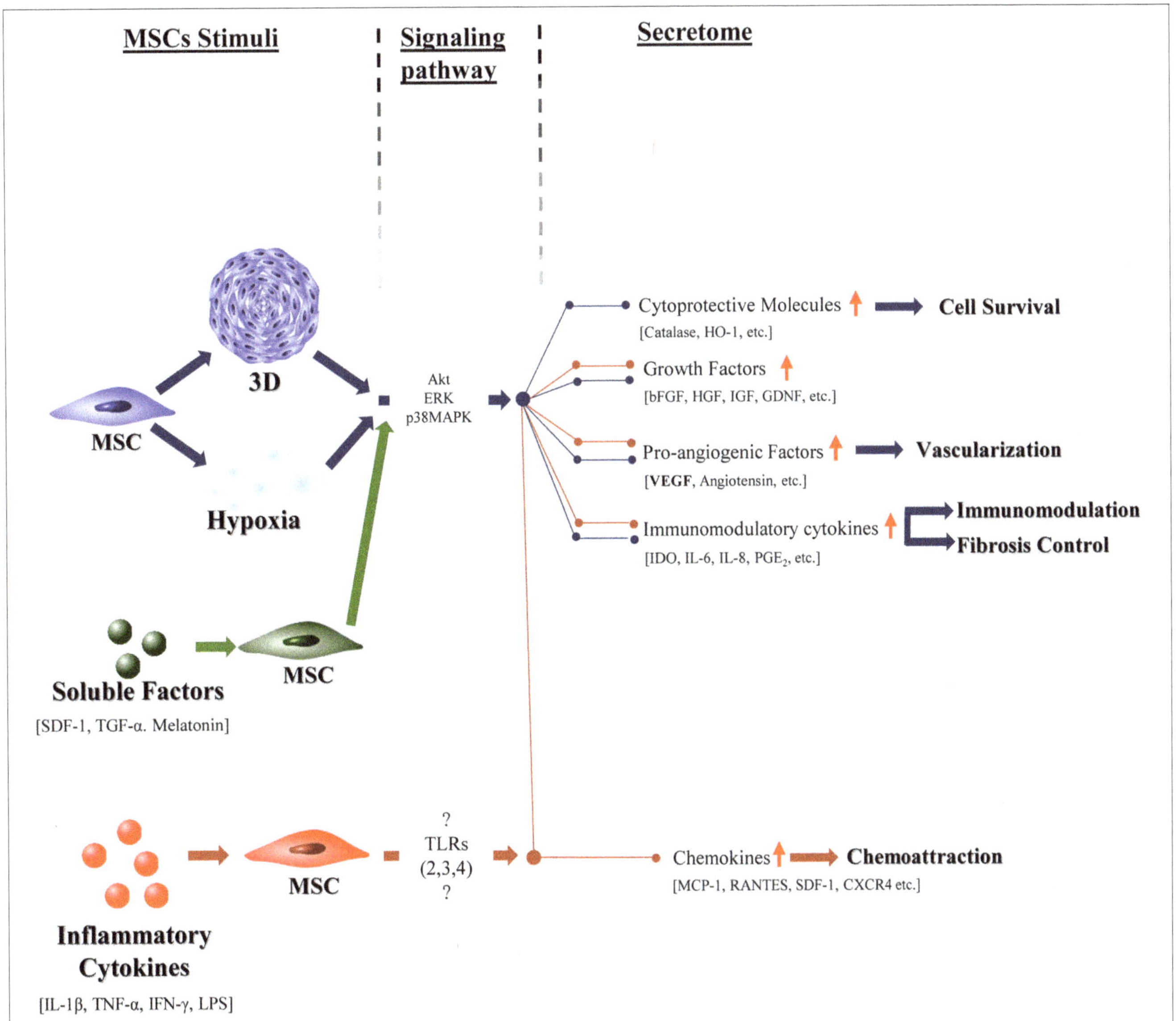

FIGURE 2 | The effect of different preconditioning stimuli in the MSCs response. Schematic representation of known effects of highly studied preconditioning factors—hypoxia (in blue), 3D culture (in blue), specific soluble factors (green), and inflammatory cytokines (red)—in the MSCs response. Blue pathway presents the effect of a hypoxic environment on the cells, which is mediated by specific signaling pathways (Akt, ERK, p38MAPK) and culminates in the stimulation of the above signaled effects. Tridimensional culture is also represented in blue. MSCs preconditioning with specific soluble factors (SDF-1, TGF-α, and melatonin) seems to stimulate the same signaling pathways as a hypoxic environment and, thus, elicit the same general response from these cells. The use of inflammatory cytokines to influence the MSC response, as represented in red, besides promoting the specific above shown effects, also stimulates the production of factors that seem to be common to all the other preconditioning factors. The pathways that mediate this activity are still to be determined.

levels of molecules associated with apoptosis (32), their immunomodulatory capacity seems secured by the diverse and abundant production of factors involved in inflammation and immune response (83, 114, 177) (**Table 3**). Accordingly, 3D MSCs-preconditioning was shown to upregulate *TSG-6* expression, as well as *SCT-1* (anti-inflammatory/anti-apoptotic protein), *LIF, IL-24, TRAIL,* and *CXCR4,* a chemokine involved in spheroid-derived-MSCs adhesion to endothelial cells. Wnt signaling cascade seemed to be involved in this effect as the expression of its inhibitor DKK1 was decreased (175). In fact, 3D pre-conditioning of MSCs demonstrated an influence in the production of, not only inflammatory cytokines, but also matrix constituents and degrading enzymes (76, 84, 85), which contributes to their anti-fibrotic capacity. Culturing MSCs as spheroids seems to further enhance their innate pro-angiogenic ability, presumably by increasing their production of angiogenic factors such as Angiogenin, bFGF, VEGF, and HGFa (85, 129). In addition, culturing MSCs in this more physiologically relevant 3D architecture was found to increase their expression of E-cadherin which, by activating the ERK/AKT signaling pathway, was responsible for the higher levels of VEGF production observed (178).

Pharmacological Agents

Pre-conditioning of MSCs with pharmacological agents may be considered as an alternative option in specific cases. For example, MSCs were pre-conditioned with atorvastatin (a statin associated with the prevention of cardiovascular disease events) (107), oxytocin (hormone) (179), Curcumin (strong anti-oxidant with anti-inflammatory properties) (180), lipopolysaccharide (LPS—endotoxin) (105), and diazoxide (used as a vasodilator in acute hypertension) (86), and their ability to treat myocardial infarction tested. All of the preconditioning protocols enhanced the survival of the transplanted MSCs *in vivo* and led to improved functional recovery and reduced infarct size (86, 105, 107, 179, 180). In general, these effects were due to increased neovascularization and reduced tissue fibrosis. All the studies, except for those using atorvastatin and oxytocin (86, 105, 180) reported a link between the observed effect and increased levels of VEGF, FGF-2, or HGF and activation of the Akt signaling pathway. In concordance, MSCs-derived EVs obtained after preconditioning with Buyang Huanwu Decoction (BYHWD), a drug that has been used for centuries for the treatment of paralysis and stroke, was shown to attenuate brain injury in a rat local cerebral ischemia model by increasing local VEGF levels (146). Atorvastatin seemed to potentiate MSCs' immunomodulatory capacity, decreasing the infiltration of inflammatory cells and the levels of TNF-α and IL-6, via a CXCR4-dependent mechanism, which the authors concluded was the primary mediator of the improvement of functional recovery and reduction of infarct size in a rat stroke model upon MSCs injection (106). Thus, it is becoming clear that MSCs' paracrine response is as dynamic as the microenvironment that surrounds them. Even minor changes in culture conditions or in the microenvironment of the injured tissue can induce dramatically different results (32).

CONCLUSION

Extensive evidence now exists to support the benefits of preconditioning of MSCs with respect to improving their capacity to induce regeneration/repair across the wide array of tissues and pathologic conditions in which these cells have been explored. Depending on the preconditioning factors used in the MSCs' culture, different signaling pathways are activated. Understanding how each different stimulus affects MSCs behavior is crucial to validate MSCs preconditioning as a tool to enhance both the safety and the disease-specific therapeutic potential of MSCs for clinical use.

Although a great deal of work to comprehend the full mechanisms of MSCs paracrine signaling upon pre-conditioning is needed, some patterns can be recognized (see **Figure 2**). In summary, pre-conditioning of MSCs with hypoxia, 3D structural organization or soluble factors as SDF-1, or TGF-β seem to activate Akt, ERK, and p38MAPK signaling pathways, that seem to increase the production of cytoprotective molecules (Catalase, HO-1, etc.), pro-regenerative (bFGF, HGF, IGF, etc.) and pro-angiogenic (VEGF) soluble factors and immunomodulatory cytokines (IDO, PGE2, IL6, etc.). On the other hand, priming of MSCs with inflammatory cytokines such as IFN-γ, IL-1β activate TLRs (namely, TLR-2/3/4) on MSCs surface which then increases the production of similar cytoprotective, pro-regenerative, pro-angiogenic and immunomodulatory molecules and further promotes chemokines secretion. Nevertheless, to-date, it has not been possible to identify one single mechanism responsible for this effect.

Moreover, data on the effects that preconditioning of MSCs exerts on the composition and therapeutic potential of their secretome is still lacking. Most studies have focused solely on the effects of hypoxia on the MSCs' secretome content and its therapeutic potential or, in alternative, on the effect that other preconditioning factors could exert on its composition. In most of the studies, there is a lack of evidence on the influence of preconditioning on both the therapeutic effect of the secretome and its composition. Perhaps most importantly, the great majority of studies exploring the clinical utility of the MSCs' secretome have tended to utilize a fairly myopic approach to study its composition, focusing on specific factors of interest in the unique pathological setting being explored, which has contributed to the difficulty in gathering a widely applicable understanding of how the secretome can be fine-tuned for maximal effect in each specific pathology.

Furthermore, due to the wide heterogeneity in MSCs employed in the research community, with respect to their tissue of origin, the health and age of the donor, protocols of cell isolation, culture and preconditioning, as well as the animal model used for testing, it becomes rather difficult to dissect the mechanistic action behind the observed effects on preconditioning of MSCs. Advances in high-throughput techniques and bioinformatic tools, in combination with a database in the area, would help to create a more comprehensive and complete understanding of the way preconditioning can be fine-tuned to increase MSCs' therapeutic utility in the future. This aspect of big data is particularly relevant when the number of studies using MSCs secretome is increasing, accomplished by an expansion of the EVs/exossomes field, in which their secretion by MSCs is being largely explored (181, 182).

Cell therapies as established so far, notwithstanding its great promise, present several obstacles concerning safety, process standardization, and practicality of the procedures needed to deliver viable cells to the hostile microenvironment often present within injured tissues (49, 183, 184). A consensus in the shifting of MSCs' therapeutic potential to their paracrine mechanism of action is being formed within the scientific community, which points to the development of guidelines to refine the experimental settings of the production of MSCs secretome, to establish more standardized protocols among the scientific community and to promote future collaborative work to close the wide gap that has existed for decades between MSCs experimental research and their clinical use.

AUTHOR CONTRIBUTIONS

JF and RG conceived the main conceptual ideas and outlined the proof. JF gathered, analyzed, and sorted the revised literature and wrote the manuscript. RG, GT, SS, GA-P, and MB revised and provided critical feedback on the manuscript. RG provided funding.

ACKNOWLEDGMENTS

We would like to acknowledge Norte Portugal Regional Operational Programme (NORTE 2020) in the framework of the project "Bioengineered Therapies for Infectious Diseases and Tissue Regeneration" (NORTE-01-0145-FEDER-000012). We also acknowledge Fundação para a Ciência e a Tecnologia (FCT) and Fundo Europeu de Desenvolvimento Regional (FEDER) funds through the COMPETE 2020-Operacional Programme for Competitiveness and Internationalization (POCI), Portugal 2020-in the framework of the project "Institute for Research and Innovation in Health Sciences" (POCI-01-0145-FEDER-007274). We also acknowledge EUROSPINE TRF for the funded project "Disc Regeneration, Immuno, and Neuro Modulation" , ref. 2017_05 . In addition, JF and RG also acknowledge FCT for funding the BiotechHealth Ph.D. fellowship (PD/BD/135486/2018) and the FCT Investigator Grant (IF/00638/2014), respectively.

REFERENCES

1. Ullah I, Subbarao RB, Rho GJ. Human mesenchymal stem cells - current trends and future prospective. *Biosci Rep*. (2015) 35:2. doi: 10.1042/BSR20150025
2. Squillaro T, Peluso G, Galderisi U. Clinical trials with mesenchymal stem cells: an update. *Cell Trans*. (2016) 25:829–48. doi: 10.3727/096368915X689622
3. Wang J, Liao L, Tan J. Mesenchymal-stem-cell-based experimental and clinical trials: current status and open questions. *Expert Opin Biol Ther*. (2011) 11:893–909. doi: 10.1517/14712598.2011.574119
4. Kim N, Cho SG. Clinical applications of mesenchymal stem cells. *Korean J Intern Med*. (2013) 28:387–402. doi: 10.3904/kjim.2013.28.4.387
5. Richardson SM, Hoyland JA, Mobasheri R, Csaki C, Shakibaei M, Mobasheri A. Mesenchymal stem cells in regenerative medicine: opportunities and challenges for articular cartilage and intervertebral disc tissue engineering. *J Cell Physiol*. (2010) 222:23–32. doi: 10.1002/jcp.21915
6. Rohban R, Pieber TR. Mesenchymal stem and progenitor cells in regeneration: tissue specificity and regenerative potential. *Stem Cells Int*. (2017) 2017:1–16. doi: 10.1155/2017/5173732
7. Christensen ME, Turner BE, Sinfield L, Kollar JK, Cullup H, Waterhouse NJ, et al. Mesenchymal stromal cells transiently alter the inflammatory milieu post-transplant to delay graft-versus-host disease. *Haematologica* (2010) 95:2102–10. doi: 10.3324/haematol.2010.028910
8. Sun L, Akiyama K, Zhang H, Yamaza T, Hou Y, Zhao S, et al. Mesenchymal stem cell transplantation reverses multi-organ dysfunction in systemic lupus erythematosus mice and humans. *Stem Cells* (2009) 27:1421–32. doi: 10.1002/stem.68
9. Liang J, Zhang H, Hua B, Wang H, Lu L, Shi S, et al. Allogenic mesenchymal stem cells transplantation in refractory systemic lupus erythematosus: a pilot clinical study. *Ann Rheum Dis*. (2010) 69:1423–9. doi: 10.1136/ard.2009.123463
10. Ciccocioppo R, Bernardo ME, Sgarella A, Maccario R, Avanzini MA, Ubezio C, et al. Autologous bone marrow-derived mesenchymal stromal cells in the treatment of fistulising Crohn's disease. *Gut* (2011) 60:788–98. doi: 10.1136/gut.2010.214841
11. Amado LC, Saliaris AP, Schuleri KH, St John M, Xie JS, Cattaneo S, et al. Cardiac repair with intramyocardial injection of allogeneic mesenchymal stem cells after myocardial infarction *Proc Natl Acad Sci USA*. (2005) 102:11474–9. doi: 10.1073/pnas.0504388102
12. Cai M, Shen R, Song L, Lu M, Wang J, Zhao S, et al. Bone marrow mesenchymal stem cells (BM-MSCs) improve heart function in swine myocardial infarction model through paracrine effects. *Sci Rep*. (2016) 6:28250. doi: 10.1038/srep28250
13. Bang OY, Lee JS, Lee PH, Lee G. Autologous mesenchymal stem cell transplantation in stroke patients. *Ann Neurol*. 57:874–82. doi: 10.1002/ana.20501
14. Mohyeddin Bonab M, Yazdanbakhsh S, Lotfi J, Alimoghaddom K, Talebian F, Hooshmand F, et al. Does mesenchymal stem cell therapy help multiple sclerosis patients? *Rep Pilot Study Iran J Immunol*. (2007) 4:50–7.
15. Karussis D, Karageorgiou C, Vaknin-Dembinsky A, Gowda-Kurkalli B, Gomori J, Kassis IM, et al. Safety and immunological effects of mesenchymal stem cell transplantation in patients with multiple sclerosis and amyotrophic lateral sclerosis. *Arch Neurol*. (2010) 67:1187–94. doi: 10.1001/archneurol.2010.248
16. Kharaziha P, Hellstrom PM, Noorinayer B, Farzaneh F, Aghajani K, Jafari F, et al. Improvement of liver function in liver cirrhosis patients after autologous mesenchymal stem cell injection: a phase I-II clinical trial. *Eur J Gastroenterol Hepatol*. (2009) 21:1199–205. doi: 10.1097/MEG.0b013e32832a1f6c
17. Shi M, Zhang Z, Xu R, Lin H, Fu J, Zou Z, et al. Human mesenchymal stem cell transfusion is safe and improves liver function in acute-on-chronic liver failure patients. *Stem Cells Transl Med*. (2012) 1:725–31. doi: 10.5966/sctm.2012-0034
18. Chang C, Wang X, Niu D, Zhang Z, Zhao H, Gong F. Mesenchymal stem cells adopt beta-cell fate upon diabetic pancreatic microenvironment. *Pancreas* (2009) 38:275–81. doi: 10.1097/MPA.0b013e318191521c
19. Holmes D. Diabetes: MSC transplant prevents [beta]-cell dysfunction. *Nat Rev Endocrinol*. (2014) 10:701. doi: 10.1038/nrendo.2014.172
20. Kursova LV, Konoplyannikov AG, Pasov VV, Ivanova IN, Poluektova MV, Konoplyannikova OA. Possibilities for the use of autologous mesenchymal stem cells in the therapy of radiation-induced lung injuries. *Bull Exp Biol Med*. (2009) 147:542–6. doi: 10.1007/s10517-009-0538-7

21. Chamberlain G, Fox J, Ashton B, Middleton J. Concise review: mesenchymal stem cells: their phenotype, differentiation capacity, immunological features, and potential for homing. *Stem Cells* (2007) 25:2739–49. doi: 10.1634/stemcells.2007-0197
22. Aggarwal S, Pittenger MF. Human mesenchymal stem cells modulate allogeneic immune cell responses. *Blood* (2005) 105:1815–22. doi: 10.1182/blood-2004-04-1559
23. Casiraghi F, Azzollini N, Cassis P, Imberti B, Morigi M, Cugini D, et al. Pretransplant infusion of mesenchymal stem cells prolongs the survival of a semiallogeneic heart transplant through the generation of regulatory T cells. *J Immunol.* (2008) 181:3933–46. doi: 10.4049/jimmunol.181.6.3933
24. Jarvinen L, Badri L, Wettlaufer S, Ohtsuka T, Standiford T, Toews GB, et al. Lung resident mesenchymal stem cells isolated from human lung allografts inhibit T cell proliferation via a soluble mediator. *J Immunol.* (2008) 181:4389–96. doi: 10.4049/jimmunol.181.6.4389
25. Chan JL, Tang KC, Patel AP, Bonilla LM, Pierobon N, Ponzio NM, et al. Antigen-presenting property of mesenchymal stem cells occurs during a narrow window at low levels of interferon-gamma. *Blood* (2006) 107:4817–24. doi: 10.1182/blood-2006-01-0057
26. Stagg J, Pommey S, Eliopoulos N, Galipeau J. Interferon-gamma-stimulated marrow stromal cells: a new type of nonhematopoietic antigen-presenting cell. *Blood* (2006) 107:2570–7. doi: 10.1182/blood-2005-07-2793
27. Caplan AI, Dennis JE. Mesenchymal stem cells as trophic mediators. *J Cell Biochem.* 98: 1076–84. doi: 10.1002/jcb.20886
28. Picinich SC, Mishra PJ, Mishra PJ, Glod J, Banerjee D. The therapeutic potential of mesenchymal stem cells. Cell- & tissue-based therapy. *Expert Opin Biol Ther.* (2007) 7:965–73. doi: 10.1517/14712598.7.7.965
29. Caplan AI. Why are MSCs therapeutic? *J Pathol.* (2009) 217:318–24. doi: 10.1002/path.2469
30. Vizoso FJ, Eiro N, Cid S, Schneider J, Perez-Fernandez R. Mesenchymal stem cell secretome: toward cell-free therapeutic strategies in regenerative medicine. *Int J Mol Sci.* (2017) 18:1852. doi: 10.3390/ijms18091852
31. Baraniak PR, McDevitt TC. Stem cell paracrine actions and tissue regeneration. *Regenerat Med.* (2010) 5:121–43. doi: 10.2217/rme.09.74
32. Madrigal M, Rao KS, Riordan NH. A review of therapeutic effects of mesenchymal stem cell secretions and induction of secretory modification by different culture methods. *J Transl Med.* (2014) 12:1–14. doi: 10.1186/s12967-014-0260-8
33. Lötvall J, Hill AF, Hochberg F, Buzás EI, Di Vizio D, Gardiner C, et al. Minimal experimental requirements for definition of extracellular vesicles and their functions: a position statement from the International Society for Extracellular Vesicles. *J Extracellu Vesicles* (2014) 3, 1 6. doi: 10.3402/jev.v3403.26913
34. Stahl PD, Raposo G. Exosomes and extracellular vesicles: the path forward. *Essays Biochem.* (2018) 62:119–24. doi: 10.1042/EBC20170088
35. Camussi G, Deregibus MC, Bruno S, Cantaluppi V, Biancone L. Exosomes/microvesicles as a mechanism of cell-to-cell communication. *Kidney Int.* (2010) 78:838–48. doi: 10.1038/ki.2010.278
36. Polacek M, Bruun JA, Elvenes J, Figenschau Y, Martinez I. The secretory profiles of cultured human articular chondrocytes and mesenchymal stem cells: implications for autologous cell transplantation strategies. *Cell Trans.* (2011) 20:1381–93. doi: 10.3727/096368910X550215
37. Kyurkchiev D, Bochev I, Ivanova-Todorova E, Mourdjeva M, Oreshkova T, Belemezova K, et al. Secretion of immunoregulatory cytokines by mesenchymal stem cells. *World J Stem Cells* (2014) 6:552–70. doi: 10.4252/wjsc.v6.i5.552
38. Linero I, Chaparro O. Paracrine effect of mesenchymal stem cells derived from human adipose tissue in bone regeneration. *PLoS ONE* (2014) 9:e107001. doi: 10.1371/journal.pone.0107001
39. Pereira T, Ivanova G, Caseiro AR, Barbosa P, Bártolo PJ, Santos JD, et al. MSCs conditioned media and umbilical cord blood plasma metabolomics and composition. *PLoS ONE* (2014) 9:e113769. doi: 10.1371/journal.pone.0113769
40. Wang M, Crisostomo PR, Herring C, Meldrum KK, Meldrum DR. Human progenitor cells from bone marrow or adipose tissue produce VEGF, HGF, and IGF-I in response to TNF by a p38 MAPK-dependent mechanism. *Am J Physiol Regul Integr Comp Physiol.* (2006) 291:R880–4. doi: 10.1152/ajpregu.00280.2006
41. Inukai T, Katagiri W, Yoshimi R, Osugi M, Kawai T, Hibi H, et al. Novel application of stem cell-derived factors for periodontal regeneration. *Biochem Biophys Res Commun.* (2013) 430:763–8. doi: 10.1016/j.bbrc.2012.11.074
42. Liu CH, Hwang S-M. Cytokine interactions in mesenchymal stem cells from cord blood. *Cytokine* (2005) 32:270–9. doi: 10.1016/j.cyto.2005.11.003
43. Osugi M, Katagiri W, Yoshimi R, Inukai T, Hibi H, Ueda M. Conditioned media from mesenchymal stem cells enhanced bone regeneration in rat calvarial bone defects. *Tissue Eng Part A* (2012) 18:1479–89. doi: 10.1089/ten.tea.2011.0325
44. Teng X, Chen L, Chen W, Yang J, Yang Z, Shen Z. Mesenchymal stem cell-derived exosomes improve the microenvironment of infarcted myocardium contributing to angiogenesis and anti-inflammation. *Cell Physiol Biochem.* (2015) 37:2415–24. doi: 10.1159/000438594
45. Li T, Yan Y, Wang B, Qian H, Zhang X, Shen L, et al. Exosomes derived from human umbilical cord mesenchymal stem cells alleviate liver fibrosis. *Stem Cells Dev.* (2012) 22:845–54. doi: 10.1089/scd.2012.0395
46. Teixeira FG, Carvalho MM, Neves-Carvalho A, Panchalingam KM, Behie LA, Pinto L, et al. Secretome of mesenchymal progenitors from the umbilical cord acts as modulator of neural/glial proliferation and differentiation. *Stem Cell Rev.* (2015) 11:288–97. doi: 10.1007/s12015-014-9576-2
47. Chen L, Xu Y, Zhao J, Zhang Z, Yang R, Xie J. Conditioned medium from hypoxic bone marrow-derived mesenchymal stem cells enhances wound healing in mice. *PLoS ONE* (2014) 9:e96161. doi: 10.1371/journal.pone.0096161
48. Ando Y, Matsubara K, Ishikawa J, Fujio M, Shohara R, Hibi H, et al. Stem cell-conditioned medium accelerates distraction osteogenesis through multiple regenerative mechanisms. *Bone* (2014) 61:82–90. doi: 10.1016/j.bone.2013.12.029
49. Parekkadan B, van Poll D, Suganuma K, Carter EA, Berthiaume F, Tilles AW, et al. Mesenchymal stem cell-derived molecules reverse fulminant hepatic failure. *PLoS ONE* (2007) 2:e941. doi: 10.1371/journal.pone.0000941
50. Imberti B, Morigi M, Tomasoni S, Rota C, Corna D, Longaretti L, et al. Insulin-like growth factor-1 sustains stem cell mediated renal repair. *J Am Soc Nephrol.* (2007) 18:2921–8. doi: 10.1681/ASN.2006121318
51. Iso Y, Spees JL, Serrano C, Bakondi B, Pochampally R, Song YH, et al. Multipotent human stromal cells improve cardiac function after myocardial infarction in mice without long-term engraftment. *Biochem Biophys Res Commun.* (2007) 354:700–6. doi: 10.1016/j.bbrc.2007.01.045
52. Javazon EH, Keswani SG, Badillo AT, Crombleholme TM, Zoltick PW, Radu AP, et al. Enhanced epithelial gap closure and increased angiogenesis in wounds of diabetic mice treated with adult murine bone marrow stromal progenitor cells *Wound Repair Regen.* (2007) 15:350–9. doi: 10.1111/j.1524-475X.2007.00237.x
53. Yang SH, Wu CC, Shih TT, Sun YH, Lin FH. *In Vitro* study on interaction between human nucleus pulposus cells and mesenchymal stem cells through paracrine stimulation. *Spine* (2008) 33:1951–7. doi: 10.1097/BRS.0b013e31817e6974
54. Brisby H, Papadimitriou N, Brantsing C, Bergh P, Lindahl A, Barreto Henriksson H. The presence of local mesenchymal progenitor cells in human degenerated intervertebral discs and possibilities to influence these *in vitro*: a descriptive study in humans. *Stem Cells Dev.* (2013) 22:804–14. doi: 10.1089/scd.2012.0179
55. Pereira CL, Teixeira GQ, Ribeiro-Machado C, Caldeira J, Costa M, Figueiredo F, et al. Mesenchymal stem/stromal cells seeded on cartilaginous endplates promote intervertebral disc regeneration through extracellular matrix remodeling. *Sci Rep.* (2016) 6:33836. doi: 10.1038/srep33836
56. Teixeira GQ, Pereira CL, Ferreira JR, Maia AF, Gomez-Lazaro M, Barbosa MA, et al. Immunomodulation of human mesenchymal stem/stromal cells in intervertebral disc degeneration: insights from a proinflammatory/degenerative *ex vivo* model. *Spine* (2017) 43:E673–82. doi: 10.1097/BRS.0000000000002494
57. Zheng ZM, Lu MM, Zhou L, Sun Y, Lv F, Leung VYL, et al. The potential of umbilical cord derived mesenchymal stem cells in intervertebral disc repair. *Global Spine J. Germany, Georg Thieme Verlag* (2014) 4:S81. doi: 10.1055/s-0034-1376649

58. Strassburg S, Hodson NW, Hill PI, Richardson SM, Hoyland JA. Bi-directional exchange of membrane components occurs during co-culture of mesenchymal stem cells and nucleus pulposus cells. *PLoS ONE* (2012) 7:e33739. doi: 10.1371/journal.pone.0033739
59. Dai W, Hale SL, Kloner RA. Role of a paracrine action of mesenchymal stem cells in the improvement of left ventricular function after coronary artery occlusion in rats. *Regenet Med.* (2007) 2:63–8. doi: 10.2217/17460751.2.1.63
60. Li Y, Chen J, Zhang CL, Wang L, Lu D, Katakowski M, et al. Gliosis and brain remodeling after treatment of stroke in rats with marrow stromal cells. *Glia* (2005) 49:407–17. doi: 10.1002/glia.20126
61. Salgado AJ, Reis RL, Sousa NJ, Gimble JM. Adipose tissue derived stem cells secretome: soluble factors and their roles in regenerative medicine. *Curr Stem Cell Res Ther.* (2010) 5:103–10. doi: 10.2174/157488810791268564
62. van Koppen A, Joles JA, van Balkom BWM, Lim SK, de Kleijn D, Giles RH, et al. Human embryonic mesenchymal stem cell-derived conditioned medium rescues kidney function in rats with established chronic kidney disease. *PLoS ONE* (2012) 7:e38746. doi: 10.1371/journal.pone.0038746
63. Silva AM, Teixeira JH, Almeida MI, Goncalves RM, Barbosa MA, Santos SG. Extracellular Vesicles: Immunomodulatory messengers in the context of tissue repair/regeneration. *Eur J Pharm Sci.* (2017) 98:86–95. doi: 10.1016/j.ejps.2016.09.017
64. Lai RC, Arslan F, Lee MM, Sze NS, Choo KA, Chen TS, et al. Exosome secreted by MSC reduces myocardial ischemia/reperfusion injury. *Stem Cell Res.* (2010) 4:214–22. doi: 10.1016/j.scr.2009.12.003
65. Doeppner TR, Herz J, Gorgens A, Schlechter J, Ludwig A, Radtke KS, et al. Extracellular vesicles improve post-stroke neuroregeneration and prevent postischemic immunosuppression. *Stem Cells Transl Med.* (2015) 4:1131–43. doi: 10.5966/sctm.2015-0078
66. Xin H, Li Y, Liu Z, Wang X, Shang X, Cui Y, et al. MiR-133b promotes neural plasticity and functional recovery after treatment of stroke with multipotent mesenchymal stromal cells in rats via transfer of exosome-enriched extracellular particles. *Stem Cells* (2013) 31:2737–46. doi: 10.1002/stem.1409
67. Ophelders DR, Wolfs TG, Jellema RK, Zwanenburg A, Andriessen P, Delhaas T, et al. Mesenchymal stromal cell-derived extracellular vesicles protect the fetal brain after hypoxia-ischemia. *Stem Cells Transl Med.* (2016) 5:754–63. doi: 10.5966/sctm.2015-0197
68. Gangadaran P, Rajendran RL, Lee HW, Kalimuthu S, Hong CM, Jeong SY, et al. Extracellular vesicles from mesenchymal stem cells activates VEGF receptors and accelerates recovery of hindlimb ischemia *J Control Rel.* (2017) 264:112–26. doi: 10.1016/j.jconrel.2017.08.022
69. Gatti S, Bruno S, Deregibus MC, Sordi A, Cantaluppi V, Tetta C, et al. Microvesicles derived from human adult mesenchymal stem cells protect against ischaemia-reperfusion-induced acute and chronic kidney injury. *Nephrol Dial Trans.* (2011) 26:1474–83. doi: 10.1093/ndt/gfr015
70. Bruno S, Grange C, Collino F, Deregibus MC, Cantaluppi V, Biancone L, et al. Microvesicles derived from mesenchymal stem cells enhance survival in a lethal model of acute kidney injury. *PLoS ONE* (2012) 7:e33115. doi: 10.1371/journal.pone.0033115
71. Zhou BR, Xu Y, Guo SL, Xu Y, Wang Y, Zhu F, et al. The effect of conditioned media of adipose-derived stem cells on wound healing after ablative fractional carbon dioxide laser resurfacing. *Biomed Res Int.* (2013) 2013:519126. doi: 10.1155/2013/519126
72. Kordelas L, Rebmann V, Ludwig AK, Radtke S, Ruesing J, Doeppner TR, et al. MSC-derived exosomes: a novel tool to treat therapy-refractory graft-versus-host disease *Leukemia* (2014) 28:970–3. doi: 10.1038/leu.2014.41
73. Fukuoka H, Suga H. Hair regeneration treatment using adipose-derived stem cell conditioned medium: follow-up with trichograms. *Eplasty* (2015) 15:e10.
74. Shin H, Ryu HH, Kwon O, Park BS, Jo SJ. Clinical use of conditioned media of adipose tissue-derived stem cells in female pattern hair loss: a retrospective case series study. *Int J Dermatol.* (2015) 54:730–5. doi: 10.1111/ijd.12650
75. Katagiri W, Osugi M, Kawai T, Hibi H. First-in-human study and clinical case reports of the alveolar bone regeneration with the secretome from human mesenchymal stem cells. *Head Face Med.* (2016) 12:5. doi: 10.1186/s13005-016-0101-5
76. Amos PJ, Kapur SK, Stapor PC, Shang H, Bekiranov S, Khurgel M, et al. Human adipose-derived stromal cells accelerate diabetic wound healing: impact of cell formulation and delivery. *Tissue Eng Part A* (2010) 16:1595–606. doi: 10.1089/ten.tea.2009.0616
77. Hemeda H, Jakob M, Ludwig AK, Giebel B, Lang S, Brandau S. Interferon-gamma and tumor necrosis factor-alpha differentially affect cytokine expression and migration properties of mesenchymal stem cells. *Stem Cells Dev.* (2010) 19:693–706. doi: 10.1089/scd.2009.0365
78. Hagmann S, Moradi B, Frank S, Dreher T, Kämmerer P, Richter WW, et al. Different culture media affect growth characteristics, surface marker distribution and chondrogenic differentiation of human bone marrow-derived mesenchymal stromal cells. *BMC Musculoskeletal Disord.* (2013) 14:223. doi: 10.1186/1471-2474-14-223
79. Lee KS, Cha SH, Kang HW, Song JY, Lee KW, Ko KB, et al. Effects of serial passage on the characteristics and chondrogenic differentiation of canine umbilical cord matrix derived mesenchymal stem cells asian-australasian *J Anim Sci.* (2013) 26:588–95. doi: 10.5713/ajas.2012.12488
80. Elahi KC, Klein G, Avci-Adali M, Sievert KD, MacNeil S, Aicher WK. Human mesenchymal stromal cells from different sources diverge in their expression of cell surface proteins and display distinct differentiation patterns. *Stem Cells Int.* (2016) 2016:5646384. doi: 10.1155/2016/5646384
81. Heathman TRJ, Rafiq QA, Chan AKC, Coopman K, Nienow AW, Kara B, et al. Characterization of human mesenchymal stem cells from multiple donors and the implications for large scale bioprocess development. *Biochem Eng J.* (2016) 108:14–23. doi: 10.1016/j.bej.2015.06.018
82. Hu X, Xu Y, Zhong Z, Wu Y, Zhao J, Wang Y, et al. A large-scale investigation of hypoxia-preconditioned allogeneic mesenchymal stem cells for myocardial repair in nonhuman primates: paracrine activity without remuscularization. *Circ Res.* (2016) 118:970–83. doi: 10.1161/CIRCRESAHA.115.307516
83. Bartosh TJ, Ylöstalo JH, Mohammadipoor A, Bazhanov N, Coble K, Claypool K, et al. Aggregation of human mesenchymal stromal cells (MSCs) into 3D spheroids enhances their antiinflammatory properties. *Proc Natl Acad Sci USA.* (2010) 107:13724–9. doi: 10.1073/pnas.1008117107
84. Suzuki S, Muneta T, Tsuji K, Ichinose S, Makino H, Umezawa A, et al. Properties and usefulness of aggregates of synovial mesenchymal stem cells as a source for cartilage regeneration. *Arthritis Res Ther.* (2012) 14:R136. doi: 10.1186/ar3869
85. Xu Y, Shi T, Xu A, Zhang L. 3D spheroid culture enhances survival and therapeutic capacities of MSCs injected into ischemic kidney. *J Cell Mol Med.* (2016) 20:203–13. doi: 10.1111/jcmm.12651
86. Cui X, Wang H, Guo H, Wang C, Ao H, Liu X, et al. Transplantation of mesenchymal stem cells preconditioned with diazoxide, a mitochondrial ATP-sensitive potassium channel opener, promotes repair of myocardial infarction in rats. *Tohoku J Exp Med.* (2010) 220:139–47. doi: 10.1620/tjem.220.139
87. Pessina A, Cocce V, Pascucci L, Bonomi A, Cavicchini L, Sisto F, et al. Mesenchymal stromal cells primed with paclitaxel attract and kill leukaemia cells, inhibit angiogenesis and improve survival of leukaemia-bearing mice. *Br J Haematol.* (2013) 160:766–78. doi: 10.1111/bjh.12196
88. Kang H, Kim KH, Lim J, Kim YS, Heo J, Choi J, et al. The therapeutic effects of human mesenchymal stem cells primed with sphingosine-1 phosphate on pulmonary artery hypertension. *Stem Cells Dev.* (2015) 24:1658–71. doi: 10.1089/scd.2014.0496
89. Pasha Z, Wang Y, Sheikh R, Zhang D, Zhao T, Ashraf M. Preconditioning enhances cell survival and differentiation of stem cells during transplantation in infarcted myocardium. *Cardiovasc Res.* (2008) 77:134–42. doi: 10.1093/cvr/cvm025
90. Herrmann JL, Wang Y, Abarbanell AM, Weil BR, Tan J, Meldrum DR. Preconditioning mesenchymal stem cells with transforming growth factor-alpha improves mesenchymal stem cell-mediated cardioprotection. *Shock* (2010) 33:24–30. doi: 10.1097/SHK.0b013e3181b7d137
91. Uemura R, Xu M, Ahmad N, Ashraf M. Bone marrow stem cells prevent left ventricular remodeling of ischemic heart through paracrine signaling. *Circ Res.* (2006) 98:1414–21. doi: 10.1161/01.RES.0000225952.61196.39
92. Leroux L, Descamps B, Tojais NF, Seguy B, Oses P, Moreau C, et al. Hypoxia preconditioned mesenchymal stem cells improve vascular and skeletal muscle fiber regeneration after ischemia through a wnt4-dependent pathway. *Mol Ther.* (2010) 18:1545–52. doi: 10.1038/mt.2010.108
93. Wei L, Fraser JL, Lu ZY, Hu X, Yu SP. Transplantation of hypoxia preconditioned bone marrow mesenchymal stem cells enhances

angiogenesis and neurogenesis after cerebral ischemia in rats. *Neurobiol Dis.* (2012) 46:635–45. doi: 10.1016/j.nbd.2012.03.002
94. Kong P, Xie X, Li F, Liu Y, Lu Y. Placenta mesenchymal stem cell accelerates wound healing by enhancing angiogenesis in diabetic Goto-Kakizaki (GK) rats. *Biochem Biophys Res Commun.* (2013) 438:410–9. doi: 10.1016/j.bbrc.2013.07.088
95. Yu J, Yin S, Zhang W, Gao F, Liu Y, Chen Z. Hypoxia preconditioned bone marrow mesenchymal stem cells promote liver regeneration in a rat massive hepatectomy model. *Stem Cell Res Ther.* (2013) 4:83. doi: 10.1186/scrt234
96. Zhang W, Liu L, Huo Y, Yang Y, Wang Y. Hypoxia-pretreated human MSCs attenuate acute kidney injury through enhanced angiogenic and antioxidative capacities. *Biomed Res Int.* (2014) 2014:462472. doi: 10.1155/2014/462472
97. Jiang CM, Liu J, Zhao JY, Xiao L, An S, Gou YC, et al. Effects of hypoxia on the immunomodulatory properties of human gingiva-derived mesenchymal stem cells *J Dent Res.* (2015) 94:69–77. doi: 10.1177/0022034514557671
98. Lan YW, Choo KB, Chen CM, Hung TH, Chen YB, Hsieh CH, et al. Hypoxia-preconditioned mesenchymal stem cells attenuate bleomycin-induced pulmonary fibrosis. *Stem Cell Res Ther.* (2015) 6:97. doi: 10.1186/s13287-015-0081-6
99. Mias C, Trouche E, Seguelas MH, Calcagno F, Dignat-George F, Sabatier F, et al. *Ex vivo* pretreatment with melatonin improves survival, proangiogenic/mitogenic activity, and efficiency of mesenchymal stem cells injected into ischemic kidney. *Stem Cells* (2008) 26:1749–57. doi: 10.1634/stemcells.2007-1000
100. Tang Y, Cai B, Yuan F, He X, Lin X, Wang J, et al. Melatonin pretreatment improves the survival and function of transplanted mesenchymal stem cells after focal cerebral ischemia. *Cell Trans.* (2014) 23:1279–91. doi: 10.3727/096368913X667510
101. Liu C, Fan Y, Zhou L, Zhu HY, Song YC, Hu L, et al. Pretreatment of mesenchymal stem cells with angiotensin II enhances paracrine effects, angiogenesis, gap junction formation and therapeutic efficacy for myocardial infarction. *Int J Cardiol.* (2015) 188:22–32. doi: 10.1016/j.ijcard.2015.03.425
102. Wang CC, Chen CH, Hwang SM, Lin WW, Huang CH, Lee WY, et al. Spherically symmetric mesenchymal stromal cell bodies inherent with endogenous extracellular matrices for cellular cardiomyoplasty. *Stem Cells* (2009) 27:724–32. doi: 10.1634/stemcells.2008-0944
103. Suenaga H, Furukawa KS, Suzuki Y, Takato T, Ushida T. Bone regeneration in calvarial defects in a rat model by implantation of human bone marrow-derived mesenchymal stromal cell spheroids. *J Mater Sci Mater Med.* (2015) 26:254. doi: 10.1007/s10856-015-5591-3
104. Zhang X, Hu M-G, Pan K, Li C-H, Liu R. 3D spheroid culture enhances the expression of antifibrotic factors in human adipose-derived mscs and improves their therapeutic effects on hepatic fibrosis. *Stem Cells Int.* (2016) 2016:4626073. doi: 10.1155/2016/4626073
105. Yao Y, Zhang F, Wang L, Zhang G, Wang Z, Chen J, et al. Lipopolysaccharide preconditioning enhances the efficacy of mesenchymal stem cells transplantation in a rat model of acute myocardial infarction. *J Biomed Sci.* (2009) 16:74. doi: 10.1186/1423-0127-16-74
106. Tsai LK, Wang Z, Munasinghe J, Leng Y, Leeds P, Chuang DM. Mesenchymal stem cells primed with valproate and lithium robustly migrate to infarcted regions and facilitate recovery in a stroke model. *Stroke* (2011) 42:2932–9. doi: 10.1161/STROKEAHA.110.612788
107. Li N, Yang YJ, Qian HY, Li Q, Zhang Q, Li XD, et al. Intravenous administration of atorvastatin-pretreated mesenchymal stem cells improves cardiac performance after acute myocardial infarction: role of CXCR(2013) 4. *Am J Trans Res.* (2015) 7:1058–70.
108. Ejtehadifar M, Shamsasenjan K, Movassaghpour A, Akbarzadehlaleh P, Dehdilani N, Abbasi P, et al. The effect of hypoxia on mesenchymal stem cell biology. *Adv Pharmaceut Bull.* (2015) 5:141–9. doi: 10.15171/apb.2015.021
109. Das R, Jahr H, van Osch GJ, Farrell E. The role of hypoxia in bone marrow-derived mesenchymal stem cells: considerations for regenerative medicine approaches. *Tissue Eng Part B Rev.* (2010) 16:159–68. doi: 10.1089/ten.teb.2009.0296
110. Hawkins KE, Sharp TV, McKay TR. The role of hypoxia in stem cell potency and differentiation. *Regenet Med.* (2013) 8:771–82. doi: 10.2217/rme.13.71
111. Berniakovich I, Giorgio M. Low oxygen tension maintains multipotency, whereas normoxia increases differentiation of mouse bone marrow stromal cells. *Int J Mol Sci.* (2013) 14:2119–34. doi: 10.3390/ijms14012119
112. Carrero R, Cerrada I, Lledo E, Dopazo J, Garcia-Garcia F, Rubio MP, et al. IL1beta induces mesenchymal stem cells migration and leucocyte chemotaxis through NF-kappaB. *Stem Cell Rev.* (2012) 8:905–16. doi: 10.1007/s12015-012-9364-9
113. Gnecchi M, He H, Noiseux N, Liang OD, Zhang L, Morello F, et al. Evidence supporting paracrine hypothesis for Akt-modified mesenchymal stem cell-mediated cardiac protection and functional improvement. *FASEB J.* (2006) 20:661–9. doi: 10.1096/fj.05-5211com
114. Xie L, Mao M, Zhou L, Zhang L, Jiang B. Signal factors secreted by 2D and spheroid mesenchymal stem cells and by cocultures of mesenchymal stem cells derived microvesicles and retinal photoreceptor neurons. *Stem Cells Int.* (2017) 2017:2730472. doi: 10.1155/2017/2730472
115. Lee MJ, Kim J, Kim MY, Bae YS, Ryu SH, Lee TG, et al. Proteomic analysis of tumor necrosis factor-alpha-induced secretome of human adipose tissue-derived mesenchymal stem cells. *J Proteome Res.* (2010) 9:1754–62. doi: 10.1021/pr900898n
116. Heo SC, Jeon ES, Lee IH, Kim HS, Kim MB, Kim JH. Tumor necrosis factor-alpha-activated human adipose tissue-derived mesenchymal stem cells accelerate cutaneous wound healing through paracrine mechanisms. *J Invest Dermatol.* (2011) 131:1559–67. doi: 10.1038/jid.2011.64
117. Jin P, Zhao Y, Liu H, Chen J, Ren J, Jin J, et al. Interferon-γ and tumor necrosis factor-α polarize bone marrow stromal cells uniformly to a Th1 phenotype. *Sci Rep.* (2016) 6:26345. doi: 10.1038/srep26345
118. Liu H, Xue W, Ge G, Luo X, Li Y, Xiang H, et al. Hypoxic preconditioning advances CXCR4 and CXCR7 expression by activating HIF-1α in MSCs. *Biochem Biophys Res Commun.* (2010) 401:509–15. doi: 10.1016/j.bbrc.2010.09.076
119. Fan H, Zhao G, Liu L, Liu F, Gong W, Liu X, et al. Pre-treatment with IL-1β enhances the efficacy of MSC transplantation in DSS-induced colitis. *Cell Mol Immunol.* (2012) 9:473–81. doi: 10.1038/cmi.2012.40
120. Kim YS, Noh MY, Cho KA, Kim H, Kwon MS, Kim KS, et al. Hypoxia/reoxygenation-preconditioned human bone marrow-derived mesenchymal stromal cells rescue ischemic rat cortical neurons by enhancing trophic factor release *Mol Neurobiol.* (2015) 52:792–803. doi: 10.1007/s12035-014-8912-5
121. Jun EK, Zhang Q, Yoon BS, Moon J, Lee G, Park G. Hypoxic conditioned medium from human amniotic fluid-derived mesenchymal stem cells accelerates skin wound healing through TGF-β/SMAD2 and PI3K/Akt pathways. *Int J Mol Sci.* (2014) 15:605–28. doi: 10.3390/ijms15010605
122. Francois M, Romieu-Mourez R, Li M, Galipeau J. Human MSC suppression correlates with cytokine induction of indoleamine 2,3-dioxygenase and bystander M2 macrophage differentiation. *Mol Ther.* (2012) 20:187–95. doi: 10.1038/mt.2011.189
123. Liang C, Chen SL, Wang M, Zhai WJ, Zhou Z, Pang AM, et al. Synergistic immunomodulatory effects of interferon-gamma and bone marrow mesenchymal stem cell]. *Zhonghua xueyexue zazhi.* (2013) 34:213–6. doi: 10.3760/cma.j.issn.0253-2727.2013.03.007
124. Noone C, Kihm A, English K, O'Dea S, Mahon BP. IFN-γ stimulated human umbilical-tissue derived cells potently suppress NK activation and resist NK mediated cytotoxicity in vitro. *Stem Cells Dev.* (2013) 15, 3003–3014. doi: 10.1089/scd.2013.0028
125. Tu Z, Li Q, Bu H, Lin F. Mesenchymal stem cells inhibit complement activation by secreting factor H. *Stem Cells Dev.* (2010) 19:1803–9. doi: 10.1089/scd.2009.0418
126. Du L, Lv R, Yang X, Cheng S, Ma T, Xu J. Hypoxic conditioned medium of placenta-derived mesenchymal stem cells protects against scar formation. *Life Sci.* (2016) 149:51–7. doi: 10.1016/j.lfs.2016.02.050
127. Lotfinia M, Lak S, Mohammadi Ghahhari N, Johari B, Maghsood F, Parsania S, et al. Hypoxia pre-conditioned embryonic mesenchymal stem cell secretome reduces il-10 production by peripheral blood mononuclear cells. *Iran Biomed J.* (2016) 21:24–31. doi: 10.6091/.21.1.24
128. Ivanova-Todorova E, Bochev I, Dimitrov R, Belemezova K, Mourdjeva M, Kyurkchiev S, et al. Conditioned medium from adipose tissue-derived mesenchymal stem cells induces CD4+FOXP3+ cells and

increases IL-10 secretion. *J Biomed Biotechnol.* (2012) 2012:295167. doi: 10.1155/2012/295167
129. Potapova IA, Gaudette GR, Brink PR, Robinson RB, Rosen MR, Cohen IS, et al. Mesenchymal stem cells support migration, extracellular matrix invasion, proliferation, and survival of endothelial cells *in vitro Stem Cells* (2007) 25:1761–8. doi: 10.1634/stemcells.2007-0022
130. Overath JM, Gauer S, Obermuller N, Schubert R, Schafer R, Geiger H, et al. Short-term preconditioning enhances the therapeutic potential of adipose-derived stromal/stem cell-conditioned medium in cisplatin-induced acute kidney injury. *Exp Cell Res.* (2016) 342:175–83. doi: 10.1016/j.yexcr.2016.03.002
131. Ma T, Grayson WL, Frohlich M, Vunjak-Novakovic G. Hypoxia and stem cell-based engineering of mesenchymal tissues. *Biotechnol Prog.* (2009) 25:32–42. doi: 10.1002/btpr.128
132. Kiani AA, Kazemi A, Halabian R, Mohammadipour M, Jahanian-Najafabadi A, Roudkenar MH. HIF-1α confers resistance to induced stress in bone marrow-derived mesenchymal stem cells. *Arch Med Res.* (2013) 44:185–93. doi: 10.1016/j.arcmed.2013.03.006
133. Deschepper M, Oudina K, David B, Myrtil V, Collet C, Bensidhoum M, et al. Survival and function of mesenchymal stem cells (MSCs) depend on glucose to overcome exposure to long-term, severe and continuous hypoxia. *J Cellu Mol Med.* (2011) 15:1505–14. doi: 10.1111/j.1582-4934.2010.01138.x
134. Cicione C, Muinos-Lopez E, Hermida-Gomez T, Fuentes-Boquete I, Diaz-Prado S, Blanco FJ. Effects of severe hypoxia on bone marrow mesenchymal stem cells differentiation potential. *Stem Cells Int.* (2013) 2013:232896. doi: 10.1155/2013/232896
135. Gnecchi M, He H, Liang OD, Melo LG, Morello F, Mu H, et al. Paracrine action accounts for marked protection of ischemic heart by Akt-modified mesenchymal stem cells *Nat Med.* (2005) 11:367–8. doi: 10.1038/nm0405-367
136. Imtiyaz HZ, Simon MC. Hypoxia-inducible factors as essential regulators of inflammation. *Curr Top Microbiol Immunol.* (2010) 345:105–20. doi: 10.1007/82_2010_74
137. Ahluwalia A, Tarnawski AS. Critical role of hypoxia sensor - HIF-1α in VEGF gene activation. Implications for angiogenesis and tissue injury healing. *Curr Med Chem.* (2005) 19:90–7. doi: 10.2174/092986712803413944
138. Bader AM, Klose K, Bieback K, Korinth D, Schneider M, Seifert M, et al. Hypoxic preconditioning increases survival and pro-angiogenic capacity of human cord blood mesenchymal stromal cells *in vitro. PLoS ONE* (2015) 10:e0138477. doi: 10.1371/journal.pone.0138477
139. Liu J, Hao H, Xia L, Ti D, Huang H, Dong L, et al. Hypoxia pretreatment of bone marrow mesenchymal stem cells facilitates angiogenesis by improving the function of endothelial cells in diabetic rats with lower ischemia. *PLoS ONE* (2015) 10:e0126715. doi: 10.1371/journal.pone.0126715
140. Bian S, Zhang L, Duan L, Wang X, Min Y, Yu H. Extracellular vesicles derived from human bone marrow mesenchymal stem cells promote angiogenesis in a rat myocardial infarction model. *J Mol Med.* (2014) 92:387–97. doi: 10.1007/s00109-013-1110-5
141. Anderson JD, Johansson HJ, Graham CS, Vesterlund M, Pham MT, Bramlett CS, et al. Comprehensive proteomic analysis of mesenchymal stem cell exosomes reveals modulation of angiogenesis via nuclear factor-kappab signaling. *Stem Cells* (2016) 34:601–13. doi: 10.1002/stem.2298
142. Park H, Park H, Mun D, Kang J, Kim H, Kim M, et al. Extracellular vesicles derived from hypoxic human mesenchymal stem cells attenuate GSK3beta expression via mirna-26a in an ischemia-reperfusion injury model. *Yonsei Med J.* (2018) 59:736–45. doi: 10.3349/ymj.2018.59.6.736
143. Han YD, Bai Y, Yan XL, Ren J, Zeng Q, Li XD, et al. Co-transplantation of exosomes derived from hypoxia-preconditioned adipose mesenchymal stem cells promotes neovascularization and graft survival in fat grafting *Biochem Biophys Res Commun.* (2018) 497:305–12. doi: 10.1016/j.bbrc.2018.02.076
144. Feng Y, Huang W, Wani M, Yu X, Ashraf M. Ischemic preconditioning potentiates the protective effect of stem cells through secretion of exosomes by targeting Mecp2 via miR-22. *PLoS ONE* (2014) 9:e88685. doi: 10.1371/journal.pone.0088685
145. Kilpinen L, Impola U, Sankkila L, Ritamo I, Aatonen M, Kilpinen S, et al. Extracellular membrane vesicles from umbilical cord blood-derived MSC protect against ischemic acute kidney injury, a feature that is lost after inflammatory conditioning. *J Extracell Vesicles* (2013) 2, 1–15. doi: 10.3402/jev.v2i0.21927
146. Yang J, Gao F, Zhang Y, Liu Y, Zhang D. Buyang huanwu decoction (BYHWD) enhances angiogenic effect of mesenchymal stem cell by upregulating vegf expression after focal cerebral ischemia. *J Mol Neurosci.* (2015) 56:898–906. doi: 10.1007/s12031-015-0539-0
147. Crisostomo PR, Wang Y, Markel TA, Wang M, Lahm T, Meldrum DR. Human mesenchymal stem cells stimulated by TNF-α, LPS, or hypoxia produce growth factors by an NFκB- but not JNK-dependent mechanism. *Am J Physiol Cell Physiol.* (2008) 294:C675–82. doi: 10.1152/ajpcell.00437.2007
148. Beegle J, Lakatos K, Kalomoiris S, Stewart H, Isseroff R, Nolta RJA, et al. Hypoxic preconditioning of mesenchymal stromal cells induces metabolic changes, enhances survival, and promotes cell retention *in vivo. Stem Cells* (2015) 33:1818–28. doi: 10.1002/stem.1976
149. Saparov A, Ogay V, Nurgozhin T, Jumabay M, Chen WCW. Preconditioning of human mesenchymal stem cells to enhance their regulation of the immune response. *Stem Cells Int.* (2016) 2016:3924858. doi: 10.1155/2016/3924858
150. Wang Y, Chen X, Cao W, Shi Y. Plasticity of mesenchymal stem cells in immunomodulation: pathological and therapeutic implications. *Nat Immunol.* (2014) 15:1009–16. doi: 10.1038/ni.3002
151. Park CW, Kim KS, Bae S, Son HK, Myung PK, Hong HJ, et al. Cytokine secretion profiling of human mesenchymal stem cells by antibody array international *J Stem Cells* (2009) 2:59–68.
152. Hsu WT, Lin CH, Chiang BL, Jui HY, Wu KK, Lee CM. Prostaglandin E2 potentiates mesenchymal stem cell-induced IL-10^{+}IFN-γ^{+}CD4^{+} regulatory T cells to control transplant arteriosclerosis. *J Immunol.* (2013) 190:2372–80. doi: 10.4049/jimmunol.1202996
153. Németh K, Leelahavanichkul A, Yuen PST, Mayer B, Parmelee A, Doi K, et al. Bone marrow stromal cells attenuate sepsis via prostaglandin E(2)-dependent reprogramming of host macrophages to increase their interleukin-10 production. *Nat Med.* (2009) 15:42–9. doi: 10.1038/nm.1905
154. Chen K, Wang D, Du WT, Han ZB, Ren H, Chi Y, et al. Human umbilical cord mesenchymal stem cells hUC-MSCs exert immunosuppressive activities through a PGE2-dependent mechanism. *Clin Immunol.* (2010) 135:448–58. doi: 10.1016/j.clim.2010.01.015
155. Maffioli E, Nonnis S, Angioni R, Santagata F, Cali B, Zanotti L, et al. Proteomic analysis of the secretome of human bone marrow-derived mesenchymal stem cells primed by pro-inflammatory cytokines. *J Proteomics* (2017) 166:115–26. doi: 10.1016/j.jprot.2017.07.012
156. Augello A, Tasso R, Negrini SM, Amateis A, Indiveri F, Cancedda R, et al. Bone marrow mesenchymal progenitor cells inhibit lymphocyte proliferation by activation of the programmed death 1 pathway. *Eur J Immunol.* (2005) 35:1482–90. doi: 10.1002/eji.200425405
157. Ren G, Zhang L, Zhao X, Xu G, Zhang Y, Roberts AI, et al. Mesenchymal stem cell-mediated immunosuppression occurs via concerted action of chemokines and nitric oxide. *Cell Stem Cell* (2008) 2:141–50. doi: 10.1016/j.stem.2007.11.014
158. Akiyama K, Chen C, Wang D, Xu X, Qu C, Yamaza T, et al. Mesenchymal-stem-cell-induced immunoregulation involves FAS-ligand-/FAS-mediated T cell apoptosis. *Cell Stem Cell* (2012) 10:544–55. doi: 10.1016/j.stem.2012.03.007
159. Di Nicola M, Carlo-Stella C, Magni M, Milanesi M, Longoni P, Matteucci DP, et al. Human bone marrow stromal cells suppress T-lymphocyte proliferation induced by cellular or nonspecific mitogenic stimuli. *Blood* (2002) 99:3838–43. doi: 10.1182/blood.V99.10.3838
160. Lee RH, Pulin AA, Seo MJ, Kota DJ, Ylostalo J, Larson BL, et al. Intravenous hMSCs improve myocardial infarction in mice because cells embolized in lung are activated to secrete the anti-inflammatory protein TSG-6. *Cell Stem Cell.* (2009) 5:54–63. doi: 10.1016/j.stem.2009.05.003
161. Roddy GW, Oh JY, Lee RH, Bartosh TJ, Ylostalo J, Coble K, et al. Action at a distance: systemically administered adult stem/progenitor cells (MSCs) reduce inflammatory damage to the cornea without engraftment and primarily by secretion of TNF-α stimulated gene/protein 6. *Stem Cells* (2011) 29:1572–9. doi: 10.1002/stem.708
162. Ryan JM, Barry F, Murphy JM, Mahon BP. Interferon-γ does not break, but promotes the immunosuppressive capacity of adult human

mesenchymal stem cells. *Clin Exp Immunol.* (2007) 149:353–63. doi: 10.1111/j.1365-2249.2007.03422.x

163. Polchert D, Sobinsky J, Douglas GW, Kidd M, Moadsiri A, Reina E, et al. IFN-γ activation of mesenchymal stem cells for treatment and prevention of graft versus host disease *Eur J Immunol.* (2008) 38:1745–55.
164. English K. Mechanisms of mesenchymal stromal cell immunomodulation. *Immunol Cell Biol.* (2013) 91:19–26. doi: 10.1038/icb.2012.56
165. Kim DS, Jang IK, Lee MW, Ko YJ, Lee D-H, Lee JW, et al. Enhanced immunosuppressive properties of human mesenchymal stem cells primed by interferon-γ. *EBio Med.* (2018) 28:261–73. doi: 10.1016/j.ebiom.2018.01.002
166. Krampera M, Cosmi L, Angeli R, Pasini A, Liotta F, Andreini A, et al. Role for interferon-gamma in the immunomodulatory activity of human bone marrow mesenchymal stem cells. *Stem Cells* (2006) 24:386–98. doi: 10.1634/stemcells.2005-0008
167. Ren G, Su J, Zhang L, Zhao X, Ling W, L'Huillie A, et al. Species variation in the mechanisms of mesenchymal stem cell-mediated immunosuppression. *Stem Cells* (2009) 27:1954–62. doi: 10.1002/stem.118
168. Lu Z, Chen Y, Dunstan C, Roohani-Esfahani S, Zreiqat H. Priming adipose stem cells with tumor necrosis factor-α preconditioning potentiates their exosome efficacy for bone regeneration. *Tissue Eng Part A* (2017) 23:1212–20. doi: 10.1089/ten.tea.2016.0548
169. Cosenza S, Toupet K, Maumus M, Luz-Crawford P, Blanc-Brude O, Jorgensen C, et al. Mesenchymal stem cells-derived exosomes are more immunosuppressive than microparticles in inflammatory arthritis. *Theranostics* (2018) 8:1399–410. doi: 10.7150/thno.21072
170. Opitz CA, Litzenburger UM, Lutz C, Lanz TV, Tritschler I, Koppel A, et al. Toll-like receptor engagement enhances the immunosuppressive properties of human bone marrow-derived mesenchymal stem cells by inducing indoleamine-2,3-dioxygenase-1 via interferon-beta and protein kinase R. *Stem Cells* (2009) 27:909–19. doi: 10.1002/stem.7
171. Sioud M, Mobergslien A, Boudabous A, Floisand Y. Evidence for the involvement of galectin-3 in mesenchymal stem cell suppression of allogeneic T-cell proliferation. *Scand J Immunol.* (2010) 71:267–74. doi: 10.1111/j.1365-3083.2010.02378.x
172. Liotta F, Angeli R, Cosmi L, Fili L, Manuelli C, Frosali F, et al. Toll-like receptors 3 and 4 are expressed by human bone marrow-derived mesenchymal stem cells and can inhibit their T-cell modulatory activity by impairing Notch signaling. *Stem Cells* (2008) 26:279–89. doi: 10.1634/stemcells.2007-0454
173. Hahn JY, Cho HJ, Kang HJ, Kim TS, Kim MH, Chung JH, et al. Pre-treatment of mesenchymal stem cells with a combination of growth factors enhances gap junction formation, cytoprotective effect on cardiomyocytes, and therapeutic efficacy for myocardial infarction. *J Am Coll Cardiol.* (2008) 51:933–43. doi: 10.1016/j.jacc.2007.11.040
174. Bai Y, Han YD, Yan XL, Ren J, Zeng Q, Li XD, et al. Adipose mesenchymal stem cell-derived exosomes stimulated by hydrogen peroxide enhanced skin flap recovery in ischemia-reperfusion injury. *Biochem Biophys Res Commun.* (2018) 500:310–7. doi: 10.1016/j.bbrc.2018.04.065
175. Potapova IA, Brink PR, Cohen IS, Doronin SV. Culturing of human mesenchymal stem cells as three-dimensional aggregates induces functional expression of CXCR4 that regulates adhesion to endothelial cells. *J Biol Chem.* (2008) 283:13100–7. doi: 10.1074/jbc.M800184200
176. Petrenko Y, Syková E, Kubinová Š. The therapeutic potential of three-dimensional multipotent mesenchymal stromal cell spheroids. *Stem Cell Res Ther.* (2017) 8:94. doi: 10.1186/s13287-017-0558-6
177. Cesarz Z, Tamama K. Spheroid culture of mesenchymal stem cells. *Stem Cells Int.* (2016) 2016:9176357. doi: 10.1155/2016/9176357
178. Lee EJ, Park SJ, Kang SK, Kim GH, Kang HJ, Lee SW, et al. Spherical bullet formation via E-cadherin promotes therapeutic potency of mesenchymal stem cells derived from human umbilical cord blood for myocardial infarction *Mol Ther.* (2012) 20:1424–33. doi: 10.1038/mt.2012.58
179. Kim YS, Ahn Y, Kwon JS, Cho YK, Jeong MH, Cho JG, et al. Priming of mesenchymal stem cells with oxytocin enhances the cardiac repair in ischemia/reperfusion injury. *Cells Tissues Organs* (2012) 195:428–42. doi: 10.1159/000329234
180. Liu J, Zhu P, Song P, Xiong W, Chen H, Peng W, et al. Pretreatment of adipose derived stem cells with curcumin facilitates myocardial recovery via antiapoptosis and angiogenesis. *Stem Cells Int.* (2015) 2015:638153. doi: 10.1155/2015/638153
181. Phan J, Kumar P, Hao D, Gao K, Farmer D, Wang A. Engineering mesenchymal stem cells to improve their exosome efficacy and yield for cell-free therapy. *J Extracell Vesicles* (2018) 7:1522236. doi: 10.1080/20013078.2018.1522236
182. Phelps J, Sanati-Nezhad A, Ungrin M, Duncan NA, Sen A. bioprocessing of mesenchymal stem cells and their derivatives: toward cell-free therapeutics. *Stem Cells Int.* (2018) 2018:9415367. doi: 10.1155/2018/9415367
183. Tolar J, Le Blanc K, Keating A, Blazar BR. Hitting the right spot with mesenchymal stromal cells (MSCs). *Stem Cells* (2010) 28:1446–55. doi: 10.1002/stem.459
184. Zhang J, Huang X, Wang H, Liu X, Zhang T, Wang Y, et al. The challenges and promises of allogeneic mesenchymal stem cells for use as a cell-based therapy. *Stem Cell Res Ther.* (2015) 6:234. doi: 10.1186/s13287-015-0240-9

9

Nature vs. Nurture: Defining the Effects of Mesenchymal Stromal Cell Isolation and Culture Conditions on Resiliency to Palmitate Challenge

Lauren K. Boland[1,2], *Anthony J. Burand*[1,2], *Devlin T. Boyt*[1,2], *Hannah Dobroski*[1,2], *Lin Di*[1,2], *Jesse N. Liszewski*[1,2], *Michael V. Schrodt*[1,2], *Maria K. Frazer*[1,2], *Donna A. Santillan*[3] and *James A. Ankrum*[1,2]*

[1] University of Iowa Fraternal Order of Eagles Diabetes Research Center, University of Iowa, Iowa City, IA, United States, [2] Roy J. Carver Department of Biomedical Engineering, University of Iowa, Iowa City, IA, United States, [3] Department of Obstetrics and Gynecology, Center for Immunology and Immune Based Diseases, Center for Hypertension Research, University of Iowa, Iowa City, IA, United States

Correspondence:
James A. Ankrum
james-ankrum@uiowa.edu

As MSC products move from early development to clinical translation, culture conditions shift from xeno- to xeno-free systems. However, the impact of isolation and culture-expansion methods on the long-term resiliency of MSCs within challenging transplant environments is not fully understood. Recent work in our lab has shown that palmitate, a saturated fatty acid elevated in the serum of patients with obesity, causes MSCs to convert from an immunosuppressive to an immunostimulatory state at moderate to high physiological levels. This demonstrated that metabolically-diseased environments, like obesity, alter the immunomodulatory efficacy of healthy donor MSCs. In addition, it highlighted the need to test MSC efficacy not only in ideal conditions, but within challenging metabolic environments. To determine how the choice of xeno- vs. xeno-free media during isolation and expansion would affect future immunosuppressive function, umbilical cord explants from seven donors were subdivided and cultured within xeno- (fetal bovine serum, FBS) or xeno-free (human platelet lysate, PLT) medias, creating 14 distinct MSC preparations. After isolation and primary expansion, umbilical cord MSCs (ucMSC) were evaluated according to the ISCT minimal criteria for MSCs. Following baseline characterization, ucMSC were exposed to physiological doses of palmitate and analyzed for metabolic health, apoptotic induction, and immunomodulatory potency in co-cultures with stimulated human peripheral blood mononuclear cells. The paired experimental design (each ucMSC donor grown in two distinct culture environments) allowed us to delineate the contribution of inherent (nature) vs. environmentally-driven (nurture) donor characteristics to the phenotypic response of ucMSC during palmitate exposure. Culturing MSCs in PLT-media led to more consistent growth characteristics during the isolation and expansion for all donors, resulting in faster doubling times and higher cell yields compared to FBS. Upon palmitate challenge, PLT-ucMSCs showed a higher susceptibility to palmitate-induced metabolic disturbance, but less susceptibility to palmitate-induced apoptosis. Most striking however, was that the PLT-ucMSCs resisted the conversion to an immunostimulatory phenotype better than their FBS counterparts. Interestingly, examining MSC suppression of PBMC proliferation at physiologic doses of palmitate magnified the differences between donors, highlighting the utility of evaluating MSC products in stress-based assays that reflect the challenges MSCs may encounter post-transplantation.

Keywords: mesenchymal stromal cells, umbilical cord mesenchymal stromal cells, obesity, palmitate, human platelet lysate, fetal bovine serum, culture media, cell therapy

INTRODUCTION

Cell bioprocessing, involving the large scale production of high volumes of consistent and efficacious cellular products, is a critical component for the translation of mesenchymal stromal cell (MSC) products from academic laboratories to clinical application (1–4). However, a mismatch in culture conditions between basic academic research and clinical translation often exists, with academic labs often relying on fetal bovine serum (FBS) supplemented media and small 2D culture systems, while clinical-grade products are grown in large batches in xeno-free medias (1, 2, 5, 6). Decisions that may once have seemed trivial or even arbitrary in the academic research environment [e.g., media composition (7–11), cell culture format (2, 12–14), or substrate-stiffness (15, 16)] become critical parameters to control in the large-scale industrial production of clinical-grade cellular therapies (17–19). Increasingly, as cellular therapies have been used to treat patients in clinical trials, it has become clear that adopting good manufacturing practices (GMP) early at the research phase can hasten the transition from the bench to the clinic (2, 4, 19, 20). With the diversification of the MSC field to include MSCs harvested from a number of tissue sources (21–24), it is critical to distinguish variance that is inherent in the cell source vs. variability that is introduced by differences in bioprocessing during initial isolation and *in vitro* expansion (25).

As findings from the lab transition to the clinic, a number of process-related changes are often needed to produce a clinical-grade product (2, 3, 18). One of the most glaring differences between MSCs studied in research labs to those produced at an industrial scale for use in patients is a switch from using animal serum as a growth supplement to xeno-free alternatives like human platelet lysate or chemically defined serum (6, 26). There is significant and valid concern surrounding the use of animal derived products in the generation of clinical-grade cell therapies (6, 27, 28). Consequently, transitioning to clinical-grade production often involves transitioning to a xeno-free culture system (1, 2). Today, a large number of xeno-free alternatives are available, many of which have associated drug master files submitted to the FDA, which can make submission of investigational new drug (IND) applications more streamlined (6, 26, 29). A number of notable advantages have been reported for xeno-free culture systems for MSCs, including enhanced cell yield (9, 10, 30), rapid growth kinetics (5, 10), elimination of xenogeneic pathogens (27), and improved genetic stability over extended culture periods (30, 31). In order to ensure that the transition from pre-clinical to clinical application is successful, it is important to understand how the process differences related to culture environments affect MSC function.

Variability in MSC phenotype due to donor age (32–35), sex (35, 36), tissue of origin (21, 37), and co-morbidities (38–40), as well as time spent in culture (41–45), has been extensively documented. A notable example of MSC donor-specific variance was the observation that certain bioactive secreted factors, namely tumor necrosis factor-inducible gene 6 (TSG6), show a sexual dimorphism with higher quantities secreted from female vs. males bone marrow donors (36). An increasingly important contributor to MSC phenotype is the presence of metabolic disease [obesity, type 2 diabetes, atherosclerosis, and metabolic syndrome (32, 34, 38, 39, 46–51)] within donors. A number of functional defects have been documented in adipose-derived MSCs from donors with metabolic disease, including blunted immunosuppressive potency against activated T cells (32, 38), reduced fibrinolytic activity (39, 46), and a diminished ability to halt the progression of neuroinflammation (40). In addition to the functional defect seen in MSCs isolated from patients with metabolic disease, recent work in our lab has demonstrated that healthy donor MSCs exposed to a "metabolically diseased" environment enriched in the saturated fatty acid, palmitate, are no longer suppressive, but stimulatory when cultured with activated peripheral blood mononuclear cells (52). Interestingly, MSCs from some donors were more sensitive to palmitate exposure than others, despite similar levels of potency in traditional palmitate-free co-culture assays. Collectively, this body of work highlights the need for cell manufacturing processes that consistently generate high quality MSCs and the need for *in vitro* potency assays capable of predicting MSC performance after transplant into challenging metabolic environments.

Although much of the variance in functional performance of MSCs has been attributed to intrinsic donor characteristics, only more recently has attention been focused on how specific process-related decisions may contribute to variance in MSC performance (1–3, 18, 20, 23, 25). While benefits have been reported for xeno-free systems, they have predominately been focused on improving the cell yield during manufacturing rather than the functional properties of the cells (17, 18). Additionally, the interaction of these systems with inherent donor variability is less well-understood (3, 53). Whether or not xeno-free culture systems produce MSCs that are more or less resilient for use in complex metabolic environments has yet to be determined.

In the present study, we investigated how early and late decisions regarding media supplementation modify the health and function of umbilical cord derived MSCs (ucMSCs). By growing ucMSCs from explant to experiment in one of two media supplements, either standard FBS or human platelet lysate, we were able to compare genetically identical cell preparations that had been acclimated to two different culture environments from isolation to analysis. Notably, this study design allowed us to delineate the contributions of donor intrinsic variability (nature) vs. process-derived variability (nurture). Additionally, we demonstrate for the first time that adapting an *in vitro* potency assay to mimic a disease environment can reveal the impact of process-related decisions on MSC performance within challenging metabolic environments.

MATERIALS AND METHODS

Umbilical Cord MSC (ucMSC) Isolation

Human umbilical cords were collected through the University of Iowa Women's Health Tissue Repository's Maternal Fetal Tissue Bank with consent from the mothers (IRB#200910784). Samples were provided to the investigators in a coded fashion and this work was deemed to not be human subjects research by the University of Iowa IRB (IRB# 201708749). After delivery, samples were placed at 4°C in PBS (Biological Industries, Cat #02-023-1A-24) supplemented with 2% Penicillin-Streptomycin solution (P/S, Biological Industries, Cat #03-031-1B) and 2% Amphotericin B (AmpB, Biological Industries, Cat #03-028-1B). Umbilical cord tissue explants were plated within 48 h of delivery. To begin, the external wall of the cord was sterilized with 70% ethanol and pharmacidal spray (Biological Industries, Cat #IC-110100-05) and placed in sterile PBS. The umbilical cords were then cut into 3 mm cross-sections with a sterile razor and subdivided into 1 mm pieces. Approximately 60 cord pieces were then divided into two dry culture dishes (Corning, Cat #430599), with Wharton's Jelly placed against the plate. Umbilical cord pieces were allowed to adhere for 10–15 min before addition of media. Warm α-minimum essential medium (Biological Industries, Cat #01-042-1A) supplemented with 1% P/S, 2% AmpB, and either 10% FBS (Biological Industries, Cat #04-121-1A-US) or 5% PLTGold Human Platelet Lysate (PLT, Biological Industries, Cat #PLTGOLD100R), was then slowly added before the dishes were placed in a 37°C incubator. Media was subsequently changed every 2–3 days. After 9 days of outgrowth the cord pieces were removed and if the cells were confluent in their colonies the cells were plated as Passage 1 (P1) in T25s. If not confluent after 9 days, media was changed, and the cells were cultured an additional 3 days before plating into T25 flasks. P1 cells were grown to 80% confluence and grown for one additional passage before cryostorage using serum-free cell freezing medium (Biological Industries, Cat #05-065-1A).

ucMSC Expansion and Growth Kinetics

ucMSCs isolated as described above were counted, plated (P1), and grown to 70–90% confluence. Cells were then harvested, counted, plated (P2), and expanded prior to cryopreservation (P3). Population doubling level at each of these splits was calculated using the formula $PDL = PDL_0 + 3.322(logP_1 - logP_0)$, where PDL_0 is the population doubling level at seeding, P_1 is the total cell yield at harvest, and P_0 is the initial cell seeding number. Cells harvested immediately after isolation were declared to have an initial PDL of 0. Time to population doubling was calculated from the initial first three cell harvests. Briefly, total cell yields at each harvest were fit using linear regression in GraphPad Prism 7 software and the linear regression equations were used to calculate the average time to population doubling for each MSC preparation.

Post-cryopreservation, MSCs were thawed at 37°C until ~1 mm^3 piece of ice remained. The cells were then transferred immediately into the appropriate pre-warmed media condition and allowed to attach overnight. Media was switched to remove cryopreservation media and cells were grown out to 70–90% confluence. Population doublings after cryostorage were calculated as before from the total cell yields after each cell harvest to confirm consistency of growth kinetics in the medias before and after cryopreservation.

BSA Preparation and Palmitate-BSA Conjugation

Bovine serum albumin (BSA) and palmitate conjugated to BSA were prepared according to a previously described protocol (52, 54, 55). Briefly, to prepare the BSA solution, 20% (w/v) fatty acid free bovine albumin (MP Biomedicals, Cat #0219477205) was gradually added to Ultra Pure water (Biological Industries, Cat #01-866-1A) at 52°C with gentle agitation until the BSA was fully dissolved, yielding a transparent yellow solution. To prepare a palmitate-BSA solution, 2.6% (w/v) of palmitic acid (Sigma-Aldrich, Cat #P0500) was added to 0.1 M NaOH (Sigma-Aldrich, Cat #221465) and heated to 70°C until fully dissolved, yielding a concentration of 100 mM. To conjugate palmitate to BSA, the 100 mM stock of palmitate in 0.1 M NaOH was added to 20% BSA at a ratio of 1:9 and heated at 55°C for 10 min. After conjugation, both the 20% (w/v) BSA solution and 10 mM Palmitate-BSA solutions were filter sterilized using a 0.45 μm PES filter (Celltreat, Cat #229709). Solutions were then aliquoted and stored at −20°C. Before use, solutions were warmed to 37°C. The total amount of BSA was held constant in all conditions by mixing ratios of fatty acid free BSA to palmitate-BSA as follows: BSA only (4:0), 0.1 mM palmitate-BSA (3:1), 0.2 mM palmitate-BSA (2:2), and 0.4 mM palmitate-BSA (0:4).

ucMSC Validation (Immunophenotyping and Differentiation)

To validate ucMSC identity via immunophenotyping, cells were stained to confirm CD105, CD73, and CD90 expression and the absence of CD45, CD34, CD11b, CD19, and HLA-DR as defined by the MSC minimal criteria (56). Positive markers were confirmed using FITC-CD105 antibody (BD Biosciences, Cat #561443), PE.Cy7-CD73 antibody (BD Biosciences, Cat #561258), and PE-CD90 antibody (BD Biosciences, Cat #A15794) with appropriate isotype controls [FITC Mouse lgG1k (BD Biosciences, Cat #556649), PE.Cy7 Mouse lgG1k (BD Biosciences, Cat #557872), and PE-CD90 Mouse lgG1 (Invitrogen, Cat #GM4993)]. A negative marker cocktail and isotype staining was performed using the PE hMSC Negative Cocktail (BD Biosciences, Cat #562530), as per manufacturer's instructions.

For trilineage differentiation (56), each preparation of ucMSCs (seven donors, maintained in either FBS-media or PLT-media) was differentiated to osteoblasts, chondroblasts, and adipocytes by culturing in osteogenic (Biological Industries, Cat #05-440-1B), chondrogenic (Biological Industries, Cat #05-220-1B-KT), or adipogenic (Biological Industries, Cat #05-330-1B-KT) differentiation medias, respectively. Briefly, for osteogenic and adipogenic differentiation, ucMSCs were plated at a density of 60,000 cells/well in 24-well plates and grown in either FBS- or PLT-media for 24 h. Cells were then visualized to ensure >80% confluence and media was exchanged to osteogenic or adipogenic differentiation medias. Cultures were maintained for 15–18 days with media exchanges every 3–4 days. Osteogenic lineage was confirmed using both bright-field imaging and Alizarin Red for mineral deposition. For adipogenic lineage confirmation, lipid droplets were identified via AdipoRed staining and fluorescent microscopy. For chondrogenic differentiation, a micromass culture (10 uL of 10 million/mL ucMSCs in chondrogenic differentiation media) was incubated for 2 h to allow for spontaneous spheroid formation in 96-well spheroid plates (Corning, Cat #4515). After 2 h, 200 uL of chondrogenic media was added for subsequent culture. Spheroid cultures were maintained for 18 days with media exchanges occurring every 3–4 days. On Day 18, chondrogenic spheroids were fixed for 5 min with 10% Formalin then mounted in Optimal Cutting Temperature (OCT) Compound. Sectioning of the chondrogenic pellets was performed by the University of Iowa Comparative Pathology Core and tissues were subsequently stained with Safranin O.

Metabolic Health and Viability Assays

To assess the metabolic function of ucMSCs exposed to palmitate, we performed an XTT assay (Cell Proliferation Kit-XTT based, Biological Industries, Cat #20-300-1000). For each media condition, 1,000 ucMSCs from every donor were plated into a 96-well plate. All wells had a total media volume of 100 μL. The ucMSCs were treated with 20% (w/v) BSA or 0.1, 0.2, or 0.4-mM palmitate-BSA as described above. After 96 h, Activation Reagent was added to warmed XTT Reagent at a ratio of 1:5,000 and 50 μL of XTT solution was added to the media of each well. The plate was read at 475 and 660 nm absorbance at 1 and 2 h time points by a microplate reader. The background reading was subtracted for all wells, and within each media condition, the background-subtracted reading was normalized to a media-only control.

To assess changes in viability, ucMSCs were incubated at 37°C with BSA, 0.1, 0.2, or 0.4 mM Palmitate-BSA for 96 h. Both detached cells in the media and adherent cells were collected and pooled for analysis. Cells were pelleted, resuspended in prepared Annexin/PI binding buffer [20 mM HEPES at pH 7.4 (Invitrogen, Cat #A14291DJ), 150 mM sodium chloride (American Bio, Cat #AB01915-10000), and 2.5 mM calcium chloride (Sigma Aldrich, Cat #C8106-500g)], and stained with 90 μg/mL FITC-Annexin V (Biolegend, Cat #640945 and 640906) and 1 mg/mL propidium iodide (Invitrogen, Cat #P3566) to final concentrations of 4.5 and 50 μg/mL, respectively, in Annexin/PI binding buffer. Samples were incubated for 30 min in the dark at room temperature. Samples were then diluted with Annexin/PI binding buffer and analyzed via flow cytometry (Accuri C6, BD Biosciences).

MSC:PBMC Co-culture With Palmitate Challenge

Peripheral blood mononuclear cells (PBMCs) from a de-identified donor were isolated from a leukapheresis reduction cone made available by the DeGowin Blood Center at the University of Iowa Hospitals and Clinics. After isolation, PBMCs were cryopreserved in a solution consisting of 50% RPMI (ThermoFisher, Cat #11875093), 40% FBS (VWR, Cat #97068085), and 10% DMSO (Fisher Scientific, Cat #D128) until use. PBMCs were thawed at 37°C 1 h prior to staining and plated in RPMI containing 10% FBS, 1% L-glutamine , and 1% P/S. In order to track generational proliferation of the PBMCs, PBMCs were stained with CFSE Cell Division Tracker Kit (Biolegend, Cat #423801) according to the manufacturer's protocol. Briefly, PBMCs resuspended in PBS at 1 million cells per mL were stained with 2 μL of 10 mM CFSE per 10 million cells to yield a final CFSE concentration of 1 μM. PBMCs were incubated at 37°C for 15 min, centrifuged at 500 g for 8 min, resuspended in RPMI to neutralize the dye, and incubated at 37°C for 30 min. Cells were centrifuged again, resuspended at 1 million per mL, and stored at 37°C until plating with the MSCs.

All 14 preparations of ucMSCs were harvested after staining the PBMCs, and 50,000 cells were plated into 24-well plates. MSCs were allowed to attach for 1 h. Prior to plating, CFSE stained PBMCs were mixed 1:1 with Human CD3/CD28 Dynabeads (Thermo Fisher, Cat #11132D) to activate T cells within the mixed population. 200,000 stimulated PBMCs were added to each MSC condition to yield an MSC to PBMC ratio of 1:4. Total media volume in each well was set at 750 μL and all co-culture conditions were either treated with BSA alone, or 0.1, 0.2, or 0.4 mM palmitate-BSA to simulate metabolic stress. PBMCs with or without Dynabeads were plated as positive and negative controls for PBMC activation and proliferation. After 5 days, PBMCs were isolated by collecting the non-adherent cell fraction from the plates. These cells were then centrifuged, the supernatants collected, and resuspended in RPMI. Supernatants were stored at −20°C for future analysis of Granzyme-B levels using a bead-based ELISA per the manufacturer's instructions (BD Biosciences, Cat #560304). PBMC proliferation was analyzed by flow cytometry (Accuri C6, BD Biosciences).

Media Exchange Experiments

Two vials of P4 ucMSC from donor 4600 isolated with FBS media, were thawed and plated in either FBS (FBS-FBS) or PLT (FBS-PLT) supplemented media. Likewise, two vials of P4 ucMSC from donor 4600 were thawed and plated in either FBS (PLT-FBS) or PLT (PLT-PLT) supplemented media. After culturing for 72 h, the ucMSCs were counted and replated to assess their growth kinetics. The cells were then cultured for another passage

(96 h), lifted, counted, and plated for either co-culture or imaging experiments. For co-culture with PBMCs, each preparation of ucMSCs were plated at a ratio 1 MSC: 4 PBMCs. PBMCs were stained with CFSE and stimulated with Dynabeads, as previously described. The co-culture was incubated for 5 days, followed by flow cytometry analysis. For morphological analysis, ucMSCs were plated at 1,000 ucMSCs/cm2 in 6-well plate and cultured for an additional 48 h. Cells were then fixed for 5 min (10% neutral buffered Formalin, Sigma-Aldrich, Cat# HT501128-4L), followed by permeabilization for 5 min (0.1% v/v Triton® X-100 in PBS, Sigma-Aldrich, Cat#T9284-500 mL), at room temperature. After permeabilization, the cells were stained with ActinRed 555 ReadyProbes Reagent (Cat# R37112, Thermo Fisher Scientific) according to the manufacturer's protocol. Briefly, for every 1 milliliter of PBS, 2 drops of reagent were added. The cells were then incubated with 1 mL of reagent in PBS at 37°C for 30 min. During the last 10 min of ActinRed staining, 1 ug/mL of Hoechst 33342 (Cat# H3570, Invitrogen) was added. Cells were then imaged using a fluorescent microscope (Leica DMI6000) at 20× magnification.

Statistical Analysis

Statistical analysis was pre-planned prior to collection of data and performed within GraphPad Prism 7. For all assays involving multiple donors, each donor was considered an independent "n." For assays involving a single donor, independent replicate experiments were considered as "n." Specific statistical tests used for each data set is listed in the caption of each figure.

RESULTS

PLT-Media Enhances ucMSC Growth Kinetics and Overall Cell Yield Compared to FBS-Media

To better understand how early choices in culture isolation and expansion affect growth rates and yield of ucMSCs, equivalent amounts of umbilical cord explants from the same donor were plated into two different media formulations ($n = 7$ umbilical cord donors), one supplemented with xeno-free 5% PLT (PLT-media) and one supplemented with traditional 10% FBS (FBS-media). Umbilical cord tissue explants were incubated to allow for ucMSC migration out of the cord and onto the underlying tissue culture plastic. After 11 days, ucMSCs were counted to determine initial cell yield (**Figure 1A**) and then culture-expanded for two additional passages prior to cryopreservation for subsequent experiments. Growing umbilical cord tissue explants in PLT-media did not enhance the initial amount of ucMSCs isolated (**Figure 1A**). Though the initial cell yield was not different between media preparations, subsequent outgrowth of ucMSCs in PLT-media resulted in a higher yield of cells (**Figure 1B**) and a significantly lower population doubling time compared to FBS-media (**Figure 1C**). To ensure that there were not large differences in total population doublings between the two different media preparations, ucMSCs were cryopreserved once 70–80% confluence was reached leading to an equivalent number of total population doublings prior to cryopreservation (**Supplemental Figure 1**). Additionally, all ucMSC preparations were characterized and found to meet ISCT minimal criteria for classification as MSCs (56), including positive staining for CD105, CD73, and CD90 and negative staining for CD45, CD34, CD19, CD11b, and HLA-DR, as well as trilineage differentiation potential (**Supplemental Figures 2–4**).

Media Supplementation Dictates a Differential Response to Palmitate-Induced Metabolic Stress and Apoptosis

In our previous study, we found that increased levels of palmitate exposure led to a slight increase in levels of apoptosis and a significant decrease in the metabolic activity of bone marrow-derived MSCs (as determined by NAD^+/NADH ratio) (52). To determine whether media supplementation led to differences in palmitate-induced apoptosis, we cultured ucMSCs in FBS- or PLT-media and exposed the cells to a range of physiological doses (57, 58) of palmitate for 96 h. Across all palmitate conditions, there was a larger overall increase in the proportion of both early (Annexin V+/Propidium Iodide-, AV+/PI-) and late (AV+/PI+) stage apoptotic cells in FBS-media (**Figures 2A,B**) compared to PLT-media. Notably, in FBS-media, ucMSC preparations showed a 2.0-fold increase in susceptibility to palmitate-induced apoptosis compared to paired PLT-media preparations as measured by late-stage apoptotic cells (**Figure 2C**). Less overall apoptosis was evident in all PLT-media donors across all tested conditions with FBS-media donors showing 2.2 and 2.6 times as many AV+/PI+ cells after 96 h of BSA and 0.2 mM Palm-BSA exposure, respectively; however, only 0.4 mM Palm-BSA achieved a significant difference in the number of late-stage apoptotic cells.

Next, to determine the effect of palmitate on the metabolic health of the ucMSCs donors, we cultured ucMSCs in increasing levels of palmitate and assessed NAD^+/NADH ratio, i.e., the redox state of the cells, by XTT. Treatment with the vehicle control (BSA) showed no difference in the baseline NAD^+/NADH ratio; however, the mean NAD^+/NADH ratio was higher in PLT-media compared to FBS-media (**Figure 3A**), which is consistent with the faster overall growth kinetics in PLT-media observed previously (**Figure 1C**). Interestingly, though FBS-media preparations showed a greater overall susceptibility to palmitate-induced apoptosis, by XTT assay, PLT-media preparations were affected more by the presence of palmitate at every dose tested (**Figure 3B**). Given that NAD^+/NADH ratio is a metric of the redox state of the cell which includes both metabolic health and proliferative ability (59), we compared the doubling time of each preparation previously calculated to the XTT absorbance at the highest dose of palmitate tested (0.4 mM Palm-BSA). As expected, there was a significant positive correlation ($R^2 = 0.5113$, *, $p = 0.0040$, Pearson correlation coefficient) between the doubling time of each preparation and the decline in XTT absorbance observed upon palmitate exposure, with lower doubling times associated with a larger decline in NAD^+/NADH ratios upon palmitate exposure (**Figure 3C**).

In addition to altering cellular metabolism, proliferating cells also increase uptake of fatty acids to produce cellular

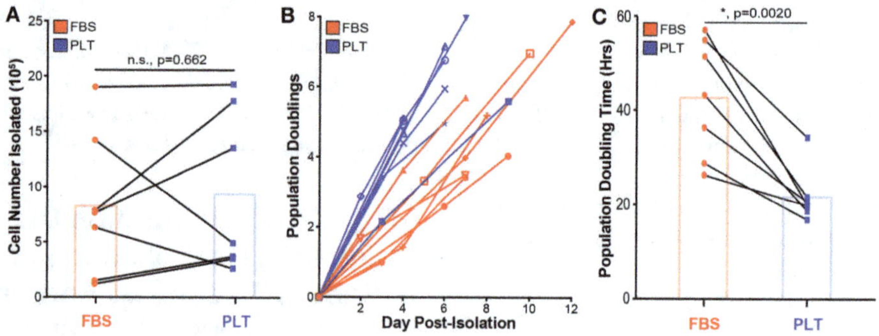

FIGURE 1 | PLT-media accelerates ucMSC growth kinetics and overall cell yield but not initial cell isolation from tissue explant. **(A)** Total cell number isolated from the initial umbilical cord tissue explant was determined via hemacytometer after 9–12 days of outgrowth. Donor paired ucMSC preparations [single donor grown in either FBS-media (red circle) or PLT-media (blue square)] are represented by connected lines. Bar graph displays the mean for all donors (paired *t*-test, $n = 7$ ucMSC donors, *n.s., p = 0.662*). **(B)** Cells isolated from the original tissue explant were plated at Day 0 Post-Isolation (P1) and allowed to grow until 70–80% confluence, before passaging. Each line represents a different preparation with each node representing the cumulative population doublings determined at P1, P2, and P3. **(C)** An average population doubling time was determined via linear regression from P1 to P3 outgrowth data for each donor. Donor paired ucMSC preparations are shown for FBS-media (red circle) and PLT-media (blue square). Bar graph displays the mean value across all donors (paired *t*-test, $n = 7$ ucMSC donors, *, $p = 0.0020$).

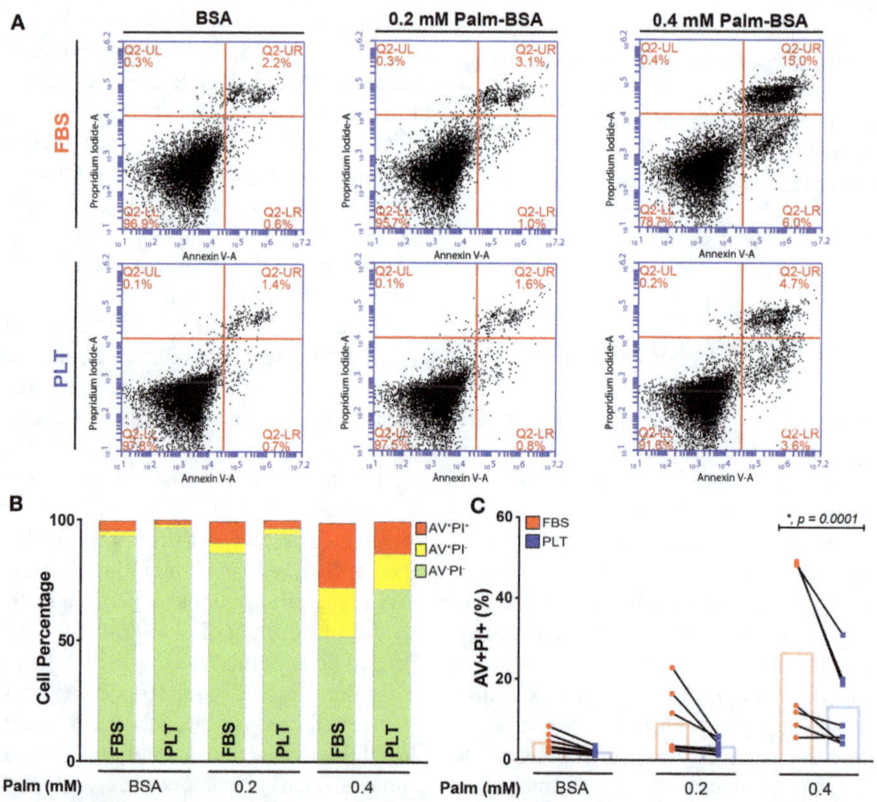

FIGURE 2 | ucMSCs are largely protected from the pro-apoptotic effects of palmitate exposure in PLT-media. **(A)** Representative flow cytometry dot plots from a single donor (3004iA) showing Annexin V (AV, x-axis) and Propidium Iodide (PI, y-axis) staining after 96 h of palmitate exposure. Top row shows the FBS-ucMSC preparation; bottom row shows the PLT-ucMSC preparation. **(B)** Summary data from FBS-ucMSC and PLT-ucMSC preparations showing average percent of viable (AV^-PI^-, green), early apoptotic (AV^+PI^-, yellow), and late apoptotic (AV^+PI^+, red) ucMSCs after 96 h of palmitate exposure. **(C)** Percent of late apoptotic cells (AV^+PI^+, red) at each concentration of palmitate. Donor paired ucMSC preparations are connected by a line between FBS-ucMSC (red circle) and PLT-ucMSC (blue square) preparation. Bar graphs represent the mean value across all donors (2-way ANOVA with Sidak Correction for multiple comparisons to FBS-ucMSC, $n = 7$ ucMSC donors, BSA: *n.s., p = 0.6518*, 0.2: *n.s., p = 0.0621*, 0.4: *, *p = 0.0001*).

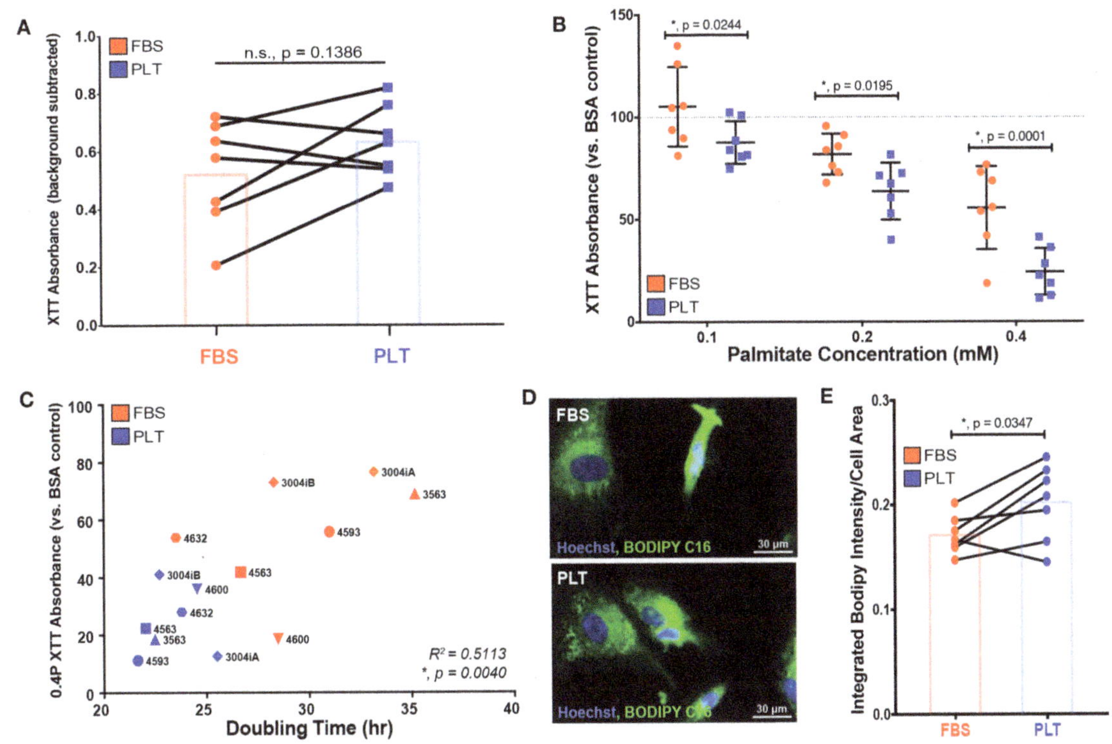

FIGURE 3 | ucMSCs grown in PLT-media show decreased overall metabolic health after palmitate exposure which correlates with higher overall proliferative rates. **(A)** Basal NAD^+/NADH ratio was determined using XTT assay after 96 h of exposure to vehicle control (BSA) in all FBS-ucMSC and PLT-ucMSC preparations (absorbance value = total absorbance – background absorbance from media only control). Paired ucMSC preparations are connected by a line between FBS-ucMSC (red circle) and PLT-ucMSC (blue square) preparations. Bar graphs represent the mean value across all donors (paired *t*-test, $n = 7$ ucMSC donors, *n.s.*, $p = 0.1386$). **(B)** The relative NAD^+/NADH ratio after 96 h of palmitate exposure was determined by normalizing the XTT absorbance to the within donor BSA control for each ucMSC preparation (BSA value is represented by the dotted line). FBS-ucMSCs (red circle) were normalized to BSA in FBS-media and PLT-ucMSCs (blue square) were normalized to BSA in PLT-media (2-way ANOVA with Sidak Correction for multiple comparisons to FBS-ucMSC, $n = 7$ ucMSC donors, 0.1: *, $p = 0.0244$, 0.2: *, $p = 0.0195$, 0.4: *, $p = 0.0001$). **(C)** Correlation plot comparing XTT absorbance at 0.4 mM Palm-BSA (y-axis) to population doubling time in the absence of palmitate (x-axis) for each donor preparation (Pearson correlation coefficient, $n = 14$ ucMSC preparations, $R^2 = 0.5113$, *, $p = 0.0040$). **(D)** Representative images of FBS-ucMSC and PLT-ucMSC preparations incubated with the fluorescent palmitate analog, BODIPY FL C16, and counterstained with Hoechst 33342 for nuclear identification, 20× magnification, Scale = 30 µm. **(E)** The total integrated intensity of fluorescent BODIPY FL C16 and cell area were determined via CellProfiler quantification of the images represented in **(D)**. Donor paired ucMSC preparations are connected by a line between FBS-ucMSC (red circle) and PLT-ucMSC (blue square) preparations. Bar graphs represent the mean value across all donors (paired *t*-test, $n = 7$ ucMSC donors, *, $p = 0.0347$).

membranes (60). To determine whether the higher proliferative rate of PLT-media donors led to increased uptake of exogenous fatty acids, all 14 ucMSC preparations were incubated with a fluorescent palmitate analog, BODIPY FL C16, and imaged to quantify BODIPY uptake on a per cell basis. The total integrated intensity of the BODIPY signal was quantified using the image processing software, CellProfiler. The total BODIPY signal was then divided by the overall cell area to account for differences in size between ucMSC grown in FBS- vs. PLT-media. All ucMSC preparations showed uptake of BODIPY FL C16 (**Figure 3D**), but the overall uptake was increased in ucMSC grown in PLT-media compared to those grown in FBS-media for nearly every donor (**Figure 3E**, *FBS-media mean* = *0.1710* ± *0.0181 vs. PLT-media mean* = *0.2019* ± *0.0365*), suggesting that PLT-media grown ucMSC preparations increase their uptake of exogenous palmitate from the environment compared to FBS-media preparations.

Immunosuppressive Potency of ucMSCs Is Critically Dependent on Media Supplementation

In our previous study, we discovered that palmitate exposure converted FBS-cultured bone marrow derived MSC from an immunosuppressive to an immunostimulatory profile in co-culture with stimulated T cells (52). Given the broad range of potential alterations precipitated by culturing ucMSCs in FBS- compared to PLT-supplemented media (10, 30, 61), we sought to determine whether culturing ucMSCs in a xeno-free, PLT based culture system would rescue the functional immunosuppressive defect previously seen in co-culture after palmitate exposure. For this study, we used a single PBMC donor in order to focus solely on the contribution of media supplementation on ucMSC potency within PBMC co-cultures.

Notably, all ucMSC preparations, regardless of previous growth in FBS- or PLT-media, suppressed the proliferation of activated PBMCs (**Figure 4A**, data points below the dotted line are immunosuppressive). Both FBS- and PLT-media preparations showed a large variation in baseline immunosuppressive potency, suppressing T cell proliferation by 5–42% in FBS-media and 3–66% in PLT-media. On average, PLT-media preparations were more suppressive at baseline (*FBS-media BSA mean* $= 71 \pm 13$, *PLT-media BSA mean* $= 59 \pm 23$); however, due to the large amount of donor-to-donor variation this value did not reach statistical significance. Interestingly, although PLT-media preparations showed less variation in growth kinetics, palmitate-induced apoptosis, and changes in $NAD^+/NADH$, they had a wider variation in baseline immunosuppressive ability compared to FBS counterparts. At low physiological levels of palmitate (0.1 mM Palm-BSA), FBS-media preparations showed a greater decline in immunosuppressive potency compared to PLT-media preparations. Transitioning from BSA to 0.1 mM Palm-BSA resulted in a more rapid loss of PBMC suppression for FBS-media MSC preparations compared to PLT-media MSC preparations (27 ± 16% increase in PBMC proliferation for FBS vs. 11 ± 6.5% increase for PLT, **Figure 4B**). At the extreme 0.4 mM Palm-BSA dose, both FBS and PLT-media preparations showed a strong immunostimulatory profile but differences between the groups did not reach statistical significance (*95% CI of difference: −85% to +4%,* $p = 0.09$). At 0.2 mM Palm-BSA, a level closer to the average serum level in patients with obesity and/or type 2 diabetes (57, 62), all FBS-media preparations and all but one PLT-media preparation converted to an immunostimulatory phenotype, which is consistent with our previously published results. However, PLT-media preparations showed a significantly less severe immunostimulatory profile compared to FBS-media preparations (**Figure 4B**, *FBS-media 0.2 Palm-BSA mean = 288 vs. PLT-media 0.2 Palm-BSA, mean = 212, 95% CI of difference: 31% to 120%,* $p = 0.0008$). Growth in PLT-media, therefore, appears to protect ucMSC immunomodulatory potential from the full extent of damage inflicted by exposure to low to moderate physiological doses of palmitate.

In order to better understand the role of the PBMCs within the co-culture system, supernatants from all of the co-cultures were collected and analyzed for Granzyme B (GRZ-B), an immune effector molecule commonly secreted by cytotoxic T cells and NK cells (63). In agreement with the PBMC proliferation data (**Figures 4A,B**), palmitate exposure led to a significant increase in GRZ-B in co-cultures with FBS-ucMSC, while co-cultures with PLT-ucMSCs only showed a slight non-significant increase (**Figure 4C**, **Supplemental Table 1**). Interestingly, this inflammatory profile seems to be dependent on both specific ucMSC donor characteristics, as well as the media supplementation used during cell expansion. In some donors, for example donor 3004iA, the preparation in FBS showed a similar change in PBMC proliferation and GRZ-B secretion in response to palmitate as the PLT preparation (FBS: 152%, 806 pg/mL GRZ-B vs. PLT: 174%, 719 pg/mL GRZ-B, **Supplemental Table 1**). However, with donor 4600, media supplementation greatly affected the performance of the donor in co-culture, with the FBS preparation showing 2.69 times higher PBMC proliferation and 18.33 times higher GRZ-B levels than the PLT preparation (**Supplemental Table 1**). This data, therefore, suggests that certain donors may be more highly affected by the choice of media supplementation during culture expansion, while other donors may behave the same regardless of choice of media supplementation.

Changing Media Supplementation Greatly Alters the Growth Kinetics of ucMSCs

In the previous series of experiments analyzing how media supplementation affected MSC immunosuppressive ability in harsh metabolic environments, we observed that there were several donors (4600, 3004iB, and 4563) that had a large change in PBMC suppression depending on their growth in FBS or PLT supplemented medias (**Supplemental Table 1**). These donors, therefore, derived the greatest benefit from isolation and expansion in PLT-media. We, therefore, sought to determine if the cells retained a "memory" of the early isolation environment or if changing the culture environment could alter the cells' behavior. To this end, we grew FBS-preparations of donor 4600 in FBS (FBS-FBS) or switched the media to PLT (FBS-PLT) and grew PLT-preparations of 4600 in PLT (PLT-PLT) or switched the media to FBS (PLT-FBS) (**Figure 5A**). In order to acclimate the cells to the new media supplementation and control for passage effects, we cultured each preparation for 7 days with one passage event prior to analysis.

Interestingly, we found that switching the media supplementation had drastic effects on the growth kinetics of the preparation. Preparations of donor 4600 that were maintained in their original media supplementation (FBS-FBS and PLT-PLT) had growth kinetics (**Figure 5B**) similar to those seen at earlier passages (**Figure 1C**). Notably, switching a FBS-preparation into PLT-media greatly improved the growth kinetics, with FBS-PLT cells growing at an average rate of 0.96 population doublings/day, a rate more similar to PLT-PLT preparations than FBS-FBS preparations (**Figure 5B**). Surprisingly, switching a PLT-preparation into FBS-media drastically reduced the proliferative ability of the cells resulting in an average growth rate of just 0.34 population doublings/day, a rate lower than cells maintained in FBS alone (**Figure 5B**). Given that the data represents growth rates after one passage and 7 total days in culture, a growth halt due to shock or adaptation delay alone does not explain the drastic difference between PLT preparations maintained in PLT compared to those same preparations transitioned to FBS-media.

Changes in Media Supplementation Drive Alterations in Cytoplasmic and Nuclear Morphological Features of ucMSCs

Recently, several groups have demonstrated the utility of analyzing complex morphological features of MSCs to demonstrate phenotypic changes precipitated by changes in priming conditions or media supplementation (10, 64–66). During the initial outgrowth of ucMSCs in new media supplementation, it appeared that ucMSCs took on the morphological features of the new media (i.e., FBS-PLT looked more similar to PLT-PLT than FBS-FBS preparations). To

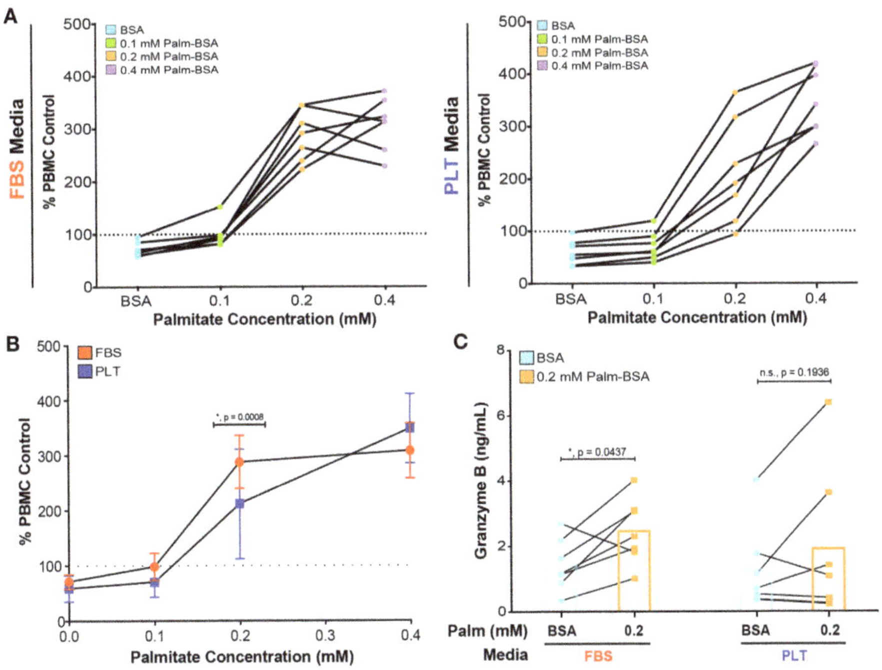

FIGURE 4 | PLT-media decreases the severity of palmitate-induced decline in immunosuppressive potency compared to FBS-media. **(A)** PBMCs, stimulated with CD3/CD28 and stained with CFSE, were exposed to increasing concentrations of palmitate in the absence (PBMC control) or presence of MSCs (1:4 ratio of MSC:PBMC). All data points in **(A)** are displayed relative to the proliferation in the PBMC control at that palmitate exposure (dotted line on graph). Each solid line represents the dose response of an individual donor, with FBS-uMSC donors (left) and PLT-ucMSC donors (right). Data points below the dotted line are immunosuppressive and data points above the line are immunostimulatory. A single PBMC donor was used. **(B)** Summary data demonstrating the dose response of all FBS-ucMSC (red) and PLT-ucMSC (blue) donors [mean ± SD, $n = 7$ ucMSC donors, 2-way ANOVA with Sidak correction for multiple comparisons to FBS-ucMSC, 0 (BSA): *n.s.*, $p = 0.9111$, 0.1: *n.s.*, $p = 0.3560$, 0.2: *, $p = 0.0008$, 0.4: *n.s.*, $p = 0.0914$]. **(C)** Protein levels of Granzyme B within the supernatant of co-cultures as determined by bead-based ELISA assay. Paired values connected by a solid line represent a single donor exposed to vehicle control (blue circles) or 0.2 mM palmitate (orange square). Bar graphs represent the mean value across all donors (2-way ANOVA with Sidak correction for multiple comparisons to BSA control, FBS: *, $p = 0.0437$, PLT: *n.s.*, $p = 0.1936$).

quantify this morphological transition, we fixed and stained cells from each of the media conditions (FBS-FBS, FBS-PLT, PLT-PLT, and PLT-FBS) with an F-actin stain to delineate the cytoplasm and Hoechst 33342 to define the nucleus (**Figure 6A**). We then analyzed the features of cells in each media condition with a modified pipeline in CellProfiler (64). Only cytoplasmic or nuclear features that showed statistically significant differences between media groups by one-way ANOVA were pursued (**Figure 6B**, **Supplemental Table 2**).

Interestingly, in line with the growth kinetic changes, the PLT-FBS transition led to the highest number of morphological changes when compared to any other media combination (**Figure 6B**). In FBS-media, PLT cells showed increased area, larger maximum and minimum feret diameters, and a larger overall perimeter in both the cytoplasm and nucleus. Although originally under bright field, FBS and PLT preparations appeared different, there were no cytoplasmic features that showed a statistically significant difference between FBS-FBS and PLT-PLT. However, at the nuclear level, PLT-PLT cells had a larger overall maximum and minimum feret diameter, as well as higher mean nuclear area and perimeter values (though area and perimeter were not statistically significant). These nuclear changes in PLT-PLT cells may be representative of more cells in S-phase of the cell cycle due to the higher overall growth rate observed (52, 67). However, PLT-FBS cells also showed a larger maximum and minimum feret diameter, area, and perimeter within the nucleus, while the overall growth rate within these cells is highly reduced compared to any other media combination. Though it would seem logical that FBS-PLT cells might take on a morphological profile between FBS-FBS and PLT-PLT cells (10), the morphological profile of FBS-PLT cells was distinct from either of the other preparations. At the cytoplasmic level, FBS-PLT cells were of a similar area, perimeter, and max/min feret diameter compared to both FBS-FBS and PLT-PLT cells; however, FBS-PLT cells showed a more rounded or spherical morphology than FBS-FBS and PLT-PLT cells with a higher extent and form factor measurement. Notably, however, at the nuclear level the mean values of most FBS-PLT measures were more similar to those of the FBS-FBS preparations, while the mean values of most PLT-FBS measures were more similar to those of the PLT-PLT cells (**Supplemental Table 2**). Cytoplasmic and nuclear changes may then be differentially regulated by transitions into new media supplementation, with cytoplasmic features being more readily plastic than their nuclear counterparts.

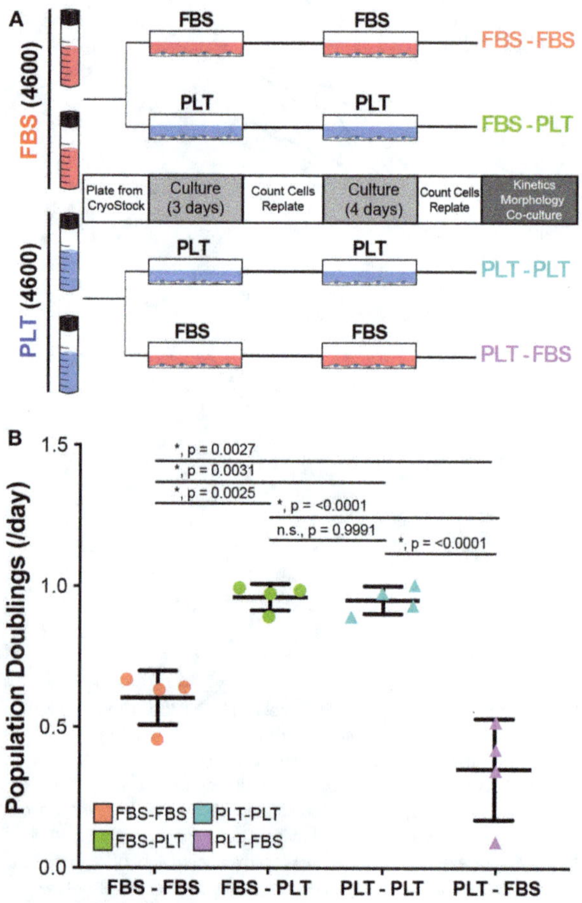

FIGURE 5 | Growth kinetics in ucMSCs are significantly altered by changes in media supplementation. **(A)** Schematic of workflow for media exchange experiments. **(B)** Daily population doublings were calculated after 7 days in maintenance media (FBS-FBS, PLT-PLT) or transition media (FBS-PLT, PLT-FBS) (1-way ANOVA, $n = 4$ independent experiments, * denotes significance $p < 0.05$). Circles represent ucMSC preparations that were initially isolated and expanded in FBS-media and triangles represent preparations originally isolated and expanded in PLT-media.

PLT-Media Improves Functional Resiliency in Challenging Metabolic Environments

Finally, we wanted to determine if transitioning media supplementation from FBS to PLT could effectively rescue the function of an ucMSC donor that had performed poorly in a co-culture potency assay when challenged with palmitate. To test the ability of media supplementation to effectively rescue a "poor performing donor," we used our previous culture scheme (**Figure 6A**) to prepare four different ucMSC preparations from the same donor (4600). These preparations of ucMSCs were then plated with activated PBMCs from three independent donors to test for immunosuppressive potency in the presence or absence of palmitate. Given the significant difference in functional performance between FBS-media and PLT-media donors previously observed in co-culture with 0.2 mM palmitate (**Figure 4B, Supplemental Table 1**), we chose this dose for the palmitate exposure condition.

In line with our previous observations, FBS-FBS cells converted from an immunosuppressive phenotype in BSA to an immunostimulatory profile when exposed to palmitate (**Figure 7**). Interestingly, at baseline, any cell preparation that had been grown in PLT-media for a period was more immunosuppressive, on average, than FBS-FBS cells (BSA, mean ± SEM: *FBS-FBS* = 65.5 ± 1.3, *FBS-PLT* = 45.1 ± 7.7, *PLT-PLT* = 41.2 ± 8.1, *PLT-FBS* = 42.2 ± 4.3). Notably, all cell preparations grown at some stage in PLT-media were also less susceptible to the palmitate-induced immunostimulatory conversion than FBS-FBS preparations from the same donor. Although there were no statistically significant differences between the immunostimulatory levels in cell preparations grown at some point in PLT-media, PLT-PLT cells showed the lowest average level of immunostimulatory behavior compared to FBS-PLT and PLT-FBS preparations (0.2P, mean ± SEM: *FBS-PLT* = 152.7 ± 31.4, *PLT-PLT* = 117.9 ± 39.7, *PLT-FBS* = 140.0 ± 38.7). Interestingly, though PLT-FBS cells showed the most striking change in morphological features and growth kinetics in response to the transition in media supplement, these changes did not translate into a poorer performance within an immunosuppressive potency assay. The lack of synchrony between changes in morphology and changes in potency performance are particularly noteworthy given the recent interest in incorporating morphological analysis into MSC potency matrices (64, 65). PLT-media, therefore, does appear to prevent some of the damage inflicted by palmitate exposure; however, it is critical to note that growth in PLT-media does not fully restore the immunosuppressive function of ucMSCs after palmitate exposure.

DISCUSSION

The choice of media supplementation for expansion of MSCs can influence a range of phenotypic characteristics including growth kinetics, morphology, and multi-lineage differentiation potential (5, 9, 10, 30, 61). In industrial scale production of therapeutic MSCs, a high priority has been placed on the ability of xeno-free supplements to amplify the overall yield of MSCs, but less focus has been placed on the functional consequences of different media compositions (3, 53). In agreement with previous findings in the field (9, 10), we observed that ucMSCs grown in xeno-free conditions showed both faster growth kinetics and greater overall yields compared to their paired counterparts grown in traditional FBS-supplemented media (**Figures 1B,C**). An interesting early feature of ucMSC preparations grown in PLT-supplemented media was a lower overall variance in performance between donors, which is also consistent with previous findings (5). Interestingly, because cells from the same initial donor tissue were isolated and grown in two different media preparations, the decrease in donor variability suggests that variance often attributed to "inherent" donor characteristics (nature) is actually heavily influenced by early bioprocessing decisions (nurture).

Based on the clear early advantage of using a xeno-free culture system, we initially hypothesized that PLT-ucMSC preparations

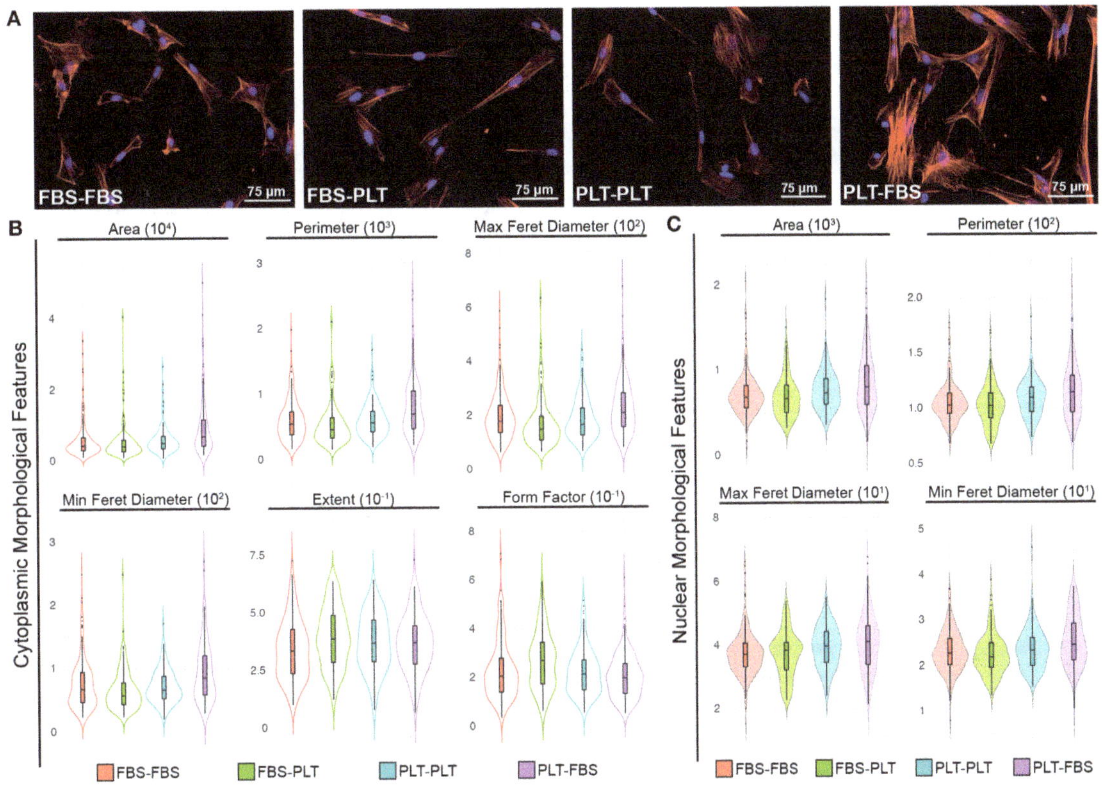

FIGURE 6 | ucMSCs respond to changes in media supplementation by modifying cytoplasmic and nuclear morphological features. **(A)** Representative images of nuclear (Hoechst 3342, blue) and cytoplasmic (ActinRed 555, red) features after 7 days of maintenance (FBS-FBS, PLT-PLT) or transition media (FBS-PLT, PLT-FBS). Each image was acquired at 20× magnification, Scale = 30 µm. **(B)** Single-cell cytoplasmic morphological features were plotted as violin plots with embedded box-plots (box: 25th, 50th, and 75th percentiles; whiskers: 10th and 90th percentiles). Only six features that were found to be significantly different between groups by 1-way ANOVA (refer to **Supplemental Table 2** for values) are represented. Approximately 10–20 whole cells were detected per image in 10 images for each media condition, yielding ~150–200 cells per media condition. **(C)** Single-cell nuclear morphological features were plotted as violin plots with embedded box-plots (10th, 25th, 50th, 75th, and 90th percentiles). Four nuclear features showed statistically significant differences between groups by 1-way ANOVA (refer to **Supplemental Table 2** for values).

would show superior performance in most, if not all, of the functional assessments. However, our assessment revealed a much more nuanced effect of growing ucMSCs in PLT-media. In comparison to FBS-ucMSCs, PLT-ucMSCs showed a decrease in apoptotic induction (**Figure 2**), but a more dramatic drop in NAD^+/NADH ratio (**Figure 3B**) in response to a metabolic stressor (palmitate exposure), while also taking up more of the fluorescent palmitate analog, BODIPY C16 (**Figures 3D,E**). A potential explanation for this finding is that although PLT-ucMSCs uptake higher levels of palmitate, the actual intracellular processing of palmitate might be distinct between the two preparations leading to the observed phenotypic differences. Future studies involving metabolic tracing of palmitate are needed to delineate if process-related decisions, like media supplementation, alter the intracellular handling of palmitate and contribute to the improved viability of PLT-ucMSCs after palmitate exposure. These findings are important for two reasons: first, in instances in which MSCs are persisting for long periods of time (e.g., bone graft or local injection), these findings identify the composition of the local metabolic environment as an important modifier of MSC viability and health (58), and second, that the functional response of MSCs can be modified through process-level decisions as simple as media supplementation.

Although MSCs are being explored for a range of clinical indications, the majority of current clinical trials are aiming to capitalize on the immunomodulatory axis of MSC function (68, 69). A growing number of studies have demonstrated that MSCs isolated from patients with metabolic disease have drastically altered immunomodulatory potential (32, 34, 38, 40, 47, 51); however, the potential corollary of the immunomodulatory performance of a healthy-donor MSC being altered in a "metabolically diseased" environment has not been well-established. The unfortunate underlying assumption is that the immunomodulatory potential of MSCs will be sustained, no matter the cues present within the transplant environment. However, several cues within serum, including complement (70) and TNF-α (71), have been shown to modulate MSC function, which challenges this assumption. Our recent work implicates metabolic cues, specifically physiologic levels of palmitate, as an additional and potent modifier of MSC immunomodulatory

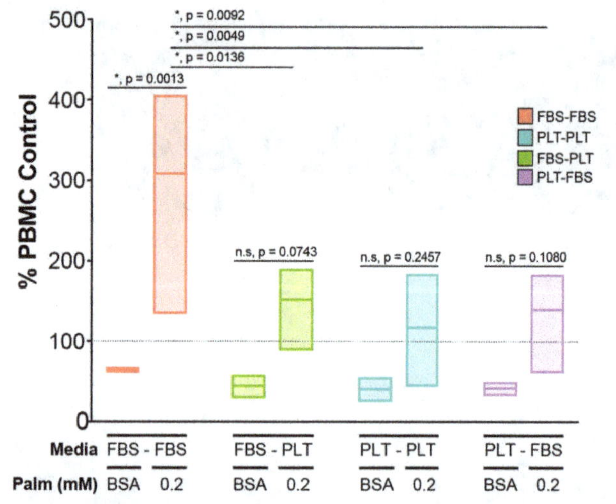

FIGURE 7 | Exposure to PLT-media partially rescues ucMSCs from the palmitate-induced loss of immunosuppressive potency. CD3/CD28-stimulated CFSE-stained PBMCs were co-cultured with ucMSCs that had been maintained in (FBS-FBS, PLT-PLT) or transitioned into a new media environment (FBS-PLT, PLT-FBS) at a 1:4 ratio of ucMSC:PBMC. A single ucMSC donor (4600) was used throughout and was co-cultured with three independent PBMC donors. The dotted line represents the PBMC only control for the vehicle (BSA) and palmitate condition (0.2 mM Palm) (2-way ANOVA with Tukey correction for multiple comparisons, $n = 3$ independent PBMC donors, * denotes significance $p < 0.05$). Box-plots show the minimum, mean, and maximum values.

performance (52). Notably, it has also recently been reported that the reliance of MSCs on specific metabolic pathways (highly glycolytic vs. oxidative) is a critical regulator of MSC immunosuppression of T cells (72). In the present study, using a broader range of donors and a different tissue source of MSCs, we have once again found that palmitate converts MSCs from an immunosuppressive to an immunostimulatory profile (**Figure 4B**) when grown in standard FBS-supplemented media. Notably, we found that although xeno-free culture does not wholly prevent palmitate-induced damage, it does lessen the severity of the damage inflicted by a metabolically toxic environment. This demonstrates that bioprocess decisions have a lasting impact on the resiliency of MSC's immunomodulatory efficacy, and could be used to tailor MSCs for use within challenging metabolic environments.

The path from cell isolation to translational application of MSCs is highly variable and incorporates a number of transition points, from the choice of isolation method to the extent of time spent in culture (2). Though industrial production of MSCs has helped to standardize transition points (18, 20), the full influence of a cell's "memory" of isolation and culture conditions remains to be uncovered. In the current study, we found that ucMSCs exhibited functional memory of their original environment in some, but not all of the aspects we profiled. Growth kinetics of ucMSCs appeared to be uniquely affected by the current growth media (**Figure 5B**), with little effect rendered by the original media supplementation. Interestingly, regarding morphological changes, nuclear features appeared to be heavily influenced by the original growth media (**Figure 6C**), while cytoplasmic features were more variable in their adaptations to new environments (**Figure 6B**). Most important to us, any cell preparation that had ever been cultured in PLT-media, regardless of the current growth media, showed an improved suppressive profile in palmitate-rich environments (**Figure 7**). Therefore, a range of phenotypic changes may occur in ucMSCs in response to environmental changes, however, these phenotypic changes do not all predict functional deficits, particularly immunosuppressive potency. This finding is a notable caution for the implementation of high-throughput strategies, like image-based morphological characterization, for predicting MSC immunomodulatory performance (64, 65). The ability of morphology to predict MSC potency appears to be heavily context dependent and in our current study, the parameters we assessed proved insufficient to predict potency within palmitate-rich environments. It appears, therefore, that ucMSCs are adaptable to culture environments, but that this adaptability is not free from the "memory" of past environments. The balance achieved between "memory" and adaptation may be a result of selective expansion of highly resilient ucMSC clones isolated early in the preparation of MSCs or the result of a lasting epigenetic imprint from the early culture environment. Future studies are needed to determine by what mechanism xeno-free growth conditions protect ucMSC immunomodulatory function in challenging metabolic environments.

In conclusion, our present study has determined that both inherent donor characteristics (nature) and process-level decisions (nurture) play critical roles in the subsequent resiliency of ucMSC function in metabolically challenging environments. Although variance between donors is apparent regardless of media supplementation, xeno-free culture systems provided more consistent performance in a number of important metrics, including overall cell yield and growth kinetics. Most importantly, culturing ucMSCs in xeno-free conditions improved viability and immunosuppressive potency in response to palmitate, a common metabolic stressor present in everyone and elevated in patients with obesity and type 2 diabetes (52, 62). As MSC therapies move into clinical use for a diverse and growingly obese patient base (73), it is more critical than ever to understand the consequence of bioprocessing decisions on the subsequent performance of MSCs in complex, pathologic *in vivo* environments. By adapting *in vitro* potency models to account for disease-relevant environmental changes, critical hidden aspects of cell health and resiliency become apparent.

AUTHOR CONTRIBUTIONS

JA and LB: conceptualization and methodology. LB and JA: formal analysis. LB, AB, DB, HD, LD, JL, MS, and MF: investigation. JA and DS: resources. LB: data curation. LB, JA,

AB, LD, HD, DB, JL, MS, and DS: writing and original draft. LB, JA, AB, HD, DB, LD, and MS: writing, review, and editing. LB: visualization. JA: supervision.

FUNDING

The majority of the chemical reagents used for this work were provided through a Biological Industries USA Research Award to JA. HD and JL received support from the Iowa Center for Research by Undergraduates at the University of Iowa. LB was supported through an NIH training grant (#5T32GM007337). AB was supported through an NIH training grant (#1T32NS045549). Additional support from the Straub Foundation awarded to JA was used to complete the project. DS was supported by the American Heart Association Strategically Focused Research Network [grant numbers 15 SFRN 23730000, 18679000, 18679001, 18679002, 18679003]. The Women's Health Tissue Repository is supported by the Carver College of Medicine, the Department of Obstetrics & Gynecology, and the National Center For Advancing Translational Sciences of the National Institutes of Health under Award Number UL1TR002537.

ACKNOWLEDGMENTS

We would like to thank the University of Iowa Women's Health Tissue Repository for supplying umbilical cords from de-identified donors in accordance with an IRB approved protocol. We would also like to thank the University of Iowa DeGowin Blood Bank and donor center for providing the leukocyte reduction cones that were used to isolate peripheral blood mononuclear cells for the immunomodulatory potency assays.

REFERENCES

1. Panchalingam KM, Jung S, Rosenberg L, Behie LA. Bioprocessing strategies for the large-scale production of human mesenchymal stem cells: a review. *Stem Cell Res Ther.* (2015) 6:225. doi: 10.1186/s13287-015-0228-5
2. Jossen V, van den Bos C, Eibl R, Eibl D. Manufacturing human mesenchymal stem cells at clinical scale: process and regulatory challenges. *Appl Microbiol Biotechnol.* (2018) 102:3981–94. doi: 10.1007/s00253-018-8912-x
3. Heathman TRJ, Rafiq QA, Chan AKC, Coopman K, Nienow AW, Kara B, et al. Characterization of human mesenchymal stem cells from multiple donors and the implications for large scale bioprocess development. *Biochem Eng J.* (2016) 108:14–23. doi: 10.1016/j.bej.2015.06.018
4. Lipsitz YY, Timmins NE, Zandstra PW. Quality cell therapy manufacturing by design. *Nat Biotechnol.* (2016) 34:393–400. doi: 10.1038/nbt.3525
5. Heathman TRJ, Stolzing A, Fabian C, Rafiq QA, Coopman K, Nienow AW, et al. Scalability and process transfer of mesenchymal stromal cell production from monolayer to microcarrier culture using human platelet lysate. *Cytotherapy.* (2016) 18:523–535. doi: 10.1016/j.jcyt.2016.01.007
6. Karnieli O, Friedner OM, Allickson JG, Zhang N, Jung S, Fiorentini D, et al. A consensus introduction to serum replacements and serum-free media for cellular therapies. *J Cytother.* (2017) 19:155–69. doi: 10.1016/j.jcyt.2016.11.011
7. Oikonomopoulos A, van Deen WK, Manansala A-R, Lacey PN, Tomakili TA, Ziman A, et al. Optimization of human mesenchymal stem cell manufacturing: the effects of animal/xeno-free media. *Sci Rep.* (2015) 5:16570. doi: 10.1038/srep16570
8. Schallmoser K, Bartmann C, Rohde E, Reinisch A, Kashofer K, Stadelmeyer E, et al. Human platelet lysate can replace fetal bovine serum for clinical-scale expansion of functional mesenchymal stromal cells. *Transfusion.* (2007) 47:1436–46. doi: 10.1111/j.1537-2995.2007.01220.x
9. Becherucci V, Piccini L, Casamassima S, Bisin S, Gori V, Gentile F, et al. Human platelet lysate in mesenchymal stromal cell expansion according to a GMP grade protocol: a cell factory experience. *Stem Cell Res Ther.* (2018) 9:124. doi: 10.1186/s13287-018-0863-8
10. Fernandez-Rebollo E, Mentrup B, Ebert R, Franzen J, Abagnale G, Sieben T, et al. Human platelet lysate versus fetal calf serum: these supplements do not select for different mesenchymal stromal cells. *Sci Rep.* (2017) 7:371–8. doi: 10.1038/s41598-017-05207-1
11. Dessels C, Potgieter M, Pepper MS. Making the switch: alternatives to fetal bovine serum for adipose-derived stromal cell expansion. *Front Cell Dev Biol.* (2016) 4:115. doi: 10.3389/fcell.2016.00115
12. Chen AK-L, Chew YK, Tan HY, Reuveny S, Oh SKW. Increasing efficiency of human mesenchymal stromal cell culture by optimization of microcarrier concentration and design of medium feed. *J Cytother.* (2015) 17:163–73. doi: 10.1016/j.jcyt.2014.08.011
13. Tavassoli H, Alhosseini SN, Tay A, Chan PPY, Oh SKW, Warkiani ME. Large-scale production of stem cells utilizing microcarriers: a biomaterials engineering perspective from academic research to commercialized products. *Biomaterials.* (2018) 181:333–46. doi: 10.1016/j.biomaterials.2018.07.016
14. Santos Dos F, Campbell A, Fernandes-Platzgummer A, Andrade PZ, Gimble JM, Wen Y, et al. A xenogeneic-free bioreactor system for the clinical-scale expansion of human mesenchymal stem/stromal cells. *Biotechnol Bioeng.* (2014) 111:1116–27. doi: 10.1002/bit.25187
15. Darnell M, Gu L, Mooney D. RNA-seq reveals diverse effects of substrate stiffness on mesenchymal stem cells. *Biomaterials.* (2018) 181:182–8. doi: 10.1016/j.biomaterials.2018.07.039
16. Li CX, Talele NP, Boo S, Koehler A, Knee-Walden E, Balestrini JL, et al. MicroRNA-21 preserves the fibrotic mechanical memory of mesenchymal stem cells. *Nat Mater.* (2016) 16:379–89. doi: 10.1038/nmat4780
17. Ménard C, Pacelli L, Bassi G, Dulong J, Bifari F, Bezier I, et al. Clinical-grade mesenchymal stromal cells produced under various good manufacturing practice processes differ in their immunomodulatory properties: standardization of immune quality controls. *Stem Cells Dev.* (2013) 22:1789–801. doi: 10.1089/scd.2012.0594
18. Trento C, Bernardo ME, Nagler A, Kuçi S, Bornhauser M, Köhl U, et al. Manufacturing mesenchymal stromal cells for the treatment of graft-versus-host disease: a survey among centers affiliated with the european society for blood and marrow transplantation. *Biol Blood Marrow Transpl.* (2018) 24:2365–70. doi: 10.1016/j.bbmt.2018.07.015
19. Liu S, de Castro LF, Jin P, Civini S, Ren J, Reems J-A, et al. Manufacturing differences affect human bone marrow stromal cell characteristics and function: comparison of production methods and products from multiple centers. *Sci Rep.* (2017) 7:46731. doi: 10.1038/srep46731
20. Mendicino M, Bailey AM, Wonnacott K, Puri RK, Bauer SR. MSC-based product characterization for clinical trials: an FDA perspective. *Cell Stem Cell.* (2014) 14:141–5. doi: 10.1016/j.stem.2014.01.013
21. Bortolotti F, Ukovich L, Razban V, Martinelli V, Ruozi G, Pelos B, et al. *In vivo* therapeutic potential of mesenchymal stromal cells depends on the source and the isolation procedure. *Stem Cell Rep.* (2015) 4:332–9. doi: 10.1016/j.stemcr.2015.01.001
22. Hass R, Kasper C, Böhm S, Jacobs R. Different populations and sources of human mesenchymal stem cells (MSC): a comparison of adult

and neonatal tissue-derived MSC. *Cell Commun Signal.* (2011) 9:12. doi: 10.1186/1478-811X-9-12
23. Phelps J, Sanati-Nezhad A, Ungrin M, Duncan NA, Sen A. Bioprocessing of mesenchymal stem cells and their derivatives: toward cell-free therapeutics. *Stem Cells Int.* (2018) 2018:1–23. doi: 10.1155/2018/9415367
24. Moll G, Ankrum JA, Kamhieh-Milz J, Bieback K, Ringdén O, Volk H-D, et al. Intravascular mesenchymal stromal/stem cell therapy product diversification: time for new clinical guidelines. *Trends Mol Med.* (2019) 25:149–63. doi: 10.1016/j.molmed.2018.12.006
25. Yin JQ, Zhu J, Ankrum JA. Manufacturing of primed mesenchymal stromal cells for therapy. *Nat Biomed Eng.* (2019) 3:90–104. doi: 10.1038/s41551-018-0325-8
26. Galipeau J, Sensebé L. Mesenchymal stromal cells: clinical challenges and therapeutic opportunities. *Cell Stem Cell.* (2018) 22:824–33. doi: 10.1016/j.stem.2018.05.004
27. Spees JL, Gregory CA, Singh H, Tucker HA, Peister A, Lynch PJ, et al. Internalized antigens must be removed to prepare hypoimmunogenic mesenchymal stem cells for cell and gene therapy. *Mol Ther.* (2004) 9:747–56. doi: 10.1016/j.ymthe.2004.02.012
28. Sundin M, Ringden O, Sundberg B, Nava S, Gotherstrom C, Le Blanc K. No alloantibodies against mesenchymal stromal cells, but presence of anti-fetal calf serum antibodies, after transplantation in allogeneic hematopoietic stem cell recipients. *Haematologica.* (2007) 92:1208–15. doi: 10.3324/haematol.11446
29. Giancola R, Bonfini T, Iacone A. Cell therapy: cGMP facilities and manufacturing. *Muscles Ligaments Tendons J.* (2012) 2:243–7. https://www.mltj.org/index.php?PAGE=articolo_dett&ID_ISSUE=632&id_article=5458
30. Crespo-Diaz R, Behfar A, Butler GW, Padley DJ, Sarr MG, Bartunek J, et al. Platelet lysate consisting of a natural repair proteome supports human mesenchymal stem cell proliferation and chromosomal stability. *Cell Transplant.* (2011) 20:797–812. doi: 10.3727/096368910X543376
31. Blázquez-Prunera A, Díez JM, Gajardo R, Grancha S. Human mesenchymal stem cells maintain their phenotype, multipotentiality, and genetic stability when cultured using a defined xeno-free human plasma fraction. *Stem Cell Res Ther.* (2017) 8:103. doi: 10.1186/s13287-017-0552-z
32. Kizilay Mancini Ö, Shum-Tim D, Stochaj U, Correa JA, Colmegna I. Age, atherosclerosis and type 2 diabetes reduce human mesenchymal stromal cell-mediated T-cell suppression. *Stem Cell Res Ther.* (2015) 6:140. doi: 10.1186/s13287-015-0127-9
33. Scruggs BA, Semon JA, Zhang X, Zhang S, Bowles AC, Pandey AC, et al. Age of the donor reduces the ability of human adipose-derived stem cells to alleviate symptoms in the experimental autoimmune encephalomyelitis mouse model. *Stem Cells Transl Med.* (2013) 2:797–807. doi: 10.5966/sctm.2013-0026
34. Kizilay Mancini Ö, Lora M, Shum-Tim D, Nadeau S, Rodier F, Colmegna I. A proinflammatory secretome mediates the impaired immunopotency of human mesenchymal stromal cells in elderly patients with atherosclerosis. *Stem Cells Transl Med.* (2017) 118:535–9. doi: 10.1002/sctm.16-0221
35. Siegel G, Kluba T, Hermanutz-Klein U, Bieback K, Northoff H, Schäfer R. Phenotype, donor age and gender affect function of human bone marrow-derived mesenchymal stromal cells. *BMC Med.* (2013) 11:146. doi: 10.1186/1741-7015-11-146
36. Lee RH, Yu JM, Foskett AM, Peltier G, Reneau JC, Bazhanov N, et al. TSG-6 as a biomarker to predict efficacy of human mesenchymal stem/progenitor cells (hMSCs) in modulating sterile inflammation *in vivo*. *Proc Natl Acad Sci USA.* (2014) 111:16766–71. doi: 10.1073/pnas.1416121111
37. Sacchetti B, Funari A, Remoli C, Giannicola G, Kogler G, Liedtke S, et al. No identical "mesenchymal stem cells" at different times and sites: human committed progenitors of distinct origin and differentiation potential are incorporated as adventitial cells in microvessels. *Stem Cell Rep.* (2016) 6:897–913. doi: 10.1016/j.stemcr.2016.05.011
38. Kizilay Mancini Ö, Lora M, Cuillerier A, Shum-Tim D, Hamdy R, Burelle Y, et al. Mitochondrial oxidative stress reduces the immunopotency of mesenchymal stromal cells in adults with coronary artery disease. *Circ Res.* (2018) 122:255–66. doi: 10.1161/CIRCRESAHA.117.311400
39. Acosta L, Hmadcha A, Escacena N, Perez-Camacho I, la Cuesta de A, Ruiz-Salmeron R, et al. Adipose mesenchymal stromal cells isolated from type 2 diabetic patients display reduced fibrinolytic activity. *Diabetes.* (2013) 62:4266–9. doi: 10.2337/db13-0896
40. Strong AL, Bowles AC, Wise RM, Morand JP, Dutreil MF, Gimble JM, et al. Human adipose stromal/stem cells from obese donors show reduced efficacy in halting disease progression in the experimental autoimmune encephalomyelitis model of multiple sclerosis. *Stem Cells.* (2016) 34:614–26. doi: 10.1002/stem.2272
41. Capasso S, Alessio N, Squillaro T, Di Bernardo G, Melone MA, Cipollaro M, et al. Changes in autophagy, proteasome activity and metabolism to determine a specific signature for acute and chronic senescent mesenchymal stromal cells. *Oncotarget.* (2015) 6:39457–68. doi: 10.18632/oncotarget.6277
42. Minieri V, Saviozzi S, Gambarotta G, Iacono Lo M, Accomasso L, Cibrario Rocchietti E, et al. Persistent DNA damage-induced premature senescence alters the functional features of human bone marrow mesenchymal stem cells. *J Cell Mol Med.* (2015) 19:734–43. doi: 10.1111/jcmm.12387
43. Jin HJ, Lee HJ, Heo J, Lim J, Kim M, Kim MK, et al. Senescence-associated MCP-1 secretion is dependent on a decline in BMI1 in human mesenchymal stromal cells. *Antioxid Redox Signal.* (2016) 24:471–85. doi: 10.1089/ars.2015.6359
44. Ankrum JA, Dastidar RG, Ong JF, Levy O, Karp JM. Performance-enhanced mesenchymal stem cells via intracellular delivery of steroids. *Sci Rep.* (2014) 4:4645. doi: 10.1038/srep04645
45. François M, Romieu-Mourez R, Li M, Galipeau J. Human MSC suppression correlates with cytokine induction of indoleamine 2,3-dioxygenase and bystander M2 macrophage differentiation. *Mol Ther.* (2012) 20:187–95. doi: 10.1038/mt.2011.189
46. Capilla-González V, López-Beas J, Escacena N, Aguilera Y, la Cuesta de A, Ruiz-Salmerón R, et al. PDGF restores the defective phenotype of adipose-derived mesenchymal stromal cells from diabetic patients. *Mol Ther.* (2018) 26:2696–709. doi: 10.1016/j.ymthe.2018.08.011
47. Serena C, Keiran N, Ceperuelo-Mallafre V, Ejarque M, Fradera R, Roche K, et al. Obesity and type 2 diabetes alters the immune properties of human adipose derived stem cells. *Stem Cells.* (2016) 34:2559–73. doi: 10.1002/stem.2429
48. Pachón-Peña G, Serena C, Ejarque M, Petriz J, Duran X, Oliva-Olivera W, et al. Obesity determines the immunophenotypic profile and functional characteristics of human mesenchymal stem cells from adipose tissue. *Stem Cells Transl Med.* (2016) 5:464–75. doi: 10.5966/sctm.2015-0161
49. Adler BJ, Kaushansky K, Rubin CT. Obesity-driven disruption of haematopoiesis and the bone marrow niche. *Nat Rev Endocrinol.* (2014) 10:737–48. doi: 10.1038/nrendo.2014.169
50. Capobianco V, Caterino M, Iaffaldano L, Nardelli C, Sirico A, Del Vecchio L, et al. Proteome analysis of human amniotic mesenchymal stem cells (hA-MSCs) reveals impaired antioxidant ability, cytoskeleton and metabolic functionality in maternal obesity. *Sci Rep.* (2016) 6:25270. doi: 10.1038/srep25270
51. Kornicka K, Houston J, Marycz K. Dysfunction of mesenchymal stem cells isolated from metabolic syndrome and type 2 diabetic patients as result of oxidative stress and autophagy may limit their potential therapeutic use. *Stem Cell Rev.* (2018) 14:337–45. doi: 10.1007/s12015-018-9809-x
52. Boland L, Burand AJ, Brown AJ, Boyt D, Lira VA, Ankrum JA. IFN-γ and TNF-α pre-licensing protects mesenchymal stromal cells from the pro-inflammatory effects of palmitate. *Mol Ther.* (2018) 26:860–73. doi: 10.1016/j.ymthe.2017.12.013
53. Heathman TR, Nienow AW, McCall MJ, Coopman K, Kara B, Hewitt CJ. The translation of cell-based therapies: clinical landscape and manufacturing challenges. *Regen Med.* (2015) 10:49–64. doi: 10.2217/rme.14.73
54. Spector AA. Structure and lipid binding properties of serum albumin. *Methods Enzymol.* (1986) 128:320–39. doi: 10.1016/0076-6879(86)28077-5
55. Cousin SP, Hügl SR, Wrede CE, Kajio H, Myers MG, Rhodes CJ. Free fatty acid-induced inhibition of glucose and insulin-like growth factor I-induced deoxyribonucleic acid synthesis in the pancreatic beta-cell line INS-1. *Endocrinology.* (2001) 142:229–40. doi: 10.1210/endo.142.1.7863
56. Dominici M, Le Blanc K, Mueller I, Slaper-Cortenbach I, Marini FC, Krause DS, et al. Minimal criteria for defining multipotent mesenchymal stromal cells. The International Society for Cellular Therapy position statement. *Cytotherapy.* (2006) 8:315–7. doi: 10.1080/14653240600855905
57. Trombetta A, Togliatto G, Rosso A, Dentelli P, Olgasi C, Cotogni P, et al. Increase of palmitic acid concentration impairs endothelial progenitor cell and bone marrow-derived progenitor cell bioavailability: role of

the STAT5/PPARγ transcriptional complex. *Diabetes.* (2013) 62:1245–57. doi: 10.2337/db12-0646

58. Gillet C, Dalla Valle A, Gaspard N, Spruyt D, Vertongen P, Lechanteur J, et al. Osteonecrosis of the femoral head: lipotoxicity exacerbation in MSC and modifications of the bone marrow fluid. *Endocrinology.* (2017) 158:490–502. doi: 10.1210/en.2016-1687
59. Hosios AM, Vander Heiden MG. The redox requirements of proliferating mammalian cells. *J Biol Chem.* (2018) 293:7490–8. doi: 10.1074/jbc.TM117.000239
60. Yao C-H, Fowle-Grider R, Mahieu NG, Liu G-Y, Chen Y-J, Wang R, et al. Exogenous fatty acids are the preferred source of membrane lipids in proliferating fibroblasts. *Cell Chem Biol.* (2016) 23:483–93. doi: 10.1016/j.chembiol.2016.03.007
61. Reis M, McDonald D, Nicholson L, Godthardt K, Knobel S, Dickinson AM, et al. Global phenotypic characterisation of human platelet lysate expanded MSCs by high-throughput flow cytometry. *Sci Rep.* (2018) 8:1579–12. doi: 10.1038/s41598-018-22326-5
62. Ma W, Wu JH, Wang Q, Lemaitre RN, Mukamal KJ, Djousse L, et al. Prospective association of fatty acids in the de novo lipogenesis pathway with risk of type 2 diabetes: the Cardiovascular Health Study. *Am J Clin Nutr.* (2014) 101:153–63. doi: 10.3945/ajcn.114.092601
63. Grossman WJ. Differential expression of granzymes A and B in human cytotoxic lymphocyte subsets and T regulatory cells. *Blood.* (2004) 104:2840–8. doi: 10.1182/blood-2004-03-0859
64. Klinker MW, Marklein RA, Surdo Lo JL, Wei C-H, Bauer SR. Morphological features of IFN-γ-stimulated mesenchymal stromal cells predict overall immunosuppressive capacity. *Proc Natl Acad Sci USA.* (2017) 114:E2598–607. doi: 10.1073/pnas.1617933114
65. Marklein RA, Klinker MW, Drake KA, Polikowsky HG, Lessey-Morillon EC, Bauer SR. Morphological profiling using machine learning reveals emergent subpopulations of interferon-γ-stimulated mesenchymal stromal cells that predict immunosuppression. *Cytotherapy.* (2018) 21:17–31. doi: 10.1016/j.jcyt.2018.10.008
66. Lam J, Bellayr IH, Marklein RA, Bauer SR, Puri RK, Sung KE. Functional profiling of chondrogenically induced multipotent stromal cell aggregates reveals transcriptomic and emergent morphological phenotypes predictive of differentiation capacity. *Stem Cells Transl Med.* (2018) 7:664–75. doi: 10.1002/sctm.18-0065
67. Roukos V, Pegoraro G, Voss TC, Misteli T. Cell cycle staging of individual cells by fluorescence microscopy. *Nat Protoc.* (2015) 10:334–48. doi: 10.1038/nprot.2015.016
68. Wang L-T, Ting C-H, Yen M-L, Liu K-J, Sytwu H-K, Wu KK, et al. Human mesenchymal stem cells (MSCs) for treatment towards immune- and inflammation-mediated diseases: review of current clinical trials. *J Biomed Sci.* (2016) 23:76. doi: 10.1186/s12929-016-0289-5
69. Ankrum J, Karp JM. Mesenchymal stem cell therapy: two steps forward, one step back. *Trends Mol Med.* (2010) 16:203–9. doi: 10.1016/j.molmed.2010.02.005
70. Moll G, Jitschin R, Bahr von L, Rasmusson-Duprez I, Sundberg B, Lönnies L, et al. Mesenchymal stromal cells engage complement and complement receptor bearing innate effector cells to modulate immune responses. *PLoS ONE.* (2011) 6:e21703. doi: 10.1371/journal.pone.0021703
71. Kota DJ, Wiggins LL, Yoon N, Lee RH. TSG-6 produced by hMSCs delays the onset of autoimmune diabetes by suppressing Th1 development and enhancing tolerogenicity. *Diabetes.* (2013) 62:2048–58. doi: 10.2337/db12-0931
72. Jitschin R, Böttcher M, Saul D, Lukassen S, Bruns H, Loschinski R, et al. Inflammation-induced glycolytic switch controls suppressivity of mesenchymal stem cells via STAT1 glycosylation. *Leukemia.* (2019) 13:392. doi: 10.1038/s41375-018-0376-6
73. Flegal KM, Kruszon-Moran D, Carroll MD, Fryar CD, Ogden CL. Trends in obesity among adults in the United States, 2005 to 2014. *JAMA.* (2016) 315:2284–8. doi: 10.1001/jama.2016.6458

10

Effects of Normothermic Machine Perfusion Conditions on Mesenchymal Stromal Cells

Jesus M. Sierra Parraga[1], Kaithlyn Rozenberg[2], Marco Eijken[3,4], Henri G. Leuvenink[5], James Hunter[2], Ana Merino[1], Cyril Moers[5], Bjarne K. Møller[3], Rutger J. Ploeg[2], Carla C. Baan[1], Bente Jespersen[4] and Martin J. Hoogduijn[1*]*

[1] Department of Internal Medicine, Erasmus Medical Center, University Medical Center Rotterdam, Rotterdam, Netherlands, [2] Nuffield Department of Surgical Sciences and Oxford Biomedical Research Centre, University of Oxford, Oxford, United Kingdom, [3] Department of Clinical Immunology, Aarhus University Hospital, Aarhus, Denmark, [4] Department of Renal Medicine, Aarhus University Hospital, Aarhus, Denmark, [5] Department of Surgery – Organ Donation and Transplantation, University Medical Center Groningen, Groningen, Netherlands

***Correspondence:**
Jesus M. Sierra Parraga
j.sierraparraga@erasmusmc.nl
Martin J. Hoogduijn
m.hoogduijn@erasmusmc.nl*

Ex-situ normothermic machine perfusion (NMP) of transplant kidneys allows assessment of kidney quality and targeted intervention to initiate repair processes prior to transplantation. Mesenchymal stromal cells (MSC) have been shown to possess the capacity to stimulate kidney repair. Therefore, the combination of NMP and MSC therapy offers potential to repair transplant kidneys. It is however unknown how NMP conditions affect MSC. In this study the effect of NMP perfusion fluid on survival, metabolism and function of thawed cryopreserved human (h)MSC and porcine (p)MSC in suspension conditions was studied. Suspension conditions reduced the viability of pMSC by 40% in both perfusion fluid and culture medium. Viability of hMSC was reduced by suspension conditions by 15% in perfusion fluid, whilst no differences were found in survival in culture medium. Under adherent conditions, survival of the cells was not affected by perfusion fluid. The perfusion fluid did not affect survival of fresh MSC in suspension compared to the control culture medium. The freeze-thawing process impaired the survival of hMSC; 95% survival of fresh hMSC compared to 70% survival of thawed hMSC. Moreover, thawed MSC showed increased levels of reactive oxygen species, which indicates elevated levels of oxidative stress, and reduced mitochondrial activity, which implies reduced metabolism. The adherence of pMSC and hMSC to endothelial cells was reduced after the thawing process, effect which was particularly profound in in the perfusion fluid. To summarize, we observed that conditions required for machine perfusion are influencing the behavior of MSC. The freeze-thawing process reduces survival and metabolism and increases oxidative stress, and diminishes their ability to adhere to endothelial cells. In addition, we found that hMSC and pMSC behaved differently, which has to be taken into consideration when translating results from animal experiments to clinical studies.

Keywords: mesenchymal stromal cells, normothermic machine perfusion, kidney repair, endothelial cells, suspension conditions, perfusion fluid, cryopreservation

INTRODUCTION

As the outcome of kidney transplantation has improved, the demand for kidney transplantation has increased and the donor organ pool to date is too small to supply the current need for transplant organs. The shortage in available donor kidneys (1) has led to the use of expanded criteria donor organs, that includes kidneys from older donors or from donors with hypertension, suboptimal kidney function or death resulting from stroke (2). This has resulted in a higher decline rate at time of offering and may also lead to a poorer outcome of the transplantation (3).

Currently, several techniques are being employed to improve the quality of expanded criteria kidneys and discarded kidneys to make them suitable for transplantation, including machine perfusion. Hypothermic machine perfusion of donor kidneys implies connection of the organ to a pump that perfuses the organ with a solution that provides the required components needed to maintain viability while also removing waste products released as a result of the metabolism and perfusion injury of the organ.

A more physiological way to assess viability of donor organs is continuous perfusion at normothermic temperature at 37°C with proper oxygenation and in the presence of necessary nutrients. A few years ago, the clinical feasibility and safety of 1 h normothermic machine perfusion (NMP) was demonstrated (4). Other groups decided to evaluate the feasibility of longer term NMP at 37°C, allowing more time to observe the kidney as well as intervene where possible. Recently, NMP has been successfully tested in a series of discarded donor kidneys for up to 24 h (5, 6). During NMP, using a bespoke red blood cell (RBC) enriched oxygenated and nutrient containing perfusate, the metabolism of the kidney resumes and allows monitoring during perfusion prior to transplantation (6, 7). Application of NMP for assessment and targeted intervention (8) to improve kidney quality is appealing and the effect of the therapy can potentially be monitored before the organ is transplanted. Thus, NMP is postulated as a promising platform to reduce kidney damage and initiate regeneration prior to transplantation.

Mesenchymal stromal cells (MSC) are multipotent cells which are found in adult tissues where they support function and repair (9). The International Society for Cellular Therapy has established the minimum criteria that a cell must meet to be considered an MSC: *in vitro* attachment to plastic, expression of several cell surface markers including CD29, CD44, CD90, the absence of endothelial and hematopoietic markers and the capacity to differentiate into cell types of mesodermal origin (10). MSC provide growth factors to progenitor cells that boost their regenerative processes (11, 12). More than 800 MSC-related studies are registered at http://clinicaltrials.gov on December 2018 and some of them have shown promising results in the treatment of kidney injury from different etiologies (13).

Abbreviations: EC, Endothelial cells; hMSC, Human mesenchymal stromal cells; HUVEC, Human umbilical cord endothelial cells; MSC, Mesenchymal stromal cells; NADH, Nicotinamide adenine dinucleotide; NMP, Normothermic machine perfusion; PAOEC, Porcine aortic endothelial cells; pMSC, Porcine mesenchymal stromal cells; ROS, Reactive oxygen species; XTT, 2,3-Bis-(2-Methoxy-4-Nitro-5-Sulfophenyl)-2H-Tetrazolium-5-Carboxanilide.

MSC are usually intravenously (IV) administered, which inevitably leads to accumulation in the lungs and poor delivery to target organs (14, 15). Obviously, delivery of MSC to the target organ while perfused on an *ex vivo* stand-alone circuit may overcome this dilemma. The idea to combine NMP and MSC therapy has generated an interest in the area of kidney transplantation (16–18). It is however unknown whether MSCs are compatible with the conditions of NMP.

In addition, a significant difference exists between preparation of therapeutic MSC for pre-clinical vs. clinical trials. In pre-clinical experiments, MSC are cultured *in vitro* and administered directly from the culture flask to laboratory animals when the cells are ready for infusion. In the human setting, large numbers of MSC are needed to treat a patient and cells are often produced at locations distant from the place of administration. This requires storage of cells in a frozen state and infusion following a delicate thawing process (13). Existing literature points out that frozen-thawed human MSC (hMSC) have an altered gene expression profile compared to cells directly retrieved from culture flasks (19), while it also has been shown that MSC immunoregulatory properties may be impaired by the freeze-thawing process (20). If the properties of MSC are impaired after the thawing process, this would mean that the results of human studies may not have the expected outcome.

It has been described that MSC from different species will exert the same actions through different mechanisms, which could affect the efficacy of MSC in various animal models (21, 22). Human and non-human primate MSC (23, 24) show marked similarities with respect to their biological properties, but it is unknown whether their therapeutic effects are comparable. Porcine models are very suitable for organ transplant and preservation studies due to the similarity in size and physiology between human and pig. It is however unknown whether MSC from human and pig behave in the same manner under NMP conditions.

With the questions above in mind, it is important to simulate conditions of NMP and assess their effect on MSC. We have evaluated the effect of the bespoke perfusate required for NMP in combination with the condition of fresh vs. frozen-thawed MSC in suspension using cells from porcine (pMSC) and from human (hMSC) origin.

MATERIALS AND METHODS

Isolation and Culture of Human and Porcine MSC and Endothelial Cells

hMSC were isolated from subcutaneous adipose tissue from healthy human kidney donors ($n = 5$) that became available during kidney donation procedures after obtaining written informed consent as approved by the Medical Ethical Committee of the Erasmus University Medical Center Rotterdam (MEC-2006-190). pMSC were isolated from subcutaneous adipose tissue ($n = 5$) collected from male pigs, which were subjected to surgery for teaching purposes, as a waste product. hMSC and pMSC were isolated as described previously (25) and phenotypically characterized by the expression of CD29, CD44, CD90 and the absence of CD31 and CD45. Human umbilical vein endothelial cells (HUVEC) were purchased from Lonza (Basel, Switzerland) and porcine aortic endothelial cells (PAOEC) were purchased from Cell Applications Inc. (San Diego, CA, USA).

Both hMSC and pMSC were cultured in minimum essential medium-α (MEM-α) (Sigma Aldrich, St. Louis, MO, USA) supplemented with penicillin (100 IU/ml), streptomycin (100 mg/ml) (1% P/S; Lonza), 2 mM L-glutamine (Lonza) and 15% fetal bovine serum (FBS; Lonza). HUVEC were cultured in endothelial growth medium 2 (PromoCell, Heidelberg, Germany). PAOEC were cultured in porcine endothelial cell media (Cell Applications, Inc.). MSC were used at passage 3-6, HUVEC were used at passage 4-8 and PAOEC at passage 3-6.

Perfusion Fluid

The perfusion fluid was a RBC-based solution with albumin as colloid adapted from NMP experiments used by several groups and allowing stable NMP of kidneys (4, 26, 27). The composition of the perfusion fluid is listed in **Table 1**.

Survival of MSC in Perfusion Fluid

MSC were trypsinized from the culture flasks at 90% confluency or thawed after cryopreservation and re-suspended either in complete culture medium or perfusion fluid at a concentration of 500,000 MSC/ml. MSC were incubated in perfusion fluid in polypropylene tubes to avoid attachment of MSC to plastic. After 30 min or 2 h in suspension, MSC were submitted to a RBC lysis process to remove the large amount of RBC present in perfusion fluid. MSC incubated in suspension with culture medium were also subjected to RBC lysis to treat both groups in the same way. Briefly, 3 ml of red blood cell lysis buffer (Invitrogen, Carlsbad, CA, USA) was added to MSC and incubated for 20 min at room temperature (RT). MSC were then washed with PBS and stained with Annexin-V (PE) and ViaProbe (PercP) to assess the number of early and late apoptotic cells. Perfusion fluid was also added to attached MSC and incubated for the same time and then trypsinized and stained as mentioned. Cells were analyzed by flow cytometry (FACS Canto II, BD Biosciences, NJ, USA) and data were analyzed using Kaluza Analysis 1.5a (Beckman Coulter, Brea, CA, USA).

TABLE 1 | Composition of perfusion fluid.

Red blood cells (Hematocrit 0.4 L/L)
Sodium (94.3 mmol/L)
Calcium (1.46 mmol/L)
Potassium (1.48 mmol/L)
Lactate (5.33 mmol/L)
Bicarbonate (26 mmol/L)
Albumin (19.1 g/L)
Glucose (2.93 mmol/L)
Mannitol (15.87 mg/L)
Creatinine (109.5 mg/L)
Amoxicillin (43.5 mmol/L)
Clavulanic Acid (16.1 mmol /L)

Mitochondrial Activity and Oxidative Stress of MSC

MSC metabolic activity was measured by a colorimetric assay based on the reduction of XTT [2,3-Bis-(2-Methoxy-4-Nitro-5-Sulfophenyl)-2H-Tetrazolium-5-Carboxanilide] by nicotinamide adenine dinucleotide (NADH) (ThermoFisher, Manhattan, NY, USA). The reagent is reduced by NADH produced during mitochondrial metabolism which results in a color change of the XTT reagent detectable by a spectrophotometer. The concentration of the reagent is measured by absorbance measured at a wavelength of 450–500 nm. This assay was performed on MSC that were in suspension for 30 and 120 min in perfusion fluid or culture medium and on attached MSC. In addition, oxidative stress of MSC was measured using CellRox reagent (ThermoFisher) according to the manufacturer's manual. CellRox is oxidized by reactive oxygen species (ROS) and emits a fluorescent signal that is measured by flow cytometry.

Proliferation of MSC

In order to assess the effects of perfusion fluid on cell proliferation, MSC were fluorescently labeled with carboxyfluorescein succinimidyl ester (CFSE) (ThermoFisher) which was added to the cells at a concentration of 5 mM. The cells were then incubated at 37°C for 15 min in the dark. Staining was stopped by the addition of twice the volume of FBS-containing culture medium and incubated for 5 min at RT. Cells were washed and exposed to perfusion fluid for 30 and 120 min at 37°C. After incubation with perfusion fluid, MSC were seeded in a 6-well-plate with regular culture medium and proliferation was measured at 24, 48, and 72 h by flow cytometry.

Release of Cytokines

A custom-made Luminex Multiplex Assay (R&D Systems, Minneapolis, MN, USA) was designed to measure the release of the following cytokines and growth factors by hMSC after incubation with perfusion fluid:

Angiopoietin-1 (ANG-1), Angiopoietin-2 (ANG-2), epidermal growth factor (EGF), hepatocyte growth factor (HGF), interferon-gamma (IFN-γ), interleukin 10 (IL-10), interleukin 6 (IL-6), monocyte chemoattractant protein 1 (MCP-1), programmed death ligand 1 (PD-L1), platelet derived growth factor AA (PDGFAA), Thrombospondin-2, tissue inhibitor of metalloproteases 1 (TIMP-1), tumor necrosis factor alpha (TNF-α), soluble tumor necrosis factor receptor 1 (sTNF-RI) and vascular endothelial growth factor (VEGF).

Subconfluent (90%) cultures of hMSC were washed and cells were incubated in perfusion fluid for 30 and 120 min. Wells were washed again and culture medium was added. After 24 h supernatants were retrieved and the array was performed according to the manufacturer's protocol. Fluorescence was measured on a Luminex 100/200 system (Luminex, Austin, TX, USA) using Xponent software. Due to the lack of pig-specific reagents this assay was only performed in with hMSC.

Adhesion of MSC to Endothelial Cells

Confluent monolayers of endothelial cells were cultured in 24-well-plates. Culture medium was replaced by either a 1:1 mix of MSC and endothelial cell (EC) medium or perfusion fluid. 200,000 MSC were fluorescently-labeled with PKH26 (Sigma) and added to each well. After 10, 30, 60, 120, and 240 min supernatant was removed and wells were washed to eliminate all non-adherent MSC. Attached cells were trypsinized and analyzed by flow cytometry. Fluorescent signal detected by flow cytometry allowed the determination of the percentage of MSC attached.

RESULTS

Cryopreserved MSC Show Reduced Survival in Perfusion Fluid and Medium

To examine the survival of MSC, fresh and cryopreserved MSC were incubated in suspension in perfusion fluid. Survival rates of freshly cultured and thawed pMSC were <40% after 30 min and decreased to 30% after 2 h in culture medium. Perfusion fluid had no negative impact on pMSC survival compared to culture medium except on thawed MSC after 30 min in perfusion (**Figure 1A**). hMSC were more resistant to suspension conditions than pMSC. Freshly cultured hMSC showed more than 95% survival in medium and in perfusion fluid. However, a significant decrease in survival to approximately 70% was observed for cryopreserved hMSC in perfusion fluid and in culture medium compared to fresh MSC. Perfusion fluid reduced survival of fresh hMSC after 2 h only minimally when compared to regular culture medium (**Figure 1B**).

MSC Show Impaired Adhesion to Endothelial Cells in Perfusion Fluid

To assess the function of surviving MSC in perfusion fluid, the capacity of MSC to adhere to EC was tested. In culture medium, fresh and thawed pMSC started to attach to PAOEC already 10 min after seeding. After 4 h, almost 100% of freshly cultured pMSC were attached to PAOEC while only 80% of thawed pMSC adhered (**Figure 2A**). Perfusion fluid strongly reduced the capacity of pMSC to attach to PAOEC regardless if they were fresh or thawed cells. Fresh hMSC adhesion to HUVEC was higher than 95% after 4 h in culture medium. In perfusion fluid, 60% of fresh hMSC were able to attach after 4 h. Cryopreserved hMSC showed 80% attachment in culture medium after 4 h and <50% of cryopreserved hMSC attached to HUVEC in perfusion fluid (**Figure 2B**). In general, thawed MSC showed a decreased capacity to attach to EC compared to fresh MSC, a difference that became more prominent in perfusion fluid.

Thawed MSC Express Higher Levels of ROS

The reduced adhesion of thawed MSC to EC could be explained by higher oxidative stress of MSC after the thawing process. The accumulation of ROS derived from the thawing process might induce damage in thawed MSC. An increased production of ROS by thawed pMSC and hMSC was observed at 30 min and 2 h after thawing both in medium and perfusion fluid (**Figure 3A**). Thawed hMSC produced a higher level of ROS than pMSC whereas ROS production in hMSC was boosted in perfusion fluid (**Figure 3B**). Thawed hMSC had elevated concentrations of ROS compared to fresh hMSC. These results indicate that freeze-thawing and perfusion fluid affects ROS production in pMSC and hMSC in the first hours after thawing.

Metabolic Activity of Mitochondria Is Reduced in Thawed MSC

ROS production leads to mitochondrial damage which results in reduced metabolic activity of cells. It is possible that MSC survive in suspension but are less metabolically active, which may explain the different capacity of MSC to adhere to EC after thawing or in perfusion fluid. To determine the effect of perfusion fluid and the freeze-thawing procedure on MSC metabolic activity, the conversion of XTT to its reduced state by mitochondria was measured in MSC. pMSC mitochondrial activity showed to be very stable and was not affected by cryopreservation (**Figure 3C**). However, perfusion fluid induced a small increase in activity after 30 min. Fresh hMSC showed a 2-fold higher mitochondrial activity than pMSC. Furthermore, fresh hMSC were more active than their thawed counterparts. After 30 min in perfusion fluid, fresh hMSC showed increased mitochondrial activity compared to culture medium (**Figure 3D**).

Freeze-Thawing Affects Proliferation of hMSC but Not pMSC

Proliferation of MSC was measured after initial incubation in suspension in perfusion fluid or medium for 30 min and 120 min, followed by 72 h culture in culture medium. Incubation in perfusion fluid did not affect pMSC proliferation (**Figures 4A,B**). Freshly cultured hMSC proliferated more than thawed hMSC after incubation in perfusion fluid or medium (**Figures 4C,D**).

Perfusion Fluid Increases the Proliferation of Attached MSC

MSC are adherent tissue cells and the suspension conditions in the previous experiments may affect their phenotype and function. To examine how adherent MSC respond to perfusion fluid, morphology, survival, metabolic activity and proliferation of attached MSC were studied. The morphology of MSC in culture was not affected after 30 min or 2 h culture in perfusion fluid (**Figures 5A–J**). Adherent pMSC and hMSC showed increased survival in perfusion fluid compared to MSC in suspension. Survival of adherent pMSC was higher than 80% in culture medium and perfusion fluid (**Figures 6A,B**) compared to a maximum of 40% survival in suspension (**Figure 1A**). hMSC showed 78% survival after 2 h in suspension in perfusion fluid, but when they were attached, survival after 2 h was 93%. No differences in survival were observed between perfusion fluid and culture medium on attached MSC (**Figures 6A,B**). No significant effect of perfusion fluid on mitochondrial activity was observed for attached MSC, although there was a trend toward a decline in activity of pMSC and hMSC over time when cultured in perfusion fluid (**Figures 6C,D**). Pre-incubation of attached pMSC in perfusion fluid increased their proliferation

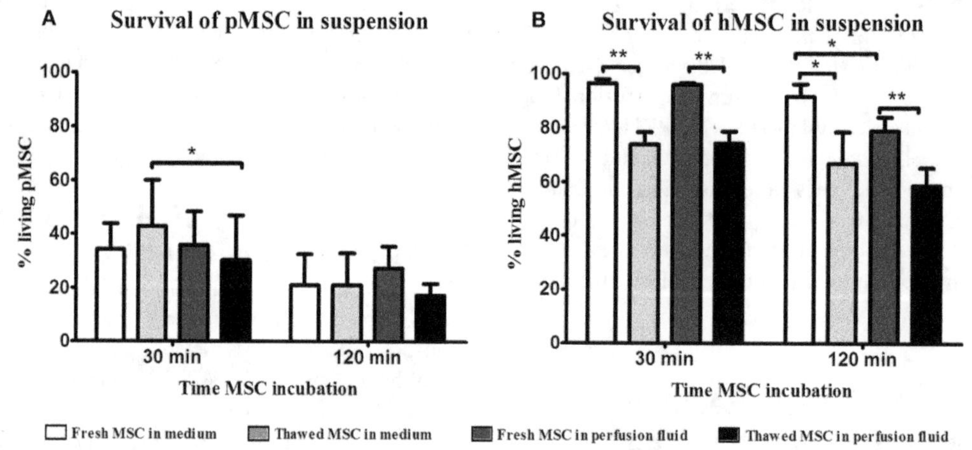

FIGURE 1 | Effect of perfusion fluid on survival of fresh and thawed pMSC and hMSC in suspension. **(A)** Perfusion fluid had minimal effect on pMSC survival compared to medium. **(B)** Perfusion fluid had minimal effects on hMSC survival. Thawed hMSC showed a lower survival in suspension compared to fresh hMSC. Perfusion fluid reduced survival of hMSC after 2 h ($n = 5$). Results are shown as means ± SD. $^{*}p < 0.05$; $^{**}p < 0.01$.

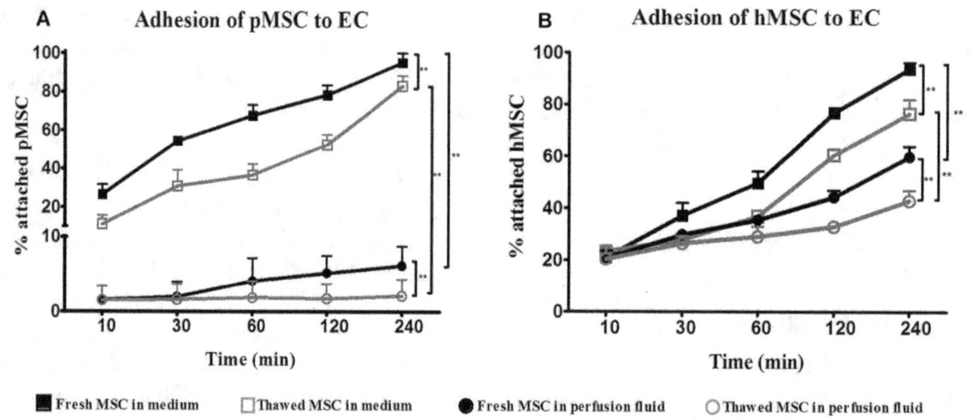

FIGURE 2 | Adhesion of fresh and cryopreserved pMSC and hMSC to EC in medium and perfusion fluid. **(A)** pMSC attachment to PAOEC over time in culture medium. pMSC show reduced binding to PAOEC in perfusion fluid. Thawed pMSC showed a further reduced ability to bind to EC in perfusion fluid. **(B)** Thawed hMSC showed reduced attachment to HUVEC compared to fresh hMSC either in culture medium or perfusion fluid ($n = 5$). Results are shown as means ± SD. $^{**}p < 0.01$.

after 24 h compared to culture medium. This effect was observed only after 120 min pre-incubation in perfusion fluid for hMSC (**Figures 6E,F**).

Secretory Profile of MSC Is Not Affected by Culture in Perfusion Fluid

The secretion of growth factors and cytokines is an important mechanism of action of MSC. To examine whether perfusion fluid would preserve the secretory profile of adherent hMSC and furthermore whether perfusion fluid induced an inflammatory response in hMSC, hMSC were incubated in perfusion fluid for 30 min or 2 h and growth factor and cytokine secretion was analyzed. We observed that the secretion of the angiogenic factors VEGF, PDGF, ANG-1, HGF and Thr2 was unaffected in perfusion fluid (**Figures 7A–E**). Inflammatory cytokines IL-6 and MCP-1 were increased 2-fold and 10-fold, respectively, in perfusion fluid (**Figures 8A,B**).

DISCUSSION

In the present work we have assessed the effect of a period of incubation in a perfusion fluid required for robust longer term *ex vivo* NMP of kidneys on MSC of both pig and human origin. Our work involved the use of cryopreserved MSC for logistic purposes, suspension conditions to deliver MSC via the renal artery using NMP and the use of an RBC-based perfusion fluid which may affect the survival and function of MSC. In order to mimic the conditions of a potential novel MSC therapy to stimulate the repair of injured kidneys while these are connected to NMP, MSC were thawed after cryopreservation and incubated in perfusion fluid in suspension. Actual infusion of MSC using NMP was not carried out as the purpose of the study was to assess the individual effect of each of the aforementioned conditions separately. In case that NMP conditions did not support MSC survival and function, future planned experimentation in the NMP setup could have been stopped, reducing economic and

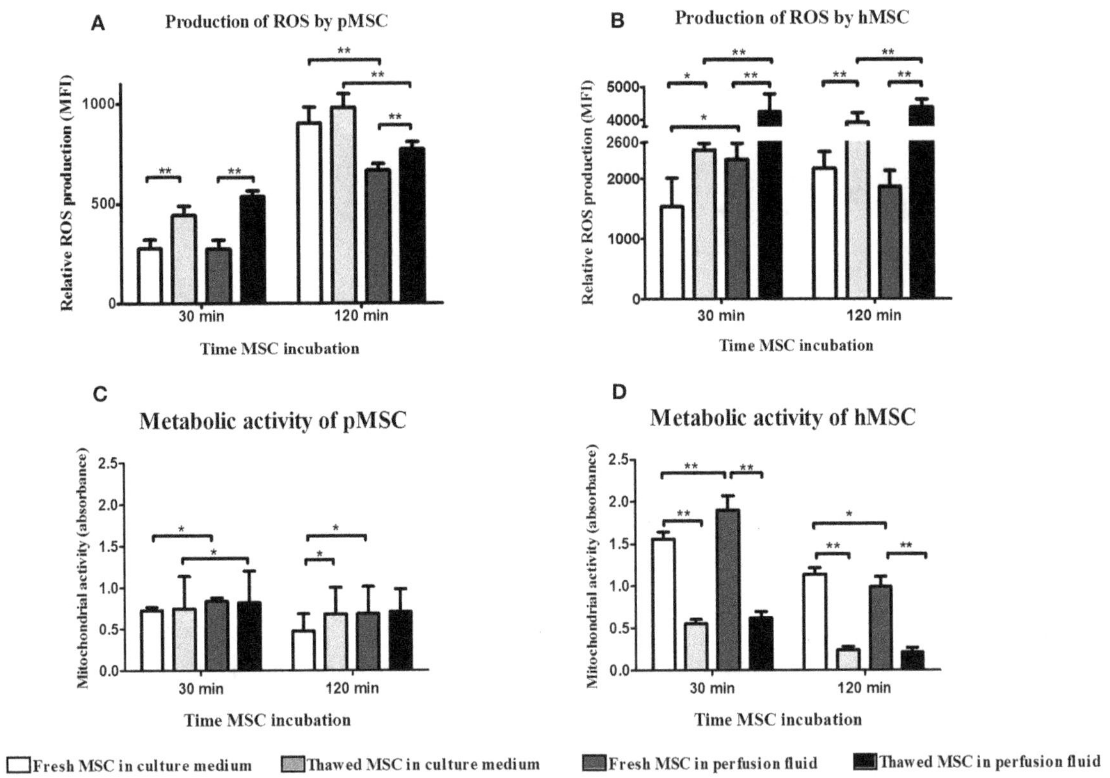

FIGURE 3 | Perfusion fluid and thawing after cryopreservation increase ROS production and reduce metabolic activity in MSC. **(A)** ROS production in fresh and frozen-thawed pMSC suspended in culture medium or in perfusion fluid. **(B)** ROS production in fresh and in frozen-thawed hMSC suspended in culture medium or in perfusion fluid. **(C)** Metabolic activity measured by XTT reduction by NADH of fresh and frozen-thawed pMSC after 30 min and 2 h incubation in perfusion fluid. **(D)** Metabolic activity measured by XTT reduction by NADH of fresh and frozen-thawed hMSC after 30 min and 2 h incubation in perfusion fluid ($n = 5$). Results are shown as means ± SD. $^{*}p < 0.05$; $^{**}p < 0.01$.

time costs. Nevertheless, the results of this study allow to take the next step and study MSC infusion through an NMP system, which is already planned to be carried out.

In the field of clinical MSC therapy, temporary cryopreservation and thawing along with vehicle solutions to deliver the therapy are important factors that can determine the success of MSC therapy (20, 28–30). The handling time until administration will have an impact on the survival and function of MSC. Survival of MSC is affected by the composition of storage media (31) and cytokine secretion profile of MSC can be altered, affecting MSC properties such as angiogenic potential (32). Our results show that the particular perfusion fluid used for our experiment is not detrimental for the secretion of angiogenic factors by hMSC.

The bespoke NMP perfusion fluid used in this study supports the survival of pig and human MSC. However, the function and metabolism of these cells are affected by the suspension conditions MSC were kept in. Specifically, perfusion fluid inhibited the adhesion of pMSC in suspension to PAOEC and hMSC to HUVEC. MSC have the capacity to adhere to endothelial cells *in vitro* under flow conditions as previously shown, especially when endothelial cells have been treated to recreate an inflammatory environment. Attachment of MSC to endothelial cells was shown to be reduced under flow compared to static conditions (33). Being in suspension in different clinically used storage solutions such as physiologic saline can influence the survival of MSC (31) and the composition of the solution that MSC are kept in modifies their metabolism and function (34, 35). Therefore, it can be deducted that the effect of MSC therapy delivered to renal grafts during *ex vivo* NMP will depend, among other factors, upon the composition of the perfusion fluid. A recent study perfused kidneys for 24 h using NMP. They infused MSC at different concentrations however, after 24 h they found 95% of infused MSC back in their perfusion solution (36), which indicate a very diminished adhesion capacity of MSC toward endothelial cells. Our results are consistent with existing literature and indicate that the composition of the perfusion fluid as well as the infusion process affects the functional properties and delivery efficiency of the final MSC product. Therefore, further knowledge need to be obtained regarding the effects of perfusion conditions on MSC delivery to the injured kidney.

Human and porcine MSC showed a negative response at several levels to cryopreservation, thawing and re-suspension in perfusion fluid. The effect of cryopreservation and thawing of hMSC has been a concern for the community as it can decrease

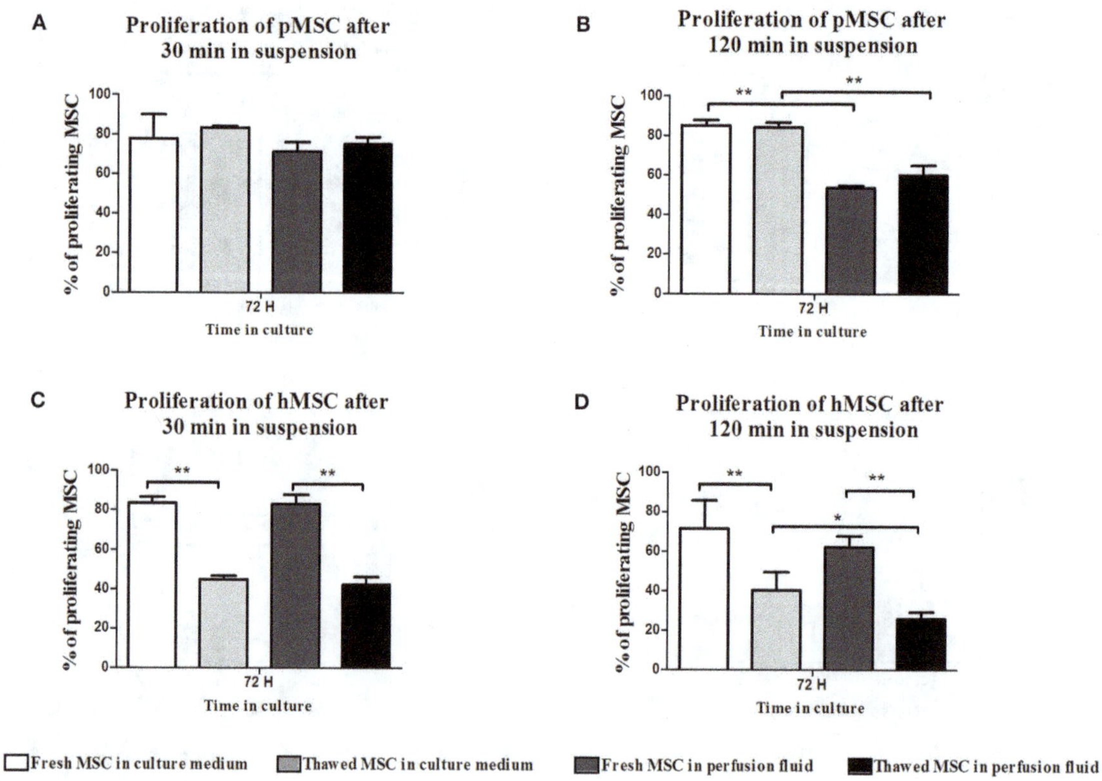

FIGURE 4 | Effect of perfusion fluid and freeze-thawing on proliferation of MSC. pMSC and hMSC were incubated in perfusion fluid for 30 and 120 min and proliferation was measured after subsequent culturing in medium after 72 h. **(A)** pMSC proliferation was not affected by incubation in perfusion fluid for 30 min. **(B)** pMSC in perfusion fluid proliferated less than pMSC in culture medium. **(C,D)** Fresh hMSC proliferated more than thawed hMSC and perfusion fluid did not have an effect on fresh hMSC proliferation. Thawed cells in perfusion fluid were the least proliferative ($n = 5$). Results are shown as means ± SD. $^*p < 0.05$; $^{**}p < 0.01$.

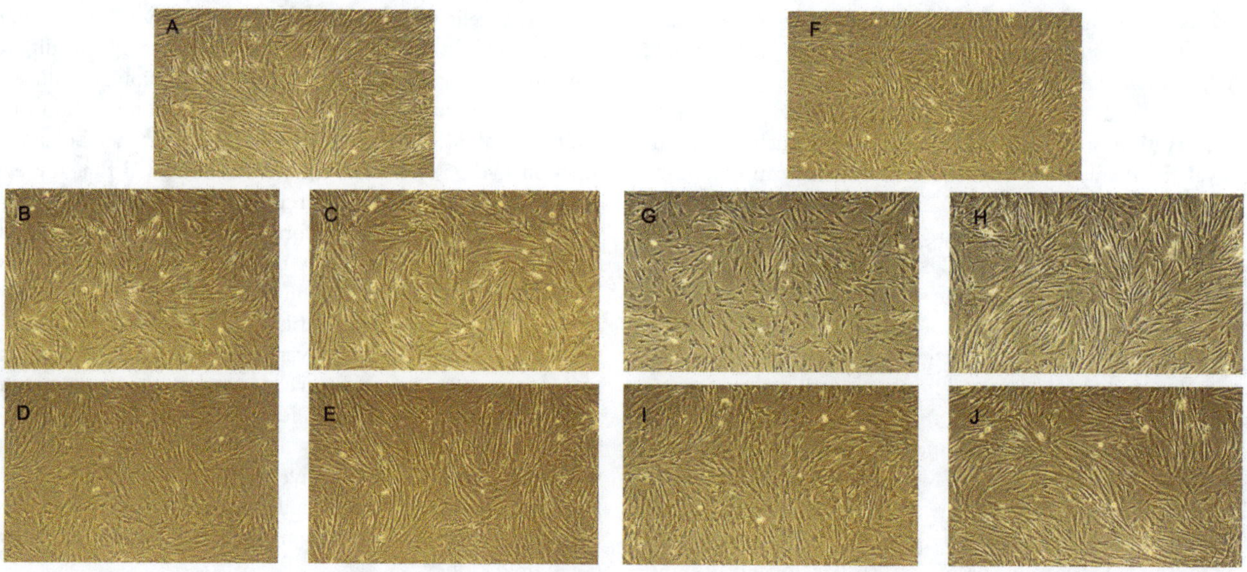

FIGURE 5 | Effect of perfusion fluid on morphology of attached MSC. **(A)** pMSC in regular culture medium. **(B,C)** pMSC in culture medium or perfusion fluid, respectively for 30 min. **(D,E)** pMSC cultured in culture medium or perfusion fluid, respectively for 120 min. **(F)** hMSC in regular culture medium. **(G,H)** hMSC cultured in culture medium or perfusion fluid, respectively for 30 min. **(I,J)** hMSC cultured in culture medium or perfusion fluid, respectively for 120 min.

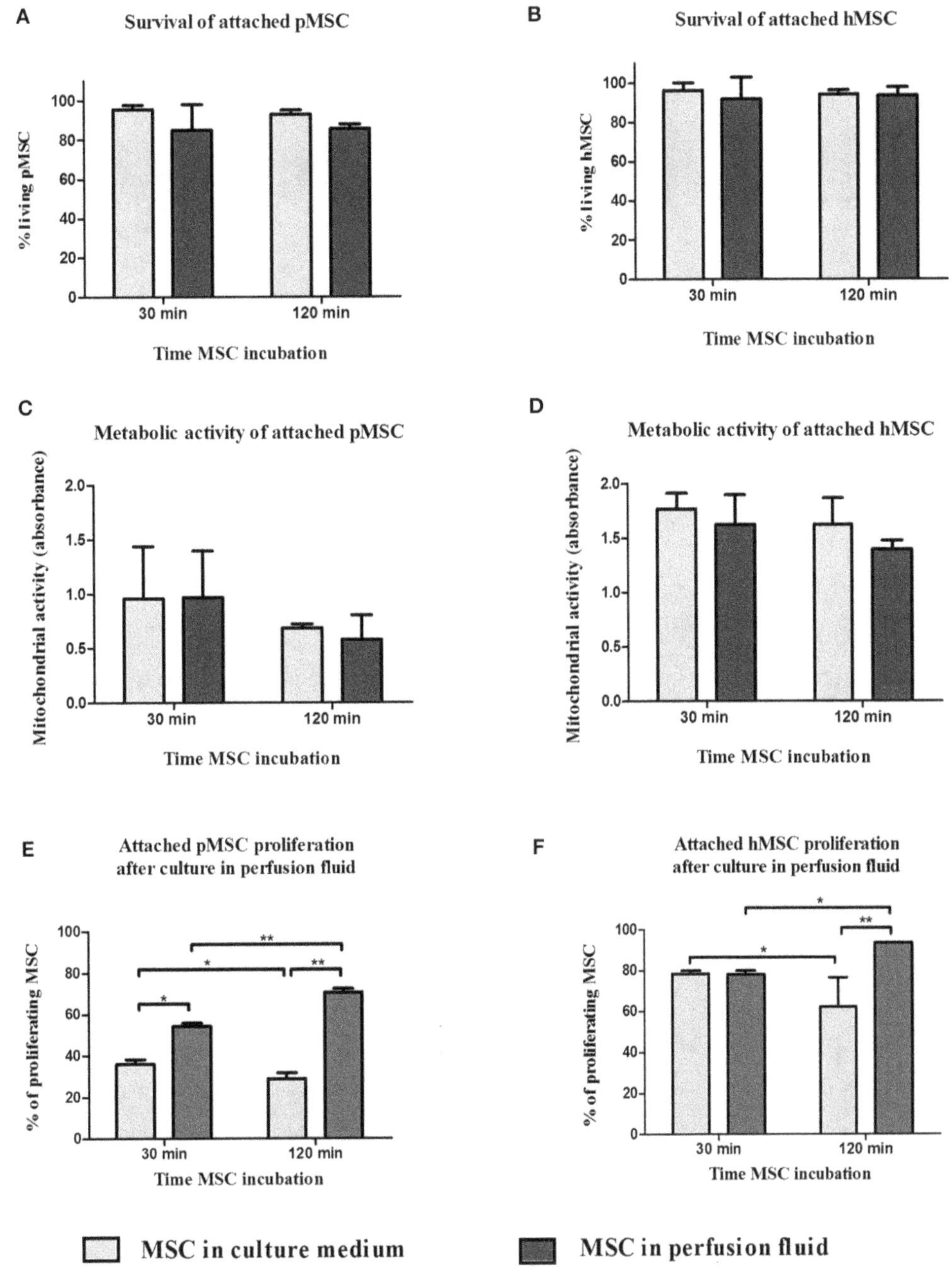

FIGURE 6 | Effect of perfusion fluid on attached MSC. Survival of attached pMSC **(A)** and hMSC **(B)** after 30 min in perfusion fluid. **(C,D)** Metabolic activity of attached pMSC **(C)** and hMSC **(D)** after 30 min in perfusion fluid measured by reduction of XTT. **(E,F)** Proliferation of attached pMSC **(E)** and hMSC **(F)** after 30 in perfusion fluid. Cells were trypsinized and re-seeded in a culture flask. Proliferation after 24 h was determined by CFSE fluorescence ($n = 5$). Results are shown as means ± SD. $^{*}p < 0.05$; $^{**}p < 0.01$.

the presumed efficacy of MSC therapy (19, 28, 30, 37). Our results confirmed this concern. It has recently been published that freezing-thawing MSC increases the production of ROS and compromises membrane stability and homeostasis in pMSC (38), which is also supported by our data. Viability of MSC is a key factor for treatments that require an active role of MSC. However, inactive MSC have been shown to retain their immunomodulatory properties (14, 39, 40). Therefore, the aim of MSC therapy dictates the required characteristics of MSC.

In vivo, MSC are tissue resident cells and are found in perivascular niches in a wide variety of tissues (41–43). *In vitro*, MSC are strictly grown as adherent cells, and for this reason MSC are cultured allowing plastic adherence to expand them (10). We demonstrated that the poor performance of MSC in

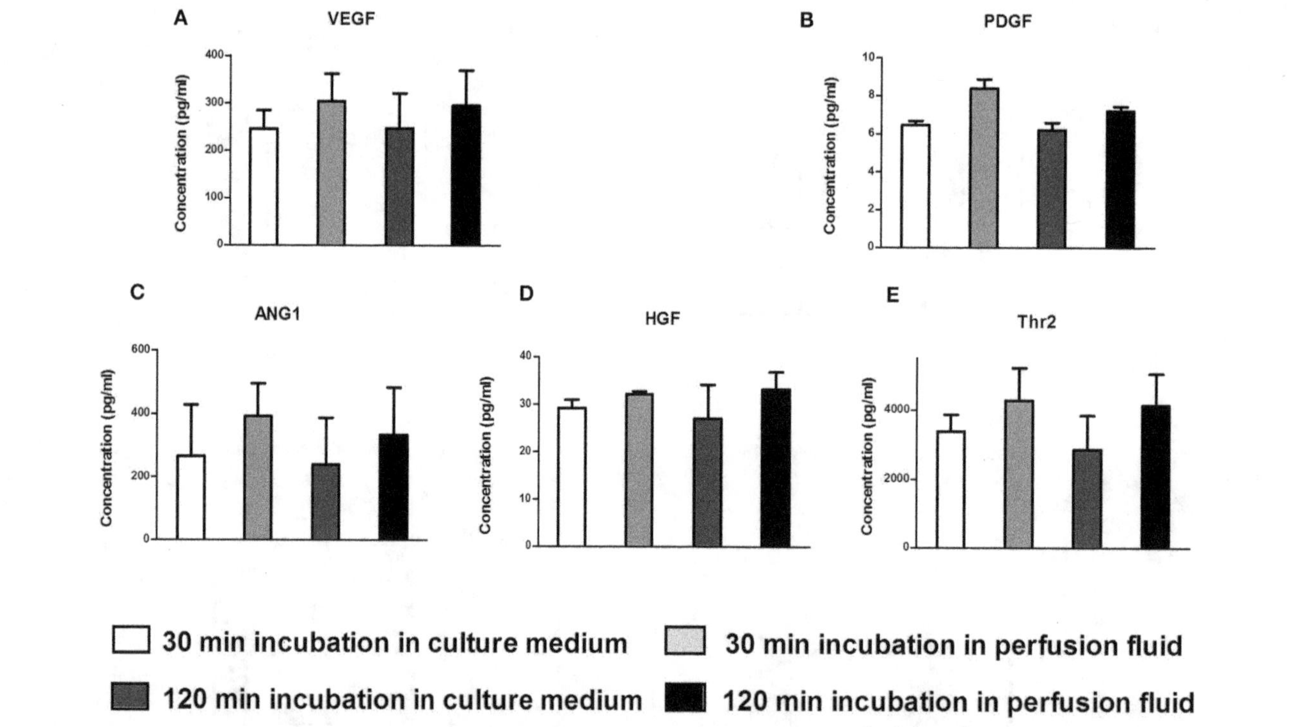

FIGURE 7 | Production of angiogenic factors by hMSC in culture medium and perfusion fluid. MSC were incubated in perfusion fluid for 30 or 120 min, washed and replaced by culture medium for 24 h. The secretion of angiogenic and growth factors was not affected by perfusion fluid. **(A–E)** Concentration of secreted VEGF, PDGF, ANG1, HGF and Thr2, respectively ($n = 5$). Results are shown as mean ± SD.

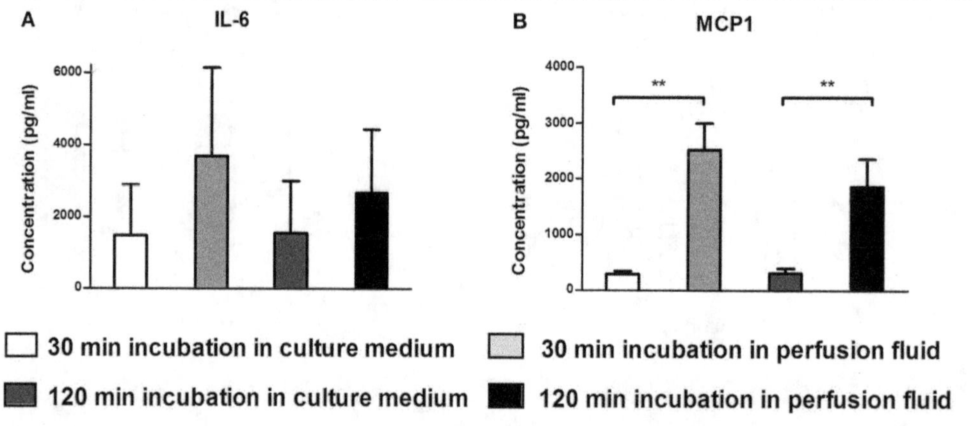

FIGURE 8 | Production of inflammatory cytokines by hMSC in culture medium and perfusion fluid. MSC were incubated in perfusion fluid for 30 or 120 min, washed and replaced by culture medium for 24 h. **(A)** The secretion of IL-6 shows tendency to increase by perfusion fluid. **(B)** The secretion of MCP-1 by MSC is increased after incubation in perfusion fluid ($n = 5$). Results are shown as means ± SD. $^{**}p < 0.01$.

perfusion fluid was primarily due to the fact that they were in suspension. Survival of both pig and human MSC cultured in adherent conditions in perfusion fluid was higher than that of MSC kept in suspension, suggesting a protective effect of adherence. Mitochondrial activity, however, was similar in MSC in suspension and attached, suggesting that MSC are active also in suspension. Proliferation of pMSC was increased after exposure to perfusion fluid which could be a response to a specific component of the perfusion fluid (31). These results suggest that the nature of MSC make them more vulnerable when they are in suspension and therefore, it should be minimized when they are administered as therapy. Being in suspension in perfusion is, however, a transient condition in the process of MSC delivery using NMP. Presumably, when MSC are delivered to the injured kidney, the damaged tissue microenvironment will help the MSC to be retained and produce regenerative factors as previously shown (44–46). We have shown that after being in contact with perfusion solution MSC can recover, proliferate and be metabolically active. In addition, the secretory profile of angiogenic factor by hMSC is not affected by perfusion fluid. Therefore, this is a promising result that hints MSC maintain their reparative potential after delivery using NMP.

Cryopreservation, thawing and suspension of MSC are inevitable conditions to infuse GMP-grade MSC to kidney transplants via NMP. These conditions are necessary to bridge differences in time and location between MSC preparation and NMP. The disadvantageous effect of these conditions on MSC has to be taken into account for the evaluation of the suitability of MSC therapy for NMP. In order to improve the viability of MSC in NMP conditions, the composition of the perfusion fluid may be adapted to provide better support for MSC survival. Another possibility is to recover MSC after thawing under favorable conditions to improve the resistance of MSC to NMP. An alternative would be to take the loss of MSC under NMP conditions into account and use higher numbers of cells, although the effect of administering large numbers of non-viable MSC to the kidney is uncertain.

Pre-clinical work in the field of transplantation is often performed using porcine models to better understand the possible behaviors of new therapies in patients. Therefore, we investigated MSC of porcine and human origin in order to determine if results can be translated. The response to the thawing process as well as to suspension conditions and perfusion fluid was quite different between MSC from both origins for the parameters studied. We are aware that differences could be related not only to species but also to age and gender, as well as site of adipose tissue harvesting from the donor. However, our results indicate that caution should be taken when interpreting *in vitro* studies with porcine cells toward the behavior of human cells. A safe translation from swine pre-clinical models to clinical studies is challenging when results are not reproducible between species.

Summarizing, NMP conditions will affect MSC but show sufficient support of their function and survival to consider MSC administration through NMP as a viable option in pursuit of a potentially beneficial cell therapy for the regeneration of injured organs. After these essential preliminary experiments, further study is now underway to determine the best way of joining these two exciting techniques in an optimal manner.

AUTHOR CONTRIBUTIONS

BJ, CB, ME, RP, and MH: conception of the study; JS, AM, and MH: designed the experiments; JS: performed the experiments and the analysis of the data; JS, AM, and MH: interpreted the data. All authors contributed to writing and reviewing the manuscript.

FUNDING

The MePEP project is funded by the Lundbeck Foundation under grant application no R198-2015-184.

REFERENCES

1. Hippen B, Ross LF, Sade RM. Saving lives is more important than abstract moral concerns: financial incentives should be used to increase organ donation. *Ann Thorac Surg.* (2009) 88:1053–61. doi: 10.1016/j.athoracsur.2009.06.087
2. Ojo AO. Expanded criteria donors: process and outcomes. *Semin Dial.* (2005) 18:463–8. doi: 10.1111/j.1525-139X.2005.00090.x
3. Nagaraja P, Roberts GW, Stephens M, Horvath S, Kaposztas Z, Chavez R, et al. Impact of expanded criteria variables on outcomes of kidney transplantation from donors after cardiac death. *Transplantation.* (2015) 99:226–31. doi: 10.1097/TP.0000000000000304
4. Nicholson ML, Hosgood SA. Renal transplantation after *ex vivo* normothermic perfusion: the first clinical study. *Am J Transplant.* (2013) 13:1246–52. doi: 10.1111/ajt.12179
5. Weissenbacher A, Lo Faro L, Boubriak O, Soares MF, Roberts IS, Hunter JP, et al. Twenty-four hour normothermic perfusion of discarded human kidneys with urine recirculation. *Am J Transplant.* (2018) 19:178–92. doi: 10.1111/ajt.14932
6. Hosgood SA, Thompson E, Moore T, Wilson CH, Nicholson ML. Normothermic machine perfusion for the assessment and transplantation of declined human kidneys from donation after circulatory death donors. *Br J Surg.* (2018) 105:388–94. doi: 10.1002/bjs.10733
7. Kaths JM, Hamar M, Echeverri J, Linares I, Urbanellis P, Cen JY, et al. Normothermic *ex vivo* kidney perfusion for graft quality assessment prior to transplantation. *Am J Transplant.* (2018) 18:580–9. doi: 10.1111/ajt.14491
8. Smith SF, Adams T, Hosgood SA, Nicholson ML. The administration of argon during *ex vivo* normothermic perfusion in an experimental model of kidney ischemia-reperfusion injury. *J Surg Res.* (2017) 218:202–8. doi: 10.1016/j.jss.2017.05.041
9. Pittenger MF, Mackay AM, Beck SC, Jaiswal RK, Douglas R, Mosca JD, et al. Multilineage potential of adult human mesenchymal stem cells. *Science.* (1999) 284:143–7. doi: 10.1126/science.284.5411.143
10. Dominici M, Le Blanc K, Mueller I, Slaper-Cortenbach I, Marini F, Krause D, et al. Minimal criteria for defining multipotent mesenchymal stromal cells. The International Society for Cellular Therapy position statement. *Cytotherapy.* (2006) 8:315–7. doi: 10.1080/14653240600855905
11. Kim HK, Lee SG, Lee SW, Oh BJ, Kim JH, Kim JA, et al. A subset of paracrine factors as efficient biomarkers for predicting vascular regenerative efficacy of mesenchymal stromal/stem cells. *Stem Cells.* (2018) 37:77–88. doi: 10.1002/stem.2920
12. Azhdari Tafti Z, Mahmoodi M, Hajizadeh MR, Ezzatizadeh V, Baharvand H, Vosough M, et al. Conditioned media derived from human adipose tissue mesenchymal stromal cells improves primary hepatocyte maintenance. *Cell J.* (2018) 20:377–87. doi: 10.22074/cellj.2018.5288
13. Reinders MEJ, van Kooten C, Rabelink TJ, de Fijter JW. Mesenchymal stromal cell therapy for solid organ transplantation. *Transplantation.* (2018) 102:35–43. doi: 10.1097/TP.0000000000001879
14. de Witte SFH, Luk F, Sierra Parraga JM, Gargesha M, Merino A, Korevaar SS, et al. Immunomodulation by therapeutic Mesenchymal Stromal Cells (MSC) is triggered through phagocytosis of MSC by monocytic cells. *Stem Cells.* (2018) 36:602–15. doi: 10.1002/stem.2779
15. Eggenhofer E, Benseler V, Kroemer A, Popp FC, Geissler EK, Schlitt HJ, et al. Mesenchymal stem cells are short-lived and do not migrate beyond the lungs after intravenous infusion. *Front Immunol.* (2012) 3:297. doi: 10.3389/fimmu.2012.00297
16. Sierra-Parraga JM, Eijken M, Hunter J, Moers C, Leuvenink H, Moller B, et al. Mesenchymal stromal cells as anti-inflammatory and regenerative mediators for donor kidneys during normothermic machine perfusion. *Stem Cells Dev.* (2017) 26:1162–70. doi: 10.1089/scd.2017.0030

17. Weissenbacher A, Hunter, J. Normothermic machine perfusion of the kidney. *Curr Opin Organ Transplant.* (2017) 22:571–6. doi: 10.1097/MOT.0000000000000470
18. Hamar M, Selzner, M. Ex-vivo machine perfusion for kidney preservation. *Curr Opin Organ Transplant.* (2018) 23:369–74. doi: 10.1097/MOT.0000000000000524
19. Hoogduijn MJ, de Witte SF, Luk F, van den Hout-van Vroonhoven MC, Ignatowicz L, Catar R, et al. Effects of freeze-thawing and intravenous infusion on mesenchymal stromal cell gene expression. *Stem Cells Dev.* (2016) 25:586–97. doi: 10.1089/scd.2015.0329
20. Moll G, Alm JJ, Davies LC, von Bahr L, Heldring N, Stenbeck-Funke L, et al. Do cryopreserved mesenchymal stromal cells display impaired immunomodulatory and therapeutic properties? *Stem Cells.* (2014) 32:2430–42. doi: 10.1002/stem.1729
21. Lohan P, Treacy O, Morcos M, Donohoe E, O'Donoghue Y, Ryan AE, et al. Interspecies incompatibilities limit the immunomodulatory effect of human mesenchymal stromal cells in the rat. *Stem Cells.* (2018) 36:1210–5. doi: 10.1002/stem.2840
22. Ling W, Zhang J, Yuan Z, Ren G, Zhang L, Chen X, et al. Mesenchymal stem cells use IDO to regulate immunity in tumor microenvironment. *Cancer Res.* (2014) 74:1576–87. doi: 10.1158/0008-5472.can-13-1656
23. Pogozhykh O, Pogozhykh D, Neehus AL, Hoffmann A, Blasczyk R, Muller, T. Molecular and cellular characteristics of human and non-human primate multipotent stromal cells from the amnion and bone marrow during long term culture. *Stem Cell Res Ther.* (2015) 6:150. doi: 10.1186/s13287-015-0146-6
24. Izadpanah R, Trygg C, Patel B, Kriedt C, Dufour J, Gimble JM, et al. Biologic properties of mesenchymal stem cells derived from bone marrow and adipose tissue. *J Cell Biochem.* (2006) 99:1285–97. doi: 10.1002/jcb.20904
25. Hoogduijn MJ, Crop MJ, Peeters AM, Van Osch GJ, Balk AH, Ijzermans JN, et al. Human heart, spleen, and perirenal fat-derived mesenchymal stem cells have immunomodulatory capacities. *Stem Cells Dev.* (2007) 16:597–604. doi: 10.1089/scd.2006.0110
26. Hosgood SA, Nicholson ML. First in man renal transplantation after *ex vivo* normothermic perfusion. *Transplantation.* (2011) 92:735–8. doi: 10.1097/TP.0b013e31822d4e04
27. Hosgood SA, Barlow AD, Hunter JP, Nicholson ML. *Ex vivo* normothermic perfusion for quality assessment of marginal donor kidney transplants. *Br J Surg.* (2015) 102:1433–40. doi: 10.1002/bjs.9894
28. Moll G, Geissler S, Catar R, Ignatowicz L, Hoogduijn MJ, Strunk D, et al. Cryopreserved or fresh mesenchymal stromal cells: only a matter of taste or key to unleash the full clinical potential of MSC therapy? *Adv Exp Med Biol.* (2016) 951:77–98. doi: 10.1007/978-3-319-45457-3_7
29. Francois M, Copland IB, Yuan S, Romieu-Mourez R, Waller EK, Galipeau, J. Cryopreserved mesenchymal stromal cells display impaired immunosuppressive properties as a result of heat-shock response and impaired interferon-gamma licensing. *Cytotherapy.* (2012) 14:147–52. doi: 10.3109/14653249.2011.623691
30. Chinnadurai R, Garcia MA, Sakurai Y, Lam WA, Kirk AD, Galipeau J, et al. Actin cytoskeletal disruption following cryopreservation alters the biodistribution of human mesenchymal stromal cells *in vivo*. *Stem Cell Reports.* (2014) 3:60–72. doi: 10.1016/j.stemcr.2014.05.003
31. Nofianti CE, Sari IN, Marlina, Novialdi, Pawitan JA. Temporary storage solution for adipose derived mesenchymal stem cells. *Stem Cell Investig.* (2018) 5:19. doi: 10.21037/sci.2018.05.04
32. Parsha K, Mir O, Satani N, Yang B, Guerrero W, Mei Z, et al. Mesenchymal stromal cell secretomes are modulated by suspension time, delivery vehicle, passage through catheter, and exposure to adjuvants. *Cytotherapy.* (2017) 19:36–46. doi: 10.1016/j.jcyt.2016.10.006
33. Ruster B, Gottig S, Ludwig RJ, Bistrian R, Muller S, Seifried E, et al. Mesenchymal stem cells display coordinated rolling and adhesion behavior on endothelial cells. *Blood.* (2006) 108:3938–44. doi: 10.1182/blood-2006-05-025098
34. Xia W, Li H, Wang Z, Xu R, Fu Y, Zhang X, et al. Human platelet lysate supports *ex vivo* expansion and enhances osteogenic differentiation of human bone marrow-derived mesenchymal stem cells. *Cell Biol Int.* (2011) 35:639–43. doi: 10.1042/CBI20100361
35. de Witte SFH, Lambert EE, Merino A, Strini T, Douben H, O'Flynn L, et al. Aging of bone marrow- and umbilical cord-derived mesenchymal stromal cells during expansion. *Cytotherapy.* (2017) 19:798–807. doi: 10.1016/j.jcyt.2017.03.071
36. Brasile L, Henry N, Orlando G, Stubenitsky, B. Potentiating Renal Regeneration using Mesenchymal Stem Cells. *Transplantation.* (2018). 103:307–13. doi: 10.1097/TP.0000000000002455
37. Chinnadurai R, Copland IB, Garcia MA, Petersen CT, Lewis CN, Waller EK, et al. Cryopreserved mesenchymal stromal cells are susceptible to T-cell mediated apoptosis which is partly rescued by ifngamma licensing. *Stem Cells.* (2016) 34:2429–42. doi: 10.1002/stem.2415
38. Gurgul A, Romanek J, Pawlina-Tyszko K, Szmatola T, Opiela, J. Evaluation of changes arising in the pig mesenchymal stromal cells transcriptome following cryopreservation and Trichostatin A treatment. *PLoS ONE.* (2018) 13:e0192147. doi: 10.1371/journal.pone.0192147
39. Luk F, de Witte S, Korevaar SS, Roemeling-van Rhijn M, Franquesa M, Strini T, et al. Inactivated mesenchymal stem cells maintain immunomodulatory capacity. *Stem Cells Dev.* (2016) 25:1342–54. doi: 10.1089/scd.2016.0068
40. Galleu A, Riffo-Vasquez Y, Trento C, Lomas C, Dolcetti L, Cheung TS, et al. Apoptosis in mesenchymal stromal cells induces *in vivo* recipient-mediated immunomodulation. *Sci Transl Med.* (2017) 9:416. doi: 10.1126/scitranslmed.aam7828
41. Fraser JK, Wulur I, Alfonso Z, Hedrick MH. Fat tissue: an underappreciated source of stem cells for biotechnology. *Trends Biotechnol.* (2006) 24:150–4. doi: 10.1016/j.tibtech.2006.01.010
42. Beltrami AP, Barlucchi L, Torella D, Baker M, Limana F, Chimenti S, et al. Adult cardiac stem cells are multipotent and support myocardial regeneration. *Cell.* (2003) 114:763–76. doi: 10.1016/S0092-8674(03)00687-1
43. Griffiths MJ, Bonnet D, Janes SM. Stem cells of the alveolar epithelium. *Lancet.* (2005) 366:249–60. doi: 10.1016/S0140-6736(05)66916-4
44. Marquez-Curtis LA, Janowska-Wieczorek, A. Enhancing the migration ability of mesenchymal stromal cells by targeting the SDF-1/CXCR4 axis. *Biomed Res Int.* (2013) 2013:561098. doi: 10.1155/2013/561098
45. Ryu CH, Park SA, Kim SM, Lim JY, Jeong CH, Jun JA, et al. Migration of human umbilical cord blood mesenchymal stem cells mediated by stromal cell-derived factor-1/CXCR4 axis via Akt, ERK, and p38 signal transduction pathways. *Biochem Biophys Res Commun.* (2010) 398:105–10. doi: 10.1016/j.bbrc.2010.06.043
46. Talkenberger K, Cavalcanti-Adam EA, Voss-Bohme A, Deutsch, A. Amoeboid-mesenchymal migration plasticity promotes invasion only in complex heterogeneous microenvironments. *Sci Rep.* (2017) 7:9237. doi: 10.1038/s41598-017-09300-3

11

Adipose-Derived Mesenchymal Stem Cells Applied in Fibrin Glue Stimulate Peripheral Nerve Regeneration

Ruslan Masgutov [1,2], Galina Masgutova [1], Adelya Mullakhmetova [1], Margarita Zhuravleva [1], Anna Shulman [3], Alexander Rogozhin [1,4], Valeriya Syromiatnikova [1], Dina Andreeva [1], Alina Zeinalova [1], Kamilla Idrisova [1], Cinzia Allegrucci [5], Andrey Kiyasov [1] and Albert Rizvanov [1]*

[1] OpenLab Gene and Cell Technologies, Institute of Fundamental Medicine and Biology, Kazan Federal University, Kazan, Russia, [2] Department of Orthopaedics, Republic Clinical Hospital, Kazan, Russia, [3] Scientific Department, Republic Clinical Hospital, Kazan, Russia, [4] Department of Neurology, Kazan State Medical Academy, Branch of Russian Medical Academy of Postgraduate Education, Kazan, Russia, [5] School of Veterinary Medicine and Science, University of Nottingham, Nottingham, United Kingdom

***Correspondence:**
Ruslan Masgutov
masgut@gmail.com

Mesenchymal stem cells (MSCs) hold a great promise for cell therapy. To date, they represent one of the best choices for the treatment of post-traumatic injuries of the peripheral nervous system. Although autologous can be easily transplanted in the injured area, clinical advances in this filed have been impaired by lack of preservation of graft cells into the injury area after transplantation. Indeed, cell viability is not retained after injection into the blood stream, and cells injected directly into the area of injury either are washed off or inhibit regeneration through scar formation and neuroma development. This study proposes a new way of MSCs delivery to the area of traumatic injury by using fibrin glue, which not only fixes cells at the site of application but also provides extracellular matrix support. Using a sciatic nerve injury model, MSC derived from adipose tissue embedded in fibrin glue were able to enter the nerve and migrate mainly retrogradely after transplantation. They also demonstrated a neuroprotective effect on DRG L5 sensory neurons and stimulated axon growth and myelination. Post-traumatic changes of the sensory neuron phenotype were also improved. Importantly, MSCs stimulated nerve angiogenesis and motor function recovery. Therefore, our data suggest that MSC therapy using fibrin glue is a safe and efficient method of cell transplantation in cases of sciatic nerve injury, and that this method of delivery of regeneration stimulants could be beneficial for the successful treatment of other central and peripheral nervous system conditions.

Keywords: rat, sciatic nerve injury, MSCs, ADSCs, fibrin glue

INTRODUCTION

Peripheral neuropathy caused by trauma and prevalent disorders is a major medical problem worldwide. Due to anatomy of the peripheral nervous system, structure damage of peripheral nerve trunks is a rather common condition. Although nerve trunk injury is not a life-threatening condition, it is accompanied by a persistent complex of nociceptive, sensory, and motor as well as trophic changes that can result in long-lasting disability, with full recovery of lost extremity functions being difficult (1, 2). Therefore, new translational research is focusing on combined methods of tissue engineering and replacement cell therapy along with the advancements in microsurgical techniques to enable efficient restoration of damaged peripheral nerves in clinical practice (3–8).

When facilitating nerve structure recovery to reduce nerve tissue injury, the main focus is to use various organic glues that join the ends of the damaged nerve (9) or are part of nerve conduit (10). The use of fibrin glue is particularly promising as it has been shown to be non-invasive and not causing inflammation or granuloma formation in post-traumatic nerve reconstruction (11–13). Fibrin glue can also induce faster re-innervation and function recovery due to reduced amount of scar tissue and directed axon growth (14, 15).

After nerve injury, the release of pro-inflammatory cytokines causes fibroblast activation and collagen hyperplasia leading to scars, fibrosis, and neuroma formation. These pro-inflammatory and nerve growth factors are necessary for axon regeneration by maintaining the balance between collagen synthesis and degradation without inducing fibroblast activation (16). Therefore, targeted delivery of potential neurorestoration stimulants is extremely important and cell therapy can be used to accelerate axon growth and help sensory neurons to survive (17, 18). To this end, mesenchymal stem cells (MSCs) have been most widely used (19) as an autologous system able to stimulate nerve tissue regeneration by neurotrophic growth factor secretion (20) and guided differentiation to the neuronal and glial lineages (21, 22).

MSCs can replace damaged tissues, secrete biologically active compounds, including homing and signaling molecules as well as myelin components necessary for structural and functional recovery of nerve fibers (23). Importantly, they do not cause immune rejection after allogeneic transplantation (24). MSCs can be easily derived from various tissues and expanded in culture. A useful source is the adipose tissue as adipose derived stem cells (ADSCs) have demonstrated MSCs properties (19, 25), together with ease of availability, derivation, and safety (26).

In order to attain efficient regeneration, transplanted cells need to survive and function in the area of injury. Cell survival after their injection is influenced by many factors, including speed of cell injection, cell density and diameter of the needle (27, 28). Regardless of the method used to assess the survival rate, cell type, and transplantation model, it is widely accepted that only a small portion of transplanted cells survive after injection (29, 30). Indeed, cells can be lost in the systemic circulation, but also local cell delivery by injection can cause additional damage and lead to neuroma development (31). Therefore, it is necessary to address a number of key issues to improve local cell retention in the area of injury before cell therapy can become a viable treatment option for peripheral nerve injury. In this study, we used fibrin glue for cell delivery. This formulation demonstrates several advantages, being minimally invasive, helping to retain cells in the area of injury, providing cells with extracellular matrix support and additional physical connection of the nerve ends (glue effect). We show that transplantation of MSCs embedded in fibrin glue promotes regeneration through a neuroprotective effect on sensory neurons and stimulation of axon growth. We also demonstrate for the first time a positive effect on recovery of motor function after injury, thus paving the way for the use of fibrin glue and MSC cell therapy in the treatment of peripheral nerve damage.

MATERIALS AND METHODS

MSC Derivation and Culture

MSC (ADSCs) were derived from the inguinal fold subcutaneous adipose tissue of healthy female Wistar rats ($n = 5$), as previously described (32). Subcutaneous fat tissue from different animals was obtained in equal volumes and immediately mixed for subsequent cultivation. Adipose tissue enzymatic disaggregation was achieved by incubation with type 1 collagenase (Biolot, Russia) at 37°C on a shaker for 1 h. Before and after the incubation the tissue was washed three times by centrifugation at 500 rpm for 5 min with DPBS solution (Dulbecco's Phosphate Buffered Saline, Paneco, Russia) containing 5% of penicillin and 5% of streptomycin. The cell pellet was resuspended and plated in culture in αMEM (Alpha Minimum Essential Medium, Invitrogen, USA) with 10% of FBS (Fetal Bovine Serum, Sigma, USA). Cultures were cultured and expanded in αMEM with 10% of FBS, 2 mM of L-glutamine (Sigma, USA) and penicillin and streptomycin (100 U/mL; 100 μg/mL) (Sigma, USA) using a MCO-15AC incubator (Sanyo, Japan) at 5% CO_2 and 37°C.

Rat ADSCs immunophenotype was evaluated using flow cytometry. Cells were trypsinized using 0.25% trypsin (Sigma, USA) and incubated in phosphate-buffered saline for 45 min with conjugated antibodies anti-CD29 (BD, USA, 556049), anti-CD44 (BioLegend, USA, 103028), anti-CD90 (SantaCruz, USA, SC-53456), anti-CD34 (SantaCruz, USA, SC-51540), and anti-CD45 (SantaCruz, USA, SC-70686). Analysis was performed using a Guava Easy Cyte 8HT flow cytometer (Millipore, United States).

After confirmation of the MSCs immunophenotype, ADSCs were transduced using a recombinant lentivirus encoding eGFP (green fluorescent protein). Transdusction efficiency was assessed 48 h later on the basis of the number of eGFP positive cells using an Axio Observer Z1 fluorescence microscope (Carl Zeiss, Germany). eGFP-expressing cells were sorted using a FACS Aria III cell sorter (BD Biosciences, USA) and expanded in cultured until transplantation. A population of cells expressing ≥90% eGFP (ADSCs-eGFP) was used for cell transplantation.

Experiments With Animals

Experiments were conducted using 62 white Wistar rats aged 4–6 months and weighting 200–300 g (GMBH "Nursery RAMTN," Moscow, Russia). For study of regeneration processes on 30 and 60th days after injury experiment was carried out on 35 male rats. Twenty two male rats used for estimation of MSCs migration on 7 and 14th days after trauma. For obtaining of MSCs we used 5 female rats additionally. All animals were acclimatized for 2 weeks before starting the experiment. Animals were kept under standard vivarium conditions with the day/night mode 12/12, with free access to feed and water. Animals were kept and used for experimental procedures in accordance with the rules accepted by Kazan Federal University and approved by the Local Ethics Committee (Permit Number 5 of May 27, 2014). Animals were used in accordance to international bioethical standards defined by the International guiding principles for biomedical research involving animals (33), the EU directive 2010/63/EC and the 3Rs principles.

The autologous nerve graft (AG) was used as an experimental nerve model. The rats were deeply anesthetized with intraperitoneal injection of chloral hydrate solution (AppliChem, Germany) at a dose of 400 mg/kg in the water for injections. Surgical approach was made to the left sciatic nerve, then a 5 mm long diastasis was formed by transecting the nerve with two parallel incisions. The ends of the nerve were joined with an autologous nerve graft, strictly maintaining the nerve fiber growth cone, and sutured without tension with 4 Prolene 10.0 (Ethicon, USA) interrupted epineural sutures. The nerve was then covered with 200 μL of Tissucol-Kit fibrin glue (Baxter AG, Austria).

Stimulation of post-traumatic regeneration was achieved through intergender allotransplantation of ADSCs-eGFP. To this end, fibrin glue (FG) containing 1 million ADSCs-eGFP was applied to the nerve after suturing. After nerve manipulations the postoperative wound was closed in layers. Animals in 4 groups were compared: (1) experimental group ($n = 5$ on 7th, $n = 5$ on 14th, $n = 5$ on 30th, and $n = 5$ on 60th days after injury)—AG+FG+ADSCs—post-traumatic recovery by regeneration stimulation using ADSCs-eGFP; (2) active control group ($n = 3$ on 7th, $n = 3$ on 14th, $n = 5$ on 30th, and $n = 5$ on 60th days after injury)—AG+FG—post-traumatic recovery by nerve coverage with fibrin glue; (3) control group ($n = 3$ on 7th, $n = 3$ on 14th, $n = 5$ on 30th, and $n = 5$ on 60th days after injury)—AG—post-traumatic recovery under natural conditions with a autologous nerve graft. The animals in experimental groups were sacrified on 7th ($n = 11$), 14th ($n = 11$), 30th ($n = 5$), and on 60th ($n = 5$) days after injury; (4) intact group ($n = 5$)—intact animals with no damage to sciatic nerve.

Antibiotics and analgesics were administered as follows: 1 mL of gentamicin (25 mg/kg, Omela, Russian Federation) was injected intramuscularly for 7 consecutive days; buprenorphine (0.5 mg/kg) was injected subcutaneously for about 7 days after surgery to minimize pain.

Assessment of Motor Function Recovery

To assess the grade of motor function recovery of the operated extremity, animals of all groups underwent a functional motor test to define the sciatic functional index (SFI) (34). The functional test was carried out once a week on day 7, 14, 21, 28, 35, 42, 49, and 56 after the injury. To this end, a run track with side walls was used (width–12 cm, length–45 cm, height–15 cm) allowing a rat to move only in one direction. Animals hind feet were covered in ink and footprints made on the run track covered with white paper were measured to calculate the SFI using the formula:

$$SFI = [((eTOF - nTOF)/nTOF) + ((nPL - ePL)/ePL) + ((eTS - nTS)/nTS) + ((eIT - nIT)/nIT)] \times 55, \text{ where}$$

eTOF is the distance from the experimental foot toes to the intact foot toes of the subsequent footprint;

nTOF is the distance from the intact foot toes to the experimental foot toes of the subsequent footprint;

PL is the length of the footprint from the heel to the third toe of the same foot;

TS is the distance between the first and the fifth toes of the same foot;

IT is the distance between the second and the fourth toes of the same foot.

All the measurements were made both for the healthy foot footprint (*n*—normal) and the operated foot footprint (e—experimental; **Supplementary Figure 1**).

Nerve Conduction Studies

Compound muscle action potentials (CMAPs) was used to assess sciatic nerve conduction of the rats both before and 30 and 60 days after the surgery. Electric muscle responses were registered using a MG-42 electromyograph (Hungary) combined with data computer analysis. After sedation, stimulant monopolar needle electrodes were inserted in the hip joint area and into the area of sciatic nerve projection. Stimulation was carried out with square-wave pulses lasting 1–2 ms. Stimulus intensity varied from 0.2 to 2 V. CMAP were registered by the monopolar needle electrodes inserted into the medial gastrocnemius: cathode was inserted into the center of the myogaster and the anode—into the tendon. Latency of the CMAP, response threshold, duration of the CMAP, and maximal amplitude were analyzed.

Vascularization Assessment

Recovery of blood flow in the distal part of the sciatic nerve was assessed by the visualization of microcirculation perfusion 14 and 30 days after the surgery using a EasyLDI laser doppler (Aimago). To this end, small operating room repetitive approach to the sciatic nerve was carried out in the animals under intraperitoneal narcosis. The device's laser beam was pointed at the distal part of the nerve and changes in microcirculation parameters over time were analyzed using the function of perfusion unit assessment in real time. In the evaluation process, the laser beam sequentially scanned the tissue of the distal stump of the sciatic nerve, while circulating red blood cells generated Doppler components in scattered light, were picked up by a photodiode, and converted into an electrical signal proportional to tissue perfusion at each measurement point. The parameters were measured in absolute perfusion units (apu) according to manufacturer's instructions.

Material Sampling and Morphological Assessment

Animals were sacrificed on day 7, 14, 30, and 60 after the surgery. Early periods were used to investigate the survival rate and migration ability of the transplanted cells. Samples collected on day 7 and 14 after the surgery were snap-frozen in liquid nitrogen with Neg50 Frozen Section Medium (Thermo Fisher Scientific, USA), and transferred and stored at −80°C. Sciatic nerve samples were cut into 6 mm thick longitudinal sections using a Cryo-Star HM560 freezing microtome (Thermo Fisher Scientific, USA). eGFP fluorescence was visualized using a LSM 780 laser scanning confocal microscope (Carl Zeiss, Germany). DAPI nuclear stain was used to assess the survival rate of transplanted cells. Hematoxylin and eosin staining used for estimation of morphological changes.

Late time points (day 30 and 60 after the injury the sciatic nerve) were used to investigate post-traumatic recovery by exposure of the L5 spinal ganglion on the operative side. Distal parts of sciatic nerves were used to assess the number of regenerating myelin fibers, whereas L5 spinal ganglia were used to assess the number of surviving neurons.

The exposed distal segments of the sciatic nerve were fixed in 2% glutaric dialdehyde, postfixed in 1% OsO_4, embedded in epoxy resin (9 ml of Epon 812, 6 ml of DDSA, 4 ml of MNA, and 0.2 ml of DHP-30) and polymerized at 60°C. The samples were sliced in 2 μm thick semifine cross sections using a LKB 2088 ultramicrotome (Leitz, Germany) and sections stained with methylene blue. The number of myelin fibers was calculated by random sample analysis of 4 nerve segments (35) using a ICC50E light microscope (Leica, Germany) with 63×100 magnification and oil objective immersion using the ImageJ 1.48v program.

Totally exposed left L5 spinal ganglia after laminectomy were fixed in 4% buffered formaline and embedded in Histomix paraffin (Biovitrum, Russia) on a Tissue-Teck® TECTM 5 paraffin embedding station (Sakura Seiki, Japan). Paraffin blocks were cut into longitudinal 7 μm thick serial sections on a HM340E rotary microtome (Thermo Fisher Scientific, USA). Deparaffinized sections were stained to assess the post-traumatic reaction of spinal ganglion neurons. Every 5th serial section was stained with 5% azure and eosin according to Romanovsky (Minimed, Russia) to visualize neuronal nuclei for cell survival analysis. To assess the response of small diameter neurons, every 5th serial section of the spinal ganglion was stained with Isolectin-B4 fluorescence cytoplasmic marker [lectin from Bandeiraea simplicifolia (BSi-B4)] (IB-4) (Sigma, L2895, 1:200) and counterstained with the nuclear dye DAPI (Sigma) before visualization.

Statistical Analysis

For all parameters of the given experimental groups, mean, and standard deviation were calculated and data presented as mean ± standard error of the mean (SEM). Student's *t*-test, one-way analysis of variance (ANOVA) with Tukey's test or two-way analysis of variance (ANOVA) were used for multiple comparisons between all experimental and control groups. The differences were considered significant if the probability threshold was <0.05% ($p < 0.05$). Data were analyzed using the Origin 7.0 SR0 Software (OriginLab, Northampton, MA, USA).

RESULTS

ADSCs isolated from rat adipose tissue had a fibroblast-like morphology and high proliferative activity. They expressed CD29, CD34, CD90 but not CD34, and CD45, confirming their MSC phenotype (**Figure 1**). Established ADSC cultures were then genetically modified to express e-GFP in order to generate a fluorescent line for cell tracing. Fluorescent green cells were detected 48 h after transduction and a population of ADSCs expressing ≥90% eGFP was obtained after sorting (**Figure 2**) Transplantation of fluorescent sorted cells allowed the visualization of MSC integration in the injured area.

Examination of the sciatic nerve longitudinal sections after transplantation demonstrated ADSCs-eGFP on the surface of epineurium as part of the FG and in the sciatic nerve at day 7 and 14 of the autologous nerve repair. Although it was not possible to count the absolute number of cells under the epineurium, the e-GFP fluorescence enabled the assessment of the migration ability of ADSCs. Cells applied in fibrin glue on the surface sciatic nerve actively migrated under the epineurium through nerve sutures and moved mainly retrogradely (**Figures 3A,B**). After 7 days, cells migrated on average 1–2 mm and cells covered 5–7 mm distance 14 days after the application. However, ADSCs were not able to move through the intact epineurium connective tissue (**Figure 3C**). No fluorescence was found in the AG+FG group where the autologous nerve graft was covered with FG without ADSCs (**Figure 3D**).

ADSCs applied together with FG on the area of the sciatic autologous nerve graft stimulated the recovery of extremity motor function. At the late stages, the parameters of motor activity in the AG+FG+ADSCs group increased on average by 26% ($p < 0.05$) as compared to the AG group and by 28% ($p < 0.05$) as compared to AG+FG group (**Figure 4**).

Assessment of muscle bioelectric activity did not reveal significant differences in CMAP threshold and CMAP latency among the compared groups on the 30th and 60th days after trauma. Initial CMAP amplitude didn't differ between the AG+FG+ADSCs group, the AG+FG-group, and control group. CMAP amplitude was only 50% from initial on the 30th day after trauma, showing axonal loss in sciatic nerve, persisting up to 60th day in both groups. The duration of CMAP in the groups with sciatic nerve injury decreased substantially 30 days after the injury as compared to the measurements in intact animals but at the same time the CMAP duration was significantly higher in the AG+FG+ADSCs group than in the AG+FG-group. Unlike the CMAP amplitude, the CMAP duration increased and reached the control group level in the 60 days after trauma. No differences in CMAP duration revealed between the AG+FG+ADSCs group and the AG+FG-group (**Figure 5**).

When assessing the restoration of blood supply by laser doppler, the vascularization parameters in the AG+FG+ADSCs group were already restored at 14 days after the injury to levels similar to those of intact animals, whereas in the AG+FG group the blood supply of the distal nerve segment was still decreased. However, no difference among the groups was found after 30 days (**Figure 6**).

Morphological examination of the sciatic nerve longitudinal sections revealed several changes associated with the injury when compared to the intact nerve (**Figure 7A**). Myelin fibers were found degenerated, with myelin breakdown in the area of the autologous nerve grafting and areas close to graft oedema. Schwann cell nuclei acquired a damaged round shape and were displaced to the periphery of the cylinder (**Figure 7B**). After transplantation, analysis of the anastomosis area revealed migration of macrophages under the epineurium and areas of hemorrhage with necrotic changes of small vessels (**Figure 7C**). Importantly, the epineurium preserved its integrity and residues

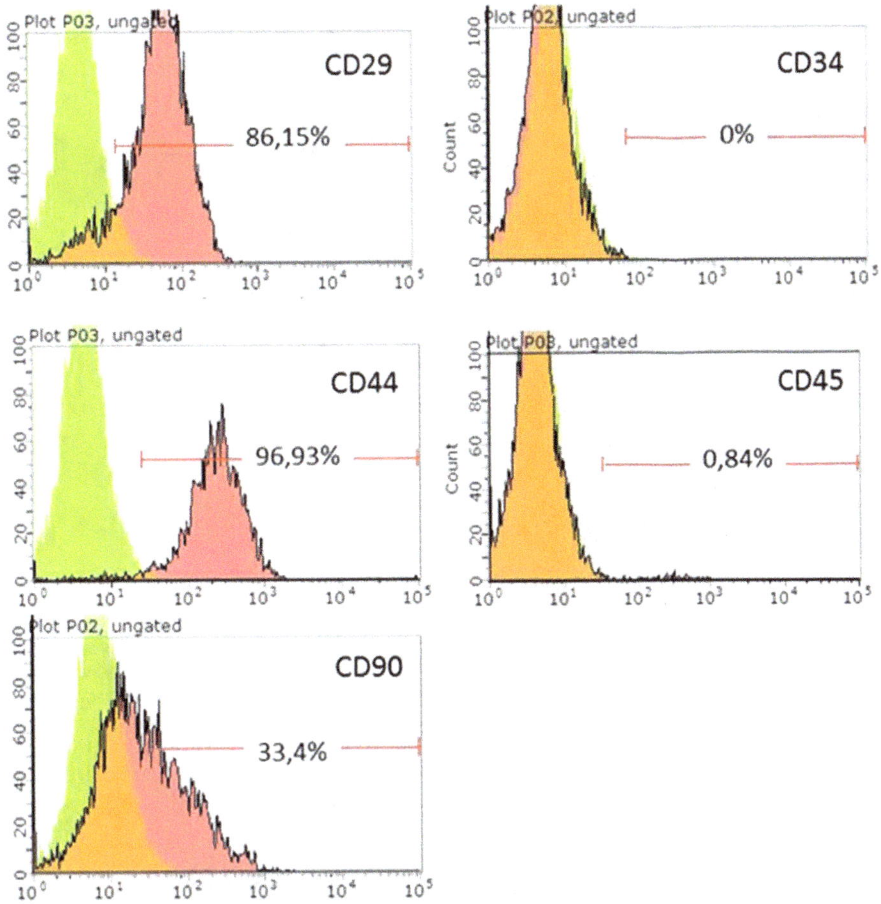

FIGURE 1 | Flow cytometry analysis of rat ADSCs. Green, isotype control; Red, stained cells.

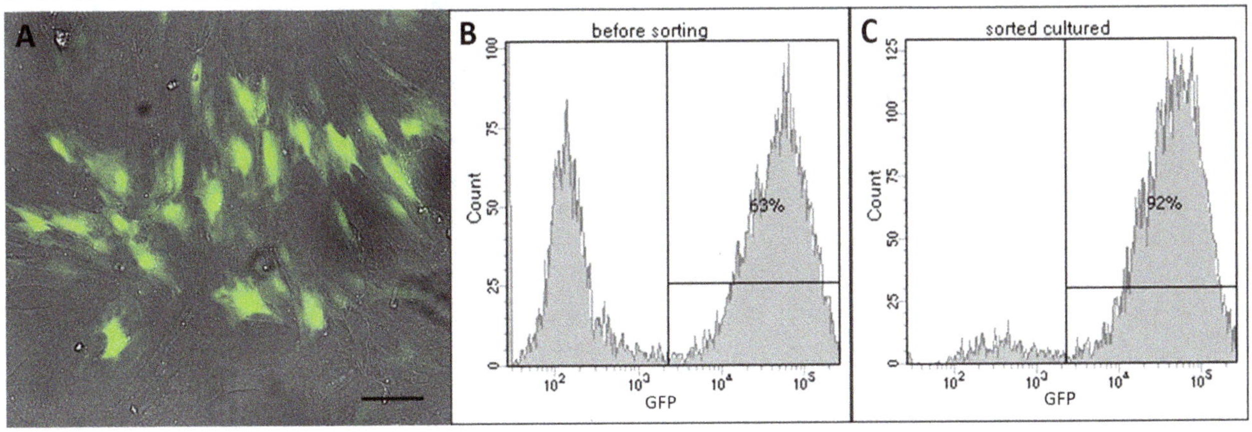

FIGURE 2 | LV-eGFP transduced rat ADSCs. **(A)** e-GFP expression 48 h after transduction. Scale bar, 100 μm. **(B)** Flow cytometry analysis of e-GFP 48 h after transduction. **(C)** Flow cytometry analysis of e-GFP-sorted cells used for transplantation.

of fibrin glue with large-sized cells and clearly visible nuclei on its surface were found in the AG+FG+ADSCs group (**Figure 7C**). Degenerative changes spread both to central and distal directions. Here, destructive changes were observed not only in myelin fibers, but also in elastic fiber that acquired a denser and more convoluted structure (**Figure 7D**). The transplantation of

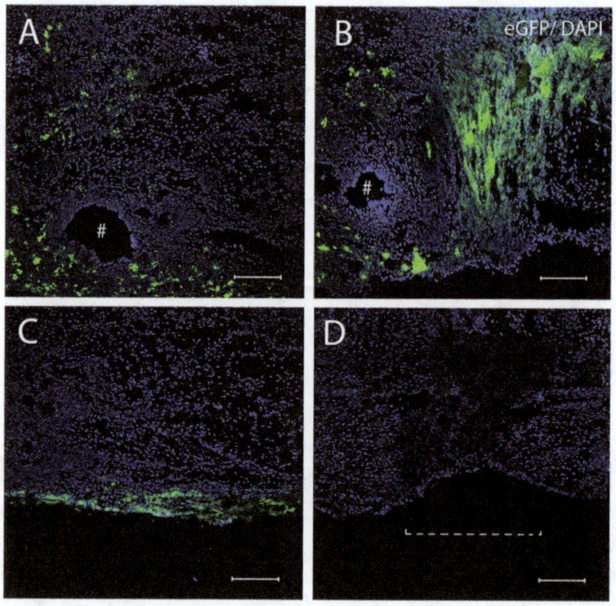

FIGURE 3 | Confocal miscopy images of longitudinal sections of the rat sciatic nerve after 7 **(A,C)** and 14 **(B)** days in the AG+FG+ADSCs group and 14 days in the AG group **(D)**. **(A,B)** The area of anastomosis of sciatic nerve after 7 days **(A)** and 14 days **(B)** the penetration of ADSCs-eGFP through the suturing area is shown. **(C)** The inability of ADSCs-eGFP invasion through the intact epineurium is shown. **(D)** Absence of fluorescence in the AG-group is shown. #, the needle hole; a dotted line, nerve anastomosis. The green fluorescence indicates transplanted cells expressing e-GFP; blue fluorescence indicates cell nuclei stained with DAPI. Scale bars: 100 μm.

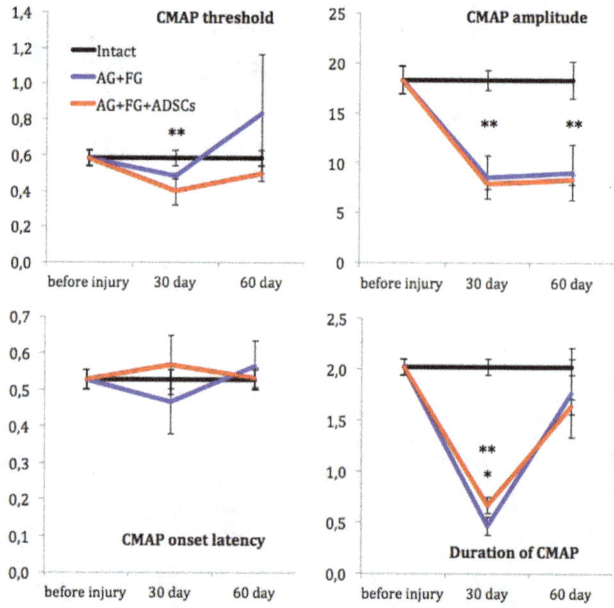

FIGURE 5 | Electroneuromyography measurements. X-axis, testing time: before injury; 30 day and 60 day after injury. Y-axis, CMAP threshold (in volts); CMAP amplitude (in millivolts); CMAP onset latency (in milliseconds) and Duration of CMAP (in milliseconds). Differences were statistically significant between the groups (*), between AG+FG+ADSCs and AG-groups; (**), between experimental and intact groups (*, **-$P < 0.05$, one-way ANOVA, Tukey's test).

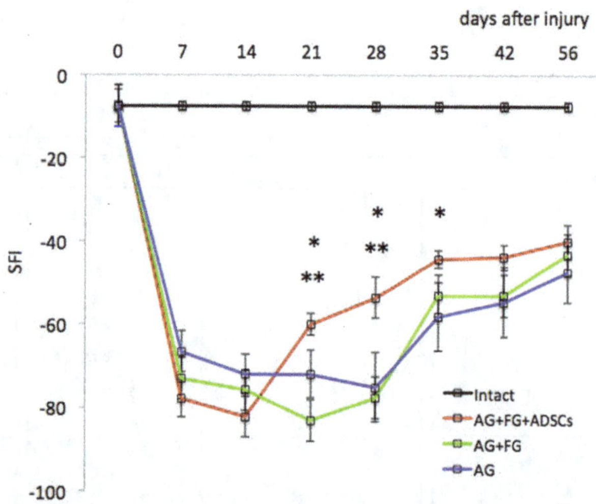

FIGURE 4 | Evaluation of motor function restoration of the hind extremity of rats (Sciatic Functional Index—test). Error bars represent standard error mean. Differences were statistically significant between the groups (*), between groups AG+FG+ADSCs and AG; (**), between groups AG+FG+ADSCs and AG+FG (*, **-$P < 0.05$, one-way ANOVA, Tukey's test).

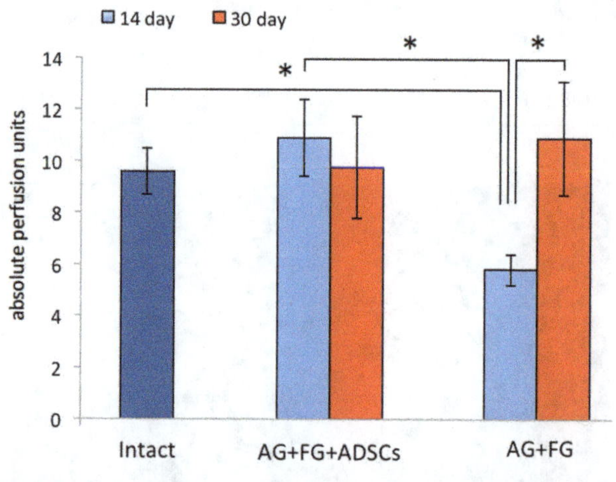

FIGURE 6 | Indicators of vascularization of the distal stump of the sciatic nerve at 14 and 30 days after AG. Differences were statistically significant between the groups *, between experimental groups (*$P < 0.05$, one-way ANOVA, Tukey's test).

ADSCs with fibrin glue significantly improved the described degenerative changes.

Myelin fibers in the distal nerve segment 30 days after the autologous nerve grafting were represented by round masses with the much smaller diameter than that of intact animals (**Figures 8A,D,G**). However, the number of myelin fibers in the nerve distal segments 30 days after the surgery in the AG+FG+ADSCs group was increased by around 18 % ($p < 0.05$) compared to the AG+FG group. The number of myelin fibers in the both groups was increased by a further 20% 60 days after the surgery compared to day 30

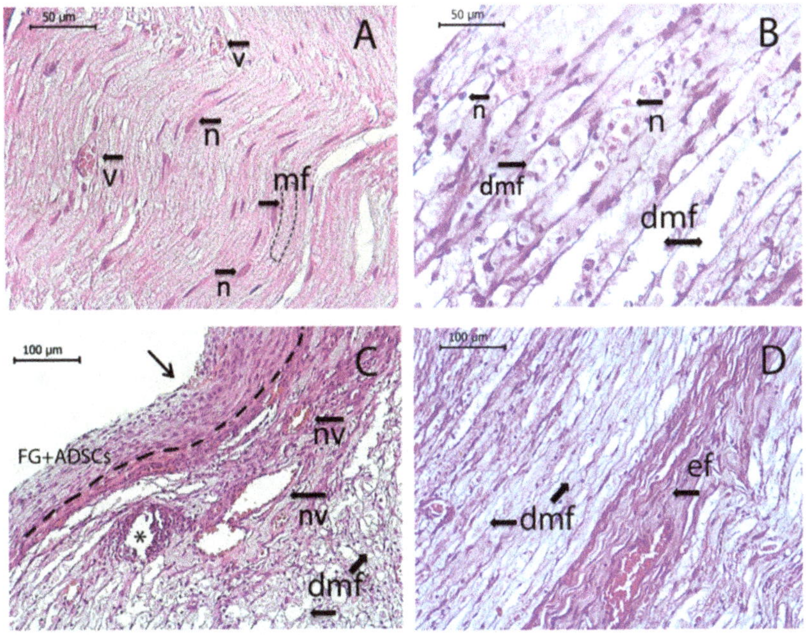

FIGURE 7 | Longitudinal sections of rat sciatic nerve. Intact nerve (*n*, nuclei of Schwann's cells; v, blood vessel; mf, normal myelin fibers) is shown in **(A)**, sciatic nerve 7 days after the operation is shown in **(B–D)**. **(B)** Autotlogus nerve graft area showing degenerated myelin fibers with oedema (dmv) with degenerated nuclei of Schwann's cells (*n*). **(C)** Location of nerve anastomosis—central part indicated by arrow, *the needle hole, a dotted line separates epineurium from the top-located fibrin glue with ADSCs, macrophage aggregation, necrotic altered blood vessels (nv) with a hemorrhage area, and degenerated myelin fibers (dmf) under the epineurium. **(D)** Distal segment of the nerve with degenerative changes of myelin fibers (dmf) and elastic fibers (ef). Hematoxylin and eosin staining. Scale bar **(A,B)** 50 μm, **(C,D)** 100 μm.

($p < 0.05$) (**Figure 8J**). The number of neurons of the L5 spinal ganglion decreases in all groups after the injury as compared to intact animals. The number of neurons decreased even further by 60 days after the surgery. Although chromatolysis was visible in the neurons 30 days after the sciatic nerve autografting (**Figures 8B,E,H**), the number of neurons of the L5 spinal ganglion was significantly higher in the AG+FG+ADSCs groups compared to AG+FG and AG. The same increase was found 60 days after the surgery, demonstrating the regenerative capacity of the grafted ADSCs (**Figure 8K**).

Assessment of the IB4+ neurons after transplantation demonstrated that IB4 expression was found not only in small (up to 30 μm in diameter) but also in medium-sized (up to 50 μm) neurons (**Figures 8C,F,I**). Although, an expected decrease in the number of IB4+ positive cells was found in injured animals, a significant increase of IB4+ neuron was found in the AG+FG+ADSCs group both at day 30 and day 60 after transplantation compared to the AG+FG control group (**Figure 8L**).

Therefore, altogether the changes observed after autologous transplantation demonstrated the enhanced regeneration capacity of ADSCs in fibrin glue that resulted in motor function recovery after injury.

DISCUSSION

Autologous nerve grafting is the gold standard procedure to ensure nerve continuity restoration in cases of diastasis after experimental injury induced to model neurodegenerative and regenerative post-traumatic processes. Indeed, the autologous conduit is biocompatibe and contains autologous Schwann cells and extracellular matrix structural proteins contributing to neuron survival and stimulating the regeneration of damaged axons (36, 37). Using functional tests, electroneurophysiology assessment, vascularization assessment, and morphometric study of the sciatic nerve and L5 spinal ganglion, in this study we were able to confirm that surgery aimed at approximating the ends of an injured nerve is not enough to achieve complete regeneration. Therefore, regeneration stimulation by means of application of autologous ADSCs to the area of nerve injury allowed a significant enhancement of regeneration through stimulation of angiogenesis and neuroprotection.

The rate and degree of nerve fiber de- and regeneration after damage is a complex process associated with inflammation, adhesion, regulation of neurotrophic factors, neurotransmitter synthesis and release, formation of the nerve growth cone, neuron survival, growth of axons and myelination, and many other factors (38, 39). At the same time, immune reactions taking place in the injury area cause the expression of pro-inflammatory transmitters around the damaged tissue and contribute to the development of scar tissue, nerve fibrosis (40), neuromas, and hyperplasia which impact on nerve damage and on conduction restoration after injury (41).

During nerve fiber regeneration the central segment of the neuronal axon interacts with glial components and grows according to the concentration gradient of chemical factors

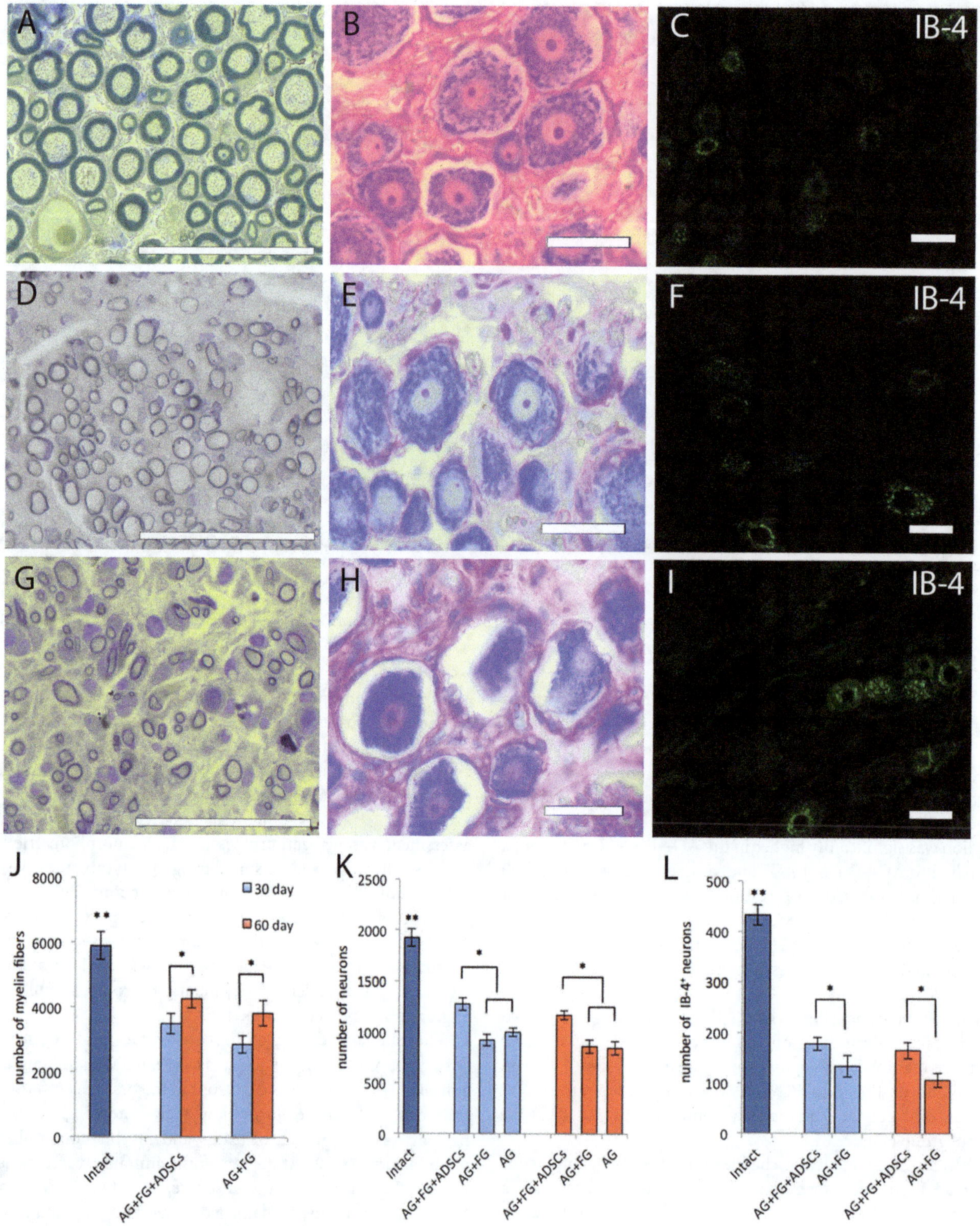

FIGURE 8 | Morphology and number of fibers and neurons in intact and transplanted animals. Top line **(A–C)**, intact animals; second line **(D–F)**, AG + FG + ADSCs group 30 days after the injury; third line **(G–I)**, AG + FG group 30 days after the injury. Scale bar scale 50 μm. **(A,D,G)** Light microscopy images of myelinated fibers fixed in osmium and stained by methylene blue. **(B,E,H)** Light microscopy images of L5 DRG neurons stained by azure-eosin. **(C,F,I)** Confocal microscopy images of IB-4+ neurons stained with the fluorescent cytoplasmic marker Isolectin-B4 [lectin from Bandeiraea simplicifolia (BSi-B4)]. **(J)** Number of myelin fibers in the distal stump of the sciatic nerve; **(K)** Number of L5 DRG neurons; **(L)** Number of IB-4+ neurons L5 DRG. Differences were statistically significant between the groups (*) between experimental groups, (**) between intact and experimental groups (*, **-$P < 0.05$, one-way ANOVA, Tukey's test).

against the background of reactive changes in the sensory neurons and degenerative processes in the peripheral segment of the nerve. The success of reinnervation depends on the ability of axons to reach their target organ. It is known that the sciatic nerve is comprised of processes of motor and sensory neurons with trophic function. If there is no contact with the axon, Schwann cells activate the synthesis of neurotrophic factors such as nerve growth factor (NGF), fibroblast growth factor (bFGF), and the ciliary neurotrophic factor (CNTF) (42, 43). These factors, together with ATP and neuregulin released from the proximal nerve ends (44, 45) promote the formation of new Schwann cells (46) and, together with acetylcholine, stimulate their further proliferation (47).

Functional stimulation of the innervated area stops when the sciatic nerve is transected, and the secretion or action of co-transmitters and trophogens is increased (48). From the survival of neurons directly depends on the speed and quality of recovery of the motor and sensitive function of the injured extremity. We demonstrate that on days 7 and 14, the recovery of motor function in the AG+FG+ADSCs and AG+FG groups is lower than in the AG group, however, it is very early to talk about any therapeutic effect on these periods. This fact at this time is due to the mechanical effect of FG on the sciatic nerve. Normally, the sciatic nerve passes freely between the muscles, and the placement of fibrin glue on its surface leads to its adhesion to the muscles, creating the effect of scarring. The result of the influence of the ADSCs observed in 14 days after the injury, when the cells that have penetrated into the nerve thickness begin to show their therapeutic effect. Neuron regenerative potential is preserved 2 to 15 days after nerve transection and up to several months after nerve crush. This interval is believed to be the most favorable for regeneration stimulation and support of neuron survival (49, 50).

Absolute difference in CMAP duration between AG+FG+ADSCs and AG+FG groups is not big despite statistical significance that rise the question of biological meaning of this observation. But they correspond with higher results of motor activity in AG+FG+ADSCs group from 21 to 35 days after trauma. Therefore, we suppose that this difference reflects higher amount of motor axons of sciatic nerve in AG+FG+ADSCs group.

Therefore, the assessment of the number of spinal ganglion neurons after the injury is essential for understanding the mechanisms of spontaneous sciatic nerve regeneration, estimation of quality of surgical procedures aimed at nerve structure restoration, and methods of stimulation of post-traumatic regeneration. L4-L5 spinal ganglia contain 98–99% of nerve cell bodies whose axons form the rat sciatic nerve (51). It is known that sciatic nerve axotomy results in the death of spinal ganglion neurons (52–55) and we demonstrated low neuron survival rate in the L5 spinal ganglion up to day 60 after the injury. When assessing the total neuron count, it was found that small diameter pain neurons are more vulnerable.

We showed that after sciatic nerve injury IB4 expression is found not only in small pain neurons but also in medium-sized neurons. IB4 is believed to be a marker of small nociceptive neurons that have unmyelinated processes in the sciatic nerve innervating the epidermis and showing lower survival both *in vitro* and *in vivo* (56). Therefore, this type of neurons is more prone to cell death as compared to medium and large diameter neurons, which is consistent with our results. Sensory IB-4+ neurons are less flexible and grow worse after axotomy. After sciatic nerve injury, IB4 neuron terminal axons can show retraction (57) and have substantially lower regeneration ability (58). It is therefore likely that IB4 expression in some medium diameter neurons after nerve injury could be due to axon growth inhibition.

Although molecular mechanisms underlying the death of spinal ganglion neurons in response to nerve axotomy are not fully understood, there is enough evidence of the role played by injury severity on neuron apoptosis and potential ability of axons to grow to respective distal receptors (59). Exogenous trophic factors can counteract the post-traumatic death of sensory neurons. Transplanted ADSCs may produce neurotrophic factors as cells transplanted into the nerve show retrograde neuroprotective effect on respective sensory neurons (52, 53). The beneficial effect of cell therapy can be seen in as little as the first week after the surgery (60). Therefore, it can be speculated that the presence in an injured nerve of exogenous progenitor cells with high paracrine activity in the first several days after the injury promotes axon growth and survival. Indeed, we did observe a significant increase in the number of myelin fibers L5 and IB4+ neurons after ADSCs transplantation, suggesting a trophic action of MSCs after injury.

In addition, morphological assessment of nerve structure at early stages after the injury revealed inflammation both in groups with cell stimulation and in control groups. The severity of inflammation was, however, much lower in the group treated with ADSCs.

In order to assess the extent of regeneration and its clinical relevance, we performed functional tests as the examination of morphological changes in the injured sciatic nerve often showed discrepancy between histopathological changes and functional observations (61).

We showed that nerve reconstruction by means of autologous nerve repair considerably reduces the ability of the operated extremity to move. It may be that nerve injuries go hand in hand with neuropathic pain (62) but we demonstrated that ADSCs can induce a significant increase in the sciatic functional index and associated motor function recovery.

Several studies demonstrated that transplanted MSCs can directly influence angiogenesis by influencing all the stages of vessel formation and maturation (63). Indeed, MSCs secrete cytokines and growth factors stimulating survival, growth, and differentiation of vascular endothelium (64), thus preventing the formation of neurotrophic ulcers caused by vessel destruction and ischemia. It is likely that the restoration of blood flow in the operated sciatic nerve could be associated with the activation of proangiogenic factor expression by transplanted ADSCs, such as fibroblast growth factor and endothelial growth factor (65, 66).

With the help of nerve conduction studies, we were able to assess the functioning of sciatic nerve motor fibers. Since fast-conducting fibers, such as axons of motor neurons (type Aα) and afferent muscle fibers (type Aβ), are thicker than slow-conducting, the process of their recovery takes more time. Unmyelinated neuron axons (type C), such as axons of the spinal ganglion sensory neurons, involved in the transmission of pain (67), temperature, and postganglionic sympathethic transmission are the slowest ones (68). Decreased amplitude and shortened CMAP duration by day 30 reflected the loss of sciatic nerve axons, most probably due to wallerian degeneration. In the period from day 30 to day 60 no increase in the CMAP amplitude was registered but the CMAP duration increased, thus demonstrating the increasing number of functionally active axons in the experimental and control groups. These changes also evidenced incomplete remyelination of regenerated axons and subsequent CMAP dispersion, which at early stages of re-innervation caused increased CMAP duration with unchanged amplitude. Significantly longer CMAP duration in the experimental group as compared to the control group by day 30 could be a sign of earlier regeneration under the influence of ADSCs, which is in line with the data we obtained for the number of myelin fibers in the distal nerve segment. The number of myelin fibers in the distal sciatic nerve segment increased in all studied groups but this parameter was higher in the group with ADSCs.

Altogether, our morphological and functional data demonstrate the beneficial effect of cell therapy with ADSCs. ADSCs transplanted with FG which facilitate the joining of the segments of the transected sciatic nerve, entered under the epineurium through the junctions of nerve segments, directionally migrated predominantly retrogradely, contributed to the sensory neurons survival through stimulating the growth of their axons, and promoted conduit vascularization by restoring the motor function of the injured extremity.

We cannot rule out a possible transdifferentiation of ADSCs into Schwann cells which could be responsible for the augmented regeneration observed; this possibility, however, warrants further investigation. Taking into account that Schwann cells play a key role in peripheral nerve survival and functioning (68), we believe that transplantation of ADSCs provides the optimal conditions for regeneration. Importantly, cell transplantation should be carried out in the acute traumatic period with minimally invasive delivery by application of potential regeneration stimulants as part of fibrin glue in order to achieve effective repair of peripheral nerve damage.

CONCLUSION

Our data suggest that MSCs transplanted in fibrin glue to the place of nerve injury have a neuroprotective effect on DRG L5 sensory neurons, stimulate axon growth, and myelination. We propose a new method of MSCs delivery to the area of traumatic injury by using fibrin glue. This method of delivery of regeneration stimulants could be beneficial for the successful treatment peripheral nerves injuries and easily translate to the clinical practice.

ETHICS STATEMENT

This study was carried out in accordance with the recommendations by the Local Ethics Committee of Kazan Federal University. The protocol was approved by the Local Ethics Committee of Kazan Federal University (Permit Number 5 of May 27, 2014).

AUTHOR CONTRIBUTIONS

RM: experimental design, collection and assembly of data, data analysis and interpretation, manuscript writing, and final approval of manuscript; GM: assembly of data, histology, data analysis and interpretation, manuscript writing; AM: work with animals, histology, functional tests, sample preparation; MZ: working with cell culture, data collection, manuscript writing, manuscript editing; AS: nerve conduction studies; ARo: nerve conduction studies, data analysis, and interpretation; VS: work with animals, acquisition of data, data processing; DA: vascularization assessment, data analysis, manuscript editing; AZ: assisted with data collection, manuscript editing; KI: work with animals, data collection; CA: conception and design, manuscript editing, and final approval of manuscript; AK: manuscript editing, and final approval of manuscript; ARi: financial support, manuscript editing, and final approval of manuscript. All authors provided critical feedback and helped shape the research, analysis and manuscript.

FUNDING

This study was supported by the Russian Government Program of Competitive Growth of Kazan Federal University. ARi was supported by state assignment 20.5175.2017/6.7 of the Ministry of Education and Science of Russian Federation. ARi was supported by Grant of the President of the Russian Federation for state support of the leading scientific schools of the Russian Federation ??-3076.2018.4. Some of the experiments were conducted using the equipment of Interdisciplinary center for collective use, Interdisciplinary Center for Analytical Microscopy and Pharmaceutical Research and Education Center, Kazan (Volga Region) Federal University, Kazan, Russia.

SUPPLEMENTARY MATERIAL

Supplementary Figure 1 | Schematic of the rat walking the track analysis; TOF (e), distance from the experimental foot toes to the intact foot toes; TOF (*n*), distance from the intact foot toes to the experimental foot toes; PL, length of the footprint from the heel to the third toe of the same foot; TS, distance between the first and the fifth toes of the same foot; IT, distance between the second and the fourth toes of the same foot, e, experimental; n, normal.

REFERENCES

1. Asplund M, Nilsson M, Jacobsson A, von Holst H. Incidence of traumatic peripheral nerve injuries and amputations in Sweden between 1998 and 2006. *Neuroepidemiology.* (2009) 32:217–28. doi: 10.1159/000197900
2. Rosberg HE, Carlsson KS, Cederlund RI, Ramel E, Dahlin LB. Costs and outcome for serious hand and arm injuries during the first year after trauma - a prospective study. *BMC Public Health.* (2013) 13:501. doi: 10.1186/1471-2458-13-501
3. di Summa PG, Kalbermatten DF, Pralong E, Raffoul W, Kingham PJ, Terenghi G, et al. Long-term *in vivo* regeneration of peripheral nerves through bioengineered nerve grafts. *Neuroscience.* (2011) 181:278–91. doi: 10.1016/j.neuroscience.2011.02.052
4. Tos P, Ronchi G, Geuna S, Battiston B. Future perspectives in nerve repair and regeneration. *Int Rev Neurobiol.* (2013) 109:165–92. doi: 10.1016/B978-0-12-420045-6.00008-0
5. Griffin JW, Hogan MV, Chhabra AB, Deal DN. Peripheral nerve repair and reconstruction. *J Bone Joint Surg Am.* (2013) 95:2144–51. doi: 10.2106/JBJS.L.00704
6. Carriel V, Alaminos M, Garzon I, Campos A, Cornelissen M. Tissue engineering of the peripheral nervous system. *Expert Rev Neurother.* (2014) 14:301–18. doi: 10.1586/14737175.2014.887444
7. Griffin M, Hindocha S, Malahias M, Saleh M, Juma A. Flap decisions and options in soft tissue coverage of the upper limb. *Open Orthop J.* (2014) 8:409–14. doi: 10.2174/1874325001408010409
8. Gu X. Progress and perspectives of neural tissue engineering. *Front Med.* (2015) 9:401–11.doi: 10.1007/s11684-015-0415-x
9. Ma S, Peng C, Wu S, Wu D, Gao C. Sciatic nerve regeneration using a nerve growth factor-containing fibrin glue membrane. *Neural Regen Res.* (2013) 8:3416–22. doi: 10.3969/j.issn.1673-5374.2013.36.007
10. Bhatnagar D, Bushman JS, Murthy NS, Merolli A, Kaplan HM, Kohn J. Fibrin glue as a stabilization strategy in peripheral nerve repair when using porous nerve guidance conduits. *J Mater Sci Mater Med.* (2017) 28:79. doi: 10.1007/s10856-017-5889-4
11. Gosk J, Knakiewicz M, Wiacek R, Reichert P. The use of the fibrin glue in the peripheral nerves reconstructions. *Polim Med.* (2006) 36:11–5.
12. Isaacs J. Major peripheral nerve injuries. *Hand Clin.* (2013) 29:371–82. doi: 10.1016/j.hcl.2013.04.006
13. Fairbairn NG, Ng-Glazier J, Meppelink AM, Randolph MA, Valerio IL, Fleming ME, et al. Light-activated sealing of nerve graft coaptation sites improves outcome following large gap peripheral nerve injury. *Plast Reconstr Surg.* (2015) 136:739–50. doi: 10.1097/PRS.0000000000001617
14. Martins RS, Siqueira MG, Da Silva CF, Plese JP. Overall assessment of regeneration in peripheral nerve lesion repair using fibrin glue, suture, or a combination of the 2 techniques in a rat model. Which is the ideal choice? *Surg Neurol.* (2005) 64(Suppl 1):S1:10–6; discussion S11:16. doi: 10.1016/j.surneu.2005.04.022
15. Tse R, Ko JH. Nerve glue for upper extremity reconstruction. *Hand Clin.* (2012) 28:529–40. doi: 10.1016/j.hcl.2012.08.006
16. Chen S, Chen ZG, Dai H, Ding J, Guo J, Han N, et al. Repair, protection and regeneration of peripheral nerve injury. *Neural Regen Res.* (2015) 10:1777–98. doi: 10.4103/1673-5374.170301
17. de Luca AC, Lacour SP, Raffoul W, di Summa PG. Extracellular matrix components in peripheral nerve repair: how to affect neural cellular response and nerve regeneration? *Neural Regen Res.* (2014) 9:1943–8. doi: 10.4103/1673-5374.145366
18. Gerth DJ, Tashiro J, Thaller SR. Clinical outcomes for Conduits and Scaffolds in peripheral nerve repair. *World J Clin Cases.* (2015) 3:141–7. doi: 10.12998/wjcc.v3.i2.141
19. Petrova ES. Injured nerve regeneration using cell-based therapies: current challenges. *Acta Naturae.* (2015) 7:38–47.
20. Lopatina T, Kalinina N, Karagyaur M, Stambolsky D, Rubina K, Revischin A, et al. Adipose-derived stem cells stimulate regeneration of peripheral nerves: BDNF secreted by these cells promotes nerve healing and axon growth *de novo*. *PLoS ONE.* (2011) 6:e17899. doi: 10.1371/journal.pone.0017899
21. Caddick J, Kingham PJ, Gardiner NJ, Wiberg M, Terenghi G. Phenotypic and functional characteristics of mesenchymal stem cells differentiated along a Schwann cell lineage. *Glia.* (2006) 54:840–9. doi: 10.1002/glia.20421
22. Gordon T. Neurotrophic factor expression in denervated motor and sensory Schwann cells: relevance to specificity of peripheral nerve regeneration. *Exp Neurol.* (2014) 254:99–108. doi: 10.1016/j.expneurol.2014.01.012
23. Cooney DS, Wimmers EG, Ibrahim Z, Grahammer J, Christensen JM, Brat GA, et al. Mesenchymal stem cells enhance nerve regeneration in a rat sciatic nerve repair and hindlimb transplant model. *Sci Rep.* (2016) 6:31306. doi: 10.1038/srep31306
24. Aggarwal S, Pittenger MF. Human mesenchymal stem cells modulate allogeneic immune cell responses. *Blood.* (2005) 105:1815–22.doi: 10.1182/blood-2004-04-1559
25. Mukhamedshina YO, Zakirova EY, Galieva LR, Kostennikov AA, Akhmetzyanova ER, Rizvanov AA. Distribution and survival of transplanted adipose-derived mesenchymal stem cells in the spinal cord injury. *BioNanoSci.* (2017) 7:608–12. doi: 10.1007/s12668-017-0440-0
26. Marconi S, Castiglione G, Turano E, Bissolotti G, Angiari S, Farinazzo A, et al. Human adipose-derived mesenchymal stem cells systemically injected promote peripheral nerve regeneration in the mouse model of sciatic crush. *Tissue Eng Part A.* (2012) 18:1264–72. doi: 10.1089/ten.TEA.2011.0491
27. Amer MH, White LJ, Shakesheff KM. The effect of injection using narrow-bore needles on mammalian cells: administration and formulation considerations for cell therapies. *J Pharm Pharmacol.* (2015) 67:640–50. doi: 10.1111/jphp.12362
28. Mukhamedshina YO, Gracheva OA, Mukhutdinova DM, Chelyshev YA, Rizvanov AA. Mesenchymal stem cells and the neuronal microenvironment in the area of spinal cord injury. *Neural Regen Res.* (2019) 14:227–37. doi: 10.4103/1673-5374.244778
29. Terrovitis JV, Smith RR, Marban E. Assessment and optimization of cell engraftment after transplantation into the heart. *Circ Res.* (2010) 106:479–94. doi: 10.1161/CIRCRESAHA.109.208991
30. Kang SK, Shin IS, Ko MS, Jo JY, Ra JC. Journey of mesenchymal stem cells for homing: strategies to enhance efficacy and safety of stem cell therapy. *Stem Cells Int.* (2012) 2012:342968. doi: 10.1155/2012/342968
31. Yarar E, Kuruoglu E, Kocabicak E, Altun A, Genc E, Ozyurek H, et al. Electrophysiological and histopathological effects of mesenchymal stem cells in treatment of experimental rat model of sciatic nerve injury. *Int J Clin Exp Med.* (2015) 8:8776–84.
32. Mukhamedshina YO, Akhmetzyanova ER, Kostennikov AA, Zakirova EY, Galieva LR, Garanina EE, et al. Adipose-derived mesenchymal stem cell application combined with fibrin matrix promotes structural and functional recovery following spinal cord injury in rats. *Front Pharmacol.* (2018) 9:343. doi: 10.3389/fphar.2018.00343
33. *International Guiding Principles for Biomedical Research Involving Animals issued by CIOMS and I. C. L. A.S.* (2012).
34. de Medinaceli L, DeRenzo E, Wyatt RJ. Rat sciatic functional index data management system with digitized input. *Comput Biomed Res.* (1984) 17:185–92.
35. Triolo D, Dina G, Taveggia C, Vaccari I, Porrello E, Rivellini C, et al. Vimentin regulates peripheral nerve myelination. *Development.* (2012) 139:1359–67. doi: 10.1242/dev.072371
36. Atlasi MA, Mehdizadeh M, Bahadori MH, Joghataei MT. Morphological identification of cell death in dorsal root ganglion neurons following peripheral nerve injury and repair in adult rat. *Iran Biomed J.* (2009) 13:65–72.
37. Masgutov R, Salafutdinov I, Bogov A, Trofimova A, Khannanova I, Mullin RI, et al. Stimulation of sciatic nerve regeneration after injury using plasmid encoding VEGF and FGF2. *Genes and Cells.* (2011) 3:1–5.
38. Angius D, Wang H, Spinner RJ, Gutierrez-Cotto Y, Yaszemski MJ, Windebank AJ. A systematic review of animal models used to study nerve regeneration in tissue-engineered scaffolds. *Biomaterials.* (2012) 33:8034–9. doi: 10.1016/j.biomaterials.2012.07.056
39. Zhou B, Yu P, Lin MY, Sun T, Chen Y, Sheng ZH. Facilitation of axon regeneration by enhancing mitochondrial transport and rescuing energy deficits. *J Cell Biol.* (2016) 214:103–19. doi: 10.1083/jcb.201605101

40. Lemke A, Penzenstadler C, Ferguson J, Lidinsky D, Hopf R, Bradl M, et al. A novel experimental rat model of peripheral nerve scarring that reliably mimics post-surgical complications and recurring adhesions. *Dis Model Mech.* (2017) 10:1015–25. doi: 10.1242/dmm.028852
41. Ngeow WC. Scar less: a review of methods of scar reduction at sites of peripheral nerve repair. *Oral Surg Oral Med Oral Pathol Oral Radiol Endod.* (2010) 109:357–66. doi: 10.1016/j.tripleo.2009.06.030
42. Ebadi M, Bashir RM, Heidrick ML, Hamada FM, Refaey HE, Hamed A, et al. Neurotrophins and their receptors in nerve injury and repair. *Neurochem Int.* (1997) 30:347–74.
43. Elfar JC, Jacobson JA, Puzas JE, Rosier RN, Zuscik MJ. Erythropoietin accelerates functional recovery after peripheral nerve injury. *J Bone Joint Surg Am.* (2008) 90:1644–53. doi: 10.2106/JBJS.G.00557
44. Birchmeier C, Nave KA. Neuregulin-1, a key axonal signal that drives Schwann cell growth and differentiation. *Glia.* (2008) 56:1491–7. doi: 10.1002/glia.20753
45. Vrbova G, Mehra N, Shanmuganathan H, Tyreman N, Schachner M, Gordon T. Chemical communication between regenerating motor axons and Schwann cells in the growth pathway. *Eur J Neurosci.* (2009) 30:366–75. doi: 10.1111/j.1460-9568.2009.06847.x
46. Stoll G, Griffin JW, Li CY, Trapp BD. Wallerian degeneration in the peripheral nervous system: participation of both Schwann cells and macrophages in myelin degradation. *J Neurocytol.* (1989) 18:671–83.
47. Pellegrino RG, Spencer PS. Schwann cell mitosis in response to regenerating peripheral axons *in vivo*. *Brain Res.* (1985) 341:16–25.
48. Ciardelli G, Chiono V. Materials for peripheral nerve regeneration. *Macromol Biosci.* (2006) 6:13–26. doi: 10.1002/mabi.200500151
49. Murakami T, Fujimoto Y, Yasunaga Y, Ishida O, Tanaka N, Ikuta Y, et al. Transplanted neuronal progenitor cells in a peripheral nerve gap promote nerve repair. *Brain Res.* (2003) 974:17–24.
50. Tohill M, Terenghi G. Stem-cell plasticity and therapy for injuries of the peripheral nervous system. *Biotechnol Appl Biochem.* (2004) 40:17–24. doi: 10.1042/BA20030173
51. Swett JE, Torigoe Y, Elie VR, Bourassa CM, Miller PG. Sensory neurons of the rat sciatic nerve. *Exp Neurol.* (1991) 114:82–103.
52. Reid AJ, Sun M, Wiberg M, Downes S, Terenghi G, Kingham PJ. Nerve repair with adipose-derived stem cells protects dorsal root ganglia neurons from apoptosis. *Neuroscience.* (2011) 199:515–22. doi: 10.1016/j.neuroscience.2011.09.064
53. Masgutov RF, Masgutova GA, Zhuravleva MN, Salafutdinov II, Mukhametshina RT, Mukhamedshina Y. O., et al. Human adipose-derived stem cells stimulate neuroregeneration. *Clin Exp Med.* (2016) 16:451–61. doi: 10.1007/s10238-015-0364-3
54. Gurgor P, Pallesen LT, Johnsen L, Ulrichsen M, de Jong IE, Vaegter CB. Neuronal death in the dorsal root ganglion after sciatic nerve injury does not depend on sortilin. *Neuroscience.* (2016) 319:1–8. doi: 10.1016/j.neuroscience.2016.01.036
55. Masgutov R, Masgutova G, Mukhametova L, Garanina E, Arkhipova SS, Zakirova E, et al. Allogenic adipose derived stem cells transplantation improved sciatic nerve regeneration in rats: autologous nerve graft model. *Front Pharmacol.* (2018) 9:86. doi: 10.3389/fphar.2018.00086
56. Duraikannu A, Martinez JA, Chandrasekhar A, Zochodne DW. Expression and manipulation of the APC-beta-catenin pathway during peripheral neuron regeneration. *Sci Rep.* (2018) 8:13197. doi: 10.1038/s41598-018-31167-1
57. Guo G, Singh V, Zochodne DW. Growth and turning properties of adult glial cell-derived neurotrophic factor coreceptor alpha1 nonpeptidergic sensory neurons. *J Neuropathol Exp Neurol.* (2014) 73:820–36. doi: 10.1097/NEN.0000000000000101
58. Tucker BA, Rahimtula M, Mearow KM. Laminin and growth factor receptor activation stimulates differential growth responses in subpopulations of adult DRG neurons. *Eur J Neurosci.* (2006) 24:676–90. doi: 10.1111/j.1460-9568.2006.04963.x
59. Gordon T, Tyreman N, Raji MA. The basis for diminished functional recovery after delayed peripheral nerve repair. *J Neurosci.* (2011) 31:5325–34. doi: 10.1523/JNEUROSCI.6156-10.2011
60. Liu GB, Cheng YX, Feng YK, Pang CJ, Li Q, Wang Y, et al. Adipose-derived stem cells promote peripheral nerve repair. *Arch Med Sci.* (2011) 7:592–6. doi: 10.5114/aoms.2011.24127
61. Zanjani LO, Firouzi M, Nabian MH, Zadegan SA, Nategh M, Erfanian R, et al. A time course study of different methods of extraneural scar induction in a rat model. *J Reconstr Microsurg.* (2016) 32:309–15. doi: 10.1055/s-0035-1571198
62. Gilron I, Baron R, Jensen T. Neuropathic pain: principles of diagnosis and treatment. *Mayo Clin Proc.* (2015) 90:532–45. doi: 10.1016/j.mayocp.2015.01.018
63. Bronckaers A, Hilkens P, Martens W, Gervois P, Ratajczak J, Struys T, et al. Mesenchymal stem/stromal cells as a pharmacological and therapeutic approach to accelerate angiogenesis. *Pharmacol Ther.* (2014) 143:181–96. doi: 10.1016/j.pharmthera.2014.02.013
64. Rubina, K., Kalinina, N., Efimenko, A., Lopatina, T., Melikhova, V., Tsokolaeva Z, et al. Adipose stromal cells stimulate angiogenesis via promoting progenitor cell differentiation, secretion of angiogenic factors, and enhancing vessel maturation. *Tissue Eng Part A.* (2009) 15:2039–50. doi: 10.1089/ten.tea.2008.0359
65. Fierro FA, Kalomoiris S, Sondergaard CS, Nolta JA. Effects on proliferation and differentiation of multipotent bone marrow stromal cells engineered to express growth factors for combined cell and gene therapy. *Stem Cells.* (2011) 29:1727–37. doi: 10.1002/stem.720
66. Pankajakshan D, Agrawal DK. Mesenchymal stem cell paracrine factors in vascular repair and regeneration. *J Biomed Technol Res:*(2014) 1:1. doi: 10.19104/jbtr.2014.107
67. Toth C, Moulin DE. *Neuropathic Pain: Causes, Management, and Understanding*. Cambridge, NY: Cambridge University Press.
68. Menorca RM, Fussell TS, Elfar JC. Nerve physiology: mechanisms of injury and recovery. *Hand Clin.* (2013) 29:317–30. doi: 10.1016/j.hcl.2013.04.002

12

A Comparison of Phenotypic and Functional Properties of Mesenchymal Stromal Cells and Multipotent Adult Progenitor Cells

Reenam S. Khan [1,2,3] *and Philip N. Newsome* [1,2,3*]

[1] National Institute for Health Research (NIHR), Birmingham Biomedical Research Centre, University Hospitals Birmingham NHS Foundation Trust, University of Birmingham, Birmingham, United Kingdom, [2] Centre for Liver Research, Institute of Immunology and Immunotherapy, University of Birmingham, Birmingham, United Kingdom, [3] Liver Unit, University Hospitals Birmingham NHS Foundation Trust, Birmingham, United Kingdom

Correspondence:
Philip N. Newsome
p.n.newsome@bham.ac.uk

Both Multipotent Adult Progenitor Cells and Mesenchymal Stromal Cells are bone-marrow derived, non-haematopoietic adherent cells, that are well-known for having immunomodulatory and pro-angiogenic properties, whilst being relatively non-immunogenic. However, they are phenotypically and functionally distinct cell types, which has implications for their efficacy in different settings. In this review we compare the phenotypic and functional properties of these two cell types, to help in determining which would be the superior cell type for different applications.

Keywords: multipotent adult progenitor cell, cellular therapy, mesenchymal stromal cell, immunomodulation, cell biology

INTRODUCTION

Cellular therapy refers to the use of cells to replace or repair damaged tissue/cells. Over the last decade there has been a tremendous development in cellular therapies for the treatment of disease. Embryonic stem cells (ESC) can potentially differentiate into cells of all three germ layers; however, research interest in ESC has been limited by ethical concerns and risk of teratoma formation. Adult cellular therapies have been widely investigated, and haematopoietic stem cell transplantation is already a well-established treatment for various malignant and non-malignant hematological disorders.

Mesenchymal stromal Cells (MSC) and Multipotent Adult Progenitor Cells (MAPC) are both non-haematopoietic cells found in bone marrow stroma, which play a role in maintenance of the haematopoietic stem cell niche (1). Following bone fracture in mice, Park et al. demonstrated mobilization of these cells, and their involvement in fracture repair (2). Great interest in these cells as a potential cellular therapy arises from evidence that they have immunomodulatory properties, can promote angiogenesis, and provide protection against apoptosis. Although MAPC and MSC co-purify, there is evidence that they are phenotypically and functionally distinct cell types.

MSCs were initially described in 1968 by Friedenstein (3), as a subtype of adult fibroblast-like cells with a high proliferative ability, capacity for self-renewal and ability to undergo tri-lineage differentiation to become osteoblasts, chondrocytes and adipocytes. Over time, it became clear that variation in isolation and culture procedures for bone marrow stromal cells contributed to generation of heterogeneous cell populations. The International Society for Cell Therapy (ISCT) subsequently published criteria for identifying MSC: (a) Bone marrow stromal cells that show plastic adherence under standard culture conditions (b) Positive for CD105, CD90, and CD73; have low levels of MHC-I; are negative for MHC-II, CD11b, CD14, CD34, CD45, and CD31 (c) Can differentiate *in vitro* into osteocytes, chondrocytes and adipocytes (4). Thirteen human MSC products have gained marketing authorization, of which nine are for allogeneic therapy and four are for autologous therapy (5), with indications including Crohn's disease, bone and adipose tissue regeneration, graft-vs.-host disease, and acute myocardial infarction.

MAPC were first described several years later, in 2001, as a novel progenitor cell in the bone marrow (6), and whilst these cells meet the ISCT criteria for MSC, they were perceived to be a more biologically primitive population than classical MSC and had greater differentiation potential. Whilst MSCs have been extensively studied, with over 900 clinical trials completed or ongoing, according to the US National Institute of Health (https://www.clinicaltrials.gov), there are fewer data published on MAPC. This review covers a summary of the key similarities and differences in the phenotypic and functional properties of these cells and the clinical data supporting their use in different settings.

SOURCING THE CELLS

Whilst MSC were originally identified as a rare population in bone marrow (BM) accounting for 0.01–0.001% of cells (7), they have also been successfully isolated from other tissues including adipose tissue (AT) (8), synovial membrane (9), skeletal muscle tissue (10), dental pulp (11), lung tissue (12), Wharton's jelly (13), umbilical cord (UC) blood (14), amniotic fluid (AF) (15), and placenta (16). Studies have compared the biological properties of MSCs isolated from different sources, and whilst some report that they have similar biological properties (13, 17, 18), others report differences in immunomodulatory activity and surface antigen expression (19–21). Furthermore, UC MSCs have been shown to have a relatively higher proliferative capacity compared to cells from other sources (22), which, has been linked to their having a more primitive phenotype. There is concurrently no consensus on which source of cells is best for clinical application. MAPC were originally isolated from the bone marrow of mice, rats and humans, but subsequently, they were also isolated from murine muscle and brain tissues (6). However, the clinical studies published on MAPC so far have all used cells obtained from human bone marrow.

CELL CULTURE AND GROWTH RATES

MAPC and MSC have distinct culture requirements (23). Whilst they are both cultured in fibronectin-coated flasks, MAPC culture medium includes the presence of growth factors (human-platelet derived growth factor, human epidermal growth factor) that are not present in many MSC culture media. Moreover, culture of MAPC takes place in conditions of relative hypoxia (5% oxygen), which is important in preventing telomerase shortening in MAPC. The consequence is that MAPC can be expanded for over 60 doublings without senescence (24), whereas for MSC, the reported population doublings range between 10 and 38 (25). Current manufacturing strategies for MAPC are capable of producing over 100,000 clinical doses from a single donor, sufficient for a clinical trial. Roobrouck et al. (26) demonstrated that the phenotypic and functional properties of the cells were influenced by culture conditions; when MAPC were cultured under MSC conditions, they acquired some of the phenotypical and functional properties of MSC and vice versa (26). Nevertheless, it is important to emphasize that MAPC and MSC are distinct cell types, rather than simply the product of different culture conditions. Following isolation and expansion, both MAPC and MSC can be cryopreserved and stored until needed, although there is evidence that upon thawing, MSCs show signs of injury even within the first 24 h, which may reduce their immunomodulatory properties and increase predisposition to immune clearance (27).

CELL PHENOTYPE AND ISSUES OF BATCH-TO-BATCH VARIATION

Phenotypically, MAPC and MSC both fulfill the ISCT criteria for identification for MSC (positive expression of CD44, CD13, CD73, CD90, and CD105, negative expression of haematopoietic (CD34, CD45, CD117), and endothelial cell markers (CD34, CD309). They are also negative for MHC class II and co-stimulatory molecules. However, MAPC do not express some of the markers expressed by MSC, such as CD140a and CD140b, for example, and this could be used to distinguish them (26). MAPC also have lower levels of MHC class I and CD44 than MSC and a higher expression of CD49d (28). MAPC and MSC also have distinct features on transcriptomic analysis, with gene signatures that correlate with their specific functional properties (26).

MAPC and MSC also have different morphology, with the former being relatively smaller cells with a trigonal shape, whereas MSC are larger cells with a "spindle"-like morphology [(29); **Figure 1**]. However, the exact size of MSC does vary according to their source, with placenta-derived MSC being relatively smaller (mean peak diameter 16 μm) than MSC from other sources (30), which are typically >20 μm in size. MSC size is also influenced by their culture conditions. For example, MSC cultured with human platelet lysate (HPL) or platelet rich plasma (PRP) can be smaller than those cultured with fetal calf serum FCS) (30, 31).

Pre-clinical studies of MAPC and MSC have involved a variety of species, including mice, rats and pigs. The genetic profile and cell secretome is slightly different between species, and this has implications for their function. For example, *in vitro*, human MSC (hMSC) proliferation is associated with a very low frequency of oncogenic formation (32), whereas murine MSC (mMSC) frequently gain chromosomal defects (33). The frequency of OCT4 gene expression, which is associated with increased expansion in culture, was found to be relatively higher in rat MAPC (rMAPC) compared with human MAPC (hMAPC) (26).

Clinical trials using purified MAPC have all sourced cells from Athersys. Comparison of MAPC products from different batches

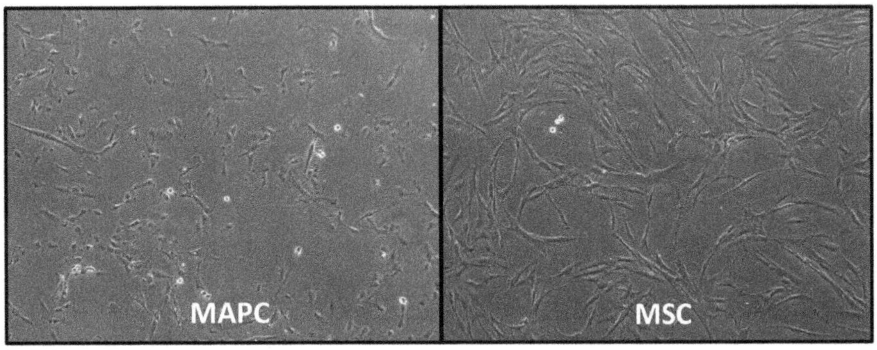

FIGURE 1 | Phase-contrast morphology of human multipotent adult progenitor cells (MAPC) and human mesenchymal stromal cells (MSC). Images courtesy of Regenesys BVBA.

has shown minimal batch-to-batch variation by way of surface antigen expression (24), growth rates, effect on suppressing T cell proliferation (34), angiogenic cytokine secretion (25–32% variance for three cytokines between 15 manufacturing runs) (35) and methylation status of 1536 CpG islands (34).

With MSC, there is no single epitope marker that can be reliably used to distinguish them, which results in heterogeneous cell populations across different studies. This makes it difficult to determine whether the effects that are seen are due to an individual cell type from the adherent cell population. Given the heterogeneity of cell types that fulfill MSC criteria, it has been suggested that one method of enhancing the efficacy of these cells would be to sort them on the basis of their expression of specific markers that are associated with favorable characteristics. For example, CD73 positivity on MSC is associated with increased capacity for self-renewal and differentiation (36). The presence of Syndecan-2 (CD-362+) on MSC has been associated with enhanced immunomodulatory properties through downregulation of CD3+ cells by degradation of the T-cell receptor (37).

MECHANISMS OF ACTION

There is evidence that MAPC and MSC can differentiate into cells of mesenchymal lineages, including bones, cartilage, fat, muscles, tendon and bone marrow. Thus, a number of trials have assessed these cells for the treatment of bone and cartilage disease, with several pre-clinical studies showing that MSCs have the potential to stimulate cartilage regeneration and delay joint destruction in osteoarthritis (38). This was supported by findings of a phase I clinical trial, in which patients with osteoarthritis also demonstrated functional improvement (39).

There is also evidence that MSC can trans-differentiate into cells from other lineages, including pancreatic islet cells (capable of producing insulin and glucagon) (40), renal tubular epithelium (41), keratinocytes (42), and hepatocytes although the biological/clinical significance of this is not clear.

MAPC appear to have greater propensity toward endothelial differentiation than MSC (26). In an *in vitro* Matrigel plug assay, human MAPC, but not MSC, could induce functional vessel formation (26). On transcriptomic analysis comparing MAPC and MSC, MAPC had over-representation of genes involved in differentiation of endothelial cells and promotion of angiogenesis, whilst MSC had over representation of genes involved in differentiation of chondrocytes and osteocytes, and in the genes involved in the development and contraction of smooth muscle and genes important for neo-vascularization (26).

However, differentiation is unlikely to explain the predominant functional role of MAPC and MSC. Firstly, experiments used labeled cells have revealed that both MAPC and MSC are rapidly cleared from the body after infusion (43–46), with <1% of MSC retained in the body at 1 week post-infusion. Secondly, upon intravenous injection, the majority of MAPC and MSC tend to get trapped in tissue capillary beds, particularly in the lungs, despite having effects in other organs (47). Thirdly, whilst cell differentiation into non-mesodermal lineages does occur, the frequency of this phenomenon is too low to completely explain the beneficial effects (48). For example, in a pig model of myocardial infarction, Wang et al. injected 50 million MAPC into the heart, and 2 weeks after infusion, only 0.55% of the cells were detectable, and of the engrafted cells, only 2% stained positive for cardiac markers (49). Fourthly, the functional properties of the cells produced through trans-differentiation of MSC and MAPC is questionable. For example, rat MSC that were induced to differentiate into "neural cells" were not able to generate normal action potentials (50), and endothelial cells generated from MSC did not express the same degree of endothelial cell markers as mature endothelial cells (51).

Some studies suggested that the rare reports on cross-germline differentiation of MSC could be ascribed to cell fusion, which may represent an alternative method by which MSC/MAPC can rescue injured cells. MSC can also communicate directly with target cells through delivery of materials, such as mitochondria, via nanotubes or connexins (52). Although there is paucity of mechanistic data on mitochondrial transfer as a putative mechanism of action of MAPC, in a porcine study of intra-cardiac MAPC infusion, treatment was associated with improvement in bio-energetic profiles (46). However,

TABLE 1 | Summary of key proteins identified in the secretome of MAPC and MSC that have therapeutic potential.

	Key components of the MAPC secretome	Key components of the MSC secretome
Chemoattraction/cell adhesion	CXCL1 (55), CXCL3 (55), CXCL5 (35), VEGF (35), sICAM1 (55), SDF1 (56), IL-8 (35)	CCL5 (57), SDF-1 (58), HGF (59), LIF (60), G-CSF (61), VEGF (60), CCL-2 (58), MCP-1 (60), ICAM1 (57), IL-8 (61)
Immunomodulation	IDO (62), TSG-6 (63), PGE_2 (63, 64), NO (63), semaphorin-7A (55)	IDO (57), PGE_2 (57), TGF-beta (65), TSG-6 (66), HGF (59), LIF (60), HLA-G (67), IL-6 (61), IL-10 (68), PD-L1 (69)
Neuroprotection	CNF (63), Galectin 1 (55), NO (63)	BDNF (70), NGF (70), GDNF (71), galectin 1 (72), NO
Anti-fibrosis	MMP1 (55), MMP2 (59), TIMP1 (59), TIMP2 (59), cathepsin B (55), bFGF (56)	MMP1 (72), MMP2 (72), MMP7 (72) MMP9, TIMP1 (72), TIMP2 (72), HGF (59), bFGF (73), Ang-1 (58)
Anti-apoptosis	bFGF (56), VEGF (35), versican (55)	VEGF (59, 60), IGF (60), HGF (59), TGFbeta, bFGF (73), GM-CSF (61), IL-6 (61)
Angiogenesis	VEGF (35), CXCL5 (35), IL-8 (35)	VEGF (59, 60), HGF (59), Ang-1 (58), bFGF (73), IGF1 (60), PDGF (73), IL-6 (61)
Anti-bacterial	Pentraxin (55), vimentin (55), lactotransferrin (55)	LL37 (74)
Proliferation	IGFBP4 (55), IGFBP5, IGFBP7, bFGF (56), VEGF (35)	FGF2 (73), VEGF (59, 60), IGFBP3 (60), IGFBP7 (60), IGDBP4 (60), PDGF (73), HGF (59), BMP (72)

Ang-1, angiopoietin 1; bFGF, basic fibroblast growth factor; BDNF, brain-derived neurotrophic factor; BMP, bone morphogenic protein; CCL2, C-C motif ligand 2; CNF, ciliary neurotrophic factor; CXCL1, C-X-C motif chemokine ligand 1; CXCL3, C-X-C motif chemokine ligand 3; CXCL5, C-X-C motif chemokine ligand 5; CCL5, C-C motif ligand 5; G-CSF, granulocyte-colony stimulating factor; GDNF, glial cell-derived neutrotrophic factor; HGF, hepatocyte growth factor; HLA-G, human leukocyte antigen G; IDO, indoleamine 2,3-dioxygenase; IGF, insulin-like growth factor; ICAM1, intercellular adhesion molecule 1; IGFBP3/4/5/7, insulin-like growth factor binding protein 3/4/5/7; IL-6/8/10, interleukin 6/8/10; LIF, leukemia inhibitory factor; LMCP-1, monocyte-chemotactic protein 1; MMP1/2/7/9, matrix metalloproteinase 1/2/7/9; TIMP 1/2, tissue inhibitor of metalloproteinase 1/2; NGF, nerve growth factor; NO, nitric oxide; PDGF, platelet-derived growth factor; PD-L1, programmed death ligand 1; PGE_2, prostaglandin E2; SDF1, stromal cell derived factor 1, sICAM1, soluble intercellular adhesion molecule 1; TGF-beta, transforming growth factor beta; TSG-6, tumor necrosis factor inducible gene 6;VEGF, vascular endothelial growth factor.

whilst in some studies cell-cell contact has been shown to be important for enhancing activity (53), in a number of studies the beneficial effects of the cell could be, at least partially, reproduced by using components from the cell secretome i.e., the set of factors/molecules released by cells into the extra-cellular space. These include exosomes (30–100 nm), generated from the endocytic pathway and release through exocytosis, whilst microvesicles (50–1,000 nm) are generated through budding from the cell surface and are released from the plasma membrane.

Evidence suggests that MSCs produce large amounts of exosomes in comparison to other cells. These exosomes may be internalized by other cells, permitting release of their contents into the cell cytoplasm (54). Whilst a thorough analysis of the MSC secretome is yet to be performed, it is clear from numerous studies that they express over a 100 proteins, many of which can regulate processes such as immune function, fibrosis, angiogenesis, and apoptosis. Burrows et al. report the only detailed proteomic analysis of the MAPC secretome, in which report identification of 97 proteins. A summary of the key components of each secretome is provided in **Table 1**.

Use of conditioned media or extra-cellular vesicles from MSCs had beneficial effects in animal models of myocardial infarction (75), colitis (76), acute liver failure (65), and Parkinson's disease (65). However, clinical experience with exosomes is currently limited. A preliminary clinical study suggested benefit of MSC exosomes in the treatment of stage 4 graft vs. host disease (77), and a clinical trial is underway to test their effect in increasing beta cell mass in patients with type I diabetes (NCT02138331). There have not yet been any clinical trials using cell-free preparations of MAPC, but it would be desirable to further explore cell-free therapy for various reasons. Firstly, it overcomes some of the problems associated with delivery of living cells, including cancer risk, potential for transmission of infections, and immune compatibility. Cell-free preparations would also be easier to store, easier to scale-up, and more cost-effective to prepare. Further, it would be possible to prepare a biological product with a high concentration of the desirable molecules.

The secretome of both MAPC and MSC is responsive to changes in their surrounding microenvironment. For example, pre-treatment with IFN-gamma was found to enhance the immunomodulatory activity of MSCs (78), and pre-treatment with TNF-alpha increases their angiogenic effect (79). Therefore, it is possible to pre-condition/prime cells *in vitro* to help in achieving the targetted effects. Pre-treatment of MSC with inflammatory molecules, including IL-1 beta, IL-23, IL-6, and IFN-gamma was found to enhance their immunomodulatory properties in a number of studies (73, 80, 81). **Figure 2** summarizes the key mechanisms by which MSC are hypothesized to act.

POTENTIAL FOR IMMUNE REJECTION

In most clinical trials of MSC, and all clinical trials using MAPC, the cells used were allogeneic and were administered without HLA matching or use of immunosuppressive medication. It is commonly believed that these cells are "immune privileged," leading to the belief that they could be used as a "one-size-fits-all" or "off-the-shelf" therapy. Indeed, cultured MSC and MAPC lack expression of MHC class II and key co-stimulatory surface molecules (CD40, CD80, and CD86), and have low levels of MHC class I molecules, which would help in protecting against a host immune response (82). Additionally, in mixed lymphocyte reactions, both MAPC and MSC do not induce a T-cell response (83).

However, recent evidence has challenged this view as MAPC and MSC are cleared rapidly from the body following infusion.

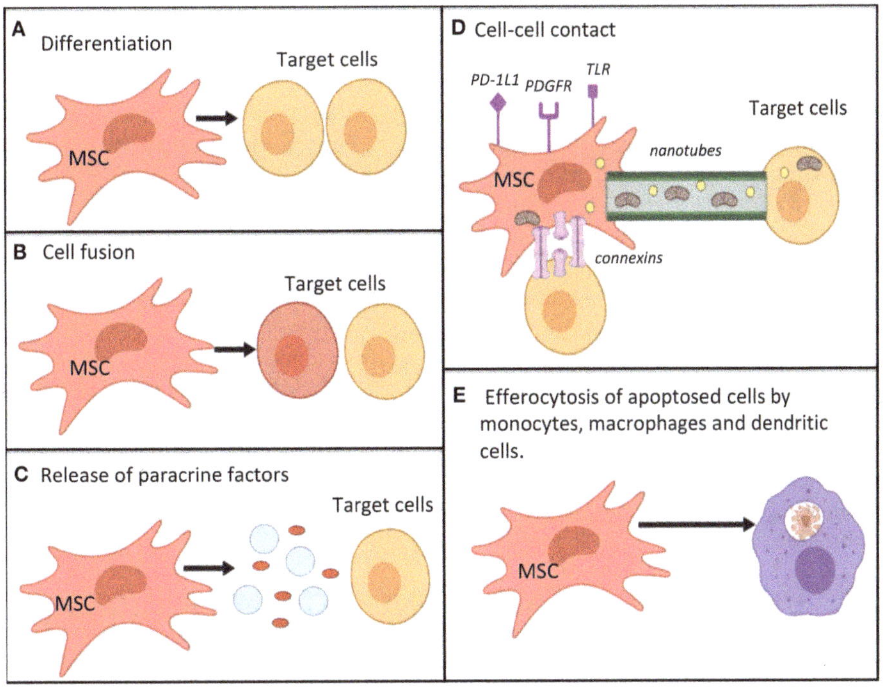

FIGURE 2 | Potential mechanisms by which mesenchymal stromal cells work for immunomodulation, restoration of cell bioenergetics and restoration of cell function; **(A)** differentiation into replacement cell types; **(B)** cell fusion with target cells for rescue of damaged or dying cells; **(C)** secretion of paracrine factors (such as growth factors, cytokines, RNA, and hormones) via micro-vesicles or exosomes. MSC autophagy may help to promote -release of cellular contents; **(D)** cell-cell contact mechanisms. MSC can interact with immune cells via various surface receptors. Transfer of organelles (e.g., mitochondria), ribonucleic acid, and chemicals may occur via nanotubes, or connections; **(E)** efferocytosis of apoptotic MSCs by monocytes, macrophages and dendritic cells. This process causes the phagocytosing cells to adopt a tolerogenic/immunomodulatory phenotype. Mechanisms **(A–E)** are not equivalent, as cell differentiation and cell fusion occur relatively infrequently. MSC, mesenchymal stromal cell; PD-1L1, programmed death ligand 1; PDGFR, platelet derived growth factor receptor; TLR, toll-like receptor.

There is evidence that upon exposure to IFN-gamma or upon differentiation, MSC upregulate MHC class I and class II expression (84). MAPC have even lower levels of MHC class I than MSC, which makes them susceptible to lysis by NK cells in their resting state (83). On exposure to IFN-gamma, MAPC upregulate MHC class I but not MHC class II expression, and therefore inflammation may promote MAPC persistence (83).

Evidence from *in vivo* studies also suggests that these cells are not truly immunologically privileged. Eliopoulos et al. administered erythropoietin-transfected MSCs from C57/BL6 mice to syngeneic or allogenic mice (85). The cells were seeded in a collagen scaffold and administered subcutaneously. Whilst the syngeneic mice had a sustained haematocrit response (in response to the erythropoietin), the allogeneic mice had a temporary spike in haematocrit before returning to baseline, suggesting clearance of the MSCs. Furthermore, allogeneic mice, but not syngeneic mice, had CD8+ and NK+ cell infiltration in the scaffolds. In another study, injection of murine luciferase-labeled MSCs to allogeneic hosts was associated with the development of memory T-cells (CD4+, CD122+, CD44+, and CD62L^{low}). Tolar et al. used luciferase labeled MAPC to demonstrate that use of T-cell and B-cell deficient mice with NK depletion were all associated with longer persistence of MAPC in mice, suggesting that all three cell types are implicated in rejecting MAPC (83).

In terms of clinical data, allogeneic MSC infusion was found to be associated with development of allo-antibodies in 13% of patients in a phase II clinical trial for GvHD (86). In a phase I clinical trial of allogeneic MAPC in patients with GvHD, infusion was associated with increased serum anti-class I titres compared to baseline, but there was no evidence of MHC class II antibody induction (87).

It has also been hypothesized that the clearance of MSC may be due to triggering of innate immunity independently of HLA-disparity, mediated by a lack of haemocompatibility (88). An "instant blood mediated inflammatory reaction" (IBMIR) has been described previously for pancreatic islet cells and hepatocytes, and is considered to be responsible for loss of up to 80% of these cells shortly after infusion (89, 90). IBMIR is triggered by exposure to host red blood cells and characterized by activation of complement/coagulation cascades, binding of activated platelets to the cells, and clot infiltration by neutrophil granulocytes and monocytes, eventually leading to cell destruction. Moll et al. demonstrated that patients infused with MSC had increased formation of blood activation markers (88). Tissue factor/CD142, which is expressed on MSC, was deemed to be the key determinant of cell haemocompatibility. Tissue factor expression was higher in cells from a higher passage, and for cells administered in higher doses, and there was a donor-to-donor variability. It was also found to vary depending on the cell source. George et al. compared the degree of tissue factor expression for MAPC derived from the bone marrow, bone marrow mononuclear cells, and MSC obtained from different sources (bone marrow, adipose tissue, amniotic fluid, and umbilical cord) (91). They found that tissue factor expression was significantly higher in MSCs originating from the adipose tissue and amniotic fluid, compared with MSC originating from bone marrow and umbilical cord, bone marrow mononuclear cells and MAPC. This would suggest that use of MAPC would be significantly advantageous for intravenous infusion because of the potential for reduced cell clearance/enhanced engraftment.

Thus, MSC are not truly "immunologically privileged," although rejection occurs slower than it does for other allogeneic cells; thus Aggarwal and Pittenger suggest a better terminology would be "immune evasive" (92). The exact timing and severity of immune rejection is likely dictated by the result of a balance between their immunogenic and immunosuppressive factors, which in turn depends on their local microenvironment. For example, in conditions where there is local immune suppression, for example due to a tumor, the immunogenic properties of the MSC may be masked (93).

Whilst an anti-donor response has been observed, it is reassuring that there have been no adverse events related to immune rejection reported in clinical trials. Furthermore, it is unclear if rejection of allogeneic cells has any impact on efficacy and this is an important area of further work. Indeed, in mixed lymphocyte reactions *in vitro*, bone marrow stromal cells were shown to suppress T-cell proliferation in in a dose-dependent manner, regardless of whether the cells were autologous or allogeneic (53). In a phase II clinical trial of patients with GvHD, there was no difference in efficacy between third-party and HLA-matched MSCs (94), but there are relatively few trials with recorded data on antibody responses. If there is formation of allo-antibodies and T-cell memory in response to allogeneic cells, repeated administration of therapy for a chronic condition may be associated with reduced efficacy.

On the other hand, recent evidence would even suggest that immune response to MSC is crucial to their function. For example, release of complement-activation products following exposure of MSC to host blood can modulate their immunomodulatory and chemotactic activity (95, 96). Contact with activated platelets, as part of the IBMIR response, can also induce extracellular matrix remodeling by MSC, which can potentially contribute to tissue repair (97). Further research is needed to determine the relative importance of IBMIR in the therapeutic efficacy of MSC/MAPC. If the therapeutic benefits on immunomodulation are outweighed by the disadvantages of increased cell clearance/reduced engraftment, it would be helpful to develop strategies to reduce/abolish tissue factor expression on the cells. In fact, given that tissue factor expression is so variable between cells and the potential impact this has on immune clearance, Moll et al. raise the question of whether haemocompatibility should be considered a release criteria for intravascular MSC therapies (98).

The term "autophagy" refers to a system of intracellular degradation that delivers cytoplasmic constituents to the lysosome. This process can be triggered by endoplasmic reticulum stress, hypoxia, and immune cell activation for example. MSC autophagy may serve to either enhance cell survival or promote cell death, depending on the surrounding micro-environment. The induction of autophagy in MSC may promote release of paracrine factors important for their immunomodulatory function (99). Further, recent work by by Galleu et al. and de Witte et al. suggests that following intravenous infusion, MSC accumulate in the lung, where they undergo apoptosis, after which they are engulfed by monocytes, which are subsequently transported elsewhere (100, 101). The process of phagocytosis was demonstrated to induce phenotypic and functional changes in monocytes which resulted in an immunomodulatory response (via release of TGF-beta and IL-10) (101, 102). In contrast, Dang et al. reported MSC autophagy was associated with dampening of their immunomodulatory efficacy (103). These conflicting findings highlight the need for additional studies to investigate this further.

POTENTIAL FOR TUMORIGENESIS

There is some evidence that links MSCs to cancer. Long term *in-vitro* culture of murine MSCs was associated with spontaneous transformation of the cells, which were then capable of promoting sarcoma formation when inoculated into immunodeficient mice (33, 104–106). Whilst there were initial reports of a similar phenomenon occurring in human MSCs in prolonged culture, some of these findings were considered to be related it to cross contamination with cancer cell-lines (106, 107). The vast majority of studies report a lack of spontaneous transformation of human MSCs despite extensive culture (108–112). Further, whilst transformation in murine MSCs can be easily induced, particularly through inactivation of *p53* and/or *Rb* genes (113, 114), in human cell lines a number of several, non-physiological, oncogenic events need to be combined for efficient induction of sarcoma (115). Of course, this does not exclude the possibility of cancer formation following infusion in patients, although the risk is low. Rodriguez et al. demonstrated that after hMSC are induced to undergo oncogenic transformation, the transformed cells lose their immunomodulatory and anti-inflammatory properties (116). For example, transformed cells do not secrete the immunomodulatory molecules prostaglandin E2 (PGE2) and PGI2, whereas they do release pro-inflammatory thromboxanes. It is reported that hMAPC remain genetically stable after prolonged culture (34); however, the effects of oncogenic transformation of hMAPC are unknown. In clinical trials, there have been no reports of cancer formation from delivery of allogeneic MSC or MAPC.

Some studies have also researched the interaction between MSC and existing tumors, particularly given that MSC have been demonstrated to home toward tumor sites (117–119). It appears that MSC may promote or inhibit tumor growth, depending on the tumor micro-environment, which as of yet remains undefined (120). The effect of MAPC on cancer growth has not been studied.

POTENTIAL FOR THROMBOSIS

Both MSC and, to a lesser degree, MAPC, express tissue factor, with the level of tissue factor expression correlating with pro-coagulant activity (91). The clinical relevance of this for VTE risk is uncertain. Results of functional coagulation assays have been shown to correlate with VTE incidence in multiple patient populations. Thousands of patients have received MSC in clinical trials and there are case reports of thromboembolic events in these patients, but the overall incidence is low (121, 122). Interestingly the case reports of thromboembolism following MSC infusion are related to umbilical-cord derived and adipose tissue derived cells rather than bone marrow derived cells. There have been no published reports of thromboembolic events in clinical trials of patients receiving MAPC infusions, although the available data is much more limited. The true incidence of thromboembolic events with cellular therapies is difficult to establish, partially because several of the patient populations receiving the MSC or MAPC received concomitant anti-coagulation (e.g., low-molecular-weight-heparin, anti-platelet agents and dextran sulfate) for conditions such as myocardial infarction, acute respiratory distress syndrome and stroke. There is also variation in tissue factor activity between cell batches, and between patients.

CELL HOMING AND BIODISTRIBUTION

One of the challenges with bone marrow stromal cell therapy has been targeting their delivery to their intended site of action. Although it is likely that MAPC and MSC exert their effects through paracrine mechanisms, localization to the target site may help in enhancing their efficacy and reducing unwanted peripheral side effects. MSCs are thought to migrate toward inflammatory cues from sites of tissue injury, and in a study of hypoxic ischaemic brain injury in rats, labeled MAPC were detected in the hippocampus regardless of whether they were administered directly into the hippocampus, or intravenously, with similar motor and neurological improvement between the groups (123).

MSC are relatively large cells in comparison to lymphocytes (diameter 15–30 μm vs. 4–12 μm, respectively) (124, 125), which means that they become easily entrapped in smaller blood vessels. The majority of MSC get trapped within the pulmonary capillary bed shortly after intravenous administration, after which they accumulate in the spleen and liver over hours to days (126–130). As MAPC are smaller than MSC, it is unsurprising that, using labeled cells injected intravenously in rats, it was demonstrated that twice as many MAPC were able to pass into the pulmonary circulation compared to MSC (131).

One potential strategy for increasing homing to target organs is intra-arterial rather than intra-venous injection, which has been shown to result in superior bio-distribution outside the lung of both MAPC and MSC (83). However, intra-arterial delivery may result in the cells becoming mechanically trapped in the microvasculature elsewhere (132). Of course, if apoptosis and phagocytosis of MSC in the lung is crucial to their mechanism of immunomodulation, as the work by Galleu et al. and de Witte et al. suggested (100, 101) then reducing pulmonary entrapment may in fact reduce efficacy.

An alternative method to encouraging homing is through cell priming. For example, pre-treatment with TNF-alpha, IFN-gamma, and IL-1 was associated with increased expression of adhesion molecules ICAM and VCAM on MSC (133), and priming with CXCL9 was associated with increased adherence of MSC to endothelial cells (134).

ROLE IN IMMUNOMODULATION

There is extensive evidence supporting the role of MSC and MAPC as modulators of immune responses (**Figure 2**), with the most well-established effects being on T-cell responses. In mixed lymphocyte reactions, bone marrow stromal cells caused a dose dependent reduction in proliferation of both CD4+ and CD8+ T-cells. When cell-cell contact was prevented, the effects persisted, but were weaker (53). Studies using purified MAPC (53, 135, 136) or MSC confirm that these cells can inhibit T-cell proliferation, with implicated factors including prostaglandin E2 (PGE2), transforming growth factor beta (TGF-beta), inducible nitric oxide synthase (iNOS) and hepHGF (53, 63, 135, 137). Both MAPC and MSC have been associated with changes in the numbers of T-cell subsets, with promotion of expression of T-reg cells (44, 138, 139). Studies also show that MSC may interfere with T-cell function, possibly through secretion of matrix metalloproteinases (MMP), such as MMP-2 and MMP-9 that can cleave CD25 from T-cells (140). The effect of MAPC on B cells has not been widely studied. MSCs have been demonstrated to inhibit B-cell proliferation, alter B cell surface antigen expression and reduce immunoglobulin production (141–143).

MAPC and MSC also affect the innate immune system. MSC have been shown to inhibit NK cell activity, as shown by reduced secretion of IL-15 and IL-2 from the NK cells, with possible mediators including PGE-2 and TGF-beta (144). Macrophages can be crudely classified as being of an M1 (pro-inflammatory) or M2 (anti-inflammatory) phenotype. Both MAPC and MSC have been associated with polarization of macrophages from an M1 (pro-inflammatory) phenotype to an M2 (anti-inflammatory) phenotype (145). This was shown in MAPC an *in vitro* model of axonal dieback (146), and in a murine model of cortical impact injury, in which the authors attributed the increase in M2: M1 ratio to be due to increased apoptosis of M1 macrophages (138). In MSC, this effect is deemed to be due to their secretion of IL-10 and arginase (147, 148). MSC can regulate dendritic cells by interfering with their differentiation to monocytes and inhibiting their activation (149). MAPC can impact immune cell infiltration. MAPC infusions were associated with a reduction of the neutrophil numbers in bronchiolar lavage samples in a sheep model of Acute Respiratory Distress Syndrome (ARDS), and in ischaemia perfusion injury of donor human lungs (150, 151).

Finally, MAPC and MSC can affect the balance of pro-inflammatory and anti-inflammatory cytokines. The secretome of MSC contains both pro-inflammatory (e.g., TNF-alpha, IFN-gamma, and IL-1B) and anti-inflammatory cytokines (e.g.,

TGF-beta 1, IL-13, and IL-18 binding protein), with the net effect likely to be the result of a balance between the two (43, 151–153). Both MAPC and MSC therapy has been associated with higher levels of a protein called TNF-alpha gene stimulated protein-6 (TSG-6), which can bind CD44 on macrophages and inhibit NfKB activation- the key controller of pro-inflammatory cytokine responses (63).

Clinical studies have investigated the utility of MAPC and MSC in a range of inflammatory and auto-immune conditions. In 2009, the first phase industry-sponsored III trial of MSCs (Prochymal) was completed to investigate their use in treating steroid-refractory GvHD (NCT00366145). The study failed to meet its' primary end-point (complete remission of GvHD 28 days after infusion). However, it was observed that response rates were higher in children, in patients who were treated early, and in patients with gut and liver GvHD. Subsequently a clinical trial was conducted of MSCs in pediatric, severe, GvHD (NCT02336230), although the trial results have not yet been published. Maziarz et al. conducted a phase I dose-escalation study of allogeneic MAPC in 36 patients undergoing myeloablative allogeneic haematopoietic stem cell transplant (87). At day 100, the overall incidences of grade II-IV graft vs. host disease (GvHD) was 37%, but in the group receiving 10 million cells per kg, incidence was 11.1%.

MSC and MAPC have also been investigated in the context of IBD. A phase III clinical trial (NCT01541579) found that allogeneic MSC sourced from adipose tissue were superior to placebo in the treatment of peri-anal fistulas associated with Crohn's disease (154). In a phase II study (NCT01240915) of MAPC in ulcerative colitis refractory to other medical treatments, no significant beneficial effect was seen.

Recent phase II trials suggest that a single MSC infusion in critically ill patients with ARDS is safe, although no impact on mortality was observed (155). Phase IIa trials are in progress to investigate whether MAPC can help in resolution of Acute Respiratory Distress Syndrome (ARDS). A press release from Athersys has reported that preliminary data show a lower mortality and a greater number of ventilator-free days in patients receiving MAPC. Clinical trials of MSC in the context of systemic lupus erythematosus (SLE) (NCT02633163) and diabetes (NCT03484741, NCT02893306, and NCT03343782) are ongoing.

In phase I studies of multiple sclerosis with amyotrophic lateral sclerosis, intrathecal and intramuscular MSC therapy was deemed safe, with possible therapeutic efficacy (156–158). The likely mechanism of benefit is a combination of anti-inflammatory properties of the MSC as well as release of neurotrophic factors.

One of the obvious concerns about using therapies with anti-inflammatory and immunomodulatory properties is the possibility that they increase incidence of infection; however, in clinical trials there has been no reported increased incidence of infection with either MAPC or MSC. To the contrary, there is evidence that MSC and MAPC have anti-microbial effects. In pre-clinical studies, MSC were found to provide protection against sepsis (159–161), whilst MAPC infusions in rats with spinal cord injury were associated with a reduced incidence of urinary tract infection (162). Possible mechanisms of MSC anti-microbial action include release of anti-bacterial peptides (74, 163), and enhancement of the phagocytic activity of neutrophils and macrophages (164). The mechanisms by which MAPC may reduce the incidence of infection are unknown, and this represents an exciting area for future research.

ROLE IN ANGIOGENESIS

Angiogenesis is the process by which new vasculature sprouts from pre-existing blood vessels. MSC can induce proliferation and migration of endothelial cells promoting tube formation. MSC have been shown to promote angiogenesis in a murine model of cardiac ischaemia reperfusion injury (165). They have also been used to promote angiogenesis in animal models of stroke, myocardial infarction, neurogenic bladder, peripheral artery disease, and stress urinary incontinence (166–168). MSC can secrete both angiogenic and anti-angiogenic factors, and the net result is likely determined by signals from the surrounding environment. For example, exposure to TGF-alpha was shown to increase levels of pro-angiogenic growth factors VEGF, hepatocyte growth factor (PDGF), IL-6 and IL-8 (169). In a murine model of acute limb ischaemic, Ryu et al. found that mice treated with MAPC had higher levels of p-selectin and recruited more $Ly6c^{lo}$ monocytes, which are pro-angiogenic (45). VEGF is an angiogenic factor, which has been identified in the MAPC secretome (55). In a study of MI in pigs, Wang et al. found that conditioned media from the MAPC had higher levels of VEGF, and levels increased further after hypoxia. The authors proposed that this could explain their finding of increased cardiac vascular density in the group treated with MAPC rather than saline (49). Medicetty et al. had similar findings in their pig model of MI (170). In models of critical limb ischaemia in mice, an increase in VEGF, bFGF, and IGF-1 were thought to be responsible for the improved blood flow in mice receiving MAPC injections (171, 172). *In vitro*, serum free conditioned media from MAPC induced endothelial tube formation, but tubes were no longer formed when CXCL5, IL-8, and VEGF were depleted, suggesting the critical role of these proteins (35).

In vitro studies suggest that the angiogenic properties of MAPC are superior to MSC. For example, expression of pro-angiogenic proteins GRO, IL-8, and VEGF were found in higher levels in MAPC than MSC, and MAPC had superior functionality in inducing formation of endothelial tubes from human umbilical vein endothelial cells (HUVEC) (26, 173).

In clinical studies, both MAPC and MSC have been investigated in the context of cardiovascular disease. Unsorted bone marrow mononuclear cells have been shown in meta-analysis to be associated with a modest but significant improvement in left ventricular ejection fraction in patients with ischaemic heart disease (174). A recently published meta-analysis including data from 950 patients (across 14 randomized placebo-controlled trials) post-myocardial infarction concluded that MSC therapy was associated with a 3.84% improvement in left ventricular ejection fraction (95% CI 2.32–5.35), and reduction in scar mass by −1.13 (95% CI −1.80 to −0.46) (175). In a phase I clinical trial, Penn et al. found that MAPC treatment in patients following ST-elevation MI was associated with improved ejection fraction (13.5%) and left ventricular stroke volume (25.4 ml), although the study was not statistically powered to detect differences in clinical outcomes (136). Differentiation into cardiomyocytes is a potential mechanism of these effects, but more likely possibilities are that the bone marrow stromal cells promote neovascularization and/or secrete molecules that promote tissue repair (176, 177).

Several clinical trials of bone marrow stromal cells in acute limb ischaemia have been reported, although these have been of limited size and are mainly not placebo-controlled (178). The studies were predominantly of autologous unsorted bone marrow mononuclear cells, rather than purified MAPC or MSC. Overall, there is some data from these (phase I/II) trials that suggests that bone marrow infusions modestly improves ankle-brachial index and pain-free walking distance, but subgroup analysis using data from placebo-controlled trials only shows no significant effect on amputation rate. There is a possibility that the limited efficacy may be related to dysfunctional angiogenesis in the autologous bone marrow cells of patients with established vascular disease (179). Whilst this has been demonstrated *in vitro* for endothelial progenitor cells and bone marrow mononuclear cells, Gremmels et al. showed no difference in angiogenic capacity *in vitro* in MSCs from patients with critical limb ischaemia vs. healthy participants (180). Further work would be needed in the way of large randomized, placebo-controlled trials to determine which type of cellular therapy (MAPC vs. MSC vs. unsorted bone marrow mononuclear cells) would be most efficacious and whether allogeneic therapy has any therapeutic advantage over autologous.

ROLE IN FIBROSIS

Fibrosis is associated with organ failure and high mortality. It is characterized by aberrant accumulation of myofibroblasts, which secrete extracellular matrix proteins like collgen and fibronectin. MSCs have been investigated for their role in reducing fibrosis in the kidney, lung, heart, skin, liver, and bone marrow.

MSCs have been shown to reduce fibrosis in a model of bleomycin induced lung fibrosis (143), and this effect could be reproduced using conditioned media from MSC (181). Cahill et al. showed that MSC promoted fibroblast migration to areas of lung injury, but also inhibited fibroblast proliferation and activation (182). Possible mechanisms include secretion of hepatocyte-growth factor by MSCs, increased levels of MMP expression, and inhibition of TGF-beta. In pre-clinical models, MSC have also been demonstrated to improve dermal fibrosis, with a reduction in alpha-sma-positive myofibroblasts and downregulation of TGF-beta, type I collagen and heat-shock protein 47 expression (183). Pre-clinical models also provide evidence of MSCs having anti-fibrotic effects in the liver, with associated reductions in TGF-beta and alpha-sma expression (184). In early phase clinical trials, there is demonstrable benefit of MSCs on liver biochemistry and MELD score, although evidence of histological benefits is lacking (185). Of course, MSCs have the potential to differentiate into fibroblasts, and so there have been concerned raised about their potential to worsen liver fibrosis; however, there is no evidence of worsening of liver fibrosis on adoptive transfer in clinical trials (185).

In a phase I clinical trial of patients with MI, MAPC were associated with reduced myocardial scarring (136), however, there is little else published clinical data of the effects of MAPC on fibrosis. *In vitro* data show that MAPC do not secrete hepatocyte-growth factor, which does play a role in the anti-fibrotic effects of MSC. However, the MAPC secretome contains a number of factors that could potentially help in reversing fibrosis. For example, MAPC secrete a number of inhibitors of TGF-beta, such as Follistatin-related proteins 1 and 3, and vasorin (55).

CYTOPROTECTIVE/ANTI-APOPTOTIC EFFECTS

Both MAPC and MSC have been shown to have anti-apoptotic effects. MSCs can protect against apoptosis by decreasing pro-apoptotic factors like Bax and cleaved caspase 3 expression, whilst increasing anti-apoptotic factors such as Bcl-2 (186).

In vitro, MSC and their exosomes have been shown to have high resistance to oxidative stress due to their constitutive expression of a number of anti-oxidant enzymes such as catalase (187). Consistent with this, MSC have been shown to protect hippocampal neurons against oxidative stress caused by amyloid oligomers in a rat model of Alzheimer's disease, which was considered to be due to release of catalase from extra-cellular vesicles, as well as secretion of IL-10, IL-6, and VEGF (188).

Traumatic brain injury usually reduces spleen size, yet rats with traumatic brain injury receiving, MAPC had preservation of their spleen size, which was associated with increased splenocyte proliferation and reduced splenocyte apoptosis (as shown by reduction in caspase 7 and caspase 12 levels on PCR) (153). Pigs receiving MAPC following induced MI had reduced cardiomyocyte apoptosis (46, 49), which was associated with reduced cytochrome C release from cells and downregulation of mitochondrial oxidative enzymes, suggestive of protection of oxidative stress. In addition there was differential expression of genes relating to metabolism and apoptosis detected on gene array (46, 49). When oligodendrocytes were exposed to sub-lethal volumes of hydrogen peroxide, subsequent co-culture with MAPC helped to prolong oligodendrocyte survival, again suggesting that they can protect against oxidative stress (63).

Conditioned media from MSC and MAPC contains numerous neurotrophic factors (189, 190). However, MAPC and MSC have distinct molecular mechanisms for neuroprotection. For example, in pre-clinical studies, it has been shown that tissue inhibitor of metalloproteinase 3 (TIMP3), released by MSCs, plays a critical role in protection against traumatic brain injury by enhancing neuronal survival and neurite outgrowth (191, 192). However, in a rat model of spinal cord injury, in which MAPC were administered with or without TIMP3, the presence of TIMP3 was actually associated with abrogation of MAPC's beneficial effects on tissue sparing and functional recovery (162).

TABLE 2 | Summary of comparison of key characteristics between multipotent adult progenitor cells and mesenchymal stromal cells.

	MAPC	MSC
Main sources of cells used in clinical studies	Bone marrow	Bone marrow, adipose tissue, umbilical cord and placenta
Size	<16 μM	>16 μM
Morphology	Smaller, triangle-shaped	Larger, spindle-shaped
Surface markers	$CD44^{low}$, CD45–, CD49d+, $MHC1^{low}$	CD44+, CD45–, CD73+, CD90+ CD105+, MHC1+, CD140+
Culture conditions	Hypoxia, with platelet-derived growth factor and epidermal growth factor	Normoxia, usually without platelet derived growth factor and epidermal growth factor
Immunogenicity	Low	Low
Limit of population doublings (whilst maintaining telomere length and cytogenetic stability)	~60	~10–38
Number of donors required for clinical dosing in trials	Single	Multiple
Haemocompatibility	Relatively high (associated with low tissue factor expression)	Relatively low, particularly for adipose-tissue derived and umbilical cord derived cells (associated with high tissue factor expression)
Potential for immunomodulation	Yes	Yes
Potential for angiogenesis	Yes (likely more than MSC)	Yes
Potential for anti-fibrotic effects	Very limited testing	Yes
Potential for anti-apoptotic effects	Yes	Yes
Safety	Yes (phase I and II clinical trials)	Yes (phase I, II, and III clinical trials)

MSC, mesenchymal stromal cells; MAPC, multipotent adult progenitor cells.

It was hypothesized that this may be because TIMP3 may interfere with MAPC migration to the site of injury (162).

In phase I clinical studies, MSC infusion was associated with a variable degree of functional improvement after spinal cord injury, associated with increased serum levels of brain-derived neurotrophic factor, glial-derived neurotrophic factor, ciliary neutrophic factor and neurotrophin 3 and 4 (193–196).

In a phase II double-blinded randomized controlled clinical trial of MAPC (Multistem) in stroke (MASTERS trial), patients with anterior circulation infarct received either Multistem ($n = 67$) or placebo ($n = 65$) (152). Whilst there was no difference in global stroke recovery at day 90, patients receiving the infusions earlier (at <36 h post-stroke) had greater improvement that those receiving the therapy at 36–48 h, and further trials are planned.

CURRENT CHALLENGES/FUTURE DIRECTIONS

Bone marrow stromal therapies are a very exciting field for research at present, with evidence showing a range of pleiotropic effects on immunomodulation, fibrosis, apoptosis, and angiogenesis. A summary of the key similarities and differences between MAPC and MSC are shown in **Table 2**.

Evidence so far from hundreds of clinical trials suggests that MAPC and MSC both have a favorable safety profile. Nevertheless, the concern remains that because the cells' activity is so dependent on surrounding stimuli, there is a possibility that they will have unpredictable side effects *in vivo*.

MSCs have been much more widely studied than MAPC. However, given the heterogeneity in cell types labeled as MSC, comparing study results is difficult. There are very few studies in which properties of MAPC and MSC have been directly compared in the same hands, using the same lab materials, such as culture media. Indeed, it may be possible that some of the trials using early-culture MSC, were in fact MAPC.

Whilst it is become clearer that MSC and MAPC are not truly immunologically privileged, their immune evasive nature gives these therapies particular advantage in acute conditions in which it may not be possible to predict the timing of the insult, and delays in delivery of cellular therapy could retract from its potential benefit. However, for use in cases such as bone transplantation, immune compatibility is more critical. A clear comparison of the efficacy of autologous vs. alloogeneic therapy in various clinical conditions is necessary. It would also be imperative to determine whether the negative impacts of IBMIR reaction can be overcome through use of low-passage cells, and what influence this has on the immunomodulatory activity of the remaining cells.

Whilst there is extensive *in vitro* and pre-clinical data supporting the efficacy of MSC and MAPC, the progress of therapy through clinical trials has been slow. There are relatively few clinical studies of MAPC, whilst hundreds of clinical trials are being performed for MSC with many of them have promising findings, the majority of these are phase I and II trials.

So far, data would suggest that the immunomodulatory and cytoprotective capacity of MAPC is equivalent to that of MSC, and that MAPC may have superior angiogenic and broader differentiation properties. In practical terms, MAPC offer the distinct advantage over classical MSC that they can be produced on a large scale, in a reproducible manner.

The use of cell-free preparations would be preferable to the administration of whole cells, and data with MSC suggests that this could be done without significant loss of efficacy. However, such data are not yet available for MAPC and in both cases, there is a need for good manufacturing practice guidelines for the large-scale production of MSC and MAPC derived products, such as exosomes.

The optimal dosing of both cellular therapies is unknown, and in clinical studies a large variation in dosing has been used. In a phase II clinical trial of patients with stroke, up to 1.2 billion cells were administered per patient (152), with no dose-related adverse effects. However, it would be important, for both cost-effectiveness of therapy and safety, to establish the minimum effective dose, which cannot be extrapolated from animal data.

AUTHOR CONTRIBUTIONS

PN had the original concept, provided intellectual input and edited the manuscript, and guarantor. RK wrote the first draft of the manuscript. All authors reviewed the final version.

FUNDING

This work is supported by an MRC grant, reference MR/S001581/1.

REFERENCES

1. Mendez-Ferrer S, Michurina TV, Ferraro F, Mazloom AR, Macarthur BD, Lira SA, et al. Mesenchymal and haematopoietic stem cells form a unique bone marrow niche. *Nature.* (2010) 466:829–34. doi: 10.1038/nature09262
2. Park D, Spencer JA, Koh BI, Kobayashi T, Fujisaki J, Clemens TL, et al. Endogenous bone marrow MSCs are dynamic, fate-restricted participants in bone maintenance and regeneration. *Cell Stem Cell.* (2012) 10:259–72. doi: 10.1016/j.stem.2012.02.003
3. Friedenstein AJ, Petrakova KV, Kurolesova AI, Frolova GP. Heterotopic of bone marrow. Analysis of precursor cells for osteogenic and hematopoietic tissues. *Transplantation.* (1968) 6:230–47. doi: 10.1097/00007890-196803000-00009
4. Dominici M, Le Blanc K, Mueller I, Slaper-Cortenbach I, Marini F, Krause D, et al. Minimal criteria for defining multipotent mesenchymal stromal cells. The International Society for Cellular Therapy position statement. *Cytotherapy.* (2006) 8:315–7. doi: 10.1080/14653240600855905
5. Jossen V, van den Bos C, Eibl R, Eibl D. Manufacturing human mesenchymal stem cells at clinical scale: process and regulatory challenges. *Appl Microbiol Biotechnol.* (2018) 102:3981–94. doi: 10.1007/s00253-018-8912-x
6. Jiang Y, Vaessen B, Lenvik T, Blackstad M, Reyes M, Verfaillie CM. Multipotent progenitor cells can be isolated from postnatal murine bone marrow, muscle, and brain. *Exp Hematol.* (2002) 30:896–904. doi: 10.1016/S0301-472X(02)00869-X
7. Gronthos S, Zannettino AC, Hay SJ, Shi S, Graves SE, Kortesidis A, et al. Molecular and cellular characterisation of highly purified stromal stem cells derived from human bone marrow. *J Cell Sci.* (2003) 116:1827–35. doi: 10.1242/jcs.00369
8. Zuk PA, Zhu M, Mizuno H, Huang J, Futrell JW, Katz AJ, et al. Multilineage cells from human adipose tissue: implications for cell-based therapies. *Tissue Eng.* (2001) 7:211–28. doi: 10.1089/107632701300062859
9. De Bari C, Dell'Accio F, Tylzanowski P, Luyten FP. Multipotent mesenchymal stem cells from adult human synovial membrane. *Arthritis Rheum.* (2001) 44:1928–42. doi: 10.1002/1529-0131(200108)44:8<1928::AID-ART331>3.0.CO;2-P
10. Young HE, Steele TA, Bray RA, Hudson J, Floyd JA, Hawkins K, et al. Human reserve pluripotent mesenchymal stem cells are present in the connective tissues of skeletal muscle and dermis derived from fetal, adult, and geriatric donors. *Anat Rec.* (2001) 264:51–62. doi: 10.1002/ar.1128
11. Perry BC, Zhou D, Wu X, Yang FC, Byers MA, Chu TM, et al. Collection, cryopreservation, and characterization of human dental pulp-derived mesenchymal stem cells for banking and clinical use. *Tissue Eng Part C Methods.* (2008) 14:149–56. doi: 10.1089/ten.tec.2008.0031
12. Gong X, Sun Z, Cui D, Xu X, Zhu H, Wang L, et al. Isolation and characterization of lung resident mesenchymal stem cells capable of differentiating into alveolar epithelial type II cells. *Cell Biol Int.* (2014) 38:405–11. doi: 10.1002/cbin.10240
13. Najar M, Raicevic G, Boufker HI, Fayyad Kazan H, De Bruyn C, Meuleman N, et al. Mesenchymal stromal cells use PGE2 to modulate activation and proliferation of lymphocyte subsets: Combined comparison of adipose tissue, Wharton's Jelly and bone marrow sources. *Cell Immunol.* (2010) 264:171–9. doi: 10.1016/j.cellimm.2010.06.006
14. Lee OK, Kuo TK, Chen WM, Lee KD, Hsieh SL, Chen TH. Isolation of multipotent mesenchymal stem cells from umbilical cord blood. *Blood.* (2004) 103:1669–75. doi: 10.1182/blood-2003-05-1670
15. Antonucci I, Stuppia L, Kaneko Y, Yu S, Tajiri N, Bae EC, et al. Amniotic fluid as a rich source of mesenchymal stromal cells for transplantation therapy. *Cell Transplant.* (2011) 20:789–95. doi: 10.3727/096368910X539074
16. Fukuchi Y, Nakajima H, Sugiyama D, Hirose I, Kitamura T, Tsuji K. Human placenta-derived cells have mesenchymal stem/progenitor cell potential. *Stem Cells.* (2004) 22:649–58. doi: 10.1634/stemcells.22-5-649
17. Yoo KH, Jang IK, Lee MW, Kim HE, Yang MS, Eom Y, et al. Comparison of immunomodulatory properties of mesenchymal stem cells derived from adult human tissues. *Cell Immunol.* (2009) 259:150–6. doi: 10.1016/j.cellimm.2009.06.010
18. Mattar P, Bieback K. Comparing the immunomodulatory properties of bone marrow, adipose tissue, and birth-associated tissue mesenchymal stromal cells. *Front Immunol.* (2015) 6:560. doi: 10.3389/fimmu.2015.00560
19. Melief SM, Zwaginga JJ, Fibbe WE, Roelofs H. Adipose tissue-derived multipotent stromal cells have a higher immunomodulatory capacity than their bone marrow-derived counterparts. *Stem Cells Transl Med.* (2013) 2:455–63. doi: 10.5966/sctm.2012-0184
20. Chao YH, Wu HP, Wu KH, Tsai YG, Peng CT, Lin KC, et al. An increase in CD3+CD4+CD25+ regulatory T cells after administration of umbilical cord-derived mesenchymal stem cells during sepsis. *PLoS ONE.* (2014) 9:e110338. doi: 10.1371/journal.pone.0110338
21. Secunda R, Vennila R, Mohanashankar AM, Rajasundari M, Jeswanth S, Surendran R. Isolation, expansion and characterisation of mesenchymal stem cells from human bone marrow, adipose tissue, umbilical cord blood and matrix: a comparative study. *Cytotechnology.* (2015) 67:793–807. doi: 10.1007/s10616-014-9718-z
22. Baksh D, Yao R, Tuan RS. Comparison of proliferative and multilineage differentiation potential of human mesenchymal stem cells derived from umbilical cord and bone marrow. *Stem Cells.* (2007) 25:1384–92. doi: 10.1634/stemcells.2006-0709
23. Vaes B, Walbers S, Gijbels K, Craeye D, Deans R, Pinxteren J, et al. Culturing protocols for human multipotent adult stem cells. *Methods Mol Biol.* (2015) 1235:49–58. doi: 10.1007/978-1-4939-1785-3_5
24. Vaes B, Van't Hof W, Deans R, Pinxteren J. Application of MultiStem((R)) allogeneic cells for immunomodulatory therapy: clinical progress and pre-clinical challenges in prophylaxis for graft versus host disease. *Front Immunol.* (2012) 3:345. doi: 10.3389/fimmu.2012.00345
25. Bara JJ, Richards RG, Alini M, Stoddart MJ. Concise review: Bone marrow-derived mesenchymal stem cells change phenotype following *in vitro* culture: implications for basic research and the clinic. *Stem Cells.* (2014) 32:1713–23. doi: 10.1002/stem.1649
26. Roobrouck VD, Clavel C, Jacobs SA, Ulloa-Montoya F, Crippa S, Sohni A, et al. Differentiation potential of human postnatal mesenchymal stem cells,

mesoangioblasts, and multipotent adult progenitor cells reflected in their transcriptome and partially influenced by the culture conditions. *Stem Cells.* (2011) 29:871–82. doi: 10.1002/stem.633

27. Moll G, Geissler S, Catar R, Ignatowicz L, Hoogduijn MJ, Strunk D, et al. Cryopreserved or fresh mesenchymal stromal cells: only a matter of taste or key to unleash the full clinical potential of MSC therapy? *Adv Exp Med Biol.* (2016) 951:77–98. doi: 10.1007/978-3-319-45457-3_7
28. Jahagirdar BN, Verfaillie CM. Multipotent adult progenitor cell and stem cell plasticity. *Stem Cell Rev.* (2005) 1:53–9. doi: 10.1385/SCR:1:1:053
29. Singh SP, Tripathy NK, Nityanand S. Comparison of phenotypic markers and neural differentiation potential of multipotent adult progenitor cells and mesenchymal stem cells. *World J Stem Cells.* (2013) 5:53–60. doi: 10.4252/wjsc.v5.i2.53
30. Moll G, Ignatowicz L, Catar R, Luecht C, Sadeghi B, Hamad O, et al. Different procoagulant activity of therapeutic mesenchymal stromal cells derived from bone marrow and placental decidua. *Stem Cells Dev.* (2015) 24:2269–79. doi: 10.1089/scd.2015.0120
31. Oeller M, Laner-Plamberger S, Hochmann S, Ketterl N, Feichtner M, Brachtl G, et al. Selection of tissue factor-deficient cell transplants as a novel strategy for improving hemocompatibility of human bone marrow stromal cells. *Theranostics.* (2018) 8:1421–34. doi: 10.7150/thno.21906
32. Prockop DJ, Brenner M, Fibbe WE, Horwitz E, Le Blanc K, Phinney DG, et al. Defining the risks of mesenchymal stromal cell therapy. *Cytotherapy.* (2010) 12:576–8. doi: 10.3109/14653249.2010.507330
33. Miura M, Miura Y, Padilla-Nash HM, Molinolo AA, Fu B, Patel V, et al. Accumulated chromosomal instability in murine bone marrow mesenchymal stem cells leads to malignant transformation. *Stem Cells.* (2006) 24:1095–103. doi: 10.1634/stemcells.2005-0403
34. Boozer S, Lehman N, Lakshmipathy U, Love B, Raber A, Maitra A, et al. Global characterization and genomic stability of human multistem, a multipotent adult progenitor cell. *J Stem Cells.* (2009) 4:17–28.
35. Lehman N, Cutrone R, Raber A, Perry R, Van't Hof W, Deans R, et al. Development of a surrogate angiogenic potency assay for clinical-grade stem cell production. *Cytotherapy.* (2012) 14:994–1004. doi: 10.3109/14653249.2012.688945
36. Suto EG, Mabuchi Y, Suzuki N, Suzuki K, Ogata Y, Taguchi M, et al. Prospectively isolated mesenchymal stem/stromal cells are enriched in the CD73(+) population and exhibit efficacy after transplantation. *Sci Rep.* (2017) 7:4838. doi: 10.1038/s41598-017-05099-1
37. Rovira-Clave X, Angulo-Ibanez M, Noguer O, Espel E, Reina M. Syndecan-2 can promote clearance of T-cell receptor/CD3 from the cell surface. *Immunology.* (2012) 137:214–25. doi: 10.1111/j.1365-2567.2012.03626.x
38. Liu Y, Wu J, Zhu Y, Han J. Therapeutic application of mesenchymal stem cells in bone and joint diseases. *Clin Exp Med.* (2014) 14:13–24. doi: 10.1007/s10238-012-0218-1
39. Orozco L, Munar A, Soler R, Alberca M, Soler F, Huguet M, et al. Treatment of knee osteoarthritis with autologous mesenchymal stem cells: a pilot study. *Transplantation.* (2013) 95:1535–41. doi: 10.1097/TP.0b013e318291a2da
40. Chen LB, Jiang XB, Yang L. Differentiation of rat marrow mesenchymal stem cells into pancreatic islet beta-cells. *World J Gastroenterol.* (2004) 10:3016–20. doi: 10.3748/wjg.v10.i20.3016
41. Li K, Han Q, Yan X, Liao L, Zhao RC. Not a process of simple vicariousness, the differentiation of human adipose-derived mesenchymal stem cells to renal tubular epithelial cells plays an important role in acute kidney injury repairing. *Stem Cells Dev.* (2010) 19:1267–75. doi: 10.1089/scd.2009.0196
42. Sasaki M, Abe R, Fujita Y, Ando S, Inokuma D, Shimizu H. Mesenchymal stem cells are recruited into wounded skin and contribute to wound repair by transdifferentiation into multiple skin cell type. *J Immunol.* (2008) 180:2581–7. doi: 10.4049/jimmunol.180.4.2581
43. Jellema RK, Ophelders DR, Zwanenburg A, Nikiforou M, Delhaas T, Andriessen P, et al. Multipotent adult progenitor cells for hypoxic-ischemic injury in the preterm brain. *J Neuroinflammation.* (2015) 12:241. doi: 10.1186/s12974-015-0459-5
44. Yang B, Hamilton JA, Valenzuela KS, Bogaerts A, Xi X, Aronowski J, et al. Multipotent adult progenitor cells enhance recovery after stroke by modulating the immune response from the spleen. *Stem Cells.* (2017) 35:1290–302. doi: 10.1002/stem.2600
45. Ryu JC, Davidson BP, Xie A, Qi Y, Zha D, Belcik JT, et al. Molecular imaging of the paracrine proangiogenic effects of progenitor cell therapy in limb ischemia. *Circulation.* (2013) 127:710–9. doi: 10.1161/CIRCULATIONAHA.112.116103
46. Jameel MN, Li Q, Mansoor A, Qiang X, Sarver A, Wang X, et al. Long-term functional improvement and gene expression changes after bone marrow-derived multipotent progenitor cell transplantation in myocardial infarction. *Am J Physiol Heart Circ Physiol.* (2010) 298:H1348–56. doi: 10.1152/ajpheart.01100.2009
47. Sohni A, Verfaillie CM. Mesenchymal stem cells migration homing and tracking. *Stem Cells Int.* (2013) 2013:130763. doi: 10.1155/2013/130763
48. Wang J, Liao L, Tan J. Mesenchymal-stem-cell-based experimental and clinical trials: current status and open questions. *Expert Opin Biol Ther.* (2011) 11:893–909. doi: 10.1517/14712598.2011.574119
49. Wang X, Jameel MN, Li Q, Mansoor A, Qiang X, Swingen C, et al. Stem cells for myocardial repair with use of a transarterial catheter. *Circulation.* (2009) 120:S238–46. doi: 10.1161/CIRCULATIONAHA.109.885236
50. Wislet-Gendebien S, Hans G, Leprince P, Rigo JM, Moonen G, Rogister B. Plasticity of cultured mesenchymal stem cells: switch from nestin-positive to excitable neuron-like phenotype. *Stem Cells.* (2005) 23:392–402. doi: 10.1634/stemcells.2004-0149
51. Volz AC, Huber B, Kluger PJ. Adipose-derived stem cell differentiation as a basic tool for vascularized adipose tissue engineering. *Differentiation.* (2016) 92:52–64. doi: 10.1016/j.diff.2016.02.003
52. Torralba D, Baixauli F, Sanchez-Madrid F. Mitochondria know no boundaries: mechanisms and functions of intercellular mitochondrial transfer. *Front Cell Dev Biol.* (2016) 4:107. doi: 10.3389/fcell.2016.00107
53. Di Nicola M, Carlo-Stella C, Magni M, Milanesi M, Longoni PD, Matteucci P, et al. Human bone marrow stromal cells suppress T-lymphocyte proliferation induced by cellular or nonspecific mitogenic stimuli. *Blood.* (2002) 99:3838–43. doi: 10.1182/blood.V99.10.3838
54. Svensson KJ, Christianson HC, Wittrup A, Bourseau-Guilmain E, Lindqvist E, Svensson LM, et al. Exosome uptake depends on ERK1/2-heat shock protein 27 signaling and lipid Raft-mediated endocytosis negatively regulated by caveolin-1. *J Biol Chem.* (2013) 288:17713–24. doi: 10.1074/jbc.M112.445403
55. Burrows GG, Van't Hof W, Newell LF, Reddy A, Wilmarth PA, David LL, et al. Dissection of the human multipotent adult progenitor cell secretome by proteomic analysis. *Stem Cells Transl Med.* (2013) 2:745–57. doi: 10.5966/sctm.2013-0031
56. Wragg A, Mellad JA, Beltran LE, Konoplyannikov M, San H, Boozer S, et al. VEGFR1/CXCR4-positive progenitor cells modulate local inflammation and augment tissue perfusion by a SDF-1-dependent mechanism. *J Mol Med.* (2008) 86:1221–32. doi: 10.1007/s00109-008-0390-7
57. Kyurkchiev D, Bochev I, Ivanova-Todorova E, Mourdjeva M, Oreshkova T, Belemezova K, et al. Secretion of immunoregulatory cytokines by mesenchymal stem cells. *World J Stem Cells.* (2014) 6:552–70. doi: 10.4252/wjsc.v6.i5.552
58. Chen L, Tredget EE, Wu PY, Wu Y. Paracrine factors of mesenchymal stem cells recruit macrophages and endothelial lineage cells and enhance wound healing. *PLoS ONE.* (2008) 3:e1886. doi: 10.1371/journal.pone.0001886
59. Wang M, Crisostomo PR, Herring C, Meldrum KK, Meldrum DR. Human progenitor cells from bone marrow or adipose tissue produce VEGF, HGF, and IGF-I in response to TNF by a p38 MAPK-dependent mechanism. *Am J Physiol Regul Integr Comp Physiol.* (2006) 291:R880–4. doi: 10.1152/ajpregu.00280.2006
60. Liu CH, Hwang SM. Cytokine interactions in mesenchymal stem cells from cord blood. *Cytokine.* (2005) 32:270–9. doi: 10.1016/j.cyto.2005.11.003
61. Pereira T, Ivanova G, Caseiro AR, Barbosa P, Bartolo PJ, Santos JD, et al. MSCs conditioned media and umbilical cord blood plasma metabolomics and composition. *PLoS ONE.* (2014) 9:e113769. doi: 10.1371/journal.pone.0113769
62. Kovacsovics-Bankowski M, Streeter PR, Mauch KA, Frey MR, Raber A, van't Hof W, et al. Clinical scale expanded adult pluripotent stem cells prevent graft-versus-host disease. *Cell Immunol.* (2009) 255:55–60. doi: 10.1016/j.cellimm.2008.10.004
63. Ravanidis S, Bogie JF, Donders R, Craeye D, Mays RW, Deans R, et al. Neuroinflammatory signals enhance the immunomodulatory and

neuroprotective properties of multipotent adult progenitor cells. *Stem Cell Res Ther.* (2015) 6:176. doi: 10.1186/s13287-015-0169-z

64. Carty F, Corbett JM, Cunha J, Reading JL, Tree TIM, Ting AE, et al. Multipotent adult progenitor cells suppress T cell activation in *in vivo* models of homeostatic proliferation in a prostaglandin E2-dependent manner. *Front Immunol.* (2018) 9:645. doi: 10.3389/fimmu.2018.00645
65. Zagoura DS, Roubelakis MG, Bitsika V, Trohatou O, Pappa KI, Kapelouzou A, et al. Therapeutic potential of a distinct population of human amniotic fluid mesenchymal stem cells and their secreted molecules in mice with acute hepatic failure. *Gut.* (2012) 61:894–906. doi: 10.1136/gutjnl-2011-300908
66. Rehman J, Traktuev D, Li J, Merfeld-Clauss S, Temm-Grove CJ, Bovenkerk JE, et al. Secretion of angiogenic and antiapoptotic factors by human adipose stromal cells. *Circulation.* (2004) 109:1292–8. doi: 10.1161/01.CIR.0000121425.42966.F1
67. Selmani Z, Naji A, Zidi I, Favier B, Gaiffe E, Obert L, et al. Human leukocyte antigen-G5 secretion by human mesenchymal stem cells is required to suppress T lymphocyte and natural killer function and to induce CD4+CD25highFOXP3+ regulatory T cells. *Stem Cells.* (2008) 26:212–22. doi: 10.1634/stemcells.2007-0554
68. Nasef A, Chapel A, Mazurier C, Bouchet S, Lopez M, Mathieu N, et al. Identification of IL-10 and TGF-beta transcripts involved in the inhibition of T-lymphocyte proliferation during cell contact with human mesenchymal stem cells. *Gene Expr.* (2007) 13:217–26. doi: 10.3727/000000006780666957
69. Augello A, Tasso R, Negrini SM, Amateis A, Indiveri F, Cancedda R, et al. Bone marrow mesenchymal progenitor cells inhibit lymphocyte proliferation by activation of the programmed death 1 pathway. *Eur J Immunol.* (2005) 35:1482–90. doi: 10.1002/eji.200425405
70. Crigler L, Robey RC, Asawachaicharn A, Gaupp D, Phinney DG. Human mesenchymal stem cell subpopulations express a variety of neuro-regulatory molecules and promote neuronal cell survival and neuritogenesis. *Exp Neurol.* (2006) 198:54–64. doi: 10.1016/j.expneurol.2005.10.029
71. Teixeira FG, Carvalho MM, Panchalingam KM, Rodrigues AJ, Mendes-Pinheiro B, Anjo S, et al. Impact of the secretome of human mesenchymal stem cells on brain structure and animal behavior in a rat model of Parkinson's disease. *Stem Cells Transl Med.* (2017) 6:634–46. doi: 10.5966/sctm.2016-0071
72. Polacek M, Bruun JA, Elvenes J, Figenschau Y, Martinez I. The secretory profiles of cultured human articular chondrocytes and mesenchymal stem cells: implications for autologous cell transplantation strategies. *Cell Transplant.* (2011) 20:1381–93. doi: 10.3727/096368910X550215
73. Linero I, Chaparro O. Paracrine effect of mesenchymal stem cells derived from human adipose tissue in bone regeneration. *PLoS ONE.* (2014) 9:e107001. doi: 10.1371/journal.pone.0107001
74. Krasnodembskaya A, Song Y, Fang X, Gupta N, Serikov V, Lee JW, et al. Antibacterial effect of human mesenchymal stem cells is mediated in part from secretion of the antimicrobial peptide LL-37. *Stem Cells.* (2010) 28:2229–38. doi: 10.1002/stem.544
75. Timmers L, Lim SK, Arslan F, Armstrong JS, Hoefer IE, Doevendans PA, et al. Reduction of myocardial infarct size by human mesenchymal stem cell conditioned medium. *Stem Cell Res.* (2007) 1:129–37. doi: 10.1016/j.scr.2008.02.002
76. Legaki E, Roubelakis MG, Theodoropoulos GE, Lazaris A, Kollia A, Karamanolis G, et al. Therapeutic potential of secreted molecules derived from human amniotic fluid mesenchymal stem/stroma cells in a mice model of colitis. *Stem Cell Rev.* (2016) 12:604–12. doi: 10.1007/s12015-016-9677-1
77. Kordelas L, Rebmann V, Ludwig AK, Radtke S, Ruesing J, Doeppner TR, et al. MSC-derived exosomes: a novel tool to treat therapy-refractory graft-versus-host disease. *Leukemia.* (2014) 28:970–3. doi: 10.1038/leu.2014.41
78. Croitoru-Lamoury J, Lamoury FM, Caristo M, Suzuki K, Walker D, Takikawa O, et al. Interferon-gamma regulates the proliferation and differentiation of mesenchymal stem cells via activation of indoleamine 2,3 dioxygenase (IDO). *PLoS ONE.* (2011) 6:e14698. doi: 10.1371/journal.pone.0014698
79. Kwon YW, Heo SC, Jeong GO, Yoon JW, Mo WM, Lee MJ, et al. Tumor necrosis factor-alpha-activated mesenchymal stem cells promote endothelial progenitor cell homing and angiogenesis. *Biochim Biophys Acta.* (2013) 1832:2136–44. doi: 10.1016/j.bbadis.2013.08.002
80. Pourgholaminejad A, Aghdami N, Baharvand H, Moazzeni SM. The effect of pro-inflammatory cytokines on immunophenotype, differentiation capacity and immunomodulatory functions of human mesenchymal stem cells. *Cytokine.* (2016) 85:51–60. doi: 10.1016/j.cyto.2016.06.003
81. Duijvestein M, Wildenberg ME, Welling MM, Hennink S, Molendijk I, van Zuylen VL, et al. Pretreatment with interferon-gamma enhances the therapeutic activity of mesenchymal stromal cells in animal models of colitis. *Stem Cells.* (2011) 29:1549–58. doi: 10.1002/stem.698
82. Ryan JM, Barry FP, Murphy JM, Mahon BP. Mesenchymal stem cells avoid allogeneic rejection. *J Inflamm.* (2005) 2:8. doi: 10.1186/1476-9255-2-8
83. Tolar J, O'Shaughnessy MJ, Panoskaltsis-Mortari A, McElmurry RT, Bell S, Riddle M, et al. Host factors that impact the biodistribution and persistence of multipotent adult progenitor cells. *Blood.* (2006) 107:4182–8. doi: 10.1182/blood-2005-08-3289
84. Le Blanc K, Tammik C, Rosendahl K, Zetterberg E, Ringden O. HLA expression and immunologic properties of differentiated and undifferentiated mesenchymal stem cells. *Exp Hematol.* (2003) 31:890–6. doi: 10.1016/S0301-472X(03)00110-3
85. Eliopoulos N, Stagg J, Lejeune L, Pommey S, Galipeau J. Allogeneic marrow stromal cells are immune rejected by MHC class I- and class II-mismatched recipient mice. *Blood.* (2005) 106:4057–65. doi: 10.1182/blood-2005-03-1004
86. Hare JM, Fishman JE, Gerstenblith G, DiFede Velazquez DL, Zambrano JP, Suncion VY, et al. Comparison of allogeneic vs autologous bone marrow-derived mesenchymal stem cells delivered by transendocardial injection in patients with ischemic cardiomyopathy: the POSEIDON randomized trial. *JAMA.* (2012) 308:2369–79. doi: 10.1001/jama.2012.25321
87. Maziarz RT, Devos T, Bachier CR, Goldstein SC, Leis JF, Devine SM, et al. Single and multiple dose MultiStem (multipotent adult progenitor cell) therapy prophylaxis of acute graft-versus-host disease in myeloablative allogeneic hematopoietic cell transplantation: a phase 1 trial. *Biol Blood Marrow Transplant.* (2015) 21:720–8. doi: 10.1016/j.bbmt.2014.12.025
88. Moll G, Rasmusson-Duprez I, von Bahr L, Connolly-Andersen AM, Elgue G, Funke L, et al. Are therapeutic human mesenchymal stromal cells compatible with human blood? *Stem Cells.* (2012) 30:1565–74. doi: 10.1002/stem.1111
89. Gustafson EK, Elgue G, Hughes RD, Mitry RR, Sanchez J, Haglund U, et al. The instant blood-mediated inflammatory reaction characterized in hepatocyte transplantation. *Transplantation.* (2011) 91:632–8. doi: 10.1097/TP.0b013e31820ae459
90. Bennet W, Sundberg B, Groth CG, Brendel MD, Brandhorst D, Brandhorst H, et al. Incompatibility between human blood and isolated islets of Langerhans: a finding with implications for clinical intraportal islet transplantation? *Diabetes.* (1999) 48:1907–14. doi: 10.2337/diabetes.48.10.1907
91. George MJ, Prabhakara K, Toledano-Furman NE, Wang YW, Gill BS, Wade CE, et al. Clinical cellular therapeutics accelerate clot formation. *Stem Cells Transl Med.* (2018) 7:731–9. doi: 10.1002/sctm.18-0015
92. Aggarwal S, Pittenger MF. Human mesenchymal stem cells modulate allogeneic immune cell responses. *Blood.* (2005) 105:1815–22. doi: 10.1182/blood-2004-04-1559
93. Dembinski JL, Wilson SM, Spaeth EL, Studeny M, Zompetta C, Samudio I, et al. Tumor stroma engraftment of gene-modified mesenchymal stem cells as anti-tumor therapy against ovarian cancer. *Cytotherapy.* (2013) 15:20–32. doi: 10.1016/j.jcyt.2012.10.003
94. Le Blanc K, Frassoni F, Ball L, Locatelli F, Roelofs H, Lewis I, et al. Mesenchymal stem cells for treatment of steroid-resistant, severe, acute graft-versus-host disease: a phase II study. *Lancet.* (2008) 371:1579–86. doi: 10.1016/S0140-6736(08)60690-X
95. Moll G, Jitschin R, von Bahr L, Rasmusson-Duprez I, Sundberg B, Lonnies L, et al. Mesenchymal stromal cells engage complement and complement receptor bearing innate effector cells to modulate immune responses. *PLoS ONE.* (2011) 6:e21703. doi: 10.1371/journal.pone.0021703
96. Schraufstatter IU, Discipio RG, Zhao M, Khaldoyanidi SK. C3a and C5a are chemotactic factors for human mesenchymal stem cells, which cause prolonged ERK1/2 phosphorylation. *J Immunol.* (2009) 182:3827–36. doi: 10.4049/jimmunol.0803055
97. Agis H, Kandler B, Fischer MB, Watzek G, Gruber R. Activated platelets increase fibrinolysis of mesenchymal progenitor cells. *J Orthop Res.* (2009) 27:972–80. doi: 10.1002/jor.20819
98. Moll G, Ankrum JA, Kamhieh-Milz J, Bieback K, Ringden O, Volk HD, et al. Intravascular mesenchymal stromal/stem cell therapy product

diversification: time for new clinical guidelines. *Trends Mol Med.* (2019) 25:149–63. doi: 10.1016/j.molmed.2018.12.006
99. An Y, Liu WJ, Xue P, Ma Y, Zhang LQ, Zhu B, et al. Autophagy promotes MSC-mediated vascularization in cutaneous wound healing via regulation of VEGF secretion. *Cell Death Dis.* (2018) 9:58. doi: 10.1038/s41419-017-0082-8
100. Galleu A, Riffo-Vasquez Y, Trento C, Lomas C, Dolcetti L, Cheung TS, et al. Apoptosis in mesenchymal stromal cells induces *in vivo* recipient-mediated immunomodulation. *Sci Transl Med.* (2017) 9:eaam7828. doi: 10.1126/scitranslmed.aam7828
101. de Witte SFH, Luk F, Sierra Parraga JM, Gargesha M, Merino A, Korevaar SS, et al. Immunomodulation by therapeutic Mesenchymal Stromal Cells (MSC) is triggered through phagocytosis of MSC by monocytic cells. *Stem Cells.* (2018) 36:602–15. doi: 10.1002/stem.2779
102. Elliott MR, Koster KM, Murphy PS. Efferocytosis signaling in the regulation of macrophage inflammatory responses. *J Immunol.* (2017) 198:1387–94. doi: 10.4049/jimmunol.1601520
103. Dang S, Yu ZM, Zhang CY, Zheng J, Li KL, Wu Y, et al. Autophagy promotes apoptosis of mesenchymal stem cells under inflammatory microenvironment. *Stem Cell Res Ther.* (2015) 6:247. doi: 10.1186/s13287-015-0245-4
104. Li H, Fan X, Kovi RC, Jo Y, Moquin B, Konz R, et al. Spontaneous expression of embryonic factors and p53 point mutations in aged mesenchymal stem cells: a model of age-related tumorigenesis in mice. *Cancer Res.* (2007) 67:10889–98. doi: 10.1158/0008-5472.CAN-07-2665
105. Mohseny AB, Szuhai K, Romeo S, Buddingh EP, Briaire-de Bruijn I, de Jong D, et al. Osteosarcoma originates from mesenchymal stem cells in consequence of aneuploidization and genomic loss of Cdkn2. *J Pathol.* (2009) 219:294–305. doi: 10.1002/path.2603
106. Tolar J, Nauta AJ, Osborn MJ, Panoskaltsis Mortari A, McElmurry RT, Bell S, et al. Sarcoma derived from cultured mesenchymal stem cells. *Stem Cells.* (2007) 25:371–9. doi: 10.1634/stemcells.2005-0620
107. Garcia S, Bernad A, Martin MC, Cigudosa JC, Garcia-Castro J, de la Fuente R. Pitfalls in spontaneous *in vitro* transformation of human mesenchymal stem cells. *Exp Cell Res.* (2010) 316:1648–50. doi: 10.1016/j.yexcr.2010.02.016
108. Bernardo ME, Zaffaroni N, Novara F, Cometa AM, Avanzini MA, Moretta A, et al. Human bone marrow derived mesenchymal stem cells do not undergo transformation after long-term *in vitro* culture and do not exhibit telomere maintenance mechanisms. *Cancer Res.* (2007) 67:9142–9. doi: 10.1158/0008-5472.CAN-06-4690
109. Choumerianou DM, Dimitriou H, Perdikogianni C, Martimianaki G, Riminucci M, Kalmanti M. Study of oncogenic transformation in *ex vivo* expanded mesenchymal cells, from paediatric bone marrow. *Cell Prolif.* (2008) 41:909–22. doi: 10.1111/j.1365-2184.2008.00559.x
110. Gou S, Wang C, Liu T, Wu H, Xiong J, Zhou F, et al. Spontaneous differentiation of murine bone marrow-derived mesenchymal stem cells into adipocytes without malignant transformation after long-term culture. *Cells Tissues Organs.* (2010) 191:185–92. doi: 10.1159/000240246
111. Grimes BR, Steiner CM, Merfeld-Clauss S, Traktuev DO, Smith D, Reese A, et al. Interphase FISH demonstrates that human adipose stromal cells maintain a high level of genomic stability in long-term culture. *Stem Cells Dev.* (2009) 18:717–24. doi: 10.1089/scd.2008.0255
112. Larson BL, Ylostalo J, Lee RH, Gregory C, Prockop DJ. Sox11 is expressed in early progenitor human multipotent stromal cells and decreases with extensive expansion of the cells. *Tissue Eng Part A.* (2010) 16:3385–94. doi: 10.1089/ten.tea.2010.0085
113. Rodriguez R, Rubio R, Masip M, Catalina P, Nieto A, de la Cueva T, et al. Loss of p53 induces tumorigenesis in p21-deficient mesenchymal stem cells. *Neoplasia.* (2009) 11:397–407. doi: 10.1593/neo.81620
114. Rubio R, Garcia-Castro J, Gutierrez-Aranda I, Paramio J, Santos M, Catalina P, et al. Deficiency in p53 but not retinoblastoma induces the transformation of mesenchymal stem cells *in vitro* and initiates leiomyosarcoma *in vivo.* *Cancer Res.* (2010) 70:4185–94. doi: 10.1158/0008-5472.CAN-09-4640
115. Li N, Yang R, Zhang W, Dorfman H, Rao P, Gorlick R. Genetically transforming human mesenchymal stem cells to sarcomas: changes in cellular phenotype and multilineage differentiation potential. *Cancer.* (2009) 115:4795–806. doi: 10.1002/cncr.24519
116. Rodriguez R, Rosu-Myles M, Arauzo-Bravo M, Horrillo A, Pan Q, Gonzalez-Rey E, et al. Human bone marrow stromal cells lose immunosuppressive and anti-inflammatory properties upon oncogenic transformation. *Stem Cell Rep.* (2014) 3:606–19. doi: 10.1016/j.stemcr.2014.08.005
117. Kucerova L, Matuskova M, Hlubinova K, Altanerova V, Altaner C. Tumor cell behaviour modulation by mesenchymal stromal cells. *Mol Cancer.* (2010) 9:129. doi: 10.1186/1476-4598-9-129
118. Karnoub AE, Dash AB, Vo AP, Sullivan A, Brooks MW, Bell GW, et al. Mesenchymal stem cells within tumour stroma promote breast cancer metastasis. *Nature.* (2007) 449:557–63. doi: 10.1038/nature06188
119. Prantl L, Muehlberg F, Navone NM, Song YH, Vykoukal J, Logothetis CJ, et al. Adipose tissue-derived stem cells promote prostate tumor growth. *Prostate.* (2010) 70:1709–15. doi: 10.1002/pros.21206
120. Stagg J. Mesenchymal stem cells in cancer. *Stem Cell Rev.* (2008) 4:119–24. doi: 10.1007/s12015-008-9030-4
121. Jung JW, Kwon M, Choi JC, Shin JW, Park IW, Choi BW, et al. Familial occurrence of pulmonary embolism after intravenous, adipose tissue-derived stem cell therapy. *Yonsei Med J.* (2013) 54:1293–6. doi: 10.3349/ymj.2013.54.5.1293
122. Lim R, Malhotra A, Tan J, Chan ST, Lau S, Zhu D, et al. First-in-human administration of allogeneic amnion cells in premature infants with bronchopulmonary dysplasia: a safety study. *Stem Cells Transl Med.* (2018) 7:628–35. doi: 10.1002/sctm.18-0079
123. Yasuhara T, Hara K, Maki M, Mays RW, Deans RJ, Hess DC, et al. Intravenous grafts recapitulate the neurorestoration afforded by intracerebrally delivered multipotent adult progenitor cells in neonatal hypoxic-ischemic rats. *J Cereb Blood Flow Metab.* (2008) 28:1804–10. doi: 10.1038/jcbfm.2008.68
124. Gao J, Dennis JE, Muzic RF, Lundberg M, Caplan AI. The dynamic *in vivo* distribution of bone marrow-derived mesenchymal stem cells after infusion. *Cells Tissues Organs.* (2001) 169:12–20. doi: 10.1159/000047856
125. Schrepfer S, Deuse T, Reichenspurner H, Fischbein MP, Robbins RC, Pelletier MP. Stem cell transplantation: the lung barrier. *Transplant Proc.* (2007) 39:573–6. doi: 10.1016/j.transproceed.2006.12.019
126. Devine SM, Cobbs C, Jennings M, Bartholomew A, Hoffman R. Mesenchymal stem cells distribute to a wide range of tissues following systemic infusion into nonhuman primates. *Blood.* (2003) 101:2999–3001. doi: 10.1182/blood-2002-06-1830
127. Koc ON, Gerson SL, Cooper BW, Dyhouse SM, Haynesworth SE, Caplan AI, et al. Rapid hematopoietic recovery after coinfusion of autologous-blood stem cells and culture-expanded marrow mesenchymal stem cells in advanced breast cancer patients receiving high-dose chemotherapy. *J Clin Oncol.* (2000) 18:307–16. doi: 10.1200/JCO.2000.18.2.307
128. Kraitchman DL, Tatsumi M, Gilson WD, Ishimori T, Kedziorek D, Walczak P, et al. Dynamic imaging of allogeneic mesenchymal stem cells trafficking to myocardial infarction. *Circulation.* (2005) 112:1451–61. doi: 10.1161/CIRCULATIONAHA.105.537480
129. Gholamrezanezhad A, Mirpour S, Bagheri M, Mohamadnejad M, Alimoghaddam K, Abdolahzadeh L, et al. *In vivo* tracking of 111In-oxine labeled mesenchymal stem cells following infusion in patients with advanced cirrhosis. *Nucl Med Biol.* (2011) 38:961–7. doi: 10.1016/j.nucmedbio.2011.03.008
130. Lee RH, Pulin AA, Seo MJ, Kota DJ, Ylostalo J, Larson BL, et al. Intravenous hMSCs improve myocardial infarction in mice because cells embolized in lung are activated to secrete the anti-inflammatory protein TSG-6. *Cell Stem Cell.* (2009) 5:54–63. doi: 10.1016/j.stem.2009.05.003
131. Fischer UM, Harting MT, Jimenez F, Monzon-Posadas WO, Xue H, Savitz SI, et al. Pulmonary passage is a major obstacle for intravenous stem cell delivery: the pulmonary first-pass effect. *Stem Cells Dev.* (2009) 18:683–92. doi: 10.1089/scd.2008.0253
132. Toma C, Wagner WR, Bowry S, Schwartz A, Villanueva F. Fate of culture-expanded mesenchymal stem cells in the microvasculature: *in vivo* observations of cell kinetics. *Circ Res.* (2009) 104:398–402. doi: 10.1161/CIRCRESAHA.108.187724
133. Ren G, Zhao X, Zhang L, Zhang J, L'Huillier A, Ling W, et al. Inflammatory cytokine-induced intercellular adhesion molecule-1 and vascular cell adhesion molecule-1 in mesenchymal stem cells are critical for immunosuppression. *J Immunol.* (2010) 184:2321–8. doi: 10.4049/jimmunol.0902023

134. Chamberlain G, Smith H, Rainger GE, Middleton J. Mesenchymal stem cells exhibit firm adhesion, crawling, spreading and transmigration across aortic endothelial cells: effects of chemokines and shear. *PLoS ONE.* (2011) 6:e25663. doi: 10.1371/journal.pone.0025663
135. Highfill SL, Kelly RM, O'Shaughnessy MJ, Zhou Q, Xia L, Panoskaltsis-Mortari A, et al. Multipotent adult progenitor cells can suppress graft-versus-host disease via prostaglandin E2 synthesis and only if localized to sites of allopriming. *Blood.* (2009) 114:693–701. doi: 10.1182/blood-2009-03-213850
136. Penn MS, Ellis S, Gandhi S, Greenbaum A, Hodes Z, Mendelsohn FO, et al. Adventitial delivery of an allogeneic bone marrow-derived adherent stem cell in acute myocardial infarction: phase I clinical study. *Circ Res.* (2012) 110:304–11. doi: 10.1161/CIRCRESAHA.111.253427
137. Meisel R, Zibert A, Laryea M, Gobel U, Daubener W, Dilloo D. Human bone marrow stromal cells inhibit allogeneic T-cell responses by indoleamine 2,3-dioxygenase-mediated tryptophan degradation. *Blood.* (2004) 103:4619–21. doi: 10.1182/blood-2003-11-3909
138. Walker PA, Bedi SS, Shah SK, Jimenez F, Xue H, Hamilton JA, et al. Intravenous multipotent adult progenitor cell therapy after traumatic brain injury: modulation of the resident microglia population. *J Neuroinflammation.* (2012) 9:228. doi: 10.1186/1742-2094-9-228
139. Eggenhofer E, Popp FC, Mendicino M, Silber P, Van't Hof W, Renner P, et al. Heart grafts tolerized through third-party multipotent adult progenitor cells can be retransplanted to secondary hosts with no immunosuppression. *Stem Cells Transl Med.* (2013) 2:595–606. doi: 10.5966/sctm.2012-0166
140. Ding Y, Xu D, Feng G, Bushell A, Muschel RJ, Wood KJ. Mesenchymal stem cells prevent the rejection of fully allogenic islet grafts by the immunosuppressive activity of matrix metalloproteinase-2 and−9. *Diabetes.* (2009) 58:1797–806. doi: 10.2337/db09-0317
141. Glennie S, Soeiro I, Dyson PJ, Lam EW, Dazzi F. Bone marrow mesenchymal stem cells induce division arrest anergy of activated T cells. *Blood.* (2005) 105:2821–7. doi: 10.1182/blood-2004-09-3696
142. Rasmusson I, Le Blanc K, Sundberg B, Ringden O. Mesenchymal stem cells stimulate antibody secretion in human B cells. *Scand J Immunol.* (2007) 65:336–43. doi: 10.1111/j.1365-3083.2007.01905.x
143. Corcione A, Benvenuto F, Ferretti E, Giunti D, Cappiello V, Cazzanti F, et al. Human mesenchymal stem cells modulate B-cell functions. *Blood.* (2006) 107:367–72. doi: 10.1182/blood-2005-07-2657
144. Klyushnenkova E, Mosca JD, Zernetkina V, Majumdar MK, Beggs KJ, Simonetti DW, et al. T cell responses to allogeneic human mesenchymal stem cells: immunogenicity, tolerance, and suppression. *J Biomed Sci.* (2005) 12:47–57. doi: 10.1007/s11373-004-8183-7
145. Vasandan AB, Jahnavi S, Shashank C, Prasad P, Kumar A, Prasanna SJ. Human Mesenchymal stem cells program macrophage plasticity by altering their metabolic status via a PGE2-dependent mechanism. *Sci Rep.* (2016) 6:38308. doi: 10.1038/srep38308
146. Busch SA, Hamilton JA, Horn KP, Cuascut FX, Cutrone R, Lehman N, et al. Multipotent adult progenitor cells prevent macrophage-mediated axonal dieback and promote regrowth after spinal cord injury. *J Neurosci.* (2011) 31:944–53. doi: 10.1523/JNEUROSCI.3566-10.2011
147. Lee KC, Lin HC, Huang YH, Hung SC. Allo-transplantation of mesenchymal stem cells attenuates hepatic injury through IL1Ra dependent macrophage switch in a mouse model of liver disease. *J Hepatol.* (2015) 63:1405–12. doi: 10.1016/j.jhep.2015.07.035
148. Anderson P, Souza-Moreira L, Morell M, Caro M, O'Valle F, Gonzalez-Rey E, et al. Adipose-derived mesenchymal stromal cells induce immunomodulatory macrophages which protect from experimental colitis and sepsis. *Gut.* (2013) 62:1131–41. doi: 10.1136/gutjnl-2012-302152
149. Beyth S, Borovsky Z, Mevorach D, Liebergall M, Gazit Z, Aslan H, et al. Human mesenchymal stem cells alter antigen-presenting cell maturation and induce T-cell unresponsiveness. *Blood.* (2005) 105:2214–9. doi: 10.1182/blood-2004-07-2921
150. Rojas M, Cardenes N, Kocyildirim E, Tedrow JR, Caceres E, Deans R, et al. Human adult bone marrow-derived stem cells decrease severity of lipopolysaccharide-induced acute respiratory distress syndrome in sheep. *Stem Cell Res Ther.* (2014) 5:42. doi: 10.1186/scrt430
151. La Francesca S, Ting AE, Sakamoto J, Rhudy J, Bonenfant NR, Borg ZD, et al. Multipotent adult progenitor cells decrease cold ischemic injury in *ex vivo* perfused human lungs: an initial pilot and feasibility study. *Transplant Res.* (2014) 3:19. doi: 10.1186/2047-1440-3-19
152. Hess DC, Wechsler LR, Clark WM, Savitz SI, Ford GA, Chiu D, et al. Safety and efficacy of multipotent adult progenitor cells in acute ischaemic stroke (MASTERS): a randomised, double-blind, placebo-controlled, phase 2 trial. *Lancet Neurol.* (2017) 16:360–8. doi: 10.1016/S1474-4422(17)30046-7
153. Walker PA, Shah SK, Jimenez F, Gerber MH, Xue H, Cutrone R, et al. Intravenous multipotent adult progenitor cell therapy for traumatic brain injury: preserving the blood brain barrier via an interaction with splenocytes. *Exp Neurol.* (2010) 225:341–52. doi: 10.1016/j.expneurol.2010.07.005
154. Panes J, Garcia-Olmo D, Van Assche G, Colombel JF, Reinisch W, Baumgart DC, et al. Expanded allogeneic adipose-derived mesenchymal stem cells (Cx601) for complex perianal fistulas in Crohn's disease: a phase 3 randomised, double-blind controlled trial. *Lancet.* (2016) 388:1281–90. doi: 10.1016/S0140-6736(16)31203-X
155. Matthay MA, Calfee CS, Zhuo H, Thompson BT, Wilson JG, Levitt JE, et al. Treatment with allogeneic mesenchymal stromal cells for moderate to severe acute respiratory distress syndrome (START study): a randomised phase 2a safety trial. *Lancet Respir Med.* (2019) 7:154–62. doi: 10.1016/S2213-2600(18)30418-1
156. Oh KW, Moon C, Kim HY, Oh SI, Park J, Lee JH, et al. Phase I trial of repeated intrathecal autologous bone marrow-derived mesenchymal stromal cells in amyotrophic lateral sclerosis. *Stem Cells Transl Med.* (2015) 4:590–7. doi: 10.5966/sctm.2014-0212
157. Petrou P, Gothelf Y, Argov Z, Gotkine M, Levy YS, Kassis I, et al. Safety and clinical effects of mesenchymal stem cells secreting neurotrophic factor transplantation in patients with amyotrophic lateral sclerosis: results of phase 1/2 and 2a clinical trials. *JAMA Neurol.* (2016) 73:337–44. doi: 10.1001/jamaneurol.2015.4321
158. Staff NP, Madigan NN, Morris J, Jentoft M, Sorenson EJ, Butler G, et al. Safety of intrathecal autologous adipose-derived mesenchymal stromal cells in patients with ALS. *Neurology.* (2016) 87:2230–4. doi: 10.1212/WNL.0000000000003359
159. Qiu G, Zheng G, Ge M, Huang L, Tong H, Chen P, et al. Adipose-derived mesenchymal stem cells modulate CD14(++)CD16(+) expression on monocytes from sepsis patients *in vitro* via prostaglandin E2. *Stem Cell Res Ther.* (2017) 8:97. doi: 10.1186/s13287-017-0546-x
160. Alcayaga-Miranda F, Cuenca J, Khoury M. Antimicrobial activity of mesenchymal stem cells: current status and new perspectives of antimicrobial peptide-based therapies. *Front Immunol.* (2017) 8:339. doi: 10.3389/fimmu.2017.00339
161. Laroye C, Gibot S, Reppel L, Bensoussan D. Concise review: mesenchymal stromal/stem cells: a new treatment for sepsis and septic shock? *Stem Cells.* (2017) 35:2331–9. doi: 10.1002/stem.2695
162. Jones JM, DePaul MA, Gregory CR, Lang BT, Xie H, Zhu M, et al. Multipotent adult progenitor cells, but not tissue inhibitor of matrix metalloproteinase-3, increase tissue sparing and reduce urological complications following spinal cord injury. *J Neurotrauma.* (2019) 36:1416–27. doi: 10.1089/neu.2018.5727
163. Gupta N, Krasnodembskaya A, Kapetanaki M, Mouded M, Tan X, Serikov V, et al. Mesenchymal stem cells enhance survival and bacterial clearance in murine *Escherichia coli* pneumonia. *Thorax.* (2012) 67:533–9. doi: 10.1136/thoraxjnl-2011-201176
164. Mei SH, Haitsma JJ, Dos Santos CC, Deng Y, Lai PF, Slutsky AS, et al. Mesenchymal stem cells reduce inflammation while enhancing bacterial clearance and improving survival in sepsis. *Am J Respir Crit Care Med.* (2010) 182:1047–57. doi: 10.1164/rccm.201001-0010OC
165. Preda MB, Ronningen T, Burlacu A, Simionescu M, Moskaug JO, Valen G. Remote transplantation of mesenchymal stem cells protects the heart against ischemia-reperfusion injury. *Stem Cells.* (2014) 32:2123–34. doi: 10.1002/stem.1687
166. Hsieh JY, Wang HW, Chang SJ, Liao KH, Lee IH, Lin WS, et al. Mesenchymal stem cells from human umbilical cord express preferentially secreted factors related to neuroprotection, neurogenesis, and angiogenesis. *PLoS ONE.* (2013) 8:e72604. doi: 10.1371/journal.pone.0072604
167. Liu G, Pareta RA, Wu R, Shi Y, Zhou X, Liu H, et al. Skeletal myogenic differentiation of urine-derived stem cells and angiogenesis using

microbeads loaded with growth factors. *Biomaterials.* (2013) 34:1311–26. doi: 10.1016/j.biomaterials.2012.10.038

168. Sharma AK, Bury MI, Fuller NJ, Marks AJ, Kollhoff DM, Rao MV, et al. Cotransplantation with specific populations of spina bifida bone marrow stem/progenitor cells enhances urinary bladder regeneration. *Proc Natl Acad Sci USA.* (2013) 110:4003–8. doi: 10.1073/pnas.1220764110
169. De Luca A, Gallo M, Aldinucci D, Ribatti D, Lamura L, D'Alessio A, et al. Role of the EGFR ligand/receptor system in the secretion of angiogenic factors in mesenchymal stem cells. *J Cell Physiol.* (2011) 226:2131–8. doi: 10.1002/jcp.22548
170. Medicetty S, Wiktor D, Lehman N, Raber A, Popovic ZB, Deans R, et al. Percutaneous adventitial delivery of allogeneic bone marrow-derived stem cells via infarct-related artery improves long-term ventricular function in acute myocardial infarction. *Cell Transplant.* (2012) 21:1109–20. doi: 10.3727/096368911X603657
171. Aranguren XL, McCue JD, Hendrickx B, Zhu XH, Du F, Chen E, et al. Multipotent adult progenitor cells sustain function of ischemic limbs in mice. *J Clin Invest.* (2008) 118:505–14. doi: 10.1172/JCI31153
172. Aranguren XL, Pelacho B, Penuelas I, Abizanda G, Uriz M, Ecay M, et al. MAPC transplantation confers a more durable benefit than AC133+ cell transplantation in severe hind limb ischemia. *Cell Transplant.* (2011) 20:259–69. doi: 10.3727/096368910X516592
173. LoGuidice A, Houlihan A, Deans R. Multipotent adult progenitor cells on an allograft scaffold facilitate the bone repair process. *J Tissue Eng.* (2016) 7:2041731416656148. doi: 10.1177/2041731416656148
174. Jeong H, Yim HW, Cho Y, Park HJ, Jeong S, Kim HB, et al. The effect of rigorous study design in the research of autologous bone marrow-derived mononuclear cell transfer in patients with acute myocardial infarction. *Stem Cell Res Ther.* (2013) 4:82. doi: 10.1186/scrt233
175. Jeong H, Yim HW, Park HJ, Cho Y, Hong H, Kim NJ, et al. Mesenchymal stem cell therapy for ischemic heart disease: systematic review and meta-analysis. *Int J Stem Cells.* (2018) 11:1–12. doi: 10.15283/ijsc17061
176. Kocher AA, Schuster MD, Szabolcs MJ, Takuma S, Burkhoff D, Wang J, et al. Neovascularization of ischemic myocardium by human bone-marrow-derived angioblasts prevents cardiomyocyte apoptosis, reduces remodeling and improves cardiac function. *Nat Med.* (2001) 7:430–6. doi: 10.1038/86498
177. Gnecchi M, Zhang Z, Ni A, Dzau VJ. Paracrine mechanisms in adult stem cell signaling and therapy. *Circ Res.* (2008) 103:1204–19. doi: 10.1161/CIRCRESAHA.108.176826
178. Teraa M, Sprengers RW, van der Graaf Y, Peters CE, Moll FL, Verhaar MC. Autologous bone marrow-derived cell therapy in patients with critical limb ischemia: a meta-analysis of randomized controlled clinical trials. *Ann Surg.* (2013) 258:922–9. doi: 10.1097/SLA.0b013e3182854cf1
179. Heeschen C, Lehmann R, Honold J, Assmus B, Aicher A, Walter DH, et al. Profoundly reduced neovascularization capacity of bone marrow mononuclear cells derived from patients with chronic ischemic heart disease. *Circulation.* (2004) 109:1615–22. doi: 10.1161/01.CIR.0000124476.32871.E3
180. Gremmels H, Teraa M, Quax PH, den Ouden K, Fledderus JO, Verhaar MC. Neovascularization capacity of mesenchymal stromal cells from critical limb ischemia patients is equivalent to healthy controls. *Mol Ther.* (2014) 22:1960–70. doi: 10.1038/mt.2014.161
181. Rathinasabapathy A, Bruce E, Espejo A, Horowitz A, Sudhan DR, Nair A, et al. Therapeutic potential of adipose stem cell-derived conditioned medium against pulmonary hypertension and lung fibrosis. *Br J Pharmacol.* (2016) 173:2859–79. doi: 10.1111/bph.13562
182. Cahill EF, Kennelly H, Carty F, Mahon BP, English K. Hepatocyte growth factor is required for mesenchymal stromal cell protection against bleomycin-induced pulmonary fibrosis. *Stem Cells Transl Med.* (2016) 5:1307–18. doi: 10.5966/sctm.2015-0337
183. Maria ATJ, Toupet K, Maumus M, Rozier P, Vozenin MC, Le Quellec A, et al. Fibrosis development in HOCl-induced systemic sclerosis: a multistage process hampered by mesenchymal stem cells. *Front Immunol.* (2018) 9:2571. doi: 10.3389/fimmu.2018.02571
184. Jang YO, Kim MY, Cho MY, Baik SK, Cho YZ, Kwon SO. Effect of bone marrow-derived mesenchymal stem cellson hepatic fibrosis in a thioacetamide-induced cirrhotic rat model. *BMC Gastroenterol.* (2014) 14:198. doi: 10.1186/s12876-014-0198-6
185. Haldar D, Henderson NC, Hirschfield G, Newsome PN. Mesenchymal stromal cells and liver fibrosis: a complicated relationship. *FASEB J.* (2016) 30:3905–28. doi: 10.1096/fj.201600433R
186. Li B, Zhang H, Zeng M, He W, Li M, Huang X, et al. Bone marrow mesenchymal stem cells protect alveolar macrophages from lipopolysaccharide-induced apoptosis partially by inhibiting the Wnt/beta-catenin pathway. *Cell Biol Int.* (2015) 39:192–200. doi: 10.1002/cbin.10359
187. Valle-Prieto A, Conget PA. Human mesenchymal stem cells efficiently manage oxidative stress. *Stem Cells Dev.* (2010) 19:1885–93. doi: 10.1089/scd.2010.0093
188. de Godoy MA, Saraiva LM, de Carvalho LRP, Vasconcelos-Dos-Santos A, Beiral HJV, Ramos AB, et al. Mesenchymal stem cells and cell-derived extracellular vesicles protect hippocampal neurons from oxidative stress and synapse damage induced by amyloid-beta oligomers. *J Biol Chem.* (2018) 293:1957–75. doi: 10.1074/jbc.M117.807180
189. Cantinieaux D, Quertainmont R, Blacher S, Rossi L, Wanet T, Noel A, et al. Conditioned medium from bone marrow-derived mesenchymal stem cells improves recovery after spinal cord injury in rats: an original strategy to avoid cell transplantation. *PLoS ONE.* (2013) 8:e69515. doi: 10.1371/journal.pone.0069515
190. Mead B, Logan A, Berry M, Leadbeater W, Scheven BA. Paracrine-mediated neuroprotection and neuritogenesis of axotomised retinal ganglion cells by human dental pulp stem cells: comparison with human bone marrow and adipose-derived mesenchymal stem cells. *PLoS ONE.* (2014) 9:e109305. doi: 10.1371/journal.pone.0109305
191. Gibb SL, Zhao Y, Potter D, Hylin MJ, Bruhn R, Baimukanova G, et al. TIMP3 attenuates the loss of neural stem cells, mature neurons and neurocognitive dysfunction in traumatic brain injury. *Stem Cells.* (2015) 33:3530–44. doi: 10.1002/stem.2189
192. Menge T, Zhao Y, Zhao J, Wataha K, Gerber M, Zhang J, et al. Mesenchymal stem cells regulate blood-brain barrier integrity through TIMP3 release after traumatic brain injury. *Sci Transl Med.* (2012) 4:161ra150. doi: 10.1126/scitranslmed.3004660
193. Satti HS, Waheed A, Ahmed P, Ahmed K, Akram Z, Aziz T, et al. Autologous mesenchymal stromal cell transplantation for spinal cord injury: a phase I pilot study. *Cytotherapy.* (2016) 18:518–22. doi: 10.1016/j.jcyt.2016.01.004
194. Vaquero J, Zurita M, Rico MA, Bonilla C, Aguayo C, Fernandez C, et al. Repeated subarachnoid administrations of autologous mesenchymal stromal cells supported in autologous plasma improve quality of life in patients suffering incomplete spinal cord injury. *Cytotherapy.* (2017) 19:349–59. doi: 10.1016/j.jcyt.2016.12.002
195. Cheng H, Liu X, Hua R, Dai G, Wang X, Gao J, et al. Clinical observation of umbilical cord mesenchymal stem cell transplantation in treatment for sequelae of thoracolumbar spinal cord injury. *J Transl Med.* (2014) 12:253. doi: 10.1186/s12967-014-0253-7
196. Mendonca MV, Larocca TF, de Freitas Souza BS, Villarreal CF, Silva LF, Matos AC, et al. Safety and neurological assessments after autologous transplantation of bone marrow mesenchymal stem cells in subjects with chronic spinal cord injury. *Stem Cell Res Ther.* (2014) 5:126. doi: 10.1186/scrt516

13

Multiplicity of Mesenchymal Stromal Cells: Finding the Right Route to Therapy

Alison Wilson [1†], Margeaux Hodgson-Garms [2†], Jessica E. Frith [2] and Paul Genever [1*]*

[1] *Department of Biology, University of York, York, United Kingdom,* [2] *Materials Science and Engineering, Monash University, Clayton, VIC, Australia*

Correspondence:
Jessica E. Frith
jessica.frith@monash.edu
Paul Genever
paul.genever@york.ac.uk

† These authors have contributed equally to this work

Over the last decade, the acceleration in the clinical use of mesenchymal stromal cells (MSCs) has been nothing short of spectacular. Perhaps most surprising is how little we know about the "MSC product." Although MSCs are being delivered to patients at an alarming rate, the regulatory requirements for MSC therapies (for example in terms of quality assurance and quality control) are nowhere near the expectations of traditional pharmaceuticals. That said, the standards that define a chemical compound or purified recombinant protein cannot be applied with the same stringency to a cell-based therapy. Biological processes are dynamic, adaptive and variable. Heterogeneity will always exist or emerge within even the most rigorously sorted clonal cell populations. With MSCs, perhaps more so than any other therapeutic cell, heterogeneity pervades at multiple levels, from the sample source to the single cell. The research and clinical communities collectively need to recognize and take steps to address this troublesome truth, to ensure that the promise of MSC-based therapies is fulfilled.

Keywords: mesenchymal stromal cell, heterogeneity, cell subpopulations, cell-based therapy, single cell technologies

INTRODUCTION

The term "MSCs" is used to describe a heterogeneous population of stromal cells, the exact nature and composition of which remains the subject of much debate. They are often characterized using criteria proposed by the International Society for Cell Therapy (ISCT) as plastic-adherent cells, expressing a distinct set of surface antigens and with the ability to differentiate *in vitro* into osteogenic, adipogenic, and chondrogenic lineages (1). This minimal definition, however, is far from definitive. MSCs exhibit unique immunomodulatory properties, support the hematopoietic niche and participate in tissue regeneration through diverse biological activities including engraftment-independent paracrine signaling. Though initially described and sourced from bone marrow we are now able to isolate MSC-like cells from a variety of tissues including adipose tissue, dental pulp, placenta, umbilical cord, and umbilical cord blood.

Although MSCs first appeared in the clinic in 1995 (2) and have since become one of the most clinically studied cell therapy platforms worldwide (3) many fundamental aspects of MSC biology remain undetermined; primarily a direct consequence of the pervasive heterogeneity that manifests itself between MSC donors, tissue sources, culture methods and individual cells within a clonal population. Furthermore, MSCs exhibit a remarkable level of plasticity over time and when presented with different microenvironments (4, 5). MSC multiplicity, and a lack of consensus in the scientific community, complicates MSC characterization and their translation into the clinic.

This review will consider the multilevel origins of heterogeneity in MSCs (see **Figure 1**) and how we should be doing more to identify, track and quantify heterogeneity in MSCs to help determine its biological importance and impact in *in vitro* and *in vivo* contexts.

CHANGE IS THE ONLY CONSTANT (*HERACLITUS, 535–475 BC*)

MSC heterogeneity has certainly obscured our understanding of MSC biology and, correctly, prompted calls to re-evaluate the use of MSCs in therapy (6–10). However, the origins of heterogeneity are complex, fascinating and a constant theme in biology. It is clear from other work, particularly in microbial systems, that heterogeneity arising in genetically identical populations can have a positive impact on overall population fitness (11–14). Stochastic fluctuations in gene expression, or "noise," can lead to phenotypic variability in clonal cell populations (11, 15) and "bet hedging" can confer survival advantages on individual cells within mixed communities when faced with environmental change (16, 17). It has been proposed that stochastic non-genetic variations (i.e., those not caused by genetic mutations) contribute to the evolution of tumors using bet hedging-like strategies (18–20) and the dynamic switching between subtly different phenotypes has been shown to influence cell fate in different adult and embryonic stem cell populations (21–23). Gene expression noise in MSCs is also likely to give rise to individual cells with different characteristics and therefore influence the aggregate function of the population. It is also clear that MSC heterogeneity is due at least in part to the existence of different subpopulations with distinct expression profiles and functional properties (24–26). It has not been determined if discrete stromal subpopulations evolve through stochastic or deterministic means, but many appear to possess properties that support general tissue maintenance [for example, immune control, vascular remodeling, hematopoiesis (25)] that are unrelated to stem cell function. Therefore, the umbrella "MSC" descriptor may actually cover a range of related but distinct cell types that are yet to be fully defined.

IMPACT OF DONOR- AND TISSUE-DEPENDENT MSC HETEROGENEITY

Cells that currently meet this broad MSC descriptor have been identified in virtually all post-natal organs and tissues (27) and while bone marrow derived MSCs (BM-MSC) are still considered the gold standard, MSCs are now frequently also isolated from adipose tissue (AT-MSCs) and umbilical cord or cord blood (UC/UCB-MSCs) (28–33). There are well-documented disparities in proliferation, differentiation potential, surface markers, transcriptional, and proteomic profile of MSCs from different sources (34–36); an overarching consensus is hard to come by. For example, prevailing MSC characteristics such as tri-lineage differentiation potential present contradictory evidence in terms of lineage preference and full tri-lineage capacity (29, 30, 32, 37). Even when derived from the same tissue of origin, MSCs demonstrate prodigious donor-to-donor variation. This may be a factor of donor health influencing MSC availability and function (38, 39). Donor age can also affect self-renewal capacity and differentiation potential, which have been reported to decline in older donors (40–43). However, differences are also apparent in healthy donors of a similar age in proliferation rate, differentiation capacity, and ultimate clinical utility (44) leading to a further addition of complexity when directly comparing samples. It is tempting to speculate that MSC heterogeneity mirrors the diversity of environments from which they may be isolated, the reality is however that our understanding of MSCs *in vivo* is still in its infancy (8).

The multiplicity of MSCs and the absence of a meaningful consensus on definitions and characterization parameters makes comparing studies within the field difficult and translating them into clinical practice even more so. Because heterogeneity is seldom accounted for, and unique cell populations used in individual research projects are rarely fully defined, many studies are not only difficult to reproduce but difficult to evaluate for comparability and impact within the field. Incomplete knowledge of the characteristics of MSCs *in vivo* and how these will relate to clinical outcomes further exacerbate the problem when considering quality control requirements for MSCs as therapeutic agents. Changes in the source materials of clinical products, e.g., a different donor, prompt regulatory authorities to require re-characterization and evidence of "comparability." In the event that comparability could not be demonstrated, product from the original and subsequent sources would be considered to be essentially different products. Thus, during clinical development, data on early product iterations could be invalidated, and post-authorization could, in the worst-case scenario, require re-authorization. In conjunction with the need for adequate cell numbers, this represents a major challenge to the acceptance of cell-based therapies as mainstream treatments; the options of extended culture or multiple donors each imply unavoidable heterogeneity. Consequently the manufacture of MSC products using processes that rely on a continuous supply of new tissue donations run the significant risk of supply constraint, interruption, and inconsistencies (10).

IN VITRO EXPANSION AND MSC HETEROGENEITY

A typical bone marrow aspirate contains just 0.01–0.001% MSCs (45) and trials for the regeneration of bone and cartilage tissue commonly use in the order of 10 million cells. The need for high levels of culture expansion adds to the challenge of generating an MSC population that retains the ability to differentiate effectively or secrete the appropriate biomolecules to induce a beneficial paracrine response. Banfi et al. investigated the growth kinetics and differentiation potential of MSCs, using fresh isolates from different donors through to passage five, and showed a dramatic decrease in MSC functionality over time (46). MSCs from the same donor and same source (iliac crest marrow aspirate) isolated at different timepoints over a

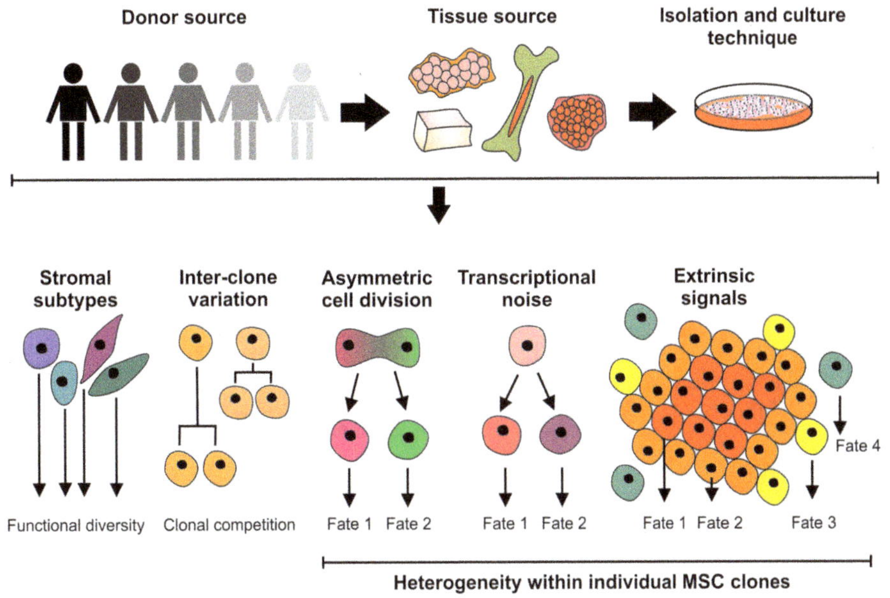

FIGURE 1 | Sources of MSC heterogeneity; considerations for the clinical application of culture-expanded MSCs. Significant variation exists in MSC cultures isolated from different donors and different tissue sites. Unrefined and non-standardized isolation and culture techniques do not select for homogeneous cell populations and are likely to give rise to a mixture of stromal cell with different functions. Differences in the growth properties of MSC clones can result in cultures being dominated by the faster-growing lines. Further levels of heterogeneity can be introduced within MSC clones through asymmetric cell division and the effects of stochastic transcriptional noise, generating cells with modified phenotypes. MSC properties will also be determined by, for example, proximity to neighboring cells and extrinsic signaling factors.

period of 6 months also show significant variation in growth rates (44). Other studies have confirmed this loss of MSC function, demonstrating reduced proliferation, colony-forming (CFU-f) efficiency, telomere length and differentiation capacity with increasing time in culture (4, 40, 47). With the mounting interest in the use of MSCs for their paracrine effect it is also noteworthy that the secreted output of MSCs has been shown to differ with number of passages (48). This reduction in therapeutic potency at the population level can mask changes within clonal MSCs. Schellenberg et al. assessed MSC clones following expansion and observed a continual decrease in CFU-f efficiency and differentiation capacity over time (49). Earlier analyses identified a complex hierarchy of MSC clones at varying stages of potency (50), so it may be that the diminishing clonal potential observed during MSC expansion is driven by subsets of cells reaching their proliferative limit or by entering the hierarchy of different stages through which cells pass during differentiation. Subsequent studies to track individual clones from MSC explant cultures showed that clonal complexity decreased markedly over 12 passages resulting in the clonal selection of a few dominant MSC clones (51).

Given the impact that culture expansion has on MSC fate, the *in vitro* environment and its influence on MSC properties is worth considering. In the majority of research laboratories, MSCs are expanded as a monolayer using standard tissue culture flasks with a plasma-treated polystyrene surface and medium containing fetal bovine serum. Surprisingly, given the detrimental effects on MSC proliferation, differentiation and paracrine activity of these basic methods, the industrial expansion of MSCs for clinical applications often still retains the same basic features (52). Scale-up can be achieved through the use of multilayered cell culture flasks (cell factories) or culture vessels specifically tailored for use with closed-box and automated systems. More advanced systems use roller bottles, hollow-fiber or stirred tank bioreactors [reviewed by (53)]. A major problem with this approach is that that these *in vitro* conditions are very different from the *in vivo* MSC microenvironment, lacking much of the complexity in terms of matrix composition, geometry, mechanical properties and interactions with other cell types. All of these microenvironmental factors are interpreted by the cell and have been shown to impact upon their behavior (54–59). At its worst, the non-physiological conditions of typical cell cultures can cause mutations or cellular defects (60) but even the best-case scenario results in cells whose behavior is markedly changed. Together, this results in loss of potential from the whole population, but MSC heterogeneity may also be driven by cells responding to local changes in the microenvironment, such as through poorly controlled substrate properties or local changes in oxygen and nutrient concentration driven by the static nature of the setup (61).

It is clear that the requirement for extended *in vitro* expansion is a major contributor to the heterogeneity of MSC populations. A deeper understanding of the impacts of different environmental cues and the mechanisms by which they drive change, will be integral to the development of technologies for the large-scale production of quality MSC populations for clinical use.

CLINICAL EXPERIENCE AND REGULATORY CONSIDERATIONS RELATED TO HETEROGENEOUS CELL THERAPY

MSC heterogeneity is multifactorial and functionally influential. Nonetheless the clinical application of MSCs does not appear to take this into account, with a selection of recent trial publications suggesting a comparatively limited assessment of cellular phenotype (**Table 1**). The criteria established for MSCs by the ISCT (1) are sometimes referenced in these studies but not necessarily met. It is of course possible that additional criteria were specified during manufacture but not published, however publication of more detail would increase our understanding of the MSC phenotypes in clinical use.

Basic requirements for all biological medicines include the necessity to define the identity, the purity and the potency of the product. The developers of cell-based medicinal products must define the "active substance"; the cell type on which the therapeutic action of the product depends. Specification limits must be established for unique identification of the active substance within the product and for quantitation of its purity. Other phenotypes present, for example those arising from a tissue biopsy or culture contaminant, and non-viable cells, are generally regarded as impurities. These impurities should be reduced as far as possible and their content in the finished product limited and defined by specifications. Cellular impurities aside, major regulatory authorities do not always require cell-based medicinal products to consist of a pure population of cells. One of the first authorized cellular therapies was the immunotherapy Provenge (Dendreon Inc), approved by the US Food and Drug Administration (FDA) in 2010 for treatment of certain prostate cancers. Provenge contains autologous peripheral blood mononuclear cells (PBMC), which are cultured with PAP-GM-CSF, a fusion protein combining granulocyte-macrophage colony-stimulating factor (GM-CSF) with a prostate cancer antigen (prostatic acid phosphatase, PAP). Antigen-presenting cells within the PBMC fraction are activated by the fusion protein, providing a tumor-directed action. The exact composition of the Provenge dose varies depending on the cellular composition of each patient's leukapheresis sample, but may contain, amongst others, T and B lymphocytes and natural killer cells so the therapy is inherently heterogeneous (77, 78). In 2015 the European Union (EU) authorized its first stem cell-based product, Holoclar (Chiesi Farmaceutici SPA, Italy). Holoclar is a population of cultured autologous human corneal epithelial cells containing limbal stem cells (LSCs) intended for treatment of ocular burns. The active substance contains only approximately 3.5% of p63bright LSCs, in a mixed population with transient amplifying meroclones and paraclones and terminally differentiated corneal epithelial cells (79). The extensive heterogeneity of the overall product, which arises from the inherent cellular variation in the patient's biopsy, was justified by evidence of relevant supportive properties provided by the non-stem majority population; these were therefore not considered to be cellular impurities (80).

In 2016, the EU approved Strimvelis [Orchard Therapeutics (Netherlands) BV], a gene therapy for treatment of adenosine deaminase (ADA) severe combined immunodeficiency (ADA-SCID), in which autologous CD34+ hematopoietic stem cells (HSCs) were transduced with ADA cDNA to provide the missing gene sequence. The active substance of Strimvelis includes not only the transduced CD34+ cells, but also the non-genetically modified CD34+ fraction, based on the fact that HSC transplantation is itself a standard treatment for ADA-SCID (81) These examples provide illustrations of the general acceptability, where justified, of heterogeneous cell populations within authorized cellular therapies. In the latter two cases, the heterogeneity specifically contributes to the overall clinical effect of the product and is not merely a consequence of the manufacturing process. The complexity associated with using fundamentally variable starting materials which are then processed, inducing further heterogeneity, implies that the purity of most cell-based products will be challenging to define. The regulators' expectation of quantitation of the population being administered in terms of identity and purity (82, 83) will be difficult to achieve definitively; it is probably more reasonable to demonstrate a degree of reproducibility across product batches and to relate the composition of each batch to those used in clinical trials than to provide exact percentages of each minor cellular component (84). The identification of relevant mechanisms of action will be of crucial importance in determining the acceptability of a degree of heterogeneity, since MSC activity in a specific clinical application should help inform selection of an ideal MSC population, whether this may be a heterogeneous preparation or a specified subset.

The inevitability of MSC heterogeneity and the consequences of culture expansion for the production of cell therapies, discussed earlier, raise key questions for developers of regenerative medicines. Whilst, as illustrated above, there is no obligation to demonstrate that a product contains only the specific cell type of interest, the challenges of definition and identification are accentuated when considering MSCs. The apparent absence of major concerns around cellular heterogeneity in whole organ and HSC transplantation is sometimes highlighted as support for a less rigorous approach to the characterization and control of cell-based therapies. However, acceptance of heterogeneity in these situations may be due in part to the fact that organ and HSC transplants are procedures which are considered to fall within the practice of medicine rather than items externally regulated as medicinal products.

FUTURE PERSPECTIVES: EMBRACING CHANGE

In order to advance the clinical utility of MSCs, it is essential that strategies to quantify heterogeneity are agreed. As a starting point, it is important to define the biological properties of the different stromal cell types within a mixed population. It is likely that stem-cell and non-stem-cell fractions are co-extracted using current protocols for MSC isolation. For regenerative therapies, it would seem logical that the stem-cell component

TABLE 1 | Sample characterization and release criteria reported in clinical trials using MSCs.

Phase	Indication	Tissue	Source	Characterization	Stated release criteria	Notes	References
I	Myocardial infarction	Bone Marrow	Allo		Positive: CD105, CD166 limits NS Negative: CD45 limits NS	"Provacel"—became Prochymal	(62)
I	Crohn's disease	Bone Marrow	Auto	HLA II (DR), CD73, CD90, CD31, CD34, CD45, CD80, CD105	CD73, CD90, and CD105 >90%		(63)
I	Graft vs. Host Disease	Bone Marrow	Allo		Positive: CD73, CD90, CD105 limit NS, Negative: CD14, CD34, CD45 limit NS		(64)
II	Graft vs. Host Disease	Bone Marrow	Allo	CD105, CD59, CD73, CD90, CD31, CD34, CD14, CD45, HLA-DR, FSP	NS		(65)
II	Multiple sclerosis	Bone Marrow	Auto	CD90, CD90, CD31, CD34, CD45	ISCT criteria	Phenotypic analysis not consistent with ISCT	(66)
I	Osteoarthritis (knee)	Bone Marrow	Auto	Positive for CD90, CD105, CD106, CD166, KDR (VEGFR2). Negative for CD34, CD45, HLA-DR	ISCT criteria	Data not presented	(67)
I	Transplant rejection	Bone Marrow	Auto	HLA II (DR), CD73, CD90, CD31, CD34, CD45, CD80, CD105	CD73, CD90, CD105 >90%		(68)
II	Kidney structure/function	Bone Marrow	Auto	HLA II (DR), CD73, CD90, CD31, CD34, CD45, CD80, CD105	CD73, CD90, CD105 >90%	Trial design, study not reported	(69)
I	Graft vs. Host Disease	Bone Marrow	Allo		CD73, CD90, CD105 >80% CD14, CD34, CD45 <10%		(70)
II	Crohn's disease	Bone Marrow	Allo		ISCT criteria	Data not presented	(71)
II	Multiple sclerosis	Bone Marrow	Auto		Positive: CD90, CD73, CD44 limits NS. Negative: CD34, CD45 limits NS		(72)
II	Myocardial infarction	Bone Marrow	Auto		Positive: CD73, CD105 >90%. Negative: CD14, CD34, CD45 <3%		(73)
I	Acute Respiratory Distress Syndrome	Bone Marrow	Allo			FC performed but no data presented	(74)
I	Osteoarthritis (knee)	Adipose	Auto	CD73, CD90, CD105, CD14, CD31, CD34, CD45, CD80, IgG1	CD14, CD45 <2% CD34<10% CD73, CD90 >90%, CD105 >80%		(75)
I/IIa	Meniscus	Bone Marrow	Auto		Positive: CD90, CD105 >80%. Negative: CD34, CD45 <10%		(76)

is the essential active ingredient, however non-differentiating stromal cells could play important supporting roles, for example in immune control; precisely why we need a full biological understanding that relates to mechanism of action. This can be achieved by exploiting techniques suitable for phenotyping individual cells, including flow cytometry, electrophysiology, microscopy (in various forms), image/morphometric analysis, lineage tracing, and powerful new single cell-omic technologies. Effective strategies will be required to ensure data are integrated, interpreted correctly and shared. The key to clinical translation will be to develop the most appropriate non-destructive biomarker identification techniques that provide functional discrimination. Reliable subtype-specific biomarkers will also support the development of treatments to target MSCs *in situ*, potentially negating the need for culture expansion. Alongside these, improved methods for MSC expansion that retain, or even promote selection of the desired MSC properties will be essential for the production of MSC products with a more defined set of characteristics and high therapeutic efficacy. Such technologies will likely incorporate biophysical as well as biochemical cues and provide platforms for scale-up culture in bioreactors. With the role of the paracrine effect of MSCs coming to the fore (85), therapies based on the MSC secretome or MSC-derived extracellular vesicles (EVs) may emerge to complement the MSC therapeutic toolkit. However, different MSC populations (or cells within that population) are still likely to produce different secretomes and so many of the fundamental challenges relating to MSC heterogeneity will remain.

Given the challenges associated with providing consistency in an MSC product from multiple tissue isolates, the generation of MSCs from pluripotent stem cell populations has garnered interest (86–92). The expansion capability of pluripotent cells means that a single clonal population can potentially be manufactured and subsequently differentiated into a virtually limitless supply of MSCs. This type of platform relieves the need for continuous tissue donations, simplifies the subject of donor-donor variation and bypasses many of the sources of MSC heterogeneity that arise when working with *ex vivo* cells. Induced pluripotent stem cells (iPSC)-derived MSCs offer the potential for large-scale production of more homogenous, off-the-shelf products with limited batch-to-batch variation that could deliver more consistent clinical outcomes. The first phase I clinical trial using iPSC-derived MSCs was completed in 2018 with promising results from Cynata Therapeutics's lead Cymerus™CYP-001 product for the treatment of graft vs. host disease (93), although the full findings have not yet been published. While the clinical use of iPSC-MSCs holds promise, an effective comparison of pluripotent cell-derived MSCs to their adult tissue counterparts is required, with appropriate safety profiling. Clonal immortalized MSC lines (both iPSC-derived and genetically modified adult MSCs) may also be developed for bulk harvesting of secreted products, proteins, and EV cargoes, which could ultimately dispense with the need for the transplantation of MSCs as a whole-cell product, however the issue of stochastic heterogeneity arising in clonal cell populations will always persist.

MSCs can offer widespread therapeutic benefits but we must balance enthusiastic demands for clinical progress against the need for better mechanistic understanding. Unraveling MSC multiplicity is the essential first step in that process.

AUTHOR CONTRIBUTIONS

AW, MH-G, JF, and PG wrote sections of the manuscript. All authors contributed to manuscript revision, read and approved the submitted version.

FUNDING

This work was supported by the Tissue Engineering and Regenerative Therapies Centre Versus Arthritis (21156).

REFERENCES

1. Dominici M, Le Blanc K, Mueller I, Slaper-Cortenbach I, Marini F, Krause D, et al. Minimal criteria for defining multipotent mesenchymal stromal cells. The International Society for Cellular Therapy position statement. *Cytotherapy.* (2006) 8:315–17. doi: 10.1080/14653240600855905
2. Lazarus HM, Haynesworth SE, Gerson SL, Rosenthal NS, Caplan AI. *Ex vivo* expansion and subsequent infusion of human bone marrow-derived stromal progenitor cells (mesenchymal progenitor cells): implications for therapeutic use. *Bone Marrow Transplant.* (1995) 16:557–64.
3. Fung M, Yuan Y, Atkins H, Shi Q, Bubela T. Responsible translation of stem cell research: an assessment of clinical trial registration and publications. *Stem Cell Rep.* (2017) 8:1190–201. doi: 10.1016/j.stemcr.2017.03.013
4. Baxter MA, Wynn RF, Jowitt SN, Wraith JE, Fairbairn LJ, Bellantuono I. Study of telomere length reveals rapid aging of human marrow stromal cells following *in vitro* expansion. *Stem Cells.* (2004) 22:675–82. doi: 10.1634/stemcells.22-5-675
5. Digirolamo CM, Stokes D, Colter D, Phinney DG, Class R, Prockop DJ. Propagation and senescence of human marrow stromal cells in culture: a simple colony-forming assay identifies samples with the greatest potential to propagate and differentiate. *Br J Haematol.* (1999) 107:275–81. doi: 10.1046/j.1365-2141.1999.01715.x
6. Bianco P, Cao X, Frenette PS, Mao JJ, Robey PG, Simmons PJ, et al. The meaning, the sense and the significance: translating the science of mesenchymal stem cells into medicine. *Nat Med.* (2013) 19:35–42. doi: 10.1038/nm.3028
7. Phinney DG. Functional heterogeneity of mesenchymal stem cells: implications for cell therapy. *J Cell Biochem.* (2012) 113:2806–12. doi: 10.1002/jcb.24166
8. McLeod CM, Mauck RL. On the origin and impact of mesenchymal stem cell heterogeneity: new insights and emerging tools for single cell analysis. *Eur Cell Mater.* (2017) 34:217–31. doi: 10.22203/eCM.v034a14
9. Pevsner-Fischer M, Levin S, Zipori D. The origins of mesenchymal stromal cell heterogeneity. *Stem Cell Rev.* (2011) 7:560–8. doi: 10.1007/s12015-011-9229-7
10. Galipeau J, Sensébé L. Mesenchymal stromal cells: clinical challenges and therapeutic opportunities. *Cell Stem Cell.* (2018) 22:824–33. doi: 10.1016/j.stem.2018.05.004
11. Elowitz MB, Levine AJ, Siggia ED, Swain PS. Stochastic gene expression in a single cell. *Science.* (2002) 297:1183–6. doi: 10.1126/science.1070919
12. Balaban NQ, Merrin J, Chait R, Kowalik L, Leibler S. Bacterial persistence as a phenotypic switch. *Science.* (2004) 305:1622–5. doi: 10.1126/science.1099390
13. Levy SF. Cellular heterogeneity: benefits besides bet-hedging. *Curr Biol.* (2016) 26:R355–7. doi: 10.1016/j.cub.2016.03.034
14. Altschuler SJ, Wu LF. Cellular heterogeneity: do differences make a difference? *Cell.* (2010) 141:559–63. doi: 10.1016/j.cell.2010.04.033
15. Pedraza JM, van Oudenaarden A. Noise propagation in gene networks. *Science.* (2005) 307:1965–9. doi: 10.1126/science.1109090
16. Veening J-W, Smits WK, Kuipers OP. Bistability, epigenetics, and bet-hedging in bacteria. *Annu Rev Microbiol.* (2008) 62:193–210. doi: 10.1146/annurev.micro.62.081307.163002
17. de Jong IG, Haccou P, Kuipers OP. Bet hedging or not? A guide to proper classification of microbial survival strategies. *Bioessays.* (2011) 33:215–23. doi: 10.1002/bies.201000127
18. Brock A, Chang H, Huang S. Non-genetic heterogeneity–a mutation-independent driving force for the somatic evolution of tumours. *Nat Rev Genet.* (2009) 10:336–42. doi: 10.1038/nrg2556
19. Gravenmier CA, Siddique M, Gatenby RA. Adaptation to stochastic temporal variations in intratumoral blood flow: the Warburg effect as a bet hedging strategy. *Bull Math Biol.* (2018) 80:954–70. doi: 10.1007/s11538-017-0261-x
20. Jolly MK, Kulkarni P, Weninger K, Orban J, Levine H. Phenotypic plasticity, bet-hedging, and androgen independence in prostate cancer: role of non-genetic heterogeneity. *Front Oncol.* (2018) 8:50. doi: 10.3389/fonc.2018.00050
21. Chang HH, Hemberg M, Barahona M, Ingber DE, Huang S. Transcriptome-wide noise controls lineage choice in mammalian progenitor cells. *Nature.* (2008) 453:544–7. doi: 10.1038/nature06965
22. Enver T, Pera M, Peterson C, Andrews PW. Stem cell states, fates, and the rules of attraction. *Cell Stem Cell.* (2009) 4:387–97. doi: 10.1016/j.stem.2009.04.011
23. Krieger T, Simons BD. Dynamic stem cell heterogeneity. *Development.* (2015) 142:1396–406. doi: 10.1242/dev.101063
24. Phinney DG, Hill K, Michelson C, DuTreil M, Hughes C, Humphries S, et al. Biological activities encoded by the murine mesenchymal stem cell transcriptome provide a basis for their developmental potential and broad therapeutic efficacy. *Stem Cells.* (2006) 24:186–98. doi: 10.1634/stemcells.2004-0236
25. Phinney DG. Biochemical heterogeneity of mesenchymal stem cell populations: clues to their therapeutic efficacy. *Cell Cycle.* (2007) 6:2884–9. doi: 10.4161/cc.6.23.5095
26. James S, Fox J, Afsari F, Lee J, Clough S, Knight C, et al. Multiparameter analysis of human bone marrow stromal cells identifies distinct immunomodulatory and differentiation-competent subtypes. *Stem Cell Rep.* (2015) 4:1004–15. doi: 10.1016/j.stemcr.2015.05.005
27. Da Silva Meirelles L, Chagastelles PC, Nardi NB. Mesenchymal stem cells reside in virtually all post-natal organs and tissues. *J Cell Sci.* (2006) 119:2204–13. doi: 10.1242/jcs.02932

28. Bieback K, Kern S, Klüter H, Eichler H. Critical parameters for the isolation of mesenchymal stem cells from umbilical cord blood. *Stem Cells.* (2004) 22:625–34. doi: 10.1634/stemcells.22-4-625
29. Erices A, Conget P, Minguell JJ. Mesenchymal progenitor cells in human umbilical cord blood. *Br J Haematol.* (2000) 109:235–42. doi: 10.1046/j.1365-2141.2000.01986.x
30. Goodwin HS, Bicknese AR, Chien SN, Bogucki BD, Quinn CO, Wall DA. Multilineage differentiation activity by cells isolated from umbilical cord blood: expression of bone, fat, and neural markers. *Biol Blood Marrow Transplant.* (2001) 7:581–8. doi: 10.1053/bbmt.2001.v7.pm11760145
31. Gronthos S, Mankani M, Brahim J, Robey PG, Shi S. Postnatal human dental pulp stem cells (DPSCs) *in vitro* and *in vivo*. *Proc Natl Acad Sci USA.* (2000) 97:13625–30. doi: 10.1073/pnas.240309797
32. Kögler G, Sensken S, Airey JA, Trapp T, Müschen M, Feldhahn N, et al. A new human somatic stem cell from placental cord blood with intrinsic pluripotent differentiation potential. *J Exp Med.* (2004) 200:123–35. doi: 10.1084/jem.20040440
33. Zuk PA, Zhu M, Mizuno H, Huang J, Futrell JW, Katz AJ, et al. Multilineage cells from human adipose tissue: implications for cell-based therapies. *Tissue Eng.* (2001) 7:211–28. doi: 10.1089/107632701300062859
34. Kern S, Eichler H, Stoeve J, Klüter H, Bieback K. Comparative analysis of mesenchymal stem cells from bone marrow, umbilical cord blood, or adipose tissue. *Stem Cells.* (2006) 24:1294–301. doi: 10.1634/stemcells.2005-0342
35. Mattar P, Bieback K. Comparing the immunomodulatory properties of bone marrow, adipose tissue, and birth-associated tissue mesenchymal stromal cells. *Front Immunol.* (2015) 6:560. doi: 10.3389/fimmu.2015.00560
36. Strioga M, Viswanathan S, Darinskas A, Slaby O, Michalek J. Same or not the same? Comparison of adipose tissue-derived versus bone marrow-derived mesenchymal stem and stromal cells. *Stem Cells Dev.* (2012) 21:2724–52. doi: 10.1089/scd.2011.0722
37. Lee OK, Kuo TK, Chen W-M, Lee K-D, Hsieh S-L, Chen T-H. Isolation of multipotent mesenchymal stem cells from umbilical cord blood. *Blood.* (2004) 103:1669–75. doi: 10.1182/blood-2003-05-1670
38. Kuznetsov SA, Mankani MH, Bianco P, Robey PG. Enumeration of the colony-forming units-fibroblast from mouse and human bone marrow in normal and pathological conditions. *Stem Cell Res.* (2009) 2:83–94. doi: 10.1016/j.scr.2008.07.007
39. Wang J, Liao L, Wang S, Tan J. Cell therapy with autologous mesenchymal stem cells-how the disease process impacts clinical considerations. *Cytotherapy.* (2013) 15:893–904. doi: 10.1016/j.jcyt.2013.01.218
40. Yang Y-HK, Ogando CR, Wang See C, Chang T-Y, Barabino GA. Changes in phenotype and differentiation potential of human mesenchymal stem cells aging *in vitro*. *Stem Cell Res Ther.* (2018) 9:131. doi: 10.1186/s13287-018-0876-3
41. D'Ippolito G, Schiller PC, Ricordi C, Roos BA, Howard GA. Age-related osteogenic potential of mesenchymal stromal stem cells from human vertebral bone marrow. *J Bone Miner Res.* (1999) 14:1115–22. doi: 10.1359/jbmr.1999.14.7.1115
42. Katsara O, Mahaira LG, Iliopoulou EG, Moustaki A, Antsaklis A, Loutradis D, et al. Effects of donor age, gender, and *in vitro* cellular aging on the phenotypic, functional, and molecular characteristics of mouse bone marrow-derived mesenchymal stem cells. *Stem Cells Dev.* (2011) 20:1549–61. doi: 10.1089/scd.2010.0280
43. Stenderup K, Justesen J, Clausen C, Kassem M. Aging is associated with decreased maximal life span and accelerated senescence of bone marrow stromal cells. *Bone.* (2003) 33:919–26. doi: 10.1016/j.bone.2003.07.005
44. Phinney DG, Kopen G, Righter W, Webster S, Tremain N, Prockop DJ. Donor variation in the growth properties and osteogenic potential of human marrow stromal cells. *J Cell Biochem.* (1999) 75:424–36.
45. Roobrouck VD, Ulloa-Montoya F, Verfaillie CM. Self-renewal and differentiation capacity of young and aged stem cells. *Exp Cell Res.* (2008) 314:1937–44. doi: 10.1016/j.yexcr.2008.03.006
46. Banfi A, Muraglia A, Dozin B, Mastrogiacomo M, Cancedda R, Quarto R. Proliferation kinetics and differentiation potential of *ex vivo* expanded human bone marrow stromal cells: implications for their use in cell therapy. *Exp Hematol.* (2000) 28:707–15. doi: 10.1016/S0301-472X(00)00160-0
47. Reiser J, Zhang X-Y, Hemenway CS, Mondal D, Pradhan L, La Russa VF. Potential of mesenchymal stem cells in gene therapy approaches for inherited and acquired diseases. *Expert Opin Biol Ther.* (2005) 5:1571–84. doi: 10.1517/14712598.5.12.1571
48. Russell AL, Lefavor R, Durand N, Glover L, Zubair AC. Modifiers of mesenchymal stem cell quantity and quality. *Transfusion.* (2018) 58:1434–40. doi: 10.1111/trf.14597
49. Schellenberg A, Stiehl T, Horn P, Joussen S, Pallua N, Ho AD, et al. Population dynamics of mesenchymal stromal cells during culture expansion. *Cytotherapy.* (2012) 14:401–11. doi: 10.3109/14653249.2011.640669
50. Russell KC, Phinney DG, Lacey MR, Barrilleaux BL, Meyertholen KE, O'Connor KC. *In vitro* high-capacity assay to quantify the clonal heterogeneity in trilineage potential of mesenchymal stem cells reveals a complex hierarchy of lineage commitment. *Stem Cells.* (2010) 28:788–98. doi: 10.1002/stem.312
51. Selich A, Daudert J, Hass R, Philipp F, von Kaisenberg C, Paul G, et al. Massive clonal selection and transiently contributing clones during expansion of mesenchymal stem cell cultures revealed by lentiviral RGB-barcode technology. *Stem Cells Transl Med.* (2016) 5:591–601. doi: 10.5966/sctm.2015-0176
52. Lechanteur C, Briquet A, Giet O, Delloye O, Baudoux E, Beguin Y. Clinical-scale expansion of mesenchymal stromal cells: a large banking experience. *J Transl Med.* (2016) 14:145. doi: 10.1186/s12967-016-0892-y
53. Mizukami A, Swiech K. Mesenchymal stromal cells: from discovery to manufacturing and commercialization. *Stem Cells Int.* (2018) 2018:4083921. doi: 10.1155/2018/4083921
54. Engler AJ, Sen S, Sweeney HL, Discher DE. Matrix elasticity directs stem cell lineage specification. *Cell.* (2006) 126:677–89. doi: 10.1016/j.cell.2006.06.044
55. Frith JE, Thomson B, Genever PG. Dynamic three-dimensional culture methods enhance mesenchymal stem cell properties and increase therapeutic potential. *Tissue Eng Part C Methods.* (2010) 16:735–49. doi: 10.1089/ten.tec.2009.0432
56. Frith JE, Mills RJ, Cooper-White JJ. Lateral spacing of adhesion peptides influences human mesenchymal stem cell behaviour. *J Cell Sci.* (2012) 125:317–27. doi: 10.1242/jcs.087916
57. Kusuma GD, Brennecke SP, O'Connor AJ, Kalionis B, Heath DE. Decellularized extracellular matrices produced from immortal cell lines derived from different parts of the placenta support primary mesenchymal stem cell expansion. *PLoS ONE.* (2017) 12:e0171488. doi: 10.1371/journal.pone.0171488
58. Etheridge L, Mason RA, Saleh F, Genever P. Cell-cell signaling pathways that regulate mesenchymal stromal cell differentiation. In: Atkinson K, editor. *The Biology and Therapeutic Application of Mesenchymal Cells.* Hoboken, NJ: John Wiley & Sons, Inc. (2016) p. 91–103.
59. Saleh FA, Genever PG. Turning round: multipotent stromal cells, a three-dimensional revolution? *Cytotherapy.* (2011) 13:903–12. doi: 10.3109/14653249.2011.586998
60. Wright WE, Shay JW. Historical claims and current interpretations of replicative aging. *Nat Biotechnol.* (2002) 20:682–8. doi: 10.1038/nbt0702-682
61. Rodrigues CAV, Fernandes TG, Diogo MM, da Silva CL, Cabral JMS. Stem cell cultivation in bioreactors. *Biotechnol Adv.* (2011) 29:815–29. doi: 10.1016/j.biotechadv.2011.06.009
62. Hare JM, Traverse JH, Henry TD, Dib N, Strumpf RK, Schulman SP, et al. A randomized, double-blind, placebo-controlled, dose-escalation study of intravenous adult human mesenchymal stem cells (prochymal) after acute myocardial infarction. *J Am Coll Cardiol.* (2009) 54:22777–86. doi: 10.1016/j.jacc.2009.06.055
63. Duijvestein M, Vos ACW, Roelofs H, Wildenberg ME, Wendrich BB, Verspaget HW, et al. Autologous bone marrow-derived mesenchymal stromal cell treatment for refractory luminal Crohn's disease: results of a phase I study. *Gut.* (2010) 59:16622–69. doi: 10.1136/gut.2010.215152
64. Herrmann R, Sturm M, Shaw K, Purtill D, Cooney J, Wright M, et al. Mesenchymal stromal cell therapy for steroid-refractory acute and chronic graft versus host disease: a phase 1 study. *Int J Hematol.* (2012) 95:182–8. doi: 10.1007/s1s2185-011-09899-2
65. Kuzmina LA, Petinati NA, Parovichnikova EN, Lubimova LS, Gribanova EO, Gaponova TV, et al. Multipotent mesenchymal stromal cells for the prophylaxis of acute graft-versus-host disease-a phase II study. *Stem Cells Int.* (2012) 2012:968213. doi: 10.1155/2012/968213
66. Bonab MM, Sahraian MA, Aghsaie A, Karvigh SA, Hosseinian SM, Nikbin B, et al. Autologous mesenchymal stem cell therapy in progressive multiple

sclerosis: an open label study. *Curr Stem Cell Res Ther.* (2012) 7:407–14. doi: 10.2174/157488812804484648
67. Orozco L, Munar A, Soler R, Alberca M, Soler F, Huguet M, et al. Treatment of knee osteoarthritis with autologous mesenchymal stem cells: a pilot study. *Transplantation.* (2013) 95:15355–41. doi: 10.1097/TP.0b00b0b13e33e3e18291a11a2ad22da
68. Reinders MEJ, de Fijter JW, Roelofs H, Bajema IM, de Vries DK, Schaapherder AF, et al. Autologous bone marrow-derived mesenchymal stromal cells for the treatment of allograft rejection after renal transplantation: results of a phase I study. *Stem Cells Transl Med.* (2013) 2:107–11. doi: 10.5966/sctm.201 22-0114
69. Reinders MEJ, Bank JR, Dreyer GJ, Roelofs H, Heidt S, Roelen DL, et al. Autologous bone marrow derived mesenchymal stromal cell therapy in combination with everolimus to preserve renal structure and function in renal transplant recipients. *J Transl Med.* (2014) 12:331. doi: 10.1186/s1s2967-014-0331-x
70. Introna M, Lucchini G, Dander E, Galimberti S, Rovelli A, Balduzzi A, et al. Treatment of graft versus host disease with mesenchymal stromal cells: a phase I study on 40 adult and pediatric patients. *Biol Blood Marrow Transplant.* (2014) 20:375–81. doi: 10.1016/j.bbmt.2013.11.033
71. Forbes GM, Sturm MJ, Leong RW, Sparrow MP, Segarajasingam D, Cummins AG, et al. A phase 2 study of allogeneic mesenchymal stromal cells for luminal Crohn's disease refractory to biologic therapy. *Clin Gastroenterol Hepatol.* (2014) 12:64–71. doi: 10.1016/j.cgh.2013.06.021
72. Llufriu S, Sepúlveda M, Blanco Y, Marín P, Moreno B, Berenguer J, et al. Randomized placebo-controlled phase II trial of autologous mesenchymal stem cells in multiple sclerosis. *PLoS ONE.* (2014) 9:e1e13936. doi: 10.1371/journal.pone.0113936
73. Lee J-W, Lee S-H, Youn Y-J, Ahn M-S, Kim J-Y, Yoo B-S, et al. A randomized, open-label, multicenter trial for the safety and efficacy of adult mesenchymal stem cells after acute myocardial infarction. *J Korean Med Sci.* (2014) 29:23–31. doi: 10.3346/jkms.2014.29.1.23
74. Wilson JG, Liu KD, Zhuo H, Caballero L, McMillan M, Fang X, et al. Mesenchymal stem (stromal) cells for treatment of ARDS: a phase 1 clinical trial. *Lancet Respir Med.* (2015) 3:24–32. doi: 10.1016/S2S213-2600(14)702911-7
75. Pers Y-M, Rackwitz L, Ferreira R, Pullig O, Delfour C, Barry F, et al. Adipose mesenchymal stromal cell-based therapy for severe osteoarthritis of the knee: a phase I dose-escalation trial. *Stem Cells Transl Med.* (2016) 5:847–56. doi: 10.5966/sctm.20155-0245
76. Whitehouse MR, Howells NR, Parry MC, Austin E, Kafienah W, Brady K, et al. Repair of torn avascular meniscal cartilage using undifferentiated autologous mesenchymal stem cells: from *in vitro* optimization to a first-in-human study. *Stem Cells Transl Med.* (2017) 6:12377–48. doi: 10.1002/sctm.16-0199
77. *Highlights of Prescribing Information.* Provenge. Available online at: https://www.fda.gov/vaccines-blood-biologics/cellular-gene-therapy-products/provenge-sipuleucel-t (accessed December 12, 2018).
78. Thara E, Dorff TB, Pinski JK, Quinn DI. Vaccine therapy with sipuleucel-T (Provenge) for prostate cancer. *Maturitas.* (2011) 69:296–303. doi: 10.1016/j.maturitas.2011.04.012
79. Holoclar European Public Assessment Report. *Ex vivo Expanded Autologous Human Corneal Epithelial Cells Containing Stem Cells.* Available online at: https://www.ema.europa.eu/documents/product-information/holoclar-epar-product-information_en.pdf (accessed December 12, 2018).
80. Flory E, Gasparini P, Jekerle V, Palomäki T, Salmikangas P. Regulatory viewpoints on the development of advanced stem cell-based medicinal products in light of the first EU-approved stem cell product. *Cell Gene Ther Insights.* (2015) 1:109–27. doi: 10.18609/cgti.2015.010
81. Strimvelis. *Common Name - Autologous CD34+ Enriched Cell Fraction That Contains CD34+ Cells Transduced With Retroviral Vector That Encodes for the Human ADA cDNA Sequence.* Available online at: https://www.ema.europa.eu/documents/assessment-report/strimvelis-epar-public-assessment-report_en.pdf (accessed December 14, 2018).
82. *Guidance for FDA Reviewers and Sponsors: Content and Review of Chemistry, Manufacturing, and Control (CMC) Information for Human Somatic Cell Therapy Investigational New Drug Applications (INDs).* Available online at: https://www.fda.gov/downloads/biologicsbloodvaccines/guidancecomplianceregulatoryinformation/guidances/xenotransplantation/ucm092705.pdf (accessed December 14, 2018).
83. European Medicines Agency. *Guideline on Human Cell-Based Medicinal Products* (EMEA/CHMP/410869/2006) (2008).
84. Carmen J, Burger SR, McCaman M, Rowley JA. Developing assays to address identity, potency, purity and safety: cell characterization in cell therapy process development. *Regen Med.* (2012) 7:85–100. doi: 10.2217/rme.11.105
85. Kusuma GD, Carthew J, Lim R, Frith JE. Effect of the microenvironment on mesenchymal stem cell paracrine signaling: opportunities to engineer the therapeutic effect. *Stem Cells Dev.* (2017) 26:617–31. doi: 10.1089/scd.2016.0349
86. Arpornmaeklong P, Brown SE, Wang Z, Krebsbach PH. Phenotypic characterization, osteoblastic differentiation, and bone regeneration capacity of human embryonic stem cell-derived mesenchymal stem cells. *Stem Cells Dev.* (2009) 18:955–68. doi: 10.1089/scd.2008.0310
87. Boyd NL, Robbins KR, Dhara SK, West FD, Stice SL. Human embryonic stem cell-derived mesoderm-like epithelium transitions to mesenchymal progenitor cells. *Tissue Eng Part A.* (2009) 15:1897–907. doi: 10.1089/ten.tea.2008.0351
88. Brown PT, Squire MW, Li W-J. Characterization and evaluation of mesenchymal stem cells derived from human embryonic stem cells and bone marrow. *Cell Tissue Res.* (2014) 358:149–64. doi: 10.1007/s00441-014-1926-5
89. De Peppo GM, Marcos-Campos I, Kahler DJ, Alsalman D, Shang L, Vunjak-Novakovic G, et al. Engineering bone tissue substitutes from human induced pluripotent stem cells. *Proc Natl Acad Sci USA.* (2013) 110:8680–5. doi: 10.1073/pnas.1301190110
90. Hynes K, Menicanin D, Mrozik K, Gronthos S, Bartold PM. Generation of functional mesenchymal stem cells from different induced pluripotent stem cell lines. *Stem Cells Dev.* (2014) 23:1084–96. doi: 10.1089/scd.2013.0111
91. Lian Q, Zhang Y, Zhang J, Zhang HK, Wu X, Zhang Y, et al. Functional mesenchymal stem cells derived from human induced pluripotent stem cells attenuate limb ischemia in mice. *Circulation.* (2010) 121:1113–23. doi: 10.1161/CIRCULATIONAHA.109.898312
92. Vodyanik MA, Yu J, Zhang X, Tian S, Stewart R, Thomson JA, et al. A mesoderm-derived precursor for mesenchymal stem and endothelial cells. *Cell Stem Cell.* (2010) 7:718–29. doi: 10.1016/j.stem.2010.11.011
93. Bloor A, Patel A, Griffin JE, Gilleece MH, Radia R, Yeung DT, et al. A phase I trial of iPSC-derived MSCs (CYP-001) in steroid-resistant acute GvHD. *Blood.* (2018) 132(Suppl. 1):4562. doi: 10.1182/blood-2018-99-110432

14

Enhanced Immunomodulation in Inflammatory Environments Favors Human Cardiac Mesenchymal Stromal-Like Cells for Allogeneic Cell Therapies

Falk Diedrichs[1,2], *Meaghan Stolk*[2], *Karsten Jürchott*[2,3], *Marion Haag*[2,4], *Michael Sittinger*[2,4] and *Martina Seifert*[2,3]*

[1] Berlin Institute of Health (BIH), Berlin, Germany, [2] BIH Center for Regenerative Therapies (BCRT), Charité-Universitätsmedizin Berlin, Corporate Member of Freie Universität Berlin, Humboldt-Universität zu Berlin, and Berlin Institute of Health, Berlin, Germany, [3] Institute of Medical Immunology, Charité-Universitätsmedizin Berlin, Corporate Member of Freie Universität Berlin, Humboldt-Universität zu Berlin, and Berlin Institute of Health, Berlin, Germany, [4] Tissue Engineering Laboratory, Charité-Universitätsmedizin Berlin, Corporate Member of Freie Universität Berlin, Humboldt-Universität zu Berlin, and Berlin Institute of Health, Berlin, Germany

***Correspondence:**
Martina Seifert
martina.seifert@charite.de

Rising numbers of patients with cardiovascular diseases and limited availability of donor hearts require new and improved therapy strategies. Human atrial appendage-derived cells (hAACs) are promising candidates for an allogeneic cell-based treatment. In this study, we evaluated their inductive and modulatory capacity regarding immune responses and underlying key mechanisms *in vitro*. For this, cryopreserved hAACs were either cultured in the presence of interferon-gamma (IFNγ) or left unstimulated. The expression of characteristic mesenchymal stromal cell markers (CD29, CD44, CD73, CD105, CD166) was revealed by flow cytometry that also highlighted a predominant negativity for CD90. A low immunogeneic phenotype in an inflammatory milieu was shown by lacking expression of co-stimulatory molecules and upregulation of the inhibitory ligands PD-L1 and PD-L2, despite *de novo* expression of HLA-DR. Co-cultures of hAACs with allogeneic peripheral blood mononuclear cells, proved their low immunogeneic state by absence of induced T cell proliferation and activation. Additionally, elevated levels of IL-1β, IL-33, and IL-10 were detectable in those cell culture supernatants. Furthermore, the immunomodulatory potential of hAACs was assessed in co-cultures with αCD3/αCD28-activated peripheral blood mononuclear cells. Here, a strong inhibition of T cell proliferation and reduction of pro-inflammatory cytokines (IFNγ, TNFα, TNFβ, IL-17A, IL-2) were observable after pre-stimulation of hAACs with IFNγ. Transwell experiments confirmed that mostly soluble factors are responsible for these suppressive effects. We were able to identify indolamin-2,3-dioxygenase (IDO) as a potential key player through a genome-wide gene expression analysis and could demonstrate its involvement in the observed immunological responses. While the application of blocking antibodies against both PD-1 ligands did not affect the immunomodulation by hAACs, 1-methyl-L-tryptophan as specific inhibitor of IDO was able to restore proliferation and to lower apoptosis of T cells. In conclusion, hAACs represent a cardiac-derived mesenchymal stromal-like cell type with a high potential for the application in an allogeneic setting, since they do not trigger T cell responses and even increase their immunomodulatory potential in inflammatory environments.

Keywords: cardiac-derived cells, immunogenicity, immunomodulation, inflammation, IFNγ, IDO

INTRODUCTION

Cardiovascular diseases are the leading cause of morbidity and mortality worldwide with ischemic heart disease alone being responsible for almost 1.8 million deaths per year in Europe (20% of all deaths; European Heart Network, 2017[1]). Even though there is a range of existing therapeutic strategies available, which have beneficial effects on the improvement of life quality and the extension of lifespan in cardiovascular patients, they often leave them with no other causal therapy option than heart transplantation (1, 2).

Numerous attempts using a variety of different cell sources were initiated over the last 20 years for development of new therapeutic treatments to induce cardiac regeneration (3). Particularly autologous cell sources, ranging from hematopoietic cells (4), over mesenchymal stromal cells (MSCs) from different tissues (5–7), to various cardiac progenitor cells (CPCs) (8–11) have been heavily investigated in this context. Specifically, the use of CPCs led to promising results: animal models of myocardial infarction demonstrated improved cardiac function after cell transplantation (12) and even first clinical studies in humans (SCIPIO and CADUCEUS trials) were able to show moderately increased regeneration of cardiac tissue (13–16).

An alternative mesenchymal-like cardiac cell type for an autologous therapeutic application in heart injury, are so called cardiac-derived adherent proliferating cells (CardAPs). This unique cell type derived from endomyocardial biopsies shares typical characteristics with MSCs but clearly distinguishes itself from all other cell types used so far in cell therapeutic application. CardAPs are positive for CD44, CD73, CD105, and CD166 but express neither the hematopoietic markers CD14, CD34 and are strikingly low for the marker CD90, which is otherwise characteristic for MSCs and fibroblasts (17). These cardiac-derived cells demonstrated increased regenerative potency by mediating angiogenesis and cardiomyogenesis, reducing cardiac hypertrophy and exhibiting immunomodulatory capacities to induce an anti-inflammatory environment (18–20). Our own immunological *in vitro* tests with these mesenchymal-like CardAPs proved their low immunogeneic status as well as the capacity to modulate the immune system toward an anti-inflammatory state (21). However, recent clinical phase-I studies with mesenchymal cell types highlighted some of the fundamental limitations of autologous cell sources (22). Manufacturing a sufficient amount of a patient-specific cell product is time consuming, thus preventing immediate availability in acute situations. Additionally, harvesting from elderly diseased patients with co-morbidities raised further concerns regarding the functional integrity and overall survival of obtained cells (23). Furthermore, it is the recent scientific consensus that every stromal cell source has to be considered as an independent entity and requires a comprehensive phenotypical and functional characterization using standardized protocols, with a particular focus on their immunological properties and immunomodulatory potency (24). This would help to identify an adequate cell source or cell subset and to promote the appropriate and safe application as a cell therapeutic or even as cell free products based on paracrine released vesicles or mediators.

For that reason, it is essential to evaluate the potential use of allogeneic cardiac-derived cells, since they can be harvested from healthy donors, have the benefit of being available at any time and can be assessed and manipulated in advance to fit the patient's needs (25). This might be important, since the transplantation of allogeneic cells or tissues always poses the risk of recognition by the recipient's immune system and induction of unwanted inflammatory responses by secretion of allo-antibodies (26, 27) or even T cell-mediated rejection responses (28, 29).

Experimental data by others with a cardiac-derived mesenchymal-like cell type indicated that those cryopreserved c-Kit$^+$ CPCs displaying low immunogeneic properties, were able to reduce local inflammatory processes and limit T cell proliferation in already ongoing immunoreactions *in vitro* (30). Additionally, the phase-I/-II CAREMI trial already proved the principal safety of allogeneic cell transplantation with previously mentioned c-Kit$^+$ selected CPCs by absence of major adverse effects after intracoronary injection (31). However, the overall benefit in cardiac improvement remains ambiguous and demands the evaluation of additional allogeneic cell sources.

Our group recently described the atrial appendage as a potential new cell source for human atrial appendage-derived cells (hAACs) that are a CD90^{low} cell product with similar pro-angiogenic characteristics compared to the endomyocardial-derived CardAPs (32). hAACs can be easily isolated from cardiac tissue and would allow allogeneic treatment for a substantial number of patients. These cells represent a mesenchymal-like cardiac-derived cell type based on the expression of the characteristic markers CD29, CD44, CD73, CD105, and CD166, but predominantly lack expression of CD90 at the same time. Precisely, this CD90^{low} phenotype could provide

[1]European Cardiovascular Disease Statistics 2017 edition. *CVD Statistics 2017* (2017). Available online at: http://www.ehnheart.org/cvd-statistics/cvd-statistics-2017.html (accessed: February 02, 2019).

Abbreviations: 1-MT, 1-methyl-L-tryptophan; CardAPs, cardiac-derived adherent proliferating cells; CD, cluster of differentiation; CFSE, carboxyfluorescein succinimidyl ester; COX-1, cyclooxygenase 1; CPCs, cardiac progenitor cells; hAACs, human atrial appendage-derived cells; HLA, human leukocyte antigen; HUVECs, human umbilical vein endothelial cells; IDO, indolamine-2,3-dioxygenase; IFNγ, interferon-gamma; IL, interleukin; MFI, mean fluorescence intensity; MSCs, mesenchymal stromal cells; PBMCs, peripheral blood mononuclear cells; PD-L, programmed death ligand; TGFβ, transforming growth factor beta; TLR3, toll-like receptor 3; TNF, tumor necrosis factor.

a beneficial tool for the enhanced repair capacity of a cell product, since it was shown that CD90 expression on cardiosphere-derived cells is negatively correlated with the scar size of injured heart tissue after cell application in myocardial infarction (33). In addition, first studies with hAACs in a mouse model of Coxsackievirus B3 (CVB3)-induced myocarditis could demonstrate, that intravenous application was able to improve the left ventricular heart function and contractility as well as to decrease tissue collagen I expression. In this experimental mouse study, immunomodulatory effects were also confirmed by detecting reduced levels of TGFβ-producing $CD68^+$ cells and regulatory T cells in the spleen of treated animals (34).

To ensure the safety and efficacy of this new hAAC product for an allogeneic transplantation in humans, it is crucial to determine whether these cells trigger immune responses in an inflammatory scenario, as seen in cell transplantation. Therefore, we aimed to assess the immunological properties of this defined cell product, test their interaction with cells from the adaptive immune system *in vitro* and gain insights into the underlying mechanism of action. First, we confirmed a mesenchymal-like surface marker expression profile after cryopreservation and assessed changes of the immune phenotype under inflammatory conditions. Second, hAACs were evaluated in immune cell co-cultures to study potential immunogeneic properties and the capacity to modulate adaptive immune responses. We could identify several potential molecules explaining the observed immune modulatory effects by a genome-wide gene expression analysis. Finally, our data revealed that indolamine-2,3-dioxygenase (IDO) is a key player of the immunomodulation by hAACs mediating the inhibition of T cell proliferation and the induction of their apoptosis.

MATERIALS AND METHODS

Isolation and Culture of Human Atrial Appendage-Derived Cells (hAACs)

Right atrial appendages, that were obtained during open-heart surgery at Deutsches Herzzentrum Berlin from eight patients, were used to generate hAACs as previously described (32). Briefly, the right atrial appendages were reduced to fragments of 1 mm^3 and cultured in Iscove's Modified Dulbecco's Medium (IMDM; Biochrom, Berlin, Germany) containing 10% allogeneic human serum (German Red Cross, Berlin, Germany), 100 U/mL penicillin and 100 µg/mL streptomycin (both from Biochrom). Outgrowing cells were harvested after about 13 days with 0.05% trypsin/0.02% EDTA (Biochrom) and then subjected to immunomagnetic sorting with CD90 microbeads (MACS; human CD90 MicroBeads kit, Miltenyi Biotec, Bergisch Gladbach, Germany). The resulting $CD90^{low}$ cell population was grown under standard culture conditions (37°C in 21% O_2 and 5% CO_2 atmosphere) at a density of 6000 cells/cm^2 in complete medium (cIDH) consisting of equal amounts of IMDM/DMEM/Ham's F12 (IDH; all Biochrom) and supplemented with 5% male heat-inactivated human AB serum (Sigma-Aldrich, St. Louis, MO, USA), 100 U/mL penicillin and 100 µg/mL streptomycin (Gibco® Life Technologies, Grand Island, NY, USA), 20 ng/mL basic fibroblast growth factor and 10 ng/mL epithelial growth factor (both from Preprotech, Hamburg, Germany) for further expansion of the purified cell product. Subsequently, hAACs were cryopreserved for at least 6 months to mimic conditions of a cell bank. After thawing, cells were routinely passaged once in cIDH medium before performing assays and were used between passages 2 and 8. Tissues were obtained according to the local guidelines of the Charité - Universitätsmedizin Berlin as well as the Declaration of Helsinki and the study was approved by the ethics committee of the Charité - Universitätsmedizin Berlin (No. 4/028/12). Human leukocyte antigen (HLA)-typing of the cells were performed in the HLA-Laboratory of the Charité - Universitätsmedizin Berlin by SSO-PCR (low) for HLA-A, HLA-B and HLA-DR. A list of all HLA-typed cells is available in the **Supplementary Table 1**.

Culture of Human Umbilical Cord-Derived Mesenchymal Stromal Cells (MSCs) and Umbilical Vein Endothelial Cells (HUVECs)

Due to their known immunomodulatory potential as previously described (35), human umbilical cord-derived MSCs were used as control cells in the immune cell co-culture experiments. Cells were kindly provided by Dirk Strunk's laboratory at the Institute of Experimental and Clinical Cell Therapy and Spinal Cord & Tissue Regeneration Center, Paracelsus Medical University (PMU) Salzburg, Austria and were obtained for human cell and tissue sample collection from the Institutional Review Board of the Medical University of Graz (protocol 19-252 ex 07/08) as described (36). Umbilical cord samples were collected from mothers that gave written informed consent after full-term pregnancies in accordance with the Declaration of Helsinki. After thawing, MSCs were grown in alpha-modified minimum essential medium (alpha-MEM; Biochrom), supplemented with 5% human male heat-inactivated AB serum (Sigma-Aldrich), 2 mM L-glutamine, 100 U/mL penicillin and 100 µg/mL streptomycin (both from Gibco® Life Technologies) at 37°C in 21% O_2 and 5% CO_2 atmosphere. HLA-Typing of the donor cells was performed by SSP PCR using Olerup SSPTM low-resolution kits (GenoVision Inc., Philadelphia, PA, USA).

HUVECs were used as positive controls in the immune cell co-culture experiments (Cascade Biologics®, Thermo Fisher Scientific, Rochester, NY, USA and Lonza, Wakersville, MD, USA). After thawing, HUVECs were cultured in EGM-2 (Lonza) with 5% human male heat-inactivated AB serum (Sigma-Aldrich), 100 U/mL penicillin and 100 µg/mL streptomycin (Gibco® Life Technologies) for further expansion.

Both cell types were passaged once before performing assays and were used between passages 2 and 8.

PBMC Isolation

Human peripheral blood mononuclear cells (PBMCs) were isolated from buffy coats (German Red Cross, Berlin, Germany; approved by the local Ethical Committee, EA1/226/14) by using a Biocoll gradient (Biochrom), as previously described (37). Briefly, following centrifugation at 800 g for 30 min without brake, PBMCs were harvested from the interphase and were washed three times with cold phosphate buffered saline solution (PBS; Biochrom). Cells were cryopreserved for later experimental use in liquid nitrogen. HLA-typing was performed in the HLA-Laboratory of the Charité - Universitätsmedizin Berlin by SSO-PCR (low) for HLA-A, HLA-B and HLA-DR.

Immunocytochemistry Analysis of hAACs

hAACs were plated on collagen I-coated (BD Biosciences, San Jose, CA, USA) 24 well dishes (Falcon, BD Biosciences). After incubation overnight, wells were washed three times with Hank's Balanced Salt Solution (HBSS; Gibco® Life Technologies) containing Mg^{2+} and Ca^{2+}, fixed with 4% paraformaldehyde (PFA; Roth, Karlsruhe, Germany) for 10 min at room temperature and washed twice with HBSS. Subsequently, the cells were incubated with 5 µg/mL wheat germ agglutinin (WGA; Biotium, Fremont, CA, USA) for 10 min at 37°C. After washing twice with HBSS, nuclei were counterstained for 15 min at room temperature with 4,6-diamidino-2-phenylindole (DAPI; Molecular probes™, Thermo Fisher). Images were taken with an Operetta® High Content Imaging System and image analysis performed by the Columbus™ Image Data Storage and Analysis System (both from Perkin Elmer, Waltham, MA, USA).

Fluorescence Staining of Cells and Flow Cytometry (FACS)

Non-adherent PBMCs were resuspended by pipetting, and adherent hAACs were first harvested using a 0.05% trypsin solution with EDTA (Gibco® Life Technologies) and transferred to 5 mL FACS tubes (Falcon, BD Biosciences). Staining procedure was performed as previously described (37). Briefly, cells were washed once with cold PBS, resuspended in a final volume of 50 µL antibody mix in cold FACS buffer [PBS supplemented with 1% fetal calve serum (FCS; both Biochrom)] and incubated for 30 min at 4°C in the dark. A list of all used antibodies and dyes as well as the respective dilution is available in the online **Supplementary Table 2**. Antibody mixes also contained the Live/Dead® violet Staining Kit (Molecular probes™, Thermo Fisher Scientific) in order to exclude dead cells from the analysis. After antibody incubation, the samples were washed with cold FACS buffer and resuspended in 1% PFA (Roth, Karlsruhe, Germany) in FACS buffer. Samples were kept at 4°C in the dark until measurement on a FACS Canto II device with FACS Diva software (Becton Dickinson, San Jose, CA, USA). Data analysis was performed using FlowJo software (TreeStar Inc., Ashland, OR, USA; RRID:SCR_008520). Gating strategies for the FACS-analysis of hAACs and PBMCs are shown in **Supplementary Figure 1**. Expression of a marker is presented either as percentage of positive cells against the unlabeled control or as geometric mean of fluorescence intensity (MFI).

Kinetic Analysis of hAAC Surface Marker Expression

hAACs were seeded on 24 well-plates (Costar®, Corning Incorporated, Kennebunk, ME, USA) at a density of 3×10^5 cells and were cultured in cIDH medium overnight. Afterwards, hAACs were either directly harvested for evaluation of constitutive MSC marker expression (CD90, CD29, CD44, CD73, CD105, CD166, CD14, CD31, CD45, c-Kit) or stimulated with 100 ng/mL of interferon-gamma (IFNγ) or a combination of 100 ng/mL IFNγ and 100 ng/mL tumor necrosis factor alpha (TNFα; both from Miltenyi Biotec) for evaluation of the immunological (HLA-ABC, HLA-E, HLA-DR, CD80, CD86, PD-L1 and PD-L2) and characteristic MSC markers (CD90, CD29, CD44, CD73, CD166). hAACs were stimulated and harvested after one, 2 and 5 days, respectively for flow cytometric analysis as described before.

hAAC/Immune Cell Co-cultures

hAACs from six different donors and control cultures with MSCs and HUVECs were seeded on rat tail collagen I-coated (BD Biosciences) 24 well plates (Costar®, Corning Incorporated) at a density of 2×10^5 cells. After attachment overnight, the adherent cells were either stimulated with 100 ng/mL IFNγ (Miltenyi Biotec) or left unstimulated for 48 h. Afterwards, the confluent monolayers were irradiated with 60 Gray using a gamma-radiation source (GSM GmbH, Leipzig, Germany) to maintain a stable cell number throughout the assay. Human HLA-mismatched PBMCs were thawed, washed three times with cold PBS (Biochrom) and labeled with 5 µM carboxyfluorescein succinimidyl ester (CFSE; Biolegend, San Diego, CA, USA) for 3 min. The staining reaction was then stopped by incubating with cold heat-inactivated human AB serum (Sigma-Aldrich) for 1 min. After washing three times with cold PBS, 3×10^5 CFSE-labeled PBMCs, that were a complete mismatch to the respective hAAC donor, were added to the hAAC, MSC and HUVEC cultures. The resulting co-cultures were maintained in 1 mL of very low endotoxin (VLE)-Roswell Park Memorial Institute (RPMI; Biochrom), supplemented with 10% human male heat-inactivated AB serum (Sigma-Aldrich), 100x L-glutamine solution, 100 U/mL penicillin and 100 µg/mL streptomycin (all from Gibco® Life Technologies). After 4 days, 250 µL of co-culture supernatant were taken for cytokine detection and 750 µL of completely supplemented VLE-RPMI were added to the cultures. Following seven days of incubation, PBMCs were harvested, stained for human immune cell defining surface markers and analyzed by flow cytometry.

Proliferation Based Immunomodulation Assay

Analogous to the hAAC/immune cell co-culture analysis, hAACs, MSCs and HUVECs were cultured on rat tail collagen I-coated (BD Biosciences) 24 well plates (Costar®, Corning Incorporated) at a density of 2×10^5 cells in the presence or absence of 100 ng/mL IFNγ (Miltenyi Biotec) for 48 h. Human PBMCs were CFSE-labeled as mentioned before and activated with a combination of 0.02 µg/mL anti-CD3 (OKT3 antibody, Janssen-Cilag, Neuss, Germany) and 0.03 µg/mL anti-CD28 (BD Biosciences). Lastly, 1×10^6 PBMCs were added to the cultures in 2 mL of completely supplemented VLE-RPMI medium. After 3 days, supernatants were taken for cytokine detection and PBMCs were harvested, stained for human immune cell defining surface markers and analyzed by flow cytometry.

Experimental settings were repeated under transwell conditions. Here, hAACs were seeded at a density of 4×10^4

cells at the bottom of rat tail collagen I-coated 24 well plates. After stimulation with IFNγ, polycarbonate transwell inserts with 0.4 μm pore size (Costar®, Corning Incorporated) were initially equilibrated for 1 h at 37°C with RPMI and subsequently 2×10^5 CFSE-labeled PBMCs were seeded into the inserts. After a co-culture time of 3 days at 37°C in a 21% O_2 and 5% CO_2 atmosphere, PBMC were harvested for flow cytometric analysis of proliferation and surface marker expression.

To selectively analyze the effects of either indoleamin-2,3-dioxygenase (IDO) or both programmed death-1 (PD-1) ligands (PD-L1 and PD-L2) in immune cell co-cultures with hAACs, 1 mM 1-methyl-L-tryptophan (1-MT; Sigma-Aldrich) was provided 2 h prior to addition of CFSE-labeled or unlabeled PBMCs and 5 μg/mL of purified anti-PD-L1 and anti-PD-L2 antibodies (both Biolegend) were added 12 h before CFSE-labeled or unlabeled PBMCs were added to the hAAC cultures.

Cytokine Detection Assays

Supernatants of mono- and co-cultures of hAACs, MSCs and HUVECs from the proliferation induction experiments were tested for IL-1β, IFNα, IFNγ, TNFα, MCP-1, IL-6, IL-8, IL-10, IL-12p70, IL-17A, IL-18, IL-23, and IL-33 using the Legendplex™ human inflammation 13-plex panel (Biolegend). The minimum detectable concentration of each cytokine is given as 0.6–2.1 pg/mL. Samples were treated following manufacturer's instructions and measured with a FACS Canto II device (Becton Dickinson).

Supernatants of hAAC co-cultures from the direct-contact immune modulation experiments were analyzed for their content of IL-1β, IL-2, IL-5, IL-10, IL-13, IL-17A, TNFα, TNFβ, IFNγ, and MDC by a multiplex assay using a Milliplex® human multi-analyte Luminex® kit (Merck KGaA, Darmstadt, Germany). Samples were treated following manufacturer's instructions and measured with a Bio-Plex® 200 multiplex analysis device (Bio-Rad®, California, USA).

Genome-Wide Gene Expression Profile

Human GeneChip U133 Plus 2.0 (Affymetrix, Santa Clara, CA, USA) was used for genome-wide gene expression profiling of hAAC samples covering over 47,000 transcripts (54,765 probes in total including double entries). RNA samples of unstimulated and IFNγ pre-stimulated hAACs were prepared with GeneChip® 3′ IVT Express Kit and GeneChip® Hybridisation, Wash and Stain Kit (Affymetrix) according to the manufacturer's instructions. In brief, 250 ng total RNA was used for cDNA synthesis and subsequent *in vitro* transcription (IVT) to amplified RNA (aRNA). 12.5 μg fragmented aRNA was used for hybridization on the chip for 16 h at 45°C. Finally, the chips were washed, stained and scanned using the Affymetrix Gene Chip Scanner 3000. Affymetrix GeneChip Operating Software (GCOS) 1.4 was used to generate CEL data files, for raw data processing and for calculation of signal intensity, signal log ratio (SLR) and *p*-value of pairwise chip comparisons AF/NP. Quality control and pre-processing was done in R[2] with the package "affy" (38). Raw data were normalized and log2-transformed using Robust Multi-array Average (RMA) algorithm implemented in this package. Thousand probe sets with the highest variances were selected in order to run a principle component analysis. Differentially expressed probe sets between the two treatment groups were selected by fitting linear models to the data and Bayesian statistics were run as implemented in the package "limma" (39). False discovery rates were used to adjust raw *p*-values for multiple testing and a minimal absolute log2-Foldchange of 1 was used for probe set selection. Mapping of differentially expressed probesets to genes and functional annotations of the DAVID database (40, 41) was done using the package "clusterProfiler" (42). Over-representation of differentially expressed genes in terms of the category "Biological Process" of the gene ontology system was done using the enrichDAVID()-function of this package. The eight top ranking results of this analysis were shown as GOcirc-plot using the "GOplot"-package (43).

RNA Extraction, cDNA Synthesis and Quantitative Polymerase Chain Reaction (qPCR)

Total RNA was extracted from unstimulated and IFNγ pre-stimulated hAACs following 48 h of incubation using the RNeasy® Mini Kit (QIAGEN, Venlo, Netherlands) according to the manufacturer's protocol. After measuring the RNA concentration with the NanoDrop 2000 spectrophotometer (Thermo Fisher Scientific), cDNA was synthesized. The reverse transcription reaction was performed using TaqMan™ Reverse Transcription Reagents Kit (Invitrogen™, Thermo Fisher Scientific). Briefly, the following components were combined to perform a 20 μL reaction volume: nuclease-free water plus total RNA (1000 ng/μL), RNase inhibitor (20 U/μL), Mg_2Cl, 10x RT Buffer, Random Hexamer Primer Mix (50 μM), dNTP Mixture (2.5 mM each dNTP) and Reverse Transcriptase (RT; 50 U/μL). Samples were incubated for 30 min at 48°C, 5 min at 95°C and subsequently cooled down at 4°C with a Thermo Flex Cycler Block (Analytik Jena, Jena, Germany). After the RT-PCR the concentration of the generated cDNA was measured with the NanoDrop 2000 to ensure a functional template for the subsequent qPCR. The qPCR was performed on a QuantStudio 6 Flex Real-Time PCR machine (Applied Biosystems, Thermo Fisher Scientific) using the SensiMix™ SYBR No-ROX kit (Bioline, London, UK). The thermal cycling conditions were comprised of a 95°C initial template denaturation for 20 s, followed by 40 cycles of PCR by applying 95°C for 15 s and 60°C for 20 s. Lastly, a final melt curve stage with 40 cycles comprising of 95°C for 15 s, 60°C for 60 s and 95°C for 15 s was performed. Three technical replicates of each sample were analyzed for gene expression of *IDO1, LGALS9, TLR3, PD-L1, PD-L2, PTGS1, HLA-G*, and *VCAM1*. All of the used primer sequences are listed in **Table 1**. The samples were normalized to the expression of the

[2]R Core Team. *R: A Language and Environment for Statistical Computing.* Vienna: R Foundation for Statistical Computing (2018). Available online at: https://www.R-project.org/

TABLE 1 | The Primer sequences of selected immunomodulatory genes.

Gene	Forward primer (5′→3′)	Reverse primer (3′→5′)
IDO1	CGGTCTGGTGTATGAAGG	CTAATGAGCACAGGAAGTTC
LGALS9	CACACATGCCTTTCCAGAAG	AAGAGGATCCCGTTCACCAT
TLR3	ATCTGTCTCATAATGGCTT	AGAAAGTTGTATTGCTGGT
PD-L1	GGCATCCAAGATACAAACTCAA	CAGAAGTTCCAATGCTGGATTA
PD-L2	GAGCTGTGGCAAGTCCTCAT	GCAATTCCAGGCTCAACATTA
PTGS1	TGTTCGGTGTCCAGTTCCAATA	ACCTTGAAGGAGTCAGGCATGAG
HLA-G	TTGGGAAGAGGAGACACGGAACA	AGGTCGCAGCCAATCATCCAC
VCAM1	CGTCTTGGTCAGCCCTTCCT	ACATTCATATACTCCCGCATCCTTC
HPRT	AGTCTGGCTTATATCCAACACTTC	GACTTTGCTTTCCTTGGTCAGG

house keeping gene *HPRT* and data were analyzed using the delta-delta Ct (ΔΔCt) method. The final results are therefore calculated as fold change of target gene expression in IFNγ pre-stimulated hAAC samples relative to the unstimulated hAAC reference samples to demonstrate upregulation of differentially expressed genes.

Raman Trapping Microscopy

Raman spectral acquisition was conducted using a BioRam® system (CellTool GmbH, Tutzing, Germany) equipped with an excitation laser wavelength of 785 nm and a laser power of 80 mW. The laser was focused through a 60x (NA 0.7) air objective. In all samples, 500 single cells were randomly selected under bright-field illumination and pinpointed for automatic spectra retrieval. Raman spectra were taken from the cytoplasm using accumulated scans of 3 × 10 s. Together with the spectra the x-y-z coordinates as well as bright-field images of each measured cell were stored. As control, 10 background measurements were taken from each sample. To assess acquired Raman spectra, multivariate data analysis was performed. Principal Component Analysis (PCA) was used for visualizing the datasets. PCA was implemented in Python 2.7, using the scikit-learn package (44). PCA score plots were used to look for clusters among the data. Circles in 2D scores plot depict 95% confidence intervals.

Statistical Analysis

Statistical Analysis and graph generation was performed with GraphPad Prism 8.0 (Graphpad Software, La Jolla, USA; RRID:SCR_002798). Statistical analyses were chosen that do rely on non-parametric distribution, since all data sets were $n \leq 10$. Statistical differences between two groups with only one variable were analyzed using the Mann-Whitney non-parametric t-test. For more than two groups with multiple variables, Kruskal Wallis one-way analysis of variance (ANOVA) with Dunn's post tests were applied. Statistical differences between two or more groups with more than two variables were analyzed using an ordinary two-way ANOVA with the Sidak's post-test. All results are shown as mean ± SEM and asterisks were assigned to the p-values in the following order: $^{*}p \leq 0.05$; $^{**}\ p \leq 0.01$; $^{***}p \leq 0.001$; $^{****}p \leq 0.0001$.

RESULTS

Cryopreserved hAACs Show Typical Characteristics for Cells of Mesenchymal Origin

It is crucial for a potential allogeneic application of hAACs to determine the general characteristics of these cells after long-term cryopreservation. hAACs from eight right atrial appendages were generated from outgrowth cultures by negative immunoselection for CD90, expanded in cell culture and cryopreserved for at least 6 months (**Figure 1A**). Thawed cells after 24 h in cell culture showed their distinctive morphology with long elongated cell bodies and fibroblast-like appearance (**Figure 1B**). Flow cytometry analysis for characteristic surface markers of mesenchymal cells confirmed the distinguishing marker profile of these cells. The hAAC cell product expressed most of the known mesenchymal stem- and progenitor markers (e.g., CD29, CD44, CD73, CD105, and CD166), while lacking the expression of the endothelial marker CD31, the hematopoietic markers CD14, CD34, and CD45 as well as the cardiac progenitor marker c-Kit (CD117) (**Figure 1C**). Yet, in contrast to classical fibroblasts and mesenchymal stromal cells, only a small proportion of cells was positive for CD90 (**Figure 1D**).

Inflammatory Priming Alters the Immune Phenotype of hAACs

To investigate the surface marker profile in an inflammatory milieu, that mimics the environmental site of cardiac injury, hAACs were stimulated with 100 ng/mL of IFNγ for 48 h. Initial experiments with stimulation by pro-inflammatory cytokines (IFNγ or a combination with TNFα) for 1, 2, and 5 days, showed an increase in surface marker expression in the majority of tested markers (HLA-ABC, HLA-DR, PD-L1) with an overall maximum of up-regulation after 2 days (**Supplementary Figure 2**). Hence, the time point of 48-h stimulation with an appropriate IFNγ concentration of 100 ng/mL was chosen to determine the relative expression levels as normalized mean fluorescence intensities (MFI) for a set of immunologically-relevant markers. Stimulation of hAACs induced similar changes in the surface marker expression of all six donors as shown in FACS histogram overlays (**Figure 2A**) of one representative donor and summarized as

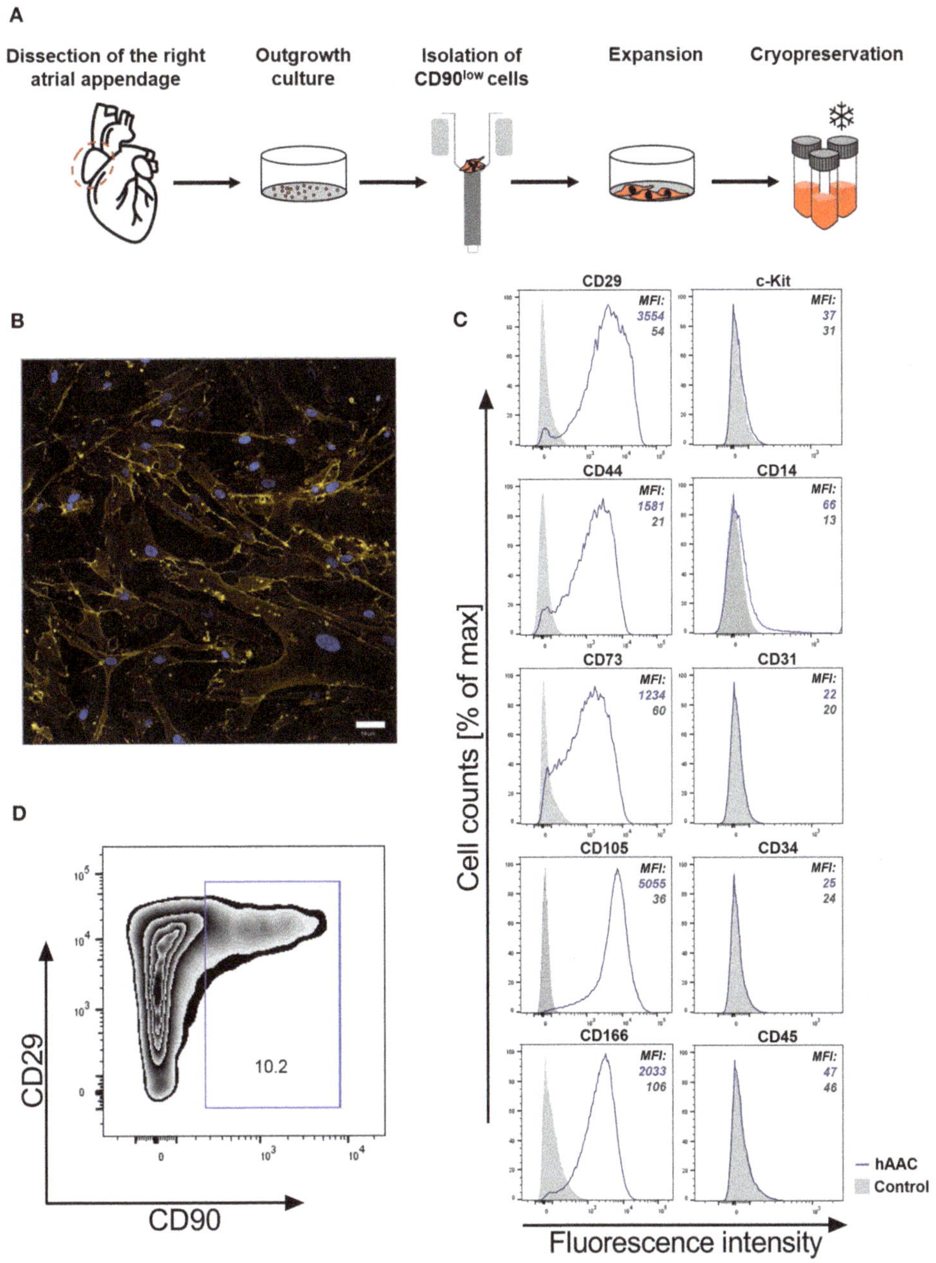

FIGURE 1 | Characteristics of human atrial appendage-derived cells (hAAC). **(A)** The general procedure to generate hAACs is shown: right atrial appendages were harvested from patients during open-heart surgery. The dissected tissue was minced into small pieces and cultured. Subsequently, the fragment's outgrowth was harvested, immuno-magnetically sorted as CD90^{low} cells and seeded again for further expansion of the desired purified hAAC product. The cells were cryopreserved and long-term stored for later experimental use. **(B)** Representative immunofluorescence images show the characteristic morphology of one hAAC donor in a cell culture plate after 12 h of incubation. Cell membranes (orange) were stained with wheat germ agglutinin (WGA) and nuclei (blue) with 4,6-diamidin-2-phenylindol (DAPI). Scale bar represents 50 μm. **(C)** Generated hAACs were harvested by treatment with trypsin after passage four and stained with human-specific antibodies against surface markers characteristic for cells of mesenchymal origin (CD29, CD44, CD73, CD105, CD166), markers for exclusion of hematopoietic contaminants (CD14, CD31, CD34, CD45) and the cardiac progenitor marker c-Kit (CD117). Representative flow cytometry histograms of one hAAC donor are shown as fluorescence intensity against cell counts for all tested markers, indicating the mean of fluorescence intensity (MFI) of positive cells (blue) compared to the unlabeled controls (gray) for each marker. **(D)** The percentage of remaining CD90^{+} cells in the hAAC cell product is shown for one representative donor in a dot plot of CD90 against CD29.

normalized MFI values (**Figure 2B**) as well as percentages of marker positive cells (**Supplementary Figure 3**). All donors expressed HLA-class I (HLA-ABC, and partly HLA-E) and low or negligible levels of HLA class II (HLA-DR) constitutively, but significantly up-regulated both HLA-molecule classes under IFNγ-stimulation. In contrast, co-stimulatory molecules such as CD80 and CD86 were completely absent, even under stimulation. Additionally, a significant increase of the

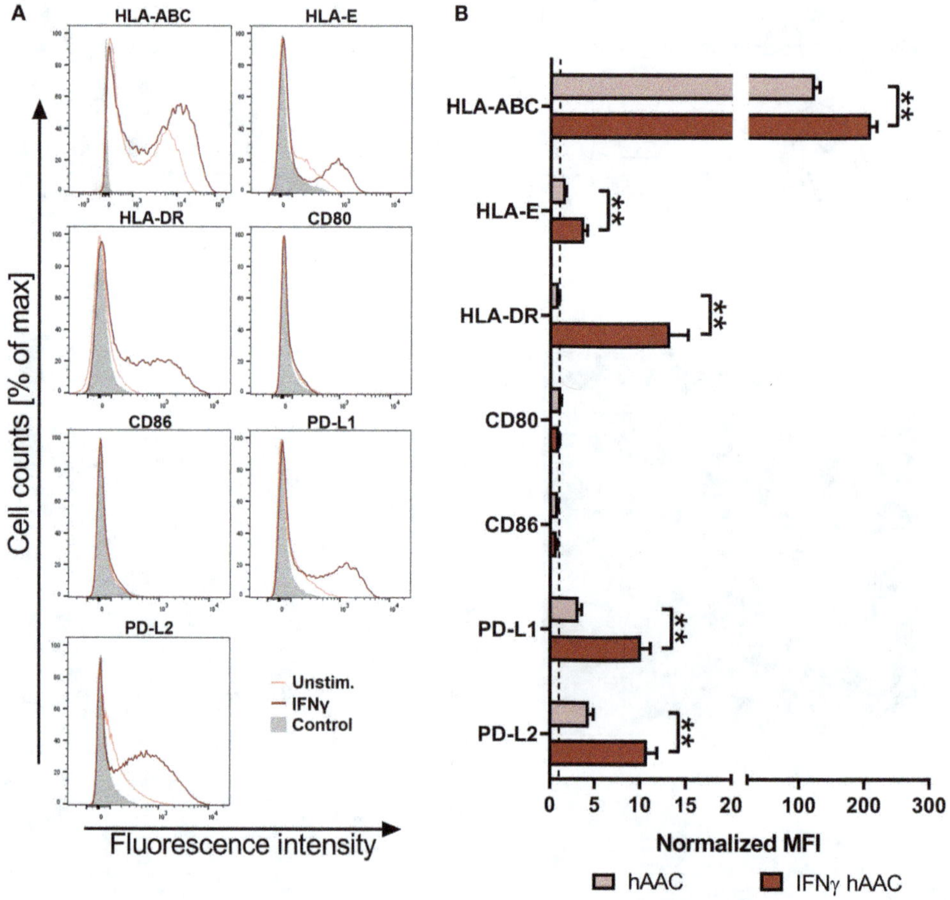

FIGURE 2 | Immune phenotype of cryopreserved hAACs under constitutive and inflammatory conditions. **(A)** Representative histogram overlays for one hAAC donor display the expression pattern for immunologically-relevant surface markers (HLA-ABC, HLA-E, HLA-DR, CD80, CD86, PD-L1, PD-L2). Cells were cultured for 48 h without additional stimulation (hAAC; light red line) or in presence of 100 ng/mL human interferon-gamma (IFNγ hAAC; dark red line). After harvest by application of trypsin, cells were stained with human-specific antibodies and analyzed by flow cytometry. Fluorescence intensity of marker expression is presented compared to the unlabeled control (Control; filled gray curve). **(B)** Summarized hAAC surface marker expression data are presented as normalized mean of fluorescence intensities (MFI), that are calculated based on the respective controls (set to one; dashed black line), and shown as mean + SEM ($n = 6$; three independent experiments with six different hAAC donors). Differences between unstimulated hAACs and IFNγ hAACs were considered significant when $^{**}p \leq 0.01$ with the Mann-Whitney *t*-test.

MFI for the immunomodulatory PD-1 ligands (PD-L1 and PD-L2) could be determined after stimulation with IFNγ (**Figure 2B**). Both markers were shown to be expressed on a considerable proportion of cells (**Supplementary Figure 3**). However, stimulation with IFNγ did not led to alterations in MFI or frequency of mesenchymal marker expression (CD29, CD44, CD73) on hAACs and even CD90 remained unchanged (**Supplementary Figure 4**).

hAACs Evade Recognition by Allogeneic T Cells *in vitro* Even in an Inflammatory Milieu

Next, we were interested in the response of T cells against allogeneic hAACs due to their key role as mediators of allo-recognition and rejection in adaptive immune responses. Accordingly, we mimicked the *in vivo* situation by co-culturing HLA-mismatched PBMCs with hAACs from six different donors and investigated the induction of allogeneic T cell responses by monitoring their activation as well as proliferation. The T cell immune responses induced by MSCs and HUVECs have been well-described in the literature and therefore both cell types were used as controls for absent or induced responses, respectively. As shown in the experimental setup (**Figure 3A**) hAACs as well as MSCs and HUVECs were cultured with or without pre-stimulation by IFNγ for 48 h, thereupon CFSE-labeled PBMCs from healthy donors were added to the cultures. After 7 days of co-culture, the surface marker expression and proliferation of T cells were analyzed by flow cytometry. Compared to the unstimulated PBMC control cultures, the presence of unstimulated HUVECs significantly induced proliferation of $CD8^+$ T cells but had no effect on the $CD4^+$ T cell compartment (**Figures 3B,C**). However, pre-stimulation with IFNγ led to highly elicited levels of both $CD4^+$ and $CD8^+$ T cell proliferation.

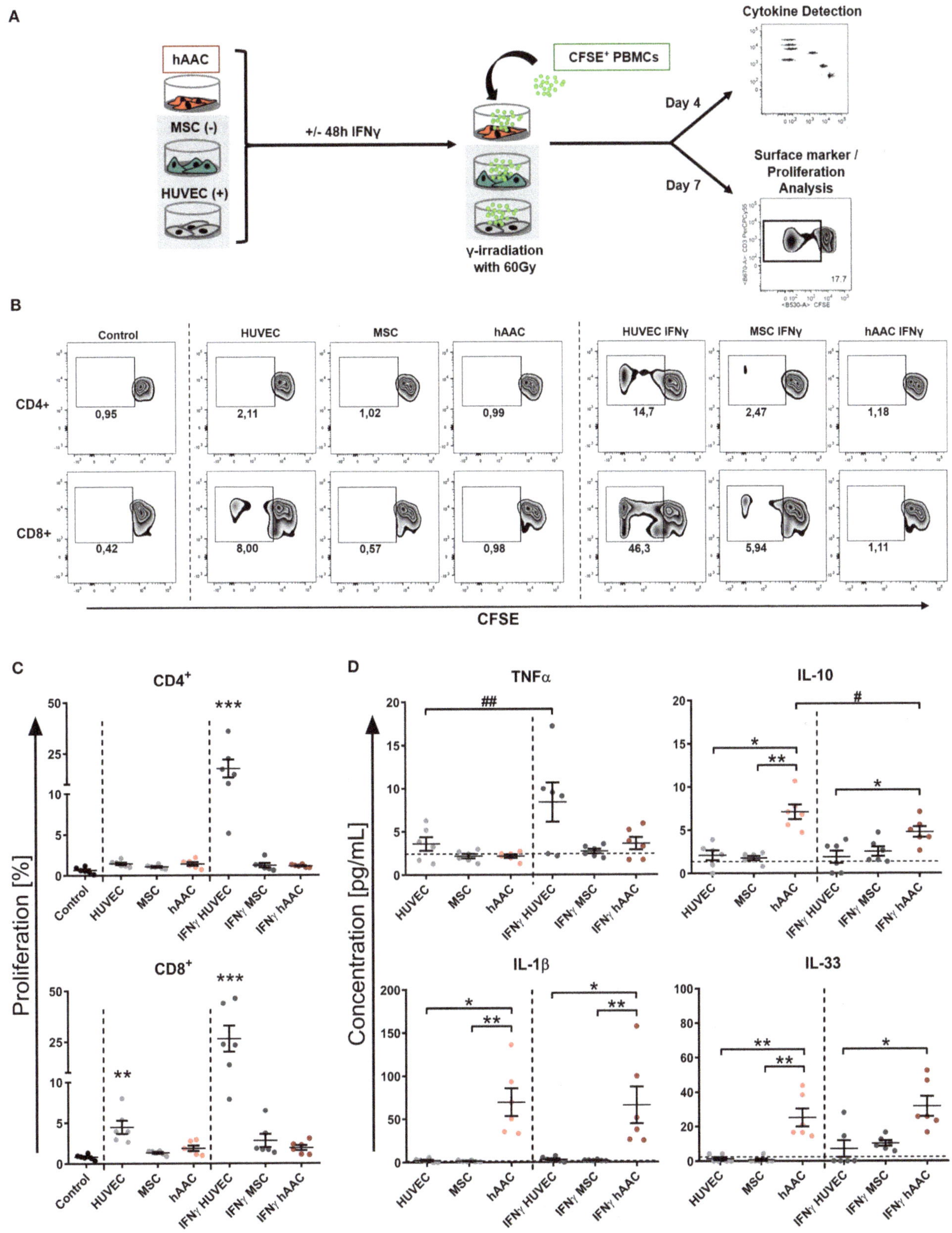

FIGURE 3 | hAACs maintain a low-immunogeneic profile even in an inflammatory environment. **(A)** The experimental setup for analyzing hAACs immunogenicity in immune cell co-cultures is illustrated: mesenchymal stromal cells from the umbilical cord (MSCs) as well as human umbilical vein endothelial cells (HUVECs) served,

(Continued)

FIGURE 3 | respectively, as cellular controls [MSC (-); HUVEC (+)] and were cultured along with hAACs for 48 h in the presence or absence of 100 ng/mL IFNγ. The cells were gamma-irradiated with 60 Gray before carboxyfluorescein succinimidyl ester (CFSE)-labeled, human leukocyte antigen (HLA)-mismatched peripheral blood mononuclear cells (PBMCs) were either added to the adherent cell cultures of all three cell types or left alone as control. After 4 days of incubation, supernatants were taken for cytokine detection using the Legendplex™ human inflammation panel and after 7 days PBMCs were harvested, stained with human immune cell specific antibodies and analyzed flow cytometrically. Levels of $CD4^+$ and $CD8^+$ T cell proliferation were detected by determining reduced CFSE signal intensity (black square). **(B)** Representative flow cytometry plots of proliferated $CD4^+$ and $CD8^+$ T cells are shown for all PBMC co-culture groups and PBMCs only (Control). **(C)** Summarized proliferation data for $CD4^+$ and $CD8^+$ T cells are presented as mean ± SEM ($n = 6$; three independent experiments with six different hAAC donor). Groups were considered significantly different compared to the Control when $^*p \leq 0.05$; $^{**}p \leq 0.01$; $^{***}p \leq 0.001$ with Kruskal Wallis ANOVA and Dunn's post-test. **(D)** Measured cytokine levels in [pg/mL] for TNFα, IL-10, IL-1β, and IL-33 are shown as mean ± SEM ($n = 6$). Cytokine levels of the PBMC control groups are depicted as dotted gray line. Groups were considered significantly different when $^*p \leq 0.05$; $^{**}p \leq 0.01$; $^{***}p \leq 0.001$ with Kruskal Wallis ANOVA and Dunn's post-test. Differences between treatments were considered significant when $\#p \leq 0.05$; $\#\#p \leq 0.01$ with ordinary two-way ANOVA and Sidak's post-test.

A similar trend was observed in the activation status that was revealed by significantly increased percentages of HLA-DR^+ T cells in the $CD4^+$ and $CD8^+$ subsets after IFNγ pre-stimulation (**Supplementary Figure 5**). hAACs on the other hand displayed low-immunogeneic properties analogous to the moderate levels of induced T cell proliferation detected in MSC co-cultures. Both, hAACs and MSCs did not induce significant changes in $CD4^+$ or $CD8^+$ T cell proliferation after IFNγ pre-stimulation as well as under unstimulated conditions (**Figures 3B,C**), neither did both cell types lead to increased expression levels of the activation marker HLA-DR on T cells (**Supplementary Figure 5**).

Additionally, supernatants were taken from co-cultures after 4 days of incubation and were evaluated for their content of various cytokines. Summarized data for TNFα, IL-10, IL-1β, and IL-33 are presented in **Figure 3D**. The amount of the pro-inflammatory cytokine TNFα significantly increased in IFNγ-stimulated HUVEC co-cultures. Contrarily, unstimulated hAAC co-cultures exclusively showed elevated levels of IL-10 release, that significantly decreased with IFNγ pre-stimulation. Significant increases of IL-1β concentrations were detected in co-cultures of unstimulated and IFNγ-stimulated hAACs. While the concentration of IL-33 also increased in unstimulated hAAC cultures, IFNγ-triggered co-cultures only showed a significant elevation compared to the cytokine levels secreted by HUVECs. Contrary to TNFα and IL-10, mono-cultures of hAACs already constitutively produced the cytokines IL-1β and IL-33 (**Supplementary Figure 6**). Other cytokines measured, like MCP-1 and IL-8, were also produced at a basal level by the adherent cells and showed no inordinate changes in PBMC co-cultures (**Supplementary Figure 7**). IFNα, IFNγ, IL-12p70, IL-17A, IL-18, and IL-23 were not detectable or only at negligible levels (data not shown).

Immunomodulatory Efficiency of hAACs Is Considerably Enhanced After IFNγ Treatment

Activation of T cells and the resulting inflammatory responses after cardiac injury largely contribute to adverse remodeling and development of chronic heart diseases in patients. We therefore examined the ability of hAACs, either unstimulated or IFNγ pre-stimulated, to modulate an already ongoing T cell response in co-cultures with αCD3/αCD28 activated PBMCs. In parallel PBMC co-cultures with both control cell types (HUVECs and MSCs) were performed. After 72 h of co-culture, proliferation rates of $CD4^+$ and $CD8^+$ T cells and cytokine release were analyzed.

As expected, the presence of HUVECs did not significantly affect the proliferation rates of T cells (**Figures 4A,B**). Contrarily, hAACs and MSCs efficiently reduced the percentage of proliferating $CD4^+$ and $CD8^+$ cells, exclusively after IFNγ triggering, below 10% (**Figures 4A,B**). A slight reduction in T cell proliferation was observed with unstimulated adherent cells, but these changes were in fact not significant. Experimental settings were repeated with hAACs under transwell conditions to test for contact-dependency of the observed immunomodulatory effects. No changes in proliferation of $CD4^+$ and $CD8^+$ T cells were detectable with unstimulated hAACs in the transwell setting. Pre-stimulation with IFNγ on the other hand caused significant reduction of proliferation levels in both T cell subsets (**Figure 4C**). The same trend was observed with a significant decrease of the activation marker CD25 on $CD4^+$ and particularly $CD8^+$ T cells after IFNγ pre-stimulation (**Supplementary Figure 8**).

Furthermore, a significant decrease of IFNγ and TNFβ concentration was measured in co-cultures with unstimulated and IFNγ-stimulated hAACs, but only the latter showed a significant reduction in the amount of released TNFα, MDC, IL-10, and IL-17A. Interestingly, co-cultures with unstimulated hAACs produced significantly more IL-1β and less IL-2 (**Figure 4D**). Other cytokines measured like IL-5 and IL-13 showed significantly lower levels in IFNγ stimulated co-cultures (data not shown).

Stimulation by IFNγ Leads to Specific Changes in the Gene Expression Profile of hAACs

Whole genome gene expression of unstimulated and IFNγ pre-stimulated hAACs were analyzed on human hgu133plus2 microarrays (Affymetrix). Data were normalized and the 1000 most variable probe sets were used in a principle component analysis. This unbiased analysis revealed a strong separation of unstimulated and IFNγ pre-stimulated samples along the first principle component reflecting the experimental design. In addition, samples were separated according to different hAAC donors in the principle component two, indicating some heterogeneity of gene expression (**Figure 5A**). Although showing these individual characteristics in gene expression, additional analyses with Raman spectroscopy revealed a

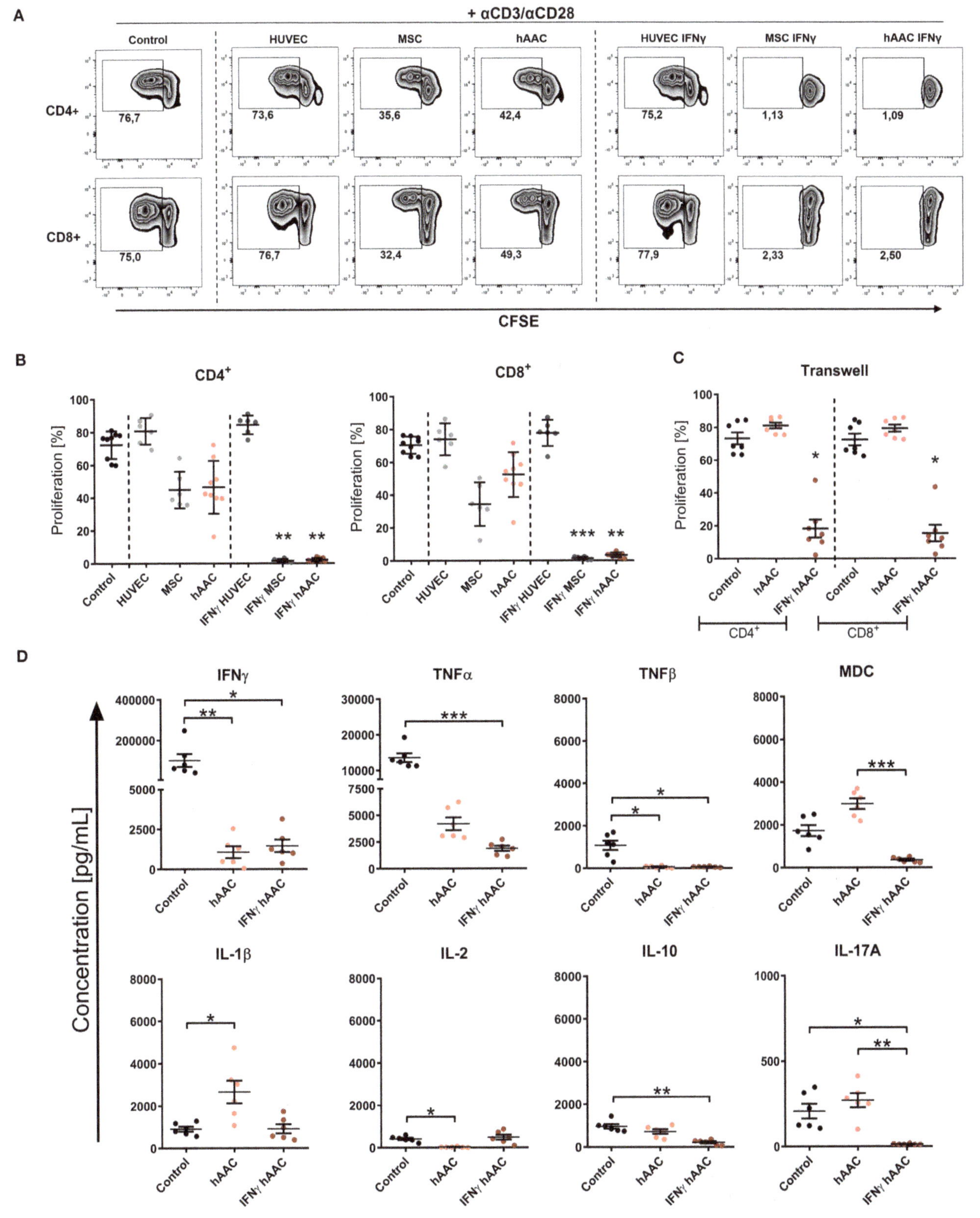

FIGURE 4 | IFNγ pre-stimulation enhances the immune-modulatory capacity of hAACs. CFSE-labeled PBMCs were activated with anti-CD3/anti-CD28 antibodies (+αCD3/αCD28) and cultured alone (Control) or in the presence of unstimulated or IFNγ-stimulated hAACs, HUVECs or MSCs for 72 h. Cells were harvested, stained

(Continued)

FIGURE 4 | with human-specific antibodies and analyzed by flow cytometry for T cell proliferation, based on reduced CFSE signal intensity. Percentages of $CD4^+$ or $CD8^+$ proliferated cells for all experimental groups are shown as representative dot plots **(A)** and as summarized data with mean ± SEM **(B)** ($n = 6–9$; four independent experiments with seven different hAAC donors). **(C)** Experimental settings were repeated with hAACs under transwell culture conditions to evaluate a contact-dependent mode of action in the observed immune-modulatory effects ($n = 7$; three independent experiments with six different hAAC donors). **(D)** Supernatants of the direct immune cell co-cultures with either unstimulated or IFNγ-treated hAACs were analyzed with a Luminex bead kit for their content of IFNγ, TNFα, TNFβ, MDC, IL-1β, IL-2, IL-10, and IL-17A. Summarized data for cytokine concentrations [pg/mL] are presented as mean ± SEM ($n = 6$; three independent experiments with six different hAAC donors). Groups were considered significantly different when $^*p \leq 0.05$; $^{**}p \leq 0.01$; $^{***}p \leq 0.001$ with Kruskal Wallis ANOVA and Dunn's post-test.

similar global phenotype of the hAAC cell product with no detectable differences in the molecular composition of cells derived from three hAAC donors (**Supplementary Figure 9**). Next, differentially expressed genes between the unstimulated and IFNγ pre-stimulated groups were determined by fitting linear models to the data and running a Bayesian statistic. Despite the genetic heterogeneity of the donors, a common response to IFNγ can be identified as shown in the heatmap (**Figure 5B**). Differentially expressed genes were subjected to an overrepresentation analysis utilizing the gene ontology system. The eight top-ranking results of the category "biological process" are shown in **Figure 5C** and were related to the immune system, cytokine-signaling and according to the experimental setup to the IFNγ-response. Remarkably, much more genes were up-regulated as down-regulated in the gene sets matched to these GO-terms. The whole results of the overrepresentation analysis are summarized in **Supplementary Table 3**. To identify a common mode of action among the hAAC donors after IFNγ pre-stimulation, the results of the genome-wide gene expression profile were checked for differential expression of known immunomodulatory genes in MSCs, including: *IDO1, PD-L1, PD-L2, NT5E, TGFB1, PTGES2, HGF, IGF, TNFAIP6, JAG1, ICOSLG, HLA-G, PGE2, IL-10, LDHB, LDHA, LGALS1, LGALS9, TLR3, ANXA1, VCAM1, PTGS1, PTGS2* and *LIF*. However, only *IDO1, LGALS9, TLR3, PD-L1, PD-L2, PTGS1, HLA-G*, and *VCAM1* were differentially expressed among all three donors and were therefore validated by qPCR. The analysis of the relative gene expression normalized to the corresponding unstimulated hAAC sample showed a very strong upregulation of *IDO1* expression in IFNγ-triggered hAACs. Yet, other immune regulatory genes like *LGALS9* (Galectin-9), *TLR3, HLA-G, PTGS1* (COX-1) as well as *PD-L1* and *PD-L2* also showed a distinct upregulation in the presence of IFNγ, but admittedly to a much lesser extent (**Figure 5D**).

IDO Predominantly Mediates Immunomodulation by IFNγ-Triggered hAACs and Acts Through Apoptosis of T Cells

Since *IDO1* was by far the most upregulated gene in IFNγ pre-stimulated hAACs (**Figure 5D**) and PD-L1/PD-L2 were already described in the literature as a key mechanism of action in cardiac-derived cells (30) and MSCs (45), we selected them as promising molecules that might be involved in the observed immunomodulatory potential of hAACs (**Figure 4B**). As illustrated in **Figure 6A**, hAACs were cultured for 48 h in the presence or absence of IFNγ. Following 2 h of pre-incubation with 1-MT as a specific inhibitor of IDO or 12 h with blocking antibodies against PD-L1 and PD-L2 (αPD1-ligand), either CFSE labeled or unlabeled PBMCs were added to the cultures and were subsequently activated with a cocktail of αCD3/αCD28 antibodies. After 72 h, PBMCs were analyzed by flow cytometry to determine proliferation and apoptosis of T cells.

hAACs without treatment by blocking agents confirmed the observed immunomodulatory effects by significantly inhibiting T cell proliferation after IFNγ pre-stimulation, shown as relative proliferation normalized to the αCD3/αCD28 PBMC control (**Figure 6B**). Additionally, significantly higher levels of late-apoptotic T cells occurred in IFNγ pre-stimulated co-cultures, shown as relative percentages of Annexin-V^+7-AAD^+ T cells normalized to the αCD3/αCD28 PBMC control (**Figure 6C**). Treatment with blocking antibodies against both PD-1 ligands neither had a significant effect on proliferation (**Figure 6B**), nor on apoptosis rates (**Figure 6C**) of $CD4^+$ and $CD8^+$ T cells when compared to the untreated control group. Notably, treatment with 1-MT, mediating the blocking of IDO, resulted in significant restoration of T cell proliferation in IFNγ-stimulated hAAC co-cultures. However, unstimulated cultures displayed rather reduced proliferation rates for both T cell subsets (**Figure 6B**). Complementary to the proliferation, treatment with 1-MT also caused significant reduction of Annexin-V^+7-AAD^+ late-apoptotic T cells for the IFNγ hAAC group relative to the untreated control group (**Figure 6C**). Yet, comparing unstimulated hAACs with IFNγ hAAC under 1-MT treatment illustrates, that levels were still elevated for apoptotic $CD4^+$ cells, but were in fact not significantly different for the $CD8^+$ subset (**Figure 6C**).

DISCUSSION

Based on the limited treatment options for cardiovascular diseases, the development of new and potent cellular therapeutics has emerged in the past decade as a promising new strategy aimed at preventing or even reversing myocardial damage.

A huge variety of cell therapy studies had been conducted with MSCs of different tissue origins, which reported only modest or no significant efficacy to improve cardiac function (46). The initial enthusiasm faded and a multitude of scientific questions remained unanswered (47). Apart from developing solutions for patient selection, timing of administration and appropriate application routes including scaffold-based approaches, efficient cell pre-conditioning or genetic manipulation (24), the most pressing task is the identification of an isolation procedure for a

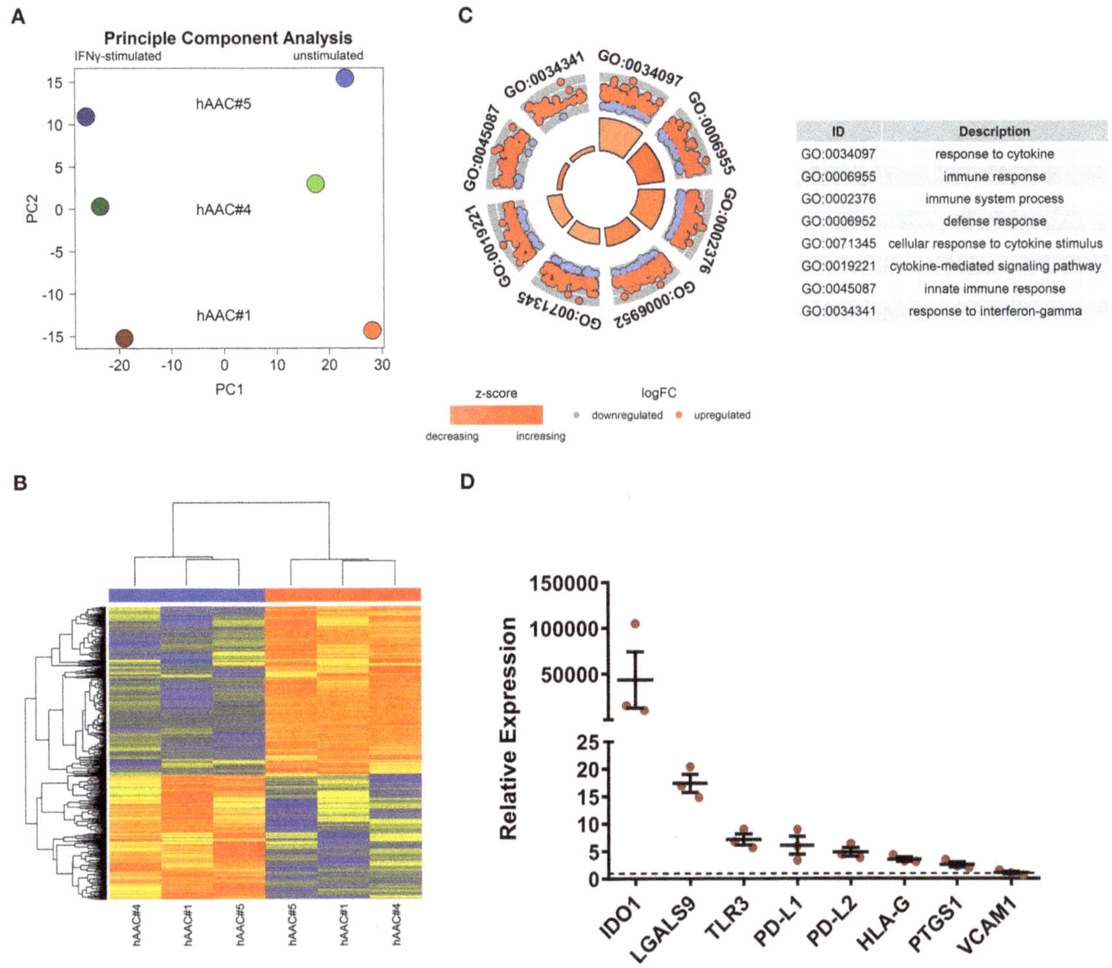

FIGURE 5 | Whole genome expression analysis and quantitative verification of immunomodulatory gene expression of unstimulated and IFNγ pre-stimulated hAACs. **(A)** Microarray expression data of unstimulated and IFNγ-stimulated hAACs were normalized and log2-transformed. Thousand probe sets with the highest variances across all samples were selected and used in a principle component analysis, showing PC1 vs. PC2 for $n = 3$ hAAC donors (#1, #4, #5). **(B)** Differentially expressed probe sets were determined by fitting linear models to the data and apply Bayesian statistics. *P*-values were adjusted for multiple testing using false discovery rate. Probe sets with an adjusted *p*-value below.05 and a minimal absolute log2-foldchange of 1 are shown in the heatmap. The blue column bar indicates the unstimulated and the red bar the IFNγ pre-stimulated hAACs of donors #1, #4, #5. **(C)** Differentially expressed probe sets were used in an overrepresentation analysis utilizing the gene ontology system. The eight top-ranking results of the category biological process are shown. **(D)** The differentially expressed immunomodulatory genes *IDO1*, *LGALS9*, *TLR3*, *PD-L1*, *PD-L2*, *HLA-G*, *PTGS1* and *VCAM1*, identified in the global microarray analysis, were validated by qPCR. Values of IFNγ pre-stimulated hAACs were normalized to the unstimulated samples (set to 1; black dotted line) by means of a ΔΔCt analysis and upregulation is shown as mean of relative expression ± SEM for $n = 3$ hAAC donors (#1, #4, #5).

suitable and effective cell type. In this context, cells isolated from close origin to the target tissue seem to be the most promising cell therapeutic strategy (48, 49). Therefore, cardiac-derived cell types excelled as a potentially powerful cell source due to their known cardio-protective and pro-angiogenic effects. Due to the lack of scientific evidence and understanding regarding these positive effects of cell-based therapies, a return to the bench-side would not only help to understand the mode of action, but also may lead to a more reliable and effective therapy for the patient (12). Moreover, large batches of healthy and instantly available cells are needed for the realization of a broad and less expensive clinical application, which clearly favors allogeneic cell sources. To exclude any unwanted immunogenicity, especially under an inflammatory and disease-related condition, it is indispensable to analyze the interaction with immune cells *in vitro*. In the present study, we addressed these issues by analyzing first the main immunological characteristics of the allogeneic hAAC cell product to estimate its immunogenicity and secondly by investigating the immunomodulation capacity focusing on T cell responses.

We could convincingly show that human allogeneic cells from the right atrial appendage (hAACs) exhibit, even in an inflammatory environment, an inherently low immunogenicity, as demonstrated by the absence of adversely induced T cell responses in immune cell co-cultures. Moreover, we could clearly present that hAACs are effective modulators of already induced

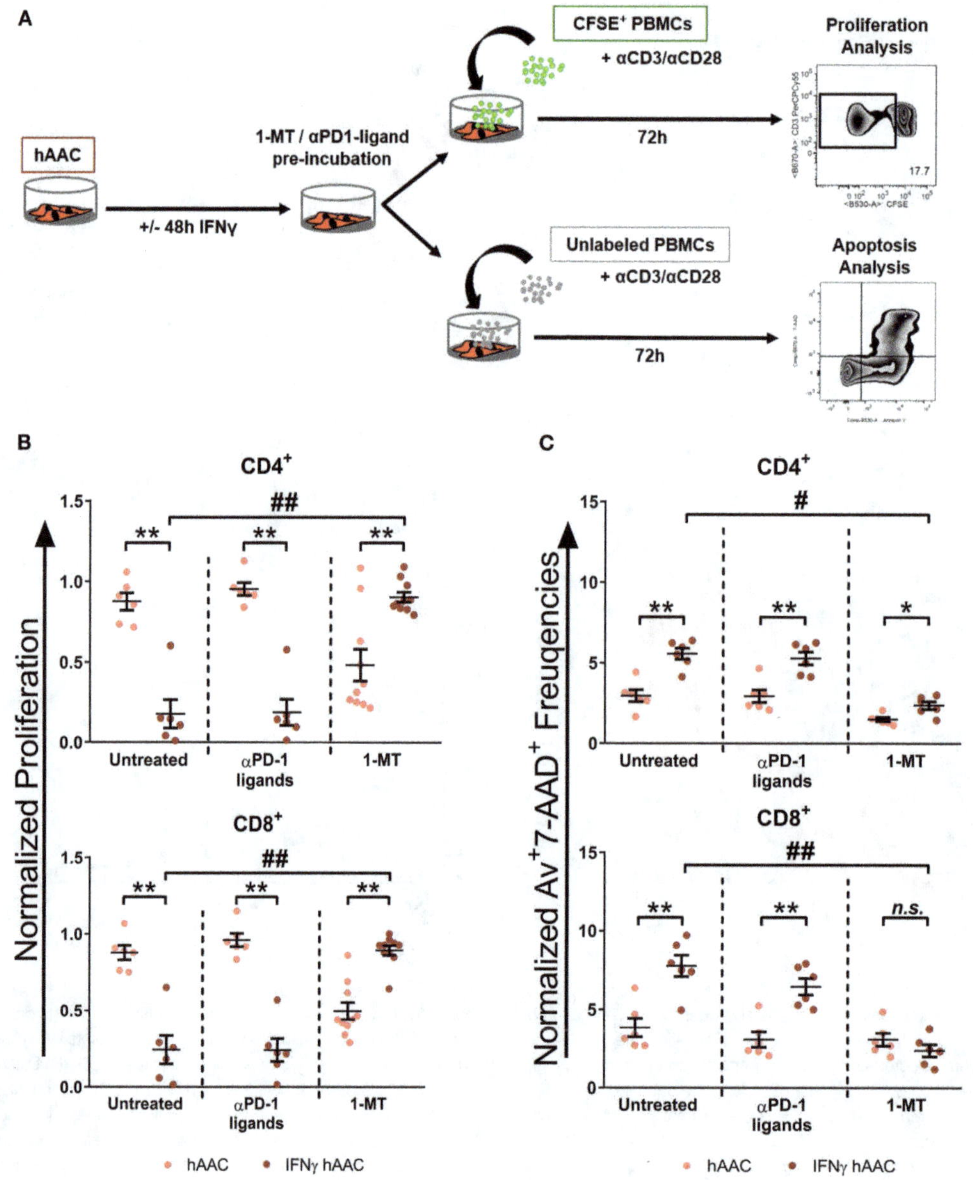

FIGURE 6 | Inhibition of T cell proliferation by hAACs is indoleamine-2,3-dioxygenase (IDO)-dependent and involves apoptosis. **(A)** The experimental setup to evaluate the potential involvement of IDO and programmed death-1 (PD-1) ligands in hAAC effects on T cells is shown schematically: hAACs were seeded and stimulated with IFNγ or left unstimulated for 48 h. Before co-cultivation with immune cells, 1-mehtyl-L-tryptophan (1-MT) as specific inhibitor of IDO or blocking antibodies against PD-L1 and PD-L2 (αPD1 ligands) were applied. Subsequently, either CFSE-labeled or unlabeled PBMCs were activated by adding a cocktail of anti-CD3/anti-CD28 (+αCD3/αCD28) antibodies and left alone as controls or cultured with hAACs for 72 h. $CD4^+$ or $CD8^+$ T cell proliferation levels as well as the percentage of Annexin-V^+ (Av+) 7-AAD^+ late apoptotic cells were determined by flow cytometry. **(B)** The normalized proliferation values for $CD4^+$ or $CD8^+$ T cell subsets were calculated based on the respective PBMC controls and are presented as mean ± SEM ($n = 6$–10; four independent experiments with seven different hAAC donors). **(C)** The normalized percentages of late-apoptotic Annexin-V^+ (Av^+) 7-AAD^+ cells within the $CD4^+$ and $CD8^+$ T cell subsets were calculated based on the respective PBMC controls and are presented as mean ± SEM (n = 6; two independent experiments with six different hAAC donors). Unstimulated (hAAC) and IFNγ-stimulated hAACs (IFNγ hAAC) within the same treatment were considered significantly different when $^*p \leq 0.05$; $^{**}p \leq 0.01$ with the Mann-Whitney t-test. Differences between treatments were considered significant when $\#p \leq 0.05$; $\#\#p \leq 0.01$ with Kruskal Wallis ANOVA and Dunn's post-test.

immune cell responses by suppressing T cell proliferation and pro-inflammatory cytokine release, especially after triggering with IFNγ. We could additionally show that IDO is one of the key mediators responsible for the observed immunomodulating features of hAACs. Our data strongly support the use of these allogeneic cardiac-derived cells for the therapeutic application in cardiovascular diseases.

So far, application of allogeneic cells for therapeutic approaches was only taken into account for mesenchymal cell types, like mesenchymal stromal cells (MSC) and cardiac progenitor cells (CPC) due to their known low immunogeneic phenotype (22, 25, 50, 51). Therefore, we checked our newly described hAAC cell product for the presence of characteristic mesenchymal cell markers and could confirm a similar expression pattern as seen in MSCs, with the only exception of a significantly reduced CD90 expression. A low CD90 expression might be an advantage for future therapeutic approaches by reducing fibrosis, as shown in a prospective analysis of a clinical trial using cardiosphere-derived cells (33). Inversely, another study using human tissue revealed that CD90 expressing fibroblasts are mainly responsible for the fibrotic thickening in peritoneal dialysis patients (52).

However, more important for the clinical application of an allogeneic cell product is to determine its potential immunogenicity (53, 54) by evaluating the expression of immunological markers under constitutive and inflammatory conditions. In the present study, we could clearly demonstrate that hAACs constitutively expressed HLA-ABC and partially HLA-E, but not HLA-DR as well as the main co-stimulatory molecules CD80 and CD86. After IFNγ stimulation, the expression profile shifted toward a further increased expression level of HLA-class I molecules and *de novo* induction of HLA-DR expression. Yet, the co-stimulation molecules remained absent. Additionally, hAACs showed expression of the immunomodulatory molecules PD-L1 and PD-L2 on a considerable proportion of cells that was further enhanced after inflammatory stimulation. Comparatively, this surface marker profile was already described for MSCs (45, 55, 56) and related cardiac cell types (30, 57).

Despite the upregulated expression of both classes of HLA-molecules after IFNγ stimulation, hAACs evaded an adaptive immune response and were thereby not able to trigger substantial $CD4^+$ and $CD8^+$ T cell proliferation, when co-cultured with human PBMCs. The in parallel tested human MSCs induced a similar proliferation response pattern in T cells. Although these results might suggest the conclusion that hAACs and MSCs share a common mode of action, they clearly differed in the spectrum of secreted cytokines within PBMC co-culture supernatants. Whereas, unstimulated and IFNγ stimulated MSCs and hAACs did not induce any TNFα release, only the cardiac mesenchymal-like hAACs showed significantly enhanced induction of the anti-inflammatory cytokine IL-10 in both conditions. However, it is striking, that exclusively IL-1ß and IL-33 levels were significantly higher in co-cultures with hAACs, which exceeded the constitutive secretion of both cytokines by hAACs alone. So far it is unclear, which effects both cytokines of the IL-1 family would have after transplantation of hAACs in a therapeutic setting. It is known that they are classical pro-inflammatory factors with an ambivalent function. On the one hand, IL-1β secretion of MSCs is involved in monocyte dependent regulation of $CD4^+$ and $CD8^+$ T cell activation by triggering the release of TGFβ (58). IL-33, on the other hand, could play an important role in cardiac tissue preservation and repair in response to myocardial injury (59–62).

Our results regarding the induction of immune cell responses with $CD90^{low}$ hACCs are comparable to data described for $CD90^+$ allogeneic human CPCs (30). Therefore, the expression or absence of CD90 on cardiac-derived cells apparently seems to be of inferior importance for the immunogenicity of the respective allogeneic cell product. However, in contrast to our study Lauden et al. demonstrated a rather low, but significant induction of $CD4^+$ T cell proliferation. Consequently, they were able to prove the induction of regulatory T cells in co-cultures of purified $CD4^+$ T cells with human CPCs.

For the potential clinical application of the hAAC cell product, not only a low immunogenicity, but also the capacity to suppress ongoing inflammatory immune responses is important, since it is widely believed to be responsible for the adverse remodeling after cardiac injury (63, 64). In co-cultures with human allogeneic PBMCs, we were able to demonstrate the capacity of hAACs as well as MSC controls to suppress $CD4^+$ and $CD8^+$ T cell proliferation nearly to the same extent. Both cell types inhibited T cell proliferation by trend if cells were cultured under unstimulated conditions and strongly enhanced their suppressive capacity after IFNγ pre-stimulation. The "licensing" effect of IFNγ was already well-described by several groups for MSCs from different sources (65–69). In contrast, no clear similarities between human CPCs and MSCs were detectable regarding the inhibitory capacity of T cell proliferation in PHA stimulated immune cell cultures as well as in mixed lymphocyte reaction settings (30). However, the experimental design of our study relied on different cell sources and a diverging setup of immune cell co-cultures, which might explain varying results of the immunomodulation. The comparable immunomodulatory efficacy of $CD90^{low}$ hAACs and the $CD90^+$ control umbilical cord MSCs argues, that CD90 expression does not play a fundamental role in our experimental setup. Although, a potential correlation was described by others for human MSCs derived from bone marrow, amnion and chorion (70), other factors seem to determine the immunosuppressive features of this specific hAAC product.

The high potency of hAACs to efficiently down-regulate already ongoing immune responses was additionally confirmed by a strong suppression of the pro-inflammatory mediators IFNγ, TNFα, and IL-17A as well as IL-2 to a minor extent. Again, these effects were further enhanced in co-cultures with IFNγ-licensed cells. The typical induction of a shift from an inflammatory toward a more anti-inflammatory secretion profile was described for human MSCs (71–74) and for CPCs (25, 75). However, presence of hAACs in triggered immune cell co-cultures rather lowered anti-inflammatory cytokines like TGFß and IL-10 instead of inducing those as described for MSCs (58, 73, 76).

Next, we wanted to get a deeper understanding of the molecular changes in hAACs after pre-stimulation with IFNγ that are likely to be responsible for the immunosuppressive or modulatory efficacy. For this, we compared unstimulated and IFNγ stimulated hAACs in a whole genome analysis by Affymetrix® microarray technology and could determine similar characteristics for all three donors between each treatment group, despite visible biological heterogeneities in their individual RNA profiles. More interestingly, the global analysis revealed, that pathways of general immune system responses (innate and adaptive), as well as cytokine signaling and IFNγ responses were

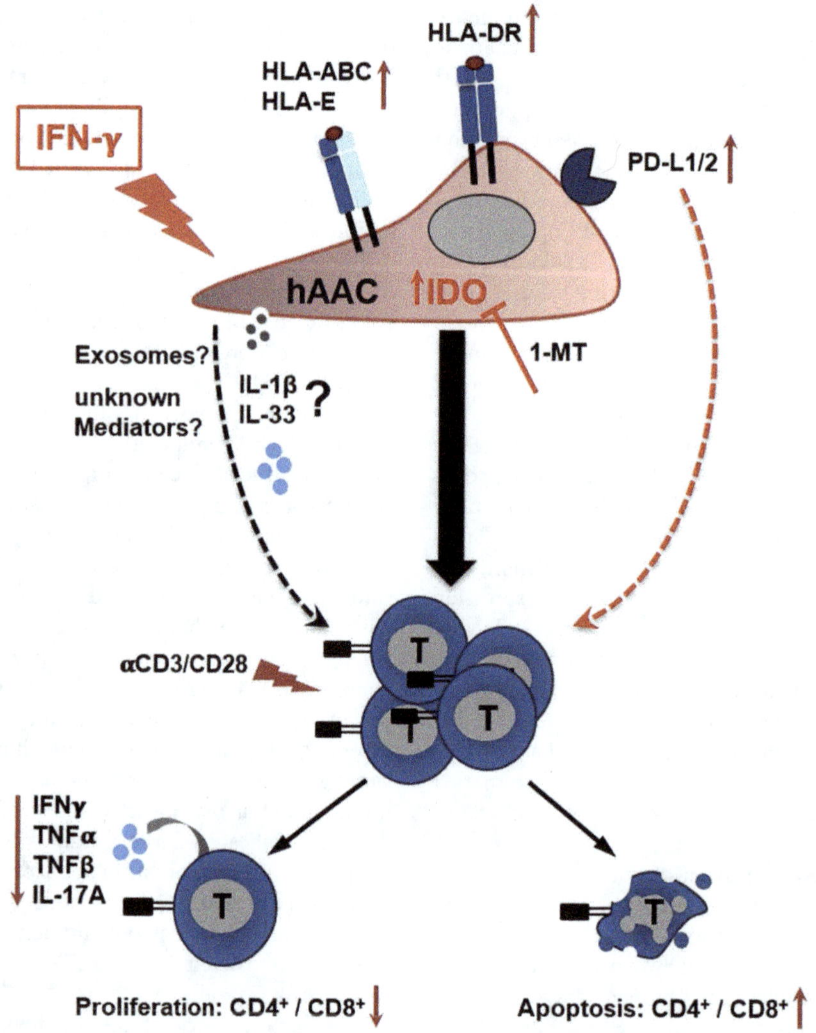

FIGURE 7 | Potential crosstalk between hAACs and T cells in an inflammatory milieu. The scheme illustrates the impact of IFNγ stimulation on CD90^{low} hAACs and its implication on the interplay with αCD3/αCD28 activated human T cells. Stimulation of hAACs alone induces upregulation of HLA-class I (HLA-ABC, HLA-E) and *de novo* expression of HLA-class II molecules (HLA-DR). Even though expression of the inhibitory co-stimulatory molecules PD-L1 and PD-L2 is enhanced on the cell surface, no specific effects in the interaction with activated T cells were detectable. Similarly, the role of the specific secretion of IL-1 family cytokines (IL-1β, IL-33) by hAACs and additional unknown mediators like exosomes remain unclear. However, a global genome-wide expression profile analysis revealed IDO as one of the strongest expressed genes in IFNγ stimulated hAACs. Specific blocking by 1-MT proved the essential involvement of IDO in the interaction with T cells. It mediates the suppression of $CD4^+$ and $CD8^+$ T cell proliferation, reduces the secretion of pro-inflammatory cytokines (IFNγ, TNFα, TNFβ, IL-17A) and induces apoptosis of both T cell subsets. We hypothesize, that this mode of hAAC interaction might contribute to the resolution of immune responses and limit thereby the effects of adverse remodeling after cardiac injury.

preferentially involved. A closer look into specific, immune-relevant molecules, that were significantly up-regulated under IFNγ stimulation, exposed *IDO1* as one of the strongest expressed genes among other interesting candidates, such as *LGALS9* (Galectin-9), *TLR3, HLA-G, PTGS1* (COX-1). The particularly high degree of *IDO1*-upregulation in IFNγ pre-stimulated hAACs was also confirmed by qPCR. IDO is often discussed to be involved in immunosuppressive effects exerted by MSCs on T cells by depletion of tryptophan and accumulation of metabolites like kynurenin (77–81). In correlation with the strong up-regulation of IDO, we could prove the involvement of this particular molecule in the suppression of T cell proliferation and the induction of their apoptosis by application of the specific inhibitor 1-MT. These observations are in accordance with a recent study, which correlated the suppressive capacity of IFNγ licensed MSCs on third party T cell proliferation to enhanced IDO and PD-L1 expression (82).

Interestingly, PD-L1 and PD-L2 were found to be up-regulated on RNA and protein level in hAACs after inflammatory stimulation. The interaction of these two molecules with PD-1 on T cells might also contribute to the observed immunosuppressive effects. For bone marrow MSCs it was recently demonstrated that the expression and secretion of both PD-1 ligands mediated suppression of $CD4^+$ T cell activation, down-regulated IL-2 secretion and induced hypo-responsiveness and cell death in T cells (45). However, in our experimental settings the application of blocking antibodies against both PD-1 ligands could neither reverse the suppression of T cell proliferation nor prevent the induction of apoptosis. We also found, that IFNγ pre-stimulated hAACs showed nearly the same suppression of αCD3/αCD28-induced T cell proliferation in direct contact as well as transwell settings. That implies that hAACs mainly mediate their immunomodulatory effects in an inflammatory milieu by soluble factors or vesicles that are able to cross the transwell membrane. A paracrine mode of action has often been suggested by others for MSCs and related mesenchymal cell types (83–86). The missing blocking effect of PD-1 ligands, which are secreted and also expressed on the cell surface, clearly discriminates hAACs from the before mentioned $CD90^+$ CPCs, that exert a more contact-dependent mode of suppression by PD-L1 involvement on ongoing immune responses (30). Conclusively, the observed increase of PD-L1 and PD-L2 expression on the cell surface seems to play only an inferior role in the immune regulation mediated by hAACs. Even though their general cellular characteristics are clearly distinguishable from fibroblasts and conventional MSCs (32), hAACs behave more like MSCs in this immunological context. The importance of the observed up-regulation of other potential immunoregulatory genes such as Galcetin-9 for the immunomodulatory capacity of hAACs has to be analyzed in more detail in future studies.

There are also still open questions remaining regarding the nature of how hAACs avoid unwanted immune responses in clinically relevant settings of cardiac injury. Apart from the proved involvement of IDO in the efficient T cell immunosuppression under IFNγ treatment, the interaction of antigen-presenting cells, like monocytes, has also to be taken into account. In this regard, it became evident that MSCs could skew monocytes toward anti-inflammatory macrophages (87–89) and induce the generation of regulatory T cells (58, 76). Additionally, it was found that MSCs could be taken up by monocytic cells and induce changes toward a non-classical monocyte phenotype with enhanced expression of PD-L1 and secretion of IL-10 that subsequently modulates adaptive immune responses (90). Furthermore, it was recently published that human CPCs attracted monocytes by a set of released cytokines and mediators and consequently changed the polarization of differentiating macrophages toward an M2-type cell (25, 91). Future studies will therefore focus on this hAAC/antigen-presenting cell interplay to gain a better understanding of the underlying mechanism of action and ensure the safe and effective translation of this hAAC cell product.

In the present study we could show, that $CD90^{low}$ hAACs isolated from human heart tissue represent a new allogeneic off-the-shelf mesenchymal-like cell product with therapeutically interesting features. Most importantly, hAACs do not trigger immunogeneic effects based on the low expression of HLA-DR and absence of co-stimulatory molecules after pro-inflammatory IFNγ stimulation. Moreover, hAACs clearly demonstrated a strong potential to inhibit ongoing immune responses even in inflammatory environments. Ultimately, we could illustrate that IDO-upregulation under IFNγ pre-treatment seems to be one of the most important players mediating the observed suppressive effects (**Figure 7**). However, the involvement of other so far not identified soluble factors cannot be excluded. This paracrine mode of action would also suggest the opportunity to isolate and use extracellular vesicles derived from hAACs for a clinical application. Our data in general would argue for a safe application of the hAAC cell product in an allogeneic setting, which also facilitates high potential to suppress already ongoing immune responses and thereby limit the progression of adverse remodeling after cardiac injury.

ETHICS STATEMENT

Human atrial appendage-derived cells (hAACs) were isolated from atrial appendages of eight donors according to the local guidelines of the Charité-Universitätsmedizin Berlin as well as the Declaration of Helsinki and the study was approved by the ethics committee of the Charité-Universitätsmedizin Berlin (No. 4/028/12). Human peripheral blood mononuclear cells (PBMCs) were isolated from buffy coats (German Red Cross, Berlin, Germany; approved by the local Ethical Committee, EA1/226/14). Umbilical cord mesenchymal stromal cells were obtained for human cell and tissue sample collection from the Institutional Review Board of the Medical University of Graz (protocol 19-252 ex 07/08). Umbilical cord samples were collected from mothers that gave written informed consent after full-term pregnancies in accordance with the Declaration of Helsinki.

AUTHOR CONTRIBUTIONS

FD led the project and was responsible for study design, execution of experiments, collection and assembly of data, data analysis, and interpretation, as well as manuscript writing. MSt established methods, performed experiments, and interpreted data. KJ designed experimental approaches, performed and interpreted the microarray analyses and revised the manuscript. MH and MSi supplied the study materials, provided administrative support, and revised the manuscript. MSe was responsible for the project conception and design, administrative support, data analysis and interpretation, manuscript writing, and final approval of the manuscript.

FUNDING

This study was supported by the BCRT [Bundesministerium für Bildung und Forschung (BMBF) grant (13GW0099)]. In part, the work was supported by the Berlin Institute of Health (BIH) with a medical doctoral research stipend and by DFG funding through the Berlin-Brandenburg School for Regenerative Therapies GSC 203 to FD.

ACKNOWLEDGMENTS

We would like to acknowledge the assistance of the BIH Core Facility Flow & Mass Cytometry—BIH FlowMassCore headed by D. Kunkel; H. Stachelscheid, and P. Mergenthaler of the Charité High Content Imaging Facility; I. Savidis and A. Levent from the Core Unit Biomarker—Immunological Study Lab (both BCRT, Charité-Universitätsmedizin Berlin) for their excellent technical assistance with the Luminex® analyses; B. Kurtze and K. Hariharan for the technical help with the qPCR measurement (both BCRT, Charité-Universitätsmedizin Berlin); A. Fleischmann and J. Ringe (both BCRT—Charité-Universitätsmedizin Berlin) for their skillful support with sample preparation and measurement of RNA microarray experiments and Laymik from the Noun Project for provision of graphical icons. We specially thank K. Schütze and H. Kremling (CellTool GmbH, Tutzing, Germany) for the extraordinary help with Raman analysis. We thank C. Stamm (Deutsches Herzzentrum Berlin) for the procurement of atrial appendages, S. Detert (BCRT—Charité-Universitätsmedizin Berlin) for the isolation of the hAAC cell product and C. Beez, M. Schneider, and N. Souidi for helpful discussions. We express our gratitude to D. Strunk (Paracelsus Medical University, Salzburg) for the supply of umbilical cord MSCs. We acknowledge support from the German Research Foundation (DFG) and the Open Access Publication Fund of the Charité-Universitätsmedizin Berlin.

REFERENCES

1. Cahill TJ, Choudhury RP, Riley PR. Heart regeneration and repair after myocardial infarction: translational opportunities for novel therapeutics. *Nat Rev Drug Discov.* (2017) 16:699–717. doi: 10.1038/nrd.2017.106
2. Owens AT, Brozena SC, Jessup M. New management strategies in heart failure. *Circul Res.* (2016) 118:480–95. doi: 10.1161/CIRCRESAHA.115.306567
3. Fernández-Avilés F, Sanz-Ruiz R, Climent AM, Badimon L, Bolli R, Charron D, et al. Global position paper on cardiovascular regenerative medicine. *Eur Heart J.* (2017) 38:2532–46. doi: 10.1093/eurheartj/ehx248
4. Schächinger V, Erbs S, Elsässer A, Haberbosch W, Hambrecht R, Hölschermann H, et al. Intracoronary bone marrow-derived progenitor cells in acute myocardial infarction. *N Engl J Med.* (2006) 355:1210–21. doi: 10.1056/NEJMoa060186
5. Houtgraaf JH, Den Dekker WK, Van Dalen BM, Springeling T, De Jong R, Van Geuns RJ, et al. First experience in humans using adipose tissue-derived regenerative cells in the treatment of patients with ST-segment elevation myocardial infarction. *J Am Coll Cardiol.* (2012) 59:539–40. doi: 10.1016/j.jacc.2011.09.065
6. Lilyanna S, Martinez EC, Vu TD, Ling LH, Gan SU, Tan AL, et al. Cord lining-mesenchymal stem cells graft supplemented with an omental flap induces myocardial revascularization and ameliorates cardiac dysfunction in a rat model of chronic ischemic heart failure. *Tissue Eng Part A.* (2013) 19:1303–15. doi: 10.1089/ten.tea.2012.0407
7. Zhao JJ, Liu XC, Kong F, Qi TG, Cheng GH, Wang J, et al. Bone marrow mesenchymal stem cells improve myocardial function in a swine model of acute myocardial infarction. *Mol Med Rep.* (2014) 10:1448–54. doi: 10.3892/mmr.2014.2378
8. Beltrami AP, Barlucchi L, Torella D, Baker M, Limana F, Chimenti S, et al. Adult cardiac stem cells are multipotent and support myocardial regeneration. *Cell.* (2003) 114:763–76. doi: 10.1016/S0092-8674(03)00687-1
9. Messina E, De Angelis L, Frati G, Morrone S, Chimenti S, Fiordaliso F, et al. Isolation and expansion of adult cardiac stem cells from human and murine heart. *Circul Res.* (2004) 95:911–21. doi: 10.1161/01.RES.0000147315.71699.51
10. Oh H, Bradfute SB, Gallardo TD, Nakamura T, Gaussin V, Mishina Y, et al. Cardiac progenitor cells from adult myocardium: Homing, differentiation, and fusion after infarction. *Proc Nat Acad Sci USA.* (2003) 100:12313–8. doi: 10.1073/pnas.2132126100
11. Smith RR, Barile L, Cho HC, Leppo MK, Hare JM, Messina E, et al. Regenerative potential of cardiosphere-derived cells expanded from percutaneous endomyocardial biopsy specimens. *Circulation.* (2007) 115:896–908. doi: 10.1161/CIRCULATIONAHA.106.655209
12. Crisostomo V, Casado JG, Baez-Diaz C, Blazquez R, Sanchez-Margallo FM. Allogeneic cardiac stem cell administration for acute myocardial infarction. *Expert Rev Cardiovasc Therapy.* (2015) 13:285–99. doi: 10.1586/14779072.2015.1011621
13. Bolli R, Chugh AR, D'Amario D, Loughran JH, Stoddard MF, Ikram S, et al. Cardiac stem cells in patients with ischaemic cardiomyopathy (SCIPIO): initial results of a randomised phase 1 trial. *Lancet.* (2011) 378:1847–57. doi: 10.1016/S0140-6736(11)61590-0
14. Chugh AR, Beache GM, Loughran JH, Mewton N, Elmore JB, Kajstura J, et al. Administration of cardiac stem cells in patients with ischemic cardiomyopathy: The SCIPIO trial: surgical aspects and interim analysis of myocardial function and viability by magnetic resonance. *Circulation.* (2012) 126(11 SUPPL.1):54–64. doi: 10.1161/CIRCULATIONAHA.112.092627
15. Makkar RR, Smith RR, Cheng K, Malliaras K, Thomson LEJ, Berman D, et al. Intracoronary cardiosphere-derived cells for heart regeneration after myocardial infarction (CADUCEUS): a prospective, randomised phase 1 trial. *Lancet.* (2012) 379:895–904. doi: 10.1016/S0140-6736(12)60195-0
16. Malliaras K, Makkar RR, Smith RR, Cheng K, Wu E, Bonow RO, et al. Intracoronary cardiosphere-derived cells after myocardial infarction: evidence of therapeutic regeneration in the final 1-year results of the CADUCEUS trial (CArdiosphere-derived aUtologous stem CElls to reverse ventricular dysfunction). *J Am Coll Cardiol.* (2014) 63:110–22. doi: 10.1016/j.jacc.2013.08.724
17. Haag M, Van Linthout S, Schröder SEA, Freymann U, Ringe J, Tschöpe C, et al. Endomyocardial biopsy derived adherent proliferating cells - a potential cell source for cardiac tissue engineering. *J Cell Biochem.* (2010) 109:564–75. doi: 10.1002/jcb.22433
18. Miteva K, Haag M, Peng J, Savvatis K, Becher PM, Seifert M, et al. Human cardiac-derived adherent proliferating cells reduce murine acute coxsackievirus B3-induced myocarditis. *PLoS ONE.* (2011) 6:e28513. doi: 10.1371/journal.pone.0028513
19. Dehne T, Adam X, Materne E-M, Reimann MC, Krüger JP, Van Linthout S, et al. A P19 and P19CL6 cell-based complementary approach to determine paracrine effects in cardiac tissue engineering. *Cells Tissues Org.* (2014) 199:24–36. doi: 10.1159/000362540
20. Miteva K, Van Linthout S, Pappritz K, Müller I, Spillmann F, Haag M, et al. Human endomyocardial biopsy specimen-derived stromal cells modulate

angiotensin II-induced cardiac remodeling. *Stem Cells Transl Med.* (2016) 5:1707–18. doi: 10.5966/sctm.2016-0031
21. Haag M, Stolk M, Ringe J, Linthout S, Van Tschöpe C, Sittinger M, et al. Immune attributes of cardiac-derived adherent proliferating (CAP) cells in cardiac therapy. *J Tissue Eng Regener Med.* (2013) 7:362–70. doi: 10.1002/term.531
22. Young PP, Schäfer R. Cell-based therapies for cardiac disease: a cellular therapist's perspective. *Transfusion.* (2015) 55:441–51. doi: 10.1111/trf.12826
23. Malliaras K, Marbán E. Cardiac cell therapy: where we've been, where we are, and where we should be headed. *Br Med Bull.* (2011) 98:161–85. doi: 10.1093/bmb/ldr018
24. Najar M, Bouhtit F, Melki R, Afif H, Hamal A, Fahmi H, et al. Mesenchymal stromal cell-based therapy: new perspectives and challenges. *J Clin Med.* (2019) 8:626. doi: 10.3390/jcm8050626
25. Al-Daccak R, Charron D. Allogenic benefit in stem cell therapy: cardiac repair and regeneration. *Tissue Antigens.* (2015) 86:155–62. doi: 10.1111/tan.12614
26. Hogen R, DiNorcia J, Dhanireddy K. Antibody-mediated rejection: what is the clinical relevance? *Curr Opin Organ Transpl.* (2017) 22:97–104. doi: 10.1097/MOT.0000000000000391
27. Terasaki PI, Cai J. Humoral theory of transplantation: further evidence. *Curr Opin Immunol.* (2005) 17:541–5. doi: 10.1016/j.coi.2005.07.018
28. Clarkson MR, Sayegh MH. T-cell costimulatory pathways in allograft rejection and tolerance. *Transplantation.* (2005) 80:555–63. doi: 10.1097/01.tp.0000168432.60022.99
29. Wood KJ, Zaitsu M, Goto R. Cell mediated rejection. *Methods Mol Biol.* (2013) 1034:71–83. doi: 10.1007/978-1-62703-493-7_3
30. Lauden L, Boukouaci W, Borlado LR, López IP, Sepúlveda P, Tamouza R, et al. Allogenicity of human cardiac stem/progenitor cells orchestrated by programmed death ligand 1. *Circul Res.* (2013) 112:451–64. doi: 10.1161/CIRCRESAHA.112.276501
31. Fernández-Avilés F, Sanz-Ruiz R, Bogaert J, Casado Plasencia A, Gilaberte I, Belmans A, et al. Safety and efficacy of intracoronary infusion of allogeneic human cardiac stem cells in patients with ST-segment elevation myocardial infarction and left ventricular dysfunction. *Circul Res.* (2018) 123:579–89. doi: 10.1161/CIRCRESAHA.118.312823
32. Detert S, Stamm C, Beez C, Diedrichs F, Ringe J, Van Linthout S, et al. The atrial appendage as a suitable source to generate cardiac-derived adherent proliferating cells for regenerative cell-based therapies. *J Tissue Eng Regener Med.* (2018) 12:e1404–17. doi: 10.1002/term.2528
33. Cheng K, Ibrahim A, Hensley MT, Shen D, Sun B, Middleton R, et al. Relative roles of CD90 and c-kit to the regenerative efficacy of cardiosphere-derived cells in humans and in a mouse model of myocardial infarction. *J Am Heart Assoc.* (2014) 3:e001260. doi: 10.1161/JAHA.114.001260
34. El-Shaffeey M, Pappritz K, Diedrichs F, Seifert M, Stamm C, Haag M, et al. Cardioprotective and immunomodulatory effects of endomyocardial biopsy- and atrial appendage-derived stromal cells in an acute model of Coxsackievirus B3-induced myocarditis. *Eur J Heart Fail.* (2018) 20: P1982.
35. Souidi N, Stolk M, Rudeck J, Strunk D, Schallmoser K, Volk H-D, et al. Stromal cells act as guardians for endothelial progenitors by reducing their immunogenicity after Co-transplantation. *Stem Cells.* (2017) 35:1233–45. doi: 10.1002/stem.2573
36. Reinisch A, Hofmann NA, Obenauf AC, Kashofer K, Rohde E, Schallmoser K, et al. Humanized large-scale expanded endothelial colony-forming cells function *in vitro* and *in vivo*. *Blood.* (2009) 113:6716–25. doi: 10.1182/BLOOD-2008-09-181362
37. Stolk M, Klatte-Schulz F, Schmock A, Minkwitz S, Wildemann B, Seifert M. New insights into tenocyte-immune cell interplay in an in vitro model of inflammation. *Sci Rep.* (2017) 7:9801. doi: 10.1038/s41598-017-09875-x
38. Gautier L, Cope L, Bolstad BM, Irizarry RA. affy–analysis of Affymetrix GeneChip data at the probe level. *Bioinformatics.* (2004) 20:307–15. doi: 10.1093/bioinformatics/btg405
39. Ritchie ME, Phipson B, Wu D, Hu Y, Law CW, Shi W, et al. limma powers differential expression analyses for RNA-sequencing and microarray studies. *Nucleic Acids Res.* (2015) 43:e47. doi: 10.1093/nar/gkv007
40. Huang DW, Sherman BT, Lempicki RA. Systematic and integrative analysis of large gene lists using DAVID bioinformatics resources. *Nat Protocols.* (2009) 4:44–57. doi: 10.1038/nprot.2008.211
41. Huang DW, Sherman BT, Lempicki RA. Bioinformatics enrichment tools: paths toward the comprehensive functional analysis of large gene lists. *Nucleic Acids Res.* (2009) 37:1–13. doi: 10.1093/nar/gkn923
42. Yu G, Wang LG, Han Y, He QY. clusterProfiler: an R package for comparing biological themes among gene clusters. *Omics J Integr Biol.* (2012) 16:284–7. doi: 10.1089/omi.2011.0118
43. Walter W, Sánchez-Cabo F, Ricote M. GOplot: an R package for visually combining expression data with functional analysis. *Bioinformatics.* (2015) 31:2912–4. doi: 10.1093/bioinformatics/btv300
44. Pedregosa F, Weiss R, Brucher M. Scikitlearn: machine learning in python. *J Mach Learn Res.* (2011) 12:2825–30. doi: 10.1007/s13398-014-0173-7.2
45. Davies LC, Heldring N, Kadri N, Le Blanc K. Mesenchymal stromal cell secretion of programmed death-1 ligands regulates T cell mediated immunosuppression. *Stem Cells.* (2017) 35:766–76. doi: 10.1002/stem.2509
46. Hausenloy DJ, Garcia-Dorado D, Bøtker HE, Davidson SM, Downey J, Engel FB, et al. Novel targets and future strategies for acute cardioprotection: position paper of the European Society of Cardiology Working Group on Cellular Biology of the Heart. *Cardiovasc Res.* (2017) 113:564–85. doi: 10.1093/cvr/cvx049
47. Nigro P, Bassetti B, Cavallotti L, Catto V, Carbucicchio C, Pompilio G. Cell therapy for heart disease after 15 years: Unmet expectations. *Pharmacol Res.* (2018) 127:77–91. doi: 10.1016/j.phrs.2017.02.015
48. Li TS, Cheng K, Malliaras K, Smith RR, Zhang Y, Sun B, et al. Direct comparison of different stem cell types and subpopulations reveals superior paracrine potency and myocardial repair efficacy with cardiosphere-derived cells. *J Am Coll Cardiol.* (2012) 59:942–53. doi: 10.1016/j.jacc.2011.11.029
49. Rossini A, Frati C, Lagrasta C, Graiani G, Scopece A, Cavalli S, et al. Human cardiac and bone marrow stromal cells exhibit distinctive properties related to their origin. *Cardiovasc Res.* (2011) 89:650–60. doi: 10.1093/cvr/cvq290
50. Karantalis V, Hare JM. Use of mesenchymal stem cells for therapy of cardiac disease. *Circul Res.* (2015) 116:1413–30. doi: 10.1161/CIRCRESAHA.116.303614
51. Ward MR, Abadeh A, Connelly KA. Concise review: rational use of mesenchymal stem cells in the treatment of ischemic heart disease. *Stem Cells Transl Med.* (2018) 7:543–50. doi: 10.1002/sctm.17-0210
52. Kawka E, Witowski J, Bartosova M, Catar R, Rudolf A, Philippe A, et al. Thy-1+/- fibroblast subsets in the human peritoneum. *Am J Physiol Renal Physiol.* (2017) 313:F1116–123. doi: 10.1152/ajprenal.00274.2017
53. Berglund AK, Fortier LA, Antczak DF, Schnabel LV. Immunoprivileged no more: measuring the immunogenicity of allogeneic adult mesenchymal stem cells. *Stem Cell Res Therap.* (2017) 8:288. doi: 10.1186/s13287-017-0742-8
54. Lohan P, Treacy O, Griffin MD, Ritter T, Ryan AE. Anti-donor immune responses elicited by allogeneic mesenchymal stem cells and their extracellular vesicles: are we still learning? *Front Immunol.* (2017) 8:1626. doi: 10.3389/fimmu.2017.01626
55. Wang WB, Yen ML, Liu KJ, Hsu PJ, Lin MH, Chen PM, et al. Interleukin-25 mediates transcriptional control of PD-L1 via STAT3 in multipotent human mesenchymal stromal cells (hMSCs) to suppress Th17 responses. *Stem Cell Rep.* (2015) 5:392–404. doi: 10.1016/j.stemcr.2015.07.013
56. Kim JY, Park M, Kim YH, Ryu KH, Lee KH, Cho KA, et al. Tonsil-derived mesenchymal stem cells (T-MSCs) prevent Th17-mediated autoimmune response via regulation of the programmed death-1/programmed death ligand-1 (PD-1/PD-L1) pathway. *J Tissue Eng Regener Med.* (2018) 12:e1022–33. doi: 10.1002/term.2423
57. Di Trapani M, Bassi G, Ricciardi M, Fontana E, Bifari F, Pacelli L, et al. Comparative study of immune regulatory properties of stem cells derived from different tissues. *Stem Cells Dev.* (2013) 22:2990–3002. doi: 10.1089/scd.2013.0204
58. Groh ME, Maitra B, Szekely E, Koç ON. Human mesenchymal stem cells require monocyte-mediated activation to suppress alloreactive T cells. *Exp Hematol.* (2005) 33:928–34. doi: 10.1016/j.exphem.2005.05.002
59. Altara R, Ghali R, Mallat Z, Cataliotti A, Booz GW, Zouein FA. Conflicting vascular and metabolic impact of the IL-33/sST2 axis. *Cardiovasc Res.* (2018) 114:1578–94. doi: 10.1093/cvr/cvy166
60. Ghali R, Altara R, Louch WE, Cataliotti A, Mallat Z, Kaplan A, et al. IL-33 (Interleukin 33)/sST2 (Soluble Suppression of Tumorigenicity 2) Axis in *Hypertension* and heart failure. *Hypertension.* (2018) 72:818–28. doi: 10.1161/HYPERTENSIONAHA.118.11157

61. Griesenauer B, Paczesny S. The ST2/IL-33 axis in immune cells during inflammatory diseases. *Front Immunol.* (2017) 8:475. doi: 10.3389/fimmu.2017.00475
62. Martin NT, Martin MU. Interleukin 33 is a guardian of barriers and a local alarmin. *Nat Immunol.* (2016) 17:122–31. doi: 10.1038/ni.3370
63. Sattler S, Fairchild P, Watt FM, Rosenthal N, Harding SE. The adaptive immune response to cardiac injury—the true roadblock to effective regenerative therapies? *NPJ Regener. Med.* (2017) 2:19. doi: 10.1038/s41536-017-0022-3
64. Prabhu SD, Frangogiannis NG. The biological basis for cardiac repair after myocardial infarction. *Circul Res.* (2016) 119:91–112. doi: 10.1161/CIRCRESAHA.116.303577
65. Crop MJ, Baan CC, Korevaar SS, IJzermans JNM, Pescatori M, Stubbs AP, et al. Inflammatory conditions affect gene expression and function of human adipose tissue-derived mesenchymal stem cells. *Clin Exp Immunol.* (2010) 162:474–86. doi: 10.1111/j.1365-2249.2010.04256.x
66. Kronsteiner B, Wolbank S, Peterbauer A, Hackl C, Redl H, van Griensven M, et al. Human mesenchymal stem cells from adipose tissue and amnion influence T-cells depending on stimulation method and presence of other immune cells. *Stem Cells Dev.* (2011) 20:2115–26. doi: 10.1089/scd.2011.0031
67. Sivanathan KN, Gronthos S, Rojas-Canales D, Thierry B, Coates PT. Interferon-gamma modification of mesenchymal stem cells: implications of autologous and allogeneic mesenchymal stem cell therapy in allotransplantation. *Stem Cell Rev Rep.* (2014) 10:351–75. doi: 10.1007/s12015-014-9495-2
68. Li X, Xu Z, Bai J, Yang S, Zhao S, Zhang Y, et al. Umbilical cord tissue-derived mesenchymal stem cells induce T lymphocyte apoptosis and cell cycle arrest by expression of indoleamine 2, 3-dioxygenase. *Stem Cells Int.* (2016) 2016:1–11. doi: 10.1155/2016/7495135
69. de Witte SFH, Merino AM, Franquesa M, Strini T, van Zoggel JAA, Korevaar SS, et al. Cytokine treatment optimises the immunotherapeutic effects of umbilical cord-derived MSC for treatment of inflammatory liver disease. *Stem Cell Res Therapy.* (2017) 8:140. doi: 10.1186/s13287-017-0590-6
70. Campioni D, Rizzo R, Stignani M, Melchiorri L, Ferrari L, Moretti S, et al. A decreased positivity for CD90 on human mesenchymal stromal cells (MSCs) is associated with a loss of immunosuppressive activity by MSCs. *Cytometry Part B Clin Cytometry.* (2009) 76:225–30. doi: 10.1002/cyto.b.20461
71. Gebler A, Zabel O, Seliger B. The immunomodulatory capacity of mesenchymal stem cells. *Trends Mol Med.* (2012) 18:128–34. doi: 10.1016/j.molmed.2011.10.004
72. Wang Y, Chen X, Cao W, Shi Y. Plasticity of mesenchymal stem cells in immunomodulation: pathological and therapeutic implications. *Nat Immunol.* (2014) 15:1009–16. doi: 10.1038/ni.3002
73. Kyurkchiev D, Bochev I, Ivanova-Todorova E, Mourdjeva M, Oreshkova T, Belemezova K, et al. Secretion of immunoregulatory cytokines by mesenchymal stem cells. *World J Stem Cells.* (2014) 6:552. doi: 10.4252/wjsc.v6.i5.552
74. Uccelli A, Moretta L, Pistoia V. Mesenchymal stem cells in health and disease. *Nat Rev Immunol.* (2008) 8:726–36. doi: 10.1038/nri2395
75. van den Akker F, Deddens JC, Doevendans PA, Sluijter JPG. Cardiac stem cell therapy to modulate inflammation upon myocardial infarction. *Biochim et Biophy Acta.* (2013) 1830:2449–58. doi: 10.1016/j.bbagen.2012.08.026
76. Melief SM, Schrama E, Brugman MH, Tiemessen MM, Hoogduijn MJ, Fibbe WE, et al. Multipotent stromal cells induce human regulatory T cells through a novel pathway involving skewing of monocytes toward anti-inflammatory macrophages. *Stem Cells.* (2013) 31:1980–91. doi: 10.1002/stem.1432
77. Tipnis S, Viswanathan C, Majumdar AS. Immunosuppressive properties of human umbilical cord-derived mesenchymal stem cells: role of B7-H1 and IDO. *Immunol Cell Biol.* (2010) 88:795–806. doi: 10.1038/icb.2010.47
78. Croitoru-Lamoury J, Lamoury FMJ, Caristo M, Suzuki K, Walker D, Takikawa O, et al. Interferon-γ regulates the proliferation and differentiation of mesenchymal stem cells via activation of indoleamine 2,3 dioxygenase (IDO). *PLoS ONE.* (2011) 6:e14698. doi: 10.1371/journal.pone.0014698
79. François M, Romieu-Mourez R, Li M, Galipeau J. Human MSC suppression correlates with cytokine induction of indoleamine 2,3-dioxygenase and bystander M2 macrophage differentiation. *Mol Therapy.* (2012) 20:187–95. doi: 10.1038/mt.2011.189
80. Hong J, Hueckelhoven A, Wang L, Schmitt A, Wuchter P, Tabarkiewicz J, et al. Indoleamine 2,3-dioxygenase mediates inhibition of virus-specific CD8(+) T cell proliferation by human mesenchymal stromal cells. *Cytotherapy.* (2016) 18:621–9. doi: 10.1016/j.jcyt.2016.01.009
81. van den Akker F, de Jager SCA, Sluijter JPG. Mesenchymal stem cell therapy for cardiac inflammation: immunomodulatory properties and the influence of toll-like receptors. *Medi Inflamm.* (2013) 2013:181020. doi: 10.1155/2013/181020
82. Guan Q, Li Y, Shpiruk T, Bhagwat S, Wall DA. Inducible indoleamine 2,3-dioxygenase 1 and programmed death ligand 1 expression as the potency marker for mesenchymal stromal cells. *Cytotherapy.* (2018) 20:639–49. doi: 10.1016/j.jcyt.2018.02.003
83. Liang X, Ding Y, Zhang Y, Tse HF, Lian Q. Paracrine mechanisms of mesenchymal stem cell-based therapy: current status and perspectives. *Cell Transpl.* (2014) 23:1045–59. doi: 10.3727/096368913X667709
84. Bruno S, Deregibus MC, Camussi G. The secretome of mesenchymal stromal cells: role of extracellular vesicles in immunomodulation. *Immunol Lett.* (2015) 168:154–8. doi: 10.1016/j.imlet.2015.06.007
85. Vizoso F, Eiro N, Cid S, Schneider J, Perez-Fernandez R. Mesenchymal stem cell secretome: toward cell-free therapeutic strategies in regenerative medicine. *Int J Mol Sci.* (2017) 18:1852. doi: 10.3390/ijms18091852
86. Zheng G, Huang R, Qiu G, Ge M, Wang J, Shu Q, et al. Mesenchymal stromal cell-derived extracellular vesicles: regenerative and immunomodulatory effects and potential applications in sepsis. *Cell Tissue Res.* (2018) 374:1–15. doi: 10.1007/s00441-018-2871-5
87. Kim J, Hematti P. Mesenchymal stem cell-educated macrophages: a novel type of alternatively activated macrophages. *Exp Hematol.* (2009) 37:1445–53. doi: 10.1016/j.exphem.2009.09.004
88. Selleri S, Bifsha P, Civini S, Pacelli C, Dieng MM, Lemieux W, et al. Human mesenchymal stromal cell-secreted lactate induces M2-macrophage differentiation by metabolic reprogramming. *Oncotarget.* (2016) 7:30193–210. doi: 10.18632/oncotarget.8623
89. Chiossone L, Conte R, Spaggiari GM, Serra M, Romei C, Bellora F, et al. Mesenchymal stromal cells induce peculiar alternatively activated macrophages capable of dampening both innate and adaptive immune responses. *Stem Cells.* (2016) 34:1909–21. doi: 10.1002/stem.2369
90. de Witte SFH, Luk F, Sierra Parraga JM, Gargesha M, Merino A, Korevaar SS, et al. Immunomodulation by therapeutic mesenchymal stromal cells (MSC) is triggered through phagocytosis of MSC by monocytic cells. *Stem Cells.* (2018) 36:602–15. doi: 10.1002/stem.2779
91. Dam N, Hocine HR, Palacios I, DelaRosa O, Menta R, Charron D, et al. Human cardiac-derived stem/progenitor cells fine-tune monocyte-derived descendants activities toward cardiac repair. *Front Immunol.* (2017) 8:1413. doi: 10.3389/fimmu.2017.01413

15

Cost-Effective, Safe and Personalized Cell Therapy for Critical Limb Ischemia in Type 2 Diabetes Mellitus

Bárbara Soria-Juan [1†], Natalia Escacena [2†], Vivian Capilla-González [2†], Yolanda Aguilera [2], Lucía Llanos [1], Juan R. Tejedo [2,3], Francisco J. Bedoya [2,3], Verónica Juan [4], Antonio De la Cuesta [5], Rafael Ruiz-Salmerón [6], Enrique Andreu [7], Lukas Grochowicz [7], Felipe Prósper [7], Fermín Sánchez-Guijo [8], Francisco S. Lozano [8], Manuel Miralles [9], Lourdes Del Río-Solá [10], Gregorio Castellanos [11], José M. Moraleda [11], Robert Sackstein [12], Mariano García-Arranz [1], Damián García-Olmo [1], Franz Martín [2,3], Abdelkrim Hmadcha [2,3‡], Bernat Soria [2,3*‡] and the Collaborative Working Group "Noma Project Team"*

[1] Fundación Jiménez Díaz Health Research Institute, Madrid, Spain, [2] Department of Regeneration and Cell Therapy, Andalusian Center for Molecular Biology and Regenerative Medicine (CABIMER), University of Pablo de Olavide-University of Seville-CSIC, Seville, Spain, [3] Spanish Biomedical Research Centre in Diabetes and Associated Metabolic Disorders (CIBERDEM), Madrid, Spain, [4] Andalusian eHealth Library, Sevilla, Spain, [5] Unidad de Isquemia Crónica de Miembros Inferiores, Hospital Victoria Eugenia de la Cruz Roja, Sevilla, Spain, [6] Servicio de Cardiología, Hospital Universitario Virgen Macarena, Sevilla, Spain, [7] Clínica Universidad de Navarra, Pamplona, Spain, [8] IBSAL-Hospital Universitario Salamanca, Salamanca, Spain, [9] Department of Surgery, University of Valencia, Valencia, Spain, [10] Cirugía Vascular, Hospital Universitario de Valladolid, Valladolid, Spain, [11] Servicio Hematología y Hemoterapia, Hospital Clínico Universitario Virgen de la Arrixaca, Murcia, Spain, [12] Herbert Wertheim College of Medicine, Florida International University, Miami, FL, United States

***Correspondence:**
Abdelkrim Hmadcha
karim.hmadcha@cabimer.es
Bernat Soria
bernat.soria@cabimer.es

† These authors share first authorship

‡ Lead authors

Cell therapy is a progressively growing field that is rapidly moving from preclinical model development to clinical application. Outcomes obtained from clinical trials reveal the therapeutic potential of stem cell-based therapy to deal with unmet medical treatment needs for several disorders with no therapeutic options. Among adult stem cells, mesenchymal stem cells (MSCs) are the leading cell type used in advanced therapies for the treatment of autoimmune, inflammatory and vascular diseases. To date, the safety and feasibility of autologous MSC-based therapy has been established; however, their indiscriminate use has resulted in mixed outcomes in preclinical and clinical studies. While MSCs derived from diverse tissues share common properties depending on the type of clinical application, they markedly differ within clinical trials in terms of efficacy, resulting in many unanswered questions regarding the application of MSCs. Additionally, our experience in clinical trials related to critical limb ischemia pathology (CLI) shows that the therapeutic efficacy of these cells in different animal models has only been partially reproduced in humans through clinical trials. Therefore, it is crucial to develop new research to identify pitfalls, to optimize procedures and to clarify the repair mechanisms used by these cells, as well as to be able to offer a next generation of stem cell that can be routinely used in a cost-effective and safe manner in stem cell-based therapies targeting CLI.

Keywords: cellular medicaments, cell-based therapy, clinical trials, diabetes, critical limb ischemia, cost-effective

INTRODUCTION

Regenerative Medicine is a new paradigm that has driven the revisiting of our understanding of biological and medical processes and suggested new treatments. According to the definition of the European Medicament Agency (EMA) and the U.S. Food and Drug Administration (FDA), Advanced Therapies include Cell and Gene Therapy and Tissue Engineering. Advanced Therapies comprise a large group of translational fields and targets in areas of unmet medical needs. Briefly, the application of cells, either alone or engineered, as a pharmacologically active substance seeks to restore the functioning of damaged tissues or organs through the protection of cellular integrity, the replacement of damaged cells, and the promotion of trophic, anti-inflammatory, and immunomodulatory effects, among others (**Figure 1**). This new therapeutic avenue also carries unknown side effects that must be deeply characterized to improve safety, feasibility, and efficacy.

In this regard, mesenchymal stem cells (MSCs) are the cell type commonly used in Regenerative Medicine due to their unique biological properties, including ease of expansion and culture. The predominant sources of stem cells are summarized in **Table 1**, namely, cells derived from the fetus, and adult tissues (1). Nine hundred forty-one studies using MSCs have been reported to date (March 2019) and registered in the database of privately and publicly funded clinical studies conducted worldwide at "*ClinicalTrials.gov.*" MSCs are multipotent non-hematopoietic progenitor cells with different degrees of stemness, derived from the mesodermal germ layer and resident in most tissues (**Table 1**). This type of cell (MSCs) can be easily expanded *in vitro* because of their fibroblastic characteristics and ability to adhere to plastic and to express specific surface marker patterns (2, 3). The Mesenchymal and Tissue Stem Cell Committee of the International Society for Cellular Therapy (ISCT) first proposed that bone marrow plastic-adherent cells generally described as "*mesenchymal stem cells*" should be defined as "*multipotent mesenchymal stromal cells,*" while the designation "*mesenchymal stem cells*" should be reticent for a subset of these cells that show stem cell activity based on clearly stated criteria (3). Since the acronym MSC may be used to define both cell populations, the combined definition "*mesenchymal stem and/or stromal cells*" is probably more appropriate, especially when the "*stemness*" of the whole MSC population has not been demonstrated (4), and it is now widely accepted that MSCs represent a heterogeneous population (5) but are considered a cellular medicament. Furthermore, MSC survival, permanent engraftment and differentiation into resident cells were initially thought to be necessary to obtain the beneficial effects of these cells, and clinical experience and several experiments have shown that one of the primary functions of MSCs, most likely their key function, is to secrete several bioactive molecules related to the environmental "*niche*" in which these cells are located. Consequently, the secretome transiently reproduces most of the effects of MSCs, and in this sense, MSCs secrete a wide variety of pro-inflammatory and anti-inflammatory cytokines, chemokines, growth factors, and prostaglandins under resting and inflammatory conditions (1, 6) (**Figure 1**).

These molecules are associated with immunomodulation [indoleamine-2,3-dioxygenase (IDO), prostaglandin-E2 (PGE-2), transforming growth factor beta (TGF-β), human leukocyte antigen-G5 (HLA-G5), and hepatocyte growth factor (HGF)], anti-apoptosis [vascular endothelial growth factor (VEGF), granulocyte-macrophage colony-stimulating factor (GM-CSF), TGF-β, stanniocalcin-1 (STC1), and insulin-like growth factor 1 (IGF-1)], angiogenesis [VEGF, monocyte chemoattractant protein 1 (MCP-1), and IGF-1), local stem and progenitor cell growth and differentiation support (SCF complex, angiopoietin-1, and stromal cell-derived factor 1 (SDF-1)], anti-fibrosis [HGF and basic fibroblast growth factor (bFGF)] and chemoattraction [chemokine (C-C motif) ligand 2 and 4 (CCL2, CCL4), and C-X-C motif chemokine 12 (CXCL12 also called SDF1)] (7). Immunomodulatory properties of MSCs and their immunoprivileged condition make these cells good candidates for use in several clinical trials related to chronic, inflammatory and autoimmune diseases, and in reducing the incidence and severity of graft-vs.-host disease (GVHD). MSCs interact with cells of the innate or adaptive immune system (T cells, B cells, NK cells, monocyte-derived dendritic cells, and neutrophils). For a cell to be recognized by the immune system, the expression of major histocompatibility complex (MHC) and costimulatory molecules is necessary. MHC class I and class II human leukocyte antigens (HLAs) are master triggers of robust immunological rejection of grafts because they present antigens to cytolytic T lymphocytes (CTL). The interaction between MSCs and immune cells provides insights into *in vivo* MSC-mediated induction of tolerance (1, 8). MSCs display a low expression level of MHC-HLA class I, while they are constitutively negative for HLA-class II; likewise, they do not express costimulatory molecules such as CD80, CD86, CD40, and CD40L (9). However, MSCs share the expression of surface markers such as vascular cell adhesion protein 1 (VCAM-1), intercellular adhesion molecule 2 (ICAM-2), and lymphocyte function-associated antigen 3 (LFA-3 or CD58) with the thymic epithelium, which are crucial for the interaction with T cells (9, 10). Whereas, MSCs remain in a quiescent state showing antiapoptotic properties and contributing to homeostasis, in an inflammatory environment (presence of IFNγ, TNFα, IL-1α, and IL-1β) they begin to exercise their immunomodulatory abilities, inhibiting the proliferation of effector cells and their cytokine production. Similarly, MSCs can block a variety of immune cell functions (1, 11) (**Figure 1**).

In addition, there is a complex "*cross-talk*" interaction between MSCs and endothelial cells. MSCs increase the proliferation and migration of endothelial cells to promote the early events of angiogenesis and to decrease the permeability of the endothelial cell monolayer. In direct cocultures of MSCs and endothelial cells, MSCs increase the persistence of preexisting blood vessels in a dose-dependent manner (12). Additionally, beneficial therapeutic effects of the use of conditioned media of MSCs have been reported, which has even been shown to be therapeutically superior to the cells themselves (13, 14) and to stimulate the proliferation of local endothelial cells (15). Likewise, in

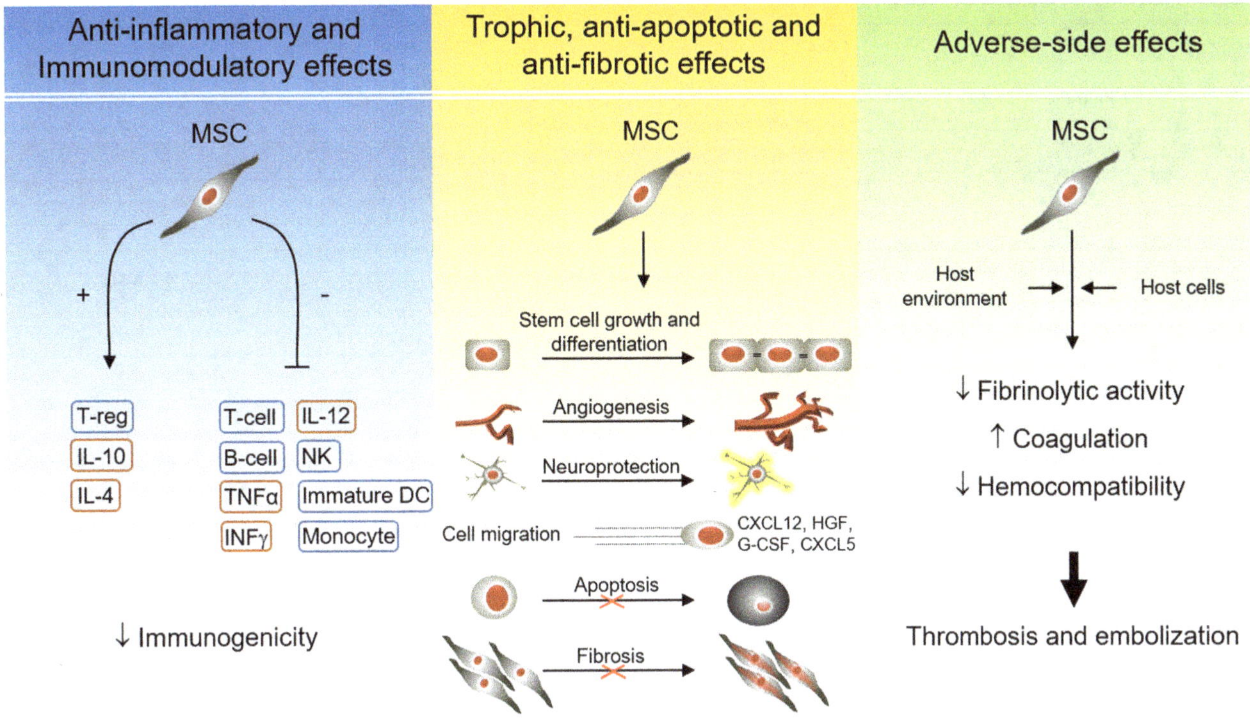

FIGURE 1 | Effects of MSCs during therapeutic application. MSCs possess a broad range of paracrine effects, including anti-inflammatory, immunomodulatory, trophic, antiapoptotic, and anti-fibrotic properties. Most of them are mediated by molecules released by MSCs, but also by direct cell-cell contacts. The paracrine properties of MSCs have beneficial effects during cell therapy for regenerative medicine. However, the interaction between MSCs and the host may result in adverse-side effects, including thrombotic events. *B-cell*, B lymphocyte; *CXCL*, C-X-C Motif Chemokine Ligand; *DC*, dendritic cell; *G-CSF*, granulocyte colony-stimulating factor; *HGF*, hepatocyte growth factor; *IL*, interleukin; *INFγ*, interferon γ; *MSC*, mesenchymal stem cells; *NK*, natural killer cells; *T-cell*, T lymphocyte; *TGFα*, transforming growth factor α; *T-reg*, regulatory T cell.

addition to direct "*cell-cell*" contact, there has been speculation of a possible transfer of mitochondria or vesicular components (secretome) that contain mRNA, microRNA and proteins (16). Noteworthy, despite the anti-inflammatory, antiapoptotic, and immunomodulatory characteristics of MSCs, and due to their ability to migrate to sites of tissue injury and inflammation, many concerns have been raised about their probable precancerous activity (17). In this regard, the functions of MSCs can be influenced by the existing microenvironment, making them acquire supportive properties toward cancer cells (8, 18). To date, no cancer has been diagnosed or has recurred in clinical trials that would originate from experimentally given MSCs. However, potential risks, related to the growth support and enhancement of undetected or "resident" cancer cells, do exist, thus the potential of tumorigenesis should be further explored and monitored to detect the possibility of tumorigenicity related to MSCs and likewise, the administration of MSCs-based therapies must be thoroughly evaluated (8, 17, 18). However, although these properties are generally attributed to all MSCs derived from different tissues, preclinical and phase I/IIa safety, and feasibility data also suggest that MSCs represent a potential therapeutic option for the treatment of Critical Limb Ischemia (CLI). Conversely, as mentioned above, evidence from different studies has suggested that MSCs from diverse sources are not identical and do not always achieve the same efficacy levels and desired outcomes. Thus, MSC effects may be influenced by the constant crosstalk between the graft and the host, which could affect the MSC fate potential. For instance, autologous MSCs from patients with inflammatory diseases (e.g., diabetes) may carry phenotypic modifications, promoting undesirable effects on the host when they come into contact with host signals (19). Here, we will provide relevant information and alternatives to possibly improve the use of MSC-based therapy to benefit type 2 diabetic patients with CLI.

CRITICAL LIMB ISCHEMIA AND DIABETES

The term CLI is used for all patients with chronic ischemic rest pain, ulcers, or gangrene in the limbs attributable to objectively proven peripheral artery disease (PAD). PAD is associated with several clinical conditions, e.g., diabetes, hypertension, cardiovascular disease, hyperlipidemia, obesity, and stroke (20). CLI is an advanced form of PAD, which is responsible for a high rate of amputations and is a major cause of morbidity and mortality worldwide. The incidence of CLI ranges from 500 to 1,000 new cases per million every year in Western Europe and North America (21, 22), and this number is expected to grow due to the aging population with a longer life expectancy and progressive increase in the incidence and prevalence of diabetes. It is estimated that more than 200 million people are living

TABLE 1 | Human stem cell sources and subtypes.

Source	Tissue	Stem cells derived and acronyms	Brief definitions
Fetal Newborn	Abortus (fetal tissues) Extra-embryonic tissues: Umbilical cord Wharton's jelly Amniotic membrane Amniotic fluid Placenta	Fetal stem cells (FSCs) Fetal structures like Wharton's jelly-derived mesenchymal stem cells (WJ-MSCs) Amniotic membrane-derived mesenchymal stem cells (Am-MSCs) Yolk sac-derived mesenchymal stem cells (YS-MSCs) Umbilical cord-derived mesenchymal stem cells (UC-MSCs) Umbilical cord blood-derived mesenchymal stem cells (UCB-MSCs) Amniotic fluid-derived mesenchymal stem cells (AF-MSCs)	Fetal stem cells are multipotent stem cells isolated from two distinct sources, the proper fetus (fetal tissues), and the supportive extra-embryonic tissues. These cells are also known as "primordial germ cells" and are isolated from tissues of 5- to 9-week fetuses obtained by therapeutic abortion. The three most reliable sources to date of abundant fetal stem cells are the placenta, amniotic fluid, and umbilical cord blood.
Adult	Bone marrow Peripheral blood	Hematopoietic stem cells (HSCs) Endothelial progenitor cells (EPCs) Bone marrow-derived mononuclear cells (BM-MNCs) Peripheral blood-derived mononuclear cells (PB-MNCs) Recombinant human granulocyte colony-stimulating factor (G-CSF)	Hematopoietic stem cells are the stem cells that give rise to other blood cells (hematopoiesis), a limited number of hematopoietic stem cells are multipotent and capable of extensive self-renewal. Endothelial progenitor cells define a group of cell population types with angiogenic activity. Endothelial progenitor cells can be obtained from the bone marrow-derived mononuclear cells fraction or from peripheral blood, and they can also be found in umbilical cord blood. Typically, endothelial progenitor cells are selected by isolation and enrichment strategies focused on the expression of surface markers CD34 and CD133.
	Bone marrow stroma Peripheral blood Adipose tissues: Fat, liposuction Others tissues: skin, gut, hair follicles, skeletal muscle, cartilage, tendon, synovium, perichondrium, cardiac tissue, oral cavity, dental pulp, salivary glands, etc.	Mesenchymal stem cells (MSCs) Bone marrow- derived mesenchymal stem cells (BM-MSCs) Peripheral blood-derived mesenchymal stem cells (PB-MSCs) Adipose tissue-derived mesenchymal stem cells (Ad-MSCs)	Mesenchymal stem cells are multipotent stem cells that can differentiate into a variety of cell types, including osteoblasts, chondrocytes, myocytes, and adipocytes.

Summary of fetal and adult stem cells subtypes. Modified from Hmadcha et al. (1).

with PAD worldwide (21), which is a common complication in patients with type 2 diabetes (23). These two factors have driven the development of a more severe degree of PAD. CLI is not a specific disease *per se*; rather, it represents a syndrome that may develop from distinct pathophysiological processes (24). Although CLI is primarily a clinic-based diagnosis, it should be confirmed objectively and early in the disease process, e.g., through the ankle-brachial index (ABI), toe-brachial index (TBI), first toe pressure (FTP), toe systolic pressure, or transcutaneous partial pressure of oxygen (tcpO_2). Further computed tomography (CT), digital subtraction angiography (DSA), Doppler echocardiography and magnetic resonance (MR) angiography are important non-invasive modalities for assessing the severity of CLI (20, 25).

PAD is a condition that is characterized by atherosclerotic occlusive disease of the lower extremities. While PAD is a major risk factor for lower extremity amputation, it is also accompanied by a high likelihood for symptomatic cardiovascular and cerebrovascular disease. Although much is known regarding PAD in the general population, the assessment and management of PAD in people with diabetes is less clear and poses some special issues. Presently, there are no established guidelines regarding the care of patients with both diabetes and PAD (26), although revascularization remains the most important therapeutic option and the main objective of CLI treatment, either by open surgery or endovascular modalities (27). Diabetic patients habitually suffer from long-segment vascular obstruction, mainly of the calf vessels (28) in type 2 diabetic patients, and the anatomic extension and distribution of atherosclerotic occlusive disease make these patients poor candidates for revascularization, resulting in continued disease progression, amputation, and death (28).

Thus, far, pharmacological treatment for these non-options patients has not been shown to be effective in the CLI course (29, 30). In fact, amputation is routinely recommended for these patients as the only option, despite its obvious dysfunctional involvement, along with the associated mortality and morbidity (27, 29). Therefore, there is a need for new effective therapeutic alternatives for a large number of patients with CLI (31).

In the diabetic patient, leg ischemia develops earlier and with greater intensity than in other diseases (vasculitis, Buerger disease). It is estimated that 15% of diabetic patients will develop CLI, and in most cases, it will lead to amputation (32). Of even greater importance, arterial lesions usually affect the more distal vessels (29). This localization of lesions makes revascularization difficult, either surgically or endovascularly.

Furthermore, the clinical presentation of diabetic patients is also different, as it entails a greater component of tissue loss and gangrene, as well as fewer clinical manifestation of pain due to the frequently associated diabetic neuropathy. Diabetic patients in particular have fewer physiological mechanisms of angiogenesis and reendothelization, and thus the course of the disease is more severe and accelerated (26). The chronic hyperglycemia present in diabetic patients results in vascular remodeling altering neovascularization (30), with deficient and/or aberrant angiogenesis (33). This phenomenon is partly due to the associated oxidative stress, underlying endothelial dysfunction, and lack of regeneration of the vascular endothelium (26, 34). In fact, neovasculogenesis is disrupted in diabetes and metabolic syndrome due to hyperglycemia and increased hemoglobin A1c (HbA1c), neuropathy, hypercholesterolemia, oxidized low-density lipoproteins (Ox-LDL), reactive oxygen species (ROS) or increased fatty acids (35–38). In healthy people, homeostatic mechanisms, by which the vascular supply increases to match metabolic demand, are activated (39). However, these mechanisms are frequently disrupted in patients with CLI, and the physiological response is not able to deliver the necessary amount of blood flow and oxygen to the affected limb, causing the arterioles of these patients to dilate to the maximum and become insensitive to provasodilator stimuli (39). This phenomenon, referred to as vasomotor paralysis, is considered the result of chronic exposure to vasorelaxation factors in patients with vascular diseases, which may explain the failure of most vasodilator therapies to improve functional capacity in individuals with PAD. Likewise, the blood vessels of patients with CLI present a decrease in wall thickness, cross-sectional area, and "*wall-to-cell wall*" ratio, among others.

The use of different types of stem cells in the treatment of CLI is aimed at stimulating neovascularization in the area of severe ischemia. The procedure consists of the administration of cells in the ischemic tissues, either intravascularly or intramuscularly (**Table 2**), to form new vascular structures and/or segregate a number of angiogenic factors that regulate the process and favor the recruitment of new cells (25, 40–51). The formation of collateral blood vessels is promoted in an effort to improve blood flow in ischemic tissue, as well as alleviate the symptoms of the disease and, in most cases, prevent amputation of the affected limb in patients who do not respond to conventional treatments. The main objective of this process is principally regenerative, restorative and anti-inflammatory (20, 25, 31). Despite the high prevalence and incidence of CLI in diabetic patients, most studies have excluded patients with diabetes or with high HbA1c (44, 47, 49, 50) (**Table 2**).

CLINICAL TRIALS FOR PAD, ISCHEMIA AND CLI

The Route of Cell Administration (Intraarterial vs. Intramuscular)

Since the publication of the Therapeutic Angiogenesis using Cell Transplantation (TACT) study (52), more than 70 reported clinical trials have been reported in patients with CLI. Only 25% of those studies included diabetic patients (25, 53). It is known that several types of stem cells, derived from different sources, have the propensity for vascular development, and could potentially be useful in the management of CLI (**Table 3**). Some of these cells have been used in preclinical models as well as clinically to treat this condition. Among them, mononuclear cells (MNCs) have been the most widely used (1, 8, 20), and despite significant steps forward in defining their potential for therapeutic purposes, further progress has been mired by unresolved questions around their definition, and mechanism of action and because of their heterogeneity (4, 54, 55).

We conducted a pilot prospective single-center study (NCT00872326), phase I/IIa, that aimed to assess the safety and efficacy of intraarterial administration of autologous bone marrow-derived MNCs (BM-MNCs) in 20 diabetic patients with CLI (25). As described by Isner and Asahara (56) and to guarantee the homing of a great number of cells, BM-MNCs were administered intraarterially into the most affected leg "*target limb*" as close as possible to the ischemic area. One year after BM-MNC infusion, there was a remarkable improvement in the clinical status of most of the target limbs. In addition, the infusion of BM-MNCs induced an unexpected benefit of an improvement in the healing process, not only for ulcers but even for minor amputations. Furthermore, early clinical benefits of cell infusion consisted of patients having a widespread perception of less limb pain, an increase in pain-free walking, and warmness in the target limb. Unlike other studies, the cell dose was 10 times smaller than the dose used by other groups. Moreover, and surprisingly, six diabetic patients reduced their need to inject insulin, probably due to a decrease in peripheral insulin resistance (25).

Similarly, the PROVASA study (Intraarterial Progenitor Cell Transplantation of Bone Marrow Mononuclear Cells for Induction of Neovascularization in Patients with Peripheral Arterial Occlusive Disease Study), a multicenter, double-blind, phase II trial (NCT00282646) with an estimated enrollment of 40 patients with ischemic rest pain or non-healing ulcers randomly assigned (1:1) to receive BM-MNC treatment or placebo, demonstrated that in patients with CLI, intraarterial administration of BM-MNCs does not increase ABI but promotes ulcer healing and reduces rest pain. Furthermore, repeated administration of functional BM-MNCs was required for successful ulcer healing associated with improved limb salvage (41). In this context, several studies have demonstrated the feasibility of intraarterial delivery of BM-MNCs, their beneficial effect on improving endothelial function (25, 57) and overall improvements of ischemic pain and ulcer healing (25, 58). However, the quality of evidence for efficacy is limited, as most studies lacked a proper placebo or sham group because of the invasive bone marrow harvesting procedure required to obtain the cells (59–61).

Although intraarterial and intramuscular injection of autologous BM-MNCs have shown similar results, and combined intraarterial and intramuscular transplantation is clinically feasible (62), to date, an overwhelming majority of clinical studies targeting PAD have relied upon intramuscular cell delivery (63). Intramuscular administration is easier and less invasive, and it results in a transient placement of cells in

TABLE 2 | Published studies using cell-based therapy to treat CLI.

Clinical trial ID (www.clinicaltrials.gov)	Phase	Cell type	Route of administration	Diabetic patients	References
NCT00872326	I/II	BM-MNCs	Intraarterial	Included	(25)
NCT00371371	I/II	BM-MNCs	Intraarterial	There is no data available	(40)
NCT00282646	I/II	BM-MNCs	Intraarterial	Included buerger disease	(41)
NCT01480414	I/II	BM-MNCs	Intramuscular	There is no data available	(42)
NCT00221143	I/II	PB-$CD34^+$	Intramuscular	There is no data available	(43)
NCT00883870	I/II	BM-MSCs	Intramuscular	Type 1 Diabetic patients were excluded	(44)
NCT01595776	I/II	Bone marrow derived- $CD133^+$	Intramuscular	There is no data available	(45)
NCT01065337	II	TRC	Intraarterial and Intramuscular	Included	(46)
NCT00392509	I/II	ALDHbr Cells	Intramuscular	Patients with HbA1c >8% were excluded	(47)
NCT00523731	I	NMPB-ACPs	Intramuscular	There is no data available	(48)
NCT00468000	II	Ixmyocel-T	Intramuscular	Patients with HbA1c >10% were excluded	(49)
NCT00533104	I/II	BM-MNC/PB-MNC	Intramuscular	Patients with HbA1c >7.5% were Excluded	(50)
NCT00721006	II	MESENDO	Intramuscular	There is no data available	(51)

ALDHbr Cells, autologous bone marrow-derived Aldehyde Dehydrogenase-bright cells; Ixmyelocel-T, BM-MNCs ($CD90^+$and $CD45^+/CD14^+$); Mesendo, combination of bone marrow-derived MSCs and EPCs; NMPB-ACPs, non-mobilized peripheral blood angiogenic cell precursors; PB-$CD34^+$, peripheral blood G-CSF-mobilized apheresis $CD34^+$ cells; TRC, tissue repair cells (expanded bone marrow cells enriched in $CD90^+$ cells).

the ischemic tissue, whereas intraarterial infusion is designed to directly inject cells into peri-ischemic areas, which are considered to have sufficient oxygen and nutrients to support cellular functions. Both delivery methods have obtained promising results in the improvement of angiogenesis (64). Klepanec et al. (65) compared the therapeutic effects of intramuscular and intraarterial delivery of BM-MNCs in a randomized manner. There were no differences among functional parameters in patients undergoing intramuscular vs. intraarterial cell supply. Preclinical (66) and clinical (67) data from our group substantiate these remarks; nevertheless, our humble experience indicates that the route of administration depends on the type and the dose of cell to be administered. Deciphering how stem cells manage the countless signals required for revascularization will improve CLI recovery, Qadura et al. (31) proposed a combination delivery of multiple cell types within supportive bioengineered matrices as a new therapeutic strategy to target CLI.

Comparison of the Cell Type

Apart from the route of administration (intramuscular, intraarterial, or combined), the most ideal cell type must be identified, and a better understanding of the effective subpopulation of stem cells is necessary as stem cells are a heterogeneous population. Some clinical trials have directly compared different cell populations (67–69) or used a combination of angiogenic stem cells. Huang et al. (68) evaluated the transplantation of peripheral blood-derived MNCs (PB-MNCs) in the treatment of diabetic patients with CLI who had received a subcutaneous injection of recombinant human granulocyte colony-stimulating factor (G-CSF) to mobilize progenitor cells, which resulted in clinical improvements including reduced limb pain and ulcers, as well as no adverse effects specifically due to cell transplantation and no lower limb amputation in the transplanted patients. Tateishi-Yuyama et al. (52) investigated the feasibility and safety of intramuscular injection of MNCs and showed that in two groups of patients, the first with unilateral ischemia infused intramuscularly with BM-MNCs in the ischemic limb and saline in the less ischemic limb, and the second with bilateral leg ischemia receiving random intramuscular injections of BM-MNCs in one leg and PB-MNCs in the other as a control, resulted in significant improvements in patients treated with BM-MNCs compared with those treated with PB-MNCs in terms of ABI and rest pain. Lu et al. (69) compared the therapeutic effect of autologous intramuscular administration of bone marrow-derived MSCs (BM-MSCs) with BM-MNCs in 20 diabetic patients with CLI and foot ulcer. The authors demonstrated that the healing rate of ulcers was significantly higher in the group treated with BM-MSCs than with BM-MNCs. Likewise, the authors concluded that BM-MSC treatment was effective and better tolerated than BM-MNCs to improve lower limb perfusion and to promote foot ulcer healing in diabetic patients with CLI. Lasala et al. (51) evaluated the efficacy and safety of autologous intramuscular administration of a combination of MSCs and endothelial progenitor cells (EPCs) in 26 patients with bilateral CLI. They found that within this phase II clinical trial (NCT00721006), the enrolled patients experienced an increase in perfusion in the treated limbs compared with the control legs and improvement in pain-free walking time and ABI after cell infusion. In this context, our group proposed a phase I/II clinical trial (NCT02287974) to study and compare the therapeutic effect of autologous BM-MNCs, autologous BM-EPCs ($CD34^+/CD133^+$ cells) and autologous adipose tissue-derived MSCs (Ad-MSCs) on inflammatory and angiogenic cytokines, resistance to insulin and a decrease in the need for insulin, as well as evaluating the safety, viability and efficiency of the intraarterial infusion of these three stem cells types in patients with type 2 diabetes with CLI. We aimed to obtain related data on the source of suitable tissue,

TABLE 3 | Registered clinical trials using cell-based therapy to treat CLI.

Clinical trial ID (www.clinicaltrials.gov)	Phase	Status	Cell type	Route of administration	Patient condition
NCT00904501	III	Completed	BM-MNC	Intramuscular	Patients with HbA1c >7.5% were Excluded
NCT01408381	II	Completed	BM-MNCs	Intraarterial	Non-diabetic
NCT00987363	I/II	Completed	BM-MNCs	Intraarterial	Diabetic
NCT01867190	I/II	Completed	BM-MNCs	Intramuscular	There is no data available
NCT00595257	I/II	Completed	BM-MNCs	There is no data available	There is no data available
NCT00498069	–	Completed	BM-MNCs	There is no data available	There is no data available
NCT01245335	III	Completed	Bone marrow-derived cells	There is no data available	Patients with HbA1c >10% were excluded
NCT00616980	I/II	Completed	$CD34^+$	Intramuscular	There is no data available
NCT01584986	II	Completed	PB-ACPs	Intramuscular	There is no data available
NCT01351610	I/II	Completed	BM-MSCs	Intravenous	Patients with HbA1c >9% were excluded
NCT01484574	II	Completed	BM-MSCs	Intramuscular	Excluded diabetic patients
NCT01824069	I/II	Completed	Ad-MSCs	Intramuscular	There is no data available
NCT01257776	I/II	Completed	Ad-MSCs	Intraarterial	Diabetic
NCT01663376	I/II	Completed	Ad-MSCs	Intramuscular	There is no data available
NCT00919958	I	Completed	PLX-PAD	Intramuscular	Patients with HbA1c >9% were excluded
NCT00951210	I	Completed	PLX-PAD	Intramuscular	There is no data available
NCT01483898	III	Completed	Ixmyocel-T	Intramuscular	Patients with HbA1c >10% were excluded
NCT00955669	I	Completed	BM-MNCs and BM-MSCs	Intramuscular	Type 2 diabetic patients
NCT00518401	I	Completed	Mesendo	Intramuscular	There is no data available
NCT00913900	I	Completed	$CD133^+$	Intramuscular	There is no data available
NCT01019681	I	Completed	UCB-MNCs	Intramuscular	There is no data available
NCT02474121	–	Available	Bone marrow-derived cells	Intramuscular	Patients with HbA1c >10% were excluded
NCT01837264	I	Active, not recruiting	BM-MNCs	There is no data available	There is no data available
NCT00956332	I/II	Active, not recruiting	MultiGene Angio	Intraarterial	There is no data available
NCT01049919	–	Ongoing, but not recruiting	BM-MNCs	Intramuscular	Patients with HbA1c >10% were excluded
NCT01745744	I/II	Ongoing, but not recruiting	Ad-MSCs	Intraarterial	Non-diabetic
NCT01386216	I	Recruiting	BM-MNCs	Intramuscular	Patients with HbA1c >10% were excluded
NCT02099500	I/II	Recruiting	AD-MSCs liposuction	Intramuscular	Patients with HbA1c >10% were excluded
NCT02915796	I	Recruiting	G-CSF $CD133^+$	Intramuscular	Diabetic
NCT02140931	II	Recruiting	PB-ACPs	Intramuscular	There is no data available
NCT01833585	III	Recruiting	PB-MNCs G-CSF	Intramuscular	There is no data available
NCT02551679	II	Recruiting	PB-ACPs	Intramuscular	There is no data available
NCT02089828	–	Recruiting	$CD34^+$ and PB-MNCs	There is no data available	There is no data available
NCT02234778	–	Recruiting	Ad-SVF cells	Intramuscular	There is no data available
NCT02805023	I/II	Recruiting	BGC101	Intramuscular	There is no data available
NCT01456819	II	Recruiting	BM-MSCs and BM-MNCs	Intramuscular	There is no data available
NCT02454231	II/III	Recruiting	EPCs and BM-MNCs	Intramuscular	Patients with HbA1c >7.5% were excluded
NCT02864654	I/II	Enrolling with invitation	ADRC from lipoaspirate	Intramuscular	There is no data available
NCT02863926	I	Not yet recruiting	Bone marrow-derived cells	Intramuscular	There is no data available
NCT02538978	III	Not yet recruiting	BM-MNCs	There is no data available	There is no data available
NCT02501018	II	Not yet recruiting	$CD34^+$	Intramuscular	There is no data available

(Continued)

TABLE 3 | Continued

Clinical trial ID (www.clinicaltrials.gov)	Phase	Status	Cell type	Route of administration	Patient condition
NCT02477540	I	Not yet recruiting	BM-MSCs	Intramuscular	Type 1 diabetic patients are excluded
NCT01686139	I	Not yet recruiting	Allogeneic BM-MSCs	Intramuscular	Type 1 and 2 diabetic patients
NCT03042572	II/III	Not yet recruiting	Allogeneic BM-MSCs	Intramuscular	There is no data available
NCT03056742	II	Not yet recruiting	Allogeneic BM-MSCs	Intramuscular	Diabetic patients are excluded
NCT02993809	I	Not yet recruiting	BM-ECs and PRPE	Intramuscular	Patients with HbA1c >7% were excluded
NCT00488020	I	Unknown	BM-MNCs	Intramuscular	There is no data available
NCT00434616	II/III	Unknown	BM-MNCs	There is no data available	There is no data available
NCT01446055	I/II	Unknown	BM-MNCs	There is no data available	Patients with HbA1c >7% were excluded
NCT01903044	I/II	Unknown	BM-MNCs	Intramuscular	There is no data available
NCT00539266	II/III	Unknown	BM-MNCs	Intramuscular	Diabetic and non-diabetic
NCT00922389	I/II	Unknown	G-CSF and PB-MNCs	Intramuscular	YES (controlled)
NCT02336646	I	Unknown	Allogeneic BM-MSC	Intramuscular	Type 1 diabetic patients are excluded
NCT01216865	I/II	Unknown	UC-MSCs	Intramuscular	Type 2 diabetic patients
NCT01558908	I/II	Unknown	ERCs	Intramuscular	Patients with HbA1c >8.5% were excluded
NCT00145262	II	Unknown	BM-MSCs and BM-MNCs	There is no data available	Patients with HbA1c >6.5% were excluded
NCT02287974	I/II	Unknown	MNCs and $CD133^+$ and Ad-MSCs	Intraarterial	There is no data available
NCT02145897	I/II	Unknown	SVF and Ad-MSCs	Intramuscular and Intravenous	Patients with HbA1c >7% were excluded

ADRCs, adipose-derived regenerative cells; Ad-SVF, adipose-derived stromal vascular fraction Cells; BGC101, mixture of cells enriched for EPCs and hematopoietic stem cells; BM-ECs, bone marrow-derived endothelial cells; ERCs, endometrial regenerative cells; Ixmyelocel-T, BM-MNCs ($CD90^+$and $CD45^+/CD14^+$); Mesendo, combination of bone marrow-derived MSCs and EPCs; MultiGeneAngio, endothelial and smooth muscle cells; PB-ACPs, autologous angiogenic cell precursors; PLX-PAD, placenta-derived MSCs; PRPE, platelets rich plasma; UC-MSCs, umbilical cord-derived MSCs.

the most appropriate cell type, optimal dose of cells, efficient and low-cost protocols, among others, to be able to offer, in the near future, a high-quality, economic, and effective therapy for those patients without current therapeutic options. Despite the extraordinary and unpaid efforts of the clinical research team and promising preliminary results obtained during the first year of follow-up showing beneficial but distinct effects of cell type treatment, for unknown reasons the sponsor of this study decided to prematurely terminate patient recruitment, and we no longer have access to the clinical data.

The Use of MSCs

Regarding the use of MSCs as a cell-based therapy for CLI, recent data suggest that the therapeutic effects of these cells in ischemic pathologies are due to the secretion of angiogenic molecules to bioactive levels and their ability to restore the microenvironment of the damaged area (70). In preclinical studies, the administration of autologous, allogeneic, and xenogeneic MSCs derived from various sources such as bone marrow, umbilical cord blood, fetal membrane and adipose tissue have been shown to be beneficial in rat and mouse models with lower limb ischemia (66). Subsequently, several phase I/IIa clinical trials have been assayed in a limited number of patients to demonstrate the safety and feasibility of MSCs obtained from different sources (71–75). MSCs isolated from healthy donors have shown uniform and consistent properties, whereas those from patients affected by degenerative and/or inflammatory disease differ in their biological and functional characteristics (8, 19, 76). In this regard, other studies using MSCs isolated from diabetic patients suggest that the hyperglycemic environment as well as other metabolic disorders associated with diabetes affect the cellular endogenous reserve and alter their proliferation, differentiation and angiogenic capacity (19, 77–79). Likewise, several groups have reported benefits of using autologous MSCs as a cell-based therapy for a wide variety of diseases, such as cardiovascular diseases (14, 31, 44, 51, 59, 61, 67, 71–75, 80–82), diabetic nephropathy (83), and diverse brain injuries including stroke, neural trauma, and heatstroke (84, 85).

In this regard, we have completed a phase I/IIa open label, randomized, dose-scalable clinical trial with 30 diabetic patients with CLI (NCT01257776) to test the safety and feasibility of the administration of autologous Ad-MSCs. A unique intraarterial cellular dose of 0.5×10^6 cells/kg (low dose) or 1×10^6 cells/kg body weight (high dose) was administered (86). Neovasculogenesis (assessed by MetaMorph® software quantification of DSA angiographies) (25) and clinical improvement (evaluated by Rutherford-Becker classification, University of Texas Diabetic Wound Scale, tcpO_2, and ABI) were compared at baseline and at follow-up. After cell infusion and after 1 year of follow-up, 40% of the diabetic patients with CLI treated with the low, and 60% of the diabetic patients with CLI treated with the high dose, respectively, experienced an increase in vascularity in the ischemic areas (**Figure 2** and **Table 4**), which correlated positively with clinical benefits and symptomatic improvement in the target limb. No patient suffered any amputation (86).

Due to the combinatorial potential for inducing angiogenesis and the immunomodulatory effects *in situ* of BM-MSCs, Gupta et al. (44) reported the results of a randomized double-blind randomized placebo controlled multicenter phase I/II study examining the efficacy and safety of intramuscular administration of allogeneic BM-MSCs in 20 patients with CLI (NCT00883870). The administration procedure of BM-MSCs at a dose of 2×10^6 cells/kg or placebo (PlasmaLyte A) was found to be feasible and safe; however, few of the efficacy parameters (such as ABI) showed significant improvements in BM-MSC arm transplant patients. Although immunogenicity may be unpredictable in cases where the administration of cells is used for a different function in the recipient than in the donor "*heterologous use*" or when injected into non-physiological sites, no evidence of harm or adverse events was detected with allogeneic administration of MSCs obtained from different source and prepared in different ways (81). This approach could result in a safe, feasible option to avoid the time involved in the process of isolation, expansion, and production of the use of autologous cells. Regardless, it is necessary to continue investigating this and other associated aspects to determine which cell type is best and feasible for a specific pathology and which type of patient health profile can benefit from this kind of cell-based therapy. Thus, improvements in cell therapy will benefit from a more precise characterization of cellular subsets in the therapeutic product.

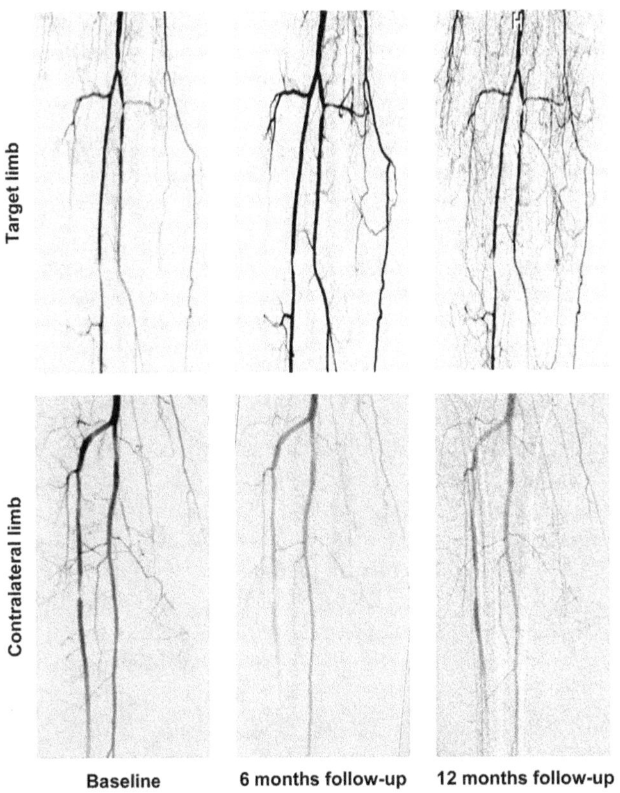

FIGURE 2 | Follow-up of diabetic patients with CLI treated with Ad-MSCs. Lower limb digital subtraction angiography (DSA) evaluation at 6- and 12-months follow-up after autologous adipose tissue-derived MSCs (Ad-MSCs) administration showing substantial neovascularization at 12 months in target limb compared to baseline and to contralateral limb.

THE IMPAIRED PROPERTIES OF MSCs COMPROMISE THEIR EFFICACY

Cell therapy is especially complex due to the nature of the product. The cellular source, isolation and expansion procedures, dose, site and procedure of administration define the cellular medicament, even without a precise knowledge of the mechanisms of action.

Immunogenicity

The administration of stem cells could interact with the host immune response (for example, in a proinflammatory environment) or have an immunomodulatory effect. Although MSCs have been considered immune-privileged in this regard, long-term exposure to the culture medium can make them more immunogenic (MSCs are isolated and expanded in medium that contains fetal bovine serum (FBS) and/or human platelet concentrate), for example, by positively regulating the normal set of histocompatibility molecules (82, 87). Thus, allogeneic use of the cells may entail a greater risk of rejection by the immune system. This rejection could lead to a loss of function of the administered cells, and consequently, their therapeutic activity could be compromised. Nevertheless, use of immunosuppressants could limit these risks, but in turn could cause adverse effects due to immunosuppressive medication.

Coagulation and Fibrinolytic Activity

MSCs have shown significant therapeutic potential due to their fibrinolytic and antithrombogenic properties (88–92). To date, several clinical trials have been conducted using autologous MSCs for the treatment of diabetes and its complications, which are presumably safer and more effective than allogenic cells. However, the therapeutic effects of MSCs have been questioned when they are derived from a diabetic milieu (19, 93–95).

MSCs isolated from healthy donors have shown uniform and consistent properties, while those from patients, such as diabetic patients, differ in their biological and functional characteristics and can reduce the beneficial therapeutic effects of autologous MSCs (19, 76, 79). In this regard, studies carried out by our group and others using Ad-MSCs of diabetic patients suggest that the hyperglycemic environment and other metabolic disorders associated with diabetes affect the cellular endogenous reserve and their proliferation, differentiation, and angiogenic capacity, among other cellular characteristics (78, 79, 96, 97). Once infused in the recipient, the cells come into direct contact with the

TABLE 4 | Improvement of clinical outcomes after implant of Ad-MSCs: Evolution of Rutherford-Becker before (baseline) and 6 and 12 months of follow-up.

Groups (10 patients/group)	Rutherford-Becker grades	Baseline cases (%)	6 months cases (%)	12 months cases (%)
Control	0	–	–	1 (10%)
	I	–	2 (25%)	2 (20%)
	II	8 (80%)	5 (62.5%)	6 (60%)
	III	2 (20%)	1 (12.5%)	1 (10%)
Low dose	0	–	3 (30%)	**4 (40%)***
	I	–	1 (10%)	1 (10%)
	II	5 (50%)	2 (20%)	2 (20%)
	III	5 (50%)	4 (40%)	3 (30%)
High dose	0	–	4 (40%)	**6 (60%)***
	I	–	3 (30%)	3 (30%)
	II	6 (60%)	–	–
	III	4 (40%)	2 (20%)	1 (10%)

Rutherford-Becker grades: Grade 0 includes: Category 0, asymptomatic and Category 1, mild claudication; Grade I: Category 2, moderate claudication and Category 3, severe claudication; Grade II: Category 4, ischemic rest pain; Grade III: Category 5, minor tissue loss and Category 6, major tissue loss, ulceration or gangrene.
**Percent in bold indicates an increase in vascularity in ischemic area of diabetic patients based on Rutherford-Becker improvement.*

tissues, bloodstream and other patient cells, and the cell-recipient interaction process still requires thorough investigation and characterization. Physiologically, MSCs reside in the perivascular compartment of almost every tissue (98, 99); however, one of the hurdles to the sustained therapeutic success of these cells is early cell loss, which is largely thought to be due to incompatibility responses after systemic infusion of the cells, a reaction termed the instant blood-mediated inflammatory reaction (IBMIR) (63, 82, 87, 97–100). This reaction suggests that the immune and inflammatory system react to cells that normally are not in contact with the blood circulation (**Figure 3**). Moreover, it has been further shown that different MSC products display varying levels of highly procoagulant tissue factor, a decrease in tissue plasminogen activator (tPA) or an increase in plasminogen activator inhibitor type 1 (PAI-1) and may adversely trigger IBMIR or microthrombosis in the target tissue (19, 97).

MSCs are considered to be safe and even to promote fibrinolysis (89, 91). Since type 2 diabetes is a systemic inflammatory disease with a prothrombotic state, intraarterial infusion of Ad-MSCs in diabetic patients was suggested and approved by the Ethical Committee and the Regulatory Agency as the appropriate treatment within a clinical trial conducted by our group (NCT01257776), but unexpectedly during the course of this clinical study, two patients developed distal microthrombosis after intraarterial Ad-MSC infusion (19). These two patients reported oppression at the infrapopliteal level and vasomotor reaction of neurogenic origin in the distal third of the target limb, accompanied by discreet pain, 10 h after the cellular infusion. MR angiography demonstrated indemnity of the arterial vessels of medium caliber. These two patients were treated with antithrombotic therapy and discharged 72 h after the symptoms disappeared (19).

Furthermore, Ad-MSCs induced an increase in expression and release of PAI-1 and reduced levels of tPA. Likewise, the quantification of D-dimer also decreased. These responses were tested with MSCs of different origins exposed to different *ex vivo* environment. Effects were more pronounced when Ad-MSCs were from type 2 diabetic patients exposed to the sera of same patients (**Figure 4**). Therefore, the efficacy of fibrinolysis decreased favoring thrombosis, and these observations were published (19) and a patent filed to identify these tentative responses for cellular medicaments (US20160161504).

Altogether, new lines of research are being promoted that focus on enhancing the therapeutic effects of stem cells by regulating their biological characteristics (76, 96, 101–105). A recent study conducted by our group has demonstrated the beneficial effects of platelet-derived growth factor-BB (PDGF-BB) in restoring the defective phenotype of Ad-MSCs derived from type 2 diabetic patients with CLI (76). In particular, this study showed that pretreatment with PDGF-BB could potentiate proliferation, migration, and homing of defective MSCs, as well as recover their impaired fibrinolytic ability (**Figure 4**). Furthermore, we demonstrated that PDGF-BB utilized the ERK-SMAD pathway to exert its beneficial effects. Therefore, pretreatment with PDGF-BB represents a suitable strategy to produce more effective MSCs for autologous therapies (76). In this context, we postulated that either the use of allogenic MSCs from healthy donors, rescue of the healthy phenotype by pretreatment of autologous MSCs with PDGF (76), expression of Akt (104) or angiotensin 1 (105), or the use of early-passage autologous MSCs (106) will be an option to improve the therapeutic effect of the cells. While we have clinical data for the first scenario, the others have only been assessed in preclinical studies; since a side effect is coagulation, pretreatment with heparin (103) may overcome this problem.

CONSIDERATIONS FOR AN ADVANCED THERAPY MEDICINAL PRODUCT

Safety, feasibility, and efficacy are mandatory to determine the viability of a clinical application for the treatment of any disease. With the exception of hematopoietic stem cell transplants, stem

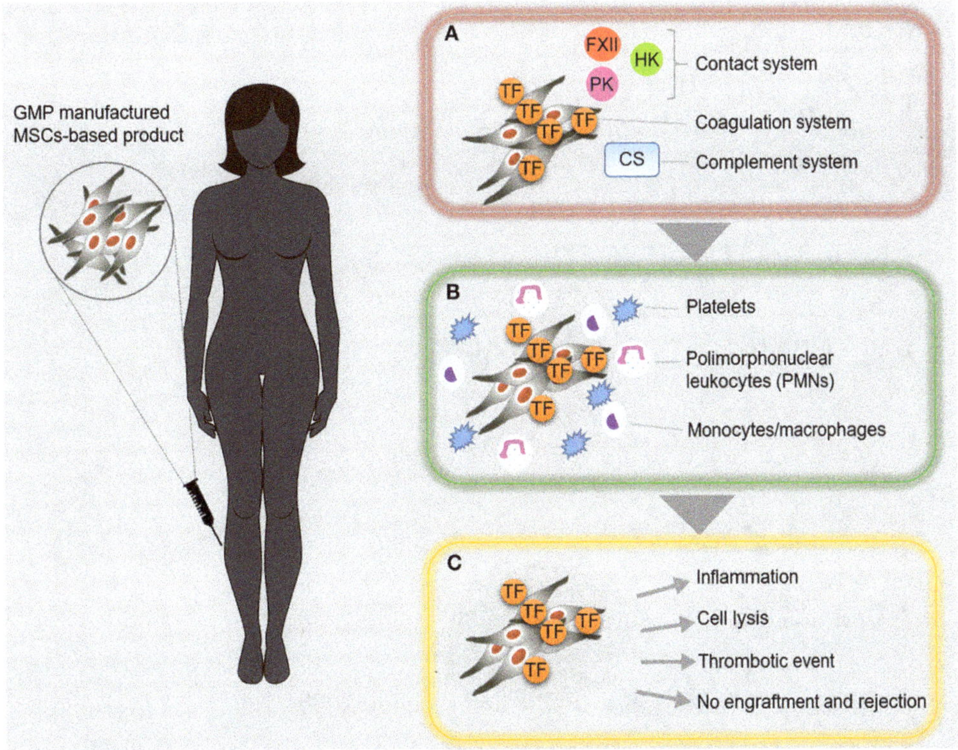

FIGURE 3 | Instant blood-mediated inflammatory reaction. **(A)** The direct contact between intravascularly infused MSC-based products and blood stream, promotes inflammatory reaction known as instant blood-mediated inflammatory reaction (IBMIR). Adipose tissue-derived MSCs (AdMSCs) express tissue factor (TF/CD142) that mediates the activation of blood coagulation cascade and complement system, leading to inflammation and thrombotic reaction. **(B)** Consequently, multiple amplification reactions result in the activation of platelets and effector cells of immune system. **(C)** IBMIR is a multifaceted phenomenon that can compromise the success of MSC-based cell therapy. *CS*, complement system; *FXII*, factor XII; *HK*, high molecular weight kininogen; *PK*, prekallikrein; *TF*, tissue factor.

cell therapies used for the treatment of any disease are considered Advanced Therapies Medicinal Products (ATMP); therefore, their development, approval and use must be in accordance with specific standards established nationally and internationally for such products. Thus, regulatory authorities warrant the safety of the studies (8, 107).

Cell expansion and culture protocol are not yet standardized (4). Currently, there is no protocol or universal definition for stem cell culture and expansion Regulatory Agencies. For example, the Spanish Agency of Medicines and Medical Devices (AEMPS), the EMA and the FDA recommend a set of standards to be followed for the production of an ATMP (EMA: Regulation (EC) No 1394/2007 provides the overall framework on ATMPs and FDA: designation described in Section 3033 of the twenty-first Century Cures Act; cell therapy medicinal products are regulated in the Code of Federal Regulations under Title 21 PART 1271, Human Cells, Tissues and Cellular, and Tissue-Based Products). The different sources of origin, as well as the different methodologies for obtaining tissue cells, make it very difficult to compare research groups in search of the fastest, most effective, economical, high-yielding, and quality-required method.

Preclinical studies have shown that cell viability after infusion is quite limited and that very few cells survive after infusion. Although the *in vivo* follow-up in humans is ethically and technically complicated, it is necessary to continue investigating in this line to determine the intrinsic mechanisms of the integration of the infused cells in the specific microenvironment. Furthermore, the ATMP dose to obtain the desired effects remains to be determined and defined. Our previous clinical data (NCT01257776, NCT01745744, and NCT02287974) show that an intraarterial dose of 1×10^6 cells/kg of body weight is more effective that a dose of 0.5×10^6 cells/kg of body weight (86). Experiments using mouse as animal models have established a minimum dose of 1×10^6 cells/kg of body weight, a quantity necessary to obtain a quantifiable but weak benefit. The dose for ATMP treatment is determined by the patient's body weight and the biodistribution of cells and paracrine factors secreted by the ATMP (MSCs) in the human recipient, and most clinical trials usually use a similar ATMP dose (108, 109). In most cases, the doses used for several clinical trials are likely not sufficient to achieve the desired outcomes and a clear therapeutic benefit. Therefore, collecting information regarding ATMP doses obtained from different sources and the influence of the host (patient recipient) medical conditions are important for proposing future clinical trials and will undoubtedly assure the safety and efficacy of ATMP-based therapies. Likewise, the frequency of administration is currently not yet determined, and the efficacy of ATMP-based therapy may be related to a precise number of repeated applications (40, 42, 110), as

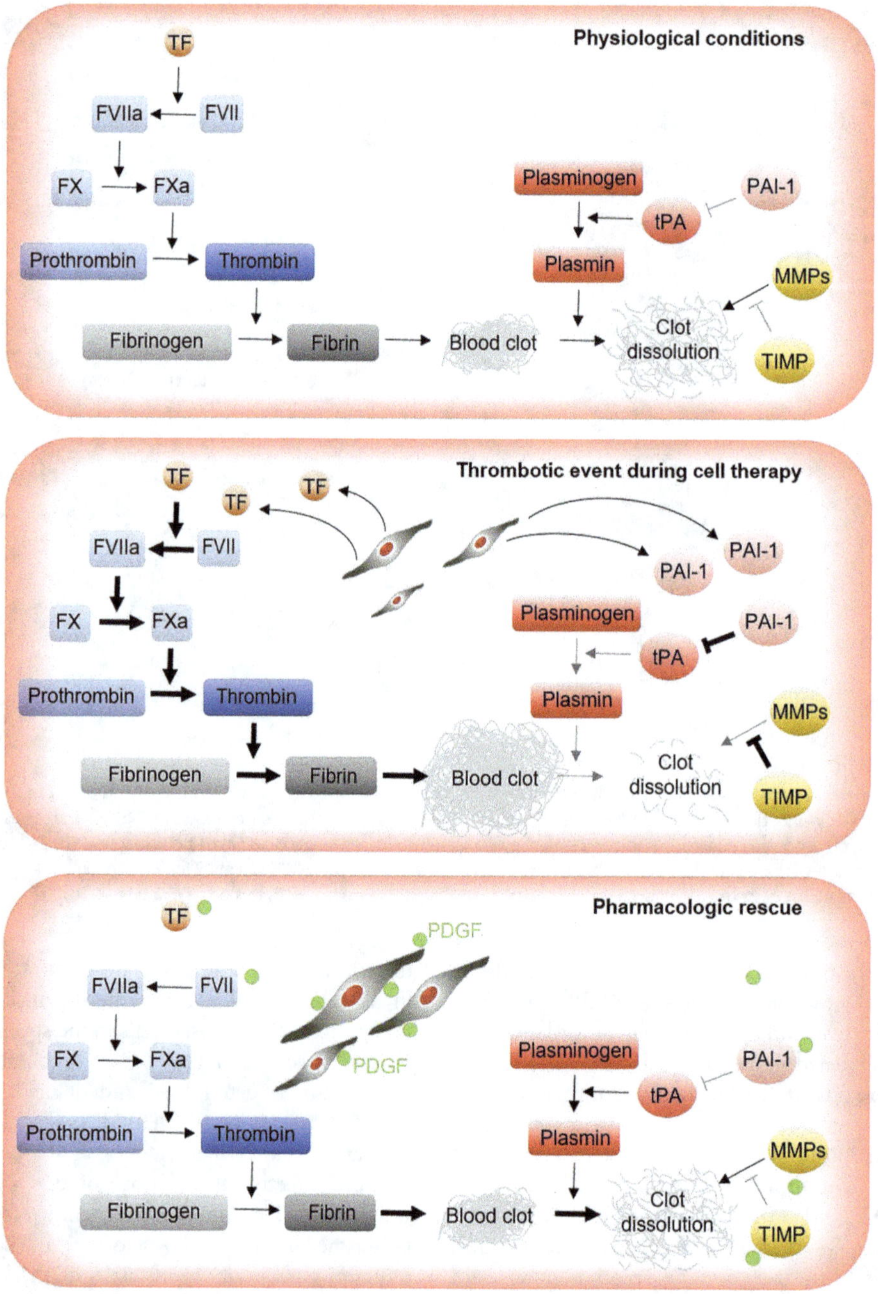

FIGURE 4 | Implications of MSCs for thrombosis risk during cell therapy. Under physiological conditions, a delicate balance between the coagulation, and fibrinolysis cascades are responsible for the effective dissolution of the clots within blood vessels. During cell therapy, transplanted MSCs may induce the expression of pro-coagulant factors (e.g., TF and PAI-1) that disrupt the coagulation/fibrinolysis balance, increasing the formation of clots and leading to thrombotic events (19). Approaches aimed to produce more safety MSC products are being investigated. For instance, the pretreatment of MSCs with PDGF results in cell products with increased fibrinolytic activity, which may help to minimize thrombotic events during cell infusion (76). *FVII*, factor VII; *FVIIa*, factor VIIa; *FX*, factor X; *FXa*, factor Xa; *MMPs*, metalloproteinases; *PAI-1*, plasminogen activator inhibitor type 1; *PDGF*, platelet derived growth factor; *TIMP*, tissue inhibitor of metalloproteinase; *TF*, tissue factor; *tPA*, tissue plasminogen activator.

determined for a conventional medicinal product. Similarly, and in addition to the doses, the duration of ATMP application remains to be determined. Thus, ATMP-based therapy should be as close as possible to conventional medicines and thus may need to be adjusted accordingly. Finally, the method and route of administration of ATMP remains inconclusive, representing another variable to be considered in future clinical trials (97). From our perspective, the most suitable ATMP (defined by the cell type, culture media and standards, doses and route of administration) for a particular disease or complication remains a challenge for Regenerative Medicine and requires further investigation.

Therefore, the desired therapeutic effect depends on many factors since the mechanism of action of an ATMP in tissue regeneration is likely to be multifaceted; ATMP potency can be determined by the ability of the injected cells to migrate, survive, integrate, differentiate, and produce functional paracrine mediator factor involved in "*cell-cell interactions.*" As mentioned above, many diseases, including diabetes, affect the phenotypic and therapeutic properties of an ATMP, and in the search for safety and efficacy, the recipient tissue must respond favorably to the administered ATMP, which would result in the activation of endogenous regeneration mechanisms (111–114). Understanding the integration of exogenous mechanisms (injected ATMP) with the endogenous recipient (host) will play a decisive role in the future clinical use of adult stem cells (8, 97).

NEW GENERATIONS OF ATMP

Advances in compliance under good manufacturing practice (GMP) standards of more sophisticated cellular products are now paving the way for new ATMP generations for use in clinical trials.

The lack of therapeutic efficacy of the generation (Generation 1) of unmodified, naïve, and wild type MSCs rated in clinical trials can be explained by the observation that, after their systemic infusion (intravenous), these cells become trapped in vascular filters (fundamentally liver and lung), with only a small percentage reaching the target tissues. Therefore, it is essential that we design strategies that favor their migration, nesting, and localization in the inflammatory and/or infectious focus to increase their effectiveness (Generation 2 MSCs) (111, 113).

Biodistribution and long-term follow-up of these cells in animal models have shown that only a few cells persist after long periods of transplantation. This phenomenon supports the idea that most of the effects of MSCs are probably based on a "*hit and run effect.*" To increase the implantation of an ATMP in the injured tissue, Sackstein et al. (113), developed a method to transiently modify the CD44 antigen present in the MSC cell membrane by enzymatic fucosylation, converting this molecule into HCELL glycoform and thus favoring the migration of MSCs to the inflamed tissues (111,

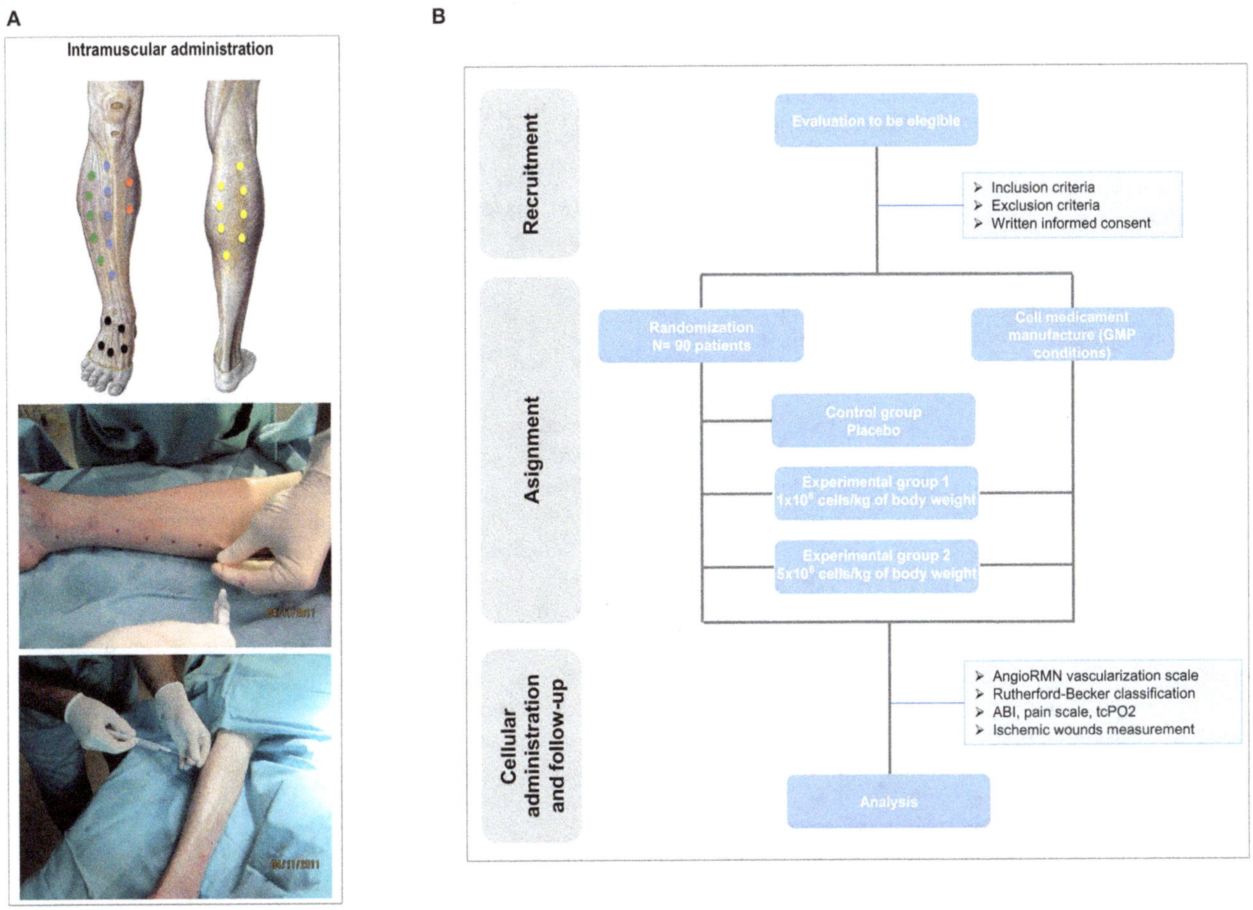

FIGURE 5 | NOMA-Project considerations. **(A)** Intramuscular infusion of stem cells directly into the muscles. Once the targeted muscular injection points of MSCs are selected (upper panel), repeated administrations of the therapeutic stem cells are directly, easily, and less invasively infused into the lower part of the limb (lower panel). **(B)** Flowchart of the phases (enrollment, intervention allocation, follow-up, and data analysis) of NOMA clinical trial.

115, 116). This method, called glycosyltransferase-programmed stereosubstitution (GPS) to custom modify cell surface glycans without affecting cell viability, has been optimized for its clinical application using an alpha-1,3-fucosyltransferase preparation and enzymatic conditions specifically designed to treat live cells and formulated to preserve the cell viability and phenotype. It has been found that this modification not only increases the adhesion of the MSCs to the endothelium but also enhances their transmigration through it by activating integrin α4β1 (VLA-4) in the absence of chemokine stimulation. Therefore, this modification by fucosylation could improve the efficacy of the treatment with MSCs by increasing their migratory capacity to the inflamed tissues after systemic administration. More detailed knowledge of the mechanisms of biodistribution, migration and specific interaction of MSCs at the damaged loci may be beneficial to design new ATMP with increased safety and efficacy (111), and in this regard, our group has collaborated in a patent to classify these tentative responses for stem cells (WO2017032612).

In this context, and as another approach, a new generation of MSCs can be engineered by increasing both cell migration and cell potency (Generation 3 MSCs). Targeting the CXCL12 and C-X-C chemokine receptor type 4 (CXCR-4) may improve the cell migration capacity of transplanted MSCs, and CXCL12 is also highly expressed in injured tissues and contributes to the recruitment of CXR4-positive cells. As a small proportion of MSCs express CXCR4 in culture, their capacity to migrate, and to respond to homing signals in damaged tissue may be reduced. Therefore, targeting CXCR4 may improve the migratory and therapeutic effects of MSCs (117). Within this context, we propose using MSCs modified to overexpress CXR4 and IL10 and/or IL7 (Generation 3 MSCs). Expression of the CXCR4 receptor will increase the migration of MSCs toward the inflammatory focus, while coexpression of the anti-inflammatory cytokine interleukin 10 (IL-10) and/or the anti-infectious cytokine interleukin 7 (IL-7) will increase the anti-inflammatory effect (IL-10) and even the anti-infective effect (IL-7).

Furthermore, we propose that the extensive use of FBS in MSC expansion media represents a clear limitation for the introduction of an ATMP at the clinical level. Currently, cell expansion is carried out in culture media supplemented with FBS. Serum use must be of a clinical grade (free of animal pathogens). Together with the growing demand for MSCs, this feature has led to a series of technical and ethical conditions for production (use of a large number of bovine fetuses) and geographic regions (zones free of prion diseases) with an associated impact on price (118–124). The substitution of FBS with human serum and platelet lysate also represents a technical limitation that is mainly related to the supply of human material and the absence of uniformity of the lots. Altogether, these considerations have enforced the development of robust processes of MSC production in chemically defined culture media free of animal and human components. These media are supplemented with recombinant proteins (albumin, insulin, TFGβ and bFGF), iron, selenium, and an antioxidant system (2-mercaptoethanol) (119, 122). The use of serum-free and xeno-free media minimizes the possible risks of contamination and adverse effects with respect to clinical application. Although several serum-free media have been described in the literature, they are chemically defined and contain known molecular components and are commercially available for cell manufacturing. We propose using our patented xeno-free and human component-free medium (WO2017021535) to expand modified MSCs.

COST OF THE PROCESS

After the introduction of CAR-T cell therapies with an actual cost of ~300,000 to 400,000 € (125) or the prices charged by PROCHYMAL (an allogeneic bone marrow-derived allogeneic MSC treatment for graft vs. host disease) or Provenge (an autologous cell therapy of dendritic cells from metastatic forms of prostate cancer), with prices between $100,000 and $200,000 US, it seems absolutely necessary to analyze the cost-effectiveness of a potential treatment to facilitate the universal coverage of healthcare. A recent survey from the International Society for Stem Cell Therapy estimates costs for a dose between 10,000 and 25,000 € (126), with additional costs from hospitalization and the endovascular department, among others, resulting in a total cost of 30,000 to 40,000 € for a single dose. This cost may be assumed for rare diseases with a low prevalence, but it seems quite difficult to extend this treatment to a highly prevalent medical condition. The only way to reduce the cost is the mass production of allogeneic doses and facilitation of administration. Intramuscular administration of allogeneic MSCs will reduce the total cost and may be as effective as the intraarterial route. The NOMA (No More Amputations) Project is aimed to address these points (**Figure 5A**). Previous cell therapy clinical trials conducted by our group have used autologous cells (BM-MNCs, BM-EPCs, and Ad-MSCs) injected intraarterially (NCT00872326, NCT01257776, NCT01745744, and NCT02287974). Given the reported adverse events, such as microthrombosis (19), clot formation (96), or IBMIR (82, 87, 97), the high cost of the treatment of complications and our preliminary data suggesting that allogeneic MSCs administered intramuscularly may be as safe and effective as intraarterial autologous MSCs, we decided to promote the NOMA Project. A description of the Project (aims, endpoint, inclusion, and exclusion criteria, follow-up, indicators and monitorization) will be published elsewhere (**Figure 5B**). The development of a safe and new generation of MSCs expanded under GMP conditions with media free of FBS and human platelet lysate (HPL) is currently under development.

We have now treated a huge number of patients with cell therapies, and the insights that we are gaining concerning the optimization of the next-generation of cell-based therapies should not be underestimated. We propose a combination among the following factors:

- A new source of healthy allogeneic donors (NOMA Project)
- A cost-effective mass production under GMP conditions (to be developed)
- A safe, friendly and less costly procedure for administration, for example, via the intramuscular route (NOMA Project)
- A new xeno-free culture medium (to be developed)

INSTITUTIONAL REVIEW BOARD AND REGULATORY COMPLIANCE

Clinical Trials

The clinical trials developed and conducted by our group were promoted by "*Fundación Progreso y Salud*" under a non-commercial investigator-driven grant, were approved by the local institutional review board (IRB), the regional Andalusian Ethics Committee, and registered in the Clinical Trials Database (www.clinicaltrials.gov), except for NOMA Clinical Trial (No-More-Amputations (NOMA); PIC18/00010; Clinical Trials of Independent Advanced Therapies).

Regulatory and ATMP Manufacturing

For the clinical trials mentioned above, the donors' source for ATMP was appropriately screened and tested for human pathogens. The therapeutic protocols were approved by the hospital ethics committee in accordance with Spanish law. All patients signed a detailed informed consent form before intervention and provided their consent for publication of the study results. These studies were conducted in accordance with the ethical standards of the Helsinki Declaration (1975). The clinical studies (NCT01257776, NCT01745744, and NCT02287974) fulfill GMP standards for application in a clinical setting. All procedures were performed in a CABIMER-GMP core facility, a certified clean room (Certificate No: ES/101I/18; inspected in accordance with Directive 2001/20/EC) that complies with the principles and guidelines of Good Manufacturing Practice laid down in Directive 2003/94/EC.

CONCLUDING REMARKS

Considering all the previously findings and despite promising results, the ATMP-based therapy applied to CLI has led to important questions regarding safety and efficacy that can be transferred to its application for other pathologies. Although MSCs display a series of properties (factor release, immunomodulation, inflammatory capacity, among others), we presently do not know how many clinical trials are necessary before a specific, safe and effective cell-based therapy can be successfully achieved to offer patients with no other therapeutic alternatives. Improvements in personalized cell production in a cost-effective, safe and effective manner, a correct diagnosis, clinical prognosis and well-defined patient profile, and the correct route of administration will undoubtedly improve this advanced cellular medicine therapy.

AUTHOR CONTRIBUTIONS

BS conceived the concept of the paper. BS-J, NE, VC-G, VJ, MG-A, DG-O, AH, and BS wrote the first draft that was circulated and all the authors contributed with different sections. All the authors (BS-J, NE, VC-G, YA, LL, JT, FB, VJ, AD, RR-S, EA, LG, FP, FS-G, FL, MM, LD, GC, JM, RS, MG-A, DG-O, FM, AH, and BS) contributed to the acquisition, analysis, interpretation of data for the work, revising it critically for important intellectual content, final approval of the version to be published, and agree to be accountable for all aspects of the work in ensuring that questions related to the accuracy or integrity of any part of the work are appropriately investigated and resolved. AH edited and submitted the final version of the manuscript.

FUNDING

The authors are supported by the Fundación Progreso y Salud, Consejería de Salud, Junta de Andalucía; FEDER co-funded grants from Instituto de Salud Carlos III (Red TerCel-Grant RD16/0011/0034 and RD16/0011/0013; PIC18/00010; PI16/00259, PI17/02104, and PI-0272-2017), Fundación Andaluza de I+D and Al-Andalus Biopharma Project (FAID-2018-1). VC-G is a recipient of a Sara Borrell contract (CD16/00118) from FEDER co-funded grants from Instituto de Salud Carlos III. The authors strongly appreciate the Non-Commercial Investigator Driven Research launched by the Ministry of Health and Consumer Affairs which allowed Spain to be at the front of Cell Therapy studies. Diabetes and Associated Metabolic Diseases Networking Biomedical Research Center (CIBERDEM) is an initiative of the Institute of Health Carlos III.

ACKNOWLEDGMENTS

Authors especially thank Nuria Mellado Damas Sanz, María Gálvez Jiménez, and Victoria Eugenia Jiménez Santos. The NOMA Project (PIC18/00010). Hospital Universitario Fundación Jiménez Díaz, Madrid: DG-O, MG-A, LL, César Aparicio and BS-J; Hospital Universitario La Paz, Madrid: Luis Riera del Moral; Clínica Universidad de Navarra, Pamplona: FP, EA, and LG; Hospital Clínico Universitario de Salamanca, Salamanca: FS-G and FL; Hospital Victoria Eugenia de la Cruz Roja, Sevilla: AD; Hospital Universitario Virgen Macarena, Sevilla: RR-S; Hospital Universitario La Fe, Valencia: MM; Hospital Clínico de Santiago de Compostela, A Coruña: Felipe Casanueva; Hospital Universitario Rio-Hortega, Valladolid: LD, Hospital Virgen de Arrixaca, Murcia: JM and GC; CABIMER, Sevilla: NE, VC-G, YA, JT, FB, FM, AH, and BS.

REFERENCES

1. Hmadcha A, Dominguez-Bendala J, Wakeman J, Arredouani M, Soria B. The immune boundaries for stem cell based therapies: problems and prospective solutions. *J Cell Mol Med.* (2009) 13:1464–75. doi: 10.1111/j.1582-4934.2009.00837.x
2. Dominici M, Le Blanc K, Mueller I, Slaper-Cortenbach I, Marini F, Krause D, et al. Minimal criteria for defining multipotent mesenchymal stromal cells. The International society for cellular therapy position statement. *Cytotherapy.* (2006) 8:315–7. doi: 10.1080/14653240600855905
3. Horwitz EM, Le Blanc K, Dominici M, Mueller I, Slaper-Cortenbach I, Marini FC, et al. Clarification of the nomenclature for MSC: the international society for cellular therapy position statement. *Cytotherapy.* (2005) 7:393–5. doi: 10.1080/14653240500319234
4. Strioga M, Viswanathan S, Darinskas A, Slaby O, Michalek J. Same or not the same? Comparison of adipose tissue-derived versus bone

marrow-derived mesenchymal stem and stromal cells. *Stem Cells Dev.* (2012) 21:2724–52. doi: 10.1089/scd.2011.0722
5. Mendez-Ferrer S, Michurina TV, Ferraro F, Mazloom AR, Macarthur BD, Lira SA, et al. Mesenchymal and haematopoietic stem cells form a unique bone marrow niche. *Nature.* (2010) 466:829–34. doi: 10.1038/nature09262
6. Hoogduijn MJ, Popp F, Verbeek R, Masoodi M, Nicolaou A, Baan C, et al. The immunomodulatory properties of mesenchymal stem cells and their use for immunotherapy. *Int Immunopharmacol.* (2010) 10:1496–500. doi: 10.1016/j.intimp.2010.06.019
7. Meirelles Lda S, Fontes AM, Covas DT, Caplan AI. Mechanisms involved in the therapeutic properties of mesenchymal stem cells. *Cytokine Growth Factor Rev.* (2009) 20:419–27. doi: 10.1016/j.cytogfr.2009.10.002
8. Escacena N, Quesada-Hernandez E, Capilla-Gonzalez V, Soria B, Hmadcha A. Bottlenecks in the efficient use of advanced therapy medicinal products based on mesenchymal stromal cells. *Stem Cells Int.* (2015) 2015:895714. doi: 10.1155/2015/895714
9. Le Blanc K. Immunomodulatory effects of fetal and adult mesenchymal stem cells. *Cytotherapy.* (2003) 5:485–9. doi: 10.1080/14653240310003611
10. Conget PA, Minguell JJ. Phenotypical and functional properties of human bone marrow mesenchymal progenitor cells. *J Cell Physiol.* (1999) 181:67–73. doi: 10.1002/(SICI)1097-4652(199910)181:1<67::AID-JCP7>3.3.CO;2-3
11. Cagliani J, Grande D, Molmenti EP, Miller EJ, Rilo HLR. Immunomodulation by mesenchymal stromal cells and their clinical applications. *J Stem Cell Regen Biol.* (2017) 3. doi: 10.15436/2471-0598.17.022
12. Duffy GP, Ahsan T, O'Brien T, Barry F, Nerem RM. Bone marrow-derived mesenchymal stem cells promote angiogenic processes in a time- and dose-dependent manner *in vitro*. *Tissue Eng Part A.* (2009) 15:2459–70. doi: 10.1089/ten.tea.2008.0341
13. Burlacu A, Grigorescu G, Rosca AM, Preda MB, Simionescu M. Factors secreted by mesenchymal stem cells and endothelial progenitor cells have complementary effects on angiogenesis *in vitro*. *Stem Cells Dev.* (2013) 22:643–53. doi: 10.1089/scd.2012.0273
14. Shrestha C, Zhao L, Chen K, He H, Mo Z. Enhanced healing of diabetic wounds by subcutaneous administration of human umbilical cord derived stem cells and their conditioned media. *Int J Endocrinol.* (2013) 2013:592454. doi: 10.1155/2013/592454
15. Potapova IA, Gaudette GR, Brink PR, Robinson RB, Rosen MR, Cohen IS, et al. Mesenchymal stem cells support migration, extracellular matrix invasion, proliferation, and survival of endothelial cells *in vitro*. *Stem Cells.* (2007) 25:1761–8. doi: 10.1634/stemcells.2007-0022
16. Aguirre A, Planell JA, Engel E. Dynamics of bone marrow-derived endothelial progenitor cell/mesenchymal stem cell interaction in co-culture and its implications in angiogenesis. *Biochem Biophys Res Commun.* (2010) 400:284–91. doi: 10.1016/j.bbrc.2010.08.073
17. Hong IS, Lee HY, Kang KS. Mesenchymal stem cells and cancer: friends or enemies? *Mutat Res.* (2014) 768:98–106. doi: 10.1016/j.mrfmmm.2014.01.006
18. Lee HY, Hong IS. Double-edged sword of mesenchymal stem cells: cancer-promoting versus therapeutic potential. *Cancer Sci.* (2017) 108:1939–46. doi: 10.1111/cas.13334
19. Acosta L, Hmadcha A, Escacena N, Perez-Camacho I, de la Cuesta A, Ruiz-Salmeron R, et al. Adipose mesenchymal stromal cells isolated from type 2 diabetic patients display reduced fibrinolytic activity. *Diabetes.* (2013) 62:4266–9. doi: 10.2337/db13-0896
20. Mamidi MK, Pal R, Dey S, Bin Abdullah BJ, Zakaria Z, Rao MS, et al. Cell therapy in critical limb ischemia: current developments and future progress. *Cytotherapy.* (2012) 14:902–16. doi: 10.3109/14653249.2012.693156
21. Fowkes FG, Rudan D, Rudan I, Aboyans V, Denenberg JO, McDermott MM, et al. Comparison of global estimates of prevalence and risk factors for peripheral artery disease in 2000 and 2010: a systematic review and analysis. *Lancet.* (2013) 382:1329–40. doi: 10.1016/S0140-6736(13)61249-0
22. Norgren L, Hiatt WR, Dormandy JA, Nehler MR, Harris KA, Fowkes FG, et al. Inter-society consensus for the management of peripheral arterial disease (TASC II). *J Vasc Surg.* (2007) 45 (Suppl. S):S5–67. doi: 10.1016/j.jvs.2006.12.037
23. Norman PE, Davis WA, Bruce DG, Davis TM. Peripheral arterial disease and risk of cardiac death in type 2 diabetes: the Fremantle Diabetes Study. *Diabetes Care.* (2006) 29:575–80. doi: 10.2337/diacare.29.03.06.dc05-1567
24. Hirsch AT. Critical limb ischemia and stem cell research: anchoring hope with informed adverse event reporting. *Circulation.* (2006) 114:2581–3. doi: 10.1161/CIRCULATIONAHA.106.666719
25. Ruiz-Salmeron R, de la Cuesta-Diaz A, Constantino-Bermejo M, Perez-Camacho I, Marcos-Sanchez F, Hmadcha A, et al. Angiographic demonstration of neoangiogenesis after intra-arterial infusion of autologous bone marrow mononuclear cells in diabetic patients with critical limb ischemia. *Cell Transplant.* (2011) 20:1629–39. doi: 10.3727/096368910X0177
26. American Diabetes A. Peripheral arterial disease in people with diabetes. *Diabetes Care.* (2003) 26:3333–41. doi: 10.2337/diacare.26.12.3333
27. Black JH III, LaMuraglia GM, Kwolek CJ, Brewster DC, Watkins MT, Cambria RP. Contemporary results of angioplasty-based infrainguinal percutaneous interventions. *J Vasc Surg.* (2005) 42:932–9. doi: 10.1016/j.jvs.2005.06.024
28. Kreitner KF, Kalden P, Neufang A, Duber C, Krummenauer F, Kustner E, et al. Diabetes and peripheral arterial occlusive disease: prospective comparison of contrast-enhanced three-dimensional MR angiography with conventional digital subtraction angiography. *AJR Am J Roentgenol.* (2000) 174:171–9. doi: 10.2214/ajr.174.1.1740171
29. Faglia E, Favales F, Quarantiello A, Calia P, Clelia P, Brambilla G, et al. Angiographic evaluation of peripheral arterial occlusive disease and its role as a prognostic determinant for major amputation in diabetic subjects with foot ulcers. *Diabetes Care.* (1998) 21:625–30. doi: 10.2337/diacare.21.4.625
30. Pries AR, Reglin B, Secomb TW. Remodeling of blood vessels: responses of diameter and wall thickness to hemodynamic and metabolic stimuli. *Hypertension.* (2005) 46:725–31. doi: 10.1161/01.HYP.0000184428.16429.be
31. Qadura M, Terenzi DC, Verma S, Al-Omran M, Hess DA. Concise review: cell therapy for critical limb ischemia: an integrated review of preclinical and clinical studies. *Stem Cells.* (2018) 36:161–71. doi: 10.1002/stem.2751
32. Steg PG, Bhatt DL, Wilson PW, D'Agostino R Sr, Ohman EM, Rother J, et al. One-year cardiovascular event rates in outpatients with atherothrombosis. *JAMA.* (2007) 297:1197–206. doi: 10.1001/jama.297.11.1197
33. Martin A, Komada MR, Sane DC. Abnormal angiogenesis in diabetes mellitus. *Med Res Rev.* (2003) 23:117–45. doi: 10.1002/med.10024
34. Keymel S, Heinen Y, Balzer J, Rassaf T, Kelm M, Lauer T, et al. Characterization of macro-and microvascular function and structure in patients with type 2 diabetes mellitus. *Am J Cardiovasc Dis.* (2011) 1:68–75.
35. Baynes JW. Role of oxidative stress in development of complications in diabetes. *Diabetes.* (1991) 40:405–12. doi: 10.2337/diabetes.40.4.405
36. Fadini GP, Sartore S, Albiero M, Baesso I, Murphy E, Menegolo M, et al. Number and function of endothelial progenitor cells as a marker of severity for diabetic vasculopathy. *Arterioscler Thromb Vasc Biol.* (2006) 26:2140–6. doi: 10.1161/01.ATV.0000237750.44469.88
37. Schatteman GC, Hanlon HD, Jiao C, Dodds SG, Christy BA. Blood-derived angioblasts accelerate blood-flow restoration in diabetic mice. *J Clin Invest.* (2000) 106:571–8. doi: 10.1172/JCI9087
38. Tepper OM, Galiano RD, Capla JM, Kalka C, Gagne PJ, Jacobowitz GR, et al. Human endothelial progenitor cells from type II diabetics exhibit impaired proliferation, adhesion, and incorporation into vascular structures. *Circulation.* (2002) 106:2781–6. doi: 10.1161/01.CIR.0000039526.42991.93
39. Coats P, Wadsworth R. Marriage of resistance and conduit arteries breeds critical limb ischemia. *Am J Physiol Heart Circ Physiol.* (2005) 288:H1044–50. doi: 10.1152/ajpheart.00773.2004
40. Teraa M, Sprengers RW, Schutgens RE, Slaper-Cortenbach IC, van der Graaf Y, Algra A, et al. Effect of repetitive intra-arterial infusion of bone marrow mononuclear cells in patients with no-option limb ischemia: the randomized, double-blind, placebo-controlled Rejuvenating Endothelial Progenitor Cells via Transcutaneous Intra-arterial Supplementation (JUVENTAS) trial. *Circulation.* (2015) 131:851–60. doi: 10.1161/CIRCULATIONAHA.114.012913
41. Walter DH, Krankenberg H, Balzer JO, Kalka C, Baumgartner I, Schluter M, et al. Intraarterial administration of bone marrow mononuclear cells in patients with critical limb ischemia: a randomized-start, placebo-controlled pilot trial (PROVASA). *Circ Cardiovasc Interv.* (2011) 4:26–37. doi: 10.1161/CIRCINTERVENTIONS.110.958348

42. Molavi B, Zafarghandi MR, Aminizadeh E, Hosseini SE, Mirzayi H, Arab L, et al. Safety and efficacy of repeated bone marrow mononuclear cell therapy in patients with critical limb ischemia in a pilot randomized controlled trial. *Arch Iran Med.* (2016) 19:388–96. doi: 10.0161906/AIM.004
43. Kawamoto A, Katayama M, Handa N, Kinoshita M, Takano H, Horii M, et al. Intramuscular transplantation of G-CSF-mobilized CD34(+) cells in patients with critical limb ischemia: a phase I/IIa, multicenter, single-blinded, dose-escalation clinical trial. *Stem Cells.* (2009) 27:2857–64. doi: 10.1002/stem.207
44. Gupta PK, Chullikana A, Parakh R, Desai S, Das A, Gottipamula S, et al. A double blind randomized placebo controlled phase I/II study assessing the safety and efficacy of allogeneic bone marrow derived mesenchymal stem cell in critical limb ischemia. *J Transl Med.* (2013) 11:143. doi: 10.1186/1479-5876-11-143
45. Arici V, Perotti C, Fabrizio C, Del Fante C, Ragni F, Alessandrino F, et al. Autologous immuno magnetically selected CD133+ stem cells in the treatment of no-option critical limb ischemia: clinical and contrast enhanced ultrasound assessed results in eight patients. *J Transl Med.* (2015) 13:342. doi: 10.1186/s12967-015-0697-4
46. Kirana S, Stratmann B, Prante C, Prohaska W, Koerperich H, Lammers D, et al. Autologous stem cell therapy in the treatment of limb ischaemia induced chronic tissue ulcers of diabetic foot patients. *Int J Clin Pract.* (2012) 66:384–93. doi: 10.1111/j.1742-1241.2011.02886.x
47. Perin EC, Silva G, Gahremanpour A, Canales J, Zheng Y, Cabreira-Hansen MG, et al. A randomized, controlled study of autologous therapy with bone marrow-derived aldehyde dehydrogenase bright cells in patients with critical limb ischemia. *Catheter Cardiovasc Interv.* (2011) 78:1060–7. doi: 10.1002/ccd.23066
48. Mutirangura P, Ruangsetakit C, Wongwanit C, Chinsakchai K, Porat Y, Belleli A, et al. Enhancing limb salvage by non-mobilized peripheral blood angiogenic cell precursors therapy in patients with critical limb ischemia. *J Med Assoc Thai.* (2009) 92:320–7.
49. Powell RJ, Marston WA, Berceli SA, Guzman R, Henry TD, Longcore AT, et al. Cellular therapy with Ixmyelocel-T to treat critical limb ischemia: the randomized, double-blind, placebo-controlled RESTORE-CLI trial. *Mol Ther.* (2012) 20:1280–6. doi: 10.1038/mt.2012.52
50. Capiod JC, Tournois C, Vitry F, Sevestre MA, Daliphard S, Reix T, et al. Characterization and comparison of bone marrow and peripheral blood mononuclear cells used for cellular therapy in critical leg ischaemia: towards a new cellular product. *Vox Sang.* (2009) 96:256–65. doi: 10.1111/j.1423-0410.2008.01138.x
51. Lasala GP, Silva JA, Minguell JJ. Therapeutic angiogenesis in patients with severe limb ischemia by transplantation of a combination stem cell product. *J Thorac Cardiovasc Surg.* (2012) 144:377–82. doi: 10.1016/j.jtcvs.2011.08.053
52. Tateishi-Yuyama E, Matsubara H, Murohara T, Ikeda U, Shintani S, Masaki H, et al. Therapeutic angiogenesis for patients with limb ischaemia by autologous transplantation of bone-marrow cells: a pilot study and a randomised controlled trial. *Lancet.* (2002) 360:427–35. doi: 10.1016/S0140-6736(02)09670-8
53. Cooke JP, Losordo DW. Modulating the vascular response to limb ischemia: angiogenic and cell therapies. *Circ Res.* (2015) 116:1561–78. doi: 10.1161/CIRCRESAHA.115.303565
54. Asahara T, Murohara T, Sullivan A, Silver M, van der Zee R, Li T, et al. Isolation of putative progenitor endothelial cells for angiogenesis. *Science.* (1997) 275:964–7. doi: 10.1126/science.275.5302.964
55. Fadini GP, Losordo D, Dimmeler S. Critical reevaluation of endothelial progenitor cell phenotypes for therapeutic and diagnostic use. *Circ Res.* (2012) 110:624–37. doi: 10.1161/CIRCRESAHA.111.243386
56. Isner JM, Asahara T. Angiogenesis and vasculogenesis as therapeutic strategies for postnatal neovascularization. *J Clin Invest.* (1999) 103:1231–6. doi: 10.1172/JCI6889
57. Saigawa T, Kato K, Ozawa T, Toba K, Makiyama Y, Minagawa S, et al. Clinical application of bone marrow implantation in patients with arteriosclerosis obliterans, and the association between efficacy and the number of implanted bone marrow cells. *Circ J.* (2004) 68:1189–93. doi: 10.1253/circj.68.1189
58. Miyamoto K, Nishigami K, Nagaya N, Akutsu K, Chiku M, Kamei M, et al. Unblinded pilot study of autologous transplantation of bone marrow mononuclear cells in patients with thromboangiitis obliterans. *Circulation.* (2006) 114:2679–84. doi: 10.1161/CIRCULATIONAHA.106.644203
59. Gremmels H, Fledderus JO, Teraa M, Verhaar MC. Mesenchymal stromal cells for the treatment of critical limb ischemia: context and perspective. *Stem Cell Res Ther.* (2013) 4:140. doi: 10.1186/scrt351
60. Teraa M, Sprengers RW, van der Graaf Y, Peters CE, Moll FL, Verhaar MC. Autologous bone marrow-derived cell therapy in patients with critical limb ischemia: a meta-analysis of randomized controlled clinical trials. *Ann Surg.* (2013) 258:922–9. doi: 10.1097/SLA.0b013e3182854cf1
61. Rigato M, Monami M, Fadini GP. Autologous cell therapy for peripheral arterial disease: systematic review and meta-analysis of randomized, nonrandomized, and noncontrolled studies. *Circ Res.* (2017) 120:1326–40. doi: 10.1161/CIRCRESAHA.116.309045
62. Bartsch T, Brehm M, Zeus T, Kogler G, Wernet P, Strauer BE. Transplantation of autologous mononuclear bone marrow stem cells in patients with peripheral arterial disease (the TAM-PAD study). *Clin Res Cardiol.* (2007) 96:891–9. doi: 10.1007/s00392-007-0569-x
63. Gupta NK, Armstrong EJ, Parikh SA. The current state of stem cell therapy for peripheral artery disease. *Curr Cardiol Rep.* (2014) 16:447. doi: 10.1007/s11886-013-0447-2
64. Amann B, Luedemann C, Ratei R, Schmidt-Lucke JA. Autologous bone marrow cell transplantation increases leg perfusion and reduces amputations in patients with advanced critical limb ischemia due to peripheral artery disease. *Cell Transplant.* (2009) 18:371–80. doi: 10.3727/096368909788534942
65. Klepanec A, Mistrik M, Altaner C, Valachovicova M, Olejarova I, Slysko R, et al. No difference in intra-arterial and intramuscular delivery of autologous bone marrow cells in patients with advanced critical limb ischemia. *Cell Transplant.* (2012) 21:1909–18. doi: 10.3727/096368912X636948
66. Alvarez Garcia J, Garcia Gomez-Heras S, Riera Del Moral L, Largo C, Garcia-Olmo D, Garcia-Arranz M. The effects of allogenic stem cells in a murine model of hind limb diabetic ischemic tissue. *PeerJ.* (2017) 5:e3664. doi: 10.7717/peerj.3664
67. Riera MLS, A. A.; Stefanov, K. S.; Tong, H.; Riera, C. L.; García-Olmo, D.; García-Arranz, M. Phase Ib Open Clinical Trial to Assess the Safety of Autologous Mesenchymal Stem Cells for the Treatment of Nonrevascularizable Critical Lower Limb Ischemia. *J Stem Cell Res Ther.* (2017) 7:391.
68. Huang P, Li S, Han M, Xiao Z, Yang R, Han ZC. Autologous transplantation of granulocyte colony-stimulating factor-mobilized peripheral blood mononuclear cells improves critical limb ischemia in diabetes. *Diabetes Care.* (2005) 28:2155–60. doi: 10.2337/diacare.28.9.2155
69. Lu D, Chen B, Liang Z, Deng W, Jiang Y, Li S, et al. Comparison of bone marrow mesenchymal stem cells with bone marrow-derived mononuclear cells for treatment of diabetic critical limb ischemia and foot ulcer: a double-blind, randomized, controlled trial. *Diabetes Res Clin Pract.* (2011) 92:26–36. doi: 10.1016/j.diabres.2010.12.010
70. Rehman J, Traktuev D, Li J, Merfeld-Clauss S, Temm-Grove CJ, Bovenkerk JE, et al. Secretion of angiogenic and antiapoptotic factors by human adipose stromal cells. *Circulation.* (2004) 109:1292–8. doi: 10.1161/01.CIR.0000121425.42966.F1
71. Bura A, Planat-Benard V, Bourin P, Silvestre JS, Gross F, Grolleau JL, et al. Phase I trial: the use of autologous cultured adipose-derived stroma/stem cells to treat patients with non-revascularizable critical limb ischemia. *Cytotherapy.* (2014) 16:245–57. doi: 10.1016/j.jcyt.2013.11.011
72. Das AK, Bin Abdullah BJ, Dhillon SS, Vijanari A, Anoop CH, Gupta PK. Intra-arterial allogeneic mesenchymal stem cells for critical limb ischemia are safe and efficacious: report of a phase I study. *World J Surg.* (2013) 37:915–22. doi: 10.1007/s00268-012-1892-6
73. Gupta PK, Krishna M, Chullikana A, Desai S, Murugesan R, Dutta S, et al. Administration of adult human bone marrow-derived, cultured, pooled, allogeneic mesenchymal stromal cells in critical limb ischemia due to buerger's disease: phase II study report suggests clinical efficacy. *Stem Cells Transl Med.* (2017) 6:689–99. doi: 10.1016/j.jvs.2017.08.009
74. Lee HC, An SG, Lee HW, Park JS, Cha KS, Hong TJ, et al. Safety and effect of adipose tissue-derived stem cell implantation in patients with critical limb ischemia: a pilot study. *Circ J.* (2012) 76:1750–60.

75. Yang SS, Kim NR, Park KB, Do YS, Roh K, Kang KS, et al. A phase I study of human cord blood-derived mesenchymal stem cell therapy in patients with peripheral arterial occlusive disease. *Int J Stem Cells.* (2013) 6:37–44. doi: 10.15283/ijsc.2013.6.1.37
76. Capilla-Gonzalez V, Lopez-Beas J, Escacena N, Aguilera Y, de la Cuesta A, Ruiz-Salmeron R, et al. PDGF restores the defective phenotype of adipose-derived mesenchymal stromal cells from diabetic patients. *Mol Ther.* (2018) 26:2696–709. doi: 10.1016/j.ymthe.2018.08.011
77. Ferrer-Lorente R, Bejar MT, Tous M, Vilahur G, Badimon L. Systems biology approach to identify alterations in the stem cell reservoir of subcutaneous adipose tissue in a rat model of diabetes: effects on differentiation potential and function. *Diabetologia.* (2014) 57:246–56. doi: 10.1007/s00125-013-3081-z
78. Minteer DM, Young MT, Lin YC, Over PJ, Rubin JP, Gerlach JC, et al. Analysis of type II diabetes mellitus adipose-derived stem cells for tissue engineering applications. *J Tissue Eng.* (2015) 6:2041731415579215. doi: 10.1177/2041731415579215
79. Rennert RC, Sorkin M, Januszyk M, Duscher D, Kosaraju R, Chung MT, et al. Diabetes impairs the angiogenic potential of adipose-derived stem cells by selectively depleting cellular subpopulations. *Stem Cell Res Ther.* (2014) 5:79. doi: 10.1186/scrt468
80. Faiella W, Atoui R. Therapeutic use of stem cells for cardiovascular disease. *Clin Transl Med.* (2016) 5:34. doi: 10.1186/s40169-016-0116-3
81. Lalu MM, McIntyre L, Pugliese C, Fergusson D, Winston BW, Marshall JC, et al. Safety of cell therapy with mesenchymal stromal cells (SafeCell): a systematic review and meta-analysis of clinical trials. *PLoS ONE.* (2012) 7:e47559. doi: 10.1371/journal.pone.0047559
82. Moll G, Jitschin R, von Bahr L, Rasmusson-Duprez I, Sundberg B, Lonnies L, et al. Mesenchymal stromal cells engage complement and complement receptor bearing innate effector cells to modulate immune responses. *PLoS ONE.* (2011) 6:e21703. doi: 10.1371/journal.pone.0021703
83. Liu Y, Tang SC. Recent progress in stem cell therapy for diabetic nephropathy. *Kidney Dis.* (2016) 2:20–7. doi: 10.1159/000441913
84. Hsuan YC, Lin CH, Chang CP, Lin MT. Mesenchymal stem cell-based treatments for stroke, neural trauma, and heat stroke. *Brain Behav.* (2016) 6:e00526. doi: 10.1002/brb3.526
85. Soria B, Martin-Montalvo A, Aguilera Y, Mellado-Damas N, Lopez-Beas J, Herrera-Herrera I, et al. Human mesenchymal stem cells prevent neurological complications of radiotherapy. *Front Cell Neurosci.* (2019) 13:204. doi: 10.3389/fncel.2019.00204
86. Escacena N. Cellular medication as a therapeutic alternative in chronic critical limb ischemia in diabetic patients without the possibility of revascularization. Dissertation Thesis. Sevilla Spain: University of Sevilla. (2016).
87. Moll G, Hult A, von Bahr L, Alm JJ, Heldring N, Hamad OA, et al. Do ABO blood group antigens hamper the therapeutic efficacy of mesenchymal stromal cells? *PLoS ONE.* (2014) 9:e85040. doi: 10.1371/journal.pone.0085040
88. Agis H, Kandler B, Fischer MB, Watzek G, Gruber R. Activated platelets increase fibrinolysis of mesenchymal progenitor cells. *J Orthop Res.* (2009) 27:972–80. doi: 10.1002/jor.20819
89. Hashi CK, Zhu Y, Yang GY, Young WL, Hsiao BS, Wang K, et al. Antithrombogenic property of bone marrow mesenchymal stem cells in nanofibrous vascular grafts. *Proc Natl Acad Sci USA.* (2007) 104:11915–20. doi: 10.1073/pnas.0704581104
90. Nakagami H, Maeda K, Morishita R, Iguchi S, Nishikawa T, Takami Y, et al. Novel autologous cell therapy in ischemic limb disease through growth factor secretion by cultured adipose tissue-derived stromal cells. *Arterioscler Thromb Vasc Biol.* (2005) 25:2542–7. doi: 10.1161/01.ATV.0000190701.92007.6d
91. Neuss S, Schneider RK, Tietze L, Knuchel R, Jahnen-Dechent W. Secretion of fibrinolytic enzymes facilitates human mesenchymal stem cell invasion into fibrin clots. *Cells Tissues Organs.* (2010) 191:36–46. doi: 10.1159/000215579
92. Prockop DJ, Oh JY. Mesenchymal stem/stromal cells (MSCs): role as guardians of inflammation. *Mol Ther.* (2012) 20:14–20. doi: 10.1038/mt.2011.211
93. Boland L, Burand AJ, Brown AJ, Boyt D, Lira VA, Ankrum JA. IFN-gamma and TNF-alpha pre-licensing protects mesenchymal stromal cells from the pro-inflammatory effects of palmitate. *Mol Ther.* (2018) 26:860–73. doi: 10.1016/j.ymthe.2017.12.013
94. Davies LC, Alm JJ, Heldring N, Moll G, Gavin C, Batsis I, et al. Type 1 diabetes mellitus donor mesenchymal stromal cells exhibit comparable potency to healthy controls *in vitro*. *Stem Cells Transl Med.* (2016) 5:1485–95. doi: 10.5966/sctm.2015-0272
95. Kim H, Han JW, Lee JY, Choi YJ, Sohn YD, Song M, et al. Diabetic mesenchymal stem cells are ineffective for improving limb ischemia due to their impaired angiogenic capability. *Cell Transplant.* (2015) 24:1571–84. doi: 10.3727/096368914X682792
96. George MJ, Prabhakara K, Toledano-Furman NE, Wang YW, Gill BS, Wade CE, et al. Clinical cellular therapeutics accelerate clot formation. *Stem Cells Transl Med.* (2018) 7:731–9. doi: 10.1002/sctm.18-0015
97. Moll G, Ankrum JA, Kamhieh-Milz J, Bieback K, Ringden O, Volk HD, et al. Intravascular mesenchymal stromal/stem cell therapy product diversification: time for new clinical guidelines. *Trends Mol Med.* (2019) 25:149–63. doi: 10.1016/j.molmed.2018.12.006
98. Bianco P, Robey PG, Simmons PJ. Mesenchymal stem cells: revisiting history, concepts, and assays. *Cell Stem Cell.* (2008) 2:313–9. doi: 10.1016/j.stem.2008.03.002
99. Crisan M, Yap S, Casteilla L, Chen CW, Corselli M, Park TS, et al. A perivascular origin for mesenchymal stem cells in multiple human organs. *Cell Stem Cell.* (2008) 3:301–13. doi: 10.1016/j.stem.2008.07.003
100. Nilsson B, Teramura Y, Ekdahl KN. The role and regulation of complement activation as part of the thromboinflammation elicited in cell therapies. *Mol Immunol.* (2014) 61:185–90. doi: 10.1016/j.molimm.2014.06.009
101. Furlani D, Ugurlucan M, Ong L, Bieback K, Pittermann E, Westien I, et al. Is the intravascular administration of mesenchymal stem cells safe? Mesenchymal stem cells and intravital microscopy. *Microvasc Res.* (2009) 77:370–6. doi: 10.1016/j.mvr.2009.02.001
102. Lee RH, Yoon N, Reneau JC, Prockop DJ. Preactivation of human MSCs with TNF-alpha enhances tumor-suppressive activity. *Cell Stem Cell.* (2012) 11:825–35. doi: 10.1016/j.stem.2012.10.001
103. Liao L, Shi B, Chang H, Su X, Zhang L, Bi C, et al. Heparin improves BMSC cell therapy: anticoagulant treatment by heparin improves the safety and therapeutic effect of bone marrow-derived mesenchymal stem cell cytotherapy. *Theranostics.* (2017) 7:106–16. doi: 10.7150/thno.16911
104. Mangi AA, Noiseux N, Kong D, He H, Rezvani M, Ingwall JS, et al. Mesenchymal stem cells modified with Akt prevent remodeling and restore performance of infarcted hearts. *Nat Med.* (2003) 9:1195–201. doi: 10.1038/nm912
105. Mei SH, McCarter SD, Deng Y, Parker CH, Liles WC, Stewart DJ. Prevention of LPS-induced acute lung injury in mice by mesenchymal stem cells overexpressing angiopoietin 1. *PLoS Med.* (2007) 4:e269. doi: 10.1371/journal.pmed.0040269
106. von Bahr L, Sundberg B, Lonnies L, Sander B, Karbach H, Hagglund H, et al. Long-term complications, immunologic effects, and role of passage for outcome in mesenchymal stromal cell therapy. *Biol Blood Marrow Transplant.* (2012) 18:557–64. doi: 10.1016/j.bbmt.2011.07.023
107. Galvez P, Clares B, Hmadcha A, Ruiz A, Soria B. Development of a cell-based medicinal product: regulatory structures in the European Union. *Br Med Bull.* (2013) 105:85–105. doi: 10.1093/bmb/lds036
108. Jiang R, Han Z, Zhuo G, Qu X, Li X, Wang X, et al. Transplantation of placenta-derived mesenchymal stem cells in type 2 diabetes: a pilot study. *Front Med.* (2011) 5:94–100. doi: 10.1007/s11684-011-0116-z
109. Tan J, Wu W, Xu X, Liao L, Zheng F, Messinger S, et al. Induction therapy with autologous mesenchymal stem cells in living-related kidney transplants: a randomized controlled trial. *JAMA.* (2012) 307:1169–77. doi: 10.1001/jama.2012.316
110. Cobellis G, Silvestroni A, Lillo S, Sica G, Botti C, Maione C, et al. Long-term effects of repeated autologous transplantation of bone marrow cells in patients affected by peripheral arterial disease. *Bone Marrow Transplant.* (2008) 42:667–72. doi: 10.1038/bmt.2008.228
111. Lopez-Lucas MD, Pachon-Pena G, Garcia-Hernandez AM, Parrado A, Sanchez-Salinas D, Garcia-Bernal D, et al. Production via good manufacturing practice of exofucosylated human mesenchymal stromal cells for clinical applications. *Cytotherapy.* (2018) 20:1110–23. doi: 10.1016/j.jcyt.2018.07.001

112. Sackstein R. The first step in adoptive cell immunotherapeutics: assuring cell delivery via glycoengineering. *Front Immunol.* (2018) 9:3084. doi: 10.3389/fimmu.2018.03084
113. Sackstein R, Merzaban JS, Cain DW, Dagia NM, Spencer JA, Lin CP, et al. *Ex vivo* glycan engineering of CD44 programs human multipotent mesenchymal stromal cell trafficking to bone. *Nat Med.* (2008) 14:181–7. doi: 10.1038/nm1703
114. Silva M, Videira PA, Sackstein R. E-selectin ligands in the human mononuclear phagocyte system: implications for infection, inflammation, and immunotherapy. *Front Immunol.* (2017) 8:1878. doi: 10.3389/fimmu.2017.01878
115. Lee T. Host tissue response in stem cell therapy. *World J Stem Cells.* (2010) 2:61–6. doi: 10.4252/wjsc.v2.i4.61
116. Pachon-Pena G, Donnelly C, Ruiz-Canada C, Katz A, Fernandez-Veledo S, Vendrell J, et al. A Glycovariant of human CD44 is characteristically expressed on human mesenchymal stem cells. *Stem Cells.* (2017) 35:1080–92. doi: 10.1002/stem.2549
117. Zheng J, Li H, He L, Huang Y, Cai J, Chen L, et al. Preconditioning of umbilical cord-derived mesenchymal stem cells by rapamycin increases cell migration and ameliorates liver ischaemia/reperfusion injury in mice via the CXCR4/CXCL12 axis. *Cell Prolif.* (2019) 52:e12546. doi: 10.1111/cpr.12546
118. Badenes SM, Fernandes TG, Cordeiro CS, Boucher S, Kuninger D, Vemuri MC, et al. Defined essential 8 medium and vitronectin efficiently support scalable xeno-free expansion of human induced pluripotent stem cells in stirred microcarrier culture systems. *PLoS ONE.* (2016) 11:e0151264. doi: 10.1371/journal.pone.0151264
119. Chase LG, Lakshmipathy U, Solchaga LA, Rao MS, Vemuri MC. A novel serum-free medium for the expansion of human mesenchymal stem cells. *Stem Cell Res Ther.* (2010) 1:8. doi: 10.1186/scrt8
120. Gottipamula S, Muttigi MS, Kolkundkar U, Seetharam RN. Serum-free media for the production of human mesenchymal stromal cells: a review. *Cell Prolif.* (2013) 46:608–27. doi: 10.1111/cpr.12063
121. Ishikawa I, Sawada R, Kato Y, Tsuji K, Shao J, Yamada T, et al. Effectivity of the novel serum-free medium STK2 for proliferating human mesenchymal stem cells. *Yakugaku Zasshi.* (2009) 129:381–4. doi: 10.1248/yakushi.129.381
122. Jayme DW, Smith SR. Media formulation options and manufacturing process controls to safeguard against introduction of animal origin contaminants in animal cell culture. *Cytotechnology.* (2000) 33:27–36. doi: 10.1023/A:1008133717035
123. Kinzebach S, Bieback K. Expansion of mesenchymal stem/stromal cells under xenogenic-free culture conditions. *Adv Biochem Eng Biotechnol.* (2013) 129:33–57. doi: 10.1007/10_2012_134
124. Wessman SJ, Levings RL. Benefits and risks due to animal serum used in cell culture production. *Dev Biol Stand.* (1999) 99:3–8.
125. June CH, Sadelain M. Chimeric antigen receptor therapy. *N Engl J Med.* (2018) 379:64–73. doi: 10.1056/NEJMra1706169
126. Hernandez I, Prasad V, Gellad WF. Total costs of chimeric antigen receptor T-cell immunotherapy. *JAMA Oncol.* (2018) 4:994–6. doi: 10.1001/jamaoncol.2018.0977

16

The Necrobiology of Mesenchymal Stromal Cells Affects Therapeutic Efficacy

Daniel J. Weiss[1], *Karen English*[2], *Anna Krasnodembskaya*[3], *Johana M. Isaza-Correa*[4], *Ian J. Hawthorne*[2] *and Bernard P. Mahon*[4*]

[1] *Department of Medicine, University of Vermont College of Medicine, Burlington, VT, United States,* [2] *Cellular Immunology Laboratory, Biology Department, Human Health Research Institute, Maynooth University, Maynooth, Ireland,* [3] *School of Medicine, Dentistry and Biomedical Sciences, Wellcome-Wolfson Institute for Experimental Medicine, Queen's University of Belfast, Belfast, United Kingdom,* [4] *Immunology & Cell Biology Laboratory, Biology Department, Human Health Research Institute, Maynooth University, Maynooth, Ireland*

Correspondence:
Bernard P. Mahon
Bernard.Mahon@mu.ie

Rapid progress is occurring in understanding the mechanisms underlying mesenchymal stromal cell (MSC)-based cell therapies (MSCT). However, the results of clinical trials, while demonstrating safety, have been varied in regard to efficacy. Recent data from different groups have shown profound and significant influences of the host inflammatory environment on MSCs delivered systemically or through organ-specific routes, for example intratracheal, with subsequent actions on potential MSC efficacies. Intriguingly in some models, it appears that dead or dying cells or subcellular particles derived from them, may contribute to therapeutic efficacy, at least in some circumstances. Thus, the broad cellular changes that accompany MSC death, autophagy, pre-apoptotic function, or indeed the host response to these processes may be essential to therapeutic efficacy. In this review, we summarize the existing literature concerning the necrobiology of MSCs and the available evidence that MSCs undergo autophagy, apoptosis, transfer mitochondria, or release subcellular particles with effector function in pathologic or inflammatory *in vivo* environments. Advances in understanding the role of immune effector cells in cell therapy, especially macrophages, suggest that the reprogramming of immunity associated with MSCT has a weighty influence on therapeutic efficacy. If correct, these data suggest novel approaches to enhancing the beneficial actions of MSCs that will vary with the inflammatory nature of different disease targets and may influence the choice between autologous or allogeneic or even xenogeneic cells as therapeutics.

Keywords: mesenchymal stromal cell, cell therapy, apoptosis, autophagy, mitochondria, extracellular vesicles, efficacy

INTRODUCTION

The efficacy of MSC administration in preclinical inflammatory models is well-documented regardless of the source of MSCs (bone marrow, adipose, placenta, other). The basic biology of MSCs, their mode of action and therapeutic efficacy in clinical studies have been reviewed extensively elsewhere (1–3). However, translation of preclinical efficacy to the clinical setting is proving difficult. A possible reason for this is a lack of understanding of the fate of MSCs when they encounter highly inflammatory microenvironments. Within this inflammatory milieu, MSCs are exposed to insults such as hypoxia and pro-inflammatory cytokines (4). What happens to MSCs during the transient period in which they are at the target site is largely unknown (5). The longstanding working hypothesis has been that viable functional MSCs are critical for efficacy. However, a number of recent studies have suggested that MSC survival in the disease milieu may not be as important as once thought. These studies elegantly demonstrate that apoptotic or dead MSCs can facilitate protection mediated by MSC administration in inflammatory microenvironments *in vivo* (6–8). However, these studies have opened up a number of questions about the processes involved in the transition from live to dead MSCs. Under what circumstances can dead MSCs substitute for viable cells? What are the limits to use? Can the pre-apoptotic cargo of extracellular vesicles (EVs) produced by MSCs or mitochondria transferred from MSCs to other cells substitute for the MSCs themselves? Is there a role for autophagy or for efferocytosis in MSCT efficacy? Does autophagy influence the soluble factors secreted by MSCs before they die? If we can better understand the fate of MSCs within the diseased microenvironment, perhaps this knowledge would lend itself to development of more optimal MSC-based cell therapies (be that live, autophagic or dead/apoptotic MSCs) and reduce the disparity between pre-clinical models and the clinical setting.

The term "necrobiology" has been used to describe the cellular processes associated with morphological, biochemical, and molecular changes which predispose, precede, and accompany cell death, as well as the consequences and tissue response to cell death (9). The observation that MSC viability and efficacy are not necessarily correlated (6, 7, 10) suggests that the necrobiology of MSCT will be a fruitful and essential area for future study. In this review we focus on key biological processes likely to affect therapeutic efficacy (**Figure 1**), summarize what is known about the questions above, and for the first time attempt to frame these disparate aspects of research within the concept of necrobiology or the biology of the dying therapeutic cell.

APOPTOTIC MSCS AND CLINICAL EFFICACY

There is relatively little data available in pre-clinical disease models in which apoptotic or dead MSCs were investigated, either as part of a direct investigation of dead/apoptotic cell actions or as part of a control group for live MSC administrations. Using pre-clinical models of respiratory diseases/critical illnesses in mice as representative examples (**Table 1**), intratracheal administration of apoptotic MSCs in models of acute lung injury or systemic administration of either fixed or heat-killed MSCs in mouse models of asthma and sepsis, respectively, did not mimic the effects of live MSC administration (11–14). Likewise the administration of other cells such as fixed fibroblasts were not beneficial, suggesting a role for MSCs that cannot be replaced by other dead cell types (11, 13). Notably, most of these studies are relatively old and did not exhaustively explore the effects of dead or apoptotic cells on immune or inflammatory cells. Whether this is a phenomenon unique to MSCs is unknown at present as there are few examples of administering other types of cells to the lung that might influence inflammatory or immune pathways. However, there are well documented anti-inflammatory bystander effects when other apoptotic cells are engulfed by macrophages and these have been recently reviewed (15). The extent to which this phenomenon is specific to lung diseases is relatively unexplored and a ripe area for further research.

In contrast, more recent studies in pre-clinical models of acute lung injury have suggested that the inflammatory environment in the lung can affect survival and subsequent efficacy of intratracheally-administered MSCs in part through activation of TRL4 signaling pathways (16). MSCs have variable effects in different mouse models of lung injuries with efficacy potentially related to the proteome profile of the BAL fluid in each respective injury (17). Another recent study demonstrated that apoptotic MSCs reduced some inflammatory endpoints in a mouse model of Th2-mediated allergic airway inflammation (7). These effects are not confined to lung disease models, a series of related studies in a rat model of cecal ligation and puncture-induced sepsis demonstrated that administration of rat adipose-derived MSCs, rendered apoptotic by 96 h culture in serum-free media, were more effective than healthy MSCs in improving survival and decreasing lung, kidney and cardiac injuries (18–21) administration of the apoptotic MSCs decreased a number of circulating and organ-specific pro-inflammatory, pro-apoptotic, and oxidative stress markers while increasing anti-apoptotic and anti-oxidant responses. The suggested mechanism(s) were that the apoptotic MSCs were more effective at dampening immune responses to the original injury, however, no specific pathways were delineated. These results suggest a more complex interaction of MSC apoptosis on efficacy in different inflammatory environments such that the inflammatory environment itself directs MSC apoptosis. Unfortunately, other more recent studies of MSC effects in a wide range of pre-clinical lung injury models have generally not included dead or apoptotic cells and thus there is opportunity for more extensive investigation (22–25).

Surprisingly, little is known about how MSCs are killed in different settings. *In vitro* studies have demonstrated the conditions for NK cell killing of MSCs (26) and this is likely to be an important mechanism for induction of MSC death *in vivo*, although few studies have examined this in detail. Similarly a role for Complement mediated killing has been proposed (27, 28). Recently, a requirement for cytotoxic $CD8^+$ T cell mediated killing (via apoptotic death) of MSCs has been shown in GvHD (7). However, the mechanisms of MSC killing (e.g., immune-mediated or as a result of exposure to microbial toxins) are likely to influence the type of death induced and the biological consequence. This might be an especially important consideration in designing cell therapeutics for lung diseases or patient subsets where there is a pathogenic microbial burden (e.g., Cystic Fibrosis).

There are even less available data on the effect of apoptotic MSCs in clinical investigations. In a notable recent example,

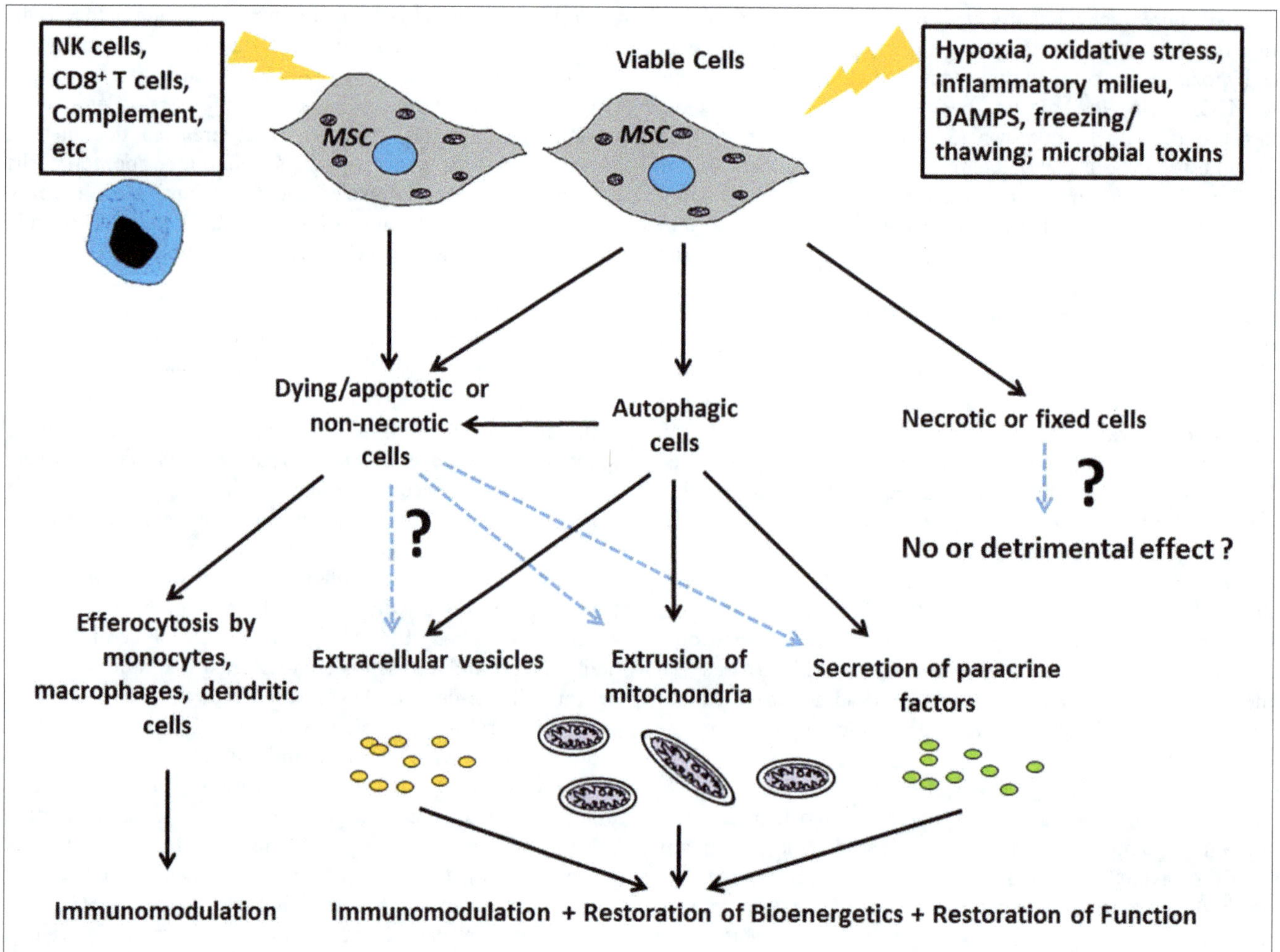

FIGURE 1 | Scheme for how the necrobiology of MSCs influences therapeutic efficacy Putative mechanisms include: as live cells through paracrine mechanisms, and through the cellular processes associated with morphological, biochemical, and molecular changes which predispose, precede, and accompany cell death. These necrobiotic processes include the response to dying and non-necrotic MSCs, the alteration of MSC biology by autophagy, and the delivery of MSC derived mitochondria or EVs to target cells and tissues.

safety but no efficacy was observed in a multi-center double-blinded randomized trial of systemic bone marrow-derived MSCs in patients with ARDS (29). In *post-hoc* analyses, the unanticipated finding was that up to 85% of the MSCs were non-viable at the time of administration. This suggests that dead MSCs may not have clinical efficacy in ARDS although there are a number of other factors to consider including timing, dose, and route of MSC administration (30). In contrast, a preliminary report of a parallel trial of bone marrow-derived MSC-like cells in ARDS patients demonstrated efficacy in major endpoints of survival, ventilator-free days and ICU stay (31). Notably, the cells utilized were fully viable at the time of administration. Therefore, viable/ live MSCs are not interchangeable therapeutically with apoptotic/dead MSC, but each have potential efficacy in different contexts and presumably by different mechanisms. In combination with the growing experience of dead/apoptotic MSCs in pre-clinical models, these clinical observations raise important and hypothesis-generating mechanistic ideas for further study.

MSC AUTOPHAGY AND CLINICAL EFFICACY

It is now known that non-necrotic cell death can be induced by diverse mechanisms and many of these are linked to the cellular processes that eliminate damaged proteins and organelles, termed autophagy (32, 33). Autophagy is a tightly regulated, complex cascade that controls the efficient delivery and fusion of damaged organelles to the autophagosome (34). Whilst this process supports cell survival and regular cell functioning, it is also associated with at least three forms of cell death- apoptosis, necroptosis, and autosis. Necroptosis is an inflammatory, caspase-independent form of cell death (33) whereas autosis is mediated by the Na^+ K^+- ATPase pump and is autophagy-gene dependent (35, 36). More broadly, autophagy is activated by microenvironmental and intracellular signals linked to ER stress, hypoxia and immune cell activation (37–39). These signals, related to tissue damage, include damage associated molecular patterns (DAMPs) and MSCs have been

TABLE 1 | Pre-clinical lung injury studies utilizing dead or apoptotic MSCs.

Injury Model	Experimental model, route, and timing of treatment	MSC Source	Outcome compared to injury effects	Potential mechanisms of MSC actions	Cell controls?	References
Acute Lung Injury	Mouse IN LPS IT MSC 4 h after LPS	Syngeneic Mouse BM Plastic Adherent	Improved survival Improved histologic inflammation and edema Decreased BALF TNF-α, MIP-2 Increased BALF and serum IL-10	None specified (soluble mediators)	Apoptotic MSC, 3T3 fibroblasts Did not mimic effects on survival or inflammation	(11)
Acute Lung Injury	Mouse IT LPS IT MSC 4 h after LPS (P 5–6); 10^6 cells/mouse	Xenogeneic Primary human umbilical cord MSC CD29+, 44+, 73+. CD34-, 45-, HLAII- osteo/adipo differentiation	Decreased mortality, histological injury (3d), BAL TNFa, MIP-2, IFNγ (3d), Th1 CD4 T cells Increased BAL IL-10 (3d), CD4/CD25/Foxp3+ Treg	Non-specified soluble mediators	Apoptotic MSCs (mitomycin C treated) Did not mimic MSC results	(12)
Asthma	Mouse ovalbumin-induced acute allergic airways inflammation Ovalbumin sensitization days 0, 7, 14 MSC IV days 7/14 (P 4–9), 5 × 10^6 cells/infusion Challenge days 25–27; Harvest d28	Allogeneic Mouse (FVB) BM Sca1+, CD44+, 106+. CD11b-, 11c-, 34-, 35-, 117- Osteo/adipo/ Chondro differentiation	Decreased histological injury, BAL total cells (especially Eosinophils & Macrophages), BAL IL-4, IL-13, splenocyte IL-4 recall Increased BAL IL-10; splenocyte IL-13, IL-10 recall	None (paracrine)	PFA-fixed MSC Did not mimic most MSC results No effect or exacerbated histology; no effect on BAL except BAL IL-13; Increased splenocyte IL-4 recall. Decreased splenocyte IL-10, IL-13 recall.	(13)
Sepsis	Mouse cecal ligation and puncture IV MSC 1 h prior, concomittant, or 24 h after surgery	Syngeneic & allogeneic Mouse BM Plastic adherent CD11b, 45 depleted	Improved survival and organ function Decreased circulating TNFα, IL-6 Increased circulating IL-10	LPS and TNFα-stimulated MSC stimulated macrophages produced IL-10 through cell-cell contact and iNOS-dependent release of PGE2	Whole bone marrow, heat-killed MSC, skin fibroblasts No effects on survival Other endpoints not assessed	(14)

shown to sense DAMPs released from dying/stressed cells (40) leading to enhanced pro-reparative and anti-inflammatory effects (40). Thus, the reparative effects of MSCs may be primed or altered by exposure to DAMPs or other stress signals that alter their interactions with other cells. During cell therapy, MSCs become exposed to such signals and autophagy is a common cellular response to such stress. Autophagy influences MSCs' therapeutic effects in at least two contrasting ways- to promote survival of the MSCs, or to induce MSC death through apoptosis, necroptosis, or autosis. The fate of the MSC is thus likely to be dependent on quantitative differences in exposure time to inflammation. Understanding the role of autophagy in MSCs at sites of inflammation could therefore inform therapeutic protocol design in the future.

The role played by autophagy as a survival mechanism to relieve stress and prevent apoptosis has been extensively studied (41). Under starvation conditions (serum deprivation, hypoxia, oxygen/glucose deprivation) or in the presence of reactive oxygen species (ROS), autophagy has been shown to promote MSC survival *in vitro* (42–44). Importantly, while sufficient levels of ROS are required to activate autophagy, excessive ROS may lead to cell death (45). This has been demonstrated where preconditioning of MSCs to serum deprivation and hypoxic conditions have prolonged survival in ischemic microenvironments through the activation of autophagic processes (46). Moreover, mitophagy in MSCs facilitates interaction with macrophages in conditions of oxidative stress whilst also preventing apoptosis (47). A number of extrinsic factors that modulate autophagy in MSCs have been identified, for example Stromal Cell Derived Factor-1β can promote MSC survival through enhanced autophagy (48). Expression of hypoxia-inducible factor 1α also protects against Oxygen-Glucose deprivation via induction of autophagy and the PI3K/AKT/mTOR signaling pathway (43), while over-expression of CPT1C in human MSCs enhances survival via an increase in autophagic flux (49). In aging mice, knockdown of insulin-like growth factor-1 enhances survival of MSCs through autophagy and prolongs MSC survival *in vivo* (50). Taken together these studies clearly show that at least in some circumstances autophagy promotes MSC survival under stress.

In addition to factors influencing the autophagic pathway in MSCs, autophagy may also lead to the production of soluble factors important for MSC's therapeutic efficacy. Vascular endothelial growth factor (VEGF) plays a key role in MSCs promotion of wound healing (51, 52), a recent study has identified that increased VEGF secretion from autophagic MSCs promoted vascularization in cutaneous wounds and improved healing (52). The induction of autophagy in MSCs may also alter their immunomodulatory function. Autophagic human bone marrow-derived MSCs can regulate $CD4^+$ T helper cell proliferation via TGF-β1 signaling (53). Activation of autophagy by rapamycin in a co-culture system enhanced MSC's ability to suppress $CD4^+$ T helper cell proliferation, whilst 3-methyladenine (3-MA), an autophagic inhibitor, reduced it. These data indicate a role for autophagy in MSCs' immunomodulatory functions of the adaptive immune response, and therefore suggest that the autophagic status of the MSCs will influence therapeutic efficacy under inflammatory conditions. However, the precise limits of this effect are unknown and there are clearly redundant and parallel mechanisms operating. For example, Chinnadurai et al. showed that while interferon gamma (IFN-γ) stimulation of MSCs upregulated the expression of autophagy genes, inhibition of autophagy via 3-MA did not affect MSCs' immunomodulatory potential (54). Furthermore, in some studies, autophagy was shown to have adverse effects on MSCs' immunomodulatory capacity. When rodent MSCs were stimulated with tumor necrosis factor (TNF) and IFN-γ, autophagy reduced MSCs' immunomodulatory effects whereas inhibition through the knockdown of Bcn1 was beneficial (55). It is difficult to compare studies that inhibit autophagy when diverse inducers and inhibitors have been used, or when different inhibitor concentrations and time points have been studied. Nevertheless, these differences are important, autophagy and indeed apoptosis are time dependent processes, and it is reasonable to assume that the activity and function of the MSCs transitioning through these processes will vary with each disease and over time. The implication for developing therapies is that future preclinical approaches will have to account more comprehensively for temporal and dose effects to be informative, but such information could well-shorten therapeutic development times if it leads to improved understanding of delivery route and dosage.

Autophagy can alter biological function following starvation or inflammation (56), and in contrast to the above, can promote autophagic cell death or autosis rather than survival. This switch in roles for autophagy is thought to be dependent on the strength of the signals present, time of treatment and availability of ATP (57). At present our understanding of the type of death induced by autophagy, tends to reflect the greater understanding of apoptosis compared to necroptosis and autosis (58). Nevertheless, autophagy-induced apoptosis has been reported as an alternative fate of MSCs exposed to an inflammatory microenvironment (59). Dang et al. demonstrated that autophagy may cause cell death in a sepsis model of inflammation. These data suggest that the cytokine cocktail presented to the MSCs from the microenvironment causes autophagy to trigger death instead of promoting cell survival. This was mediated via the interaction with the ROS/ERK pathway resulting in the downregulation of Bcl-2. Inhibition of autophagy in MSCs led to increased production of prostaglandin E2 (55) and enhanced immunoregulation in pre-clinical models of EAE (55) and sepsis (59). Dang et al. (55) also reported that the induction of apoptosis reduced the therapeutic effect of MSCs, however, it has recently been demonstrated in a GvHD model that apoptotic MSCs are still immunosuppressive (7). Galleu et al. recorded that apoptotic MSCs (apoMSC) could reduce effector cell number in the lung and spleen of GvHD mice (7). Importantly, phagocytes producing indolamine 2,3-dioxygenase were required for the protection associated with apoMSC when administered intraperitoneally but not intravenously (60). These and other studies from the Hoogduijn group (5, 6) suggest that the therapeutic effects of both live and apoMSC are dependent on interactions with specific phagocytic cell populations. These observations also highlight the important interaction between

MSCs and macrophages and the contribution of innate immune modulation to therapeutic efficacy (61). Given the contrasting data surrounding the effects on efficacy of autophagy in MSCs it is clear that further study is needed, especially of dose and temporal responses. Nevertheless, it is possible to state that the inflammatory environment plays an important role in the MSC fate of survival or death, that autophagic processes are involved in this fate decision, and that subsequent interaction of MSCs with innate cells such as monocytes/macrophages influence therapeutic efficacy. From the above studies, it seems likely that whereas pro-survival processes are likely to be linked in part to therapeutic effects through MSCs' production of paracrine factors (e.g., VEGF, etc.), necrobiological-related efficacy is more likely to operate through the interaction between MSCs and the innate immune cells such as monocytes/macrophages (6).

MITOCHONDRIAL TRANSFER BY MSCS AND CLINICAL EFFICACY

Cell death, oxidative stress, and autophagy are all linked to mitochondrial function (62), so it understandable that the mitochondrion has a role in MSC efficacy. More surprising have been the now well-documented reports that reprogramming of host cells by MSCs is significantly mediated by their ability to transfer functionally active mitochondria to somatic recipient cells. Mitochondrial transfer has been found to play a critical role in therapeutic effect of MSCs in the pre-clinical models of multiple diseases including brain injury, cardiac myopathies, muscle sepsis, and acute (ARDS) (63, 64) and chronic respiratory disorders (asthma and COPD) (65, 66). Mitochondria are transferred between cells via tunneling nanotubules (TNTs), cell fusion, and can also be contained in secreted extracellular vesicles (EV) (67). These mitochondria are functionally active and their transfer results in the enhancement of oxidative phosphorylation coupled with alleviation of oxidative stress in recipient cells leading to restoration of impaired functional activity (e.g., surfactant secretion, phagocytosis and wound healing) and cytoprotective effects. The consequences and mechanisms of mitochondrial transfer have been comprehensively reviewed previously (67–69). As mitochondrial dysfunction contributes to pathophysiology of various diseases, strategies aiming to protect mitochondria from injury or to increase biogenesis are being increasingly explored as promising therapeutic opportunities. Replacement of damaged mitochondria through donation from MSCs is a faster and physiologically more economical route for the recipient cell at the site of injury as compared to the mitochondrial biogenesis and therefore, appears to be an efficient means for disease attenuation (70).

In addition to protective effects due to improved bioenergetics, there is evidence of the involvement of mitochondrial transfer in cellular rejuvenation and transcriptional reprogramming (71). Studies by Acquistapace et al. demonstrated a key beneficial role of MSC mitochondria for reprogramming of post-mitotic murine cardiomyocytes toward proliferating cardiac progenitor-like cells through spontaneous cell fusion (72).

Although the precise mechanisms regulating mitochondrial extrusion from MSCs as well as their uptake by recipient cells remain to be investigated, it is clear that the injury microenvironment will have an impact on the rate and efficiency of this process. Thus, we have recently demonstrated that hypercapnia, a condition often associated with low tidal volume ventilation in ARDS, induces mitochondrial dysfunction and although the rate of mitochondrial transfer from MSCs to recipient cells is not changed, these dysfunctional mitochondria are not able to improve recipient cell bioenergetics and promote capacity of the lung epithelial cells to wound closure. This is in good agreement with the finding of Paliwal et al. demonstrating that mitochondria from MSCs with higher mitochondrial respiration capacities are more effective in suppression of mtROS in stressed recipient cells (73). Li et al. have demonstrated that pre-treatment with anti-oxidants such as N-acetyl-L-cysteine and L-ascorbic acid 2-phosphate enhanced mitochondrial transfer from the anti-oxidant treated population of the bone marrow derived MSCs to the untreated population of MSCs injured by H_2O_2 (74).

A key study by Mahrouf-Yorgov et al. reports that mitochondria released from dying cells at the site of injury are an important environmental cue that controls the cytoprotective function of MSCs and regulates their capacity for mitochondrial transfer (75). It was shown that upon oxidative stress, somatic cells (cardiomyocytes and endothelial cells) release mitochondria which are engulfed and degraded by MSCs, leading to induction of heme oxygenase-1 (HO-1) and stimulation of autophagy and mitochondrial biogenesis. As a result, the ability of MSCs to donate their mitochondria to injured cells to alleviate oxidative stress injury was enhanced (75). Reactive oxygen species and inflammatory cytokines (e.g., TNF-α) have also been postulated to play a role in the regulation of mitochondrial donation (67–69, 76). Our unpublished data suggest that mitochondrial transfer from MSCs to lung epithelial cells is enhanced in inflammatory environments. Taken together these data strongly suggest that mitochondrial transfer and cell death are related and relevant to clinical efficacy.

The processes of MSC mitochondrial transfer and autophagy are intrinsically interdependent. Phinney et al. have demonstrated that MSCs extrude their mitochondria in EVs which express autophagosomal markers, suggesting that this phenomenon is a result of incomplete autophagy (47). Physiologically, MSCs reside in the low oxygen environment of the bone marrow stem cell niche and the authors observed that conventional culture of MSCs in normoxia- (21% oxygen) induced oxidative stress, thereby promoting MSC mitophagy, however instead of degradation, mitochondria were directed outside of the cells (47). Ghanta et al. then demonstrated the importance of autophagy in maintaining healthy mitochondrial function and promoting survival in MSCs during oxidative stress (45).

In the view of accumulating evidence that after *in vivo* administration, MSCs undergo apoptosis and fragmentation and subsequent elimination by phagocytes (6, 7, 77), it is plausible to hypothesize that mitochondria could be released during fragmentation and taken up by surrounding somatic cells and particularly macrophages. We have previously demonstrated that mitochondrial transfer from healthy MSCs through extracellular vesicles results in macrophage metabolic reprogramming toward M2-like phenotype with enhanced phagocytic activity (63, 64) however whether or not mitochondria released from dying MSCs retain the same properties and exert similar effects remains to be determined.

MSC-DERIVED EXTRACELLULAR VESICLES, AND CLINICAL EFFICACY

If there is a cell-specific therapeutic benefit from using MSCs as opposed to any apoptotic cell, then cell-specific characteristics and mechanisms need further exploration. Apoptotic cells maintain some biological activities as they begin the orderly process of disassembly and death. In particular apoptotic cells can produce a range of EVs and apoptotic bodies that can influence their microenvironment (78, 79). There has been an explosion in the literature describing how exosomes and other EVs can act as biological modulators. Healthy, viable MSCs are well-characterized producers of a wide range of EVs with different cargos. These can include microvesicles, including those bearing mitochondria (see above), as well as exosomes (66) that are now recognized as powerful mediators of intercellular communication locally and systemically. Exosomes, and presumably their cargo, can activate or suppress aspects of immunity such as cytokine secretion, immune cell differentiation and polarization, and T cell activation (47, 80–82). In addition, processes such as angiogenesis, proliferation, oncogenesis, and microenvironmental conditioning can all be affected by exosomes. MSC derived exosomes (even in the absence of their viable MSC producer) can thus have detectable therapeutic influences in human systems (**Table 2**). The influences of exosomes are largely defined by their cargo, which can include cytosolic and membrane proteins, mRNA and non-coding RNA including miRNA (miR), and the nature of the EV cargo of MSCs is influenced by the extracellular environment (**Table 3**). Several studies have linked treatment with MSC-derived exosomes to improvement in models of liver, kidney, heart, skin, lung and other diseases (90–93). The influence of MSC-derived exosomes on lung injury is especially important to studies of clinical efficacy given that the lung is a major site of MSC accumulation in the early period after delivery (94, 95). MSC-derived exosomes regulate vascular remodeling and reduce hypoxic pulmonary hypertension in rodent models. These exosomes reduced the activation of the hypoxic transcription factor STAT3 and the expression of the miRNA-17 superfamily but restored miRNA-204 in lung (normally reduced in human pulmonary hypertension) (83). In an acute respiratory distress syndrome (ARDS) model, alveolar macrophages treated with MSC-derived CD44$^+$ EVs also reduced lung injury (64). As discussed above, EVs promoted mitochondrial transfer to the macrophages increasing their phagocytic capacity and inducing an anti-inflammatory response. These findings suggest that intravenously delivered MSC therapies that see an accumulation of viable pre-apoptotic MSCs in the lung vasculature, have the potential to produce EV with extensive biological effects.

Perhaps the most profound influence of MSC-derived exosomes and EVs is the regulation of innate immune responses. Phinney et al. showed that MSC-derived exosomes with miRNA cargo inhibited macrophage activation by modulating Toll-like receptor signaling (47, 96). Macrophages treated with MSC-derived exosomes activated NF-kB and changed the expression of 50 of the 84 TLR-associated proteins evaluated, including IL-1β, COX2, IL-10, CCL2, TNF, MyD88, TLR 1,4,5,7,8 and 9, IRAK1, and TRAF6 (47). The breadth of biological processes downstream of these factors is very extensive and hints at the potential scale of effects that might be influenced by exosomes produced in the early, pre-apoptotic, phase of MSC therapy. In adaptive immunity (and hence of relevance to cell therapy for autoimmune disease and transplantation), EVs from bone marrow-derived MSCs increased production of immunosuppressive IL-10 and the proliferation of regulatory T cells in peripheral blood mononuclear cell cultures stimulated with anti-CD3/CD28 beads (81). In this research, treatment with MSC-EVs alone resulted in apoptosis of T cell populations. Interestingly, in other studies, exosomes secreted by HIF-1α-overexpressing donor MSCs were enriched for the Notch ligand Jagged-1 (97). Subcutaneous injection of these exosomes in a Matrigel plug assay induced angiogenesis (97). This builds on earlier work showing that Jagged-1 was an important contact dependent signal by which MSCs induced tolerogenic dendritic cells (DC) (98). Given that DCs have an antigen acquisition sentinel function throughout the body, and that these are key cells in shaping adaptive immunity, exosomal cargos of Notch ligands might prove an important modulator of immunity in multiple cell therapies. The immunomodulatory properties of MSC-derived subcellular particles indicate their potential as a novel cell-free therapy for treatment of immunological disorders, especially through interaction with antigen presenting cells (61). This is borne out by a recent study showing differential effects of membrane derived particles from MSCs either untreated or pre-treated with IFN-γ. Whilst both particle types decreased the frequency of CD14$^+$ CD16$^+$ inflammatory monocytes, the particles derived from IFN-γ treated cells also promoted anti-inflammatory PD-L1 expressing monocytes (10). This provides a mechanistic basis for earlier work showing that IFN-γ does not break but enhances the immunosuppressive capacities of MSCs and MSC-like cells (26, 99, 100). In the context of necrobiology, these data indicate that pre-apoptotic MSCs used as therapies in inflammatory microenvironments could be responsible for a switch toward an anti-inflammatory response through subcellular particles through their intra-vesicular or surface cargo (10).

The second aspect of MSC necrobiology that could affect therapeutic efficacy is the very recent observation from tumor biology that apoptotic cancer cells produce EVs with the characteristics of exosomes (78, 101). Apoptotic cell-derived extracellular vesicles (apoEVs) appear to be enriched with snRNA and spliceosomal proteins that can alter mRNA splicing in recipient cells (101). This finding is consistent with other studies showing that EVs produced during apoptosis are not

TABLE 2 | Selected studies of MSC-derived exosomes in human models.

Experimental System	MSC	Cargo	Method of exosome isolation	Effect	Study
PBMC co-culture	Bone marrow (healthy donors)	ND	Ultracentrifugation and precipitation	Suppressed TNF-a & IL-1b but increased anti-inflammatory factor TGF-b *in vitro*	(80)
PBMC co-culture	Bone marrow (healthy donors)	ND	Ultracentrifugation	Increased Treg/Teff ratio and IL-10 concentration in culture medium	(81)
Monocyte-derived macrophages	Bone marrow (healthy donors)	ND	Ultracentrifugation	Suppressed pro-inflammatory cytokine production, increased M2 macrophage marker expression, and augmented phagocytic capacity of human monocyte derived macrophages in non-contact cultures	(64)
Isolated human pulmonary artery endothelial cells	Umbilical cord	ND	S200 size-exclusion chromatography, differential centrifugation and ultracentrifugation	Regulated STAT3-mediated signaling	(83)
Human umbilical cord vein endo- thelial cells (HUVECs)	Bone marrow (healthy donors)	1,927 proteins identified	Differential centrifugation, filtration and ultracentrifugation	Proteomic analysis of proteins contained in exosomes released by MSC under ischemic like conditions. Mostly proteins such as platelet, epidermal or fibroblast derived growth factors, as well as proteins from nuclear factor-kappaB (NFkB) signaling pathway	(84)
HUVEC & human breast carcinoma-derived cell lines	Bone marrow (healthy donors)	miRNA-100	Differential centrifugation, filtration and ultracentrifugation	Decreased expression of VEGF in breast cancer-derived cells by modulating the mTOR/HIF-1α signaling axis	(85)
Comparative study	Bone marrow (healthy donors)	730 proteins identified in microvesicles	Sucrose cushion centrifugation & ultracentrifugation	Proteomic analysis identified proteins involved in cell proliferation, adhesion, migration, and morphogenesis	(86)

TABLE 3 | Studies showing the influence of inflammatory environment on human MSC-derived exosome cargo.

Stimulation	MSC	Cargo	Isolation/Treatment	Effect compared to control	References
TNF-α + IFN-γ overnight	Human bone marrow derived MSC	ICAM 1, CXCL12, and CCL5. 11 miRNAs with direct or indirect immunomodulatory function	Tangential flow filtration	Stimulated MSC EVs increased anti-inflammatory response through COX2/PGE2 pathway modulation	(87)
TNF-α + IFN-γ overnight	Human bone marrow derived MSC	ND	Tangential flow filtration	Improved mechanical sensitivity in rat spinal cord injury model	(88)
TGF-β, IFN-γ, or TGF-β + IFN-γ for 72 h	Human umbilical cord derived MSC	Exosomes from MSC treated with TGF-β and IFN-γ contained more IFN-γ, IL-10, and IDO	Centrifugation and PEG6000	EV from MSCs treated with TGF-β and IFN-γ induced Tregs differentiation	(89)

simply debris but have important immune regulatory roles in autoimmunity, infection and cancer (78, 79). Thus, apoEVs including those with exosome characteristics are the conduit of intercellular communication in physiologic and pathologic contexts. In this regard, it is important to note that there has yet to be a comprehensive description of exosomes produced by apoptotic MSCs. We do not know the degree to which apoptotic MSCs produce apoEVs, nor how this is affected by MSC history, stimulation, source, or apoptotic stage. Nevertheless, it is clear that MSC-derived subcellular particles' contents are not static but vary by tissue origin, MSC activity, and the cellular environment of the MSCs (96). It remains reasonable to assume that apoptotic MSCs produce apoEVs with potential to modify target cells.

Overall, the important implication is that MSCs (whether viable or non-viable) delivered to a patient are likely to be accompanied by or result in EV with diverse cargo produced prior to or after therapeutic deployment. However, it is also worth remembering that broader animal studies of MSCs that do not consider exosome function, could be unwittingly measuring a confounding effect of bovine exosomes present in the serum constituents of culture medium. This is usually well-controlled for in studies designed to discover exosome effects, but less often in studies of the MSC function itself. The range of such effects are extensive and could be influencing multiple disease models (**Table 2**). Nonetheless, the beneficial effects of exosomes derived from various sources has led to over 100 human phase I/II clinical trials, although to date there are very few reports of trials involving human MSC derived exosomes (www.clinicaltrials.gov). Those that have been registered target pancreatic cancer, macular holes, cerebrovascular disorders and diabetes, but most seem to be in the recruitment phase at present.

CONCLUSION

A common aspect of all the above aspects of MSC necrobiology is the significant role played by innate immune cells to counter the pathologic processes. Thus, efferocytosis or the processes linked to removing apoptotic MSCs are likely to contribute to the therapeutic benefit in studies where MSC viability is not essential (2, 102). This is likely to extend beyond simple uptake of apoptotic MSCs by macrophages, dendritic or other cells (61), and extend to the range of EVs, mitochondria, and other signals produced by dying MSCs and which profoundly alter the tissue microenvironment and innate immune cells (8, 102). The importance of these processes in the regular homeostatic function of endogenous MSCs is not known. Nevertheless, in the context of cell therapy, the efficacy of MSC can be attributed to either live/viable or dying/dead MSCs in different disease contexts, and these benefits are attributable to downstream effects linked to: a) the biological activity to (or evoked by) the intended therapeutic component (the viable MSC itself or its derivatives) and/or b) the recipient's response to MSCs that are in the process of dying (**Figure 1**). Without this understanding, and a greater appreciation of the complex necrobiology of MSCs, we are unlikely to understand the mechanisms of cell therapy action or rationally design improvements. Thus, the necrobiology of the mesenchymal stromal cell is likely to be a fruitful area for improving the efficacy or removing confounding influences on cell therapy.

AUTHOR CONTRIBUTIONS

DW has performed literature research, designed the review layout, wrote, and revised the review. KE has performed literature research, designed the review layout, wrote, and revised the review. AK has performed literature research, designed the review layout, wrote, and revised the review. JI-C has performed the literature research and contributed to the section on extracellular vesicles. IH has performed the literature research and contributed to the section on autophagy. BM has performed literature research, designed the review layout, wrote, and revised the review. All authors agree to be accountable for the content of the work.

FUNDING

DW is funded by the National Institutes of Health (HL127144-01, EB024329), the Cystic Fibrosis Foundation, and the Medical Technology Consortium. KE is supported by Science Foundation Ireland (SFI) under Grant Number 13/SIRG/2172 through a Starting Investigator Research Grant award and the Irish Research Council Laureate award IRCLA/2017/288. AK is supported by UK Medical Research Council Research Awards (MRC MR/R025096/1 and MR/S009426/1).

REFERENCES

1. English K. Mechanisms of mesenchymal stromal cell immunomodulation. *Immunol Cell Biol.* (2013) 91:19–26. doi: 10.1038/icb.2012.56
2. Galipeau J, Sensebe L. Mesenchymal stromal cells: clinical challenges and therapeutic opportunities. *Cell Stem Cell.* (2018) 22:824–33. doi: 10.1016/j.stem.2018.05.004
3. Weiss DJ. Concise review: current status of stem cells and regenerative medicine in lung biology and diseases. *Stem Cells.* (2014) 32:16–25. doi: 10.1002/stem.1506
4. Greco SJ, Rameshwar P. Microenvironmental considerations in the application of human mesenchymal stem cells in regenerative therapies. *Biologics Targets Ther.* (2008) 2:699–705. doi: 10.2147/BTT.S2765
5. Eggenhofer E, Benseler V, Kroemer A, Popp FC, Geissler EK, Schlitt HJ, et al. Mesenchymal stem cells are short-lived and do not migrate beyond the lungs after intravenous infusion. *Front Immunol.* (2012) 3:297. doi: 10.3389/fimmu.2012.00297
6. de Witte SFH, Luk F, Sierra Parraga JM, Gargesha M, Merino A, Korevaar SS, et al. Immunomodulation by therapeutic mesenchymal stromal cells (MSC) is triggered through phagocytosis of MSC by monocytic cells. *Stem Cells.* (2018) 36:602–15. doi: 10.1002/stem.2779
7. Galleu A, Riffo-Vasquez Y, Trento C, Lomas C, Dolcetti L, Cheung TS, et al. Apoptosis in mesenchymal stromal cells induces *in vivo* recipient-mediated immunomodulation. *Sci Transl Med.* (2017) 9:eaam7828. doi: 10.1126/scitranslmed.aam7828
8. Luk F, de Witte SF, Korevaar SS, Roemeling-van Rhijn M, Franquesa M, Strini T, et al. Inactivated mesenchymal stem cells maintain immunomodulatory capacity. *Stem Cells Dev.* (2016) 25:1342–54. doi: 10.1089/scd.2016.0068
9. Darzynkiewicz Z, Juan G, Li X, Gorczyca W, Murakami T, Traganos F. *Cytometry* in cell necrobiology: analysis of apoptosis and accidental cell death (necrosis). *Cytometry.* (1997) 27:1–20.
10. Goncalves FDC, Luk F, Korevaar SS, Bouzid R, Paz AH, López-Iglesias C, et al. Membrane particles generated from mesenchymal stromal cells modulate immune responses by selective targeting of pro-inflammatory monocytes. *Sci Rep.* (2017) 7:12100. doi: 10.1038/s41598-017-12121-z
11. Gupta N, Su X, Popov B, Lee JW, Serikov V, Matthay MA. Intrapulmonary delivery of bone marrow-derived mesenchymal stem cells improves survival and attenuates endotoxin-induced acute lung injury in mice. *J Immunol.* (2007) 179:1855–63. doi: 10.4049/jimmunol.179.3.1855
12. Sun J, Han ZB, Liao W, Yang SG, Yang Z, Yu J, et al. Intrapulmonary delivery of human umbilical cord mesenchymal stem cells attenuates acute lung injury by expanding $CD4^+CD25^+$ Forkhead Boxp3 $(FOXP3)^+$ regulatory T cells and balancing anti- and pro-inflammatory factors. *Cell Physiol Biochem.* (2011) 27:587–96. doi: 10.1159/000329980
13. Kavanagh H, Mahon BP. Allogeneic mesenchymal stem cells prevent allergic airway inflammation by inducing murine regulatory T cells. *Allergy.* (2011) 66:523–31. doi: 10.1111/j.1398-9995.2010.02509.x
14. Nemeth K, Leelahavanichkul A, Yuen PS, Mayer B, Parmelee A, Doi K, et al. Bone marrow stromal cells attenuate sepsis via prostaglandin E(2)-dependent reprogramming of host macrophages to increase their interleukin-10 production. *Nat Med.* (2009) 15:42–9. doi: 10.1038/nm.1905
15. Saas P, Daguindau E, Perruche S. Concise review: apoptotic cell-based therapies-rationale, preclinical results and future clinical developments. *Stem Cells.* (2016) 34:1464–73. doi: 10.1002/stem.2361
16. Gupta N, Sinha R, Krasnodembskaya A, Xu X, Nizet V, Matthay MA, et al. The TLR4-PAR1 axis regulates bone marrow mesenchymal stromal cell survival and therapeutic capacity in experimental bacterial pneumonia. *Stem Cells.* (2018) 36:796–806. doi: 10.1002/stem.2796

17. Islam D, Huang Y, Fanelli V, Delsedime L, Wu S, Khang J, et al. Identification and modulation of microenvironment is crucial for effective MSC therapy in acute lung injury. *Am J Respir Crit Care Med.* (2018) 199:1214–24. doi: 10.1164/rccm.201802-0356OC
18. Chang CL, Leu S, Sung HC, Zhen YY, Cho CL, Chen A, et al. Impact of apoptotic adipose-derived mesenchymal stem cells on attenuating organ damage and reducing mortality in rat sepsis syndrome induced by cecal puncture and ligation. *J Transl Med.* (2012) 10:244. doi: 10.1186/1479-5876-10-244
19. Chen HH, Chang CL, Lin KC, Sung PH, Chai HT, Zhen YY, et al. Melatonin augments apoptotic adipose-derived mesenchymal stem cell treatment against sepsis-induced acute lung injury. *Am J Transl Res.* (2014) 6:439–58.
20. Chen HH, Lin KC, Wallace CG, Chen YT, Yang CC, Leu S, et al. Additional benefit of combined therapy with melatonin and apoptotic adipose-derived mesenchymal stem cell against sepsis-induced kidney injury. *J Pineal Res.* (2014) 57:16–32. doi: 10.1111/jpi.12140
21. Sung PH, Chang CL, Tsai TH, Chang LT, Leu S, Chen YL, et al. Apoptotic adipose-derived mesenchymal stem cell therapy protects against lung and kidney injury in sepsis syndrome caused by cecal ligation puncture in rats. *Stem Cell Res Ther.* (2013) 4:155. doi: 10.1186/scrt385
22. Augustine S, Avey MT, Harrison B, Locke T, Ghannad M, Moher D, et al. Mesenchymal stromal cell therapy in bronchopulmonary dysplasia: systematic review and meta-analysis of preclinical studies. *Stem Cells Transl Med.* (2017) 6:2079–93. doi: 10.1002/sctm.17-0126
23. Mei SH, Dos Santos CC, Stewart DJ. Advances in stem cell and cell-based gene therapy approaches for experimental acute lung injury: a review of preclinical studies. *Hum Gene Ther.* (2016) 27:802–12. doi: 10.1089/hum.2016.063
24. Serrano-Mollar A. Cell therapy in idiopathic pulmonary fibrosis(dagger). *Med Sci.* (2018) 6:64. doi: 10.3390/medsci6030064
25. Zhang LB, He M. Effect of mesenchymal stromal (stem) cell (MSC) transplantation in asthmatic animal models: a systematic review and meta-analysis. *Pulmonary Pharmacol Therapeut.* (2019) 54:39–52. doi: 10.1016/j.pupt.2018.11.007
26. Noone C, Kihm A, English K, O'Dea S, Mahon BP. IFN-gamma stimulated human umbilical-tissue-derived cells potently suppress NK activation and resist NK-mediated cytotoxicity *in vitro*. *Stem Cells Dev.* (2013) 22:3003–14. doi: 10.1089/scd.2013.0028
27. Li Y, Fung J, Lin F. Local inhibition of complement improves mesenchymal stem cell viability and function after administration. *Mol Ther.* (2016) 24:1665–74. doi: 10.1038/mt.2016.142
28. Li Y, Lin F. Mesenchymal stem cells are injured by complement after their contact with serum. *Blood.* (2012) 120:3436–43. doi: 10.1182/blood-2012-03-420612
29. Matthay MA, Calfee CS, Zhuo H, Thompson BT, Wilson JG, Levitt JE, et al. Treatment with allogeneic mesenchymal stromal cells for moderate to severe acute respiratory distress syndrome (START study): a randomised phase 2a safety trial. *Lancet Respiratory Med.* (2019) 7:154–62. doi: 10.1016/S2213-2600(18)30418-1
30. Weiss DJ. Cell-based therapies for acute respiratory distress syndrome. *Lancet Respiratory Med.* (2019) 7:105–6. doi: 10.1016/S2213-2600(18)30477-6
31. Bellingan G, Jacono F, Bannard-Smith J, Brealey D, Meyer N, Thickett D, et al. Ting, primary analysis of a phase 1/2 study to assess MultiStem® cell therapy, a regenerative Advanced Therapy Medicinal Product (ATMP), in Acute Respiratory Distress Syndrome (MUST-ARDS). *B14. Late Break Clin Trials.* (2019) 199:A7353. doi: 10.1164/ajrccm-conference.2019.199.1_MeetingAbstracts.A7353
32. Dunn WA Jr. Autophagy and related mechanisms of lysosome-mediated protein degradation. *Trends Cell Biol.* (1994) 4:139–43. doi: 10.1016/0962-8924(94)90069-8
33. Goodall ML, Fitzwalter BE, Zahedi S, Wu M, Rodriguez D, Mulcahy-Levy JM, et al. The autophagy machinery controls cell death switching between apoptosis and necroptosis. *Dev Cell.* (2016) 37:337–49. doi: 10.1016/j.devcel.2016.04.018
34. Kirisako T, Baba M, Ishihara N, Miyazawa K, Ohsumi M, Yoshimori T, et al. Formation process of autophagosome is traced with Apg8/Aut7p in yeast. *J Cell Biol.* (1999) 147:435–46. doi: 10.1083/jcb.147.2.435
35. Liu Y, Levine B. Autosis and autophagic cell death: the dark side of autophagy. *Cell Death Differentiation.* (2014) 22:367. doi: 10.1038/cdd.2014.143
36. Liu Y, Shoji-Kawata S, Sumpter RM Jr, Wei Y, Ginet V, Zhang L, et al. Autosis is a Na^+,K^+-ATPase-regulated form of cell death triggered by autophagy-inducing peptides, starvation, and hypoxia-ischemia. *Proc Natl Acad Sci USA.* (2013) 110:20364–71. doi: 10.1073/pnas.1319661110
37. Fang Y, Tan J, Zhang Q. Signaling pathways and mechanisms of hypoxia-induced autophagy in the animal cells. *Cell Biol Int.* (2015) 39:891–8. doi: 10.1002/cbin.10463
38. Fougeray S, Pallet N. Mechanisms and biological functions of autophagy in diseased and ageing kidneys. *Nat Rev Nephrol.* (2015) 11:34–45. doi: 10.1038/nrneph.2014.201
39. Kouroku Y, Fujita E, Tanida I, Ueno T, Isoai A, Kumagai H, et al. ER stress (PERK/eIF2alpha phosphorylation) mediates the polyglutamine-induced LC3 conversion, an essential step for autophagy formation. *Cell Death Differentiation.* (2007) 14:230–9. doi: 10.1038/sj.cdd.4401984
40. Auletta JJ, Deans RJ, Bartholomew AM. Emerging roles for multipotent, bone marrow-derived stromal cells in host defense. *Blood.* (2012) 119:1801–9. doi: 10.1182/blood-2011-10-384354
41. Sommermann TG, Mack HI, Cahir-McFarland E. *Autophagy* prolongs survival after NFkappaB inhibition in B-cell lymphomas. *Autophagy.* (2012) 8:265–7. doi: 10.4161/auto.8.2.18763
42. Liu J, Hao H, Huang H, Tong C, Ti D, Dong L, et al. Hypoxia regulates the therapeutic potential of mesenchymal stem cells through enhanced autophagy. *Int J Lower Extremity Wounds.* (2015) 14:63–72. doi: 10.1177/1534734615573660
43. Lv B, Hua T, Li F, Han J, Fang J, Xu L, et al. Hypoxia-inducible factor 1 alpha protects mesenchymal stem cells against oxygen-glucose deprivation-induced injury via autophagy induction and PI3K/AKT/mTOR signaling pathway. *Am J Transl Res.* (2017) 9:2492–9.
44. Zhang Q, Yang YJ, Wang H, Dong QT, Wang TJ, Qian HY, et al. Autophagy activation: a novel mechanism of atorvastatin to protect mesenchymal stem cells from hypoxia and serum deprivation via AMP-activated protein kinase/mammalian target of rapamycin pathway. *Stem Cells Dev.* (2012) 21:1321–32. doi: 10.1089/scd.2011.0684
45. Ghanta S, Tsoyi K, Liu X, Nakahira K, Ith B, Coronata AA, et al. Mesenchymal stromal cells deficient in autophagy proteins are susceptible to oxidative injury and mitochondrial dysfunction. *Am J Respir Cell Mol Biol.* (2017) 56:300–9. doi: 10.1165/rcmb.2016-0061OC
46. Moya A, Larochette N, Paquet J, Deschepper M, Bensidhoum M, Izzo V, et al. Quiescence preconditioned human multipotent stromal cells adopt a metabolic profile favorable for enhanced survival under ischemia. *Stem Cells.* (2017) 35:181–96. doi: 10.1002/stem.2493
47. Phinney DG, Di Giuseppe M, Njah J, Sala E, Shiva S, St. Croix CM, et al. Mesenchymal stem cells use extracellular vesicles to outsource mitophagy and shuttle microRNAs. *Nat Commun.* (2015) 6:8472. doi: 10.1038/ncomms9472
48. Herberg S, Shi X, Johnson MH, Hamrick MW, Isales CM, Hill WD. Stromal cell-derived factor-1beta mediates cell survival through enhancing autophagy in bone marrow-derived mesenchymal stem cells. *PLoS ONE.* (2013) 8:e58207. doi: 10.1371/journal.pone.0058207
49. Roa-Mansergas X, Fado R, Atari M, Mir JF, Muley H, Serra D, et al. CPT1C promotes human mesenchymal stem cells survival under glucose deprivation through the modulation of autophagy. *Sci Rep.* (2018) 8:6997. doi: 10.1038/s41598-018-25485-7
50. Yang M, Wen T, Chen H, Deng J, Yang C, Zhang Z. Knockdown of insulin-like growth factor 1 exerts a protective effect on hypoxic injury of aged BM-MSCs: role of autophagy. *Stem Cell Res Ther.* (2018) 9:284. doi: 10.1186/s13287-018-1028-5
51. Wu Y, Chen L, Scott PG, Tredget EE. Mesenchymal stem cells enhance wound healing through differentiation and angiogenesis. *Stem Cells.* (2007) 25:2648–59. doi: 10.1634/stemcells.2007-0226

52. An Y, Liu WJ, Xue P, Ma Y, Zhang LQ, Zhu B, et al. Autophagy promotes MSC-mediated vascularization in cutaneous wound healing via regulation of VEGF secretion. *Cell Death Dis.* (2018) 9:58. doi: 10.1038/s41419-017-0082-8
53. Gao L, Cen S, Wang P, Xie Z, Liu Z, Deng W, et al. Autophagy improves the immunosuppression of $CD4^+$ T cells by mesenchymal stem cells through transforming growth factor-beta1. *Stem Cells Transl Med.* (2016) 5:1496–505. doi: 10.5966/sctm.2015-0420
54. Chinnadurai R, Copland IB, Ng S, Garcia M, Prasad M, Arafat D, et al. Mesenchymal stromal cells derived from Crohn's patients deploy indoleamine 2,3-dioxygenase-mediated immune suppression, independent of autophagy. *Mol Ther.* (2015) 23:1248–61. doi: 10.1038/mt.2015.67
55. Dang S, Xu H, Xu C, Cai W, Li Q, Cheng Y, et al. *Autophagy* regulates the therapeutic potential of mesenchymal stem cells in experimental autoimmune encephalomyelitis. *Autophagy.* (2014) 10:1301–15. doi: 10.4161/auto.28771
56. Marino G, Niso-Santano M, Baehrecke EH, Kroemer G. Self-consumption: the interplay of autophagy and apoptosis. *Nat Rev Mol Cell Biol.* (2014) 15:81–94. doi: 10.1038/nrm3735
57. Munoz-Pinedo C, Martin SJ. Autosis: a new addition to the cell death Tower of Babel. *Cell Death Dis.* (2014) 5:e1319. doi: 10.1038/cddis.2014.246
58. Kroemer G, Levine B. Autophagic cell death: the story of a misnomer. *Nat Rev Mol Cell Biol.* (2008) 9:1004–10. doi: 10.1038/nrm2529
59. Dang S, Yu ZM, Zhang CY, Zheng J, Li KL, Wu Y, et al. Autophagy promotes apoptosis of mesenchymal stem cells under inflammatory microenvironment. *Stem Cell Res Ther.* (2015) 6:247. doi: 10.1186/s13287-015-0245-4
60. Cheung TS, Galleu A, von Bonin M, Bornhauser M, Dazzi F. Apoptotic mesenchymal stromal cells induce prostaglandin E2 in monocytes: implications for the monitoring of mesenchymal stromal cells activity. *Haematologica.* (2019). doi: 10.3324/haematol.2018.214767. [Epub ahead of print].
61. Carty F, Mahon BP, English K. The influence of macrophages on mesenchymal stromal cell therapy: passive or aggressive agents? *Clin Exp Immunol.* (2017) 188:1–11. doi: 10.1111/cei.12929
62. Scherz-Shouval R, Elazar Z. ROS, mitochondria and the regulation of autophagy. *Trends Cell Biol.* (2007) 17:422–7. doi: 10.1016/j.tcb.2007.07.009
63. Jackson MV, Morrison TJ, Doherty DF, McAuley DF, Matthay MA, Kissenpfennig A, et al. Mitochondrial transfer via tunneling nanotubes is an important mechanism by which mesenchymal stem cells enhance macrophage phagocytosis in the *in vitro* and *in vivo* models of ARDS. *Stem Cells.* (2016) 34:2210–23. doi: 10.1002/stem.2372
64. Morrison TJ, Jackson MV, Cunningham EK, Kissenpfennig A, McAuley DF, O'Kane CM, et al. Mesenchymal stromal cells modulate macrophages in clinically relevant lung injury models by extracellular vesicle mitochondrial transfer. *Am J Respir Crit Care Med.* (2017) 196:1275–86. doi: 10.1164/rccm.201701-0170OC
65. Ahmad T, Mukherjee S, Pattnaik B, Kumar M, Singh S, Kumar M, et al. Miro1 regulates intercellular mitochondrial transport & enhances mesenchymal stem cell rescue efficacy. *EMBO J.* (2014) 33:994–1010. doi: 10.1002/embj.201386030
66. Li X, Zhang Y, Yeung SC, Liang Y, Liang X, Ding Y, et al. Mitochondrial transfer of induced pluripotent stem cell-derived mesenchymal stem cells to airway epithelial cells attenuates cigarette smoke-induced damage. *Am J Respir Cell Mol Biol.* (2014) 51:455–65. doi: 10.1165/rcmb.2013-0529OC
67. Torralba D, Baixauli F, Sanchez-Madrid F. Mitochondria know no boundaries: mechanisms and functions of intercellular mitochondrial transfer. *Front Cell Dev Biol.* (2016) 4:107. doi: 10.3389/fcell.2016.00107
68. Murray LMA, Krasnodembskaya AD. Concise review: intercellular communication via organelle transfer in the biology and therapeutic applications of stem cells. *Stem Cells.* (2019) 37:14–25. doi: 10.1002/stem.2922
69. Paliwal S, Chaudhuri R, Agrawal A, Mohanty S. Regenerative abilities of mesenchymal stem cells through mitochondrial transfer. *J Biomed Sci.* (2018) 25:31. doi: 10.1186/s12929-018-0429-1
70. Agrawal A, Mabalirajan U. Rejuvenating cellular respiration for optimizing respiratory function: targeting mitochondria. *Am J Physiol Lung Cell Mol Physiol.* (2016) 310:L103–L113. doi: 10.1152/ajplung.00320.2015
71. Fergie N, Todd N, McClements L, McAuley D, O'Kane C, Krasnodembskaya A. Hypercapnic acidosis induces mitochondrial dysfunction and impairs the ability of mesenchymal stem cells to promote distal lung epithelial repair. *FASEB J.* (2019) 33:5585–98. doi: 10.1096/fj.201802056R
72. Acquistapace A, Bru T, Lesault PF, Figeac F, Coudert AE, le Coz O, et al. Human mesenchymal stem cells reprogram adult cardiomyocytes toward a progenitor-like state through partial cell fusion and mitochondria transfer. *Stem Cells.* (2011) 29:812–24. doi: 10.1002/stem.632
73. Paliwal S, Chaudhuri R, Agrawal A, Mohanty S. Human tissue-specific MSCs demonstrate differential mitochondria transfer abilities that may determine their regenerative abilities. *Stem Cell Res Ther.* (2018) 9:298. doi: 10.1186/s13287-018-1012-0
74. Li CJ, Chen PK, Sun LY, Pang CY. Enhancement of mitochondrial transfer by antioxidants in human mesenchymal stem cells. *Oxid Med Cell Longevity.* (2017) 2017:8510805. doi: 10.1155/2017/8510805
75. Mahrouf-Yorgov M, Augeul L, Da Silva CC, Jourdan M, Rigolet M, Manin S, et al. Mesenchymal stem cells sense mitochondria released from damaged cells as danger signals to activate their rescue properties. *Cell Death Differentiation.* (2017) 24:1224–38. doi: 10.1038/cdd.2017.51
76. Zhang Y, Yu Z, Jiang D, Liang X, Liao S, Zhang Z, et al. iPSC-MSCs with high intrinsic MIRO1 and sensitivity to TNF-alpha yield efficacious mitochondrial transfer to rescue anthracycline-induced cardiomyopathy. *Stem Cell Rep.* (2016) 7:749–63. doi: 10.1016/j.stemcr.2016.08.009
77. Braza F, Dirou S, Forest V, Sauzeau V, Hassoun D, Chesne J, et al. Mesenchymal stem cells induce suppressive macrophages through phagocytosis in a mouse model of asthma. *Stem Cells.* (2016) 34:1836–45. doi: 10.1002/stem.2344
78. Caruso S, Poon IKH. Apoptotic cell-derived extracellular vesicles: more than just debris. *Front Immunol.* (2018) 9:1486. doi: 10.3389/fimmu.2018.01486
79. Lynch C, Panagopoulou M, Gregory CD. Extracellular vesicles arising from apoptotic cells in tumors: roles in cancer pathogenesis and potential clinical applications. *Front Immunol.* (2017) 8:1174. doi: 10.3389/fimmu.2017.01174
80. Chen W, Huang Y, Han J, Yu L, Li Y, Lu Z, et al. Immunomodulatory effects of mesenchymal stromal cells-derived exosome. *Immunol Res.* (2016) 64:831–40. doi: 10.1007/s12026-016-8798-6
81. Del Fattore A, Luciano R, Pascucci L, Goffredo BM, Giorda E, Scapaticci M, et al. Immunoregulatory effects of mesenchymal stem cell-derived extracellular vesicles on T lymphocytes. *Cell Transplant.* (2015) 24:2615–27. doi: 10.3727/096368915X687543
82. Del Fattore A, Luciano R, Saracino R, Battafarano G, Rizzo C, Pascucci L, et al. Differential effects of extracellular vesicles secreted by mesenchymal stem cells from different sources on glioblastoma cells. *Expert Opin Biol Ther.* (2015) 15:495–504. doi: 10.1517/14712598.2015.997706
83. Lee C, Mitsialis SA, Aslam M, Vitali SH, Vergadi E, Konstantinou G, et al. Exosomes mediate the cytoprotective action of mesenchymal stromal cells on hypoxia-induced pulmonary hypertension. *Circulation.* (2012) 126:2601–11. doi: 10.1161/CIRCULATIONAHA.112.114173
84. Anderson JD, Johansson HJ, Graham CS, Vesterlund M, Pham MT, Bramlett CS, et al. Comprehensive proteomic analysis of mesenchymal stem cell exosomes reveals modulation of angiogenesis via nuclear factor-kappaB signaling. *Stem Cells.* (2016) 34:601–13. doi: 10.1002/stem.2298
85. Pakravan K, Babashah S, Sadeghizadeh M, Mowla SJ, Mossahebi-Mohammadi M, Ataei F, et al. MicroRNA-100 shuttled by mesenchymal stem cell-derived exosomes suppresses *in vitro* angiogenesis through modulating the mTOR/HIF-1alpha/VEGF signaling axis in breast cancer cells. *Cell Oncol.* (2017) 40:457–70. doi: 10.1007/s13402-017-0335-7
86. Kim HS, Choi DY, Yun SJ, Choi SM, Kang JW, Jung JW, et al. Proteomic analysis of microvesicles derived from human mesenchymal stem cells. *J Proteome Res.* (2012) 11:839–49. doi: 10.1021/pr200682z
87. Harting MT, Srivastava AK, Zhaorigetu S, Bair H, Prabhakara KS, Toledano Furman NE, et al. Inflammation-stimulated mesenchymal stromal cell-derived extracellular vesicles attenuate inflammation. *Stem Cells.* (2018) 36:79–90. doi: 10.1002/stem.2730
88. Ruppert KA, Nguyen TT, Prabhakara KS, Toledano Furman NE, Srivastava AK, Harting MT, et al. Human mesenchymal stromal cell-derived extracellular vesicles modify microglial response and improve clinical outcomes in experimental spinal cord injury. *Sci Rep.* (2018) 8:480. doi: 10.1038/s41598-017-18867-w

89. Zhang Q, Fu L, Liang Y, Guo Z, Wang L, Ma C, et al. Exosomes originating from MSCs stimulated with TGF-beta and IFN-gamma promote Treg differentiation. *J Cell Physiol.* (2018) 233:6832–40. doi: 10.1002/jcp.26436
90. Collino F, Pomatto M, Bruno S, Lindoso RS, Tapparo M, Sicheng W, et al. Exosome and microvesicle-enriched fractions isolated from mesenchymal stem cells by gradient separation showed different molecular signatures and functions on renal tubular epithelial cells. *Stem Cell Rev.* (2017) 13:226–43. doi: 10.1007/s12015-016-9713-1
91. Lai RC, Arslan F, Lee MM, Sze NS, Choo A, Chen TS, et al. Exosome secreted by MSC reduces myocardial ischemia/reperfusion injury. *Stem Cell Res.* (2010) 4:214–22. doi: 10.1016/j.scr.2009.12.003
92. Lou G, Chen Z, Zheng M, Liu Y. Mesenchymal stem cell-derived exosomes as a new therapeutic strategy for liver diseases. *Exp Mol Med.* (2017) 49:e346. doi: 10.1038/emm.2017.63
93. Wu P, Zhang B, Shi H, Qian H, Xu W. MSC-exosome: a novel cell-free therapy for cutaneous regeneration. *Cytotherapy.* (2018) 20:291–301. doi: 10.1016/j.jcyt.2017.11.002
94. Cahill EF, Kennelly H, Carty F, Mahon BP, English K. Hepatocyte growth factor is required for mesenchymal stromal cell protection against bleomycin-induced pulmonary fibrosis. *Stem Cells Transl Med.* (2016) 5:1307–18. doi: 10.5966/sctm.2015-0337
95. Kennelly H, Mahon BP, English K. Human mesenchymal stromal cells exert HGF dependent cytoprotective effects in a human relevant pre-clinical model of COPD. *Sci Rep.* (2016) 6:38207. doi: 10.1038/srep38207
96. Phinney DG, Pittenger MF. Concise review: MSC-derived exosomes for cell-free therapy. *Stem Cells.* (2017) 35:851–8. doi: 10.1002/stem.2575
97. Gonzalez-King H, Garcia NA, Ontoria-Oviedo I, Ciria M, Montero JA, Sepulveda P. Hypoxia inducible factor-1alpha potentiates jagged 1-mediated angiogenesis by mesenchymal stem cell-derived exosomes. *Stem Cells.* (2017) 35:1747–59. doi: 10.1002/stem.2618
98. Cahill EF, Tobin LM, Carty F, Mahon BP, English K. Jagged-1 is required for the expansion of $CD4^+$ $CD25^+$ $FoxP3^+$ regulatory T cells and tolerogenic dendritic cells by murine mesenchymal stromal cells. *Stem Cell Res Ther.* (2015) 6:19. doi: 10.1186/s13287-015-0021-5
99. English K, Barry FP, Field-Corbett CP, Mahon BP. IFN-gamma and TNF-alpha differentially regulate immunomodulation by murine mesenchymal stem cells. *Immunol Lett.* (2007) 110:91–100. doi: 10.1016/j.imlet.2007.04.001
100. Ryan JM, Barry F, Murphy JM, Mahon BP. Interferon-gamma does not break, but promotes the immunosuppressive capacity of adult human mesenchymal stem cells. *Clin Exp Immunol.* (2007) 149:353–63. doi: 10.1111/j.1365-2249.2007.03422.x
101. Pavlyukov MS, Yu H, Bastola S, Minata M, Shender VO, Lee Y, et al. Apoptotic cell-derived extracellular vesicles promote malignancy of glioblastoma via intercellular transfer of splicing factors. *Cancer Cell.* (2018) 34:119–35.e10. doi: 10.1016/j.ccell.2018.05.012
102. Ghahremani Piraghaj M, Soudi S, Ghanbarian H, Bolandi Z, Namaki S, Hashemi SM. Effect of efferocytosis of apoptotic mesenchymal stem cells (MSCs) on C57BL/6 peritoneal macrophages function. *Life Sci.* (2018) 212:203–12. doi: 10.1016/j.lfs.2018.09.052

17

The Utilization of Freezing Steps in Mesenchymal Stromal Cell (MSC) Manufacturing: Potential Impact on Quality and Cell Functionality Attributes

Sofia Oja, Tanja Kaartinen, Marja Ahti, Matti Korhonen, Anita Laitinen and Johanna Nystedt*

Advanced Cell Therapy Centre, Finnish Red Cross Blood Service, Helsinki, Finland

Correspondence:
Johanna Nystedt
johanna.nystedt@gmail.com

Some recent reports suggest that cryopreserved and thawed mesenchymal stromal cells (MSCs) may have impaired functional properties as compared to freshly harvested MSCs from continuous cultures. A cryopreservation step in the manufacturing process brings important benefits, since it enables immediate off-the-shelf access to the products and a completion of all quality testing before batch release and administration to the patient. Cryopreservation is also inevitable in MSC banking strategies. In this study, we present the results from the MSC stability testing program of our in-house manufactured clinical-grade allogeneic bone marrow-derived MSC product that is expanded in platelet lysate and frozen in passage 2. The current manufacturing protocol contains only one freezing step and the frozen MSC product is thawed bed-side at the clinic. We can conclude superior viability and cell recovery of the frozen and thawed MSC product utilizing the validated freezing and thawing protocols we have developed. The MSC phenotype and differentiation potential was generally found to be unaltered after thawing, but the thawed cells exhibited a 50% reduced, but not completely abolished, performance in an *in vitro* immunosuppression assay. The *in vitro* immunosuppression assay results should, however, be interpreted with caution, since the chosen assay mainly measures one specific immunosuppressive mechanism of MSCs to suppress T-cell proliferation. Since at least two freezing steps are usually necessary in MSC banking strategies, we went on to investigate the impact of repeated freezing on MSC quality attributes. We can conclude that two freezing steps with a preceding cell culture phase of at least one passage before freezing is feasible and does not substantially affect basic cell manufacturing parameters or quality attributes of the final frozen and thawed product. Our results suggest, however, that an exhaustive number of freezing steps (≥4) may induce earlier senescence. In conclusion, our results support the utilization of frozen MSC products and MSC banking strategies, but emphasize the need of always performing detailed studies on also the cryopreserved MSC counterpart and to carefully report the cryopreservation and thawing protocols.

Keywords: cryopreservation, freezing, MSC, mesenchymal stromal cell, immunosuppression, cell banking, master cell bank

HIGHLIGHTS

- 1-2 freezing steps for MSCs in early passage is feasible and preserves most of the *in vitro* functional properties.Interim freezing steps are not reflected in standard manufacturing parameters.
- the *in vitro* immunosuppressive performance of frozen and thawed MSCs may be different from their fresh counterparts with a reduced, but not abolished *in vitro* performance specific for the IDO pathway.
- *in vitro* immunosuppression assay results ought to be interpreted with caution.
- cryopreserved and thawed MSCs may be different from their fresh counterparts, but that does not necessarily translate to reduced clinical efficacy.

INTRODUCTION

Mesenchymal stromal cells (MSCs) are being widely studied as potential cell therapy medicinal products due to their immunomodulatory properties, which have been established by *in vitro* studies and in several clinical trials (1, 2). Within this context, MSC therapy may hold substantial promise particularly in the treatment of inflammatory and autoimmune conditions and MSCs have therefore been widely employed as a salvage treatment option in refractory graft-vs.-host disease (GvHD) in its acute form (3–6). It is, however, becoming evident that albeit some patients with severe acute GvHD markedly benefit from the MSC treatment, some patients experience no improvement of the symptoms (7, 8). Based on numerous published patient cohorts and thousands of treated patients, the safety of MSC therapy appears clear, but less certain is the efficacy of the MSC therapy. It is currently evident that the overwhelming positive results reported from *in vitro* and preclinical animal studies have largely not yet translated into full clinical efficacy.

Clearly, there is still much to be learned and optimized with regards to the *in vivo* interactions of MSCs in human pathological states. It has been thought that allogeneic MSCs do not provoke an overt immune reaction in the host even when the host and donor are not human leukocyte antigen (HLA) matched. This concept has been challenged recently, but fortunately not from a safety point of view. Allogeneic MSCs are obviously not as hypoimmunogenic as originally thought and an immune activation of host cytotoxic T-cells and cytotoxic activity against MSCs is actually critical for effective immunosuppression through phagocytosis of apoptotic MSCs and subsequent macrophage polarization (9, 10). The apoptosis-based MSC immunomodulation mechanism has significantly improved our understanding on the mechanistic properties of MSCs, but we also need to clarify how the functional properties of MSCs may be affected by differences in the manufacturing strategies and culturing conditions.

Clinical MSC preparations can either be fresh, meaning the cells are detached from the cell cultures just before administration to patients, or the cells can be cryopreserved and thawed bedside just before administration. Naturally, every step of the clinical manufacturing process need to fulfill local legislation, such as the Advanced Therapy Medicinal Product (ATMP) legislation in all EU countries, and comply with GMP requirements specific for the area. MSCs are usually manufactured from third-party healthy donors and administered in a completely HLA-mismatched allogeneic setting. Not unsurprisingly, it is therefore more common to utilize cryopreserved MSCs in the clinical setting, since cryopreservation includes several obvious benefits: (1) rapid access to the cell product in acute conditions, (2) convenient logistics to the clinical site, (3) minimal reconstitution activities of the cell product bed-side, (4) substantial quality benefits with all quality testing completed before batch release and administration to the patient, (5) large-scale manufacturing and master cell bank opportunities, and (6) enablement of treatment protocols with numerous identical cell doses. The effect of cryopreservation on MSC therapeutic properties has, however, become highly controversial. Some recent reports suggest that cryopreserved MSCs may have impaired functional properties when compared with freshly harvested MSCs from continuous cultures (11–13). On the contrary, some recent conflicting studies have shown that the functionality and cell characteristics of cryopreserved MSCs are comparable to fresh MSCs (14–16). Cryopreserved MSCs have been explored in several clinical studies for graft-vs.-host disease by us and others with partially encouraging results (7, 17–21). These conflicting results warrant further studies to elucidate the impact of a cryopreservation step in the manufacture of clinical-grade MSCs.

In this study, we performed an in-depth analysis of our in-house manufactured and cryopreserved platelet lysate-expanded bone marrow-derived MSC product, which is provided frozen in passage 2 (p2) for clinical use. We report superior viability and cell recovery of our cryopreserved MSC product utilizing the freezing and thawing protocols we have established. The MSC phenotype and differentiation potential was generally unaltered after thawing, but we report a reduced, but not abolished, *in vitro* immunosuppression capacity directly after thawing. Since our current manufacturing protocol contains only one freezing step in p2 and multiple freezing steps are usually necessary in MSC banking strategies, we went on to investigate the impact of repeated freezing. We can conclude that two freezing steps with preceding cell culture of at least one passage before freezing is feasible and does not substantially affect basic cell manufacturing parameters (cell yield, growth kinetics, population doubling (PD) number) or the basic quality attributes of the final frozen and thawed product. Our results indicate that freezing of MSCs is feasible and preserves most of the functional properties of the product, keeping in mind that the *in vitro* immunosuppressive potency assay results with thawed MSCs could indicate a reduced IDO-dependent immunomodulatory capacity. Our results might also suggest that an exhaustive amount of freezing steps (≥4) may accelerate the induction of senescence.

MATERIALS AND METHODS

MSC Starting Material

Healthy, voluntary bone marrow (BM) donors were recruited specifically for donating starting material for clinical-grade MSC manufacturing. The donor recruitment, eligibility assessment and bone marrow aspiration protocol was approved by the local ethics committee and the BM collection procedure was authorized by the Finnish Medicines Agency under a tissue establishment license. Forty milliliters of bone marrow was aspirated under local anesthesia from the posterior iliac crest into heparinized syringes after a written informed consent. Donor eligibility was assessed by an extensive health questionnaire and testing for infectious diseases (NAT tests: HIV-1, HBV, and HCV by Ultrio NAT, B19 (Parvo), HAV; Serological tests: HBsAg, HCVAb, HIVAb/Ag, HBcAb, TrpaAb (Syphilis), HTLVAb and CMVAbG). The donors had an average age of 25 years (range 18–35 years) and the male/female ratio was 30/70%, respectively.

Bone Marrow-Derived MSC (BM-MSC) Culture Protocol

The clinical-grade MSC culture protocol layout is presented in **Figure 1** and has been previously published together with clinical data (7, 8). Cultured MSCs are classified as ATMPs in the EU and the clinical-grade MSC products were manufactured in the GMP facility of the Finnish Red Cross Blood Service in Helsinki under a national ATMP hospital exemption license authorized by the Finnish Medicines Agency. The BM sample was processed within 2 h of aspiration and was diluted with DPBS (CTS Life Technologies, Thermo Fischer Scientific, MA, USA) containing 0.02% Versene (EDTA, Lonza, Switzerland) before density gradient centrifugation (Ficoll-Paque Premium, GE Healthcare Biosciences, Sweden). Primary cultures (p0) were initiated by plating isolated BM mononuclear cells (MNCs) at 400 000 cells/cm^2. All culturing steps were performed at +37°C and in humidified atmospheric oxygen with 5% CO2. The animal serum-free, antibiotic-free basal culture medium consisted of D-MEM low glucose (Life Technologies) supplemented with 40 IU/ml heparin (Heparin Leo 5000 IU/ml, Leo Pharma, Sweden) and 10% pooled platelet lysate (Finnish Red Cross Blood Service, Helsinki, Finland; for more details see (22, 23). Culture medium was replaced twice a week and the cells were detached with TrypLE Select CTS (Life Technologies). The cells were plated at a density of 1000 cells/cm^2 from p1 and onwards. Culture confluency and cell morphology were monitored by phase contrast microscopy. Cell numbers and viability were determined using the NucleoCounter NC-100 (ChemoMetec A/S, Denmark). Research-grade MSC cultures (from p3 onwards) were cultured outside the GMP facility and the cultures were additionally supplemented with 100 IU/ml Penicillin and Streptomycin (Gibco, Life Technologies).

Freezing and Thawing of MSCs

Clinical-grade MSCs in passage 2 were detached from cultures with TrypLE Select CTS (Life Technologies), washed with DPBS and suspended in a pre-chilled freezing medium consisting of 10% dimethylsulfoxide (DMSO, CryoSure, Wak-Chemie Medical GmbH, Germany) and 90% human serum albumin (Albunorm 200 g/l, Octapharma, Switzerland). Clinical-grade cells were dispensed in freezing bags with either 50 × 10^6 or 100 × 10^6 cells/bag (CryoMACS®, Miltenyi Biotech GmbH, Germany) as described previously (7). Research-grade cultures were frozen in smaller cryovials (Nunc CryoTubes®, Nalge Nunc, NY, USA), but with comparable cell density. Both freezing bags and cryovials were slowly chilled at −70°C before transferring the bags or vials to vapor (clinical-grade) or liquid (research-grade) nitrogen tanks at ≤ −140°C. Frozen bags or cryovials were thawed at +37°C water bath according to a standardized operating procedure. The exact thawing time was always documented and never exceeded 3 min. Clinical-grade cells in freezing bags were always diluted in saline to 2 × 10^6 cells/ml immediately after thawing, after which the thawed cells were filtered through a 200 μm clinical filter. The cells were allowed to rest at room temperature for 5 to 15 min before analysis or further processing.

Colony Forming Units and Population Doubling Numbers

Original MSC amounts in the bone marrow aspirate was determined by the colony forming units-fibroblast (CFU-F) assay (24). Four Lakhs isolated BM-MNCs/cm^2 were plated in replicate on 6-well plates in basal culture medium and were let to adhere for 72 h. Non-adherent cells were removed, the wells were washed four times with DPBS and culturing was continued for 5–10 days until individual colonies consisting of at least 20 cells were clearly visible. The cells were fixed with methanol (Methanol EMSURE® ACS, Merck KgAa, Germany), stained with Giemsa stain (Giemsa's Azur Eosin Methylene Blue, Merck, KgAa) and the number of colonies in each well was calculated. The CFU-F colony numbers were correlated to the plated MNC numbers and culture area cm^2. Population doubling (PD) numbers were calculated for each passage according to the equation [PD = log^2 (NH/N1)], where N1 is the number of seeded cells/cm^2, and NH is the number of harvested cells/cm2. The CFU-F number of the p0 culture was used as the first N1 (22). The cumulative PD number thus also include the amount of population doublings that occur during the initial p0 culture stage.

Cell cultures were considered fully senescent when the confluency of the culture remained under 30% after 2 weeks of culturing without passaging combined with manifested senescence-associated morphological characteristics such as enlarged cell size and granularity (25).

Cell Surface Phenotype by Flow Cytometry

The MSC immunophenotype was determined by flow cytomety using the Navios Cytometer (Beckmann Coulter, IN, USA) and FlowJo analysis software (version 7.4.1. TreeStar Inc., CA, USA). The cell surface antigens CD44, CD49e, CD13, CD90, CD73, CD29, CD86, CD105, HLA-ABC, CD14, CD19, CD34, CD45, CD80, and HLA-DR were analyzed and at least 5000 cells were included in the analysis. All antibodies were purchased from BD Biosciences (CA, USA) except CD80 (Beckmann Coulter) and HLA-DR (Abcam, UK). The proportion of non-viable cells were determined during analysis by staining the cells with propidium iodide (50 μg/ml in PBS, BD Biosciences).

Differentiation Assays

Osteogenic Differentiation Assay

MSCs were plated at a density of 1000 cells/cm^2 in 6-well-plates and cultured in platelet lysate supplemented basal culture

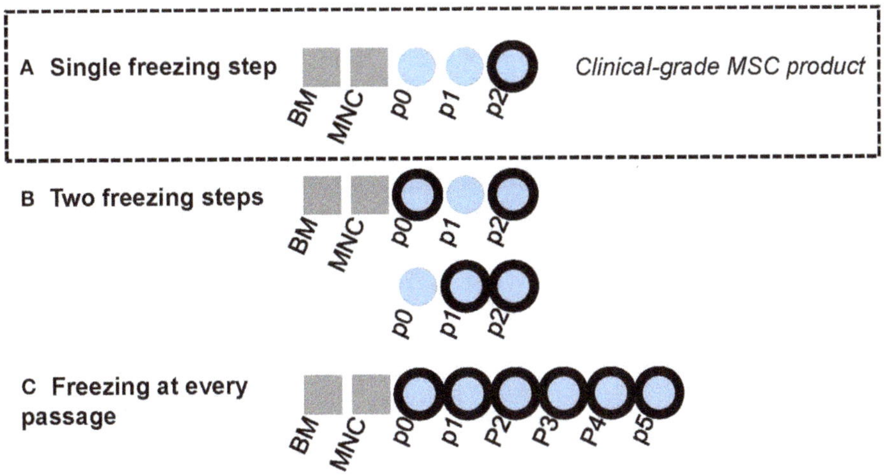

FIGURE 1 | Graphical layout of experimental design. Bold black outliners indicate freezing steps. **(A)** In-house BM-MSC manufacturing protocol for clinical-grade MSC product. The final product in passage two is cryopreserved. **(B)** The effect of adding one interim freezing step at either passage 0 or passage 1 in addition the final freezing step at passage 2 was explored in this study. **(C)** Some MSC lines were cultured until passage 5 in this study with freezing at every passage.

medium with 100 IU/ml penicillin and streptomycin until 70% confluency. Osteogenic differentiation was induced with 0.1 μM dexamethasone (Dexamethasone, BioXtra, Sigma, MO, USA), 50 μM ascorbic acid (Ascorbic acid 2-phosphate, Sigma) and 10 mM β-glyserophosphate (β-Glyserophosphate disodium salt pentahydrate, AppliChem, Germany). The MSCs were maintained in the induction medium until the formation of visible calcium phosphate precipitate. The cells were fixed with 4% paraformaldehyde (PFA) and the calcium phosphate precipitate was stained using von Kossa staining.

Adipogenic Differentiation Assay

MSCs were plated at density 1000 cells/cm^2 in 6-well plates and cultured in the platelet lysate supplemented basal culture medium with 100 IU/ml penicillin and streptomycin until 70–100% confluent. Adipogenic differentiation was induced for 3–4 days by an induction medium consisting of αMEM GlutaMax (Gibco, Life Technologies), 10% inactivated fetal bovine serum (Gibco, Life Technologies), 20 mM Hepes (Gibco, Life Technologies), 100 U/ml penicillin and streptomycin (Gibco, Life Technologies) supplemented with the induction cocktail of 0.1 mM indomethacin (Sigma), 44 μg/ml 3-isobutyl-methyl-xanthine (IBMX-22*), 0.5 μg/ml insulin (Insulin-0.25*) and 0.4 μg/ml dexamethasone (DM-200*) (*Preadipocyte Differentiation Medium Supplement Pack, PromoCell, Italy). Control cells were only maintained in induction medium without the induction cocktail. Differentiation was finalized by culturing the cells in a terminal differentiation medium consisting of induction medium supplemented with 0.1 mM indomethacin (Sigma), 0.5 μg/ml insulin (Insulin-0.25*) and 3.0 μg/ml ciglitazone (Ciglitazone-1.5*) (*Preadipocyte Differentiation Medium Supplement Pack) for 2–4 weeks until visible lipid droplets could be observed. The cells were fixed with 4% PFA and were stained using Sudan III.

Immunosuppression Assays

Basic T-Cell Proliferation Assay

Peripheral blood mononuclear cells (PBMNCs) from healthy voluntary blood donors were used as responder cells and were isolated from buffy coats (Finnish Red Cross Blood Service, Helsinki, Finland) by density gradient centrifugation (Ficoll-Paque Premium, GE Healthcare). Isolated PBMNCs were cryopreserved in 10% DMSO (CryoSure DMSO, Wak-Chemie Medical GmbH) and 90% inactivated fetal bovine serum (Gibco, Life Technologies). Responder cells from at least two individual donors were used in each experiment. The responder cells were thawed in +37°C water bath, were let to rest at room temperature for at least 30 min and were subsequently filtered through a 30 μm sterile filter and labeled with 5 μM 5(6)-Carboxyfluorescein succinimidyl ester (CFSE mixed isomers, Molecular Probes, Life Technologies). Co-cultures with MSCs were initiated by first allowing the MSCs to adhere for 3–4 h (max 24 h) before adding the labeled responder cells to the co-cultures together with the activation cocktail composing of 6.5 μg/ml CD3 (anti-human CD3, clone Hit3a, Bio Legend, CA, USA) and 6.5 μg/ml CD28 (anti-human CD28, clone CD28.2, Bio Legend) antibodies. All co-cultures were performed with a T-cell culture medium consisting of RPMI 1640 Medium with L-glutamine and HEPES buffer, 100 U/ml penicillin-streptomycin and 5% inactivated fetal bovine serum (all from Gibco, Life Technologies). Every experiment was performed with replicate samples, at least two different responder cells and also including different responder:MSC cell ratios (1:10, 1:20 and 1:50). Each experiment also contained an IDO inhibition verification with 1.0 mM of the IDO inhibitor 1-Methyl-L-Tryptophan (L-1MT, Sigma). The responder:MSC cell ratio was always 1:10 in the IDO inhibitor tests. The co-cultures were incubated for 4–5 days before flow cytometric analysis of T-cell proliferation (Navios, Beckmann Coulter or BD FACS Aria IIu, Becton Dickinson, NH, USA).

Alternative Set-Ups

To elucidate the impact of assay design in the final read-out of the immunosuppression assay, three alternative set-ups were also utilized: the MSCs were either (1) combined directly after thawing with the labeled responder cells and activation cocktail or (2) allowed to adhere to the assay plates for exactly 24 h before adding the labeled responder cells and the activation cocktail (24 h culture rescue), or (3) the responder cells were pre-activated with the CD3/CD28 antibodies for 24 h before adding fresh MSCs from continuous cultures to the assay ($n = 2$).

Analysis of Mean Cell Area

A detailed protocol for image acquisition and analysis is described in Oja et al. (25). Briefly, MSCs were plated at density 3000 cells/cm^2 in 6-well culture plates and were allowed to adhere and proliferate for 48 h before fixation with 4% PFA (Sigma). Fixed MSCs were permeabilized with 0.1% Triton X-100/PBS (Sigma) and the nuclei were stained with 0.125 μg/ml DAPI and the cytoplasm with 1 μg/ml Cell Mask Deep Red Stain (both from Life Technologies). The images were acquired using the Cell Insight high content screening platform (Thermo Scientific, IL, USA). Signals from DAPI and Cell Mask stains were recorded on separate channels, using filters 386 and 630 nm, respectively. Images were acquired from 6 wells resulting in 1998 images per channel on each run and mean cell area in different passages were calculated.

Statistical Methods

Statistical analyses were performed using the GraphPad Prism (version 7.02) software (GraphPad Software Inc., La Jolla, California). A two-tailed paired t-test with a Mann-Whitney test was used for the comparison of two groups. 1way ANOVA with either Bonferroni's multiple comparisons test or Brown-Forsythe test (**Figure 6C**) was utilized when comparing multiple groups. Differences were considered statistically significant when $p < 0.05$.

RESULTS

Excellent Viability and Cell Recovery and Restored Basic MSC Characteristics After a Single Freezing Step at Passage 2

As part of the stability testing program of our in-house manufactured clinical-grade bone marrow-derived MSC product, which is frozen in p2 (7, 22), we investigated the quality and functional properties of the product before and after the single freezing step at p2. The outlines of the clinical MSC culture protocol is presented in **Figure 1A**. The cells were frozen in doses of either 50 or 100 × 10^6 cells in 50 ml freezing bags and were thawed and diluted in 0.9% NaCl according to a validated protocol in use bedside in the clinics, which also includes forcing the cells through a syringe and 200 μm filter (see Materials and Methods section for more details). The viability of 26 individual clinical MSC batches in p2 before freezing is presented in **Figure 2A**. The mean viability of these batches was 94.8% (min 91.5%, max 96.7%) before dispensing the cells in freezing bags and initiating the freezing. The viability and recovery was studied in 8–9 individual bags (derived from four individual MSC batches) after freezing and thawing. As presented in **Figure 2A**, the mean viability and cell recovery after thawing was 94.2% (min 92.7%, max 95.4%) and 96.4% (min 88.2%, max 100%), respectively. We can therefore conclude a superior viability and cell recovery after one freezing step in p2 utilizing the freezing and thawing protocols we have developed and validated for large cell amounts (50 or 100 × 10^6 cells) in 50 ml freezing bags and even after forcing the thawed cells through standard infusion materials composing of a syringe, 200 μm filter and infusion lines.

The osteogenic and adipogenic differentiation potential of cells from p2 was examined before and after freezing. We compared the differentiation potential in at least three individual clinical-grade MSC batches and we always observed a preserved differentiation potential in all of these batches (**Figure 2B**). The basic MSC phenotype, as studied by flow cytometry, was also compared before and after freezing in the exact same MSC batches (**Figure 2C**). The generally accepted positive MSC markers CD90, CD73, CD105, CD44, CD49e, CD13, CD29, and HLA-ABC remained unaltered in the flow cytometry analysis directly after thawing (**Figure 2C**, left panel) and clearly fulfilled the ISCT minimal criteria of ≥95% positive expression (26). The expression of the negative MSC markers CD14, CD19, CD34, and CD45 (**Figure 2C**, right panel) also fulfilled the ISCT minimal criteria (26) and were consistently ≤2% with the exception of HLA-DR (**Figure 2C**, right panel). The HLA-DR cell surface expression in pooled platelet lysate cultured BM-MSC has previously been shown to be routinely ≥2% (22). The mean HLA-DR cell surface expression in the 26 individual MSC batches analyzed in this study was 16.2% at p2 before freezing (min 0.9%, max 56.5%) (data not shown). The HLA-DR expression was above the ≥2% expression limit also after freezing and thawing as presented in **Figure 2C** (right panel), but it is interesting that there was a drop of about 8% in the HLA-DR cell surface expression as compared to the expression levels in the same cell batches before freezing. It is also noteworthy that we observed a substantial increase in PI+ positive cells in the post thaw cell samples (**Figure 2C**, right panel), indicating that the FACS staining and analysis procedure is harsh on newly thawed cells, since the cell viability (as determined with the NucleoCounter NC-100 method that also measures PI penetration) was concluded to be excellent directly after thawing as presented in **Figure 2A**.

The *in vitro* immunosuppression capacity of 16 individual clinical MSC batches in p2 before freezing is presented in **Figure 2D** utilizing a T-cell proliferation assay. The proliferative response was normalized to the proliferation control set as 100% in each assay. Every MSC batch consistently inhibited the proliferation of the activated responder T-cells to a mean of 31.06% of the original proliferation rate (min 13.75%, max 59.93%). Addition of the IDO inhibitor 1-Methyl-L-Tryptophan (L-1MT) significantly inhibited the immunosuppression capacity of the MSCs by approx. Fifty percent resulting in a mean T-cell proliferation rate of 69.52%

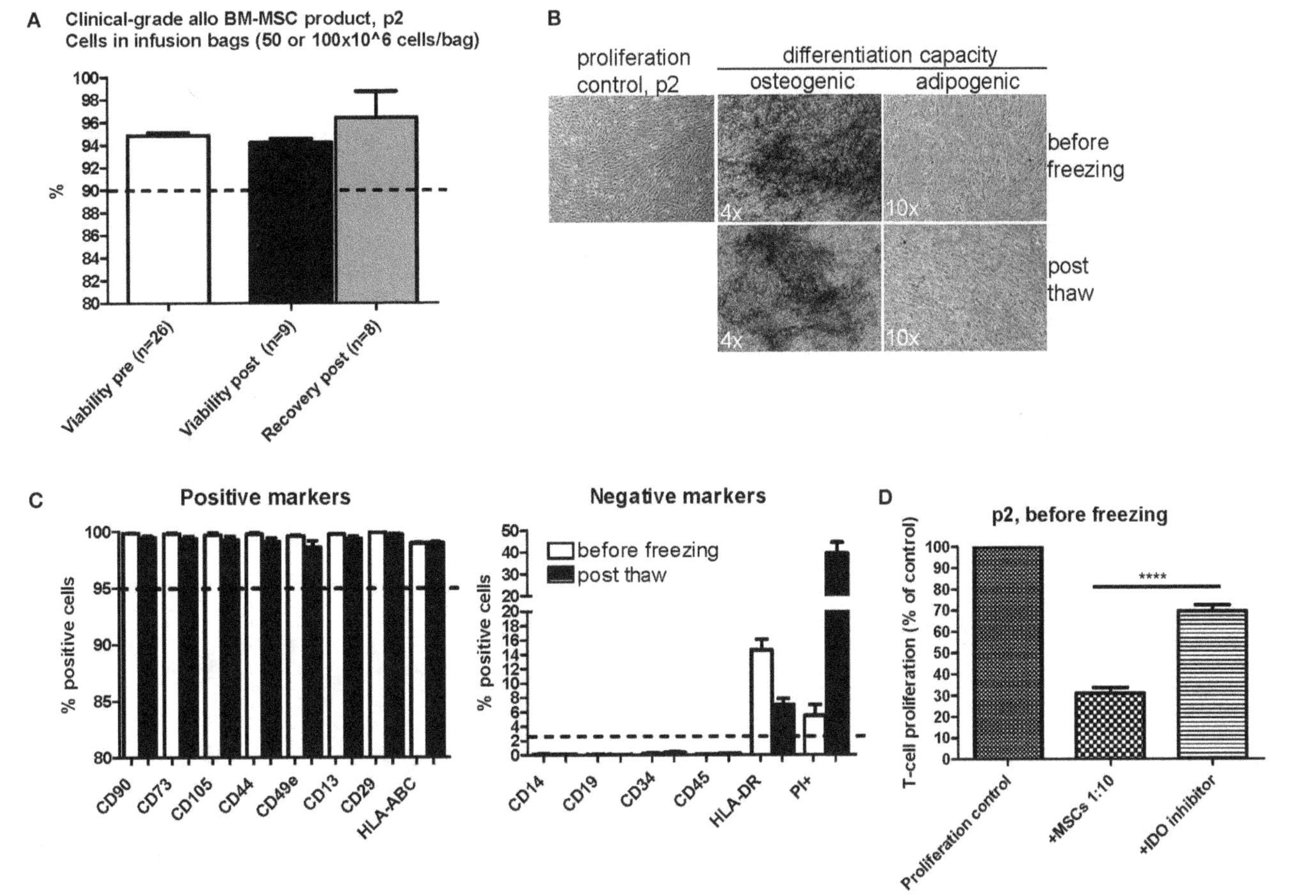

FIGURE 2 | Characterization of clinical-grade allogeneic BM-MSC product in passage 2 before and after one freezing step. **(A)** Viability before freezing (white) and after thawing (black) and cell recovery after thawing (gray). The cells were frozen at passage 2 in freezing bags with either 50 or 100 × 10^6 cells/bag. Results are presented as mean ± SEM. The viability differences were non-significant (p-value 0.1716 with a non-parametric two-tailed t-test and 95% confidence interval). **(B)** Osteogenic and adipogenic differentiation capacity of MSCs before and after freezing. Representative phase contrast microscope pictures are presented (n = 3). Enlargement is indicated in each picture separately. **(C)** MSC phenotype before (white) and after (black) freezing as studied by flow cytometry. Results are presented as mean ± SEM (n = 3). **(D)** T-cell proliferation assay results of clinical-grade MSCs in passage 2 before freezing with IDO inhibition. Results are presented as mean ± SEM (n = 16). 1-Methyl-L-Tryptophan (L-1MT) was utilized as an IDO inhibitor. ****p < 0.0001.

of the proliferation control (min 37.58%, max 92.50%) (P-value <0.0001 with a two-tailed paired t-test) (**Figure 2D**). The immunosuppression capacity of frozen and thawed MSCs is presented in **Figures 4, 5**.

Introducing an Additional Interim Freezing Step Does Not Affect Basic MSC Manufacturing Parameters

We next wanted to explore the possible impact of introducing additional interim freezing steps at either p0 or p1 in addition to the final freezing step at p2 (**Figure 1B**). No differences were seen in the cumulative culture time (**Figure 3A**), the cumulative population doubling (PD) number (**Figure 3B**) or the cell yield/cm^2 (**Figure 3C**) until reaching p2 when comparing unfrozen MSCs initiated from fresh bone marrow to MSCs frozen at p0 or p1 and subsequently subcultured until p2. Unfrozen and interim frozen cells from the same MSC batches were compared in these experiments and the experiments were repeated with four individual MSC batches. No statistically significant differences could be observed in either the p2 cell viability or the recovery after thawing utilizing only one freezing step (control) or two freezing steps either' at p0 + p2 or p1 + p2 (**Figure 3D**). The osteogenic and adipogenic differentiation capacity was also clearly not affected by an additional freezing step and the differentiation results of p2 post thaw cells after two freezing steps (**Figure 3E**) were completely comparable to the differentiation potential after one freezing step and, importantly, directly comparable to the differentiation capacity of unfrozen MSCs (**Figure 2B**).

Impaired *in vitro* Immunosuppression Activity of Thawed MSCs

The immunosuppression capacity of MSCs before freezing and after one freezing step at p2 or two freezing steps (p0

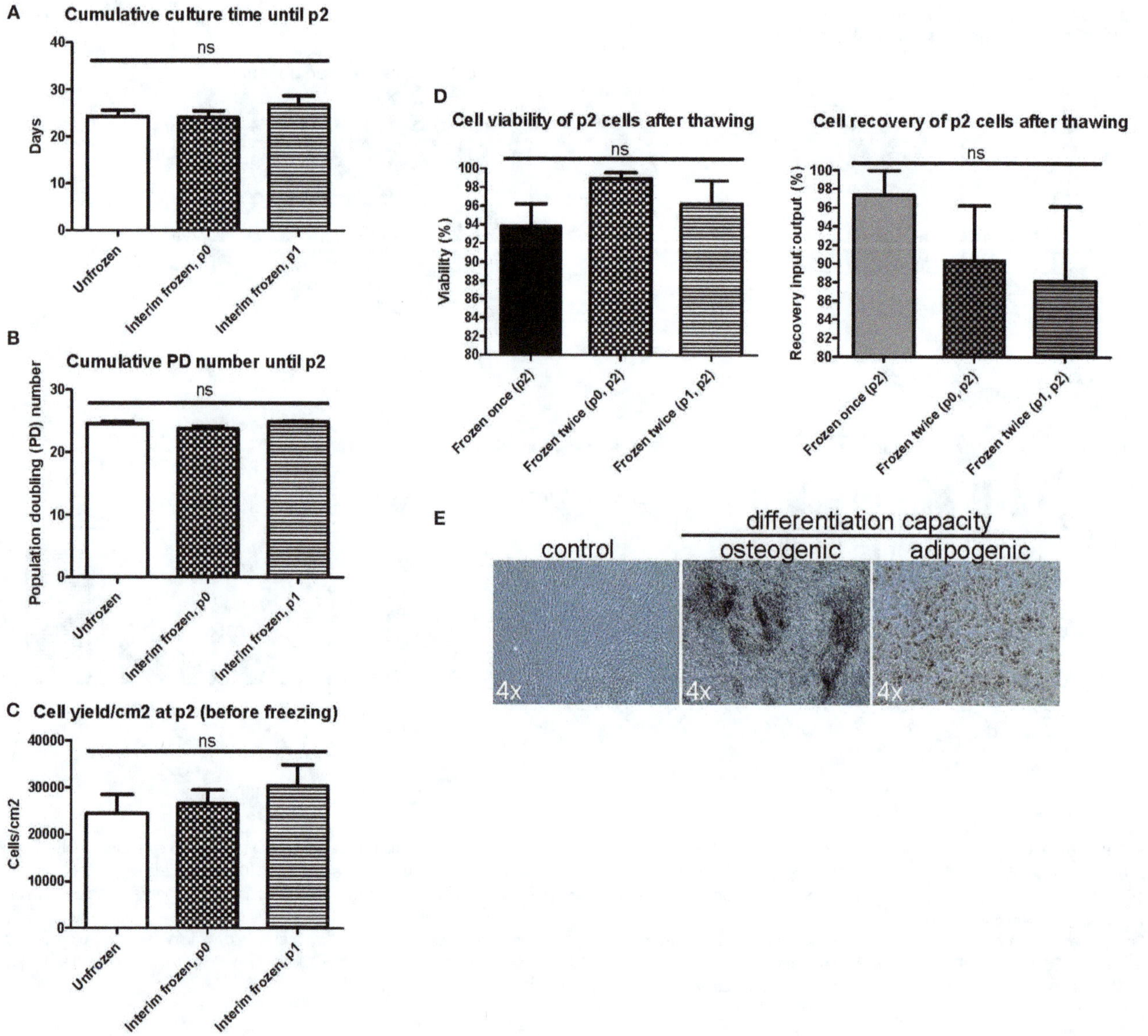

FIGURE 3 | Manufacturing parameters and characterization of BM-MSCs in passage 2 after two freezing steps. MSC products in passage 2 (p2) are compared after no preceding freezing steps (unfrozen), one freezing step at passage 0 (p0) or one freezing step at passage 1 (p1). **(A)** Cumulative culture time at p2. **(B)** Cumulative population doubling (PD) time. **(C)** Cell yield per cm^2 at p2 before freezing. **(D)** Cell viability (left) and cell recovery (right) at p2 after freezing and thawing. **(E)** Osteogenic and adipogenic differentiation capacity of MSCs frozen twice in either p0 or p1 and p2. Representative phase contrast microscope pictures presented from $n = 3$. Enlargement indicated for each picture separately. Results in **(A–D)** are presented as mean ± SEM ($n = 4$).

+ p2 or p1 + p2) was studied primarily using a T-cell proliferation assay (**Figures 4A,B**). The results indicate that the immunosuppression potential of frozen and thawed MSCs is reduced by approx. Fifty percent ($p < 0.0001$, $n = 20$) as compared to the fresh counterpart, at least in this particular *in vitro* assay design for immunosuppression (**Figure 4B**). It is, however, evident that additional interim freezing steps at p0 or p1 do not affect the immunosuppression potential of the final product at p2 (**Figure 4A**) and, interestingly, the difference in immunosuppression before and after freezing is not as evident and even not statistically significant in cells after two freezing steps. It should be pointed out that the MSCs were introduced to the FCS containing immunosuppression assay medium when initiating the assay, which is also the situation for the thawed MSCs. It should also be pointed out that the immunosuppression potential of frozen and thawed MSCs was not completely abolished, but was merely reduced by approx. Fifty percent to approximately 66% of the proliferation control (**Figure 4B**). Interestingly, the immunosuppression potential of frozen and thawed MSCs is comparable to the immunosuppression assays with fresh MSC where the IDO inhibitor 1-Methyl-L-Tryptophan (L-1MT) was added as presented in **Figure 2D**.

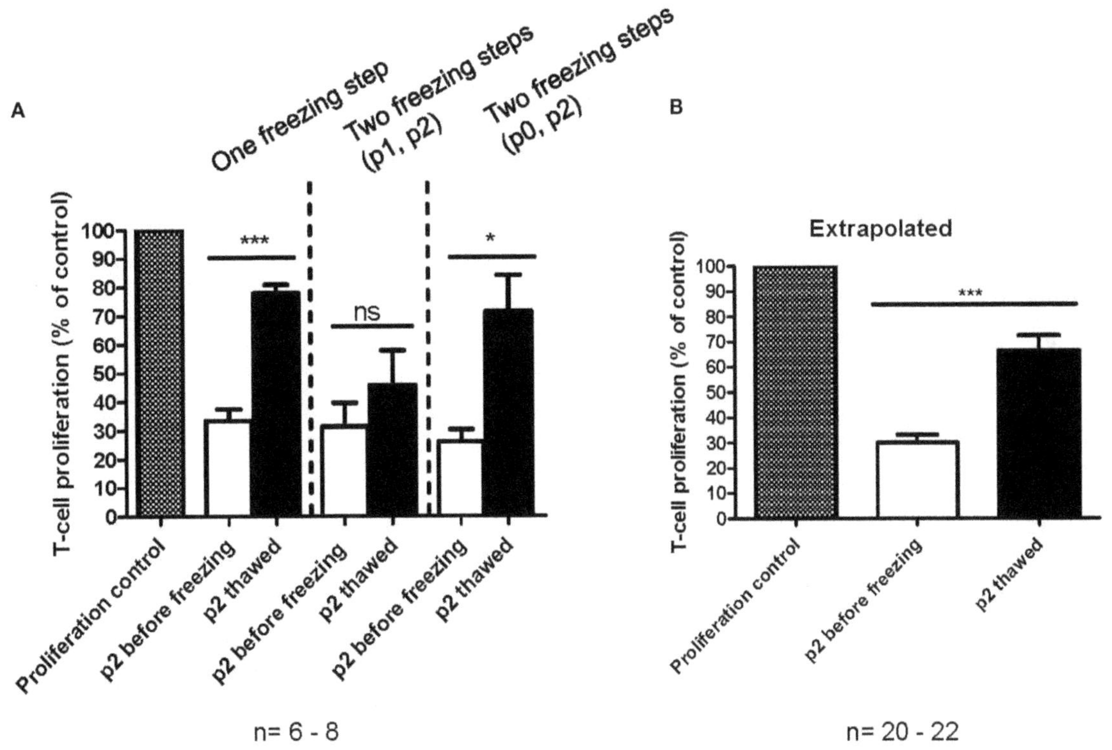

FIGURE 4 | *In vitro* immunosuppression capacity of cryopreserved MSCs as measured by a T-cell proliferation assay. **(A)** Immunosuppression capacity of MSCs before (white) and after freezing and thawing (black) with one freezing step at passage 2 (p2) or two freezing steps at either p0 or p1 and p2. The results are presented as mean ± SEM (n = 6-8). **(B)** Comparison of fresh (white) and frozen and thawed (black) MSCs in p2. The data in extrapolated from all the data presented in **(A)**, regardless of the number of freezing steps. The results are presented as mean ± SEM (n = 20–22). $^*p < 0.05$, $^{***}p < 0.001$.

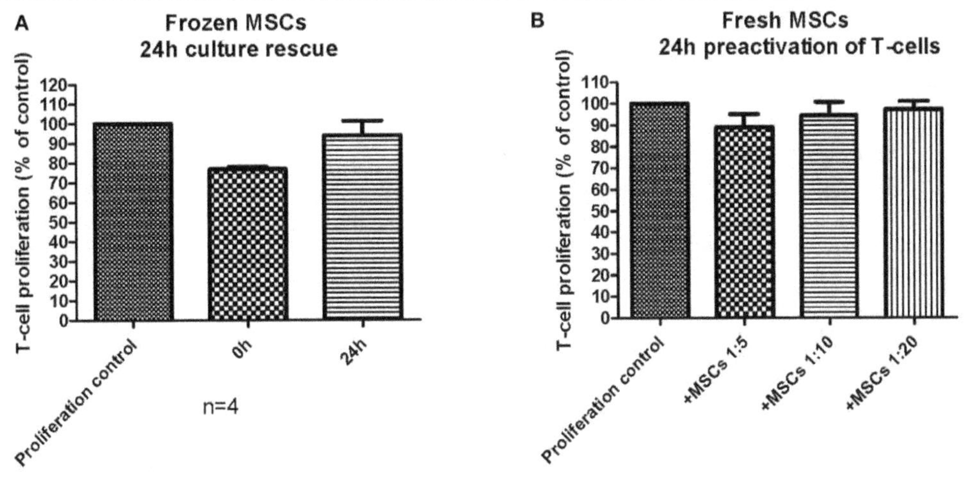

FIGURE 5 | The impact of assay design and the performance of an *in vitro* immunosuppression assay. **(A)** Frozen and thawed MSCs were either combined directly (0 h) with activated T-cells or pre-plated (culture rescued) for 24 h before combined with the activated T-cells (n = 4). **(B)** Fresh MSCs were combined with T-cells, which had been pre-activated for 24 h before combining with the MSCs. The results are presented as mean ± SEM.

Immunosuppression *in vitro* Assay Design and Impact On Readout

The chosen primary immunosuppression assay used in this study was a T-cell proliferation assay, where the MSCs were always first plated alone and allowed to adhere (usually 3–4 h and up to 24 h with fresh MSCs), after which the labeled responder MNCs were added to the assay together with the T-cell activator molecules CD3 and CD28. Fresh MSCs consistently inhibited

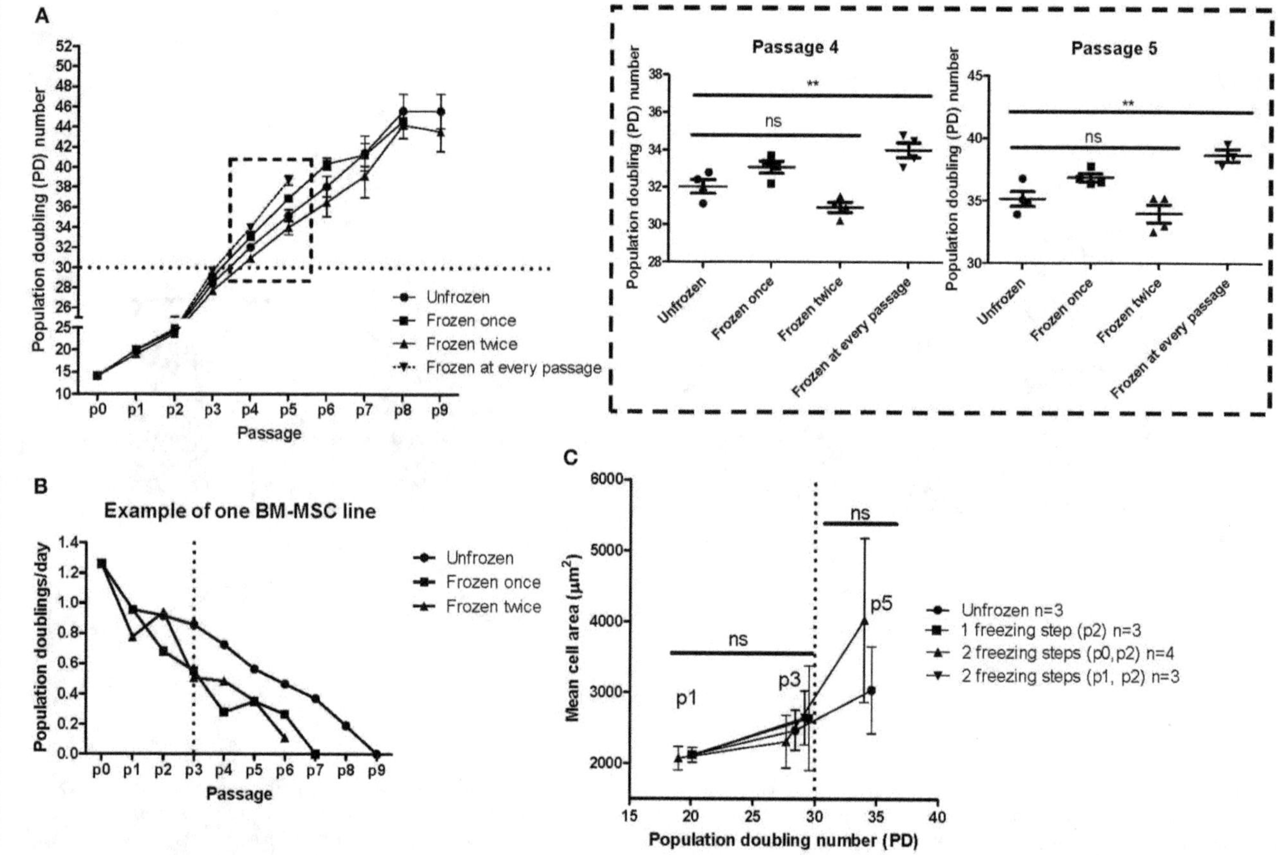

FIGURE 6 | MSC freezing and impact on replicative senescence. **(A)** Cumulative population doubling (PD) numbers at each passage of the same MSC line without any freezing steps, one freezing step, two freezing steps or freezing at every passage until reaching replicative senescence. The results are presented as mean ± SEM of four individual MSC lines. The box with the dashed boarder lines containing the results for passage 4 and passage 5 is enlarged on the right. **(B)** Example of the growth kinetics of one of the four individual MSC lines presented in **Figure 6A**. **(C)** Mean cell area compared to population doubling (PD) numbers of either unfrozen MSCs or after either one or two freezing steps. $^{**}p < 0.01$.

the T-cell proliferation with all studied MSC batches in this assay format with a variable initial MSC adhering time from 3–24 h (**Figure 4**). Since the frozen and thawed MSCs were not allowed to adhere for more than 9 h, we next wanted to confirm that the reduced immunosuppression of the frozen and thawed MSCs (**Figure 4**) were not due to a shorter adhering time. In **Figure 5A**, frozen and thawed MSCs were either combined with the responder cells directly (0 h) after thawing or after a 24 h adherence time. This 24 h adhering period can alternatively be named a 24 h culture rescue period as used by others (11). As can be seen in **Figure 5A**, a 24 h rescue period did not improve the *in vitro* immunosuppression response and, on the contrary, almost completely diminished the *in vitro* immunosuppression response of thawed MSCs. The immunosuppression of thawed MSCs immediately combined with the responder cells (0 h) was completely comparable to the frozen and thawed MSCs after 3–9 h of preplating as presented in **Figure 4**. In our hands and with the chosen assay format, we could not see a benefit of a 24 h culture rescue period for the frozen and thawed MSCs.

We further wanted to explore how small changes in the immunosuppression assay design could possibly impact the final result. In the basic immunosuppression assay format, we activate the responder MNCs simultaneously when added to the MSCs. We therefore explored the impact of utilizing pre-activated responder MNCs with fresh MSCs (**Figure 5B**). The result was strikingly different: the fresh MSCs, which always perform well in the *in vitro* immunosuppression assay, produced only a very modest immunosuppression, even in higher cell:responder ratios, and the immunosuppression was even worse than for the frozen and thawed MSCs presented in **Figure 4**. This experiment demonstrates well how easily an immunosuppression readout can be modified by small, but critical, changes in the assay design.

The Impact of Freezing on Replicative Senescence

Four individual MSC lines derived from individual bone marrow donors were cultured from fresh bone marrow and with continuous passaging until reaching complete replicative senescence (**Figure 6A**). During these culturing experiments, aliquots were taken during passaging and were cryopreserved (see experiment layout in **Figure 1C**). Frozen interim samples were subsequently thawed and cultured until senescence. We finally compared the culture kinetics of the same individual MSC line without any freezing steps, one freezing step, two freezing steps or freezing at every passage until reaching p5 (**Figure 6A**). We chose to analyze the data by comparing the population doubling (PD) numbers at each passage vs. the passage numbers. A strikingly similar culture kinetics with no statistically significant differences at any passage number was observed between the unfrozen cultures vs. cells that had been frozen once (at p1) or twice (at p0 and p2). We can, however, conclude that there is more variability in the culture kinetics after higher passaging numbers >p3. Interestingly, some significant differences were revealed in the cultures where the MSCs were frozen at every passage from p0 until p5, where the cells that were frequently frozen and thawed clearly had a higher PD number compared to cells at the same passage number which had either never been frozen or had been frozen once or twice during culturing (**Figures 1C, 6A**, blow up). This was, however, evident only at the higher passage numbers 4 and 5 and was not still evident at passage 2 or 3. This might indicate that freezing and thawing affects the senescence mechanisms, but only after an exhaustive number of freezing steps. Furthermore, the possible impact of freezing steps might be very individual for a particular MSC line, since one of the four analyzed MSC lines clearly demonstrated a different culture kinetics after one or two interim freezing steps as compared to the unfrozen culture (**Figure 6B**). In this particular case, it looked like this individual MSC line reached senescence (defined as when the number of PD's/day reached zero) two passages earlier (**Figure 6B**). It is also evident that the PD capacity per day decreases as early as from p2-3, but markedly after p3 (**Figure 6B**). We finally looked at the mean cell area correlated to both passage and PD numbers as an indicator of early senescence (**Figure 6C**). The cell area is clearly unaffected after one to two freezing steps as compared to their unfrozen counterparts during early passage up to p3 and PD numbers up to 30 (**Figure 6C**). At later passage numbers the variability in cell size increases, but the cell size differences between cells that have been frozen twice and their unfrozen counterpart is unsignificant (**Figure 6C**), which indicates that a moderate number of freezing steps does not significantly affect early onset of senescence of MSCs.

DISCUSSION

The conflicting results on the potential functional consequences of MSC cryopreservation led us to thoroughly investigate the functionality and quality attributes of our clinical-grade MSC product, mainly by utilizing a set of standard quality control and batch release assays used in clinical MSC manufacturing by us and others. Our in-house GMP manufacturing protocol for an allogeneic BM-derived platelet lysate-expanded MSC product contains only one freezing step at p2 (**Figure 1A**). The cryopreserved MSC products in p2 are subsequently provided frozen to the clinics when needed for the treatment of refractory acute GvHD and are thawed bed-side immediately before i.v. administration according to a published standard operating procedure (7, 22). In favor of MSC cryopreservation, we report in this study (and as part of our MSC product stability program) a 94.2% viability and 96.4% recovery for the frozen and thawed MSC product in p2, which to our knowledge is superior to other published reports (11, 12, 27, 28). We can also conclude that the basic MSC cell phenotype for platelet lysate-expanded MSCs and the osteogenic and adipogenic differentiation potentials are unaltered after freezing and thawing. Furthermore, introducing additional interim freezing steps at p0 and p1 does not alter the viability, recovery, cumulative PD number, cell yield, cell phenotype or differentiation potential of the frozen and thawed MSC product in p2, which is in line with an earlier study performed with research-grade MSCs (29). When comparing PD numbers between different studies it should be noted that our PD numbers, unlike most other studies, are calculated starting from the original CFU-F numbers which thus also includes the first cell divisions of the original MSCs in the primary culture. Our cumulative PD numbers demonstrate that on average 14 population doublings actually take place during the first cell culture stage in p0, which is usually disregarded in most other studies.

An additional conclusion made from this study, and in favor of MSC cryopreservation, is that we cannot see a clear correlation between a modest number of cryopreservation steps and earlier induction of senescence. Our data, however, give reason to believe that a) the induction of senescence is highly heterogenous between individual MSC batches (derived from different donors) and that b) exhaustive numbers of freezing steps ($\geq$4) might induce earlier onset of senescence. We made a choice to mainly utilize a set of standard quality control and batch release assays for clinical MSC products. We must, therefore, point out that we did not analyze the impact of MSC cryopreservation on a molecular level, where the effect of a stressful cell handling procedure (which cryopreservation and thawing most certainly is) may be more evident (12, 16). It is nevertheless unclear whether transient changes on a molecular level will ever translate to a functional level that really ultimately affect cell functionality or cell performance. We cannot, however, emphasize enough the importance of utilizing process developed and fully validated protocols for MSC freezing and thawing combined with fixed, unambiguous standard operating procedures and documentation of every thawing procedure also at the clinical site. This may in the end be the most relevant parameter for the quality of frozen and thawed MSCs.

The crucial question however still remains: does cryopreservation negatively affect the immunomodulatory potential of MSCs as demonstrated in some studies (11, 12)? We tried to address this question by an *in vitro* immunosuppression assay currently in use as a potency assay by us in our in-house clinical-grade MSC batch release. Our immunosuppression assay utilizes PBMNCs derived from allogeneic blood donors as responder cells. A single well in this assay only contains PBMNCs from a single donor, as opposed to a mixed lymphocyte reaction (MLR). The T-cells in the PBMNC fraction are activated with CD3 and CD28 antibodies and the MSCs are not irradiated, as usually is the case in an MLR setup. Our assay format consistently displays a potent immunosuppressive *in vitro* behavior of fresh

MSCs with a 70% inhibition of T-cell proliferation at a 1:10 ratio (**Figure 2D**). It is also evident that this particular assay measures immunosuppression linked to the IDO pathway, since the addition of the IDO inhibitor L-1MT reduces the MSC-induced T-cell proliferation inhibition by over 50% (**Figure 2D**). It is, however, clear that frozen and thawed MSCs do not perform as well as their unfrozen counterparts in this assay (**Figure 4**). When extrapolating all the immunosuppression data we have from frozen and thawed MSCs, regardless of the number of freezing steps before p2, we can conclude that the immunosuppressive potential is reduced by ∼50% as compared to the fresh counterpart (**Figure 4B**). Strikingly, the result is almost identical to the IDO inhibitor L-1MT result (**Figure 2D**). Since we made an observation that the assay performance was somewhat inconsistent in the experiments comparing fresh and frozen cells after one or two freezing steps (**Figure 4A**), where we even saw some frozen MSCs performing almost as well as fresh cells, we were challenged to critically evaluate the design of our assay and in particular the assay compatibility with frozen and thawed MSCs. As we exemplify in **Figure 5**, the assay readout is highly dependent on what might be considered "insignificant experimental details" such as plating times of the MSCs and the activation protocol of the T-cells. It is striking that changing the activation protocol to a 24 h pre-activation of the T-cells before MSC addition dramatically changes the outcome and even fresh MSCs did not perform in our assay anymore (**Figure 5B**). It is also interesting that we could not restore the immunosuppressive potential with a 24 h rescue period/plating period after thawing in **Figure 5A** as demonstrated by François et al. (11).

Our results ultimately forces one to ask the question: what does our *in vitro* immunosuppression assay measure? Our assay format is standardized, but it differs significantly from an MLR assay, where PBMNCs from two unrelated donors are mixed to induce the allostimulation of T-cells. This assay can to some degree measure the proliferation of activated CD3+ T-cells and an IDO-pathway related immunosuppression, but it is clear it does not directly address the effect of MSCs on other cellular components of PBMNCs such as B cells, regulatory T-cells, dendritic cells and monocytes and macrophages (30, 31). The assay does not either measure other alternative MSC immunosuppressive pathways such as the adenosinergic pathway by ectonucleotidases CD73 and CD39 (32). The adenosinergic immunosuppressive pathway has interestingly been shown to be an important immunoregulatory mechanism of MSCs in situations where extracellular ATP is available in excess, such as in tissue injury (32). Furthermore, the exciting MSC immunomodulation paradigm shift recently presented by Galleu et al. (9) demonstrates that MSCs undergo extensive caspase activation and apoptosis after infusion in the presence of cytotoxic cells. After infusion, recipient phagocytes engulf apoptotic MSCs and subsequently produce IDO after polarization to M2 macrophages, which is ultimately necessary for effecting immunosuppression. de Witte et al. furthermore demonstrated that both CD8+ T-cell and CD56+ natural killer (NK) populations are responsible for initiating MSC apoptosis (10). This mechanism has also been supported by clinical data, whereby only those GvHD patients with high cytotoxic activity against MSCs responded to MSC infusion, and those GvHD patients with low activity did not respond to MSC therapy (9). This new apoptosis-based MSC immunomodulatory mechanism challenges the design of *in vitro* immunosuppression assays even further, since the assay should also contain functional cell types to induce MSC apoptosis and, importantly, functional monocytes that can polarize to M2 macrophages. It has also been recently presented that an optimal *in vitro* immunosuppression assay should not allow MSCs to adhere to plastic to avoid baseline activation of monocytes adhering to the plastic and for instance polypropylene tubes should be utilized instead in order to more closely resemble the *in vivo* setting (10). Another evident weakness in the performance of an *in vitro* immunosuppression assay might be the utilization of fetal calf serum (FCS) in the T-cell expansion media. In our case, platelet lysate-expanded MSCs encounter FCS for the first time in the immunosuppression assay. We have thus far been unsuccessful in finding a xenofree substitute for FCS with equal performance in this assay format. Utilizing FCS in the immunosuppression assay does not cause obvious evident problems for fresh platelet lysate-expanded MSCs. The xenogeneic stress might be tolerable for fresh MSCs, but when a cell is simultaneously recovering from a thawing procedure, the xenogeneic stress caused by FCS might impact the performance of the MSCs in the assay. Luetzkendorf et al. demonstrated preserved immunomodulatory functions of their platelet lysate expanded cryopreserved MSCs in an FCS-based assay, but by utilizing mitogen (phytohemagglutinin; PHA)-stimulated PBMNCs and a readout based on bromodeoxyuridine (BrdU) incorporation (27). Furthermore, Kuci et al. convincingly demonstrated a clear *in vitro* immunosuppressive result for their frozen and thawed clinical-grade MSC product MSC-FFM, but by a two-way MLR assay (20). It is, however, also evident that the Frankfurt platelet lysate-expanded MSC-FFM product is manufactured in a very different manner from ours utilizing pooled BMNCs from multiple donors to produce the MSCs (20, 21), which might significantly impact the immunomodulatory performance, also *in vitro*.

As discussed above, the weakness of our study may lie in the design of the immunosuppression assay, or more precisely, in the utilization of only one assay format (31). Furthermore, we did not measure cytokines in the assay media (31). It is, however, an inevitable fact that *in vitro* functionality data can only answer questions concerning efficacy to a limited degree. In the light of our immunosuppression results, it is evident that the cryopreserved and thawed MSCs perform differently from the fresh counterpart, but that does not necessarily translate to reduced immunomodulatory efficacy *in vivo*. The ultimate answers therefore lie within the clinical data and unless the MSC product is rigorously evaluated and demonstrated to correlate with clinical outcomes, the *in vitro* immunosuppression assay can only be used as a research tool. Our clinical-grade MSC product has been in clinical use since 2013 and the results of the first aGvHD patient cohort consisting of 26 adult and pediatric patients, most with grade III-IV GvHD, have been published earlier (7). We can conclude an overall response rate of 62% at day 28, but with a disappointing overall survival of only 22% after a median follow-up time of 767 days (range

74-1270 days) (7). Our results also presented a markedly different overall long-term survival rate between adult and pediatric patients and between complete responders and non-responders. Thus far, we have been unable to find molecular markers that can predict the outcome or find a mechanistic explanation to the variable degree of response (8). Several other MSC GvHD patient cohorts have been published lately utilizing a similar MSC manufacturing strategy including cryopreservation and supplementing the culture media with pooled platelet lysate (21, 33–37). When comparing our published clinical cohort with the six other published MSC GVHD cohorts utilizing platelet lysate-expanded and cryopreserved MSCs, it can be concluded that although the overall response rate at day 28 is to some extent comparable (ranging from 47–83%), there is a large difference in the reported overall survival rates ranging from 16.6–67% (follow-up times from 6 months to >2 years). The very variable follow-up times in the published patient cohorts so far and the mixed degree of adult vs. pediatric patients complicates the study comparisons. The obvious limitation in all these cohorts is the seemingly small number of patients included and the lack of control groups and randomization. The very recent publication by the Frankfurt am Main group presents the largest patient cohort of 69 patients, but mostly consisting of pediatric patients (74%) (21). The Frankfurt cohort presents superior overall response rates of 83% at day 28 and a very promising overall survival rate of 67%, but it needs to be pointed out that the mean follow-up time was only 8.1 months and the clinical monitoring needs to be continued. Nevertheless, the Frankfurt overall response rates and preliminary survival data is superior to the other MSC GvHD cohorts with frozen, platelet lysate-expanded MSCs. The Frankfurt MSC-FFM manufacturing strategy differs significantly from the other studies by utilizing a pooled BM-MNC bank derived from eight allogeneic donors as a source for bulk production of the clinical-grade MSC products (20), which naturally might result in a more immunosuppressive product, but the MSC-FFM product is cryopreserved and the manufacturing strategy includes at least two freezing steps. It is therefore tempting to speculate, in light with the new apoptosis-based theory (9), that cryopreserved MSCs might in fact cause a stronger immunosuppression since it has been demonstrated that freeze-thawed MSCs are more prone to activate the instant blood mediated inflammatory reaction (IBMIR) and display increased sensitivity to complement lysis, which leads to increased MSC apoptosis (12). This is in-line with another study demonstrating that cryopreserved MSCs are significantly more susceptible to contact-dependent apoptosis when co-cultured with activated T-cells (13). The long-term survival data of patient cohorts utilizing cryopreserved MSC products will provide more insights into what parameters might cause differences in the outcome, but the available published clinical data suggest that the functional consequences of cryopreservation might be inconsequential as compared to other MSC manufacturing and administration details.

CONCLUSIONS

Our results support the utilization of frozen MSC products and MSC banking strategies and suggest that 1-2 freezing steps for MSCs in early passage is feasible and preserves most of the *in vitro* functional properties. Interim freezing steps are not reflected in standard manufacturing parameters. The *in vitro* immunosuppressive potential of frozen and thawed MSCs can be interpreted as different from the fresh counterpart, with a reduced, but definitely not abolished *in vitro* performance specific for the IDO pathway, but *in vitro* immunosuppression assay results ought to be interpreted with caution to avoid false conclusions. Our study emphasizes the need of always performing detailed studies on also the cryopreserved MSC counterpart and to thoroughly validate both the freezing and thawing procedures and to monitor the thawing procedures with unambiguous standard operating procedures. All MSC products are not equal, and frozen-thawed MSCs might differ to some extent to their fresh counterpart. This, however, does not mean that cryopreserved MSCs are not efficacious, they just might work differently from fresh MSCs. In light of available clinical data, it is tempting to speculate that the ultimate parameter affecting efficacy might not actually be cryopreservation, but essential differences in MSC manufacturing strategies.

AUTHOR CONTRIBUTIONS

SO planned and performed the majority of the experiments and assisted in the results analysis and manuscript writing. TK and AL performed some of the experiments, assisted in the results analysis and contributed to the interpretation of the results. MA performed some of the experiments and assisted in results analysis. MK provided study materials, supported the supervision of the study and contributed to the interpretation of the results. JN conceived the original idea, planned the experiments, supervised the study, analyzed the results, and wrote the manuscript. All authors have commented and accepted the manuscript.

FUNDING

This study was partly funded by a State Research Funding Finland grant (VTR) granted to the Finnish Red Cross Blood Service.

ACKNOWLEDGMENTS

The authors sincerely thank Satu Happonen, Sirkka Hirschovits, Johanna Matikainen, Birgitta Rantala, Susanna Räsänen, and Iris Tallus for expert technical assistance.

REFERENCES

1. Galipeau J, Sensébé L. Mesenchymal stromal cells: clinical challenges and therapeutic opportunities. *Cell Stem Cell.* (2018) 22:824–33. doi: 10.1016/j.stem.2018.05.004
2. Ankrum JA, Ong JF, Karp JM. Mesenchymal stem cells: immune evasive, not immune privileged. *Nat Biotechnol.* (2014) 32:252–60. doi: 10.1038/nbt.2816
3. Introna M, Rambaldi A. Mesenchymal stromal cells for prevention and treatment of graft-versus-host disease: successes and hurdles. *Curr Opin Organ Transplant.* (2015) 20:72–8. doi: 10.1097/MOT.0000000000000158
4. Hashmi S, Ahmed M, Murad MH, Litzow MR, Adams RH, Ball LM, et al. Survival after mesenchymal stromal cell therapy in steroid-refractory acute graft-versus-host disease: systematic review and meta-analysis. *Lancet Haematol.* (2016) 3:e45–52. doi: 10.1016/S2352-3026(15)00224-0
5. Chen X, Wang C, Yin J, Xu J, Wei J, Zhang Y. Efficacy of mesenchymal stem cell therapy for steroid-refractory acute graft-versus-host disease following allogeneic hematopoietic stem cell transplantation: a systematic review and meta-analysis. *PLoS ONE.* (2015) 10:e0136991. doi: 10.1371/journal.pone.0136991
6. Amorin B, Alegretti AP, Valim V, Pezzi A, Laureano AM, da Silva MAL, et al. Mesenchymal stem cell therapy and acute graft-versus-host disease: a review. *Hum Cell.* (2014) 27:137–50. doi: 10.1007/s13577-014-0095-x
7. Salmenniemi U, Itala-Remes M, Nystedt J, Putkonen M, Niittyvuopio R, Vettenranta K, et al. Good responses but high TRM in adult patients after MSC therapy for GvHD. *Bone Marrow Transplant.* (2017) 52:606–8. doi: 10.1038/bmt.2016.317
8. Keto J, Kaartinen T, Salmenniemi U, Castren J, Partanen J, Hanninen A, et al. Immunomonitoring of MSC-treated GvHD patients reveals only moderate potential for response prediction but indicates treatment safety. *Mol Ther Methods Clin Dev.* (2018) 9:109–18. doi: 10.1016/j.omtm.2018.02.001
9. Galleu A, Riffo-Vasquez Y, Trento C, Lomas C, Dolcetti L, Cheung TS, et al. Apoptosis in mesenchymal stromal cells induces *in vivo* recipient-mediated immunomodulation. *Sci Transl Med.* (2017) 9:eaam7828. doi: 10.1126/scitranslmed.aam7828
10. de Witte SFH, Luk F, Sierra Parraga JM, Gargesha M, Merino A, Korevaar SS, et al. Immunomodulation by therapeutic mesenchymal stromal cells (MSC) is triggered through phagocytosis of MSC by monocytic cells. *Stem Cells.* (2018) 36:602–15. doi: 10.1002/stem.2779
11. François M, Copland IB, Yuan S, Romieu-Mourez R, Waller EK, Galipeau J. Cryopreserved mesenchymal stromal cells display impaired immunosuppressive properties as a result of heat-shock response and impaired interferon-γ licensing. *Cytotherapy.* (2012) 14:147–52. doi: 10.3109/14653249.2011.623691
12. Moll G, Alm JJ, Davies LC, Von Bahr L, Heldring N, Stenbeck-Funke L, et al. Do cryopreserved mesenchymal stromal cells display impaired immunomodulatory and therapeutic properties? *Stem Cells.* (2014) 32:2430–42. doi: 10.1002/stem.1729
13. Chinnadurai R, Copland IB, Garcia MA, Petersen CT, Lewis CN, Waller EK, et al. Cryopreserved mesenchymal stromal cells are susceptible to T-cell mediated apoptosis which is partly rescued by IFNγ licensing. *Stem Cells.* (2016) 34:2429–42. doi: 10.1002/stem.2415
14. Gramlich OW, Burand AJ, Brown AJ, Deutsch RJ, Kuehn MH, Ankrum JA. Cryopreserved mesenchymal stromal cells maintain potency in a retinal ischemia/reperfusion injury model: toward an off-the-shelf therapy. *Sci Rep.* (2016) 6:26463. doi: 10.1038/srep26463
15. Cruz FF, Borg ZD, Goodwin M, Sokocevic D, Wagner D, McKenna DH, et al. Freshly thawed and continuously cultured human bone marrow-derived mesenchymal stromal cells comparably ameliorate allergic airways inflammation in immunocompetent mice. *Stem Cells Transl Med.* (2015) 4:615–24. doi: 10.5966/sctm.2014-0268
16. Hoogduijn MJ, de Witte SFH, Luk F, van den Hout-van Vroonhoven MCGN, Ignatowicz L, Catar R, et al. Effects of freeze–thawing and intravenous infusion on mesenchymal stromal cell gene expression. *Stem Cells Dev.* (2016) 25:586–97. doi: 10.1089/scd.2015.0329
17. Resnick IB, Barkats C, Shapira MY, Stepensky P, Bloom AI, Shimoni A, et al. Treatment of severe steroid resistant acute GVHD with mesenchymal stromal cells (MSC). *Am J Blood Res.* (2013) 3:225–38.
18. Le Blanc K, Frassoni F, Ball L, Locatelli F, Roelofs H, Lewis I, et al. Mesenchymal stem cells for treatment of steroid-resistant, severe, acute graft-versus-host disease: a phase II study. *Lancet.* (2008) 371:1579–86. doi: 10.1016/S0140-6736(08)60690-X
19. Kebriaei P, Isola L, Bahceci E, Holland K, Rowley S, McGuirk J, et al. Adult human mesenchymal stem cells added to corticosteroid therapy for the treatment of acute graft-versus-host disease. *Biol Blood Marrow Transplant.* (2009) 15:804–11. doi: 10.1016/j.bbmt.2008.03.012
20. Kuci Z, Bonig H, Kreyenberg H, Bunos M, Jauch A, Janssen JWG, et al. Mesenchymal stromal cells from pooled mononuclear cells of multiple bone marrow donors as rescue therapy in pediatric severe steroid-refractory graft-versus-host disease: a multicenter survey. *Haematologica.* (2016) 101:985–94. doi: 10.3324/haematol.2015.140368
21. Bader P, Kuci Z, Bakhtiar S, Basu O, Bug G, Dennis M, et al. Effective treatment of steroid and therapy-refractory acute graft-versus-host disease with a novel mesenchymal stromal cell product (MSC-FFM). *Bone Marrow Transplant.* (2018) 53:852–62. doi: 10.1038/s41409-018-0102-z
22. Laitinen A, Oja S, Kilpinen L, Kaartinen T, Moller J, Laitinen S, et al. A robust and reproducible animal serum-free culture method for clinical-grade bone marrow-derived mesenchymal stromal cells. *Cytotechnology.* (2016) 68:891–906. doi: 10.1007/s10616-014-9841-x
23. Strunk D, Lozano M, Marks DC, Loh YS, Gstraunthaler G, Schennach H, et al. International forum on GMP-grade human platelet lysate for cell propagation. *Vox Sang.* (2018) 113:e1–e25. doi: 10.1111/vox.12593
24. Castro-Malaspina H, Gay RE, Resnick G, Kapoor N, Meyers P, Chiarieri D, et al. Characterization of human bone marrow fibroblast colony-forming cells (CFU-F) and their progeny. *Blood.* (1980) 56:289–301.
25. Oja S, Komulainen P, Penttila A, Nystedt J, Korhonen M. Automated image analysis detects aging in clinical-grade mesenchymal stromal cell cultures. *Stem Cell Res Ther.* (2018) 9:6. doi: 10.1186/s13287-017-0740-x
26. Dominici M, Le Blanc K, Mueller I, Slaper-Cortenbach I, Marini F, Krause D, et al. Minimal criteria for defining multipotent mesenchymal stromal cells. The international society for cellular therapy position statement. *Cytotherapy.* (2006) 8:315–7. doi: 10.1080/14653240600855905
27. Luetzkendorf J, Nerger K, Hering J, Moegel A, Hoffmann K, Hoefers C, et al. Cryopreservation does not alter main characteristics of good manufacturing process-grade human multipotent mesenchymal stromal cells including immunomodulating potential and lack of malignant transformation. *Cytotherapy.* (2015) 17:186–98. doi: 10.1016/j.jcyt.2014.10.018
28. Lechanteur C, Briquet A, Giet O, Delloye O, Baudoux E, Beguin Y. Clinical-scale expansion of mesenchymal stromal cells: a large banking experience. *J Transl Med.* (2016) 14:1–15. doi: 10.1186/s12967-016-0892-y
29. Mamidi MK, Nathan KG, Singh G, Thrichelvam ST, Mohd Yusof NAN, Fakharuzi NA, et al. Comparative cellular and molecular analyses of pooled bone marrow multipotent mesenchymal stromal cells during continuous passaging and after successive cryopreservation. *J Cell Biochem.* (2012) 113:3153–64. doi: 10.1002/jcb.24193
30. Bloom DD, Centanni JM, Bhatia N, Emler CA, Drier D, Leverson GE, et al. A reproducible immunopotency assay to measure mesenchymal stromal cell-mediated T-cell suppression. *Cytotherapy.* (2015) 17:140–51. doi: 10.1016/j.jcyt.2014.10.002
31. Galipeau J, Krampera M, Barrett J, Dazzi F, Deans RJ, DeBruijn J, et al. International society for cellular therapy perspective on immune functional assays for mesenchymal stromal cells as potency release criterion for advanced phase clinical trials. *Cytotherapy.* (2016) 18:151–9. doi: 10.1016/j.jcyt.2015.11.008
32. Kerkela E, Laitinen A, Rabina J, Valkonen S, Takatalo M, Larjo A, et al. Adenosinergic immunosuppression by human mesenchymal stromal cells requires co-operation with T cells. *Stem Cells.* (2016) 34:781–90. doi: 10.1002/stem.2280
33. von Dalowski F, Kramer M, Wermke M, Wehner R, Rollig C, Alakel N, et al. Mesenchymal stromal cells for treatment of acute steroid-refractory graft versus host disease: clinical responses and long-term outcome. *Stem Cells.* (2016) 34:357–66. doi: 10.1002/stem.2224
34. Introna M, Lucchini G, Dander E, Galimberti S, Rovelli A, Balduzzi A, et al. Treatment of graft versus host disease with mesenchymal stromal cells: a phase I study on 40 adult and pediatric patients. *Biol Blood Marrow Transplant.* (2014) 20:375–81. doi: 10.1016/j.bbmt.2013.11.033

35. Lucchini G, Introna M, Dander E, Rovelli A, Balduzzi A, Bonanomi S, et al. Platelet-lysate-expanded mesenchymal stromal cells as a salvage therapy for severe resistant graft-versus-host disease in a pediatric population. *Biol Blood Marrow Transplant.* (2010) 16:1293–301. doi: 10.1016/j.bbmt.2010.03.017
36. Te Boome LCJ, Mansilla C, van der Wagen LE, Lindemans CA, Petersen EJ, Spierings E, et al. Biomarker profiling of steroid-resistant acute GVHD in patients after infusion of mesenchymal stromal cells. *Leukemia.* (2015) 29:1839–46. doi: 10.1038/leu.2015.89
37. Dotoli GM, De Santis GC, Orellana MD, de Lima Prata K, Caruso SR, Fernandes TR, et al. Mesenchymal stromal cell infusion to treat steroid-refractory acute GvHD III/IV after hematopoietic stem cell transplantation. *Bone Marrow Transplant.* (2017) 52:859–62. doi: 10.1038/bmt.2017.35

Permissions

All chapters in this book were first published by Frontiers; hereby published with permission under the Creative Commons Attribution License or equivalent. Every chapter published in this book has been scrutinized by our experts. Their significance has been extensively debated. The topics covered herein carry significant findings which will fuel the growth of the discipline. They may even be implemented as practical applications or may be referred to as a beginning point for another development.

The contributors of this book come from diverse backgrounds, making this book a truly international effort. This book will bring forth new frontiers with its revolutionizing research information and detailed analysis of the nascent developments around the world.

We would like to thank all the contributing authors for lending their expertise to make the book truly unique. They have played a crucial role in the development of this book. Without their invaluable contributions this book wouldn't have been possible. They have made vital efforts to compile up to date information on the varied aspects of this subject to make this book a valuable addition to the collection of many professionals and students.

This book was conceptualized with the vision of imparting up-to-date information and advanced data in this field. To ensure the same, a matchless editorial board was set up. Every individual on the board went through rigorous rounds of assessment to prove their worth. After which they invested a large part of their time researching and compiling the most relevant data for our readers.

The editorial board has been involved in producing this book since its inception. They have spent rigorous hours researching and exploring the diverse topics which have resulted in the successful publishing of this book. They have passed on their knowledge of decades through this book. To expedite this challenging task, the publisher supported the team at every step. A small team of assistant editors was also appointed to further simplify the editing procedure and attain best results for the readers.

Apart from the editorial board, the designing team has also invested a significant amount of their time in understanding the subject and creating the most relevant covers. They scrutinized every image to scout for the most suitable representation of the subject and create an appropriate cover for the book.

The publishing team has been an ardent support to the editorial, designing and production team. Their endless efforts to recruit the best for this project, has resulted in the accomplishment of this book. They are a veteran in the field of academics and their pool of knowledge is as vast as their experience in printing. Their expertise and guidance has proved useful at every step. Their uncompromising quality standards have made this book an exceptional effort. Their encouragement from time to time has been an inspiration for everyone.

The publisher and the editorial board hope that this book will prove to be a valuable piece of knowledge for researchers, students, practitioners and scholars across the globe.

List of Contributors

Giuseppe Remuzzi and Federica Casiraghi
Department of Molecular Medicine, Istituto di Ricerche Farmacologiche Mario Negri IRCCS, Bergamo, Italy

Manuel Alfredo Podestà
Department of Molecular Medicine, Istituto di Ricerche Farmacologiche Mario Negri IRCCS, Bergamo, Italy
Department of Health Sciences, Università degli Studi di Milano, Milan, Italy

Behnam Sadeghi, Gianluca Moretti, Bita Kohein and Olle Ringdén
Translational Cell Therapy Research (TCR), Department of Clinical Science Intervention and Technology (CLINTEC), Karolinska Institutet and Karolinska University Hospital, Stockholm, Sweden

Fabian Arnberg and Staffan Holmin
Department of Clinical Neuroscience, Karolinska Institutet and Karolinska University Hospital, Stockholm, Sweden
Department of Neuroradiology, Karolinska Institutet and Karolinska University Hospital, Stockholm, Sweden

Erik Samén
Department of Clinical Neuroscience, Karolinska Institutet and Karolinska University Hospital, Stockholm, Sweden
Department of Neuroradiology, Karolinska Institutet and Karolinska University Hospital, Stockholm, Sweden
Department of Radiopharmacy, Karolinska Institutet and Karolinska University Hospital, Stockholm, Sweden

Rusan Catar
Department of Nephrology and Internal Intensive Care Medicine, Charité-Universitätsmedizin Berlin, Corporate Member of Freie-Universität Berlin, Humboldt-Universität zu Berlin, and Berlin Institute of Health (BIH), Berlin, Germany
Berlin Institute of Health (BIH), Berlin, Germany

Sven Geissler
BIH Center for Regenerative Therapies (BCRT), Charité-Universitätsmedizin Berlin, Corporate Member of Freie-Universität Berlin, Humboldt-Universität zu Berlin, and Berlin Institute of Health (BIH), Berlin, Germany
Julius Wolff Institute (JWI), Charité-Universitätsmedizin Berlin, Corporate Member of Freie-Universität Berlin, Humboldt-Universität zu Berlin, and Berlin Institute of Health (BIH), Berlin, Germany

Guido Moll
Department of Nephrology and Internal Intensive Care Medicine, Charité-Universitätsmedizin Berlin, Corporate Member of Freie-Universität Berlin, Humboldt-Universität zu Berlin, and Berlin Institute of Health (BIH), Berlin, Germany
BIH Center for Regenerative Therapies (BCRT), Charité-Universitätsmedizin Berlin, Corporate Member of Freie-Universität Berlin, Humboldt-Universität zu Berlin, and Berlin Institute of Health (BIH), Berlin, Germany
Berlin-Brandenburg School for Regenerative Therapies (BSRT), Charité-Universitätsmedizin Berlin, Corporate Member of Freie-Universität Berlin, Humboldt-Universität zu Berlin, and Berlin Institute of Health (BIH), Berlin, Germany

Henry Caplan, Scott D. Olson, Akshita Kumar, Mitchell George, Karthik S. Prabhakara, Pamela Wenzel, Supinder Bedi, Naama E. Toledano-Furman, Fabio Triolo and Charles S. Cox Jr.
Department of Pediatric Surgery, McGovern Medical School, University of Texas Health Science Center at Houston, Houston, TX, United States

Julian Kamhieh-Milz
Department of Transfusion Medicine, Charité Universitätsmedizin Berlin, Corporate Member of Freie Universität Berlin, Humboldt-Universität zu Berlin, and Berlin Institute of Health (BIH), Berlin, Germany

Galina Masgutova, Adelya Mullakhmetova, Margarita Zhuravleva, Valeriya Syromiatnikova, Dina Andreeva, Alina Zeinalova, Kamilla Idrisova, Andrey Kiyasov and Albert Rizvanov
OpenLab Gene and Cell Technologies, Institute of Fundamental Medicine and Biology, Kazan Federal University, Kazan, Russia

Ruslan Masgutov
OpenLab Gene and Cell Technologies, Institute of Fundamental Medicine and Biology, Kazan Federal University, Kazan, Russia
Department of Orthopaedics, Republic Clinical Hospital, Kazan, Russia

Cinzia Allegrucci
School of Veterinary Medicine and Science, University of Nottingham, Nottingham, United Kingdom

Alexander Rogozhin
OpenLab Gene and Cell Technologies, Institute of Fundamental Medicine and Biology, Kazan Federal University, Kazan, Russia
Department of Neurology, Kazan State Medical Academy, Branch of Russian Medical Academy of Postgraduate Education, Kazan, Russia

Bárbara Soria-Juan, Lucía Llanos, Mariano García-Arranz and Damián García-Olmo
Fundación Jiménez Díaz Health Research Institute, Madrid, Spain

Natalia Escacena, Vivian Capilla-González and Yolanda Aguilera
Department of Regeneration and Cell Therapy, Andalusian Center for Molecular Biology and Regenerative Medicine (CABIMER), University of Pablo de Olavide-University of Seville-CSIC, Seville, Spain

Juan R. Tejedo, Francisco J. Bedoya, Franz Martín, Abdelkrim Hmadcha and Bernat Soria
Department of Regeneration and Cell Therapy, Andalusian Center for Molecular Biology and Regenerative Medicine (CABIMER), University of Pablo de Olavide-University of Seville-CSIC, Seville, Spain
Spanish Biomedical Research Centre in Diabetes and Associated Metabolic Disorders (CIBERDEM), Madrid, Spain

Verónica Juan
Andalusian eHealth Library, Sevilla, Spain

Antonio De la Cuesta
Unidad de Isquemia Crónica de Miembros Inferiores, Hospital Victoria Eugenia de la Cruz Roja, Sevilla, Spain

Rafael Ruiz-Salmerón
Servicio de Cardiología, Hospital Universitario Virgen Macarena, Sevilla, Spain

Enrique Andreu, Lukas Grochowicz and Felipe Prósper
Clínica Universidad de Navarra, Pamplona, Spain

Fermín Sánchez-Guijo and Francisco S. Lozano
IBSAL-Hospital Universitario Salamanca, Salamanca, Spain

Manuel Miralles
Department of Surgery, University of Valencia, Valencia, Spain

Lourdes Del Río-Solá
Cirugía Vascular, Hospital Universitario de Valladolid, Valladolid, Spain

Gregorio Castellanos and José M. Moraleda
Servicio Hematología y Hemoterapia, Hospital Clínico Universitario Virgen de la Arrixaca, Murcia, Spain

Robert Sackstein
Herbert Wertheim College of Medicine, Florida International University, Miami, FL, United States

Alison Wilson and Paul Genever
Department of Biology, University of York, York, United Kingdom

Margeaux Hodgson-Garms and Jessica E. Frith
Materials Science and Engineering, Monash University, Clayton, VIC, Australia

Ramón Menta, Pablo Mancheño-Corvo, Maitane Ortiz-Virumbrales, Alvaro Avivar-Valderas, Cristina Ramírez, Borja Del Río, Ángel Herrero-Méndez, Itziar Palacios, Eleuterio Lombardo and Olga DelaRosa
Takeda Madrid, Cell Therapy Technology Center—Cell Therapies, Madrid, Spain

Cristina Martín-Martín and José Luís Castañer
Department of Immunology, University Hospital Ramon y Cajal, Madrid, Spain

Julián Panés
Department of Gastroenterology, Hospital Clínic, IDIBAPS, CIBERehd, Barcelona, Spain

Damián García-Olmo
Department of Surgery, Hospital U. Fundación Jiménez Díaz, Madrid, Spain

Wilfried Dalemans
TiGenix NV, Leuven, Belgium

Graciosa Q. Teixeira and Susana G. Santos
Instituto de Investigação e Inovação em Saúde (i3S), Universidade do Porto, Porto, Portugal
Instituto de Engenharia Biomédica, Universidade do Porto, Porto, Portugal

Joana R. Ferreira, Raquel M. Gonçalves and Mário A. Barbosa
Instituto de Investigação e Inovação em Saúde (i3S), Universidade do Porto, Porto, Portugal
Instituto de Engenharia Biomédica, Universidade do Porto, Porto, Portugal
Instituto de Ciências Biomédicas Abel Salazar, Universidade do Porto, Porto, Portugal

Graça Almeida-Porada
Wake Forest Institute for Regenerative Medicine, Winston-Salem, NC, United States

Lauren K. Boland, Anthony J. Burand, Devlin T. Boyt Hannah Dobroski, Lin Di, Jesse N. Liszewski, Michael V. Schrodt, Maria K. Frazer and James A. Ankrum
University of Iowa Fraternal Order of Eagles Diabetes Research Center, University of Iowa, Iowa City, IA, United States
Roy J. Carver Department of Biomedical Engineering, University of Iowa, Iowa City, IA, United States

Donna A. Santillan
Department of Obstetrics and Gynecology, Center for Immunology and Immune Based Diseases, Center for Hypertension Research, University of Iowa, Iowa City, IA, United States

Reenam S. Khan and Philip N. Newsome
National Institute for Health Research (NIHR), Birmingham Biomedical Research Centre, University Hospitals Birmingham NHS Foundation Trust, University of Birmingham, Birmingham, United Kingdom
Centre for Liver Research, Institute of Immunology and Immunotherapy, University of Birmingham, Birmingham, United Kingdom
Liver Unit, University Hospitals Birmingham NHS Foundation Trust, Birmingham, United Kingdom

Andreas Robert Rudolf Weiss
Department of Surgical Sciences, Dunedin School of Medicine, University of Otago, Dunedin, New Zealand

Marc Hendrik Dahlke
Department of Surgery, Robert Bosch Hospital, Stuttgart, Germany

Daniel J. Weiss
Department of Medicine, University of Vermont College of Medicine, Burlington, VT, United States

Karen English and Ian J. Hawthorne
Cellular Immunology Laboratory, Biology Department, Human Health Research Institute, Maynooth University, Maynooth, Ireland

Anna Krasnodembskaya
School of Medicine, Dentistry and Biomedical Sciences, Wellcome-Wolfson Institute for Experimental Medicine, Queen's University of Belfast, Belfast, United Kingdom

Johana M. Isaza-Correa and Bernard P. Mahon
Immunology & Cell Biology Laboratory, Biology Department, Human Health Research Institute, Maynooth University Maynooth, Ireland

Karsten Jürchott
BIH Center for Regenerative Therapies (BCRT), Charité Universitätsmedizin Berlin, Corporate Member of Freie Universität Berlin, Humboldt-Universität zu Berlin, Berlin Institute of Health (BIH), Berlin, Germany

Anastazja Andrzejewska, Janosch Schoon, Taimoor Hasan Qazi, Frauke Andrea Sass, Georg Duda, and Sven Geissler
BIH Center for Regenerative Therapies (BCRT), Charité Universitätsmedizin Berlin, Corporate Member of Freie Universität Berlin, Humboldt-Universität zu Berlin, Berlin Institute of Health (BIH), Berlin, Germany
Berlin-Brandenburg School for Regenerative Therapies, Charité Universitätsmedizin Berlin, Corporate Member of Freie Universität Berlin, Humboldt-Universität zu Berlin, BIH, Berlin, Germany
Julius Wolff Institute, Charité Universitätsmedizin Berlin, Corporate Member of Freie Universität Berlin, Humboldt-Universität zu Berlin, BIH, Berlin, Germany

Dorit Jacobi, Antje Blankenstein and Simon Reinke
BIH Center for Regenerative Therapies (BCRT), Charité Universitätsmedizin Berlin, Corporate Member of Freie Universität Berlin, Humboldt-Universität zu Berlin, Berlin Institute of Health (BIH), Berlin, Germany
Julius Wolff Institute, Charité Universitätsmedizin Berlin, Corporate Member of Freie Universität Berlin, Humboldt-Universität zu Berlin, BIH, Berlin, Germany

Guido Moll
BIH Center for Regenerative Therapies (BCRT), Charité Universitätsmedizin Berlin, Corporate Member of Freie Universität Berlin, Humboldt-Universität zu Berlin, Berlin Institute of Health (BIH), Berlin, Germany
Berlin-Brandenburg School for Regenerative Therapies, Charité Universitätsmedizin Berlin, Corporate Member of Freie Universität Berlin, Humboldt-Universität zu Berlin, BIH, Berlin, Germany
Julius Wolff Institute, Charité Universitätsmedizin Berlin, Corporate Member of Freie Universität Berlin, Humboldt-Universität zu Berlin, BIH, Berlin, Germany
Department of Nephrology and Internal Intensive Care Medicine, Charité Universitätsmedizin Berlin, Corporate Member of Freie Universität Berlin, Humboldt-Universität zu Berlin, BIH, Berlin, Germany

Rusan Catar
BIH Center for Regenerative Therapies (BCRT), Charité Universitätsmedizin Berlin, Corporate Member of Freie Universität Berlin, Humboldt-Universität zu Berlin, Berlin Institute of Health (BIH), Berlin, Germany
Department of Nephrology and Internal Intensive Care Medicine, Charité Universitätsmedizin Berlin, Corporate Member of Freie Universität Berlin, Humboldt-Universität zu Berlin, BIH, Berlin, Germany

David Krüger
Julius Wolff Institute, Charité Universitätsmedizin Berlin, Corporate Member of Freie Universität Berlin, Humboldt-Universität zu Berlin, BIH, Berlin, Germany

Mathias Streitz, Stephan Schlickeiser and Sarina Richter
BIH Center for Regenerative Therapies (BCRT), Charité Universitätsmedizin Berlin, Corporate Member of Freie Universität Berlin, Humboldt-Universität zu Berlin, Berlin Institute of Health (BIH), Berlin, Germany
Institute of Medical Immunology, Charité Universitätsmedizin Berlin, Corporate Member of Freie Universität Berlin, Humboldt-Universität zu Berlin, BIH, Berlin, Germany

Naima Souidi and Christien Beez
BIH Center for Regenerative Therapies (BCRT), Charité Universitätsmedizin Berlin, Corporate Member of Freie Universität Berlin, Humboldt-Universität zu Berlin, Berlin Institute of Health (BIH), Berlin, Germany
Berlin-Brandenburg School for Regenerative Therapies, Charité Universitätsmedizin Berlin, Corporate Member of Freie Universität Berlin, Humboldt-Universität zu Berlin, BIH, Berlin, Germany
Institute of Medical Immunology, Charité Universitätsmedizin Berlin, Corporate Member of Freie Universität Berlin, Humboldt-Universität zu Berlin, BIH, Berlin, Germany

Ulrike Krüger and Tomasz Zemojtel
BIH Core Unit Genomics Charité Universitätsmedizin Berlin, Corporate Member of Freie Universität Berlin, Humboldt-Universität zu Berlin, Berlin Institute of Health, Berlin, Germany

Dirk Strunk
Berlin Center for Advanced Therapies, Charité Universitätsmedizin Berlin, Corporate Member of Freie Universität Berlin, Humboldt-Universität zu Berlin, BIH, Berlin, Germany

Petra Reinke
BIH Center for Regenerative Therapies (BCRT), Charité Universitätsmedizin Berlin, Corporate Member of Freie Universität Berlin, Humboldt-Universität zu Berlin, Berlin Institute of Health (BIH), Berlin, Germany
Institute of Medical Immunology, Charité Universitätsmedizin Berlin, Corporate Member of Freie Universität Berlin, Humboldt-Universität zu Berlin, BIH, Berlin, Germany
Spinal Cord Injury and Tissue Regeneration Center, Experimental and Clinical Cell Therapy Institute, Paracelsus Medical University, Salzburg, Austria

Falk Diedrichs
Berlin Institute of Health (BIH), Berlin, Germany
BIH Center for Regenerative Therapies (BCRT), Charité-Universitätsmedizin Berlin, Corporate Member of Freie Universität Berlin, Humboldt-Universität zu Berlin, and Berlin Institute of Health, Berlin, Germany

Meaghan Stolk
BIH Center for Regenerative Therapies (BCRT), Charité-Universitätsmedizin Berlin, Corporate Member of Freie Universität Berlin, Humboldt-Universität zu Berlin, and Berlin Institute of Health, Berlin, Germany

Karsten Jürchott and Martina Seifert
BIH Center for Regenerative Therapies (BCRT), Charité-Universitätsmedizin Berlin, Corporate Member of Freie Universität Berlin, Humboldt-Universität zu Berlin, and Berlin Institute of Health, Berlin, Germany
Institute of Medical Immunology, Charité-Universitätsmedizin Berlin, Corporate Member of Freie Universität Berlin, Humboldt-Universität zu Berlin, and Berlin Institute of Health, Berlin, Germany

Michael Sittinger and Marion Haag
BIH Center for Regenerative Therapies (BCRT), Charité-Universitätsmedizin Berlin, Corporate Member of Freie Universität Berlin, Humboldt-Universität zu Berlin, and Berlin Institute of Health, Berlin, Germany
Tissue Engineering Laboratory, Charité-Universitätsmedizin Berlin, Corporate Member of Freie Universität Berlin, Humboldt-Universität zu Berlin, and Berlin Institute of Health, Berlin, Germany

Xuegang Yuan
Department of Chemical and Biomedical Engineering, Florida State University, Tallahassee, FL, United States

Teng Ma
Institute of Molecular Biophysics, Florida State University, Tallahassee, FL, United States

Timothy M. Logan
Department of Chemistry and Biochemistry, Florida State University, Tallahassee, FL, United States

Jesus M. Sierra Parraga, Carla C. Baan, Ana Merino, and Martin J. Hoogduijn
Department of Internal Medicine, Erasmus Medical Center, University Medical Center Rotterdam, Rotterdam, Netherlands

Kaithlyn Rozenberg, James Hunter and Rutger J. Ploeg
Nuffield Department of Surgical Sciences and Oxford Biomedical Research Centre, University of Oxford, Oxford, United Kingdom

Bjarne K. Møller
Department of Clinical Immunology, Aarhus University Hospital, Aarhus, Denmark

Marco Eijken
Department of Clinical Immunology, Aarhus University Hospital, Aarhus, Denmark
Department of Renal Medicine, Aarhus University Hospital, Aarhus, Denmark

Bente Jespersen
Department of Renal Medicine, Aarhus University Hospital, Aarhus, Denmark

Henri G. Leuvenink and Cyril Moers
Department of Surgery - Organ Donation and Transplantation, University Medical Center Groningen, Groningen, Netherlands

Sofia Oja, Tanja Kaartinen, Marja Ahti, Matti Korhonen, Anita Laitinen and Johanna Nystedt
Advanced Cell Therapy Centre, Finnish Red Cross Blood Service, Helsinki, Finland

Index

www.ingramcontent.com/pod-product-compliance
Lightning Source LLC
LaVergne TN
LVHW080927210926
848485LV00004B/742